UNDER THE SPELL OF THE GAUGE PRINCIPLE

ADVANCED SERIES IN MATHEMATICAL PHYSICS

Published

Vol. 1: Mathematical Aspects of String Theory
edited by S-T Yau

Vol. 2: Bombay Lectures on Highest Weight Representations of Infinite Dimensional Lie Algebras
by V G Kac and A K Raina

Vol. 3: Kac-Moody and Virasoro Algebras: A Reprint Volume for Physicists
edited by P Goddard and D Olive

Vol. 4: Harmonic Mappings, Twistors and σ-Models
edited by P Gauduchon

Vol. 5: Geometric Phases in Physics
edited by A Shapere and F Wilczek

Vol. 7: Infinite Dimensional Lie Algebras and Groups
edited by V Kac

Vol. 8: Introduction to String Field Theory
by W Siegel

Vol. 9: Braid Group, Knot Theory and Statistical Mechanics
edited by C N Yang and M L Ge

Vol. 10: Yang-Baxter Equations in Integrable Systems
edited by M Jimbo

Vol. 11: New Developments in the Theory of Knots
edited by T Kohno

Vol. 12: Soliton Equations and Hamiltonian Systems
by L A Dickey

Vol. 14: Form Factors in Completely Integrable Models of Quantum Field Theory
by F A Smirnov

Vol. 15: Non-Perturbative Quantum Field Theory – Mathematical Aspects and Applications
Selected Papers of Jürg Fröhlich

Vol. 16: Infinite Analysis – Proceedings of the RIMS Research Project 1991
edited by A Tsuchiya, T Eguchi and M Jimbo

Vol. 17: Braid Group, Knot Theory and Statistical Mechanics (II)
edited by C N Yang and M L Ge

Vol. 18: Exactly Solvable Models and Strongly Correlated Electrons
edited by V Korepin and F H L Ebler

Vol. 20: State of Matter
edited by M Aizenman and H Araki

Advanced Series in Mathematical Physics
Vol. 19

UNDER THE SPELL OF THE GAUGE PRINCIPLE

G. 't Hooft

Institute for Theoretical Physics
University of Utrecht
The Netherlands

World Scientific
Singapore • New Jersey • London • Hong Kong

Published by

World Scientific Publishing Co. Pte. Ltd.

P O Box 128, Farrer Road, Singapore 9128

USA office: Suite 1B, 1060 Main Street, River Edge, NJ 07661

UK office: 73 Lynton Mead, Totteridge, London N20 8DH

The author and publisher would like to thank the following publishers of the various journals and books for their assistance and permission to reproduce the selected reprints found in this volume:

The American Physical Society (*Phys. Rev. D*); Springer-Verlag (*Trends in Elementary Particle Theory*); Uppsala University, Faculty of Science (*Reports of the CELSIUS/LINNÉ Lectures*).

While every effort has been made to contact the publishers of reprinted papers prior to publication, we have not been successful in some cases. Where we could not contact the publishers, we have acknowledged the source of the material. Proper credit will be accorded to these publishers in future editions of this work after permission is granted.

UNDER THE SPELL OF THE GAUGE PRINCIPLE

ISBN: 981-02-1308-5
ISBN: 981-02-1309-3 (pbk)

Printed in Singapore by Utopia Press.

To my mother

To the memory of my father

PREFACE

Our view of the world that we live in has changed. And it continues to change, as we learn more. What became abundantly clear in the course of the twentieth century is that the terminology most suited to describe the Laws of Physics is no longer that of everyday life. Instead, we have learned to create a new language, the language of mathematics, to describe elementary particles and their constituents. To facilitate this work we often introduce familiar words from the ordinary world to describe some law, principle or quantity, such as "energy", "force", "matter". But then these are given a meaning that is much more accurately defined than in our social lives. Sometimes the relation with the conventional meaning is remote, such as in "potential energy" and "action". And sometimes they mean something altogether different, for instance "color", or "charm".

The notion of "gauge invariance" also originated from a concept in daily life: instruments can be "gauged" by comparing them with other instruments, but the way they work is independent from the way they are gauged. The notion has been redefined with mathematical precision for theories in physics. Gauge symmetries, both "global" (i.e. space-time independent) and "local" (space-time dependent) became key issues in elementary particle physics. This book contains a selection of my work of over twenty years of research. The guiding principles in this work were symmetry and elegance of the magnificent edifice that we call our universe. And the most important symmetry was gauge symmetry.

Tremendous successes have been achieved in the recent past by the application of advanced mathematics to physics. However, some of the papers published in recent physics journals seem to be doing nothing but preparing and extending fragile mathematical constructions without the slightest indication as to how these structures should be used to build a theory of physics. I think it is important always to keep in mind that the mathematics we use in trying to understand our world is just a tool, a language sharp and precise enough to help us where ordinary words fail. Mathematics can never *replace* a theory. And, although I love mathematics, none of my papers were intended to improve our understanding of pure mathematics.

What I have always sought to do is quite the opposite. Using mathematics as a tool I have tried to identify the most urgent problems, the most baffling questions that are standing in our way towards a better understanding of our physical world. And then the art is to select out of those the ones that are worth being further pursued by a theoretical physicist. More often than not I end up immersed in mathematical equations. The next question is then always: "How will these equations help me answer the question I started off with? How do I *interpret* my mathematical expressions?" In this light one has to read and understand the articles reproduced in this book.

This book is a collection of what I consider to be the more salient chapters of my work. It is by no means meant to be complete. There are several important subjects in quantum field theory on which I have also made investigations, in particular monopoles and statistics, instanton solutions, lattice theory, classical and quantum gravity, and fundamental issues in quantum mechanics. All these subjects were either too technical or too incomplete to be included in a review book such as this, but that does not mean that I would consider them to be less important. I also could have included more work that I did with my co-authors, Martinus Veltman, Bernard de Wit, Peter Hasenfratz, Tevian Dray, Stanley Deser, Roman Jackiw, Karl Isler, Stilyan Kalitzin and others. In any case, if this book was supposed to be my "collected works" I sincerely hope it to be incomplete because I plan to continue my investigations.

It is impossible to produce important contributions to science without the insight, advice, help and support of numerous colleagues and friends. Among them, of course, the co-authors I just mentioned. Before Veltman became my teacher I learned a lot from N. G. van Kampen, who attempted to afflict me with his passion for precise arguments, discontent with half explanations, and total dedication to theoretical physics. Later I benefitted from so many inspiring discussions with equal-minded colleagues that I find it impossible to print out all their names. Often the importance of a discussion only became manifest much later at a time that I had forgotten who it was who told me.

I thank them all.

CONTENTS

CHAPTER 1
INTRODUCTION

When I began my graduate work, renormalization was considered by many physicists to be a dirty word. We had to learn about it because, inspite of its ugliness, renormalization seemed to work, at least in one limited area of particle physics, that of quantum electrodynamics. Renormalization was thought to be ugly because the procedure was ill-formulated and apparently an *ad hoc* and imprecise cure for a fundamental shortcoming in the quantum field theories of the day. But for those who analyzed the situation sufficiently carefully the physical reasoning behind renormalization wasn't that ugly at all. Indeed it seemed to be a quite natural fact that the fundamental interaction constants one puts into a theory do not need to be identical to the charges and masses one actually measures. If the theory is formulated on a very fine but discretized lattice instead of a continuum of space and time, the relation between the input parameters and the measured quantities will be quite direct. But now it so happens that if we take the limit where the mesh size approaches to zero, the input parameters needed to reproduce the measured quantities walk off the scale. This seems to be no disaster as long as we can't measure these input, or "bare", parameters directly.

Nowadays we know very well that such theories are *not* infinitely precise from a mathematical point of view, except possibly when the bare coupling strengths run to zero instead of infinity (such as in quantum chromodynamics). This latter, rare, property was not yet known about at the time, so in some sense the critics were right: our theories were mathematically empty. It is here that one needs more than just mathematical skill, namely insight into the requirements for our theories as needed to answer our questions. It would be too much to ask for infinite mathematical precision. If, in lieu of that, one could construct a theory that allows for an infinite *asymptotic* expansion in terms of some parameter that is measured by our experimental friends to be small, what may be achieved is a theory that is more than sufficiently precise to meet all our purposes.

My advisor and colleague, Tini Veltman, had an even more pragmatic view. He had reached the conclusion that Yang–Mills gauge field theories should be used in describing elementary particles by elimination: all the alternatives were uglier, and indeed quite a few aspects of the *weak* interactions pointed towards a renormalizable Yang–Mills theory for them. In any case, it was his perseverence that kept us on this track. Even though there was a moment that he was practically certain to have proven Yang–Mills theories to be nonrenormalizable (he had been studying the pure massive case and was not interested in the Higgs mechanism), he kept pointing towards the facts experimenters had told us about the weak interactions: Yang–Mills theories *had* to be right!

How could gauge invariance have anything to do with renormalization? To analyze this question it is crucial to understand the physical requirements of a particle theory at short distance scales. Suppose we want a theory that describes all known particles at large distance scales, and essentially nothing else at small distances. If many new particle types would arise at very small distances, that is, if we would have a large series of increasingly massive particles, then we would be stuck with a large, in general infinite number of uncontrollable parameters. Such a theory would have zero predictive power. Thus, as measured at a short distance scale, what we want is a theory containing only very light particles. It is all a question of economy; we are dealing here with in some sense the most efficient theories of nature, carrying the least possible amount of structure at small distances.

Ideally, we would like to have a system in which besides the mass parameters also all coupling parameters have the dimension of a negative power of a length, so that at small distance scales the particles are nearly free. In a renormalizable theory we settle for the next best thing, namely dimensionless couplings. One can then prove a very powerful theorem: *the only perturbative field theories with massless particles and dimensionless couplings are theories featuring fundamental particles with spin* 0, 1 *or* $\frac{1}{2}$. Furthermore, the spin 1 fields have an invisible component: the longitudinal part decouples. All for the better, because the longitudinal parts carry essentially no kinetic terms, so they behave as *ghosts*, particles that would contribute in a disastrous way to the S matrix if they were real; unitarity and positivity of the energy would be destroyed.

Now suppose we do want the particles at large distance scales to have a finite mass. Massive particles with spin 1 have a larger number of physical degrees of freedom than massless ones. For massless spin 1 particles only two of the four components of their vector field $A_\mu(x)$ are independently observable (the two possible helicities of the radiation field); massive particles on the other hand carry three independent degrees of freedom: the values $m_z = -1$, 0 and $+1$. If we would simply add a tiny mass term to the field equations we would give the bad ghost component of the vector fields at all distance scales the status of a physically observable field. The value of this field component is ill-controlled at short distances. Its wild oscillations would effectively produce strong interactions, more precisely, interactions with a negative mass dimension, or in other words: such a theory is nonrenormalizable.

The only mechanism that can protect the ghost components from becoming observable under all circumstances is a complete local gauge invariance. This invariance must be *exact* at small distance scale, and therefore should also survive at large distance scale. The great discovery of Peter Higgs was that local gauge invariance is not at all incompatible with finite masses for the vector particles at large distance scales. All one needs to do is add physical degrees of freedom to the model that can play the role of the needed transverse field components. If these fields behave as scalar fields at short distances they don't give rise to any ghost problem there. At large distances they conspire with the vector field to describe the three independent components of a massive spin 1 particle.

The trick used is that although the local gauge invariance has to be exact, the gauge transformation rule can be more general than in a pure Yang–Mills system. It may also involve the scalar field, even if the energetically preferred value of this scalar field is substantially different from zero. Now if we had been dealing with merely a *global* gauge symmetry this would have been an example of spontaneous symmetry breaking: the vacuum, with its nonvanishing scalar field, is not gauge invariant. It was tempting to refer to the present situation as "spontaneous breakdown of local gauge symmetry", but this would be somewhat inaccurate. In our theories the vacuum does *not* break local gauge invariance. Any state in Hilbert space that fails to be invariant under local gauge transformations is an unphysical state. Strictly speaking, the vacuum is entirely gauge invariant here. The reason why we can do our calculations in practice as if the symmetry were spontaneously broken is that one can choose the gauge condition such that the scalar field points in a (more or less) fixed direction. In view of the above, we prefer not to use the phrase "spontaneous local symmetry breakdown", but rather the phrase "Higgs mechanism" to characterize this realization of the theory where the vector particles are massive.

Space and time are continuous. This is how it has to be in all our theories, because it is the only way known to implement the experimentally established fact that we have exact Lorentz invariance. It is also the reason why we must restrict ourselves to renormalizable quantum field theories for elementary particles. As a consequence we can consider unlimited scale transformations and study the behavior of our theories at all scales (Chapter 3). This behavior is important and turns out to be highly nontrivial. The fundamental physical parameters such as masses and coupling constants undergo an effective change if we study a theory at a different length and time scale, even the ones that had been introduced as being dimensionless. The reason for this is that the renormalization procedure that relates these constants to physically observable particle properties depends explicity on the mass and length scales used.

It was proposed by A. Peterman and E. C. G. Stueckelberg[1], back in 1953, that the freedom to choose one's renormalization subtraction points can be seen as an invariance group for a renormalizable theory. They called this group the "renormalization group". In 1954 Murray Gell-Mann and Francis Low observed that

[1]E. C. G. Stueckelberg and A. Peterman, *Helv. Phys. Acta* **26** (1953) 499.

the optimal choice of the subtraction depends on the energy and length scales at which one studies the system. Consequently it turned out that the most important subgroup of the renormalization group corresponds to the group of scale transformations. Later, Curtis G. Callan and independently Kurt Symanzik derived from this invariance partial differential equations for the amplitudes. The coefficients in these equations depend directly on the subtraction terms for the renormalized interaction parameters.

The subtraction terms depend to some extent on the details of the subtraction scheme used. For gauge field theories Veltman and the author had introduced the so-called dimensional renormalization procedure. It turns out that if the subtraction terms obtained from this procedure are used for the renormalization group equations these equations simplify. Furthermore there is a purely algebraic relation between the dimensional subtraction terms and the original parameters of the theory. This enables us to express the scaling properties of the most general renormalizable theory directly in terms of all interaction constants via an algebraic master equation. In deriving this equation one can make maximal use of gauge invariance. One can extend this master equation in order to derive counter terms for nonrenormalizable theories such as perturbative quantum gravity, but it would be incorrect to relate these terms to the scaling behavior of this theory, because here the canonical dimension of the interaction parameter, Newton's gravitational constant, does not vanish but is that of an inverse mass-squared.

In the 1960's and early 1970's it was thought that all renormalizable field theories scale in such a way that the effective interaction strengths *increase* at smaller distance scales. Indeed, there existed some theorems that suggest a general law here, which were based on unitarity and positivity for the Feynman propagators. I had difficulties understanding these theorems, but only later realized why. I had made some preliminary investigations concerning scaling behavior, and had taken those theories I understood best: the gauge theories. Here the scaling behavior seemed to be quite different. Now that we have the complete algebra for the scaling behavior of all renormalizable theories we know that non-Abelian gauge theories are the *only* renormalizable field theories that may scale in such a way that all interactions at small distances become weak. The reason why they violate the earlier mentioned theorems is that in the renormalized formulation of the propagators the ghost particles play a fundamental role and the positivity arguments are invalid. This new development was of extreme importance because it enabled us to define theories with strong interactions at large distance scales in terms of a rapidly convergent perturbative formulation at small distances.

The new theory for the strong interactions based on this principle was called "quantum chromodynamics". According to this theory all hadronic subatomic particles are built from more elementary constituent particles called "quarks". These quarks are bound together by a non-Abelian gauge force, whose quanta are called "gluons". The crucial assumption was that quarks and gluons behave nearly as free particles as long as they stay close together, but attract each other with strong binding forces if they are far apart. This fits with the renormalization group properties of the system, but the fact that complete separation of the quarks from each other and the gluons is impossible under any circumstance could not be understood from arguments based on perturbative formulations of the theory. Only a nonperturbative approach could possibly explain this.

Quantum chromodynamics is one of the very few renormalizable field theories in four space-time dimensions that one can hope to understand nonperturbatively. But how could one possibly explain the confinement phenomenon?

Permanent quark confinement must have everything to do with gauge invariance. In his early searches for an explanation the author hit upon a feature in gauge theories that at first sight is entirely unrelated to this problem: the existence of magnetic monopoles. Purely magnetically charged particles had been considered before in quantum field theories, notably by Paul A. M. Dirac in 1934. He had derived that, in natural units, the product of the magnetic charge unit g and the electric charge unit e had to be an integer multiple of 2π. This implies that one can never apply perturbative methods to the interactions of magnetic monopoles because they interact super strongly. What was discovered, also independently by Alexander M. Polyakov[2] in Moscow, was that certain extended solutions of the non-Abelian field equations carry a magnetic charge that obeys Dirac's quantization condition. These solutions behave like classical particles in the perturbative limit and the strong interactions among them are classical (i.e. unquantized) interactions.

These monopoles are interesting in their own right, since they may be a feature of certain grand unified schemes for the fundamental particles and may play a role in cosmological theories. But also they may indeed provide for a quark confinement mechanism. Quark confinement is now generally considered to be a consequence of superconductivity of the vacuum state with respect to monopoles. It is related to the Higgs mechanism if one exchanges electric with magnetic forces. Magnetic monopoles were a consequence of the *topological* properties of gauge theories. Are they the only consequences? Algebraic topology turns out to be a very rich subject, and indeed there is much more, as explained in Chapter 4. Magnetic monopoles are stable objects in three dimensions. If one searches for objects stable in one dimension one obtains domain walls. Stability in two dimensions is enjoyed by "vortices". One may now also ask for topologically stable field structures in four dimensions rather than three. In a space-time diagram such structures would show up as a special kind of "events". An example of such a configuration that looked interesting had been considered by Alexander Belavin, Albert Schwartz, Yuri Tyupkin and A. Polyakov.

But what was the *physical* interpretation of such "events"? Would they change and enrich the theory as much as magnetic monopoles? They certainly would, as seen in Chapter 5. Events of this sort could be calculated to be extremely rare in most theories, except when the interactions are strong, or if the temperature is very high. In every respect they represented some kind of "tunneling transition". But tunneling from what to where? At first sight the tunneling was merely from some gauge field configuration to a gauge-rotated field configuration. What was special about that? One answer is: the side effects. The event goes associated with

[2] A. M. Polyakov, *JETP Lett.* **20** (1974) 194.

an unusual transition between energy levels in the Dirac sea. The consequence of this transition is that in most of our standard theories some apparent conservation laws are violated in these events. For the weak interaction theories the violated conservation law is that of baryon number and lepton number. In the strong interaction the result is nonconservation of chiral $U(1)$ charge. There are still some hopes that baryon number violation will be seen in future laboratory experiments (although I do not share this optimism), but chiral charge nonconservation in the strong interactions had already been observed for some time, causing confusion and embarassment among theorists who failed to understand what was going on. The new, "topological" events, which we called "instantons", explain this nonconservation in a satisfactory way. It was finally understood why the strong interactions render the η particle so much heavier than the pions.

The mechanism that keeps quarks permanently bound together in quantum chromodynamics would be a lot easier to understand if we had a soluble version of QCD. Ordinary perturbation expansion, which had served us so well in the previous quantum field theories, was useless here because our question concerns the region of strong couplings. Does there not exist a simplified version of QCD that is exactly soluble, or better even, is the coupling constant expansion the only expansion we can perform, or does there exist some other asymptotic expansion? At first sight the coupling constant g seems to be the only variable available to expand in, but that is not true. There are at least two others.[3] One variable is the dimension D of space-time. In 2 space-time dimensions the theory becomes substantially simpler; however it is not exactly soluble in 2 dimensions, and an expansion in $D-2$ does not seem to be very enlightening. A very interesting expansion parameter is $1/N$, where N is a parameter if we replace the color gauge group $SU(3)$ by $SU(N)$. At first sight the theory does not seem to simplify at all as $N \to \infty$, since there still is an infinite class of highly complicated Feynman graphs.

But it does, in a very special way. If we keep $g^2N = \tilde{g}^2$ fixed we obtain a theory which, again, as expanded in powers of $\tilde{g}^2$ produces an infinite series of Feynman diagrams, and they are too complicated to sum up, even in the limit. But the simplification that does take place is a topological one: all surviving diagrams in the $N \to \infty$ limit are "planar", which means that they can be drawn on a 2-dimensional plane without any crossings of the lines. The two-dimensional structures one obtains this way remind one very much of string theory, such that the strings connect every quark with an antiquark. This is the topic of Chapter 6.

But what is the use of this observation if one still cannot sum the required diagrams? First of all, the series of diagrams *can* be summed in 2 dimensions. In the exact theory for $D = 2$, $N = \infty$, one obtains a beautiful spectrum of hadrons consisting of permanently confined quarks. It would be nice if now the expansion with respect to $D-2$ could be performed, but that is not easy. There are two other reasons for further studying QCD in 4 dimensions in the $N \to \infty$ limit. One is that one may consider the theory in a lightcone gauge. The stringy nature of the interactions is then fairly apparent and one may hope to obtain a sensible description

[3]A third involves the topological parameter θ, connected with instantons.

of hadrons this way.[4] Another has to do with more rigorous mathematical aspects of the theory. One may expect namely that the diagrammatic expansion of the theory in terms of powers of $\tilde{g}^2$ converges better than that of the full theory. If one counts the total *number* of diagrams one finds this to diverge only geometrically with the order, whereas this number diverges factorially in the full theory.

Now if any single diagram would only give a contribution to the amplitudes that is strictly bounded by a universal limit, then one would have an expansion that converges within some circle of convergence in the complex $\tilde{g}$ plane. But we are not that fortunate. The contribution of a single diagram is not bounded. This is because there were infinities that had to be subtracted. For each individual diagram the renormalization procedure does what it is supposed to do, namely produce a finite, hence useful expression. But when it comes to summing all those diagrams up, the result of these subtractions is a new divergence. Fortunately, the ultraviolet divergent diagrams only form a small subset of all diagrams, and maybe there is a different way to get these under control. Just because the total number of diagrams is much better behaved than in the full theory one can try to perform resummation tricks here. What we were able to prove is that, if some Higgs mechanism provides for masses so that also the infrared divergences are tamed, and if furthermore the coupling constant is small enough, then the resummation procedure named after Borel[5] is applicable.

Originally we did this in Euclidean space, but analytic continuation towards Minkowski space does not produce serious new problems. The only drawback is that the proof is only rigorous if the coupling constant $\tilde{g}$ is so small that the theory is utterly trivial. From a physical point of view nothing new is added to what we already knew or suspected from ordinary perturbation theory. One does not obtain the hadron spectrum or even a confinement mechanism this way. It is rather from the more formal, mathematical point of view that this result is interesting, because it shows that this limit can be constructed in all mathematical rigor. A procedure to remove the constraint that there should be masses, or, equivalently, that $\tilde{g}$ should be kept small, is still lacking. Indeed, to solve that problem a better understanding of the vacuum state is needed. Atempts to Borel resum such theories bounce off against the fact that we do not know the vacuum expectation values of an infinite class of composite operators.

Then, in Chapter 7, we finally attack the confinement problem directly. Some details of this mechanism were first exposed when Kenneth G. Wilson[6] produced the $1/g$ expansion of a gauge theory on a lattice. A linearly rising electric potential energy between two quarks then emerges naturally. As indicated earlier, confinement has to do with superconductivity of magnetic monopoles. But it is also directly related to gauge-invariance. It had been argued by V. Gribov[7] that the

[4]There was a recent report of substantial progress in lightcone QCD. See S. J. Brodsky and G. P. Lepage, in *Perturbative Quantum Chromodynamics*, ed. A. H. Müller (World Scientific, Singapore, 1989), p. 93.

[5]Émile Borel, Leçons sur les séries divergentes, Paris, Gauthier-Villars, 1901.

[6]K. G. Wilson, *Phys. Rev.* **D10** (1974) 2445.

[7]V. N. Gribov, *Nucl. Phys.* **B139** (1978) 1.

usual procedure for fixing the gauge freedom in non-Abelian gauge theories is not unambiguous. His claim that this ambiguity may have something to do with quark confinement was originally greeted with skepticism. But it turned out to be true. If one meticulously fixes the gauge in an unambiguous manner one finds a new phenomenon, namely a sea of color-magnetic monopoles and antimonopoles. It is these monopoles that may undergo Bose condensation. Whether Bose condensation actually occurs or not is not something one can establish from topological arguments alone, because the ordinary Higgs mechanism would be impossible to exclude, even if one has no fundamental scalar field at hand; composite scalar fields could also be used for the Higgs mechanism. So one has to look at dynamical properties of the theory, such as asymptotic freedom.

What could be established was a precise description of the confinement mechanism in these terms, as well as alternative modes a theory such as QCD could condense into. Since these alternatives could not be ruled out we have not actually given a "proof" that QCD explains quark confinement. But the logical structure of the confinement mode is now so well understood that no basic mystery seems to be left.

And thus elementary particle physics ran out of mysteries. The problem of uniting special relativity with quantum mechanics has been solved. The solution is called "renormalizable quantum field theory", which in general includes gauge theories. The small distance divergence is in its most essential form only logarithmic, and the solution of that should be looked for by postulating hierarchies of field theories, each hierarchy valid at its own characteristic distance scale.

But this does not solve the biggest mystery of all: why do these theories work? From a mathematical point of view they work because they are the most economical constructions in terms of physical degrees of freedom: we postulated the least amount of structure possible at small distance scales. That postulate is what makes these theories so unique. But why should we postulate optimal economy? And what determines which structures exist at all at small distance scales? How does our chain of hierarchies get started? We all think we know where to look for the answer to such questions: the gravitational force. Gravity adds a new fundamental constant to our physical world, namely Newton's constant. It is incredibly small, indicating that gravity dominates the natural forces at a tremendously tiny distance scale. At this distance scale, called the Planck scale, everything in physics will have to be reconsidered. It is here namely that all our presently known techniques fall short.

Numerous attempts have been made to attack this problem. Basically all we want is unite *general* relativity with quantum mechanics. Now general relativity even without quantum mechanics is a highly complex theory, and the roles played by the concepts "energy" and "time" in general relativity are on the one hand equally crucial as, and on the other hand fundamentally different from the ones they play in quantum mechanics. Well-known are the approaches where a new kind of symmetry is introduced, called "supersymmetry". The resulting "supergravity theory" was potentially interesting because it seemed to be not as divergent as gravity alone. But I never gave this approach by itself much chance of success because it did not address the most fundamental aspect of the difficulty, which is the fact that Newton's constant has the dimensionality of a length squared, so that at distance scales shorter than the Planck length the gravitational force, supersymmetric or not, runs out of control.

The same objection, though with a little more hesitation, can be brought against the "superstring" approach to quantizing gravity. The space-time in which the superstring moves is a continuous space-time, and yet we have a distance scale at which a smooth metric becomes meaningless. On the other hand a flat background metric is usually required at ultrashort distances, even in string theories. It is my conviction that a much more drastic approach is inevitable. Space-time ceases to make sense at distances shorter than the Planck length. Here again I reject a purely mathematical attack, particularly when the math is impressive for its stunning complexity, yet too straightforward to be credible. The point here is also that our problem is not only a mathematical one but more essentially physical as well: what is it precisely that we want to know, and what do we know already?

There is a different, more "physical" way to see what goes wrong at small distances. Suppose we want to describe any physical phenomenon that is localized at a distance scale smaller than the Planck scale. According to an established paradigm in quantum mechanics this system will contain momenta that will be spread over more than one Planck unit of momentum, and its energy will be of the same order of magnitude. But then it is easy to see that this energy would be confined to within its Schwarzschild horizon, which will stretch beyond one Planck radius. Hence gravitational collapse must already have occurred: such a system is a black hole, and its size will be larger than a Planck unit. This proves that confining any system to within a Planck unit is impossible.

Indeed, black holes form a natural barrier. Obviously they must be a central theme in any theory with gravity and quantum mechanics that boasts to be complete. This discussion is conspicuously missing in any superstring theory of quantum gravity, which is why I don't believe these theories can be complete. In my attempts towards a better understanding of this problem I begin with black holes. How can their presence be reconciled with the laws of quantum mechanics? A brilliant discovery had been made before by Stephen Hawking[8]: quantum field theory gives black holes a behavior fundamentally different from unquantized general relativity: they radiate. Upon studying this system further one stumbles upon more surprises. It is a fascinating subject. One can imagine thought experiments that may ultimately lead to the resolution of our most fundamental questions. This is Chapter 8.

One of the weaknesses a scientist may fall victim to later in his career is that he (or she) may begin to ponder about the deeper significance of his theories in a wider context, the direction they are heading to, the expectations they may hold for the more distant future. What would the most natural, the most satisfactory, the most complete results look like? Could there be an *ultimate* theory? Somewhat surprising, perhaps, is that our present insights do indeed suggest that there may

[8]S. W. Hawking, *Commun. Math. Phys.* **43** (1975) 199.

be such a thing as an ultimate description of all physical degrees of freedom and the laws according to which these should evolve. As argued above, space-time itself at the Planck length does not seem to allow for any further subdivision. Could an ultimate physical theory be formulated around that length scale? Our last chapter deals with such questions.

CHAPTER 2
RENORMALIZATION OF GAUGE THEORIES

CHAPTER 2
RENORMALIZATION OF GAUGE THEORIES

Introduction to Gauge Field Theory [2.1]

The pioneering paper of Chen Ning Yang and Robert Mills[9] in 1954 has inspired many physicists in the late 1960's to construct theories for the weak and the strong interactions. In the following papers it is assumed that the reader is somewhat familiar with this fundamental idea, namely to consider field equations that are invariant under symmetry transformations that vary from point to point in space-time. It is basically just a generalization of the Maxwell equations which indeed have such a symmetry:

$$\Psi(\mathbf{x},t) \to \exp ie\Lambda(\mathbf{x},t)\,\Psi(\mathbf{x},t)\,,$$

$$A_\mu(\mathbf{x},t) \to A_\mu(\mathbf{x},t) - \partial_\mu\Lambda(\mathbf{x},t)\,,$$

or a space-time dependent rotation in the complex plane. Yang and Mills replaced these by higher dimensional rotations, which in general are non-Abelian.

It was clear however that the Yang–Mills equations would describe *massless* spin one particles, just as the Maxwell equations, and that these would interact as if they were electrically charged. And it was evident that such particles do not exist in the real world. This leads us to ask two questions:

1. Can one add a mass term to the Yang–Mills equations, such that they become physically more plausible?

2. Since power counting suggests renormalizability, can one renormalize the quantum version of this theory?

As explained in the previous chapter, the answer to both these questions is yes, provided that one invokes what is now known as the Higgs mechanism. Without the Higgs mechanism the power counting argument would fail because the longitudinal components of the vector fields (which decouple in the massless case) have no kinetic part in their propagator and therefore interact nonrenormalizably.

[9] C. N. Yang and R. L. Mills, *Phys. Rev.* **96** (1954) 191.

But this had to be proved. The problem Veltman was studying in the late 1960's was that the contributions of the ghost particles, all in the longitudinal sector, did not seem to add up properly to give a unitary theory. The ghost problem itself had been studied by Richard Feynman for the massive case, and Bryce DeWitt, Ludwig Faddeev and Victor Popov in the massless case. It was not immediately realized that the massless theory is not simply the limit of a massive one where the mass is sent to zero[10] because the transverse component would continue to contribute in loop diagrams: it can be pair-produced.

The papers by Faddeev and Popov,[11] mostly in Russian, were not immediately available to us, but a short paper by them in *Physics Letters*[12], in which they summarized their arguments, made their procedure quite clear to me. Basically the message was that the S-matrix amplitudes, just like all quantum mechanical amplitudes, are functional integral expressions. Like always in integrals, when one performs a symmetry transformation in the integrand, one has to keep track of the Jacobian factors. These are usually just big determinants. I decided to rewrite these determinants as Gaussian integrals, because then one can more easily read off what the Feynman rules for these are. These Feynman rules are just as if there are some extra complex scalar particles, but because of a sign switch we found that that these scalar particles had to be seen as fermions rather than bosons. And then the trick was that one had to combine the Faddeev–Popov prescription with the Higgs–Kibble mechanism. So we were dealing with *two* kinds of ghost fields rather than one.

In papers (2.1) and (2.2) we consider this procedure as given. But we did not wish to trust the details of the functional integral approach, because the way it dealt with the renormalization infinities did not seem to be rational. Instead, we consider the perturbative formulation of the theory. We then *prove* that, order by order in the perturbation expansion and after renormalization, this theory indeed obeys all physically relevant requirements such as unitarity and causality. That it was wise to be this careful soon became clear: there may sometimes be *anomalies*, and in that case the procedure does not work. A theory with anomalies in the local gauge sector is inconsistent.

The first paper is an introduction to the second, illustrating the prescriptions in a simple example. In this example the three gauge bosons $B^{1,2,3}$ all get the same mass M, see eq. (2.11). That the masses are all the same is not a consequence of the local gauge symmetry but of a somewhat hidden *global* symmetry (2.7). This symmetry is explicitly broken by the λ_2 term, so that higher order corrections may cause a relative mass shift. We first see this mass shift in the fermions. The explicit expression for $\mathcal{L}^{break}$ is:

$$\mathcal{L}^{break} = \lambda_2 F(\bar{\chi}_1 \psi_1 + \bar{\psi}_1 \chi_1) + \mathcal{L}^{break,int} ,$$

to be obtained by substituting (2.10) in the last part of (2.5).

[10]H. van Dam and M. Veltman, *Nucl. Phys.* **B21** (1970) 288.

[11]L. D. Faddeev, *Theor. Math. Phys.* **1** (1969) 3 (in Russian); *Theor. Math. Phys.* **1** (1969) 1 (English translation).

[12]L. D. Faddeev and V. N. Popov, *Phys. Lett.* **25B** (1967) 29.

This symmetry structure is very special. It is one of the few systems with computable mass splittings for some of the elementary fields. This was the reason for my interest in this model.

Introduction to DIAGRAMMAR [2.2]

Perturbation expansions with respect to small coupling constants are often looked upon as ugly but necessary tools, and repeatedly physicists attempt to avoid them altogether. It cannot be sufficiently emphasized however that perturbation expansions are an absolutely essential ingredient in quantized gauge field theories. Many of our cherished particle theories can only be *defined* perturbatively. This means that their treatment can only be considered as being mathematically rigorous if we consider all observable quantities as formal power series in terms of a small parameter. This is sometimes referred to as "nonstandard analysis": we extend the field of numbers to the field of power expansions. The strength of this modification of our mathematics lies in the fact that the expansions need not have a finite radius of convergence (in general they don't). If we substitute a finite number, say g^2 for the coupling constants we can trust the series to be meaningful only as long as the next term is a smaller correction than the previous one. Later we will see that in most cases the series will behave as $C^N(g^2)^N N!$. This then implies that observables cannot be computed more accurately than with margins of the order of e^{-1/Cg^2}, in many cases more than good enough!

It should be noted that shifts of vacuum expectation values are not difficult to deal with within this philosophy of nonstandard analysis. Most importantly, one can observe that the perturbation expansion for a quantum field theory is equivalent to dividing the amplitudes up in so-called Feynman diagrams. Feynman diagrams are nothing more than book-keeping devices for the various contributions to the amplitudes. They are central in the next paper, "DIAGRAMMAR", written together with M. Veltman. With this word we intended to indicate that our prescriptions are nothing but the "grammatical rules" for working with diagrams. It also means "diagrams" in Danish.

There are circumstances where one hopes to be able to do better than perturbation expansion. In asymptotically free theories one might be able to replace the perturbation parameter g^2 by the much smaller number at an arbitrarily small distance scale, so that our margin should be tightened significantly, perhaps all the way to zero. Thus a theory such as QCD can perhaps be given a completely rigorous mathematical basis. This has never been proven, but it indeed seems to be plausible. Another point is that one expects phenomena that are themselves of the order of e^{-1/Cg^2}, as explained in Chapter 5. These cannot be handled with Feynman diagrams.

In "DIAGRAMMAR" we first show how to work with the diagrams, then how transformations in the field variables correspond to combinatorial manipulations of the diagrams, exactly as in expressions for functional integrals. If a series of diagrams is geometrically divergent (for instance if we sum the propagator insertions, then *unitarity* tells us exactly that the analytic sum is to be taken. We show how the Faddeev–Popov procedure is translated in diagrammatic language. Most importantly, it is now seen how and why these tricks continue to work when the theory is renormalized. The infinities cancel if the anomalies cancel.

Then dimensional regularization and renormalization are explained. There are three steps to be taken. First we have to define what it means to have a non-integer number of dimensions. Fortunately this is unambiguous at the level of the diagrams — though not at all so "beyond" the perturbation expansion! Secondly we have to deal with integrations that diverge even at non-rational space-time dimension near four; they are easily tamed by analytic continuation from the regions where the integrals do converge. In practice one does this by partial integrations. Thirdly we observe that precisely at *integer* dimensions some infinities are not tamed by partial integrations: the logarithmic ones. These show up as powers of $1/\epsilon$, where $\epsilon = 4 - n$, n is the number of space-time dimensions. They have to be cancelled out by inserting the proper counterterms into the Lagrangian. One then ends up with finite, physically meaningful expressions.

In Section 11 of "DIAGRAMMAR" a transformation is described that had been introduced before as "Bell–Treiman transformation", by my co-author M. Veltman. The name is something of a joke. There is no reference to either Bell or Treiman. It would have been more appropriate to call them "Veltman transformations".

The Slavnov–Taylor identities play a crucial role in the renormalization procedure of gauge theories. They get the attention they deserve in DIAGRAMMAR. The proofs here are as they were first derived. Nowadays a more elegant method exists: the Becchi–Rouet–Stora–Tyupkin quantization procedure. These authors observed that our identities follow from a global symmetry.

Now I had tried such symmetries myself long before DIAGRAMMAR was written, but without success. The crucial, and brilliant, ingredient invented by BRST was that the symmetry generators had to be anticommuting numbers.

Introduction to Gauge Theories with Unified Weak, Electromagnetic and Strong Interactions [2.3]

The last paper in this chapter needs little further introduction. It displays the enormous progress made in just a few years of renormalizable gauge theory. The J/ψ particle, the last crucial ingredient needed to render credibility to what was to become the standard model, had just been found. I was a little too optimistic in applying asymptotic freedom to understand the J/ψ energy levels, hence my underestimation of the splitting between the ortho and the para levels in Table 1. But these levels, at that time not yet seen, would soon be discovered and everything fell into place. Of course most of the more exotic models speculated about in this paper were ruled out during the years that followed. What remained was to be called the "Standard Model". The solution to the eta problems, Section 10 was basically correct, but would be understood better in 1975, with the discovery of the instanton effects, see Chapter 5.

CHAPTER 2.1

GAUGE FIELD THEORY*

G. 't Hooft
CERN — Geneva
On leave from the University of Utrecht, Netherlands

1. INTRODUCTION

There are several possible approaches to quantum field theory. One may start with a classical system of fields, interacting through non-linear equations of motion which are subsequently "quantized". Alternatively, one could take the physically observed particles as a starting point; then define a Hilbert space, local operator fields, and an interaction Hamiltonian. More ambitious, perhaps, is the functional integral approach, which has the advantage of being obviously Lorentz covariant.

All these approaches have one unpleasant and one pleasant feature in common. The unpleasant one is that in deriving the S-matrix for the theory, one encounters infinities of different types. In order to get rid of these, one has to invoke a rather ad hoc "renormalization procedure", thus changing and undermining the theory halfway. The pleasant feature, on the other hand, is that one always ends up with a simple calculus for the S matrix: the Feynman rules. Few physicists object nowadays to the idea that these Feynman diagrams contain more truth than the underlying formalism, and it seems only rational to abandon the aforementioned principles and use the diagrammatic rules as a starting point.

It is this diagrammatic approach to quantum field theory which we wish to advertise. The short-circuiting has several advantages. Besides the fact that it implies a considerable simplification, in particular in the case of gauge theories, one can simply superimpose the renormalization prescriptions on the Feynman rules. As for unitarity and causality, the situation has now been reversed: we shall have to investigate under which conditions these Feynman rules describe a unitary and causal theory. Within such a scheme many more or less doubtful or complicated theorems from the other approaches can be proved completely rigorously.

Clearly, Feynman diagrams merely describe an asymptotic expansion of a theory for the coupling constants going to zero, and strict ly speaking, nothing is known about the theory with finite coupling constants. But the other approaches are not better in this respect, if it comes to calculations of physical effects. A really rigorous formulation of quantum field theory with finite interactions has not yet been given and we are of the opinion that attempts at such a for-

Proceedings of the Adriatic Summer Meeting on Particle Physics,
Edited by M. Martinis *et al.* (North Holland, 1973).

mulation can only succeed if the perturbation expansion, formulated in the simplest possible way (Feynman diagrams) is well understood.

In these notes, which must be considered as an introduction to the CERN report called "DIAGRAMMAR" |1|, we shall outline the basic steps that have to be taken in order to formulate and understand a gauge field theory. We shall illustrate our arguments with a simple example of a non-Abelian gauge model (which is not realistic physically).

In Section 2 we give a review of the construction process of a model in general, showing the Brout-Englert-Higgs-Kibble phenomenon.

In Section 3 the quantization problem is formulated and the essential steps in the proof of renormalizability are indicated.

2. CONSTRUCTION OF A MODEL

In the construction of a model one must try to combine the experimental observations with the theoretical requirements. There is of course no logical prescription how to do that, and therefore we shall here only consider the theoretical principles |2| and leave it to the reader to alter our little example in such a way as to obtain finally physically more interesting results.

a) First we choose the GAUGE GROUP. This is a group of internal symmetry transformations that depend on space and time |3|. In our example this group will be SU(2)×U(1) (at each space-time point). Let us denote an infinitesimal gauge transformation as e^T, which is generated by an infinitesimal function of space-time, $\Lambda^a(x)$, a = 1,2,3,0. The condition that the infinitesimal gauge transformations generate a group is

$$e^{T_{(1)}}e^{T_{(2)}} = e^{T_{(2)}}e^{T_{(1)}}e^{[T_{(1)},T_{(2)}]+O(T^2_{(1)})+O(T^2_{(2)})} , \tag{2.1}$$

where $[T_{(1)}, T_{(2)}]$ is again an infinitesimal gauge transformation, generated by

$$\Lambda^a_{(3)}(x) = f_{abc}\Lambda^b_{(1)}(x)\Lambda^c_{(2)}(x) . \tag{2.2}$$

Here the indices between the parentheses denote different choices for Λ, and f_{abc} are structure constants of the group. In our example,

$$\begin{aligned} f_{abc} &= \varepsilon_{abc} \quad \text{for } a,b,c \neq 0 \\ &= 0 \quad \text{otherwise.} \end{aligned}$$

(Often we shall write $g\Lambda^a$ instead of Λ^a, where g is some coupling constant.)

By choosing the gauge group we also fix the set of vector particles: one for each generator T. In our example:

$$B_\mu^a(x), \qquad a = 1,2,3 \quad \text{and} \quad A_\mu(x) \ .$$

b) Now we choose the other fields. They all must be representations of the gauge group. Example: A Bose field ϕ_i with "isospin" $\frac{1}{2}$ (a two-component representation of SU(2)) and "charge" 0 (a scalar for U(1) transformations). A Fermi field ψ_i with "isospin" $\frac{1}{2}$ and "charge" 1; a Fermi field X with "isospin" 0 and "charge" 1.

c) The GAUGE TRANSFORMATION LAW must satisfy the commutation rules (2.1), (2.2). There is one way to satisfy the requirements. In our example the infinitesimal transformations are

$$B_\mu^{a\prime} = B_\mu^a - \partial_\mu \Lambda^a + g_1 \varepsilon_{abc} \Lambda^b B_\mu^c \ ;$$

$$\phi' = \phi - \frac{1}{2} i g_1 \sum_{a=1}^{3} \tau^a \Lambda^a \phi \ ,$$

$$\psi' = \psi - \frac{1}{2} i g_1 \sum_{a=1}^{3} \tau^a \Lambda^a \psi + i g_2 \Lambda^o \psi \ ,$$

$$X' = X + i g_2 \Lambda^o X \ . \tag{2.3}$$

d) Next we write down a Lagrange density, which we shall call the SYMMETRIC LAGRANGIAN, $\mathcal{L}^{\text{inv}}(x)$. It is a function of the fields and their space-time derivatives at the point x. The corresponding Lagrange equations

$$\frac{\delta \mathcal{L}}{\delta A_i(x)} - \partial_\mu \frac{\delta \mathcal{L}}{\delta\, \partial_\mu A_i(x)} = 0 \ , \tag{2.4}$$

where $A_i(x)$ is any of the fields B_μ^a, A_μ, ϕ, ψ_i or X will describe to a first approximation the propagation of these fields, the quanta of which are the physical particles. The fact that we have a quantum theory will necessitate "higher order" corrections (the loops in the Feynman graphs).

This Lagrangian must be <u>invariant</u> under the gauge transformations (2.3). In our example we take

$$\mathcal{L}^{\text{inv}} = -\frac{1}{4} \sum_{a=1}^{3} G_{\mu\nu}^a G_{\mu\nu}^a - \frac{1}{4} F_{\mu\nu} F_{\mu\nu} - \frac{1}{2} (D_\mu \phi)^* \cdot D_\mu \phi - \frac{1}{2} \mu_o^2 \phi^* \cdot \phi$$

$$- \frac{1}{2} \lambda_1 (\phi^* \cdot \phi)^2 - \overline{X}(\gamma_\mu D_\mu + m_X) X - \overline{\psi} \cdot (\gamma_\mu D_\mu + m_\psi) \psi$$

$$+ \lambda_2 (\overline{X} \phi^* \cdot \psi + \overline{\psi} \cdot \phi\, X) \ , \tag{2.5}$$

in which

$$G_{\mu\nu}^a = \partial_\mu B_\nu^a - \partial_\nu B_\mu^a + g_1 \varepsilon_{abc} B_\mu^b B_\nu^c \ ,$$

$$F_{\mu\nu} = \partial_\mu A_\nu - \partial_\nu A_\mu ,$$

$$D_\mu \phi = \partial_\mu \phi - \frac{1}{2} i g_1 \sum_{a=1}^{3} \tau^a B_\mu^a \phi ,$$

$$D_\mu \psi = \partial_\mu \psi - \frac{1}{2} i g_1 \sum_{a=1}^{3} \tau^a B_\mu^a \psi + i g_2 A_\mu \psi ,$$

$$D_\mu X = \partial_\mu X + i g_2 A_\mu X . \qquad (2.6)$$

All terms in the Lagrangian must have dimension 4 or less. Dimension is counted as follows: a derivative has dimension one, a Bose field has dimension one and a Fermi field has dimension $\frac{3}{2}$. Given that Bose propagators behave as $\frac{1}{k^2}$ and Fermi propagators as $\frac{1}{k}$ for $|k| \to \infty$ this requirement will enable us to estimate the maximal "overall" divergence of an integral in a diagram, if only the external lines are known. For instance, the overall divergence of the diagram in fig. 1 is linear, independent of its internal topology (in fig. 1 there are 11 Fermion and 8 Boson propagators,

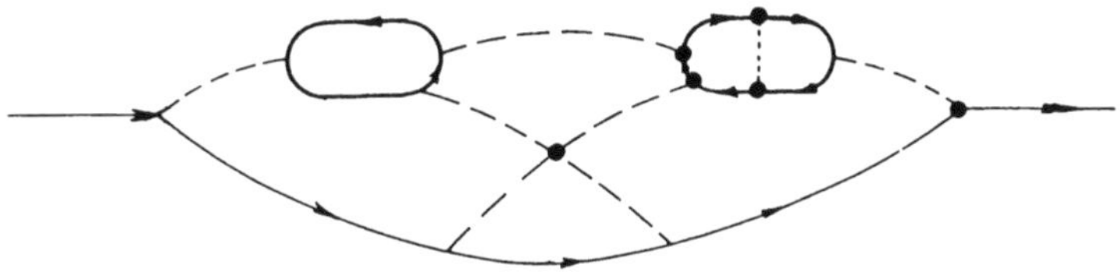

———— fermion ($\sim \frac{1}{k}$) - - - - boson ($\sim \frac{1}{k^2}$)

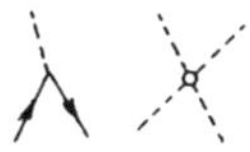
vertices from $\bar{\psi}\phi\chi$ and $(\phi^*\phi)^2$, respectively, (for instance).

Fig. 1

and 7 loops for integration, $7 \cdot 4 - 11 \cdot 1 - 8 \cdot 2 = 1$.)

e) Examine the GLOBAL SYMMETRIES that is, all symmetries that are independent of space time. In the example: Parity (P), Charge conjugation (C), Time reversal (T), and, if $\lambda_2 \to 0$, an unexpected SU(2) symmetry:

$$\phi_i' = \phi_i + \Lambda_1 \varepsilon_{ij} \phi_j^* + \Lambda_2 i \varepsilon_{ij} \phi_j^* + \Lambda_3 i \phi_i , \qquad (2.7)$$

Λ_i space-time independent.

The importance of the global symmetries is that all possible interaction terms that do not violate these symmetries nor the local gauge symmetry, must be present in the Lagrangian (2.5). This is why the term with λ_1 must be present.

If μ_o^2, m_χ and $m_\psi>0$, then the model contains massive Fermions and ϕ particles. But note that our Lagrangian does not contain a mass term for the vector particles, simply because no gauge invariant mass term exists. So these spin-one particles are massless and they interact with each other. As a consequence there are huge infrared divergencies of a type that cannot be cured as in quantum electrodynamics. It is expected that the eventual solution of the quantized equations will be drastically different from the classical ones (2.4) and completely governed by these infrared effects (conjecture: strongly interacting Regge particles), but nobody knows how to solve this problem.†

Nevertheless, the model is interesting, and that is because we can take the variable μ_o^2 to be negative. If we then assume for a moment that the fields in $\mathcal{L}$ are classical, then the Hamilton density $\mathcal{H}$, obtained from $\mathcal{L}$, contains a term

$$\frac{1}{2}\mu_o^2\phi^*\cdot\phi + \frac{1}{2}\lambda_1(\phi^*\cdot\phi)^2 , \qquad \mu_o^2 < 0 . \tag{2.8}$$

The energy is minimal not if $\phi=0$, but if

$$|\phi| \equiv F = \left(\frac{|\mu_o^2|}{\lambda}\right)^{1/2}. \tag{2.9}$$

Because of gauge invariance, we can always rotate until ϕ_i is parallel to the spinor (1,0). Small fluctuations of the various fields near this equilibrium state are now described by massive equations of motion, as we shall see in f.

We can now proceed in two ways. Either we

I) first quantize the "symmetric" theory and then construct the new vacuum, corresponding to this "equilibrium state", or we

II) first define new fields ϕ, with the equilibrium value (2.9) subtracted, and then quantize this "asymmetric" theory.

We choose possibility no. II. Therefore:

f) Shift those fields which have a vacuum expectation value |4|. In the example

$$\phi_i = F\binom{1}{0} + \frac{1}{\sqrt{2}}\begin{pmatrix} Z + iY_3 \\ -Y_2 + iY_1 \end{pmatrix} . \tag{2.10}$$

Here the number F satisfies, up to possible quantum corrections, eq.

(2.9), and Z and Y_a are the new field variables (real). The Lagrangian in terms of the new fields is the NON-SYMMETRIC LAGRANGIAN. In our example:

$$\begin{aligned}\mathcal{L}^{\text{inv}} = & -\tfrac{1}{4} G^a_{\mu\nu}G^a_{\mu\nu} - \tfrac{1}{2} M^2 (B^a_\mu)^2 - \tfrac{1}{4} F_{\mu\nu}F_{\mu\nu} \\ & -\tfrac{1}{2}(\partial_\mu Z)^2 - \tfrac{1}{2} M_Z^2 Z^2 - \tfrac{1}{2}(\partial_\mu Y_a)^2 \\ & -\bar{\psi}\cdot(\gamma_\mu D_\mu + m_\psi)\psi - \bar{X}(\gamma_\mu D_\mu + m_\chi)X \\ & + \mathcal{L}^{\text{int}} + \mathcal{L}^{\text{break}} - \beta\left[\tfrac{1}{2}(Z^2+Y_a^2) + \frac{2M}{g_1} Z\right] - MY_a \partial_\mu B^a_\mu ,\end{aligned} \tag{2.11}$$

where

$$\begin{aligned} M^2 &= \tfrac{1}{2} g_1^2 F^2 \ ; \quad M_Z^2 = 2\lambda_1 F^2 \quad \text{and} \\ \beta &= \mu_o^2 + \lambda_1 F^2 . \end{aligned} \tag{2.12}$$

The interaction term is rather complicated now and not so relevant for the discussion. In our example it is

$$\begin{aligned}\mathcal{L}^{\text{int}} = & -\tfrac{1}{2} g_1 \partial_\mu Y_a \varepsilon_{abc} B^b_\mu Y_c + \tfrac{1}{2} g B^a_\mu (Z\partial_\mu Y_a - Y_a \partial_\mu Z) \\ & -\tfrac{1}{8} g^2 B_\mu^2 (Z^2 + Y_a^2) - \tfrac{1}{2} g M B_\mu^2 Z \\ & -\frac{M_Z^2}{4M} g Z(Z^2 + Y_a^2) - \frac{M_Z^2}{32 M^2}(Z^2 + Y_a^2)^2 .\end{aligned} \tag{2.13}$$

The term $\mathcal{L}^{\text{break}}$ arises from the λ_2 term and breaks the global SU(2) symmetry (the Fermion masses are also split).

Note that there is a subgroup of $SU(2)_{\text{local}} \times SU(2)_{\text{global}}$ that leaves the spinor $\binom{1}{0}$ in (2.10) invariant. This is the new symmetry group, broken by the λ_2 term.

As we do not insist on eq. (2.9) for F, our shift is free, and we get a new "free parameter" β in (2.11). In general, we shall require that all diagrams where the Z particle vanishes into the vacuum cancel (fig. 2). In lowest order this corresponds to $\beta=0$.

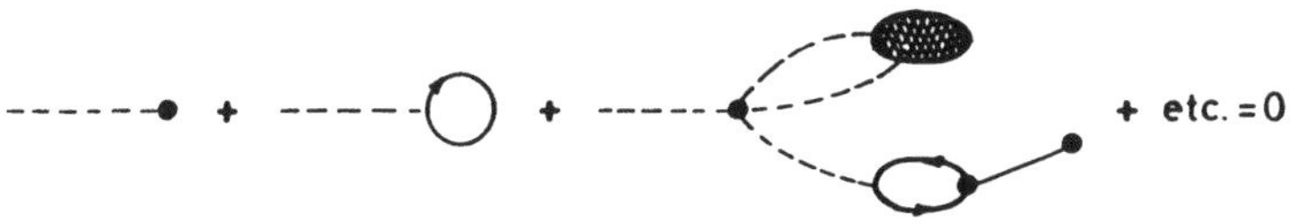

Fig. 2
The condition for the parameter β.

g) The obtained Lagrangian does not at all look invariant under the local gauge transformations, in particular the vector-mass term and the λ_2 term. (Of course, the U(1) invariance here is obvious). But we can rewrite the original gauge transformation law (2.3) in terms of the new fields. We get the NEW GAUGE TRANSFORMATION LAW:

$$Y_a' = Y_a + \frac{1}{2} g_1 \varepsilon_{abc} \Lambda^b Y_c - \frac{1}{2} g_1 \Lambda^a Z - M\Lambda^a ,$$

$$Z' = Z + \frac{1}{2} g_1 \Lambda^a Y_a . \tag{2.14}$$

No change in:

$$B_\mu^{a\prime} = B_\mu^a - \partial_\mu \Lambda^a + g_1 \varepsilon_{abc} \Lambda^b B_\mu^c .$$

In the general case,

$$A_i' = A_i + \hat{t}_i^a \Lambda^a + g\, s_{ij}^a \Lambda^a A_j , \tag{2.15}$$

where $\hat{t}_i^a$ are either coefficients with the dimension of a mass, or the derivative ∂_μ for the vector fields.

Under these gauge transformations the Lagrangian is <u>still invariant</u>. Its nonsymmetric form is merely a consequence of the practically arbitrary, nonsymmetric new coefficients t_i^a.

Although we have much more bookkeeping to do now, the general principle, the invariance under a gauge transformation, is not changed by the shift. Only the transformation law is a little bit different.

Note that the new t coefficients are of <u>zeroth</u> order in g_1.

h) The model contains PHYSICAL PARTICLES and GHOSTS. Ghosts correspond to those fields that can be turned away independently by gauge transformations. To determine which fields are ghosts it is sufficient to consider only the lowest-order parts of the gauge transformation, described by the t coefficients. In our example one can take $\partial_\mu A_\mu = 0$, and then either turn the Y_a fields away or one of the spin-components of the vector fields W_μ^a. The first choice is obviously the most convenient one. Compare (2.14); we choose $\Lambda^a = \frac{1}{M} Y_a +$ + perturbation expansion in g_1. Our model evidently contains three massive spin-one particles B_μ^a, a massless photon A_μ (with only two possible helicities), one massive spinless particle Z, and three massive Fermions X and ψ_i (of which the latter form a doublet).

i) It is instructive to consider first the classical field equations. Just as in electrodynamics one must choose a gauge condition in order to remove the gauge freedom. In general, we can write this condition in the form

$$C^a(x) = 0 \, , \tag{2.16}$$

where $C^a(x)$ is some function of the fields that must not be gauge invariant. The number of components of C^a must be identical to the number of generators in the gauge group, here four. In subsection h) we would have

$$\begin{aligned} C^o &= \partial_\mu A_\mu \\ C^a &= Y_a \end{aligned} \qquad \text{or} \qquad \begin{aligned} C^o &= \partial_\mu A_\mu \\ C^a &= \partial_\mu B_\mu \end{aligned}$$

One may impose such a gauge condition in an elegant way by replacing the original Lagrangian $\mathcal{L}^{inv}$ by

$$\mathcal{L} = \mathcal{L}^{inv} - \frac{1}{2} (C^a)^2 \, . \tag{2.17}$$

The equations of motion corresponding to this Lagrangian are fulfilled if any small variation of the fields does not change $\int \mathcal{L}\, d^4x$. Let us choose as a small variation an infinitesimal gauge transformation, described by $\Lambda^b(x)$:

$$\mathcal{L} \to \mathcal{L} - C^a \cdot \frac{\delta C^a}{\delta \Lambda^b} \cdot \Lambda^b \, , \tag{2.18}$$

where $\frac{\delta C^a}{\delta \Lambda^a}$ stands formally for the variation of C^a under a gauge trans formation, and must be unequal to zero. Hence C^a must be zero. Applying other variations to the Lagrangian, we get back the original equations of motion, now with the gauge condition $C^a = 0$.

In our example it is convenient to choose

$$\begin{aligned} C^a_{(\lambda)}(x) &= \frac{1}{\lambda} \partial_\mu B^a_\mu - \lambda M Y_a \, , \\ C^o_{(\lambda)}(x) &= \partial_\mu A_\mu , \end{aligned} \tag{2.19}$$

where λ is a free parameter.

Consequently, after a gauge transformation,

$$C^{a\prime}_{(\lambda)} = C^a_{(\lambda)} + \frac{\delta C^a_{(\lambda)}}{\delta \Lambda^b} \Lambda^b \, ,$$

more explicitly,

$$\begin{aligned} C^{a\prime}_{(\lambda)} &= C^a_{(\lambda)} - \frac{1}{\lambda} \partial^2 \Lambda^a + \frac{g_1}{\lambda} \varepsilon_{abc} \partial_\mu (\Lambda^b B^c_\mu) \\ &\quad + \lambda M^2 \Lambda^a - \lambda M (\tfrac{1}{2} g_1 \varepsilon_{abc} \Lambda^b Y_c - \tfrac{1}{2} g_1 \Lambda^a Z) \, ; \\ C^{o\prime}_{(\lambda)} &= C^o_{(\lambda)} - \partial^2 \Lambda^o \, . \end{aligned} \tag{2.20}$$

The reason why this particular choice for $C^a(x)$ is convenient is that the bilinear interaction

$$-M\, Y_a \partial_\mu B^a_\mu$$

in $\mathcal{L}^{inv}$, eq. (2.11), is cancelled, so that the B_μ and Y propagators are decoupled. The importance of the gauge parameter λ will be made clear in the next section.

The model of our example becomes more interesting if we assign a U(1) charge not to X but to the field ϕ. We then get something closely resembling the Weinberg model |5|. Its gauge structure is a bit more complicated.

3. THE MATHEMATICAL STRUCTURE OF GAUGE FIELD THEORY

The models constructed along the lines given in the previous section are renormalizable (under one additional condition, see ii) in this section). For the proof of this statement several steps must be made. We indicate those here without going into any of the technical details (which can all be found in "DIAGRAMMAR"). We give the steps in an order which is logical in the mathematical sense. This is not the order in which the theorems have been derived, but the order in which a final proof should be given. Historically, the Feynman rules were first found by functional integral methods but the proof was heuristic and not very rigorous. As our first step we shall

i) postulate the Feynman rules. These are always uniquely defined by a Lagrangian. As in the classical case, one must choose a gauge function $C^a_{(\lambda)}(x)$, for which we shall again take (2.19) in our example. The Lagrangian however is not the one of the previous section, but

$$\mathcal{L} = \mathcal{L}^{inv} - \frac{1}{2}(C^a)^2 + \mathcal{L}^{F.-P.} \,, \tag{3.1}$$

where $\mathcal{L}^{F.-P.}$ describes a new "ghost particle", called Faddeev-Popov ghost, occurring only in closed loops, and it depends on the choice of $C^a(x)$:

$$\mathcal{L}^{F.-P.} = \phi^{*a}(x)\, \frac{\delta C^a(x)}{\delta \Lambda^b(y)}\, \phi^b(y) \,, \tag{3.2}$$

where ϕ^a is a new complex field. In our example,

$$\mathcal{L}^{F.-P.} = \phi^{*a}(-\frac{1}{\lambda}\partial^2 + \lambda M^2)\phi^a \tag{3.3}$$

$$+ \phi^{*a}\left[\frac{g_1}{\lambda}\varepsilon_{abc}\,\partial_\mu(\phi^b B^c_\mu) - \lambda M(\frac{1}{2} g_1 \varepsilon_{abc}\phi^b Y_c - \frac{1}{2} g_1 \phi^a Z)\right] - \phi^{*o}\partial^2\phi^o .$$

The ϕ particle may not occur among the external lines and because the vertices only connect two ϕ lines, it only occurs in single loops. A further prescription is that there must be one more minus sign for each ϕ loop ("wrong statistics").

Note that the ϕ corresponding to the Abelian group U(1) is a free particle and therefore drops out. The mass of the SU(2)-ϕ depends on the gauge parameter λ.

ii) Regularize the theory, that is, we must modify the theory slightly in terms of a small parameter, say ε, such that all divergent integrals become finite and the physical situation is attained in the limit $\varepsilon \to 0$. The modification must be as gentle as possible and should not destroy gauge invariance so that we can still use all our theorems. The most elegant method |7| is the dimensional regularization procedure, in which $\varepsilon = 4-d$, d is the "number of space-time dimensions". What we mean by this can be formulated completely rigorously as long as we confine ourselves to diagrams (but that we have already decided). For some theories, however, containing $\gamma^5 = \gamma^1\gamma^2\gamma^3\gamma^4$, or the tensor $\varepsilon_{\alpha\beta\gamma\delta}$, the extension to arbitrary dimensions cannot be made, and the method does not work. Indeed, such theories suffer, more often than not, from the so-called "Bell-Jackiw" anomalies |6|. They are not renormalizable.

iii) Now that all Green's functions are finite, we prove that the S matrix is independent of the choice of $C^a(x)$. For this we must consider non-local field transformations ("canonical transformations") and prove the Slavnov identities. These identities are essential and from them the gauge independence of the S matrix follows. It is in the proof of the Slavnov identities where the "group property", eqs. (2.1) and (2.2), enters.

iv) Find a set of gauge functions $C^a_\lambda(x)$, such that

a) the theory is "renormalizable by power counting" for all $\lambda < \infty$ (compare subsection 2d and fig. 1);

b) the S matrix is unitary in the space of states with energy*) $E < \lambda$. This can be established by means of the "cutting rules";

c) the gauge functions must stay Lorentz-invariant.

---

*) Note that the ghosts in our example all have mass λM.

(One could also require renormalizability and complete unitarity but give up Lorentz invariance for a while; this seems not very practical.)

Now we must consider the limit where the regulator parameter ε approaches its physical value zero.

v) Renormalize, according to some well-defined prescription. The most secure prescription is to add "local" counter terms to the Lagrangian that cancel one by one the divergencies in all possible diagrams. In the dimensional procedure they have the form of ordinary terms but with coefficients $\frac{g^2}{\varepsilon}$, $\frac{g^4}{\varepsilon^2}$, etc. It must be proved that only local counter terms are sufficient to remove all divergencies, otherwise the cutting rules, e.g. unitarity are violated. Here we make use of the causality-dispersion relations (also derived from cutting rules).

vi) Evidently, we altered the Lagrangian. Does this not spoil gauge invariance? We show that the new, "renormalized" Lagrangian is invariant under new, "renormalized" gauge transformation laws. We must convince ourselves that the gauge group (in our example $SU(2)\times U(1)$ is not changed into another one (for example, $U(1)\times U(1)\times U(1)\times U(1)$). In fact, the new laws are equivalent to the old ones for "renormalized" fields. It then follows that the theory is "multiplicitatively" renormalizable. We could only prove this if the gauge function $C^a(x)$ has been chosen without bilinear field combinations, such as

$$C^a(x) = \partial_\mu B^a_\mu + \alpha (B^a_\mu)^2 .$$

vii) Finally show that not only the regularized but also the renormalized S matrix is independent of the choice of $C^a(x)$, apart from the obvious fact that a variation of $C^a(x)$ may have to be accompanied by a slight change of the original variables g_1, M, etc. It is this latter remark which makes this point far from easy to deal with, but it follows from the "multiplicative" renormalizability in the case of linear C.

REFERENCES

|1| G. 't Hooft and M. Veltman, "DIAGRAMMAR", CERN 73-9 (1973), Chapter 2.2 of this book.

|2| See also: B.W. Lee, Phys. Rev. D5 (1972) 823;
B.W. Lee and J. Zinn Justin, Phys. Rev. D5 (1972) 3121, 3137, 3155.

|3| C.N. Yang and R.L. Mills, Phys. Rev. 96 (1954) 191.

|4| F. Englert and R. Brout, Phys. Rev. Letters 13 (1964) 321;
P.W. Higgs, Phys. Letters 12 (1964) 132; Phys. Rev. Letters 13 (1964) 508; Phys. Rev. 145 (1966) 1156;
G.S. Guralnik, C.R. Hagen and T.W.B. Kibble, Phys. Rev. Letters 13 (1964) 585;
T.W.B. Kibble, Phys. Rev. 155 (1967) 627.

|5| S. Weinberg, Phys. Rev. Letters 19 (1967) 1264.

|6| J.S. Bell and R. Jackiw, Nuovo Cimento 60A (1969) 47.

|7| See however also the approach of A. Slavnov, Theor. and Math. Phys. 13 (1972) 174.

*Reprinted from *Proceedings of the Adriatic Summer Meeting on Particle Physics*, Rovinj, Yugoslavia, September 23 - October 5, 1973.

†Note added: This paper was written in 1973, before "asymptotic freedom" led to the general acceptance of QCD. The conjecture, which refers to gauge theories without Higgs mechanism, is now generally believed to be correct.

CHAPTER 2.2

DIAGRAMMAR

G. 't Hooft and M. Veltman
CERN — European Organization for Nuclear Research — Geneva

Reprint of CERN Yellow Report 73-9 :
Diagrammar by G. 't Hooft and M. Veltman
with the kind permission of CERN

1. INTRODUCTION

With the advent of gauge theories it became necessary to reconsider many well-established ideas in quantum field theories. The canonical formalism, formerly regarded as the most conventional and rigorous approach, has now been abandoned by many authors. The path-integral concept cannot replace the canonical formalism in defining a theory, since path integrals in four dimensions are meaningless without additional and rather ad hoc renormalization prescriptions.

Whatever approach is used, the result is always that the S-matrix is expressed in terms of a certain set of Feynman diagrams. Few physicists object nowadays to the idea that diagrams contain more truth than the underlying formalism, and it seems only rational to take the final step and abandon operator formalism and path integrals as instruments of analysis.

Yet it would be very shortsighted to turn away completely from these methods. Many useful relations have been derived, and many more may be in the future. What must be done is to put them on a solid footing. The situation must be reversed: diagrams form the basis from which everything must be derived. They define the operational rules, and tell us when to worry about Schwinger terms, subtractions, and whatever other mythological objects need to be introduced.

The development of gauge theories owes much to path integrals and it is tempting to attach more than a heuristic value to path integral derivations. Although we do not rely on path integrals in this paper, one may think of expanding the exponent of the interaction Lagrangian in a Taylor series, so that the algebra of the Gaussian integrals becomes exactly identical to the scheme of manipulations with Feynman diagrams. That would leave us with the problems of giving the correct $i\epsilon$ prescription in the propagators, and to find a decent renormalization scheme.

There is another aspect that needs emphasis. From the outset the canonical operator formalism is not a pertubation theory, while diagrams certainly are perturbative objects. Using diagrams as a starting point seems therefore to be a capitulation in the struggle to go beyond perturbation theory. It is unthinkable to accept as a final goal a perturbation theory, and it is not our purpose to forward such a notion. On the contrary, it becomes more and more clear that perturbation theory is a very useful device to discover equations and properties that may hold true even if the perturbation expansion fails. There are already several examples of this mechanism: on the simplest level there is for instance the treatment of unstable particles, while if it comes to unfathomed depths the Callan-Symanzik equation may be quoted. All such treatments have in common that global properties are established for diagrams and then extrapolated beyond perturbation theory. Global properties are those that hold in arbitrary order of perturbation theory for the grand total of all diagrams entering at any given order. It is here that very naturally the concept of the global diagram enters: it is for a given order of perturbation theory, for a given number of external lines the sum of all contributing diagrams. This object, very often presented as a blob, an empty circle, in the following pages, is supposed to have a significance beyond perturbation theory. Practically all equations of the canonical formalism can be rewritten in terms of such global diagrams, thereby opening up the arsenal of the canonical formalism for this approach.

A further deficiency is related to the divergencies of the perturbation series. Traditionally it was possible to make the theory finite within the context of the canonical formalism. For instance quantum electrodynamics can be made finite by means of Pauli-Villars regulator fields, representing heavy particles with wrong metric or wrong statistics. Judicial choice of masses and coupling constants makes everything finite and gauge invariant and turns the canonical formalism into a reasonably well-behaving machine, free of objects such as $\delta(0)$, to name one. Unfortunately this is not the situation in the case of gauge theories. There the most suitable regulator method, the dimensional regularization scheme, is defined exclusively for diagrams, and up to now nobody has seen a way to introduce a dimensional canonical formalism or path integral. The very concept of a field, and the notion of a Hilbert space are too rigid to allow such generalizations.

The treatment outlined in the following pages is not supposed to be complete, but rather meant as a first, more or less pedagogical attempt to implement the above point of view featuring global diagrams as primary objects. The most important properties of the canonical as well as path integral formalism are rederived: unitarity, causality, Faddeev-Popov determinants, etc. The starting point is always a set of Feynamn rules succinctly given by means of a Lagrangian. No derivation of these rules is given: corresponding to any Lagrangian (with very few limitations concerning its form) the rules are simply defined.

Subsequently, Green's functions, a Hilbert space and an S-matrix are defined in terms of diagrams. Next we examine properties like unitarity and causality of the resulting theory. The basic tool for that are the cutting equations derived in the text. The use of these equations relates very closely to the classical work of Bogoliubov, and Bogoliubov's definition of causality is seen to hold. The equations remain true within the framework of the continuous dimension method: renormalization can therefore be treated à la Bogoliubov.

Of course, the cutting equations will tell us in general that the theory is not unitary, unless the Lagrangian from which we started satisfies certain relations. In a gauge theory moreover, the S-matrix is only unitary in a "physical" Hilbert space, which is a subspace of the original Hilbert space (the one that was suggested by the form of the Lagrangian).

To illustrate in detail the complications of gauge theory we have turned to good old quantum electrodynamics. Even if this theory lacks some of the complications that may arise in the general case it turns out to be sufficiently structured to show how everything works.

The metric used throughout the paper is

$$x_\mu = (\vec{x},\ ict), \quad k^2 = \vec{k}^2 + k_4^2, \quad k_4 = ik_0 = \frac{iE}{c}.$$

The factors i in the fourth components are only there for ease of notation, and should not be reversed when taking the complex conjugate of a four-vector

$$j_\mu = (\vec{j},\ ij_0),$$

$$j_\mu^* = (\vec{j}^*,\ ij_0^*) = (\vec{j}^*,\ -j_4^*).$$

In our Feynman rules we have explicitly denoted the relevant factors $(2\pi)^4 i$, but often omitted the δ functions for energy-momentum conservation.

2. DEFINITIONS

2.1. Definition of the Feynman Rules

The purpose of this section is to spell out the precise form of the Feynman rules for a given Lagrangian. In principle, this is very straightforward: the propagators are defined by the quadratic part of the Lagrangian, and the rest is represented by vertices. As is well known, the propagators are minus the inverse of the operator found in the quadratic term, for example

$$\mathcal{L} = \frac{1}{2}\phi(\partial^2 - m^2)\phi \rightarrow (k^2 + m^2 - i\varepsilon)^{-1},$$

$$\mathcal{L} = -\bar{\psi}(\gamma^\mu\partial_\mu + m)\psi \rightarrow (i\gamma k + m)^{-1} = \frac{-i\gamma k + m}{k^2 + m^2 - i\varepsilon}.$$

Customarily, one derives this using commutation rules of the fields, etc. We will simply skip the derivation and define the propagator, including the $i\varepsilon$ prescription for the pole.

Similarly, vertices arise. For instance, if the interaction Lagrangian contains a term providing for the interaction of fermions and a scalar field one has

$$\mathcal{L}_I = g(\bar{\psi}\psi)\phi$$

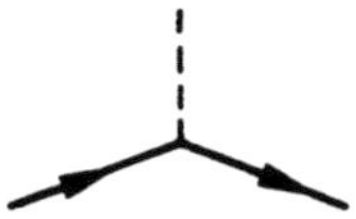

g

In this and similar cases there is no difficulty in deriving the rules by the usual canonical formalism. If however derivatives, or worse non-local terms, occur in $\mathcal{L}$, then complications arise. Again we will short-circuit all difficulties and define our vertices, including non-local vertices, directly from the Lagrangian. Furthermore, we will allow sources that can absorb or emit particles. They are an important tool in the analysis. In the rest of this section we will try to define precisely the Feynman rules for the general case, including factors π, etc. Basically the recipe is the straightforward generalization of the simple cases shown above.

The most general Lagrangian to be discussed here is

$$\mathcal{L}(x) = \psi_i^*(x)V_{ij}\psi_j(x) + \frac{1}{2}\phi_i(x)W_{ij}\phi_j(x) + \mathcal{L}_I(\psi^*,\psi,\phi). \tag{2.1}$$

The ψ_i and ϕ_i denote sets of complex and real fields that may be scalar, spinor, vector, tensor, etc., fields. The index i stands for any spinor, Lorentz, isospin, etc., index. V and W are matrix operators that may contain derivatives, and whose Fourier transform must have an inverse. Furthermore, these inverses must satisfy the Källén-Lehmann representation, to be discussed later. The interaction Lagrangian $\mathcal{L}_I(\psi^*,\psi,\phi)$ is any polynomial in certain coupling constants g as well as the fields. This interaction Lagrangian is allowed to be non-local, i.e. not only depend on fields in the point x, but also on fields at other space time points x', x", ... The coefficients in the polynomial expansion may be functions of x. The explicit form of a general term in $\mathcal{L}_I(x)$ is

$$\int d_4x_1 d_4x_2, \ldots, \alpha_{i_1 i_2 \ldots}(x, x_1, x_2, \ldots) \qquad (2.2)$$

$$\psi^*_{i_1}(x_1), \ldots, \psi_{i_m}(x_m), \ldots, \phi_{i_n}(x_n), \ldots .$$

The α may contain any number of differential operators working on the various fields.

Roughly speaking propagators are defined to be minus the inverse of the Fourier transforms of V and W, and vertices as the Fourier transforms of the coefficients α in $\mathcal{L}_I$.

The action S is defined by

$$iS = i \int d_4x \mathcal{L}(x). \qquad (2.3)$$

In $\mathcal{L}$ we make the replacement

$$\psi_i(x) = \int d_4k \; a_i(k) e^{ikx} ,$$

$$\psi^*_i(x) = \int d_4k \; b_i(k) e^{ikx} ,$$

$$\phi_i(x) = \int d_4k \; c_i(k) e^{ikx} ,$$

$$\alpha_{i_1 i_2 \ldots}(x, x_1, x_2, \ldots) =$$

$$= \int d_4k\, d_4k_1\, d_4k_2, \ldots, e^{ikx+ik_1(x-x_1)+ik_2(x-x_2)+\ldots}\,\bar{\alpha}_{i_1\ldots}(k,k_1,k_2,\ldots).$$

The action times i takes the form

$$(2\pi)^4 i b(k)\bar{V}_{ij}(k)a_j(k) + \frac{1}{2}(2\pi)^4 i c_i(k)\bar{W}_{ij}(k)c_j(k)$$

$$+ \ldots + (2\pi)^4 i\delta_4(k + k_1 + \ldots)\bar{\alpha}_i(k, k_1, k_2, \ldots) \times$$

$$\times\, b_{i_1}(k_1), \ldots, a_{i_m}(k_m), \ldots, c_{i_n}(k_n) + \ldots ,$$

(2.4)

each term integrated over the momenta involved. The $\bar{V}$ and $\bar{W}$ contain a factor ik_n (or $-ik_n$) for every derivative $\partial/\partial x_\mu$ acting to the right (left) in V and W, respectively. The α contain a factor $ik_{j\mu}$ for every derivative $\partial/\partial x_{j\mu}$ acting on a field with argument x_j.

The propagators are defined to be:

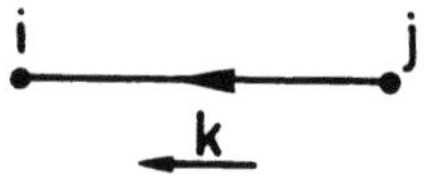

$$\Delta_{Fij}(k) = -\frac{1}{(2\pi)^4 i}\,[\bar{V}^{-1}(k)]_{ij},$$

$$\Delta_{Fij}(k) = -\frac{1}{(2\pi)^4 i}\left[\frac{1}{2}\bar{W}(k) + \frac{1}{2}\tilde{\bar{W}}(-k)\right]^{-1}_{ij} \qquad (2.5)$$

Here $\tilde{\bar{W}}$ is $\bar{W}$ reflected, i.e. $\tilde{\bar{W}}_{ij} = \bar{W}_{ji}$. In the rare case of real fermions the propagator must be minus the inverse of the anti-symmetric part of W. Furthermore, there is the usual $i\varepsilon$ prescription for the poles of these propagators. The momentum k in Eqs. (2.5) is the momentum flow in the direction of the arrow.

The definition of the vertices is:

$$(2\pi)^4 i \sum_{\{1\ldots m-1\}}\ \sum_{\{m\ldots n-1\}}\ \sum_{\{n\ldots\}} (-1)^P \times$$

$$\times\, \alpha_{i_1\ldots}(k, k_1, k_2, \ldots)\delta_4(k + k_1 + \ldots). \qquad (2.6)$$

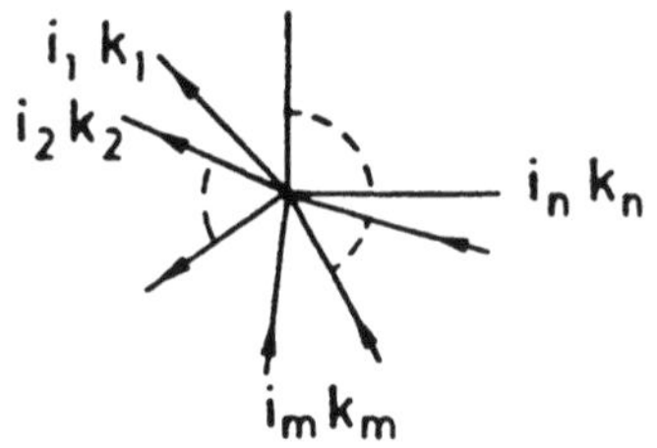

The summation is over all permutations of the indices and momenta indicated. The momenta are taken to flow inwards. Any field ψ^* corresponds to a line with an arrow pointing outwards; a field ψ gives an opposite arrow. The ϕ fields give arrow-less lines. The factor $(-1)^P$ is only of importance if several fermion field occur. All fermion fields are taken to anticommute with all other fermion fields. There is a factor -1 for every permutation exchanging two fermion fields.

The coefficients α will often be constants. Then the sum over permutations results simply in a factor. It is convenient to include such factors already in $\mathcal{L}_I$; for instance

$$\mathcal{L}_I(x) = \frac{1}{3!6!} g\psi^*(x)^3 \psi(x)^6$$

gives as vertex simply the constant g.

As indicated, the coefficients α may be functions of x, corresponding to some arbitrary dependence on the momentum k in (2.6). This momentum is not associated with any of the lines of the vertex. If we have such a k dependence, i.e. the coefficient α is non-zero for some non-zero value of k, then this vertex will be called a <u>source</u>. Sources will be indicated by a cross or other convenient notation as the need arises.

A diagram is obtained by connecting vertices and sources by means of propagators in accordance with the arrow notations. Any diagram is provided with a combinatorial factor that corrects for double counting in case identical particles occur. The computation of these factors is somewhat cumbersome; the recipe is given in Appendix A.

Further, if fermions occur, diagrams are provided with a sign. The rule is as follows:

i) there is a minus sign for every closed fermion loop;
ii) diagrams that are related to each other by the omission or addition of boson lines have the same sign;

iii) diagrams related by the exchange of two fermion lines, internal or external, have a relative minus sign.

EXAMPLE 2.1.1 Electron-electron scattering in quantum-electrodynamics. Some diagrams and their relative sign are given in the figure. The fourth and fifth diagrams are really the same diagram and should not be counted separately:

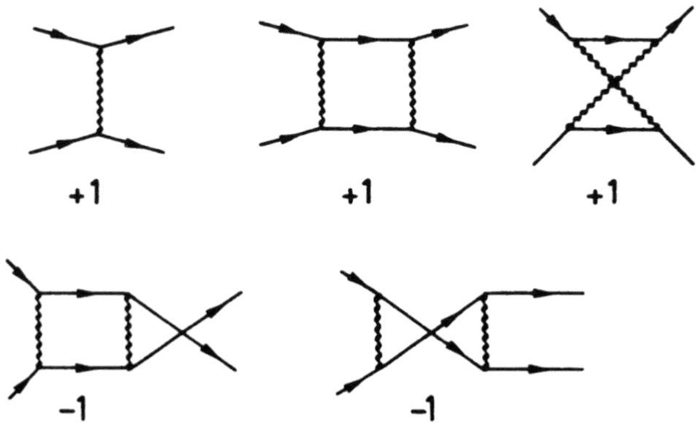

This raises the question of the recognition of topologically identical diagrams. In Appendix A some considerations pertinent to the topology of quantum electrodynamics are presented.

2.2 A Simple Theorem ; Some Examples

THEOREM Diagrams are invariant for the replacement

$$\psi_i^* \to \psi_k^* X_{ki}$$

in the Lagrangian (1.1), where X is any matrix that may include derivatives but must have an inverse.

The theorem is trivial to prove. Any oriented propagator obtains a factor X^{-1} that cancels against the factor X occurring in the vertices. It is left to the reader to generalize this theorem to include transformations of the ψ's and ϕ's.

Some examples illustrating our definitions are in order.

EXAMPLE 2.2.1 Charged scalar particles with ψ^4 interaction

$$\mathcal{L}(x) = \psi^*(\partial^2 - m^2)\psi + \frac{g}{2!2!}(\psi^*\psi)^2 + J^*(x)\psi + \psi^* J(x). \tag{2.7}$$

Only $\psi^*(x)$ and $\psi(x)$ in the same point x occur, and in accordance with the description given at the beginning of the previous section we have here a local Lagrangian. The functions J(x) and $J^*(x)$ are source functions.

Propagator:

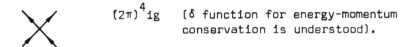

$$\Delta_F = \frac{1}{(2\pi)^4 i} \frac{1}{k^2 + m^2 - i\varepsilon}.$$

Vertex:

$(2\pi)^4 ig$ (δ function for energy-momentum conservation is understood).

Sources:

$(2\pi)^4 iJ^*(k)$

$(2\pi)^4 iJ(k)$

The functions J(k), $J^*(k)$ are the Fourier transforms of J(x) and $J^*(x)$

$$J(x) = \int d_4k e^{-ikx} J(k).$$

Note that k is the momentum flowing from the line into the source.

Some diagrams:

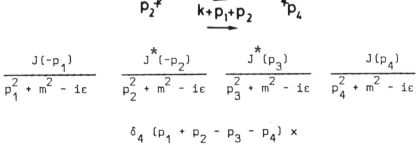

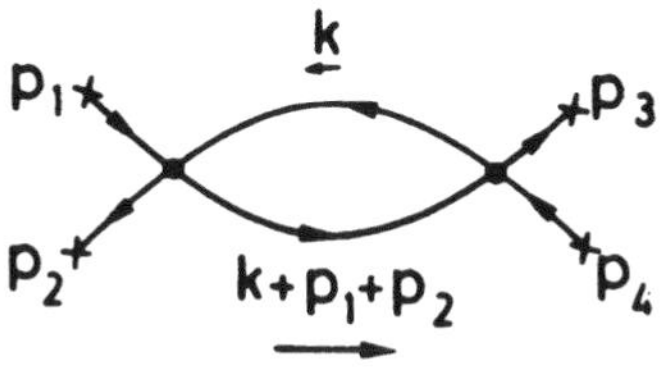

$$\frac{J(-p_1)}{p_1^2 + m^2 - i\varepsilon} \quad \frac{J^*(-p_2)}{p_2^2 + m^2 - i\varepsilon} \quad \frac{J^*(p_3)}{p_3^2 + m^2 - i\varepsilon} \quad \frac{J(p_4)}{p_4^2 + m^2 - i\varepsilon}$$

$$\delta_4 (p_1 + p_2 - p_3 - p_4) \times$$

$$\times g^2 \int d_4k \frac{1}{k^2 + m^2 - i\varepsilon} \frac{1}{(k + p_1 + p_2)^2 + m^2 - i\varepsilon}.$$

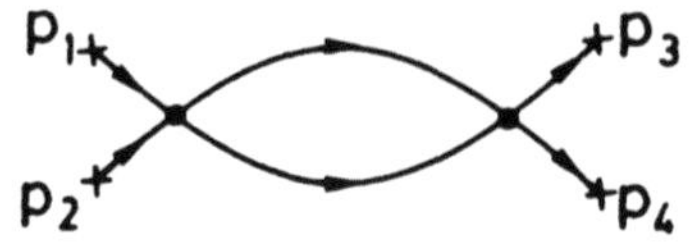

$$\frac{J(-p_1)}{p_1^2 + m^2 - i\varepsilon} \frac{J(-p_2)}{p_2^2 + m^2 - i\varepsilon} \frac{J^*(p_3)}{p_3^2 + m^2 - i\varepsilon} \frac{J^*(p_4)}{p_4^2 + m^2 - i\varepsilon}$$

$$\delta_4 (p_1 + p_2 - p_3 - p_4) \times$$

$$\times \frac{1}{2} g^2 \int d_4k \frac{1}{k^2 + m^2 - i\varepsilon} \frac{1}{(k + p_1 + p_2)^2 + m^2 - i\varepsilon}.$$

The factor 1/2 in the second case is the combinatorial factor occurring because of the identical particles in the intermediate state.

EXAMPLE 2.2.2 Free real vector mesons

$$\mathcal{L} = -\frac{1}{4} (\partial_\mu W_\nu - \partial_\nu W_\mu)^2 - \frac{1}{2} m^2 W_\mu^2. \tag{2.8}$$

In terms of four real fields $\phi_\alpha \equiv W_\alpha$, $\alpha = 1, \ldots, 4$,

$$\mathcal{L} = \frac{1}{2}\phi_\alpha [-(\overleftarrow{\partial}_\mu \overrightarrow{\partial}_\mu + m^2)\delta_{\alpha\beta} + \overleftarrow{\partial}_\alpha \overrightarrow{\partial}_\beta]\phi_\beta \equiv \frac{1}{2} \phi_\alpha V_{\alpha\beta} \phi_\beta.$$

The matrix V in momentum space is

$$-\delta_{\alpha\beta}(k^2 + m^2) + k_\alpha k_\beta.$$

The vector meson propagator becomes:

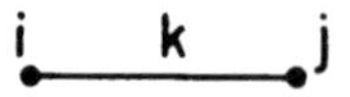

$$\Delta_{F\alpha\beta} = -\frac{1}{(2\pi)^4 i} (V^{-1})\alpha\beta = \frac{1}{(2\pi)^4 i} \frac{\delta_{\alpha\beta} + k_\alpha k_\beta/m^2}{k^2 + m^2 - i\varepsilon} \tag{2.9}$$

Indeed

$$[\delta_{\alpha\beta}\,(k^2 + m^2) - k_\alpha k_\beta]\;\frac{\delta_{\beta\lambda} + k_\beta k_\lambda/m^2}{k^2 + m^2 - i\varepsilon} =$$

$$= \frac{1}{k^2 + m^2 - i\varepsilon}\;[\delta_{\alpha\lambda}\,(k^2 + m^2) - k_\alpha k_\lambda$$

$$+ k_\alpha k_\lambda\,\frac{k^2 + m^2}{m^2} - k_\alpha k_\lambda\,\frac{k^2}{m^2}\;] = \delta_{\alpha\lambda}.$$

<u>EXAMPLE 2.2.3</u> Electron in an external electromagnetic field A_μ

$$\mathcal{L} = -\bar{\psi}(\gamma^\mu \partial_\mu + m)\psi + ieA_\mu \bar{\psi}\gamma^\mu \psi\ . \qquad (2.10)$$

Note that $\bar{\psi} = \psi^* \gamma^4$. Because of our little theorem the matrix γ^4 can be omitted in giving rules for the diagrams. Then:

$$\Delta_F = \frac{1}{(2\pi)^4 i}\,\frac{-i\gamma k + m}{k^2 + m^2 - i\varepsilon}\ . \qquad (2.11)$$

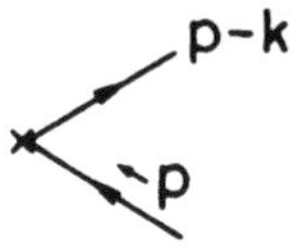

$$-(2\pi)^4\ e\ \gamma^\mu A_\mu(k)\ .$$

As a further application of our theorem we may substitue $\psi \rightarrow (-\gamma^\nu \partial_\nu + m)\psi$ to give

$$\mathcal{L} = \bar{\psi}(\partial^2 - m^2)\psi + ieA_\mu \bar{\psi}\gamma^\mu(-\gamma^\nu \partial_\nu + m)\psi\ .$$

This gives the equivalent rules:

$$\Delta_F = \frac{1}{(2\pi)^4 i} \frac{1}{k^2 + m^2 - i\varepsilon}$$

$$- (2\pi)^4 e A_\mu(k)\gamma^\mu(-i\gamma^\nu p_\nu + m) .$$

2.3 Internal consistency

Two points need to be investigated. First, the separation in real and complex fields is really arbitrary, because for any complex field ϕ one can always write

$$\phi = \frac{1}{\sqrt{2}}(A + iB) , \qquad \phi^* = \frac{1}{\sqrt{2}}(A - iB) ,$$

where A and B are real fields. The question is whether the results will be independent of the representation chosen. This indeed is the case, and may be best explained by considering an example:

$$\mathcal{L} = \phi_i^* V_{ij} \phi_j + J_i^* \phi_i + \phi_i^* J_i$$

$$- \frac{1}{(2\pi)^4 i} (V^{-1})_{ij}$$

$$(2\pi)^4 i J_i^*$$

$$(2\pi)^4 i J_i .$$

The diagram containing two sources is:

$$-(2\pi)^4 i J_i^* (V^{-1})_{ij} J_j .$$

On the other hand, let us write $\phi = (1/\sqrt{2})(A + iB)$

$$\mathcal{L} = \frac{1}{2} A_i V_{ij} A_j + \frac{1}{2} B_i V_{ij} B_j + \frac{i}{2} A_i V_{ij} B_i - \frac{i}{2} B_i V_{ij} A_j +$$

$$+ \frac{1}{\sqrt{2}} (J_i^* A_i + i J_i^* B_i + A_i J_i - i B_i J_i) .$$

Defining the real field X_i

$$X = \begin{pmatrix} A \\ B \end{pmatrix},$$

we have

$$\mathcal{L} = \frac{1}{2} X_i W_{ij} X_j + F_i^* X_i + F_i X_i$$

with

$$W = \begin{pmatrix} V^s & iV^a \\ -iV^a & V^s \end{pmatrix}, \quad F^* = \frac{1}{\sqrt{2}} \begin{pmatrix} J_i^* \\ iJ_i^* \end{pmatrix}, \quad F = \frac{1}{\sqrt{2}} \begin{pmatrix} J_i \\ -iJ_i \end{pmatrix},$$

where the superscripts s and a denote the symmetrical and the antisymmetrical part, respectively.

Let Y now be the inverse of V. Thus

$$VY = 1, \quad YV = 1.$$

Writing $V = V^s + V^a$, $Y = Y^s + Y^a$ one finds, comparing the reflected VY with YV

$$V^sY^s + V^aY^a = 1,$$

$$V^sY^a + V^aY^s = Y^aV^s + Y^sV^a = 0.$$

The inverse of the matrix W is therefore

$$W^{-1} = \begin{pmatrix} Y^s & iY^a \\ -iY^a & Y^s \end{pmatrix},$$

where $Y = V^{-1}$. The source-source diagram becomes

$$-(2\pi)^4 iF^* W^{-1} F = -(2\pi)^4 i \frac{1}{2} 2J^* (Y^s + Y^a) J = -(2\pi)^4 iJ^* V^{-1} J$$

as before. Clearly the separation into real and imaginary parts amounts to separation into symmetric and antisymmetric parts of the propagator.

The second point to be investigated is the question of separation of the quadratic, i.e. propagator defining, part in $\mathcal{L}$. Thus let there be given the Lagrangian

$$\mathcal{L} = \overset{*}{\phi}_i V_{ij} \phi_i + \overset{*}{\phi}_i V'_{ij} \phi_j \ .$$

One can either say that one has a propagator $-(V + V')^{-1}$ for the ϕ-field, or alternatively a propagator $-(V)^{-1}$ and a vertex:

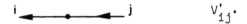

V'_{ij}.

However, these two cases give the same result. Summing over all possible insertions of the vertex V' one finds:

$$-\frac{1}{V} + [-\frac{1}{V}]V' [-\frac{1}{V}] + \ \dots \ = -\frac{1}{V}\frac{1}{1 + V'/V} = -\frac{1}{V + V'} .$$

This demonstrates the internal consistency of our scheme of definitions. We leave it as an exercise to the reader to verify that combinational factors check if one makes the replacement $\phi \to (A + iB)/\sqrt{2}$, for instance for the diagrams:

$\frac{1}{2}$

2.4 Definition of the Green's Functions

Let there be given a general Lagrangian of the form (2.1). Corresponding to any field we introduce source terms (we will need many, but write only one for each field)

$$\mathcal{L} \to \mathcal{L} + \overset{*}{J}_i \psi_i + \overset{*}{\psi}_i J_i + \phi_i K_i \ .$$

According to the previous rules such terms give rise to the following vertices:

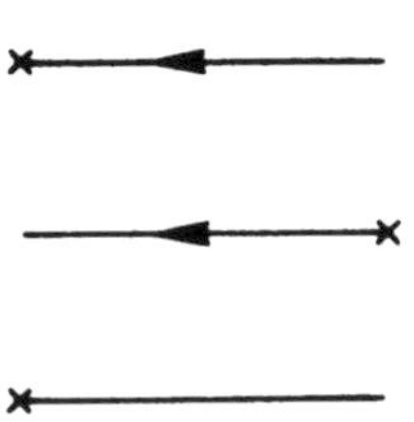

$(2\pi)^4 i J^*(k)$;

$(2\pi)^4 i J(k)$,

$(2\pi)^4 i K(k)$,

where the k dependence implies Fourier transformation.

Remember now that the Lagrangian was a polynomial in some coupling constants. For a given order in these coupling constants we may consider the sum of all diagrams connecting n sources. All n sources are to be taken different, because we want to be able

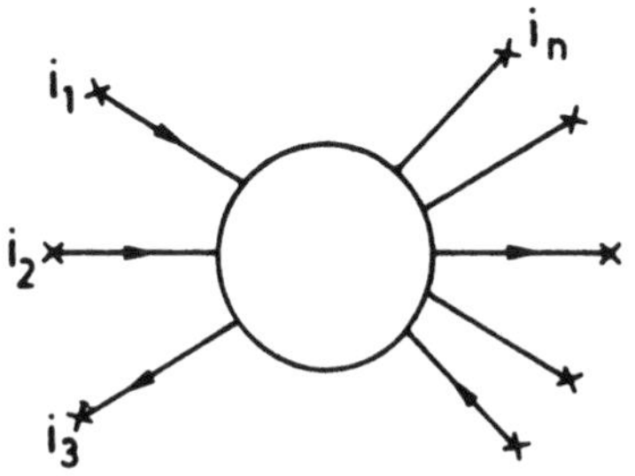

to vary the momenta independently. Each of these diagrams, and a fortiori their sum, will be of the form

$$J_{i_1}(k_1) J'_{i_2}(k_2) \dots K_{i_n}(k_n) G_{i_1 \dots i_n}(k_1, \dots, k_n) .$$

The function G_i will be called the n-point Green's function for the given external line configuration for the order in the coupling constants specified. Factors $(2\pi)^4 i$ from the source vertices are included. The k_i denote the momenta flowing from the sources into the diagrams. The propagators that connect to the sources are included in this definition.

The first example of Section 2.2 shows some diagrams contributing to the second-order four-point Green's functions.

2.5 Definition of the S-Matrix ; Some Examples

Roughly speaking the S-matrix obtains from the Green's functions in two steps: (i) the momenta of the external line are put on mass shell, and (ii) the sources are normalized such that they correspond to emission or absorption of one particle. Both these statements are somewhat vague, and we must precise them, but they reflect the essential physical content of the reasoning below.

Consider the diagrams connecting two sources:

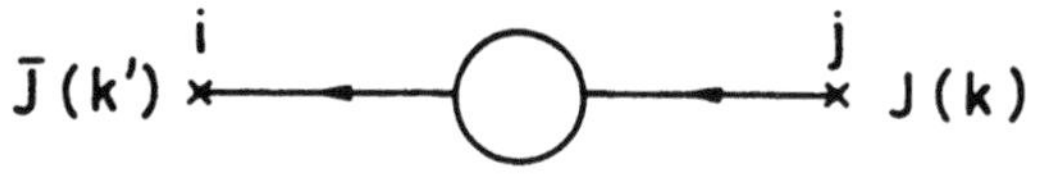

The corresponding expression is

$$\bar{J}_i(k')G_{ij}(k, k')J_j(k). \tag{2.12}$$

The two-point Green's function will in general have a pole (or possibly many single poles) at some value $-M^2$ of the squared four-momentum k_ν. If there is no pole there will be no corresponding S-matrix element; such will be the case if a particle becomes unstable because of the interactions. At the pole the Green's function will be of the form

$$G_{ij}(k, k') = (2\pi)^4 i\delta_4(k + k')\frac{K_{ij}(k)}{k^2 + M^2} \text{ at } k^2 \to -M^2.$$

The matrix residue K_{ij} can be a function of the components k_μ, with the restriction that $k^2 = -M^2$.

First we will treat the currents for emission of a particle, corresponding to incoming particles of the S-matrix. Define a new set of currents $J_i^{(a)}$, one for every non-zero eigenvalue of K, which are mutually orthogonal and eigenstates of the matrix K(k)

$$J_i^{(a)*}J_i^{(b)} = 0 \quad \text{if} \quad a \neq b\,,$$

$$K_{ij}(k)J_j^{(a)}(k) = f^a(k)J_i^{(a)}(k)\,, \tag{2.13}$$

and normalized such that

$$[J_j^{(a)}(k)]^* K_{ji}(k)J_j^{(a)}(k) = \begin{cases} 1 & \text{for integer spin} \\ \dfrac{k_0}{m} & \text{for half-integer spin}\,. \end{cases} \tag{2.14}$$

This is possible only if all eigenvalues of K are positive. In the case of negative eigenvalues normalization is done with minus the right-hand side of Eq. (2.14).

The sources thus defined are the properly normalized sources for emission of a particle or antiparticle (the latter follows from considering $\tilde{K}(-k)$). The properly normalized sources for absorption of a particle or antiparticle follow by considering Eqs. (2.13) and (2.14) but with k replaced by -k in K.

The above procedure defines the currents up to a phase factor. We must take care that the phase factor for the emission of a certain particle agrees with that for absorption of that same particle. This is fixed by requiring that the two-point Green's function provided with such sources has precisely the residue 1 (or k_0/m) for $k^2 = -m^2$.

The matrix elements of the matrix S' (almost, but not exactly the S-matrix, see below) for n ingoing particles and m outgoing particles are defined by

$$<p_1 b_1, \ldots, p_m b_m | S' | k_1 a_1, \ldots, k_n a_n> \equiv \prod_{r=1}^{n} \lim_{k_r^2 = -M_r^2} J_{i_r}^{(a_r)}(k_r)$$

$$\times (k_r^2 + M_r^2) \prod_{s=1}^{m} \lim_{p_s^2 = -M_s^2} J_{j_s}^{(b_s)}(p_s)(p_s^2 + M_s^2)$$

$$\times G_{i_1 \ldots j_m}(k_1, \ldots, k_n, -p_1, \ldots, -p_m). \qquad (2.15)$$

The energies $k_{10}, \ldots, k_{n0}$ and $p_{10}, \ldots, p_{m0}$ are all positive. The minus sign for the momenta p in the Green's function appears because in the matrix element the momenta of the outgoing particles are taken to be flowing out, while in our Green's functions the convention was that all momenta flow inwards.

EXAMPLE 2.5.1 Scalar particles. Near the physical mass pole the two-point Green's function will be such that

$$\frac{G(k, k')}{(2\pi)^4 i\delta_4(k - k')} = \frac{1}{Z^2(k^2 + M^2)}. \qquad (2.16)$$

For any k the properly normalized external current is $\tilde{J} = Z$, and the prescription to find S'-matrix elements with external scalar particles is

$$\langle p_1, \ldots, p_m | S' | k_1, \ldots, k_n \rangle = \prod_{r=1}^{n} \lim_{k_r^2 = -M^2} Z(k_r^2 + M^2) \times$$
$$\times \prod_{s=1}^{m} \lim_{p_s^2 = -M^2} Z(p_s^2 + M^2) G(k_1, \ldots, k_n, -p_1, \ldots, -p_m). \tag{2.17}$$

EXAMPLE 2.5.2 Fermions as in QED. Near the physical mass pole the two-point Green's function will be such that

$$\frac{G_{ij}(k, k')}{(2\pi)^4 i\delta_4(k - k')} = \frac{1}{Z^2(i\gamma k + M)} = \frac{K_{ij}(k)}{k^2 + M^2}, \tag{2.18}$$

with

$$K_{ij}(k) = \frac{1}{Z^2} (-i\gamma k + M)_{ij} . \tag{2.19}$$

A set of eigenstates of this matrix is provided by the solutions $u^a(k)$ of the Dirac equation

$$(i\gamma k + M)u = 0 , \quad u^{a*} u^a = 1 , \quad a = 1, 2 . \tag{2.20}$$

Because of the normalization condition Eq. (2.14), we must take for the currents $J^{(a)}(k)$

$$J^{(a)}(k) = u^a(k) \frac{Z}{2M} \sqrt{2k_0} , \tag{2.21}$$

with the Dirac spinors normalized to 1, see Eq. (2.20). Note that the factor 1/2M cancels upon multiplication of this source with the propagator numerator (2.19).

The antiparticle emission currents are obtained by considering $\tilde{K}(-k)$. The solutions are the antiparticle spinors

$$\bar{u}^a(k) , \quad a = 3, 4.$$

For outgoing antiparticles and particles $K(-k)$ and $\tilde{K}(k)$ must be considered to obtain the proper absorption currents. The solutions are the spinors $u^a(k)$, $a = 3, 4$ and $\bar{u}^a(k)$, $a = 1, 2$.

The phase condition $\bar{u}^a K u^a = 2M$ dictates a minus sign for the source for emission of an antiparticle (see Appendix A).

Let us now complete the S-matrix definition. The prescription given above results in zero when applied to the two-point function, because there will be two factors $K^2 + M^2$ (one for every source). Thus we evidently do not obtain exactly the S-matrix that one has in such cases. This discrepancy is related to the treatment of the over-all δ function of energy-momentum conservation, when passing from S-matrix elements to transition probabilities. Anyway, to get the S-matrix we must also allow lines where particles go through without any interaction, and associate a factor 1 with such lines. These particles must, of course, have a mass as given by the pole of their propagator.

Matrix elements of the S-matrix, including possible lines going straight through, will be denoted by graphs with external lines that have no terminating cross. The convention is : left are incoming particles, right outgoing. Energy flows from left to right. The direction of the arrows is of course not related to the direction of the energy flow.

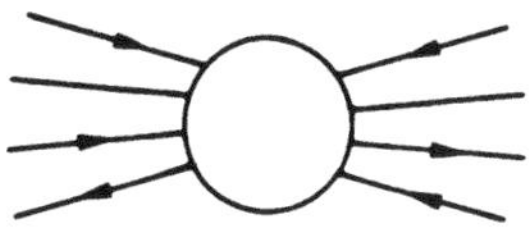

We emphasize again that the above S-matrix elements include diagrams containing interaction free lines. For instance, the diagram shown is included in the 3-particle-in/3-particle-out S-matrix element.

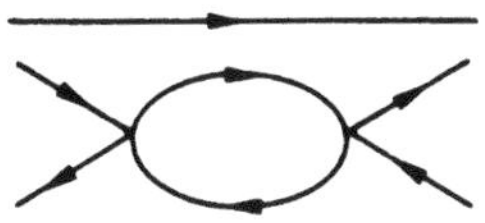

The definition of the S-matrix given above applies if there is no gauge symmetry. For gauge theories some of the propagators correspond to "ghost" particles that are assumed not to have physical relevance. In defining the S-matrix, the sources must be restricted to emit or absorb only physical particles. Such sources will be called <u>physical sources</u> and have to be defined in the precise context of the gauge symmetry. To show in such cases that the S-matrix is unitary requires then special effort.

2.6 Definition of $S^\dagger$

The matrix elements of the matrix $S^\dagger$ are defined as usual by

$$\langle \alpha | S^\dagger | \beta \rangle = \langle \beta | S | \alpha \rangle^* , \qquad (2.22)$$

where the complex conjugation implies also the replacement $i\varepsilon \to -i\varepsilon$ in the propagators.

The matrix elements of $S^\dagger$ can also be obtained in another way. In addition to the Lagrangian $\mathcal{L}$ defining S, consider the conjugated Lagrangian $\mathcal{L}^\dagger$. It is obtained from $\mathcal{L}$ by complex conjugation and reversal of the order of the fields. The latter is only relevant for fermions. This $\mathcal{L}^\dagger$ may be used to define another S-matrix; let $\bar{S}$ denote the matrix obtained in the usual way from $\mathcal{L}^\dagger$, however with the opposite sign for the $i\varepsilon$ in the propagators and also the replacement $i \to -i$ in the notorious factors $(2\pi)^4 i$. We claim that the matrix elements of $\bar{S}$ are equal to those of $S^\dagger$. In formula we get

$$\langle \alpha | S(\mathcal{L}, i)^\dagger | \beta \rangle = \langle \alpha | S(\mathcal{L}^\dagger, -i) | \beta \rangle .$$

The proof rests mainly on the observation that an incoming particle source is obtained by considering (notation of Section 2.5, Eq. (2.13)):

$$K_{ij}(k) J_j(k)$$

and the complex conjugate of an outgoing particle source by study of:

$$J_i(-k) K^*_{ij}(k) ,$$

or, equivalently,

$$\tilde{K}^*_{ij}(-k) J_j(k) ,$$

where $\tilde{K}_{ij} = K_{ji}$. This is indeed what corresponds to complex conjugation of the propagator defining part of $\mathcal{L}^\dagger$

$$[\phi^*_i V_{ij}(\partial) \phi_i]^* = \phi^*_i V^\dagger_{ij}(-\partial) \phi .$$

Note the change of sign of the derivative: what worked to the right works now to the left, which implies a minus sign. $V^\dagger$ is obtained by transposition and complex conjugation of V. Clearly complex conjugation and exchange of incoming and outgoing states corresponds to the use of $V^\dagger$ instead of V.

EXAMPLE 2.6.1 Fermion coupled to complex scalar field

$$\mathcal{L} = -\bar{\psi}(\gamma\partial + m)\psi + \phi^*(\partial^2 - m^2)\phi + g\bar{\psi}(1 + \gamma^5)\psi\phi^* .$$

The lowest order S-matrix element is:

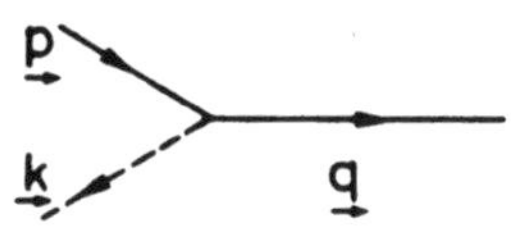

$$< q|S|p,\ k> = (2\pi)^4 ig\delta_4(p + k - q) \times$$
$$\times\ \bar{u}(q)(1 + \gamma^5)u(p)\ \sqrt{4q_o p_o}$$

According to Eq. (2.22)

$$< p,\ k|S^+|q >= -(2\pi)^4 ig^* \delta_4(p + k - q)\bar{u}(p)(1 - \gamma^5)u(q)\sqrt{4q_o p_o}\ .$$

There is a minus sign because $\gamma^5\gamma^4 = -\gamma^4\gamma^5$.

Consider now $\mathcal{L}^\dagger$

$$\mathcal{L}^\dagger = -\bar{\psi}(\gamma\partial + m)\psi + \phi^*(\partial^2 - m^2)\phi + g^*\bar{\psi}(1 - \gamma^5)\psi\phi\ .$$

There are minus signs because of γ^4, γ^5 exchange and $\partial \to -\partial$ changes except $\partial_4 \to \partial_4$. Including the sign change for $(2\pi)^4 i$, we obtain for $\bar{S}$:

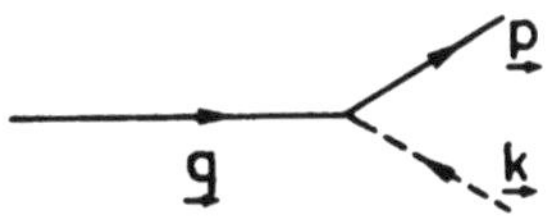

$$< p,\ k|\bar{S}|q > = -(2\pi)^4\ ig^*\delta^4(p+k-q)\times$$
$$\times\ \bar{u}(p)\ (1 - \gamma^5)u(q)\ \sqrt{4p_o q_o}\ ,$$

which equals the result for $S^\dagger$ found above.

To summarize, the matrix elements of $S^\dagger$ can be obtained either directly from their definition (2.22), or by the use of different Feynman rules. These new rules can be obtained from the old ones by reversing all arrows in vertices and propagators and replacing all vertex functions and propagators by their complex conjugate (for the propagators this means using the Hermitian conjugate propagators). Also, the factors $(2\pi)^4 i$ and the $i\epsilon$ terms in the propagators are to be complex conjugated. The in- and out-state source functions are defined by the usual procedure, involving now the Hermitian conjugate propagators.

Let us finally, for the sake of clarity, formulate the definition of ingoing and outgoing states in the diagram language for both the old and the new rules:

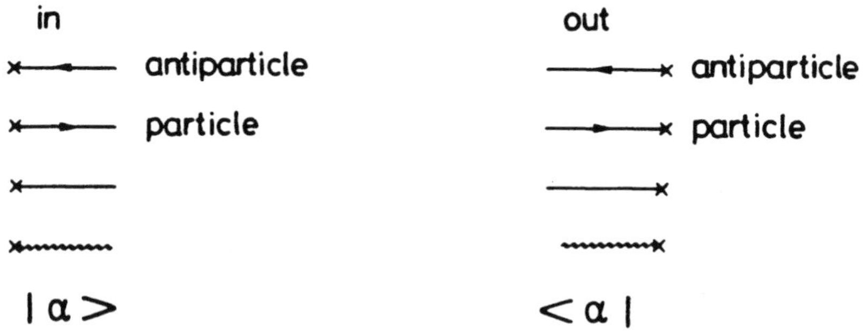

2.7 Definition of Transition Probabilities ; Cross-Sections and Lifetimes

The S-matrix elements are the transition amplitudes of the theory. The probability amplitudes are defined by the absolute value squared of these amplitudes. Conservation of probability requires that the S-matrix be unitary

$$\sum_\beta |\langle\beta|S|\alpha\rangle|^2 = \sum_\beta \langle \alpha|S^\dagger|\beta\rangle \langle\beta|S|\alpha\rangle = 1 \,. \tag{2.23}$$

This property will be true only if the diagrams satisfy certain conditions, and we will investigate this in Section 6.

In the usual way, lifetimes and cross-sections can be deduced from the transition probabilities.

Consider the decay of particle α into particles 1, 2, 3, ..., n. The decay width Γ (= inverse lifetime τ) is

$$\frac{1}{\tau} = \Gamma = \int \frac{d_3p_1}{2p_{10}(2\pi)^3}, \ldots, \frac{d_3p_n}{2p_{no}(2\pi)^3} \frac{\delta_4(p - p_1 - p_2 \cdots - p_n)}{2p_o(2\pi)^4} \times$$

$$\times \langle \alpha|M^\dagger|1, 2, \ldots, n\rangle \langle 1, 2, \ldots, n|M|\alpha\rangle, \tag{2.24}$$

where the matrix elements of M (and $M^\dagger$) are those of S (and $S^\dagger$) without the energy-momentum conservation δ function

$$\langle \beta|M|\alpha\rangle = \frac{\langle \beta|S|\alpha\rangle}{\delta_4(p_\alpha - p_\beta)}. \tag{2.25}$$

The $p_{\alpha o}$, ..., p_{no} are the energies of particles α, 1, 2, ..., n. Dividing Γ by $\hbar = 6.587 \times 10^{22}$ MeV • sec gives Γ in $\sec^{-1}$, provided all was computed in natural units ($\hbar = c = 1$) with the MeV as the only unit left.

Next, consider the scattering of a particle α on a particle β at rest giving rise to a final state with n particles. The cross-section σ for this process is

$$\sigma = \int \frac{d_3 p_1}{2p_{10}(2\pi)^3}, \ldots, \frac{d_3 p_n}{2p_{no}(2\pi)^3} \frac{p_{\alpha o}}{|\vec{p}_\alpha|} \frac{\delta_4(p_\alpha + p_\beta - p_1 - p_2 \cdots - p_n)}{4 p_{\alpha o} p_{\beta o} (2\pi)^4} \times$$

$$\times \langle \alpha, \beta | M^\dagger | 1, 2, \ldots, n \rangle \langle 1, 2, \ldots, n | M | \alpha, \beta \rangle. \tag{2.26}$$

Here $\vec{p}_\alpha$ is the three-momentum of the incoming particle α in the rest system of the β particle (target particle). Multiplying by $(\hbar c)^2 = (1.9732 \times 10^{-11}$ MeV • cm$)^2$, one obtains σ in cm^2.

3. DIAGRAMS AND FUNCTIONAL INTEGRALS

In all proofs we will rely only on combinatorics of diagrams. However, the rules and definitions may look somewhat ad hoc, and the purpose of this section is to show a formal equivalence with certain integral expressions.

Imagine a world with only a finite number of space-time points x^a_μ, $a = 1, \ldots, N$; $\mu = 1, \ldots, 4$. For simplicity, we consider only the case of real boson fields. The action S is now

$$S = \frac{1}{2} \sum_{a,b} \phi_i(x^a) W^{ab}_{ij} \phi_j(x^b) + \sum_a \mathcal{L}_I(x_a, \phi) \,, \qquad (3.1)$$

where W is taken to be symmetric. Derivatives occur as differences. The diagram rules corresponding to this action are as given in Section 2.

Suppose now that S is real if ϕ is real. The following theorem holds.

The rules defined in Section 2 for connected diagrams are precisely the same as the mathematical rules to obtain the function Γ defined by

$$e^{i\Gamma} = C \int e^{iS(\phi(x))} \prod_a \prod_i d\phi_i(x^a) \,, \qquad (3.2)$$

where the right-hand side is understood as a series expansion in terms of the coefficients in $\mathcal{L}_I$. Here C is an arbitrary constant not depending on the sources in $\mathcal{L}_I$. The set of all diagrams including disconnected ones are obtained by expanding $e^{i\Gamma}$.

Instead of Eq. (3.2) we will use the condensed notation

$$e^{i\Gamma} = C \int \mathcal{D}\phi e^{iS(\phi)} \,. \qquad (3.3)$$

The theorem is easy to prove. First calculate the integral for free particles, using for the action the expression

$$S_0(\phi, J) = \frac{1}{2} \sum \phi_i(x^a) W^{ab}_{ij} \phi_j(x^b) + \sum J_i(x^a) \phi_i(x^a) \qquad (3.4)$$

with arbitrary source functions J. Define

$$e^{i\Gamma_o(J)} = C_o \int \mathcal{D}\phi\, e^{iS_o(\phi,J)} . \tag{3.5}$$

Γ_o can be found by making a shift in the integration variables

$$\phi_i(x^a) = \phi_i'(x^a) - \sum_b (W^{-1})^{ab}_{ij} J_j(x^b) , \tag{3.6}$$

so that

$$S_o(\phi,J) = \frac{1}{2}\sum \phi' W \phi' - \frac{1}{2}\sum J_i(x^a)(W^{-1})^{ab}_{ij} J_j(x^b) . \tag{3.7}$$

The primed fields are not coupled to the J! We find

$$\Gamma_o(J) = -\frac{1}{2}\sum J W^{-1} J , \tag{3.8}$$

because the integral over the ϕ' is now a source independent constant and can be absorbed in the constant C in Eq. (3.3). Note the factor 1/2 because of identical sources.

We see that Γ^o is nothing but the free particle propagator. Even the $i\varepsilon$ prescription can be introduced correctly if we introduce a smooth cut-off for the integral for large values of the fields ϕ

$$e^{\frac{1}{2}i\phi W\phi} \rightarrow e^{\frac{1}{2}i\phi W\phi} e^{-\frac{1}{2}\varepsilon\phi^2} . \tag{3.9}$$

This makes the integral well defined even in directions where $\phi W\phi = 0$. The result is the usual $i\varepsilon$ addition to W.

Let us now consider also interaction terms. The perturbation expansion is

$$e^{i\sum_a \mathcal{L}_I(x^a)} = 1 + i\sum_a \mathcal{L}_I(x^a,\phi) + \frac{i^2}{2!}\sum_{a,b} \mathcal{L}_I(x^a,\phi)\mathcal{L}_I(x^b,\phi)\ldots \tag{3.10}$$

or

$$e^{iS} = e^{iS_o}\left[1 + i\sum \mathcal{L}_I + \ldots\right], \tag{3.11}$$

where the sources are included in S_o. From the definition of S_o we have

$$[i\phi_i(x^a)i\phi_j(x^b)\ \ldots]\ e^{iS_0(\phi,J)} = \left[\frac{\partial}{\partial J_i(x^a)}\frac{\partial}{\partial J_j(x^b)}\cdots\right] e^{iS_0(\phi,J)} . \tag{3.12}$$

Consequently we can replace everywhere the fields $\phi_i(x^a)$ by the derivative $\delta/\delta J_i(x^a)$

$$\int \mathcal{D}\phi e^{iS(\phi)} = C\left[1 + i \sum_a \mathcal{L}_I\left(x^a, \frac{\partial}{\partial J(x)}\right) + \ldots\right] \int \mathcal{D}\phi e^{iS_0(\phi,J)} . \tag{3.13}$$

The remaining integral is precisely the one computed before and is equal to exp $(i\Gamma_0)$ with Γ_0 given in Eq. (3.8). Expanding this exponential

$$e^{i\Gamma_0} = 1 - \frac{1}{2} i \sum JW^{-1}J + \frac{1}{8} i^2(JW^{-1}J)^2 \ldots , \tag{3.14}$$

we see that every term in Eq. (3.13) corresponds to a diagram with vertices and propagators as defined in Section 2. Diagrams not coupled to sources are called vacuum renormalizations and must be absorbed in the constant C. Note the combinatorial factors due to the occurrence of identical sources.

The functional integral notation (3.3) for the amplitude in the presence of sources is elegant and compact, and it is very tempting to write the amplitudes of relativistic field theories in this way. Many theorists can indeed not resist this temptation. Let us see what is involved. In our definition (3.3) we restricted ourselves to a finite number of space-time points. In any realistic theory this number is infinite and summations are to be replaced by integrations. So a suitable limiting procedure must be defined, but unfortunately these definitions cannot always be given in a manner free of ambiguities. Generally speaking, difficulties set in about at the same point that difficulties appear in the usual operator formalism, in particular we mention higher order derivatives or worse non-localities in the Lagrangian. We shall, therefore, in this report not try to formulate such a definition but attach to the result of manipulations with functional integrals a heuristic value. Everything is to be verified explicitly by combinatorics of diagrams. But having verified certain basic algebraic properties we can happily manipulate these "path" integrals. One of the most interesting manipulations used with great advantage in connection with gauge theories is to change of field variables. Both local and non-local canonical transformations turn out to be correctly described by the path integral formulae, as we shall see later.

We may finally mention that the various sign prescriptions, in the case of fermion fields, cannot be obtained in a simple way. They can be corrected for by hand afterwards, or in a more sophisticated way an algebra of anticommuting variables can be introduced. Such work can be found in the literature and is quite straightforward.

4. KALLEN-LEHMANN REPRESENTATION

The quantities $-V^{-1}$ and $-W^{-1}$ (see Eqs. (2.5)) that are defined as propagators in the theory are more precisely called the <u>bare</u> propagators. This is contrast to the two-point Green's function that when divided by $-(2\pi)^8\delta_4(k + k')$ is called the <u>dressed</u> propagator. Both the bare and dressed propagators are required to satisfy the Kallén-Lehmann representation, but it will turn out that the dressed propagators satisfy this automatically if the bare propagators do.

Consider any propagator and decompose it into invariant functions. For instance, for vector mesons

$$(2\pi)^4 i\Delta_{F\mu\nu}(k) = \delta_{\mu\nu} f_1(k^2) + k_\mu k_\nu f_2(k^2) . \tag{4.1}$$

The Kallén-Lehmann representation for the invariant functions is

$$f(-s) = \int_{a\geq 0}^{\infty} \frac{\rho(s')}{s' - s - i\varepsilon} ds' . \tag{4.2}$$

The functions $\rho(s')$ must be real. For bare propagators we will insist that the $\rho(s')$ are a sum of δ functions

$$\rho(s') = \sum_i a_i \delta(s' - m_i^2) \qquad \text{for bare propagators} \tag{4.3}$$

with real coefficients a_i, and real positive m_i^2.

Now introduce the Fourier transform of the propagators

$$\Delta_{Fij}(x) = \int d_4k e^{ikx} \Delta_{Fij}(k) . \tag{4.4}$$

Corresponding to the decomposition of $\Delta_F(k)$ in invariant functions we will have a decomposition involving derivatives. For instance, for vector mesons

$$\Delta_{F\mu\nu}(x) = \delta_{\mu\nu} \Delta_1(x) - \partial_\mu\partial_\nu \Delta_2(x) , \tag{4.5}$$

where Δ_1 and Δ_2 are the Fourier transforms of f_1 and f_2 above. The statement that any function $f(x)$ satisfies the Kallén-Lehmann representation is equivalent to the statement

$$f(x) = \theta(x_0)f^+(x) + \theta(-x_0)f^-(x) , \tag{4.6}$$

where $f^+(f^-)$ is a positive (negative) energy function

$$f^{\pm}(x) = \frac{1}{(2\pi)^3}\int_{a\geq 0}^{\infty} ds'\rho(s') \int d_4k e^{ikx}\theta(\pm k_0)\delta(k^2 + s') . \tag{4.7}$$

The proof is very simple. Using the Fourier representation

$$\theta(x_0) = \frac{1}{2\pi i}\int_{-\infty}^{\infty} d\tau \frac{e^{i\tau x_0}}{\tau - i\varepsilon}, \tag{4.8}$$

one derives immediately the result. Of course this derivation is only correct if the integral in Eq. (4.2) exists, and this is in general the case. If not one must introduce regulators, to be discussed below. However, we need the representation (4.6) not only for the invariant functions, but also for the complete propagator such as in Eq. (4.5). Whether this is true depends on the convergence of the dispersion integrals. If the functions f^+ and f^- go for $x_0 = 0$ sufficiently smoothly over into one another, then the expression (4.6) has no particular singularity at $x_0 = 0$ and one may ignore the action of derivatives on the θ functions. Now from the above equation (4.7) it is clear that $f^+(x_0) = f^-(-x_0)$, so in any case $f^+(0) = f^-(0)$. This is enough to treat the case of one derivative such as occurring in the case of fermion propagators. Consider now

$$\partial_0 f^{\pm} = \frac{1}{(2\pi)^3}\int ds'\rho(s') \int d_4k e^{ikx}(-ik_0)\theta(\pm k_0)\delta(k^2 + s'). \tag{4.9}$$

For $x_0 = 0$ we can do the k_0 integral

$$\partial_0 f^{\pm}\Big|_{x_0=0} = \frac{1}{(2\pi)^3}\int d_3k e^{i\vec{k}\vec{x}} \int ds'\rho(s')\left(\mp\frac{i}{2}\right). \tag{4.10}$$

Clearly $\partial_0 f^+ = \partial_0 f^-$ in $x_0 = 0$ only if the dispersion integral is superconvergent

$$\int ds'\rho(s') = 0 . \tag{4.11}$$

For bare propagators, see Eq. (4.3), this can only be true if some of the coefficients a_i are negative, some posotive. This implies the existence of negative metric particles, which in turn may lead to unphysical results, such as negative lifetimes or cross-sections, or lack of unitarity depending on how one defines things.

Let us now assume that the superconvergence equation (4.11) holds. Then one obtains indeed

$$\partial_\mu \partial_\nu (\theta(x_0) f^+ + \theta(-x_0) f^-) = \theta(x_0) \partial_\mu \partial_\nu f^+ + \theta(-x_0) \partial_\mu \partial_\nu f^- , \tag{4.12}$$

using

$$\delta(x_0) \partial_0 f^+ - \delta(x_0) \partial_0 f^- = 0 ,$$

as well as

$$\delta'(x_0) f^+ - \delta'(x_0) f^- = 0 .$$

5. INTERIM CUT-OFF METHOD: UNITARY REGULATORS

Because of divergencies, the definition of diagrams given in Section 2 may be meaningless. Moreover, the propagators may not satisfy the Kallén-Lehmann representation because of lack of superconvergence (see Section 4). To avoid such difficulties we introduce what we will call unitary regulators. This regulator method works for any theory in the sense that it allows a proper definition of all diagrams and is moreover very suitable in connection with the proof of unitary, causality, etc. It fails in the case of Lagrangians invariant under a gauge group, for which we will later introduce a more sophisticated method.

The prescription is exceedingly simple : construct things so that any propagator is replaced by a sum of propagators. The extra propagators correspond to heavy particles, and coefficients (the a_i in Eq. (4.3)) are chosen such that the high momentum behaviour is very good. This is achieved as follows. Introduce in the original Lagrangian $\mathcal{L}$, Eq. (2.1), sets of regulator fields ψ_i^λ and ϕ_i^λ in the following way:

$$\mathcal{L}^r = \psi_i^* V_{ij} \psi_j + \sum_\lambda a_\lambda \psi_i^{\lambda *} (V + M^\lambda)_{ij} \psi_j^\lambda + \phi_i W_{ij} \phi_j + $$

$$+ \sum_\lambda b_\lambda \phi_i^\lambda (W + m^\lambda)_{ij} \phi_j^\lambda + \mathcal{L}_I \left(\psi^* + \sum \psi^{\lambda *}, \psi + \sum \psi^\lambda, \phi + \sum \phi^\lambda \right). \tag{5.1}$$

The coefficients a_λ and b_λ and the mass matrices M^λ and m^λ must be chosen so as to assure the proper high momentum behaviour.

The propagators for the regulating fields ψ^λ are

$$\Delta_{Fij}^{R\lambda} = -a_\lambda^{-1} (V + M^\lambda)_{ij}^{-1} \ . \tag{5.2}$$

Because in the interaction Lagrangian ψ is everywhere replaced by $\psi + \Sigma\, \psi^\lambda$ (similarly ψ^* and ϕ), only the following combination of propagators will occur in the diagrams,

$$-\frac{1}{V} - \sum_\lambda \frac{1}{a_\lambda} \frac{1}{V + M^\lambda} \ . \tag{5.3}$$

Choosing the appropriate coefficients and mass matrices all diagrams will become finite. If every eigenvalue of the mass matrix is made very large, diagrams that were finite before the introduction of regulators will converge to their unregulated value.

EXAMPLE 5.1 Charged scalar particles with ϕ^4 interaction

$$\mathcal{L}^R = \psi^* (\partial^2 - m^2)\psi - \psi^{1*}(\partial^2 - m^2 - M^2)\psi^1 + \frac{g}{4}(\psi^* + \psi^{1*})^2 \times$$

$$\times (\psi + \psi^1)^2 . \qquad (5.4)$$

$$\psi \text{ propagator:} \quad \frac{1}{(2\pi)^4 i} \frac{1}{k^2 + m^2} . \qquad (5.5)$$

$$\psi^1 \text{ propagator:} \quad - \frac{1}{(2\pi)^4 i} \frac{1}{k^2 + m^2 + M^2} . \qquad (5.6)$$

The combination occurring in the diagrams is

$$\frac{1}{(2\pi)^4 i} \frac{M^2}{(k^2 + m^2)(k^2 + m^2 + M^2)} \qquad (5.7)$$

This behaves as k^{-4} at high momenta. The diagrams shown in Example 2.2.1 are now finite for large but finite M^2.

6. CUTTING EQUATIONS

6.1 Preliminaries

In order to keep the work of this section transparent, we will suppress indices, derivatives, etc. In particular, for vertices we retain only a factor $(2\pi)^4 ig$ in momentum space, that is ig in coordinate space. There is no difficulty whatsoever in reintroducing the necessary details.

It is assumed that diagrams are sufficiently regulated, so that no divergencies occur.

The starting point is the decomposition of the propagator into positive and negative energy parts

$$\Delta_{ij}(x) = \theta(x_0)\Delta^+_{ij}(x) + \theta(-x_0)\Delta^-_{ij}(x) \ , \qquad (6.1)$$

$$\Delta^{\pm}_{ij}(x) = \frac{1}{(2\pi)^3} \int d_4k e^{ikx}\theta(\pm k_0)\rho(k^2) \ , \qquad (6.2)$$

with $x = x_i - x_j$. Here we used the notation $\Delta_{ij}(x) = \Delta_{Fij}(x_i - x_j)$.

In view of the reality of the spectral functions ρ we have $\Delta^{\pm}_{ij} = (\Delta^{\mp}_{ij})^*$. Also

$$\Delta^{\pm}_{ij} = \Delta^{\mp}_{ji} \ . \qquad (6.3)$$

Consequently

$$\Delta^{*}_{ij} = \theta(x_i - x_j)\Delta^-_{ij} + \theta(x_j - x_i)\Delta^+_{ij} \ . \qquad (6.4)$$

As usual

$$\theta(x) = \frac{1}{2\pi i} \int_{-\infty}^{\infty} d\tau \frac{e^{i\tau x}}{\tau - i\varepsilon} = \begin{cases} 1 & \text{if } x_0 > 0 \\ 0 & \text{if } x_0 < 0 \end{cases} \qquad (6.5)$$

$$\theta(x) + \theta(-x) = 1 \ .$$

The representation (6.5) can be verified by choosing the integration contour with a large semi-circle in the upper (lower) complex plane for x_0 positive (negative).

Consider now a diagram with n vertices. Such a diagram represents in coordinate space an expression containing many propagators depending on arguments $x_1, \ldots, x_n$. We will denote such an expression by

$$F(x_1, x_2, \ldots, x_n).$$

For example, the triangle diagram represents the function:

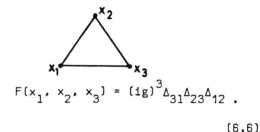

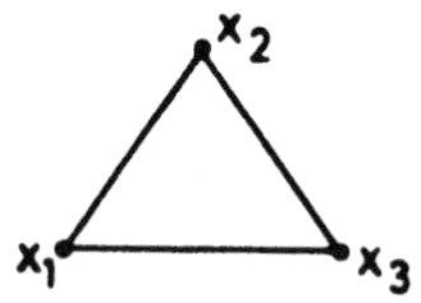

$$F(x_1, x_2, x_3) = (ig)^3 \Delta_{31}\Delta_{23}\Delta_{12} . \tag{6.6}$$

Every diagram, when multiplied by the appropriate source functions and integrated over all x contributes to the S-matrix. The contribution to the T-matrix, defined by

$$S = 1 + iT \tag{6.7}$$

is obtained by multiplying by a factor -i. Unitarity of the S-matrix implies an equation for the imaginary part of the so defined T matrix

$$T - T^\dagger = iT^\dagger T . \tag{6.8}$$

The T-matrix, or rather the diagrams, are also constrained by the requirement of causality. As yet nobody has found a definition of causality that corresponds directly to the intuitive notions; instead formulations have been proposed involving the off-mass-shell Green's functions. We will employ the causality requirement in the form proposed by Bogoliubov that has at least some intuitive appeal and is most suitable in connection with a diagrammatic analysis. Roughly speaking Bogoliubov's condition can be put as follows: if a space-time point x_1 is in the future with respect to some other space-time point x_2, then the diagrams involving x_1 and x_2 can be rewritten in terms of functions that involve positive energy flow from x_2 to x_1 only.

The trouble with this definition is that space-time points

cannot be accurately pinpointed with relativistic wave packets corresponding to particles on mass-shell. Therefore this definition cannot be formulated as an S-matrix constraint. It can only be used for the Green's functions.

Other definitions refer to the properties of the fields. In particular there is the proposal of Lehmann, Symanzik and Zimmermann that the fields commute outside the light cone. Defining fields in terms of diagrams, this definition can be shown to reduce to Bogoliubov's definition. The formulation of Bogoliubov causality in terms of cutting rules for diagrams will be given in Section 6.4.

6.2 The Largest Time Equation

Let us now consider a function $F(x_1, x_2, \ldots, x_n)$ corresponding to some diagram. We define new functions F, where one or more of the variables $x_1, \ldots, x_n$ are underlined. Consider

$$F(x_1, x_2, \ldots, \underline{x_i}, \ldots, \underline{x_j}, \ldots, x_n) . \tag{6.9}$$

This function is obtained from the original function F by the following:

i) A propagator Δ_{ki} is unchanged if neither x_k nor x_i are underlined.
ii) A propagator Δ_{ki} is replaced by Δ^+_{ki} if x_k but not x_i is underlined.
iii) A propagator Δ_{ki} is replaced by Δ^-_{ki} if x_i but not x_k is underlined.
iv) A propagator Δ_{ki} is replaced by Δ^*_{ki} if x_k and x_i are underlined.
v) For any underlined x replace one factor i by -i. Apart from that, the rules for the vertices remain unchanged.

(6.10)

Equations (6.1) and (6.4) lead trivially to an important equation, the largest time equation. Suppose the time x_{io} is larger than any other time component. Then any function F in which x_i is not underlined equals minus the same function but with x_i now underlined

$$F(x_1, \ldots, x_i, \ldots, \underline{x_j}, \ldots, x_n) = -F(x_1, \ldots, \underline{x_i}, \ldots, \underline{x_j}, \ldots, x_n). \tag{6.11}$$

The minus sign is a consequence of point (v). In view of what follows it is useful to invent a diagrammatic representation of the newly-defined functions:

> Any function F is represented by a diagram where any vertex corresponding to an underlined variable is provided with a circle.

EXAMPLE 6.2.1 If $F(x_1, x_2, x_3)$ is given by Eq. (6.6) then

$$F(x_1, \underline{x_2}, \underline{x_3}) = (ig)^3 \Delta^+_{31} \Delta^*_{23} \Delta^-_{12} \; . \qquad (6.12)$$

The corresponding diagram is:

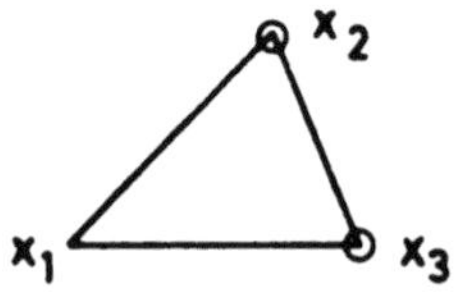

If the time component of x_3 is largest we have, for instance:

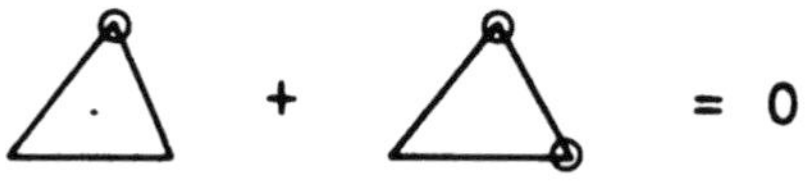

From such a diagram it is impossible to see if a given line connecting a circled to an uncircled vertex corresponds to a Δ^+ or Δ^- function. But due to Eq. (6.2) the result is the same anyway. The important fact is that energy always flows from the uncircled to the circled vertex, because of the θ function in Eq. (6.2). Of course there is no restriction on the sign of energy flow for lines connecting two circled or two uncircled vertices.

6.3 Absorptive Part

To obtain the contribution of a diagram to the S-matrix the corresponding function $F(x_1, \ldots, x_n)$ must be multiplied with the appropriate source functions for the ingoing and outgoing lines and integrated over all x_i. For instance, the function $F(x_1, \ldots, x_6)$ corresponding to the diagram:

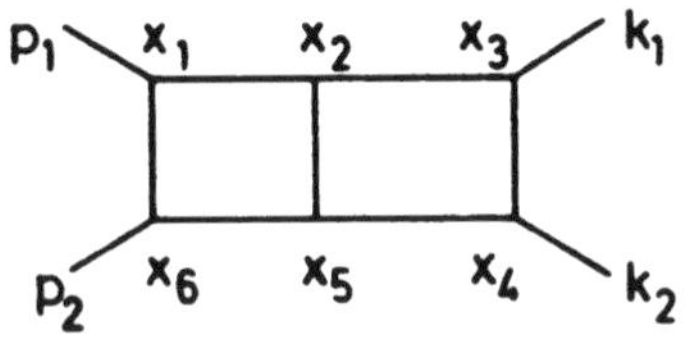

must be multiplied by

$$e^{ip_1x_1} e^{ip_2x_6} e^{-ik_1x_3} e^{-ik_2x_4}$$

and subsequently integrated over $x_1, \dots x_6$.

The restriction that Eq. (6.11) only holds if x_{io} is larger than the other time components makes it impossible to write down the analogue of Eq. (6.11) in momentum space.

We will now write down an equation which follows directly from the largest time equation, but holds whatever the time ordering of the various x_i. Thus consider a function $F(x_1, \dots, x_n)$ corresponding to some diagram. We have

$$\sum_{\text{underlinings}} F(x_1, \dots, \underline{x_i}, \dots, \underline{x_j}, \dots, x_n) = 0. \tag{6.13}$$

The summation goes over all possible ways that the variables may be underlined. For instance, there is one term without underlined variables, n terms with one underlined variable, etc. There is also one term, the last, where all variables are underlined. Under certain conditions, to be discussed in Section 7, that term is related to the first term

$$F(\underline{x_1}, \underline{x_2}, \dots, \underline{x_n}) = F(x_1, x_2, \dots, x_n)^* . \tag{6.14}$$

The proof of Eq. (6.13) is trivial. Let one of the x, say x_i, have the largest time component. Then on the left-hand side of Eq. (6.13) any diagram with x_i not underlined cancels against the term where in addition x_i is underlined, by virtue of the largest time equation.

Now we can multiply Eq. (6.13) by the appropriate source factors and integrate over all x. We obtain a set of functions $\bar{F}$ depending on the various external momenta and further internal momenta (loop momenta). Here, the bar on F denotes the Fourier transform. The functions $\bar{F}$ are composed of $\Delta^{\pm}$, Δ and Δ^* functions

$$\Delta^{\pm}(k) = \frac{1}{(2\pi)^3}\,\theta(\pm k_o)\int_{a\geq o}^{\infty} ds'\rho(s')\delta(k^2 + s')\ , \qquad (6.15)$$

$$\Delta(k) = \frac{1}{(2\pi)^4 i}\int_{a\geq o}^{\infty} ds'\rho(s')\,\frac{1}{k^2 + s' - i\varepsilon}\ . \qquad (6.16)$$

We now observe that in the resulting equation many terms will be zero, due to conflicting energy θ functions. Take, for example, a diagram*) with one underlined point which is not connected to an outgoing line:

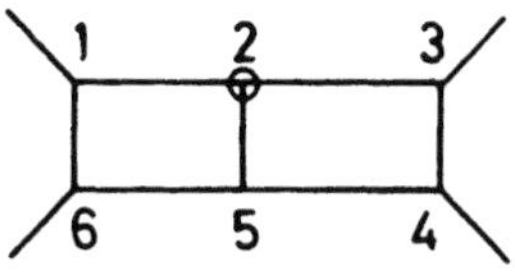

Because of the θ functions in the $\Delta^{\pm}$ (see Eq. (6.15)), energy is forced to flow towards the circled vertex. Since energy conservation holds in that vertex this is impossible and we conclude that this diagram is zero. The same is true if vertex 5 is circled instead. Further, if the momenta p_1 and p_2 represent incoming particles, which implies energy flowing from the outside into vertices 1 and 6, then also the diagrams with vertex 1 and/or vertex 6 circled are zero. Also the diagram with 2 and 5 circled is zero, even if now energy may flow in either direction between 2 and 5, because all other lines ending in 2 or 5 force energy flow towards these vertices. We thus come to the following result.

A diagram containing circled vertices gives rise to a non-zero contribution if the circled vertices form connected regions that contain one or more outgoing lines. And also the uncircled vertices must form connected regions involving incoming lines.

Thus, for example,

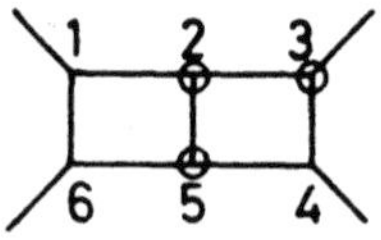

*) In this and in the following diagrams, incoming particles are at the left, outgoing at the right (energy flows in at the left, out at the right).

is zero because in vertex 4 we have a conflicting situation. Examples of non-zero diagrams are:

Note that an ingoing line may be attached to a circled region.

Since the circled vertices form connected regions we may drop the circles and indicate the region with the help of a boundary line:

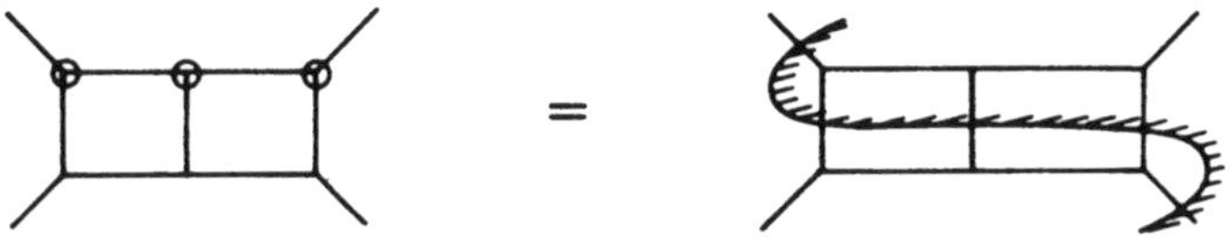

Here is an example of another diagram:

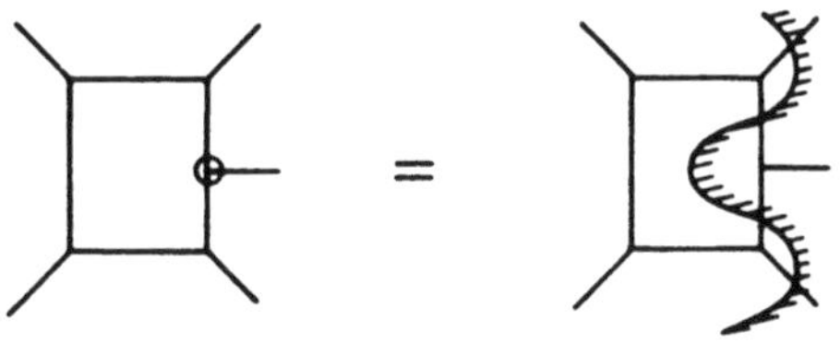

Note that no special significance is attached to the cutting of an external line.

Taking the above into account, Eq. (6.13) now reduces to

$$\bar{F}(k_1, \ldots, k_n) + \bar{\underline{F}}(k_1, \ldots, k_n) = - \sum_{\text{cuttings}} \bar{F}_c(k_1, \ldots, k_n) \ . \tag{6.17}$$

Here $\bar{F}$ is the Fourier transform of the function F without underlinings, $\bar{\underline{F}}$ the Fourier transform of the function F with all variables underlined. The functions F_c correspond to all non-zero diagrams containing both circled and uncircled vertices. They correspond to all possible cuttings of the original diagram with the prescription that for a cut line the propagator function $\Delta(k)$ must be replaced by $\Delta^{\pm}(k)$ with the sign such that energy is forced to flow towards the shaded region. Equation (6.17) is Cutkosky's cutting rule.

Remembering that the T-matrix is obtained by multiplying by -i, we see that Eq. (6.17) is of exactly the same structure as the unitarity equation (6.8). There is one notable difference: Eq. (6.17) holds for a single diagram, while unitarity is a property true for a transition amplitude, that is for the sum of diagrams contributing to a given process.

Equation (6.17) holds for any theory described by a Lagrangian, whether it is unitary or not. The Feynman rules for $\bar{\bar{F}}$ are, however, different from those for $T^{\dagger}$ (Section 2.6). Therefore, if Eq. (6.17) is truly to imply unitarity a number of properties must hold. This will be discussed later.

6.4 Causality

Again consider any diagram, that represents a function $F(x_1, \ldots, x_n)$. Take any two variables, say x_i and x_j. Let us suppose that the time component of x_j is larger than x_{io}. The following equation holds independently of the time ordering of the other time components

$$\sum_{\substack{\text{underlinings} \\ \text{except } x_i}} F(x_1, \ldots, \underline{x_k}, \ldots, x_n) = 0 \quad \text{if} \quad x_{io} < x_{jo} . \tag{6.18}$$

Again terms cancel in pairs. We do not need the diagrams where x_i is underlined, because we know for sure that x_{io} is never the largest time.

Equation (6.18), when multiplied by the appropriate source functions and integrated over all x except x_i and x_j, is the single diagram version of Bogoliubov's causality condition. His notation is

$$\frac{\delta^2 S}{\delta g(x_i)\delta g(x_j)} \hat{S} + \frac{\delta S}{\delta g(x_i)} \frac{\delta \hat{S}}{\delta g(x_j)} = 0 \quad \text{if} \quad x_{io} < x_{jo} . \tag{6.19}$$

Here the first term describes cut diagrams (including the case of no cut at all -- the unit part of $\hat{S}$) with x_i and x_j not circled, and the second term denotes diagrams with x_j but not x_i circled. $\hat{S}$ is the S-matrix obtained from the conjugate Feynman rules (i.e. all vertices underlined), and will often be equal to $S^{\dagger}$. Further g(x) is the coupling constant, made into a function of space-time.

Similarly we can consider the case where $x_{io} > x_{jo}$. Then we have an equation where now x_j is never to be underlined. Separating off the term with no variable underlined we may combine equations, with the result

$$F(x_1, x_2, \ldots, x_n) = -\theta(x_{jo} - x_{io}) \sum_i{}' F(x_1, \ldots, \underline{x_k}, \ldots, x_n) - \theta(x_{io} - x_{jo}) \sum_j{}' F(x_1, \ldots, \underline{x_k}, \ldots, x_n). \tag{6.20}$$

The prime indicates absence of the term without underlined variables. The index i implies absence of diagrams with x_i underlined.

The summations in Eq. (6.20) still contain many identical terms, namely those where neither x_i nor x_j are underlined. Also these may be taken together to give

$$F(x_1, \ldots, x_n) = -\sum_{ij}{}' F(x_1, \ldots, x_n) - \theta(x_{jo} - x_{io}) \times \sum_{\substack{j \text{ underlined} \\ i \text{ not}}} F(x_1, \ldots, x_n) - \theta(x_{io} - x_{jo}) \times \sum_{\substack{i \text{ underlined} \\ j \text{ not}}} F(x_1, \ldots, x_n) . \tag{6.21}$$

The first term on the right-hand side of Eq. (6.21) is a set of cut diagrams, with x_i and x_j always in the unshaded region. They represent the product $\bar{S}S$ with the restriction that x_i and x_j are vertices of S. We can now apply the same equation, with the same points x_i and x_j, to the diagrams for S in this product. Doing this as many times as necessary, the right-hand side of Eq. (6.21) can be reduced entirely to the sum of two terms, one containing a function $\theta(x_{io} - x_{jo})$ multiplying a function whose Fourier transforms contains θ functions forcing energy flow from i to j, the other containing the opposite combination. This is precisely of the form indicated in Section 6.1.

Let us now return to Eq. (6.21). Introducing for θ the Fourier representation Eq. (6.5), we can see θ as another kind of propagator connecting the points x_i and x_j. Multiplying by the appropriate source function and integrating over all x_i we obtain the following diagrammatic equation:

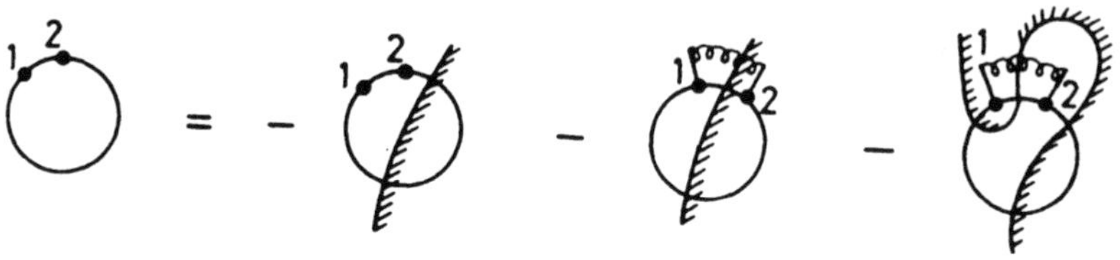

The blob stands for any diagram or collection of diagrams. The points 1 and 2 indicate two arbitrarily selected vertices. The "self inductance" is the contribution due to the θ function, and is obviously non-covariant:

$$\frac{1}{(2\pi)i}\,\frac{1}{-k_0 - i\varepsilon}\,\delta_3(\vec{k})\ . \qquad (6.23)$$

Of course, in the diagrams on the right-hand side summation over all cuts with the points 1 and 2 in the position shown is intended.

This is perhaps the right moment to summarize the Feynman rules for the cut diagrams. As an example we take the simple scalar theory:

Propagator in unshadowed region:

$$\frac{1}{(2\pi)^4 i}\,\frac{1}{k^2 + m^2 - i\varepsilon}$$

Propagator in shadowed region:

$$-\frac{1}{(2\pi)^4 i}\,\frac{1}{k^2 + m^2 + i\varepsilon} \qquad (6.24)$$

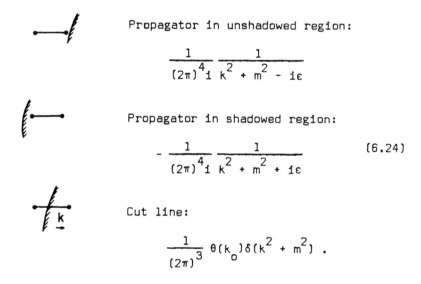

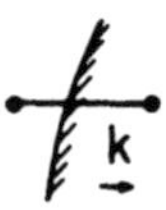

Cut line:

$$\frac{1}{(2\pi)^3}\,\theta(k_0)\delta(k^2 + m^2)\ .$$

Vertex in unshadowed region: $(2\pi)^4 ig$.

Vertex in shadowed region: $-(2\pi)^4 ig$.

For a spin-1/2 particle everything obtains by multiplying with the factor $-i\gamma k + m$:

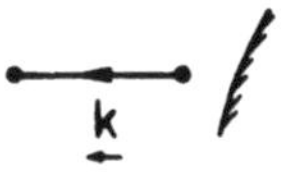

$$\frac{1}{(2\pi)^4 i} \frac{-i\gamma k + m}{k^2 + m^2 - i\varepsilon}$$

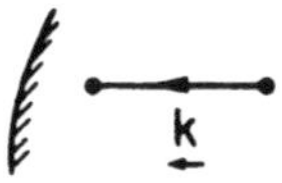

$$-\frac{1}{(2\pi)^4 i} \frac{-i\gamma k + m}{k^2 + m^2 + i\varepsilon} \tag{6.25}$$

$$\frac{1}{(2\pi)^3} (-i\gamma k + m)\theta(-k_0)\delta(k^2 + m^2)$$

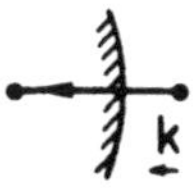

$$\frac{1}{(2\pi)^3} (-i\gamma k + m)\theta(k_0)\delta(k^2 + m^2).$$

The most simple application concerns the case of only one propagator connecting two sources. We will let these sources emit and absorb energy, but we will not put anything on mass-shell. Indeed, nowhere have mass-shell conditions been used in the derivations.

Thus consider:

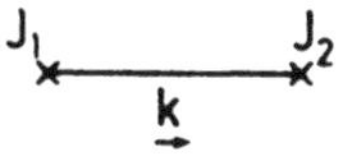

$$(2\pi)^4 i J_1(k) \frac{1}{k^2 + m^2 - i\varepsilon} J_2(k)$$

with J_1 and J_2 non-zero only if $k_0 > 0$. The unitarity equation (6.17) reads:

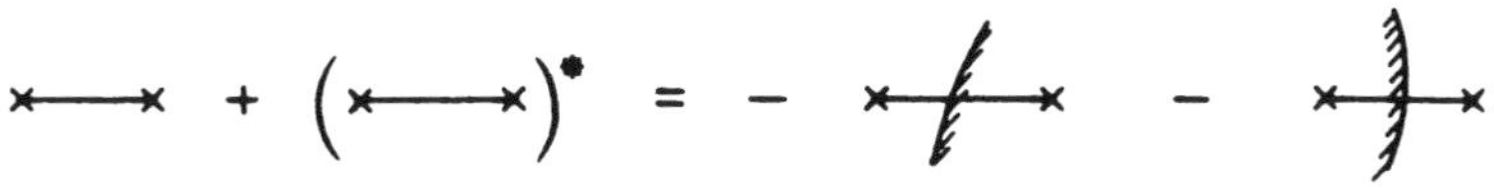

The complex conjugation does apply to everything except the sources J. The second term on the right-hand side is zero, because of the condition $k_0 > 0$. The equation becomes

$$J\left[(2\pi)^4 i \frac{1}{k^2 + m^2 - i\varepsilon} - (2\pi)^4 i \frac{1}{k^2 + m^2 + i\varepsilon}\right]J =$$

$$= J[+ \frac{i^2(2\pi)^8}{(2\pi)^3} \theta(k_0)\delta(k^2 + m^2)]J.$$

Note that the vertex in the shadowed region gives a factor $-(2\pi)^4 i$. With

$$\frac{1}{a - i\varepsilon} = P\left(\frac{1}{a}\right) + i\pi\delta(a) ,$$

we see the equation holds true.

Also Eq. (6.22) can be verified:

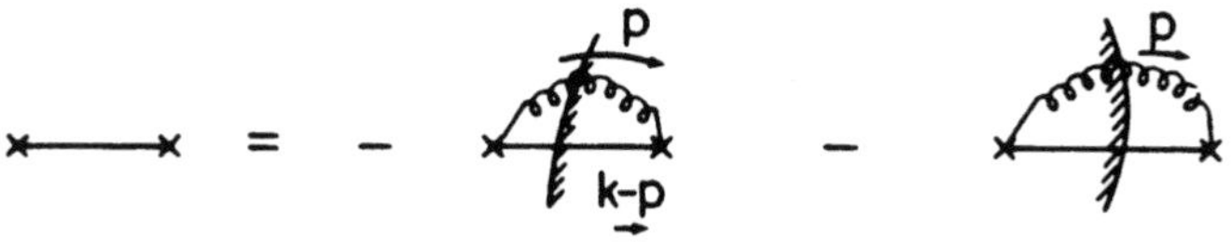

We now obtain (note the minus sign for vertex in shadowed region)

$$(2\pi)^4 i \frac{1}{k^2 + m^2 - i\varepsilon} = \frac{(2\pi)^8 i^2}{(2\pi)^3 2\pi i} \times$$

$$\times \int_{-\infty}^{\infty} dp_0 \left(\frac{1}{-p_0 - i\varepsilon} \theta(k_0 - p_0)\delta[(k - p)^2 + m^2] + \right.$$

$$\left. + \frac{1}{p_0 - i\varepsilon} \theta(-k_0 + p_0)\delta[(k - p)^2 + m^2] \right)_{\vec{p}=0} .$$

The four-vector p_0 has zero space components (see expression (6.23)). The p_0 integration is trivial and gives the desired result.

6.5 Dispersion Relations

Equations (6.22) are nothing but dispersion relations, valid for any diagram. Let τ be the fourth component of the momentum flowing through the self-inductance. Let $f^+(\tau)$ be the function corresponding to the cut diagrams excluding the τ-line in the second term on the right-hand side of Eq. (6.22), with τ directed from 1 to 2. Similarly for $f^-(\tau)$. If f and f' represent the left-hand side and the first term on the right-hand side, respectively, we have

$$f = -f' - \frac{1}{2\pi i} \int_{-\infty}^{\infty} \frac{d\tau}{-\tau - i\varepsilon} f^{+}(\tau) - \frac{1}{2\pi i} \int_{-\infty}^{\infty} \frac{d\tau}{\tau - i\varepsilon} f^{-}(\tau) \; . \tag{6.26}$$

Of course, all the functions f depend on the various external momenta. The function $f^{+}(\tau)$ will be zero for large positive τ, namely as soon as τ becomes larger than the total amount of energy flowing into the diagram in the unshadowed region. Similarly $f^{-}(\tau)$ is zero for large negative τ.

The dispersion relations Eq. (6.26) are very important in connection with renormalization. If all subdivergencies of a diagram have been removed by suitable counter terms, then all cut diagrams will be finite (involving products of subdiagrams with certain finite phase-space integrals). According to Eq. (6.26) the infinities in the diagram must then arise because of non-converging dispersion integrals. Suitable subtractions, i.e. counter terms, will make the integrals finite.†

It may finally be noted that our dispersion relations are very different from those usually advertised. We do not disperse in some external Lorentz invariant, such as the centre-of-mass energy or momentum transfer.

†Note added: Indeed, a rigorous proof of multiplicative renormalizability of many theories, at the expense of exclusively local counter terms at all orders, can be constructed without too much trouble from these dispersion relations.

7. UNITARITY

If the cutting equation (6.17), diagrammatically represented as:

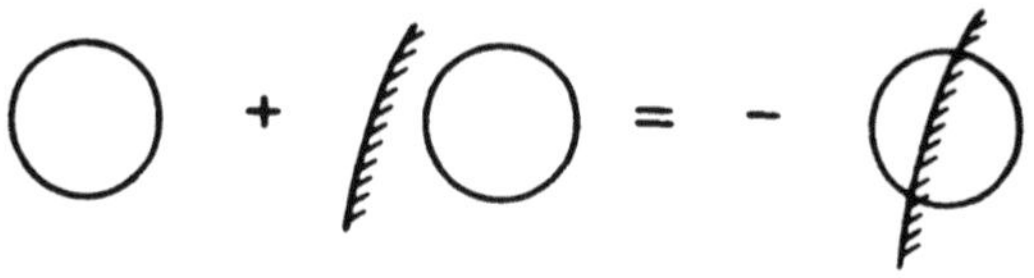

corresponding to $T - T^\dagger = iT^\dagger T$, is to imply unitarity, the following must hold:

i) The diagrams in the shadowed region must be those that occur in $S^\dagger$;
ii) The Δ^+ functions must be equal to what is obtained when summing over intermediate states.

Referring to our discussion of the matrix $S^\dagger$, in Section 2.6, we note that point (i) will be true if the Lagrangian generating the S-matrix is its own conjugate.

Point (ii) amounts to the following. The two-point Green's function, on which the definition of the S-matrix sources was based, contained a matrix K_{ij} (see Eq. (2.12) and following). In considering $S^\dagger S$ one will encounter (particle-out of S, particle-in of $S^\dagger$):

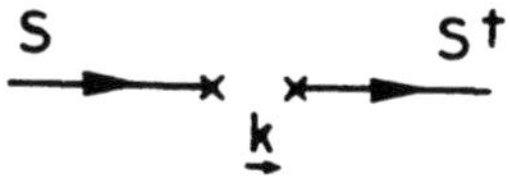

$$\sum_a K^\dagger_{ij}(-k) J_j^{*(a)}(k) J_\ell^a(k) K_{\ell m}(-k) \qquad (7.1)$$

in the sum over intermediate states. The K are from the propagators attached to the sources. Because of $\mathcal{L} = (\mathcal{L})_{conjugate}$ we have $K^\dagger_{\ell m}(-k) = K_{\ell m}(k)$. Also if $J(k)K(-k) \sim J(k)$ then $K^\dagger(-k)J^* \sim J^*$, showing that J and J^* are the appropriate eigen currents of S and $S^\dagger$. If unitarity is to be true we require that this sum (7.1) occurring in $S^\dagger S$ equals the matrix K_{im} occurring when cutting a propagator.

The proof of this is trivial. Suppose K_{ij} is diagonal with diagonal elements λ_i. The current defining equations (2.13) and (2.14) imply that the currents are of the form

$$J^{(a)} = \begin{pmatrix} 0 \\ \cdot \\ \cdot \\ \cdot \\ 0 \\ 1/\sqrt{\lambda_a} \\ 0 \\ \cdot \\ \cdot \\ \cdot \\ 0 \end{pmatrix}$$

There are no currents corresponding to zero eigenvalues. Obviously

$$\sum_a J^{(a)*} J^{(a)} = K^{-1} , \tag{7.2}$$

and this remains true if one provides the currents with phase factors, etc.

For spin-1/2 particles things are slightly more complicated, because of γ^4 manipulations. For instance, one will have

$$K^{\dagger}(-k)\gamma^4 = \gamma^4 K(k) . \tag{7.3}$$

Also the normalization of the currents is different. One finds the correct expression when summing up particle-out/particle-in states, but a minus sign extra for antiparticle-out/antiparticle-in states. This factor is found back in the prescription -1 for every fermion loop. A few examples are perhaps useful.

EXAMPLE 7.1

$$\mathcal{L} = -\bar{\psi}(\gamma\partial + m)\psi + \frac{1}{2}\phi(\partial^2 - m^2)\phi + g\bar{\psi}\psi\phi .$$

Propagators and vertices have been given before. The appropriate source functions and related equations are given in Appendix A.

There are four two-point Green's functions:

$$\bar{u}^a(k)\,\frac{-i\gamma k + m}{k^2 + m^2}\,u^a(k)\,\frac{2k_o}{4m^2} \;;\; a = 1, 2 ,$$

$$-\bar{u}^a(k)\,\frac{i\gamma k + m}{k^2 + m^2}\,u^a(k)\,\frac{2k_0}{4m^2}\;;\; a = 3,\,4\,.$$

Note the minus sign for the incoming antiparticle wave-function.

Scalar particle self-energy (we write also δ functions):

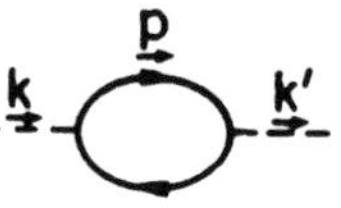

$$-g^2\delta_4(k - k')\int d_4p\,\frac{-i\gamma p + m}{p^2 + m^2 - i\varepsilon}\;\frac{-i\gamma(p - k) + m}{(p - k)^2 + m^2 - i\varepsilon}\,.$$

Note the minus sign for the closed fermion loop. Cut diagram (remember $-(2\pi)^4 i$ for vertex in shadowed region):

$$-(2\pi)^2 g^2\delta_4(k - k')\int d_4p(-i\gamma p + m)\theta(p_0)\delta(p^2 + m^2)\;\times$$

$$\times\;(-i\gamma(p - k) + m)\theta(k_0 - p_0)\delta\left[(p - k)^2 + m^2\right].$$

Decay of scalar into two fermions:

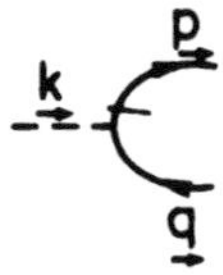

$$(2\pi)^4 i g\,\sqrt{4p_0q_0}\;\bar{u}(p)u^\alpha(q)\delta_4(k - p - q)\,.$$

The superscript α now indicates antiparticle spinor. The complex conjugate, but with k' instead of k, is

$$-(2\pi)^4 i g\sqrt{4p_0q_0}\;\bar{u}^\alpha(q)u(p)\delta_4(k' - p - q)\,.$$

The product of the two summed over intermediate states is

$$(2\pi)^8 g^2 4p_0q_0\int\frac{d_3p}{(2\pi)^3 2p_0}\,\frac{d_3q}{(2\pi)^3 2q_0}\,\delta_4(k - p - q)\delta_4(k' - p - q)\times$$

$$\times \frac{1}{2p_0} (-i\gamma p + m) \frac{-1}{2q_0} (i\gamma q + m) \ .$$

Note the minus sign for the q-spinor sum.

Since $p_0 = \sqrt{\vec{p}^2 + m^2}$ we have

$$\int \frac{d_3 p}{2p_0} = \int d_4 p \theta(p_0) \delta(p^2 + m^2)$$

and similarly for q. The q integration can be performed

$$-(2\pi)^2 g^2 \delta_4(k - k') \int d_4 p \theta(p_0) \delta(p^2 + m^2) \theta(k_0 - p_0) \times$$

$$\times [-i\gamma(p - k) + m] \delta[(p - k)^2 + m^2] (-i\gamma p + m) \ ,$$

which indeed equals the result for the cut diagram. The minus sign for the closed fermion loop appears here as a minus sign in front of the antiparticle spinor summation. One may wonder what happens in the case where an antiparticle line is cut, but when there is no closed fermion loop. An example is provided by the antiparticle self-energy as compared to particle self-energy:

Somehow there must also be an extra minus sign for the first diagram. Indeed it is there, because the first diagram contains an incoming antiparticle whose wave-function has a minus sign.

All this demonstrates a tight interplay between statistics (minus sign for fermion loops) and the transformation properties under Lorentz transormations of the spinors. The latter requires the normalization to energy divided by mass as given before, and also relates to the minus sign for antiparticle source summation.

If an integer spin field is assigned Fermi statistics in the form of minus signs for line interchanges then unitarity will be violated.

8. INDEFINITE METRIC

If the numerator K of a propagator has a negative eigenvalue at the pole then unitarity cannot hold because Eq. (7.2) cannot hold. Unitarity can be restored if we introduce the convention that a minus sign is to be attached whenever such a state appears. It is said that the state has negative norm, and transition probabilities as well as cross-sections and lifetimes can now take negative values. This is, of course, physically uncacceptable, and particles corresponding to these states are called ghosts. In theories with ghosts special mechanisms must be present to assure absence of unphysical effects. In gauge theories negative metric ghosts occur simultaneously with certain other particles with positive metric, in such a way that the transition probabilities cancel. Also the second type of particle, although completely decent, is called a ghost and has no physical significance.

9. DRESSED PROPAGATORS

The perturbation series as formulated up to now will in general be divergent in a certain region. Consider the case of a scalar particle interacting with itself and possibly other particles. Let $\delta_4(k - k')\Gamma(k^2)$ denote the contribution of all self-energy diagrams that cannot be separated into two pieces by cutting one line. These diagrams are called irreducible self-energy diagrams. The two-point Green's function for this scalar particle is of the form

$$-(2\pi)^8\delta_4(k - k')\bar{\Delta}_F(k). \tag{9.1}$$

The function $\bar{\Delta}_F$ is called the dressed propagator. This in contrast to the propagator of the scalar particle, called the bare propagator. If we denote this bare propagator by Δ_F we find

$$\bar{\Delta}_F = \Delta_F + \Delta_F\Gamma\Delta_F + \Delta_F\Gamma\Delta_F\Gamma\Delta_F + \dots . \tag{9.2}$$

Summing this series

$$\bar{\Delta}_F = \frac{\Delta_F}{1 - \Gamma\Delta_F} \tag{9.3}$$

corresponding to the diagrams:

where the hatched blobs stand for the irreducible self-energy diagrams.

The function Γ is proportional to the coupling constant of the theory. It is clear that the perturbation series converges only if $\Gamma\Delta_F < 1$. But if Δ_F has a pole for a certain value of the four-momentum then this series will certainly not converge near this pole, unless Γ happens to be zero there. And if we remember that the definition of the S-matrix involves precisely the behaviour of the propagators at the poles we see that this problem needs discussion.

There are two possible solutions to this difficulty. The simplest solution is to arrange things in such a way that indeed Γ

is zero at the pole. This can be done by introducing a suitable vertex in the Lagrangian. This vertex contains two scalar fields and equals minus the value of Γ at the pole. For instance, suppose

$$\Delta_F = \frac{1}{(2\pi)^4 i} \frac{1}{k^2 + m^2 - i\varepsilon} . \qquad (9.4)$$

The function $\Gamma(k^2)$ can be expanded at the point $k^2 + m^2 = 0$

$$\Gamma(k) = \Gamma_0 + (k^2 + m^2)\Gamma_1 + \Gamma_2(k). \qquad (9.5)$$

Γ_0 and Γ_1 are constants, Γ_2 is of order $(k^2 + m^2)^2$. Introduce in the Lagrangian a term that leads to the vertex:

$$-[\Gamma_0 + (k^2 + m^2)\Gamma_1]. \qquad (9.6)$$

Instead of Γ we will now have a function Γ':

Γ' will be of order $(k^2 + m^2)^2$ and the series (9.2) converges at the pole. Actually this reasoning is correct only in lowest order because the new vertex can also appear inside the hatched blob. Through the introduction of further higher order vertices of the type (9.6) the required result can be made accurate to arbitrary order.

This procedure, involving mass and wave-function renormalization corresponding to Γ_0 and Γ_1 type terms, respectively, embodies certain inconveniences. First of all, it may be that Γ_0 and Γ_1 contain imaginary parts. This occurs if the particle becomes unstable because of the interaction. Such truly physical effects are part of the content of the theory and the above neutralizing procedure cannot be carried through. Furthermore, in the case of gauge theories, the freedom in the choice of terms in the Lagrangian is limited by gauge invariance, and it is not sure that the procedure can be carried through without gauge invariance violation.

An alternative solution is to use directly the summed expression (9.3) for the propagators in the diagrams. These diagrams must then, of course, contain no further internal self-energy parts, that is they must be skeleton diagrams. The function Γ occurring in the propagator must then be calculated with a certain accuracy in the coupling constant. For instance, in lowest order for any Green's function the recipe is to compute tree diagrams using bare propagators. In the next order there are diagrams with one closed loop (no self-energy loops) and bare propagators, and tree diagrams with dressed propagators where Γ is computed by considering one-closed-loop self-energy diagrams.

It is clear that this recipe leads to all kinds of complications, which however do not appear to be very profound. Mainly, kinematics and the perturbation expansion have to be considered together. With respect to renormalization the complications are trivial, because then only the behaviour of the propagator for very large k^2 is of importance. But then the series expansion Eq. (9.2) is permitted.

For the rest of this section we will consider the implications of the use of dressed propagators for cutting rules.

As a first step we note that the dressed propagator satisfies the Kallén-Lehmann representation. This follows from the causality relation Eq. (6.21), or in picture Eq. (6.22) for the two-point Green's function, where x_1 and x_2 are taken to be the in-source and out-source vertices, repectively. The first term on the right-hand side of these equations is then zero. Indeed, this gives precisely Eq. (6.1), the decomposition of the (dressed)propagator into positive and negative frequency parts. The derivation of the cutting relations, which is based on this decomposition goes through unchanged.

However, with respect to unitarity there is a further sublety. Using dressed propagators the prescription is to use only diagrams without self-energy insertions, that is skeleton diagrams. Now the Δ^+ function corresponding to a dressed propagator is evidently obtained by cutting a dressed propagator. The dressed propagator is a series, and cutting gives the result (we use again the example of a scalar particle)

$$\mathrm{Re}\,\bar{\Delta}_F = \bar{\Delta}_F(-\mathrm{Re}\,\Gamma)\bar{\Delta}_F^* + \text{pole part} \tag{9.7}$$

or

$$\mathrm{Re}\,\bar{\Delta}_F = \mathrm{Re}\,\frac{1}{(2\pi)^4 i[k^2+m^2+i\Gamma/(2\pi)^4]} =$$

$$= \frac{-\mathrm{Re}\,\Gamma}{(2\pi)^8|k^2+m^2+i\Gamma/(2\pi)^4|^2} + \frac{\delta(k^2+M^2)}{2(2\pi)^3 Z}\,, \tag{9.8}$$

where we used

$$k^2 + m^2 + i\Gamma/(2\pi)^4 \sim Z(k^2 + M^2 - i\varepsilon) + O[(k^2 + M^2)^2] \tag{9.9}$$

near the pole. Now Re Γ is obtained when cutting the irreducible self-energy diagrams. In diagrammatic form we have:

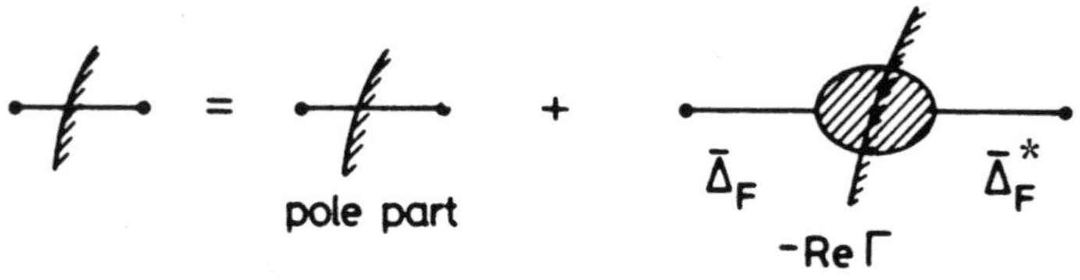

Note the occurrence of dressed propagators in the second term on the right.

The subtlety hinted at above is the following. The matrix S contains skeleton diagrams with dressed propagators. Cutting these diagrams apparently results in expressions obtained when cutting self-energy diagrams. Indeed, these arise automatically in $S^\dagger S$. If S contains skeleton diagrams of the type:

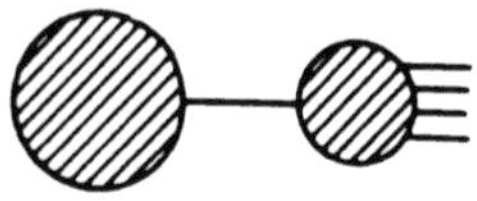

then $S^\dagger S$ contains the type:

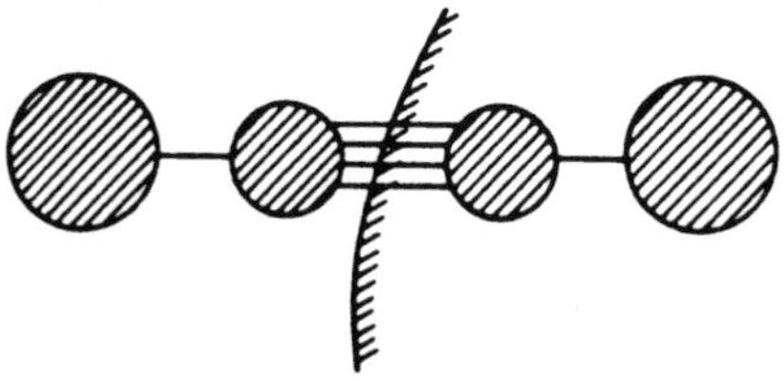

Even if S and $S^\dagger$ contain no reducible diagrams, the product $S^\dagger S$ nevertheless has self-energy structures. They correspond to what is obtained by cutting a dressed propagator.

10. CANONICAL TRANSFORMATIONS

10.1 Introduction

In this section we study the behaviour of the theory under field transformations. Fields by themselves are not very relevant quantities, from the physical point of view. The S-matrix is supposed to describe the physical content of the theory, and there is no direct relation between S-matrix and fields. Given the Green's functions fields may be defined; the Green's functions, however, can be considered as a rather arbitrary extension of the S-matrix to off-mass-shell values of the external momenta. Within the framework of perturbation theory it is possible, up to a point, to define the fields by considering those Green's functions that behave as smoothly as possible when going off mass-shell. For gauge theories this is still insufficient to fix the theory, there being many choices of fields (and Green's functions) that give the same physics (S-matrix) with equally smooth behaviour. For gravitation the situation is even more bewildering, up to the point of frustration.

In the study of field transformations path integrals have been of great heuristic value. The essential characteristics will be shown in the next section.

10.2 Path Integrals

A few simple equations form the basis of all path integral manipulations, and will be listed here.

Let α be a complex number with a positive non-zero imaginary part. Furthermore, $z = x + iy$ is a complex variable. We have

$$\iint dz e^{i\alpha z^* z} = \int_{-\infty}^{\infty} dx \int_{-\infty}^{\infty} dy e^{i\alpha(x^2+y^2)} . \qquad (10.1)$$

Introducing polar coordinates

$$= 2\pi \int_0^{\infty} e^{i\alpha r^2} r\, dr = \frac{i\pi}{\alpha} . \qquad (10.2)$$

Incidentally, realizing that the expression (10.1) is a pure square

$$\int_{-\infty}^{\infty} dx e^{i\alpha x^2} = \sqrt{\frac{i\pi}{\alpha}} . \qquad (10.3)$$

Let now z be a complex n-component vector, and A a complex

n x n matrix. The generalization of Eqs. (10.1), (10.2) is

$$\int dz_1 \ldots dz_n e^{i(z^*,Az)} = \frac{\pi^n i^n}{\det(A)} , \qquad (10.4)$$

where

$$\int dz_j = \int_{-\infty}^{\infty} dx_j \int_{-\infty}^{\infty} dy_j , \quad z_j = x_j + iy_j . \qquad (10.5)$$

Equation (10.4) follows trivially from Eqs. (10.1), (10.2) in the case where A is diagonal. Next note that the integration measure is invariant for unitary transformations U. To see this write

$$U = A + iB ,$$

where A and B are real matrices. The fact that $UU^\dagger = 1$ implies

$$A\tilde{A} + B\tilde{B} = 1 , \quad B\tilde{A} - A\tilde{B} = 0 , \qquad (10.6)$$

where the wiggle denotes reflection. If now $z' = Uz$ then $x' = Ax - By$ and $y' = Ay + Bx$. That is, the 2n dimensional vector x, y transforms as

$$\begin{pmatrix} x' \\ y' \end{pmatrix} = \begin{pmatrix} A & -B \\ B & A \end{pmatrix} \begin{pmatrix} x \\ y \end{pmatrix} .$$

The determinant of this 2n x 2n transformation matrix is 1. In fact, the matrix is orthogonal, because multiplication with its transpose gives 1, on account of the identities (10.6).

Because of the invariance of both the integration measure as well as the complex scalar product under unitary transformations, we conclude that Eq. (10.5) holds for any complex matrix A that can be diagonalized by means of a unitary transformation and where all diagonal matrix elements have a positive non-zero imaginary part.

Consider now a path integral involving a Lagrangian depending on a set of real fields A_i (see Section3)

$$\int \mathcal{D}A_i e^{iS(A)} \qquad (10.7)$$

where the action S is given by

$$S(A) = \int d_4x\mathcal{L}(A) \ . \tag{10.8}$$

Suppose we want to use other fields B_i that are related to the fields A_i as follows

$$A_i(x) = B_i(x) + f_i(x, B) \ . \tag{10.9}$$

The f_i are arbitrary functions (apart from the fact that Eq. (10.9) must be invertible). They may depend on the fields B at any sapce-time point including the space-time point x.

According to well-known rules

$$\mathcal{D}A_i = \det\left(\frac{\partial A_i}{\partial B_j}\right) B_j \tag{10.10}$$

or

$$\mathcal{D}A_i = \det\left(\delta_{ij} + \frac{\delta f_i(B)}{\delta B_j}\right)\mathcal{D}B_j \ . \tag{10.11}$$

The determinant is simply the Jacobian of this transformation. It is very clumsy to work with this determinant, especially if we realize that it involves the fields at every space-time point separately. Fortunately there exits a nice method that makes things easy. Let ϕ be a complex field. According to Eq. (10.4) we have

$$\frac{1}{\det\,(X_{ij})} = C\int\mathcal{D}\phi e^{iS(\phi)} \ , \tag{10.12}$$

where C is an irrelevant numerical factor, and

$$S(\phi) = \int d_4x\phi_i^* X_{ij}\phi_j \ . \tag{10.13}$$

However, Eq. (10.11) involves a determinant, while Eq. (10.12) is for the inverse of a determinant. We must invert Eq. (10.12).

The expression on the right-hand side of Eq. (10.12) is a path integral of the type considered in Section 3. It involves a complex field ϕ as well as fields B_i that appear simply as sources. The "action" is (see Eq. (10.11))

$$S(\phi) = \int d_4x\phi_i^*(x)\phi_i(x) + \int d_4x\; d_4x'\phi_i^*(x)Y_{ij}(x,\, x',\, B)\phi_j(x') \tag{10.14}$$

with

$$Y_{ij}(x,\, x',\, B) = \delta f_i(x,\, B)/\delta B_j(x')\ . \tag{10.15}$$

The diagrams corresponding to Eq. (10.12) involve a ϕ-propagator that is simply 1. There are only vertices involving two ϕ-lines:

$$-\frac{1}{(2\pi)^4 i}\,\delta_{ij}\ ,$$

$$(2\pi)^4 iY(k',\, k,\, B)\ .$$

The blob contains B-fields, but we have not indicated them explicitly. The only diagrams that can occur are closed ϕ-loops involving one, two or more Y-vertices. We will write down the first few:

Zeroth order in Y: 1

First order : 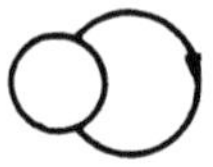

Second order : $+\frac{1}{2}$ 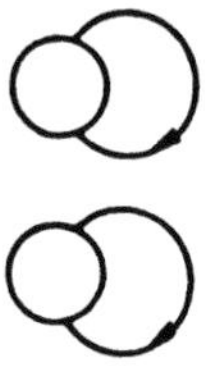 $+\frac{1}{2}$ 

The explicitly written factor 1/2 is a combinatorial factor. It follows also by treating the path integral along the lines indicated in Section 3. In fact, it is easily established that in n^{th} order there will be a contribution of n times the first-order loop with a factor 1/n! Similarly for the two-Y loop. In this way, it is seen that the whole series adds up to an exponential, with in the exponent only single closed loops

$$\int \mathcal{D}\phi e^{iS(\phi)} = C \exp\,(\Gamma) \tag{10.16}$$

$\Gamma =$ 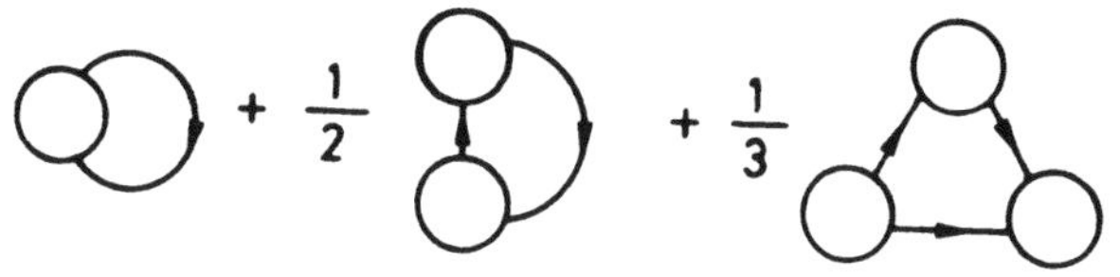 $+ \cdots$

This can be easily inverted simply by replacing Γ by $-\Gamma$, that is by the prescription that every closed ϕ-loop must be given a minus sign. We so arrive at the equation

$$\int \mathcal{D} A_i \, e^{iS(A)} = \int \mathcal{D} B_i \int \mathcal{D} \phi_i \, e^{iS(\phi)+iS'(B)} \tag{10.17}$$

with the additional rule that every closed ϕ-loop must be given a minus sign. $S(\phi)$ is given in Eq. (10.14). Finally $S'(B)$ follows by substituting the transformation Eq. (10.9) into the action for the A-field, Eq. (10.8)

$$S'(B) = S[B + f(B)] \; . \tag{10.18}$$

Summarizing, the theory remains unchanged if a field transformation is performed, provided closed loops of ghost particles (with a minus sign/loop) are also introduced. The vertices in the ghost loop are determined by the transformation law.

In the following sections we will derive this same result, however without the use of path integrals.

10.3 Diagrams and Field Transformations

In this section we will consider field transformations of the very simplest type. This we do in order to make the mechanism transparent.

Consider the Lagrangian

$$\mathcal{L}(\phi, \phi^*, A) = \frac{1}{2} \phi_i V_{ij} \phi_j + \frac{1}{2} A_i (\partial^2 - m^2) A_i + \mathcal{L}_I(\phi) + J_i \phi_i + H_i A_i . \tag{10.19}$$

We assume V to be symmetric. $\mathcal{L}_I$ is any interaction Lagrangian not involving the A-fields; the latter are evidently free fields. The diagrams corresponding to this Lagrangian are well-defined; if necessary use can be made of the previously-given regularization procedure. The ϕ- and A-propagators are:

$$\frac{1}{(2\pi)^4 i}\,(-V^{-1})_{ij}\ ,$$

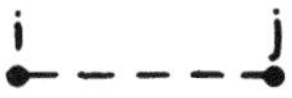

$$\frac{1}{(2\pi)^4 i}\,\frac{\delta_{ij}}{k^2 + m^2 - i\varepsilon}\ . \qquad (10.20)$$

The vertices stemming from $\mathcal{L}_I$ involve ϕ-lines only.

We want to study another Lagrangian $\mathcal{L}'$ obtained from by the replacement

$$\phi_i \to \phi_i + \alpha_{ij}A_j\ , \qquad (10.21)$$

where α is any matrix. The A remain unchanged.

<u>THEOREM</u> The Green's functions of the new Lagrangian $\mathcal{L}'$ are equal to those of the original Lagrangian. In particular, A remains a free field, that is all Green's functions involving an A as external leg are zero, except for A-propagators connecting directly two A-sources.

The proof of this theorem is very simple, and consists mainly of expanding the quadratic part of $\mathcal{L}'$. We have (dropping all indices)

$$\mathcal{L}' = \frac{1}{2}\,\phi V\phi + \phi V\alpha A + \frac{1}{2}\,A\tilde{\alpha}V\alpha A + \frac{1}{2}\,A(\partial^2 - m^2)A +$$

$$+ \mathcal{L}_I(\phi + \alpha A) + J(\phi + \alpha A) + HA\ . \qquad (10.22)$$

Here the wiggle denotes reflection.

Rather then trying to invert the complete matrix in the quadratic part we will treat as the propagator part only the terms $\phi V\phi$ and $A(\partial^2 - m^2)A$. The remaining terms are treated as interaction terms. We obtain the following A-ϕ vertices:

$$(2\pi)^4 i(V\alpha)_{ij}\ ,$$

$$(2\pi)^4 i(\tilde{\alpha}V\alpha)_{ij}\ . \qquad (10.23)$$

They will be called "special vertices". In addition to these there will be many vertices involving A-lines, because of the replacement (10.21) performed in the interaction Lagrangian.

Consider now any vertex containing an A-field (excluding the A-sources H). Because the A-line arises due to the substitution (10.21) there always exists a similar vertex with the A-line replaced by a ϕ-line. But the A-line can be connected to that vertex also after transformation to a ϕ-line via one of the special vertices. The sum of the two possibilities is zero.

EXAMPLE 10.3.1 In the transformed Lagrangian $\mathcal{L}'$, Eq. (10.22) we have a new vertex

$(2\pi)^4 iJ\alpha$

connecting an A-line with the source of the ϕ-field. In addition, we have the original vertex:

$(2\pi)^4 iJ$

and then also the diagram involving one special vertex. The factor V in the vertex cancels:

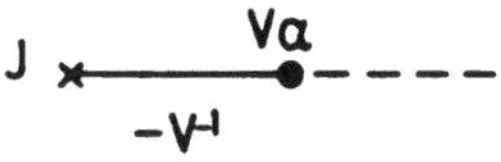

against the propagator, leaving a minus sign. The diagram cancels against the previous one.

EXAMPLE 10.3.2 For the two-point Green's function at the zero loop level we have:

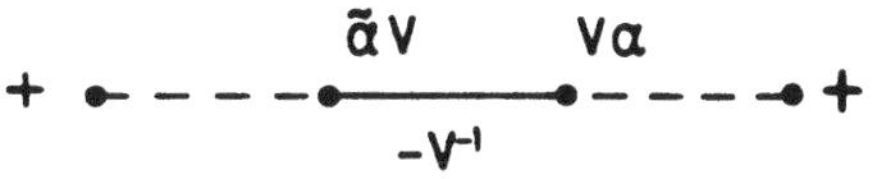

is with α^4, etc.

Only the first term survives; all others cancel in pairs.

Diagrams with loops involving A-lines also cancel since they involve vertices already discussed.

The above theorem can easily be generalized to the case of more complicated transformations, such as

$$\phi \to \phi + f(A)$$

with f any function of the A fields.

EXAMPLE 10.3.3 Consider the transformation

$$\phi_i \to \phi_i + \alpha_{ijk\ell} A_j A_k A_\ell \; .$$

The following special vertices are generated (we leave the factors $(2\pi)^4 i$ as understood):

Again, by the same trivial mechanism as before we have, for example:

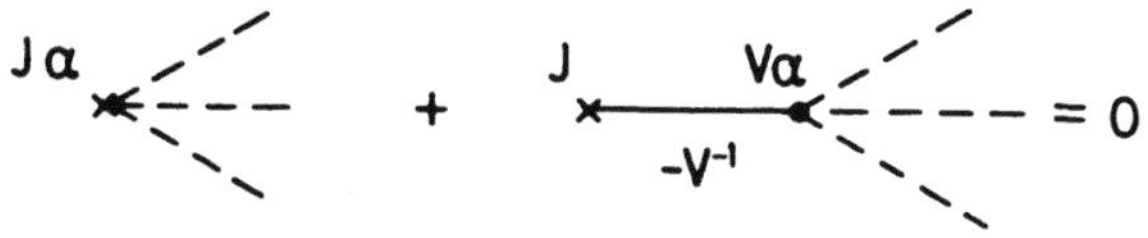

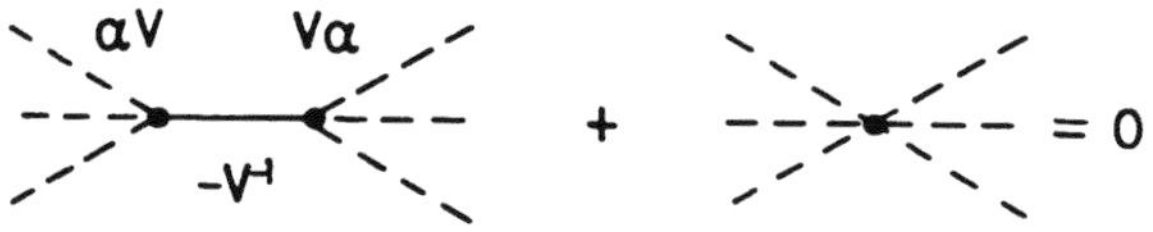

The situation becomes more complicated if we allow the function f also to depend on the ϕ-fields, even in a non-local way.

10.4 Local and Non-local Transformations

Starting again from the Lagrangian (10.19) we consider now the general substitution

$$\phi \to \phi + f(A, \phi) \; .$$

We will now have special vertices involving the unspecified function f. The function f will be represented as a blob with a certain number of A- and ϕ-lines of which we will indicate only three explicitly. The propagator matrix V is always there as a factor and is indicated by a dot. The ϕ-line attached at that point will be called the original ϕ-line of the special vertex.

Now this original ϕ-line may be connected to any of the old vertices of the theory or to another original ϕ-line. Again the cancellation mechanism works.

EXAMPLE 10.4.1

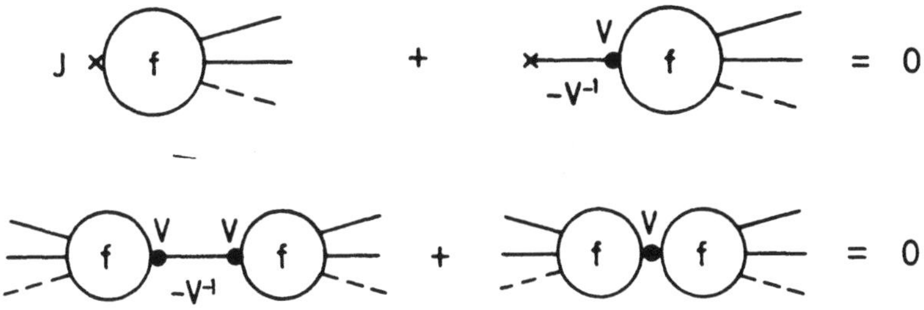

No new feature arises in these cases. However, if an original ϕ-line is connected to any of the other ϕ-lines of a special vertex (except the fVf vertex that is cancelled out already as shown above) no cancellation occurs. For example, there is nothing that cancels the following construction:

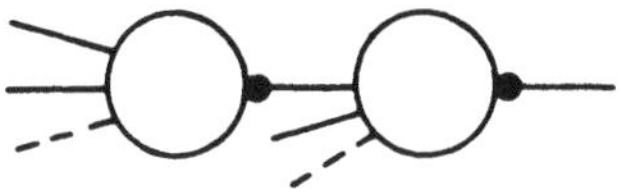

Thus we get a non-zero extra contribution only if the original ϕ-line is connected to another ϕ-line of the special vertex. All this means that the new Lagrangian contains all the contributions of the original Lagrangian plus a new kind of diagram where all the original ϕ-lines are connected to one of the other -lines of the special vertex. Such diagrams contain at least one closed loop of special vertices with "wings" of old vertices as well as special vertices.

EXAMPLE 10.4.2 Some new diagrams:

Considering these diagrams one immediately notices that the factors V in the special vertices are always multiplied with the propagators $-V^{-1}$. This then gives as net result the simple propa-

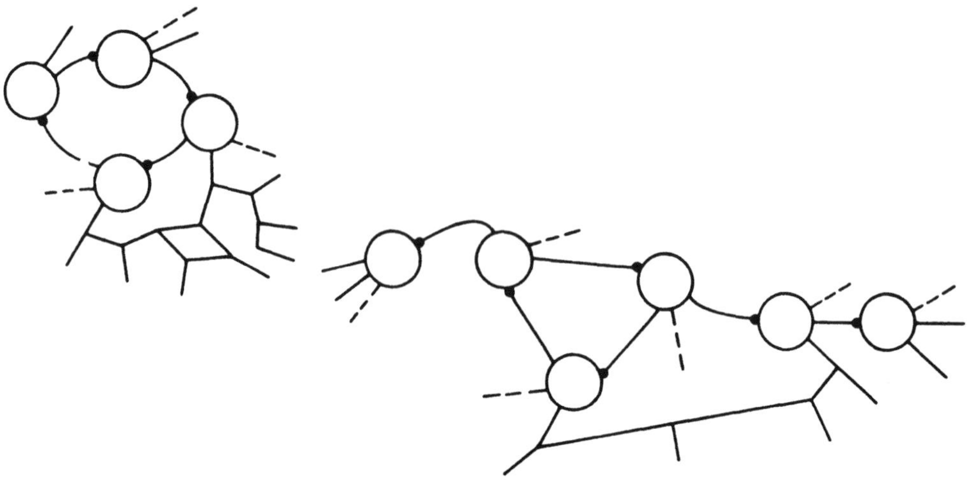

gator factor -1 for the "ghost" connecting the f vertices. See Eq. (10.14).There is no momentum dependence in that propagator. If now the special vertices are local, i.e. only polynomial dependence on the momenta and no factors such as $1/k^2$, then these vertices can be represented by a point and the closed loop momentum integration becomes the integral over a polynomial. Within the regularization scheme to be introduced later such integrals are zero, and nothing survives.

If, however, the field transformation is non-local these new diagrams survive and give an additional contribution with respect to what we had from the original Lagrangian.

How can we get back to the original Green's functions? At first sight it seems that we must simply subtract these diagrams, i.e. introduce new vertices in the transformed Lagrangian that produce precisely such diagrams, but with the opposite sign. This however ignores the possibility that the original ϕ-line of a special vertex connects to any of these new vertices. The correct solution is quite simple: introduce vertices that reproduce the closed loops only, without the "wings", and give each of those closed loops a minus sign. Thus at the Lagrangian level the extra terms are as depicted below:

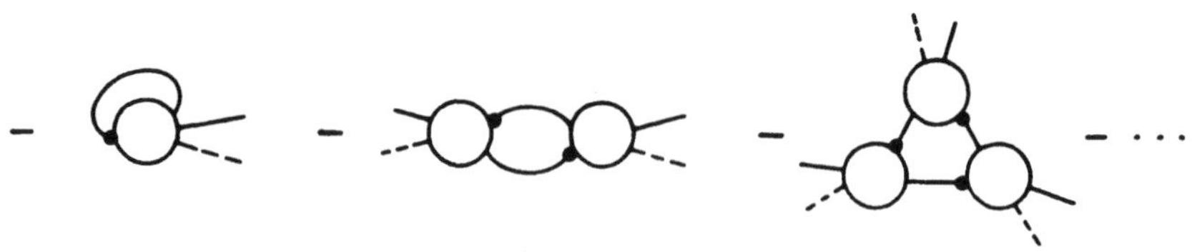

Now there is also the possibility that the original ϕ-line of a special vertex connects to these counter loops. In this way counter "winged" diagrams arise automatically and need not to be introduced by hand.

<u>EXAMPLE 10.4.3</u>. The following cancellation occurs in case of a Lagrangian including counter closed loops:

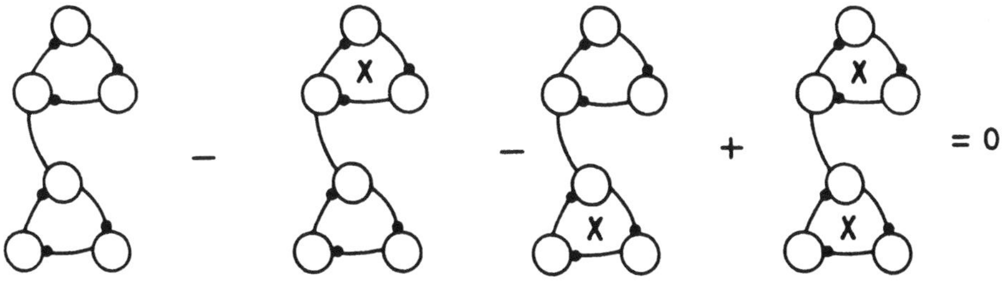

The crosses denote counter closed loops.

The solution may thus be summarized as follows. Start from a Lagrangian $\mathcal{L}(\phi)$. Perform the transformation $\phi_i(x) \rightarrow \phi_i(x) + f_i(x, \phi)$. Add a ghost Lagrangian $\mathcal{L}_{ghost}$. This ghost Lagrangian must be such that it gives rise to "wingless" closed loops of functions f, with one of the external lines removed, at which point another (or the same) function f is attached. The connection is by means of a propagator $-1/(2\pi)^4 i$. Of course, every function also carries a factor $(2\pi)^4 i$. Further there is a minus sign for every such closed loop.

As indicated, the vertices in such loops are the f with one line taken off. Symbolically

$$\frac{\delta f_i(x, \phi)}{\delta \phi_i(x')} \; .$$

Pictorially

$f(x, \phi)$

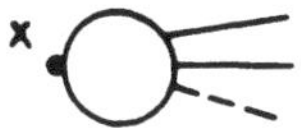

$\dfrac{\delta f(x, \phi)}{\delta \phi(x')}$

This agrees indeed precisely with the result found with the help of path integrals.

It may be that the reader is somewhat worried about combinatorial factors in these cancellations. A well-known theorem states that combinatorial factors are impossible to explain; everyone must convince himself that the above ghost loop prescription leads exactly to the required cancellations. There is really nothing difficult here; we do not want to suggest that there is. A good guideline is always given by the path integral formulation. Another way is to convince oneself that, for every possibility of special lines and vertices connecting up, there is a similar possibility arising from the ghost Lagrangian. That is, the precise factor in front of any possibility is not relevant, as long as it is known that it is the same as found in the counterpart.

10.5 Concerning the Rigour of the Derivations

There is nothing mysterious about the previous derivations -- provided we remember that we are working within the context of perturbation theory. Thus the transformations must be such that the new Lagrangian is of the type as described at the beginning. In particular the quadratic part of the new Lagrangian must be such that propagators exist. For instance, a substitution of the form $\phi \to \phi - \phi$ is clearly illegal. In short, the canonical transformation must be invertible.

There is a further problem connected with the $i\epsilon$ prescription. The propagators are not strictly $-V^{-1}$, in the notation of the previous section, but have been modified through the $i\epsilon$ addition. The factors V appearing in the special vertices must therefore be modified also such that the key-relation $(-V^{-1})V = -1$ remains strictly true. In connection with gauge field theory the functions f themselves contain often factors V^{-1}; careless handling of such factors may lead to errors. As we will see the correct $i\epsilon$ prescription for Faddeev-Popov ghosts can be established by precise consideration of these circumstances.

11. THE ELECTROMAGNETIC FIELD

11.1 Lorentz Gauge; Bell-Treiman Transformations and Ward Identities

The theorem about transformations of fields proved in the foregoing section applies to any Lagrangian $\mathcal{L}$ and is quite general. We will now exploit its consequences in the case of the electromagnetic field Lagrangian, in order to derive in particular the so-called "generalized Ward identities" for the Green's functions of the theory.

We start considering the free electromagnetic field case, since it contains all the main features of the problems we want to study. This includes the case of interactions with other particles (e.g. electrons) provided these interactions are introduced in a gauge-invariant way.

The Lagrangian giving rise to the Maxwell equations is

$$\mathcal{L} = -\frac{1}{4}(\partial_\mu A_\nu - \partial_\nu A_\mu)^2 + A_\mu J_\mu \ . \tag{11.1}$$

As is well known, in trying to apply the canonical formalism to the quantization of this Lagrangian, many difficulties are encountered. Such difficulties are essentially due to some redundancy of the lectromagnetic field variables, meaning that some combination of them is actually decoupled from the other ones and themselves.

Within the scheme defined in Section2, to get rules for diagrams and then a theory from a given Lagrangian, such difficulties manifest themselves in the fact that the matrix V_{ij} that defines the propagator has no inverse in the case of the Lagrangian (11.1).

To avoid such problems conventionally one adds to the Lagrangian (11.1), with some motivations, a term $-1/2(\partial_\mu A_\mu)^2$. Then one obtains

$$\mathcal{L}' = \mathcal{L} - \frac{1}{2}(\partial_\mu A_\mu)^2 = -\frac{1}{2}(\partial_\mu A_\nu)^2 + A_\mu J_\mu \ . \tag{11.2}$$

Now the propagator is well defined:

$$\frac{1}{(2\pi)^4 i}\,\frac{\delta_{\mu\nu}}{k^2 - i\varepsilon}$$

and we have the vertex:

$$(2\pi)^4 i\tilde{J}_\mu(\kappa) .$$

This certainly defines a theory, but in fact we have no idea if the physical content is as that described by the Maxwell equations. This is mainly the subject we want to study now. In doing that we will encounter many other somewhat related problems which must be solved when constructing a theory for the electromagnetic field. The key to all these problems is provided by the generalized Ward identities.

In order to derive these identities we add to the Lagrangian (11.2) a free real scalar field B, coupled to a source J_B. We get

$$\mathcal{L}''(A, B) = -\frac{1}{2}(\partial_\mu A_\nu)^2 + J_\mu A_\mu + \frac{1}{2}B(\partial^2 - \mu^2)B + J_B B . \tag{11.3}$$

We will now perform a Bell-Treiman transformation. This is a canonical transformation that is in form also a gauge transformation. Here

$$A_\mu \to A_\mu - \varepsilon\partial_\mu B \tag{11.4}$$

depending on a parameter ε. The replacement (11.4) in $\mathcal{L}''$, Eq. (11.3), gives up to first order in this parameter

$$\mathcal{L}''(A_\mu - \varepsilon\partial_\mu B, B) = \mathcal{L}''(A, B) + \varepsilon(\partial_\mu A_\nu)\partial_\mu\partial_\nu B - \varepsilon\partial_\mu B J_\mu . \tag{11.5}$$

As stated, the transformation (11.4) is in form just a gauge transformation

$$A_\mu \to A_\mu + \partial_\mu\Lambda , \tag{11.8}$$

with Λ any function of x. Note that the first Lagrangian, Eq. (11.1), apart from the source term is invariant under such a transformation. On the contrary, the second Lagrangian (11.2), and a fortiori Eq. (11.3), is not gauge invariant because the added term $-1/2(\partial_\mu A_\mu)^2$ is not. Precisely the source term and this $-1/2(\partial_\mu A_\mu)^2$ term are responsible for the difference, in Eq. (11.5), between the original Lagrangian $\mathcal{L}''(A_\mu, B)$ and the transformed one.

Let us come back to our local canonical transformation (11.4). The difference between the original and the new Lagrangian gives rise to the following extra vertices:

$$-\varepsilon(2\pi)^4 k^2 k_\mu \,,$$

$$\varepsilon(2\pi)^4 k_\mu \tilde{J}_\mu \,. \qquad (11.7)$$

Here the dotted lines denote the B-field, and the short double line at the sources denotes the gauge factor ∂_μ appearing in the gauge transformation (11.6) in front of Λ.

(Note: A derivative ∂_μ, acting on a field in an interaction term of the Lagrangian, gives i times the momentum of the field flowing towards the vertex.)

Of course the vertex term $\varepsilon(2\pi)^4 k_\mu J_\mu$ would give no contribution if the source of the electromagnetic field J_μ were a "gauge-invariant source", namely if $\partial_\mu J_\mu = 0$. In the discussion which follows we want, however, to allow in general also "non-gauge-invariant sources", whose four-divergence is different from zero.

As a consequence of the general theorem proved in the foregoing section, the B-field remains a free field, just as in the theory described by the Lagrangian $\mathcal{L}''$. To first order in ε, considering the n-point Green's function with one B source we have:

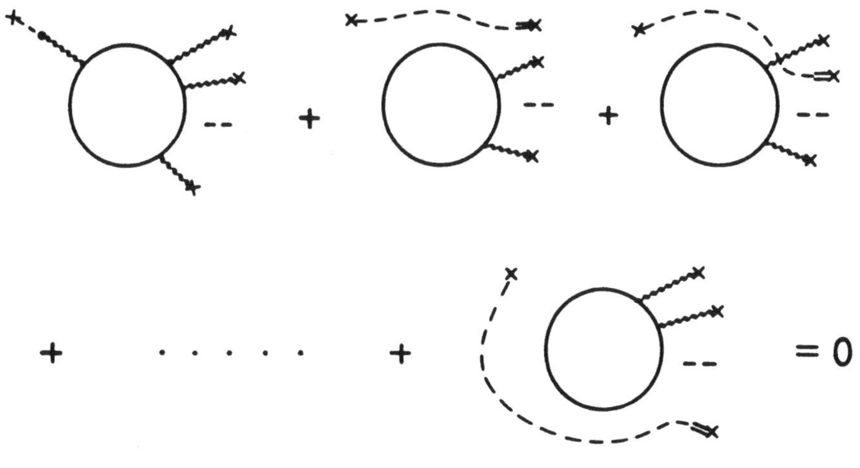

In the first diagram the B-A_μ vertex is followed by a photon propagator. One has:

$$-\varepsilon(2\pi)^4 k^2 k_\mu \frac{1}{(2\pi)^4 \cdot i}\,\frac{1}{k^2} = i\varepsilon k_\mu \ .$$

Apart from a sign this is precisely the same factor as occurring in the $\partial_\mu B J_\mu$ vertex ($\varepsilon i k_\mu (2\pi)^4 i J_\mu$) and we may use the same notation for it:

The resulting Ward identities are then:

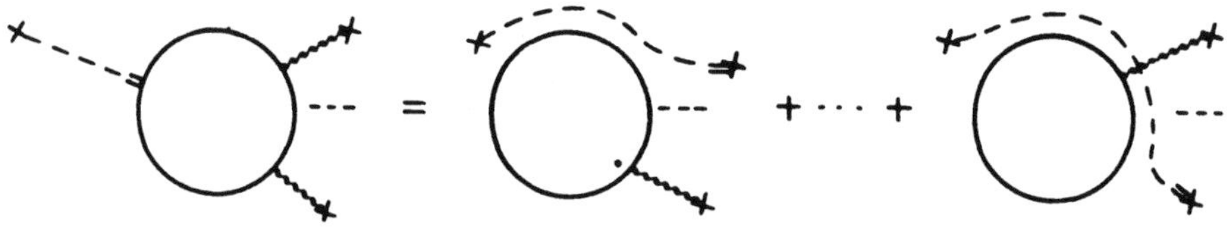

(11.8)

This result looks perhaps a little strange, since usually one tends to forget the lines that go straight through without interaction. As a matter of fact, this is allowed only in the case of gauge-invariant sources, for which, since then $k_\mu J_\mu = 0$, we have:

Equation (11.8) is an equation for Green's functions. Of course in defining Green's functions one should not employ particular properties of sources, such as $\partial_\mu J_\mu = 0$. As we will see, the S-matrix will be defined using gauge-invariant sources for the electromagnetic field, and for such cases the right-hand side of Eq. (11.8) is zero.

We remark that the Ward identity, Eq. (11.8), is trivially true as it stands in the case of no interaction that we are considering. It is in fact sufficient to realize that here, for example for the four-point Green's function, we have:

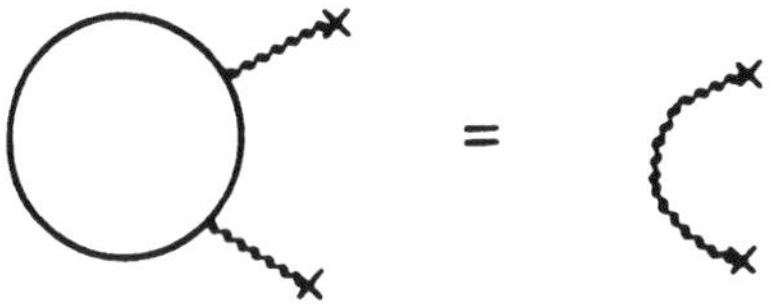

It is worth while to note that the above derivation goes through also if electrons, or any other particles, are present, provided the photon is coupled to them in a gauge-invariant way. Let us consider in some details the introduction of the electrons into the theory. The Lagrangian to start from is

$$\mathcal{L}(A_\mu,\ \psi,\ \bar\psi,\ B) = \mathcal{L}''(A_\mu,\ B) - \bar\psi(\gamma\partial + m)\psi +$$

$$+\ ie\bar\psi\gamma^\mu\psi A_\mu + \bar J_e\psi + \bar\psi J_e\ ,$$

where $\mathcal{L}''(A_\mu, B)$ is given in Eq. (11.3). The new pieces of this Lagrangian involving the electron field, the source terms $\bar J_e\psi + \bar\psi J_e$ excepted, are invariant under the transformation (11.4) if also the electron field is transformed

$$\psi \to e^{-i\varepsilon eB}\psi \simeq (1 - i\varepsilon eB)\psi + O(\varepsilon^2)\ ,$$

$$\bar\psi \to \bar\psi e^{+i\varepsilon eB} \simeq \bar\psi(1 + i\varepsilon eB) + O(\varepsilon^2)\ .$$

Then, up to first order in ε, we have

$$\mathcal{L}\,[A_\mu - \varepsilon\partial_\mu B,\ (1 - i\varepsilon eB)\psi,\ \bar\psi(1 + i\varepsilon eB),\ B] = \mathcal{L}(A_\mu,\ \psi,\ \bar\psi,\ B) +$$

$$+\ \varepsilon(\partial_\mu A_\nu)\partial_\mu\partial_\nu B - \varepsilon\partial_\mu BJ_\mu - i\varepsilon eB\bar J_e\psi + i\varepsilon eB\bar\psi J_e\ .$$

The extra vertices of the transformed Lagrangian are then the same as those of the free field case, given in (11.7), together with the following ones involving the electron sources:

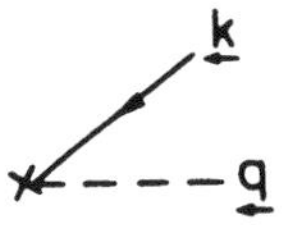

$$-(2\pi)^4 i\bar J_e(q + k)i\varepsilon e\ ,$$

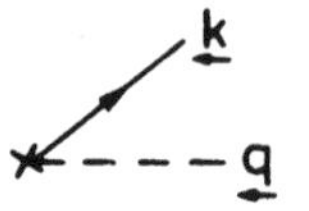

$$+(2\pi)^4 iJ_e(q + k)i\varepsilon e \ .$$

Therefore, for Green's functions involving no external electron lines, the Ward identities are exactly the same as in Eq. (11.8), even if in this case the bubbles contain any number of closed electron loops. When external electron lines are present, the Ward identities receive additional contributions of the form:

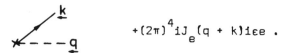

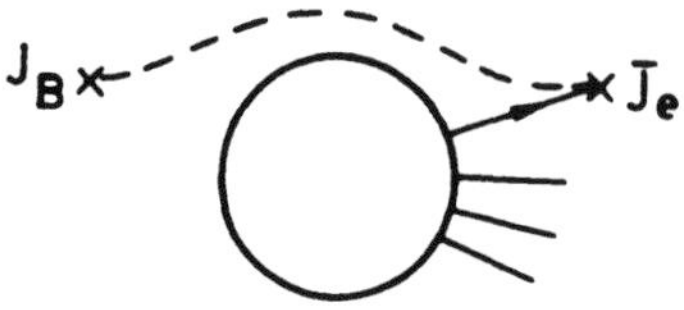

11.2 Lorentz Gauge: S-Matrix and Unitarity

Let us now investigate the S-matrix, keeping in mind for simplicity the free electromagnetic field Lagrangian $\mathcal{L}$, Eq. (11.2). The matrix K (see Sections 2 and 7) is here

$$K_{\mu\nu} = \delta_{\mu\nu} \ .$$

From the cutting equation, Section 7, we derive the fact that unitarity would hold if the sources were normalized according to Eq. (7.2). However, the complex conjugate of a four-vector is defined to have an additional minus sign in its fourth component (see the Introduction), so we are forced to attribute to the fourth component of the vector particle a negative metric (Section 8). The sources $J_\mu^{(a)}$, $a = 1, \ldots, 4$ are chosen such that

$$\delta_{\mu\nu} J_\nu^{(a)} = f^{(a)} J_\mu^{(a)} \ ,$$

$$J_\mu^{*(a)} \delta_{\mu\nu} J_\nu^{(a)} = \eta(a) = 1 \quad \text{for} \quad a = 1, 2, 3 \ , \qquad (11.9)$$

$$= -1 \quad \text{if} \quad a = 4 \ .$$

On the other hand, it is well known that due to gauge invariance we do have a positive metric theory, but with only two photon polarizations. In the system where $k_\mu = (0, 0, \kappa, ik_0)$, we label the sources as follows

$$J_\mu^{(1)} = (1, 0, 0, 0) ,$$

$$J_\mu^{(2)} = (0, 1, 0, 0) ,$$

$$J_\mu^{(3)} = (0, 0, 1, 0) ,$$

$$J_\mu^{(4)} = (0, 0, 0, i) .$$

We now postulate that only the first two of these sources emit physical photons. In terms of a non-covariant object $z_\mu = (0, 0, -\kappa, i\kappa_0)$ obtained from k_μ by space-reflection, we have

$$\sum_{a=1,2} J_\mu^{(a)} J_\nu^{(a)*} = \delta_{\mu\nu} + \frac{k_\mu z_\nu + z_\mu k_\nu}{(kz)} ,$$

whereas

$$K_{\mu\nu}^{-1} = \delta_{\mu\nu} .$$

Considering certain sets of cut diagrams we can apply the Ward identities to the left- and the right-hand side, to see that the terms proportional to k_μ and k_ν, respectively cancel among themselves. For instance, at the left-hand side of a cut diagram one can apply the Ward identity:

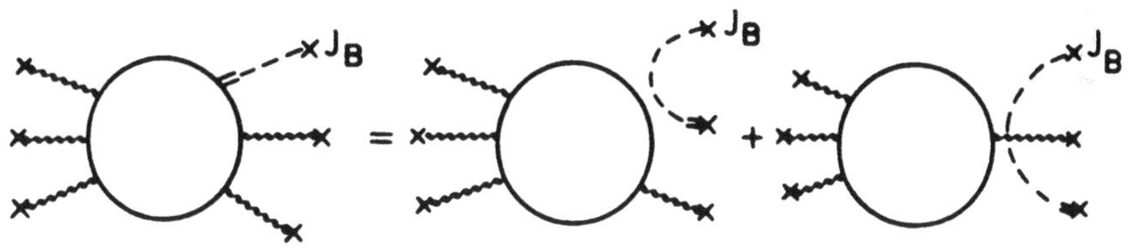

As before the double line indicates a factor ik_μ. Note that all polarizations can occur at the intermediate states, i.e. the out-states for S and the in-states for $S^\dagger$, but the other lines are physical.

In our case the right-hand side of this Ward identity is zero because the sources on the right-hand side <u>absorb</u> energy only, and therefore:

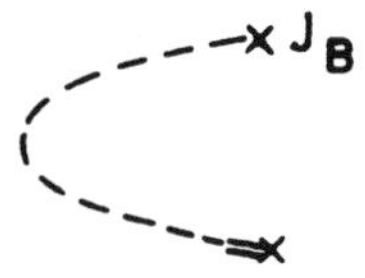

So, due to the Ward identities, one may replace the factor $K^{-1}_{\mu\nu}$ in the intermediate states, found from the cutting equation, by $J^{(1)}_\mu J^{(1)}_\nu + J^{(2)}_\mu J^{(2)}_\nu$, which implies unitarity in a Hilbert space with only transverse photons.

We must make a slight distinction between physical sources and gauge-invariant sources. The combination

$$J_\mu = k_o J^{(3)}_\mu + \kappa J^{(4)}_\mu = (0, 0, k_o, i\kappa)$$

is gauge invariant because $\partial_\mu J_\mu = 0$, but on mass shell this source is proportional to k_μ and it gives no contribution due to the Ward identities, so despite its gauge invariance, it is unphysical, in the sense that it emits nothing at all, not even ghosts.

11.3 Other Gauges: The Faddeev-Popov Ghost

Before going on, let us come back for a moment and try to interpret the Ward identity we have proved. Starting from a gauge-invariant Lagrangian, Eq. (11.1), we added the term

$$-\frac{1}{2}(\partial_\mu A_\mu)^2$$

in order to define a propagator. The Ward identities follow by performing a Bell-Treiman transformation, i.e. a canonical transformation that is also a gauge transformation. As we already noted, only the above term (and the source term) gives rise to a contribution, namely $\varepsilon(\partial_\mu A_\mu)(\partial^2 B)$. This is the only coupling of the B-field formally appearing in the transformed Lagrangian. Now the fact that B remains a free field, i.e. gives zero when part of a Green's function, implies that $\partial_\mu A_\mu$ is also free. This is actually the content of the Ward identity. Therefore we conclude that the addition of the term $-1/2(\partial_\mu A_\mu)^2$ does not change the physics of the theory.

This defines our starting point. If, due to a gauge invariance, a Lagrangian is singular, i.e. the propagators do not exist, then a "good" Lagrangian can be obtained by adding a term

$-\frac{1}{2}C^2$, where C behaves under gauge transformations as $C \to C + \hat{t}\Lambda$. Here $\hat{t}$ is any field-independent quantity that may contain derivatives. The argument given above, showing that the addition of the term $-1/2(\partial_\mu A_\mu)^2$ does not modify the physical content of the theory, can equally well be applied here. C will appear to be a free field, as can be seen by performing a Bell-Treiman transformation. However, for this simple recipe to be correct, as in the case explicitly considered $C = \partial_\mu A_\mu$, one needs $\hat{t}$, defined by the gauge transformation $C \to C + t\Lambda$, not to depend on any fields.

The difficulty with Yang-Mills fields is that the gauge transformations are more complicated, so much so that no simple C with the required properties exists. (Actually there exists a choice of C for Yang-Mills fields that is acceptable from this point of view, namely $C = \alpha A_3$, $\alpha \to \infty$. See R.L. Arnowitt and S.I. Fickler, Phys. Rev. 127, 1821 (1962).) For this reason we will now study quantum electrodynamics using a gauge function C that has more complicated properties under gauge transformations.

We take for understood the theory corresponding to the Lagrangian

$$\mathcal{L}(A_\mu) = \mathcal{L}_{inv}(A_\mu) - \frac{1}{2}C^2 , \quad C = \partial_\mu A_\mu . \qquad (11.10)$$

The sources will be taken to be gauge invariant and are included in $\mathcal{L}_{inv}$. Later we will consider also non-gauge-invariant sources.

Let us now suppose that we would like to have $\partial_\mu A_\mu + \lambda A_\mu^2$ instead of $\partial_\mu A_\mu$ for the function C. We can go from the above Lagrangian with $C = \partial_\mu A_\mu$ to the case of $C = \partial_\mu A_\mu + \lambda A_\mu^2$ by means of a non-local Bell-Treiman transformation

$$A_\mu \to A_\mu + \lambda\partial^{-2}\partial_\mu(A_\nu)^2 ,$$

or, more explicitly,

$$A_\mu \to A_\mu - i\lambda\partial_\mu \int d_4x'\Delta(x - x')A_\nu^2(x') , \qquad (11.11)$$

with

$$\Delta(x - x') = \frac{1}{(2\pi)^4 i} \int d_4k e^{ik(x-x')} \frac{1}{k^2} .$$

Here we have a situation as described in Subsection 10.5. For the subsequent manipulations to be true we must in fact supply a $-i\varepsilon$ to the denominator. Note that

$$\partial_x^2 \Delta(x - x') = i\delta_4(x - x') .$$

$\mathcal{L}_{inv}$ is unchanged under this somewhat strange gauge transformation, and

$$\partial_\mu A_\mu \to \partial_\mu A_\mu + \lambda A_\mu^2 .$$

In view of the structure of our canonical transformation we may expect that a ghost Lagrangian must be added.

Performing the transformation (11.11) one gets to first order in ε the following special vertex (see Section 10):

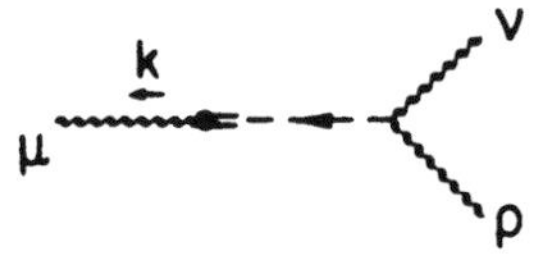

$$-2\lambda(-1)(2\pi)^4 ik^2 ik_\mu \frac{1}{k^2} \delta_{\nu\rho} .$$

Here, as usual, the dot indicates the factor V_{ij} which is minus the inverse photon propagator, and the short double line a factor ik_μ. The dotted line represents the function $\Delta(x - x')$, which is just like a scalar massless particle propagator. The theorem proved in Section 10 shows that the theory described by the transformed Lagrangian remains unchanged if we also provide for ghost loops, constructed by connecting the original photon lines to one of the other photon lines of the special vertex. For example, we have:

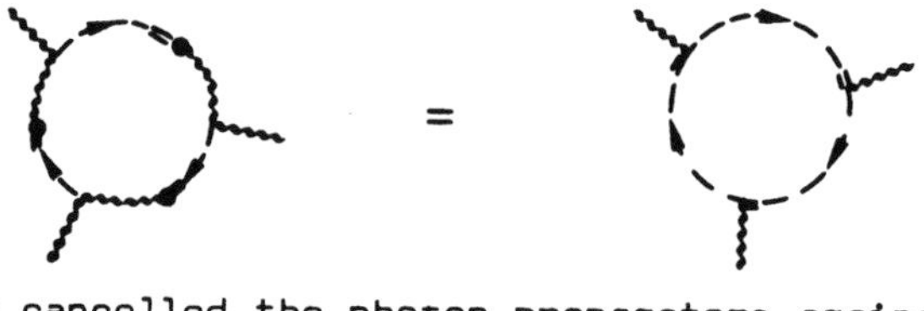

Here we have cancelled the photon propagators against the dots, so that the new vertex:

$$(2\pi)^4 i2\lambda ik_\mu$$

appears which can be formed by introducing a massless complex field ϕ interacting with the photon via the interaction Lagrangian $\mathcal{L}_I = 2\lambda\phi\, A_\mu \partial_\mu \phi$. All the closed loops constructed this way must have a minus sign in front.

We come to the conclusion that the Lagrangian

$$\mathcal{L} = \mathcal{L}_{inv}(A_\mu) - \frac{1}{2}(\partial_\mu A_\mu + \lambda A_\mu^2)^2 + \phi^* \partial^2 \phi + \qquad (11.12)$$

$$+ 2\lambda\phi^* A_\mu \partial_\mu \phi \quad ,$$

with the prescription that every closed ϕ loop gets a minus sign, reproduces the same Green's functions as before, when we had $C = \partial_\mu A_\mu$ and no ghost particles. The ϕ particle is called the Faddeev-Popov ghost. The $i\varepsilon$ prescription for its propagator is the usual one. It is perhaps noteworthy that the Faddeev-Popov ghost is <u>not</u> the ghost with propagator -1 of Section 10. The F-P ghost is the internal structure of the transformation function, see transformation (11.11).

<u>EXAMPLE 11.3.1</u> The Feynman rules from the Lagrangian (11.12) are:

$$\frac{1}{(2\pi)^4 i} \frac{\delta_{\mu\nu}}{k^2 - i\varepsilon} \, ,$$

$$\frac{1}{(2\pi)^4 i} \frac{1}{k^2 - i\varepsilon}$$ (-1 for every closed loop) ,

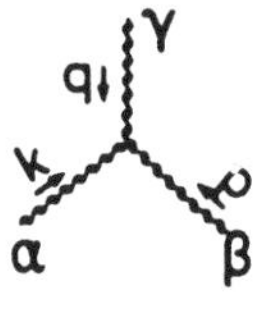

$$2(2\pi)^4 \lambda (\delta_{\alpha\beta} q_\gamma + \delta_{\alpha\gamma} p_\beta + \delta_{\beta\gamma} k_\alpha) \, ,$$

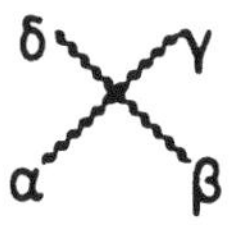

$$-4(2\pi)^4 i \lambda^2 (\delta_{\alpha\beta}\delta_{\gamma\delta} + \delta_{\alpha\gamma}\delta_{\beta\delta} + \delta_{\alpha\delta}\delta_{\beta\gamma}) \, ,$$

$$-(2\pi)^4 2\lambda k_\mu \, .$$

Let us consider photon-photon scattering in the zero loop approximation. The following diagrams:

a) b) c) d)

must give zero. This is indeed the case

$$a) \rightarrow (2\pi)^4 2\lambda\delta_{\alpha\beta}(-k_\lambda) \frac{1}{(2\pi)^4 i} \frac{1}{k^2} (2\pi)^4 2\lambda k_\lambda \delta_{\gamma\delta} = 4(2\pi)^4 i\lambda^2 \delta_{\alpha\beta}\delta_{\gamma\delta} \, ,$$

$$b) \rightarrow 4(2\pi)^4 i\lambda^2 \delta_{\beta\delta}\delta_{\alpha\gamma} \, ,$$

$$c) \rightarrow 4(2\pi)^4 i\lambda^2 \delta_{\alpha\delta}\delta_{\beta\gamma} \, ,$$

$$d) \rightarrow -4(2\pi)^4 i\lambda^2 (\delta_{\alpha\beta}\delta_{\gamma\delta} + \delta_{\alpha\gamma}\delta_{\beta\delta} + \delta_{\alpha\delta}\delta_{\beta\gamma}) \, .$$

Note that in the contributions from diagrams (a), (b), (c), many terms vanish, because our external sources are taken to be gauge invariant.

Of course, to get a contribution from the Faddeev-Popov ghost, one has to consider examples of diagrams containing loops. Indeed, one notes for instance:

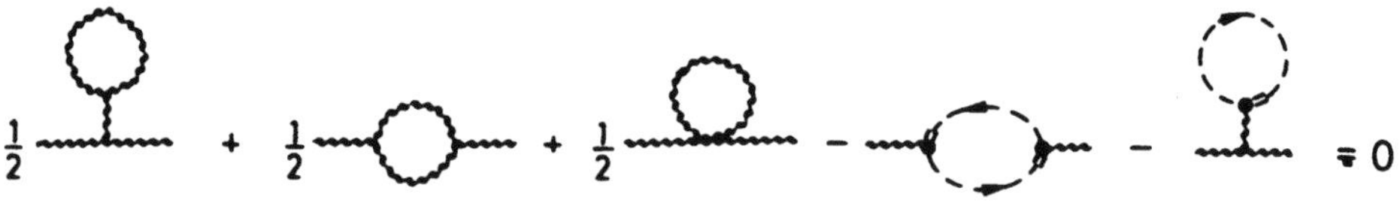

Actually to prove this cancellation we need a gauge-invariant regularization method, which will be provided in the following.

We want to understand now whether a general prescription can be given which, in a general gauge defined by the function C, allows us to write down immediately the Lagrangian for the ghost particle ϕ.

In the case considered of $C = \partial_\mu A_\mu + \lambda A_\mu^2$, it is easy to see that the ghost Lagrangian in (11.12) comes out by the following prescription.

Take the function C. Under a gauge transformation one has to first order in Λ

$$C \to C + \mathcal{M}\Lambda$$

with $\mathcal{M}$ some operator, which in the actual case depends on the fields A_μ. Then the ghost Lagrangian is simply

$$\phi^* \mathcal{M} \phi \ .$$

Here

$$A_\mu \to A_\mu + \partial_\mu \Lambda \ ,$$

$$C = \partial_\mu A_\mu + \lambda A_\mu^2 \to C + \partial^2 \Lambda + 2\lambda A_\mu \partial_\mu \Lambda + O(\Lambda^2) \ , \qquad (11.13)$$

from which (11.12) follows.

The generality of this prescription can be understood by looking at our manipulations. The non-local transformation (11.11) was taken such that $\partial_\mu A \to \partial_\mu A_\mu + \lambda A_\mu^2$. If we consider a gauge transformation

$$\partial_\mu A_\mu \to \partial_\mu A_\mu + \partial^2 \Lambda \ ,$$

then we must solve the equation

$$\partial^2 \Lambda = \varepsilon A_\mu^2 \ .$$

The fact that the propagator $1/k^2$ appeared in our loops is thus simply related to the behaviour $\partial_\mu A_\mu \to \partial_\mu A_\mu + \partial^2\Lambda$. Secondly, the special vertices are

$$A_\mu V_{\mu\nu} \partial_\nu \Lambda \ ,$$

where $V_{\mu\nu}$ is in the usual way related to the photon propagator. Then the vertices appearing in the ghost loops arise from the prescription: remove one factor A_μ from the expression λA_μ^2 and replace it by $\partial_\mu\Lambda$. Diagrammatically:

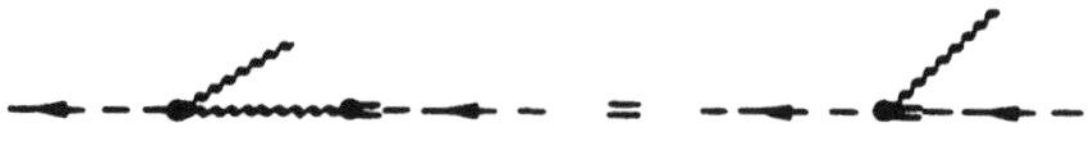

A factor of 2 provides for the two possible ways of removing an A_μ-field from λA_μ^2. This is indeed precisely what one obtains to first order in Λ by submitting λA_μ^2 to a gauge transformation, see Eq. (11.13).

The beauty of this recipe to find the ghost Lagrangian is that there is no reference to the function C that we started from. To construct the ghost Lagrangian one needs only the C actually required. Also we need only infinitesimal gauge transformations.

EXAMPLE 11.3.2 The above prescription works also in the case of $C = \partial_\mu A_\mu$ that we started from. In fact we have

$$\partial_\mu A_\mu \to \partial_\mu A_\mu + \partial^2\Lambda \ ,$$

meaning that the ghost Lagrangian should be

$$\mathcal{L}_\varphi = \phi^* \partial^2 \phi$$

with no coupling however of the ϕ-field to the electromagnetic field. Therefore, as we already know, no ghost loop is to be added in this case.

Even if the above manipulations are quite solid one may have some doubts concerning the final prescription for the ghost Lagrangian. To establish correctness of the theory, as we did in the case of $C = \partial_\mu A_\mu$, Ward identities are needed. In the general case such identities are more complicated than those given before and we will call them Slavnov-Taylor identities.

11.4 The Slavnov-Taylor Identities

In this section we will derive the Slavnov-Taylor identities, using only local canonical transformations. One can always keep in mind the explicit case of $C = \partial_\mu A_\mu + \lambda A_\mu^2$, but we now start adapting a general notation anticipating the general case.

Let there be given a function C, that behaves under a general gauge transformation $A_\mu \to A_\mu + \partial_\mu \Lambda$ as follows

$$C \to C + (\hat{m} + \hat{\ell})\Lambda + O(\Lambda^2) . \qquad (11.14)$$

Here we split up the earlier factor $\mathcal{M} = (\hat{m} + \hat{\ell})$ with the part $\hat{\ell}$ depending on the field A_μ. For $C = \partial_\mu A_\mu + \lambda A_\mu^2$, we have $\hat{m} = \partial^2$ and $\hat{\ell} = 2\lambda A_\mu \partial_\mu$.

Our starting point will be the Lagrangian

$$\mathcal{L}(A_\mu) = \mathcal{L}_{inv}(A_\mu) - \frac{1}{2} C^2 + \phi^* \hat{m}\phi + \phi^* \hat{\ell}\phi + J_\mu A_\mu \qquad (11.15)$$

with the -1/loop prescription for ghost loops. As before, we add to this Lagrangian a piece describing a free field B

$$\frac{1}{2} B(\partial^2 - \mu^2)B + J_B B$$

and perform the local Bell-Treiman transformation

$$A_\mu \to A_\mu + \hat{t}_\mu B$$

with $\hat{t}_\mu = \partial_\mu$.

Working to first order in the field B

$$\mathcal{L}(A_\mu + \partial_\mu B) = \mathcal{L}(A_\mu) - C(\hat{m} + \hat{\ell})B + \phi^* (\hat{e} + \hat{d})B\phi + J_\mu \partial_\mu B . \qquad (11.16)$$

Here we used the transformation

$$\hat{\ell} \to \hat{\ell} + (\hat{e} + \hat{d})B + O(B^2) .$$

Again we distinguish the field-independent part $\hat{e}$ and the field-dependent $\hat{d}$. In the actual case of $\hat{\ell} = \lambda A_\mu \partial_\mu$ we have

$$\hat{e}B = 2\lambda \partial_\mu B \partial_\mu , \qquad \hat{d} = 0 .$$

The vertices involving the B-field in the transformed Lagrangian (11.16) are (the over-all factor $(2\pi)^4 i$ is always left understood):

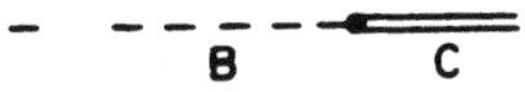

$-\hat{m}$ from $-C\hat{m}B$,

$-\hat{\ell}$ from $-C\hat{\ell}B$,

$\hat{e}$ from $\phi^*\hat{e}B\phi$,

$\hat{t}_\mu \tilde{J}_\mu$ from $\partial_\mu BJ_\mu$.

The vertex corresponding to $\phi\, \hat{d}B\phi$ is not drawn since in the actual case ($C = \partial_\mu A_\mu + \lambda A_\mu^2$) $\hat{d}$ is zero. The vertices $C\hat{m}B$ and $C\hat{\ell}B$ appear with a minus sign in the Lagrangian: this sign is not included in the vertex definition and thus must be provided separately. The statement that B remains a free field becomes then to first order in the B-vertices:

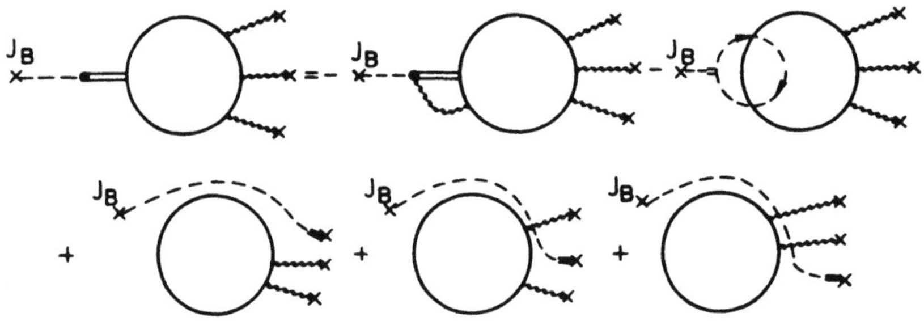

In the second term at the right-hand side we exhibited explicitly the whole loop to which the B-line is attached, together with the associated minus sign. The blobs are built up from diagrams containing propagators and vertices as given in the previous section. The ϕ particles go around in loops only, and such loops are included in the blobs.

In the above equation there is one crucial point. Consider the very simplest case, no interaction and only one photon source. Since $C = \partial_\mu A_\mu + \lambda A_\mu^2$, one has to zeroth order in λ:

The first diagram (similarly to the second one) contains the B propagator pole, a photon propagator pole, the vertex factor $(2\pi)^4 i\hat{m}$ and the factor $\hat{t}_\mu$. The last diagram has only the B pole and a factor $\hat{t}_\mu = \partial_\mu$. The equation can be true only if $\hat{m}$ is zero precisely on the photon pole. We had $\hat{m} = \partial^2 \rightarrow -k^2$, and we must therefore always take

$$(2\pi)^4 i\hat{m} = (2\pi)^4 i(\partial^2 + i\varepsilon) \rightarrow -(2\pi)^4 i(k^2 - i\varepsilon).$$

As will be seen, this $\hat{m}$ is minus the inverse of the ghost propagator. In the general case, considering lowest order identities, one must check on the i in the propagators in the manner described here.

We will perform some manipulations on Eq. (11.17). Consider the first diagram on the right-hand side. Using:

the contribution of this diagram becomes:

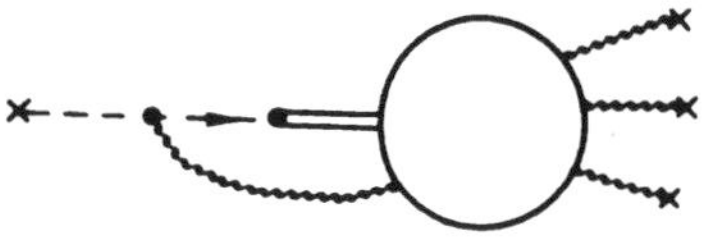

with a new vertex:

Treating this first vertex as a photon source we see that we can iterate, getting:

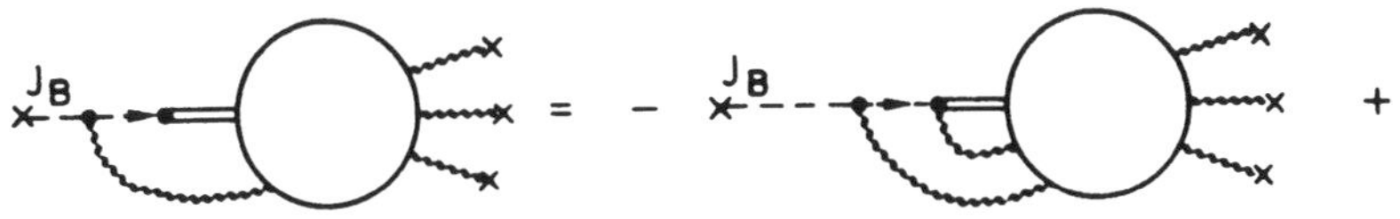

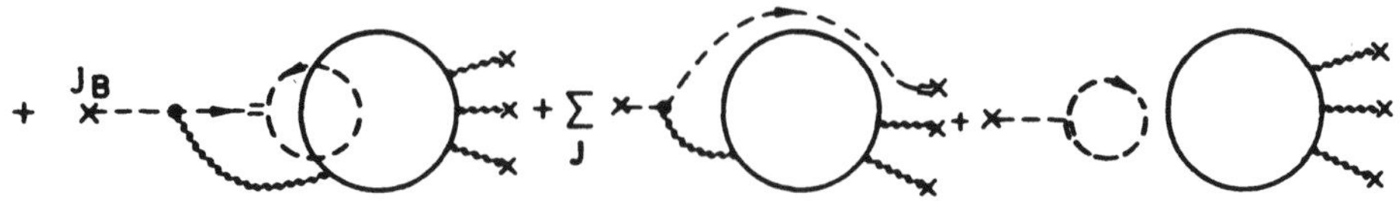

Σ_J means sum over all the photon sources. The last diagram arises because the first vertex is treated as a source. Some new vertices enter, whose meaning is unambigously defined. For example, we have:

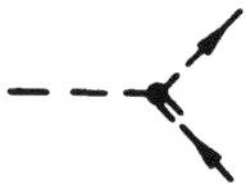

which correspond precisely to a vertex $(2\pi)^4 i\hat{e}$ or to an interaction term $\phi^* \partial_\mu B \partial_\mu \phi$. In particular, since this term is invariant under the interchange $B \to \phi$, we have:

This equality, trivial in the case of quantum electrodynamics, becomes much more tricky in the case of non-Abelian gauge symmetries. Then a similar identity follows from the group structure, and is closely related to the Jacobi identity for the structure constants of the group.

The right-hand side of Eq. (11.18) can now be inserted in place of the first diagram on the right-hand side of the original Eq. (11.17). We note that some cancellations occur. The second diagram on the right-hand side of Eq. (11.17) contains also the configuration where the ghost has no further interaction:

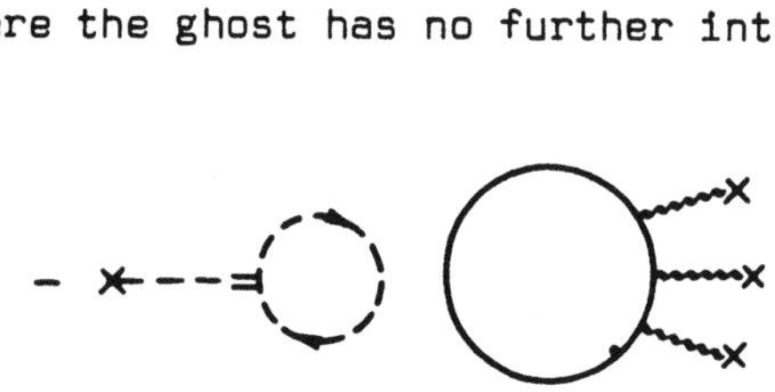

This cancels precisely the last diagram of the iterated equation, after insertion in Eq. (11.17).

This process can be repeated indefinitely and one finally arrives at the equation:

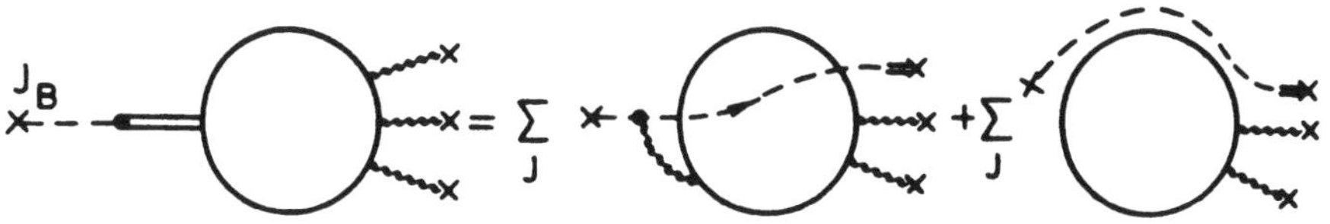

The ghost line going through the diagram may have zero, one, two, etc., vertices of the type:

$\hat{I}$

If one notes now that the mass of the B-field is an absolutely free parameter in our discussion, we can also give now the B-field the propagator $-(2\pi)^4 i\hat{m}^{-1}$. In this way we can simply reduce the first diagram on the right-hand side of Eq. (11.19) to:

and also include in it the last diagram of Eq. (11.19). Observing that:

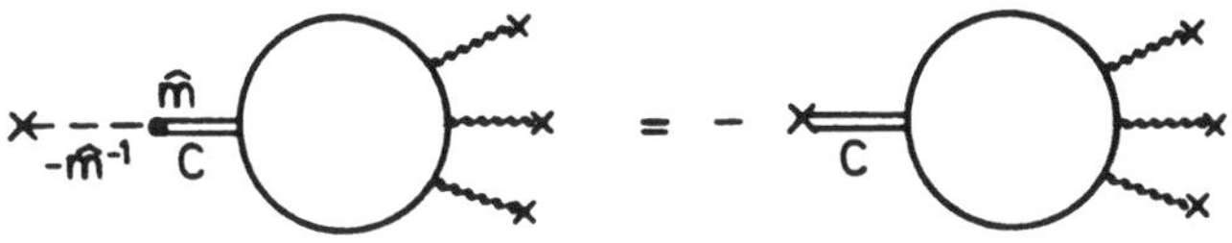

we get the identity:

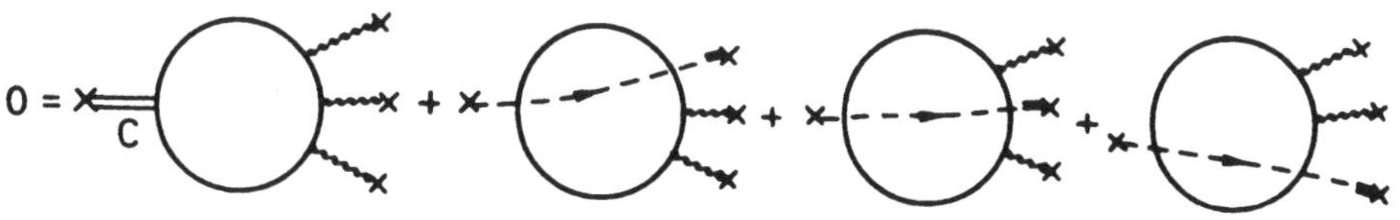

Note: In the case of $C = \partial_\mu A_\mu$ this equation coincides with the Ward identity (11.8). One has indeed:

 $= (2\pi)^4 i\tilde{J}_B \dfrac{1}{(2\pi)^4 i k^2} (-ik_\mu) =$ $-$ 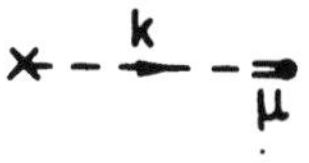

so that the first diagram of Eq. (11.20) can also be drawn as:

$-$ 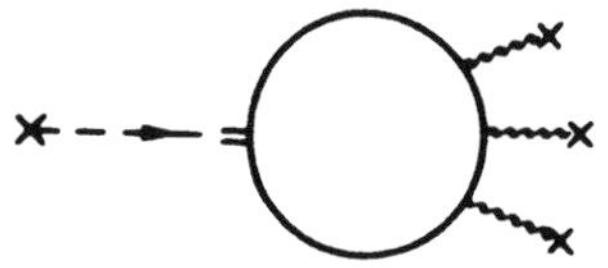

Furthermore, in this case of $C = \partial_\mu A_\mu$ the ghost does not interact and then:

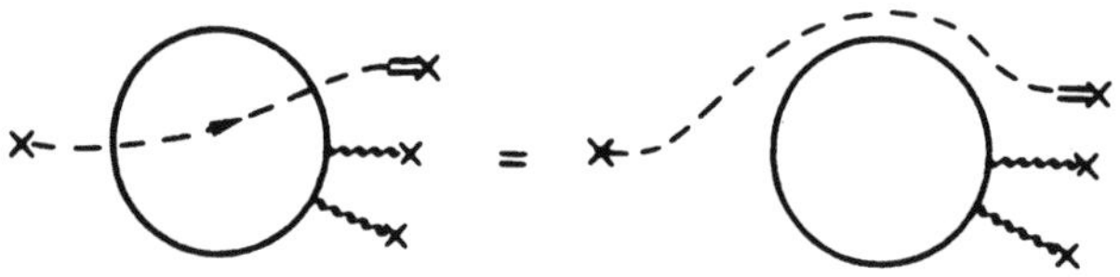

Equation (11.20) can be generalized as follows. Suppose that a source is coupled to several electromagnetic fields, for instance

$$JA_\mu^2 \quad \text{or} \quad JA_\mu^4$$

or even more general to a local but otherwise arbitrary function R(A) of the A_μ. Example:

$$J(\partial_\mu A_\mu + \lambda A_\mu^2 + \kappa A_\mu^4, \ldots) \; .$$

The only way this comes into the above derivation is through the behaviour of this whole term under gauge transformations. Suppose one has in the Lagrangian (11.15) also the coupling

$$JR(A_\mu) \; .$$

Furthermore, under a gauge transformation

$$A_\mu \to A_\mu + \partial_\mu \Lambda ,$$

one has

$$R \to R + \hat{r}\Lambda + \hat{\rho}\Lambda .$$

$\hat{r}$ is the part independent of the A_μ and $\hat{\rho}$ contains the A_μ dependent parts. Both $\hat{r}$ and $\hat{\rho}$ may contain derivatives.

Now, performing the Bell-Treiman transformation with the B-field one obtains the vertices:

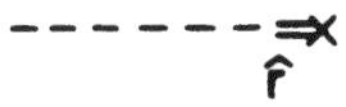

from $J\hat{r}B$,

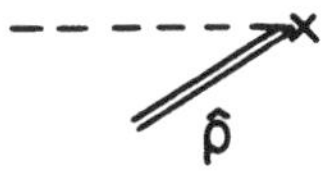

from $J\hat{\rho}B$,

where the double line denotes a collection of possibilities, including at least one photon line. The whole derivation can be carried through unchanged, and given many sources J^a coupled to many field functions R_a we get:

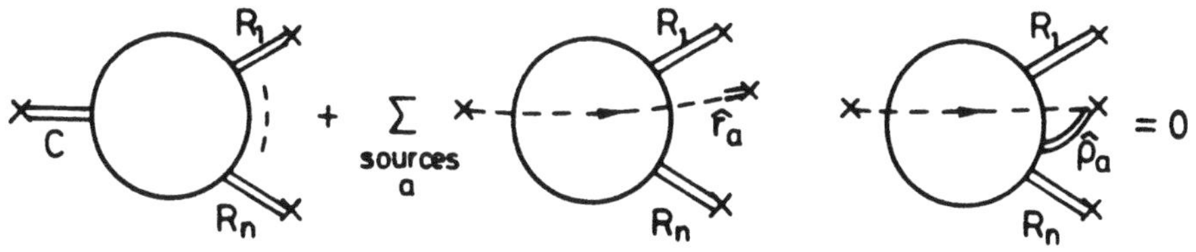

where $\hat{r}_a$ and $\hat{\rho}_a$ are defined by the behaviour of the functions R_a under gauge transformations

$$\text{if} \quad A_\mu \to A_\mu + \partial_\mu \Lambda , \quad R_a \to R_a + \hat{r}_a\Lambda + \hat{\rho}_a(A_\mu)\Lambda .$$

The above identities are the Slavnov-Taylor identities.

11.5 Equivalence of Gauges

From the preceding sections we obtain the following prescription for handling a theory with gauge invariance. First choose a non-gauge-invariant function C and add $-1/2C^2$ to the Lagrangian. Next consider the properties of C under infinitesimal gauge transformations, for instance

$$A_\mu \to A_\mu + \partial_\mu \Lambda ,$$

$$C \to C + (\hat{m} + \hat{\ell})\Lambda .$$

A ghost part must also be added to the Lagrangian.

$$\mathcal{L} = \mathcal{L}_{inv} - \frac{1}{2} C^2 + \varphi^* (\hat{m} + \hat{\ell})\varphi ,$$

with the prescription of providing a factor -1 for every closed ϕ loop. Clearly the choice of C is milited by the fact that the operator $\hat{m}$, defining the ghost propagator, must have an inverse. Moreover C must together with $\mathcal{L}_{inv}$ define non-singular A propagators, but this is automatic if C breaks the gauge invariance.

Suppose we had taken a slightly different C

$$C' = C + \varepsilon R .$$

Now under a gauge transformation

$$C' \to C' + (\hat{m} + \hat{\ell})\Lambda + \varepsilon(\hat{r} + \hat{\rho})\Lambda ,$$

where $\hat{r}$ and $\hat{\rho}$ are, respectively, the field-independent and field-dependent parts resulting from a gauge transformation of R. For example, if $R = A_\mu^2$ then $R \to R + 2A_\mu \partial_\mu \Lambda$ and we have $\hat{r} = 0$ and $\hat{\rho} = 2A_\mu \partial_\mu$.

The ghost Lagrangian must be changed accordingly, and we get

$$\mathcal{L}' = \mathcal{L}_{inv} - \frac{1}{2} (C + \varepsilon R)^2 + \phi^* (\hat{m} + \hat{\ell} + \varepsilon\hat{r} + \varepsilon\hat{\rho})\phi . \tag{11.22}$$

We now prove that to first order in ε the S-matrix generated by $\mathcal{L}'$ equals the S-matrix generated by $\mathcal{L}$. This is clearly sufficient to have equivalence of any two gauges that can be connected by a series of infinitesimal steps.

Let us compare the Green's functions of $\mathcal{L}'$ with those of $\mathcal{L}$. The difference is given by Green's functions containing one ε-vertex. From Eq. (11.22) these vertices are:

The difference between an $\mathcal{L}'$ and Green's function with the same external legs is then:

We exhibited explicitly the minus sign associated with the ghost loops. Opening up the top vertex this difference is:

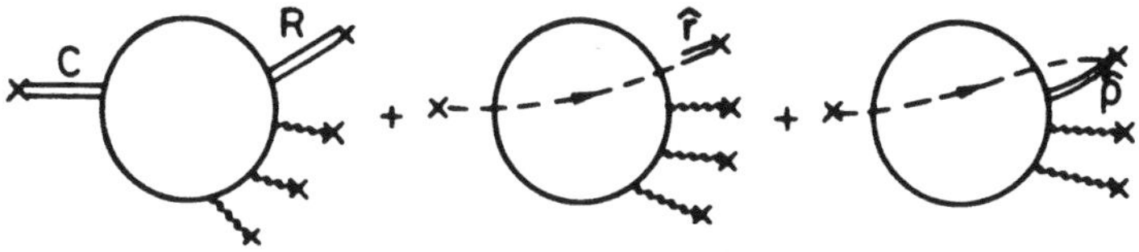

If now all the original sources are gauge invariant, namely $\partial_\mu J_\mu = 0$, we see that this difference is zero as a consequence of the Slavnov-Taylor identities, Eq. (11.21). The diagrams in Eq. (11.21), where the ghost line is attached to such currents, give no contribution, since for these currents we have $\hat{r}_\mu = \partial_\mu$ (and $\hat{r}_\mu J_\mu = 0$) and $\hat{\rho} = 0$. As the S-matrix is defined on the basis of gauge-invariant sources, we see that the S-matrix is invariant under a change of gauge as given above.

11.6 Inclusion of Electrons; Wave-Function Renormalization

The preceding discussion has been carried out in such a way that the inclusion of electrons changes practically nothing. The main difference is that we now must introduce sources that emit or absorb electrons and such sources are not gauge invariant. This complicates somewhat the discussion of the equivalence of gauges.

Thus consider electron source terms

$$J_e\psi + \bar{\psi}J_e \ .$$

Under gauge transformations $(A_\mu \to A_\mu + \partial_\mu\Lambda)$

$$\psi \to e^{ie\Lambda}\psi \simeq \psi + ie\Lambda\psi \ ,$$

$$\bar{\psi} \to \bar{\psi} - ie\bar{\psi}\Lambda \ .$$

Let us consider the difference of the Green's functions of two different gauges in the presence of an electron source. For simplicity we will only draw one of them explicitly. The Slavnov-Taylor identity is:

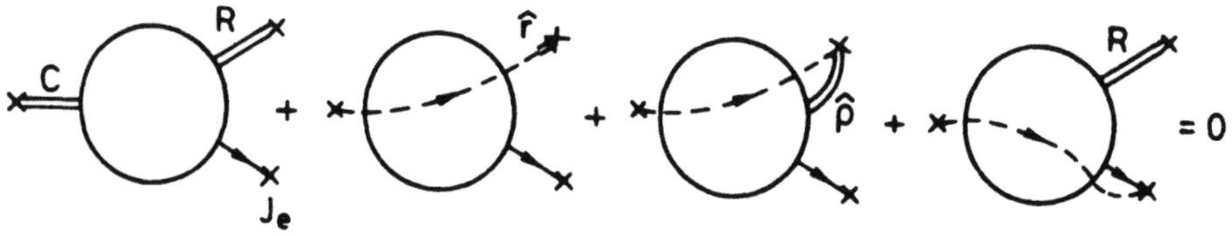

The first three terms are precisely those found in considering the difference of the Green's functions, or more precisely of objects obtained when unfolding a vertex in the difference of the Green's function (see preceding section).

Folding back the C and R source to obtain again the true difference of the Green's function and using the Slavnov-Taylor identity we get:

$$\mathcal{G}_{\mathcal{L}'} - \mathcal{G}_{\mathcal{L}} =$$

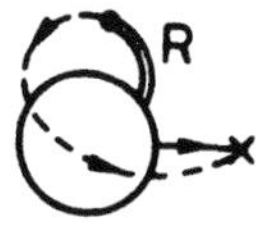

In general, such a type of diagram will have no pole as one goes to the electron mass-shell. Thus passing to the S-matrix the contribution of most diagrams will disappear. But included in the above set are also diagrams of the type:

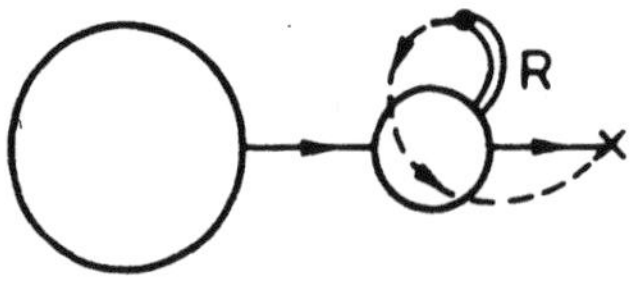

which do have a pole. However, we must not forget that the S-matrix definition is based on the two-point function. The sources must be such that the residue of the two-point function including the sources is one.

Consider the two-point function for the electron:

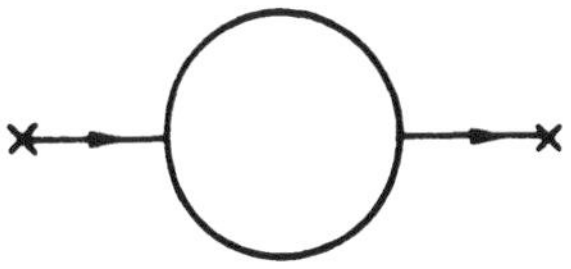

Also this function will change, and for any of the two sources we will have precisely the same change as above:

In accordance with our definition of the S-matrix we must redefine our sources such that the residue of the two-point function remains one; that is

$$J' = J - \delta ,$$

where

δ = value at the pole of 

It is seen that including the redefinition of the sources the S-matrix is unchanged under a change of gauge.

In conventional language, the above shows that the electron wave-function renormalization is gauge dependent, but that is of no consequence for the S-matrix.

12. COMBINATORIAL METHODS

There are essentially three levels of sophistication with which one can do combinatorics. On the first level one simply uses identities of vertices that are a consequence of gauge invariance. Example: electrons interacting with photons. The part of the Lagrangian containing electrons is

$$-\bar{\psi}(\gamma\partial + m)\psi + ieA_\mu\bar{\psi}\gamma^\mu\psi \ .$$

Now write

$$\psi \to \psi + ieB\psi \ , \qquad \bar{\psi} \to \bar{\psi} - ieB\bar{\psi} \ , \qquad A_\mu \to A_\mu + \partial_\mu B \ .$$

Of course this Lagrangian remains invariant under these transformations, but we want to understand this in terms of diagrams. One has, not cancelling anything, as extra terms

$$-ie\bar{\psi}(\gamma\partial + m)B\psi + ie\bar{\psi}B(\gamma\partial + m)\psi + ie\partial_\mu B\bar{\psi}\gamma^\mu\psi \ .$$

These are simply vertices:

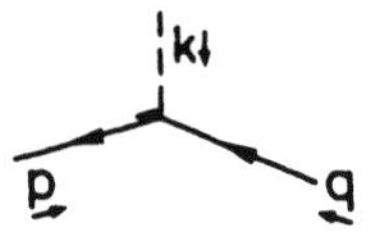

$$-ie(2\pi)^4 i(-i\gamma p + m) \ ,$$

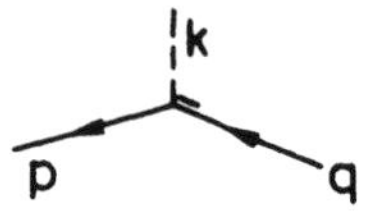

$$ie(2\pi)^4 i(i\gamma q + m) \ ,$$

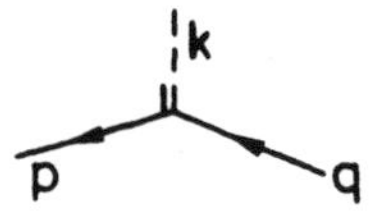

$$ie(2\pi)^4 i(i\gamma k) \ .$$

How does the cancellation manifest itself? In the Lagrangian one must write

$$\partial_\mu(B\psi) = \partial_\mu B\psi + B\partial_\mu\psi$$

to see it. Here we take the first vertex and write

$$p = -q - k.$$

Then one sees explicitly how the three vertices together give zero

$$-ie(2\pi)^4 i\{i\gamma q + i\gamma k + m - i\gamma q - m - i\gamma k\} = 0 ,$$

and this remains true if we replace m everywhere by m - iε. Only then can one say that the vertices contain factors which are inverse propagators; now we know for sure that the following Green's function identity holds (in lowest order):

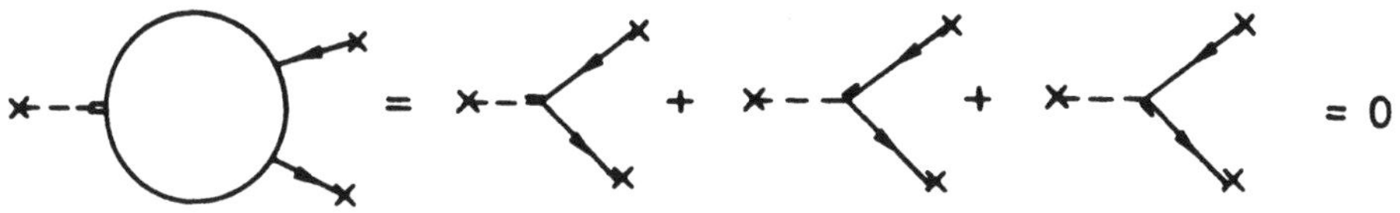

where the short double line indicates the inverted propagator including iε. Exploiting this fact we obtain:

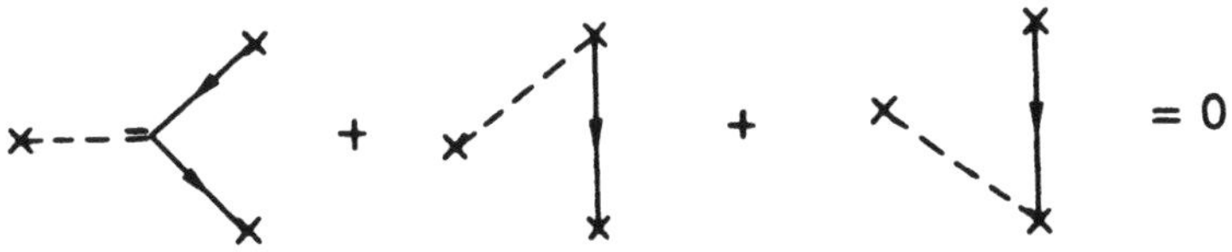

which is a Ward identity given before.

The above makes clear how the very first level combinatorics is the building block for the second-level combinatorics. This second-level combinatorics uses the fact that all terms cancel when one has gauge invariance. But this fact must be verified explicitly by means of first-level combinatorics to ascertain that the iε prescription is consistent with the gauge invariance. It is just by such a type of reasoning that the iε in the ghost propagator is fixed, as shown before.

The third level of combinatorics is that when one uses non-local canonical transformations. These can be used to derive the Slavnov-Taylor identities directly, as was done by Slavnov. One must be very careful about the iε prescription; our procedure whereby this identity was derived by means of first- and second-level combinatorics shows that the usual -iε prescription for the ghost propagator is the correct one.

13. REGULARIZATION AND RENORMALIZATION

13.1 General Remarks

As noted before, the regulation scheme introduced in the beginning is not gauge invariant. We need a better scheme that can also be used in case of gauge theories, Abelian or non-Abelian. An elegant method is the dimension regularization scheme which we will discuss now.

A good regularization scheme must be such that for certain values of some parameters (the masses Λ in the unitary cut-off method) the theory is finite and well defined. The physical theory obtains in a certain limit (masses becoming very big), and one requires that quantities that were already finite before regularization was introduced remain unchanged in this limit.

In order to obtain a finite physical theory it will be necessary to introduce counter-terms in the Lagrangian. A theory is said to be renormalizable if by addition of a finite number of kinds of counterterms a finite physical theory results. This physical theory must of course not only be finite, but also unitary, causal, etc. With regard to a regularization and a renormalization procedure for a gauge theory, some problems arise which are peculiar to this kind of theory. As demonstrated before, the Ward identities (or more generally the Slavnov-Taylor identities) must hold, because they play a crucial role in proving unitarity. Since regularization implies the introduction of new rules for diagrams, and because new vertices, corresponding to the renormalization counterterms are added, it becomes problematic as to whether our final renormalized theory satisfies Ward identities. As far as the regularization procedure is concerned, a first step is to provide a scheme in which the Ward identities are satisfied for any value of the regularization parameters.

This goal is obtained in the dimensional regularization scheme. It must be stressed, however, that this does not guarantee that the counterterms satisfy Ward identities and indeed, in general, one has considerable difficulty in proving Ward identities for the renormalized theory. The point is this: in the unrenormalized theory there exists Ward identities, and an invariant cut-off procedure guarantees that the counterterms satisfy certain relations. From these relations one can derive new Ward identities satisfied by the renormalized theory. However, these Ward identities turn out to be different: one may speak of renormalized Ward identities. That is, one can prove that the renormalized theory has a certain symmetry structure (giving rise to certain Ward identities), but one has to show also that this symmetry is the same as that of the unrenormalized theory. Let us remark once more that Ward identities are needed for the renormalized theory because they are needed in proving unitarity. Indeed, Ward identities have nothing to do with renormalizability but everything to do with unitarity.

Another problem to be considered is the following one. If one admits the possibility that the renormalized symmetry is different from the unrenormalized one, then formally the following can happen. If one carries through a renormalization program, one must first make a choice of gauge in the unrenormalized theory. Perhaps then the symmetry of the renormalized Lagrangian depends on the initial choice of gauge. Given a gauge theory one must show explicitly that this is not the case, and that the various renormalized Lagrangians belonging to different gauge choices in the unrenormalized theory are related by a change of gauge with respect to the renormalized symmetry. In quantum electrodynamics this problem is stated as the gauge independence of the renormalized theory.

13.2 Dimensional Regularization Method: One-Loop Diagrams

In the dimensional regularization scheme a parameter n is introduced that in some sense can be visualized as the dimension of space time. For $n \neq 4$ a finite theory results; the physical theory obtains in the limit $n = 4$.

As a first step, we define the procedure for one-loop diagrams. The example we will treat makes clear explicitly that the dimensional method in no way depends on the use of Feynman parameters. Actually, since for two or more closed loops ultraviolet divergencies may also be transferred from the momentum integrations to the Feynman parameter integrations the use of Feynman parameters in connection with the dimensional regularization scheme must be avoided, or at least be done very judiciously. The procedure for multiloop diagrams will be defined in subsequent subsections.

Consider a self-energy diagram with two scalar intermediate particles in n dimensions:

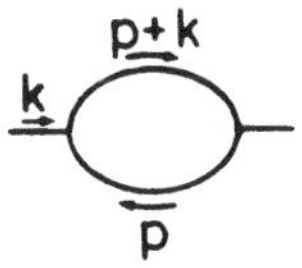

$$I_n = \int d_n p \frac{1}{(p^2 + m^2 - i\varepsilon)[(p + k)^2 + M^2 - i\varepsilon]} . \quad (13.1)$$

In the integrand the loop momentum p is an n component vector.

This expression makes sense in one-, two-, and three-dimensional space; in four dimensions the integral is logarithmically divergent. To evaluate this integral we can go to the k rest-frame $(k = 0, 0, 0, i\mu)$. Next we can introduce polar coordinates in the remaining space dimensions

$$I_n = \int_{-\infty}^{\infty} dp_0 \int_0^{\infty} \omega^{n-2}\, d\omega \int_0^{2\pi} d\theta_1 \int_0^{\pi} d\theta_2 \sin\theta_2 \int_0^{\pi} d\theta_3 \sin^2\theta_3 \ldots$$

$$\ldots \int_0^{\pi} d\theta_{n-2} \sin^{n-3}\theta_{n-2} \times$$

$$\times \frac{1}{(-p_0^2 + \omega^2 + m^2 - i\varepsilon)[-(p_0 + \mu)^2 + \omega^2 + M^2 - i\varepsilon]} . \tag{13.2}$$

Here ω is the length of the vector p in the n-1 dimensional subspace. The integrand has no dependence on the angles $\theta_1, \ldots, \theta_{n-2}$, and one can integrate using

$$\int_0^{\pi} \sin^m\theta\, d\theta = \sqrt{\pi}\, \frac{\Gamma\left(\frac{m}{2} + \frac{1}{2}\right)}{\Gamma\left(\frac{m}{2} + 1\right)} . \tag{13.3}$$

These and other useful formulae will be given in Appendix B. The result is

$$I_n = \int_{-\infty}^{\infty} dp_0 \int_0^{\infty} \omega^{n-2}\, d\omega\, \frac{2\pi^{(n-1)/2}}{\Gamma\left(\frac{n-1}{2}\right)} \times$$

$$\times \left(\frac{1}{(-p_0^2 + \omega^2 + m^2)[-(p_0 + \mu)^2 + \omega^2 + M^2]} \right) . \tag{13.4}$$

This integral makes sense also for non-integer, in fact also for complex n. We can use this equation to define I_n in the region where the integral exists. And outside that region we define I_n as the analytic continuation in n of this expression.

Apparently this expression becomes meaningless for $n \leq 1$, for then the ω integral diverges near $\omega = 0$, or for $n \geq 4$ due to the ultraviolet behaviour. The lower limit divergence is not very serious and mainly a consequence of our procedure. Actually for $n \to 1$ not only does the integral diverge, but also the Γ

function in the denominator, so that I_1 from Eq. (13.4) is an undetermined form ∞/∞. Let us first fix n to be in the region where the above expression exists, for instance $1.5 < n < 1.75$. Next we perform a partial integration with respect to ω^2

$$d\omega\, \omega^{n-2} = \frac{1}{2}\, d\omega^2 (\omega^2)^{(n-3)/2} = \frac{1}{2}\, d\omega^2 \frac{2}{n-1} \frac{d}{d\omega^2} (\omega^2)^{(n-1)/2} .$$

For n in the given domain the surface terms are zero. Using $z\Gamma(z) = \Gamma(z+1)$ and repeating this operation λ times we obtain

$$I_n = \frac{2\pi^{(n-1)/2}}{\Gamma\left(\frac{n-1}{2} + \lambda\right)} \int_{-\infty}^{\infty} dp_0 \int_0^{\infty} d\omega\, \omega^{n-2+2\lambda} \left(-\frac{\partial}{\partial\omega^2}\right)^{\lambda} \{\cdots\} . \tag{13.5}$$

We have derived this equation for $1 < n < 4$. However it is meaningful also for $1 - 2\lambda < n < 4$. The ultraviolet behaviour is unchanged, but the divergence near $\omega = 0$ is seen to cancel agianst the pole of the Γ function. Note that

$$\Gamma(z) = \frac{\Gamma(z+1)}{z} ;$$

thus for $z \to 0$ this behaves as $1/z$.

The last equation is analytic in n for $1 - 2\lambda < n < 4$. Since it coincides with the original I_n for $1 < n < 4$ it must be equal to the analytic continuation of I_n outside $1 < n < 4$. Clearly, we now have an explicit expression for I_n for arbitrarily small values of n (taking λ sufficiently large). It is equally obvious that if the original expression which we started from had been convergent, then its value would have been equal to I_n for $n \to 4$. Moreover, cutting equations can be derived using only time and energy components, and as long as the last given expression for I_n exists, and p_0 and ω integrations can be exchanged (i.e. for $n < 4$) these cutting equations can be established.

Let us now see what happens for $n \geq 4$. Again we will use the method of partial integrations to perform the analytic continuation. For simplicity we set $\lambda = 0$. First fix n in the region $1 < n < 4$. Next we insert

$$1 = \frac{1}{2}\left(\frac{dp_0}{dp_0} + \frac{d\omega}{d\omega}\right) .$$

Next we perform partial integration with respect to p_0 and . Again in the given domain the surface terms are zero. Further

$$\frac{1}{2}\left\{-p_0\frac{d}{dp_0} - \omega\frac{d}{d\omega}\right\}\omega^{n-2}\{\ldots\} =$$

$$= \frac{1}{2}\left\{-n + 2 + \frac{-2p_0^2 + 2\omega^2}{(-p_0^2 + \omega^2 + m^2)} + \frac{-2p_0(p_0 + \mu) + 2\omega^2}{[-(p_0 + \mu)^2 + \omega^2 + M^2]}\right\}\omega^{n-2} \times$$

$$\times \{\ldots\} =$$

$$= \frac{1}{2}\left\{-n + 6 + \frac{-2m^2}{(-p_0^2 + \omega^2 + m^2)} + \frac{2p_0\mu + 2\mu^2 - 2M^2}{[-(p_0 + \mu)^2 + \omega^2 + M^2]}\right\}\omega^{n-2} \times$$

$$\times \{\ldots\} .$$

Inserting this in the right-hand side of Eq. (13.4) gives

$$I_n = \frac{-n + 6}{2} I_n - I'_n \tag{13.6}$$

or

$$I_n = -\frac{2}{n-4} I'_n$$

with

$$I'_n = \frac{\pi^{(n-1)/2}}{\Gamma\left(\frac{n-1}{2}\right)} \int_{-\infty}^{\infty} dp_0 \int_0^{\infty} \omega^{n-2}\, d\omega \times$$

$$\times \left\{\frac{2m^2}{(-p_0^2 + \omega^2 + m^2)^2[-(p_0 + \mu)^2 + \omega^2 + M^2]} + \right.$$

$$\left. + \frac{-2p_0\mu - 2\mu^2 + 2M^2}{(-p_0^2 + \omega^2 + m^2)\,[-(p_0 + \mu)^2 + \omega^2 + M^2]^2}\right\} . \tag{13.7}$$

The integral I'_n is convergent for $1 < n < 5$. Now the I_n given before and this expression are equal for $1 < n < 4$; since the last expression is analytic for $n < 5$ with a simple pole at $n = 4$, it must be equal to the analytic continuation of I_n.

The above procedure may be repeated indefinitely. One finds that I_n is of the form

$$I_n = \Gamma\left(\frac{4-n}{2}\right) f(n, \mu, m, M) , \tag{13.8}$$

where the function f is well-behaved for arbitrarily large n. The Γ function shows simple poles at $n = 4, 6, 8, \ldots$.

We see now why the limit $n \to 4$ cannot be taken: there is a pole for $n = 4$. It is very tempting to say that one must introduce a counterterm equal to minus the pole and its residue. But if unitarity is to be maintained this counterterm may not have an imaginary part, i.e. it must be a polynomial in μ, m and M. Thus we must find the form of the residue of the pole. It will turn out to be of the required form. To show that it is of the polynomial form in the general case of many loops necessitates use of the cutting equations. For the one-loop case at hand we will simply compute I_n suing Feynman parameters. One has

$$I_n = \int_0^1 dx \int d_n p \frac{1}{[p^2 + 2pkx + k^2 x + M^2 x + m^2(1-x)]^2} .$$

Shifting integration variables ($p' = p + kx$), making the Wick rotation, and introducing n-dimensional polar coordinates, one computes

$$I_n = \frac{i\pi^{n/2}\,\Gamma\left(2-\frac{n}{2}\right)}{\Gamma(2)} \int_0^1 dx \frac{1}{[M^2 x + m^2(1-x) + k^2 x(1-x)]^{2-n/2}} .$$

In this way we have explicitly I_n in the form of Eq. (13.8). For $n = 4, 6, 8, \ldots$, the integrand is a simple polynomial, for $n = 4$ the integral gives simply 1. Thus the pole term is

$$PP(I_n) = \frac{i\pi^2}{1} \frac{2}{4-n} , \tag{13.9}$$

where PP stands for "pole part". Using the equation

$$a^\varepsilon = e^{\varepsilon \ln a} = 1 + \varepsilon \ln a + O(\varepsilon^2) ,$$

one may compute

$$\lim_{n=4} [I_n - PP(I_n)] = -i\pi^2 \int_0^1 dx \ln \left[M^2 x + m^2(1 - x) + \right.$$

$$\left. + k^2 x(1 - x)\right] + C.$$

Here C is a constant related to the n dependence other than in the exponent of the denominator, containing for instance ln π, from $\pi^{n/2}$. In general C is a polynomial just as the pole part of I_n. Since we could have taken as our starting point an I_n multiplied by b^{n-4}, where b is any constant, we see that C is undetermined. This is the arbitrariness that always occurs in connection with renormalization.

We now must do some work that will facilitate the treatment of the multiloop case. Let us consider the cutting equation. One has:

$$= f^+(k^2) = -(2\pi)^2 \int d_n p \theta(-p_0)\delta(p^2 + m^2) \times$$

$$\times \theta(p_0 + k_0)\delta[(p + k)^2 + M^2] .$$

In the rest frame

$$f^+(\mu) = -(2\pi)^2 \frac{2\pi^{(n-1)/2}}{\Gamma\left(\frac{n-1}{2} + \lambda\right)} \int_0^\infty d\omega \omega^{n-2+2\lambda} \left(-\frac{\partial}{\partial\omega^2}\right)^\lambda \times$$

$$\times \int_{-\infty}^{\infty} dp_0 \theta(-p_0)\delta(-p_0^2 + \omega^2 + m^2)\theta(p_0 + \mu)\delta(-2p_0\mu + M^2 - m^2 - \mu^2). \qquad (13.10)$$

The function $f^-(k^2)$ can be obtained by changing the signs of the arguments of the θ functions.

In coordinate space one has

$$f(x) = \theta(x_0)f^+(x) + \theta(-x_0)f^-(x) .$$

The Fourier transform of this statement is

$$f(\kappa) = \frac{1}{2\pi i} \int_{-\infty}^{\infty} d\tau \, \frac{1}{\tau - i\varepsilon} \, f^{+}(k + \tau) \; +$$

$$+ \frac{1}{2\pi i} \int_{-\infty}^{\infty} d\tau \, \frac{1}{\tau - i\varepsilon} \, f^{-}(k - \tau) \; . \qquad (13.11)$$

τ can be considered as an n vector with all components zero except the energy component. As long as the ω integral is sufficiently convergent (i.e. $n < 4$) one may exchange freely the ω and τ integration. Doing the p_0 and τ integrations one obtains of course the old result for I_n, which is not very interesting. Let us therefore leave the τ integration in front of the ω integral and compute the p_0 integral. One finds for the p_0 integral in f^+

$$\int dp_0 \ldots = \frac{1}{2|\mu|} \, \theta(\mu - M - m)\delta(\omega^2 - \kappa^2) \; ,$$

where

$$\kappa = \sqrt{\frac{(\mu^2 - m^2 - M^2)^2 - 4m^2M^2}{4\mu^2}} \; .$$

The θ function expresses the fact that μ must be positive and furthermore that κ^2 must be positive in order for the ω integration to give a non-zero result. For f^- one obtains the same, except for the change $\theta(\mu - M - m) \to \theta(-\mu - M - m)$.

Also the ω integration can be done, and is of course independent of λ (this follows as usual by considering the integral for $1 < n < 4$ and doing the necessary partial integrations). For $\lambda = 0$ one finds

$$f^{+}(\mu) = -(2\pi)^2 \, \frac{2\pi^{(n-1)/2}}{\Gamma\left(\frac{n-1}{2}\right)} \, \frac{\kappa^{n-3}}{4\mu} \, \theta(\mu - M - m) \; .$$

The complete function $f(k) = I_n$ obtains as given above. We must study

$$\int d\tau \, \frac{1}{\tau - i\varepsilon} \, f^{+}(\mu + \tau) \longrightarrow \int_{M+m}^{\infty} \frac{d\tau}{\tau - \mu - i\varepsilon} \, \frac{\kappa^{n-3}}{4|\tau|\Gamma\left(\frac{n-1}{2}\right)} \equiv g^{+}(\mu) \; ,$$

where

$$\kappa = \sqrt{\frac{(\tau^2 - \mu^2 - M^2)^2 - 4m^2M^2}{4\tau^2}} .$$

Since

$$\kappa^2(\tau, m, M) = (\tau - m - M)\sigma(\tau, m, M) , \qquad (13.12)$$

where σ is positive and finite at threshold $\tau \to m + M$, this integral is not well defined for $n \leq 1$. This is how the divergence at $\omega = 0$, previously found in Eq. (13.4), manifests itself here. But this is again no problem, and really due to the fact that our derivation is correct only for $n > 1$. We will come back to that below. And there is no trouble in constructing the analytic continuation to smaller values of n. This can be done by performing partial integrations with respect to the factor $(\tau - M - m)$ in Eq. (13.12).

For $n \geq 4$, however, the integral diverges for large values of $|\tau|$. This can be handled as follows. The expression (13.12) for $g^+(\mu)$ is nothing but a dispersion relation, and we may perform a subtraction

$$g^+(\mu) = g(0) + \mu \int_{M+m}^{\infty} \frac{d\tau}{\tau(\tau - \mu - i\varepsilon} \frac{\kappa^{n-3}}{4|\tau|\Gamma\left(\frac{n-1}{2}\right)} ,$$

$$g(0) = \int_{m+M}^{\infty} \frac{d\tau}{4\tau^2} \frac{\kappa^{n-3}}{\Gamma\left(\frac{n-1}{2}\right)} .$$

As before, by inserting $d\tau/d\tau$ it may be shown that $g(0)$ has a pole at $n = 4$. The remainder of g^+, however, is perfectly well-behaved for $n < 5$. Again we see that the pole terms (in $n = 4$) have the proper polynomial behaviour; they are like subtractions in a dispersion relation.

We must now clear up a final point, namely the question of the behaviour of the τ integral near threshold. Consider as an example the function

$$f(\tau) = \frac{(\tau - 1)^\alpha}{\tau^4}$$

In the complex τ plane we have a pole at $\tau = 0$ and for non-integer α a cut along the real axis from $\tau = 1$ to $\tau = \infty$. Multiply this function with $(\tau - \mu - i\varepsilon)^{-1}$ and integrate over a small circle around the point $\tau = \mu$. One obtains

$$2\pi i f(\mu) \ .$$

On the other hand the contour may be enlarged; we get

$$f(\mu) = \frac{1}{2\pi i} \int_C \frac{d\tau}{\tau - \mu - i\varepsilon} \frac{(\tau - 1)^\alpha}{\tau^4} + \text{contribution of the origin,}$$

where C is as in the following diagram:

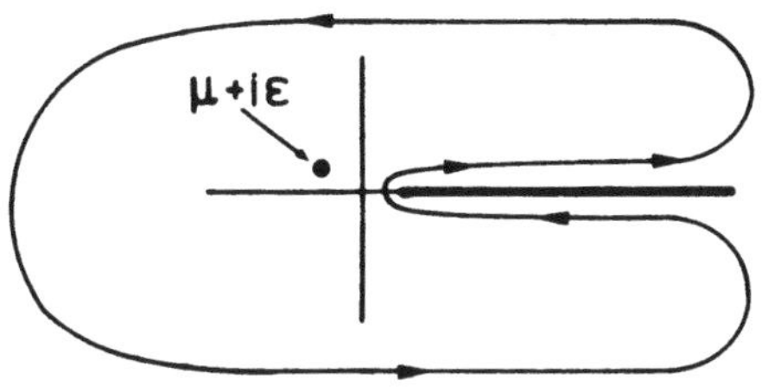

The circle at infinity may be ignored provided $\alpha < 4$. Since now the integrand has a quite singular behaviour at $\tau = 1$, this point must be treated carefully. The contour may be divided into a contribution of a small circle with radius ε around this point, and the rest.

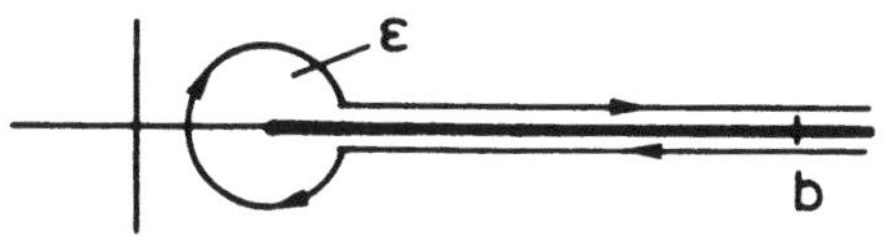

In considering the integral over the circle, τ may be set to one except in the factor $(\tau - 1)^\alpha$. Moreover, we may introduce the change of variable $\tau = \tau - 1$. Writing $\tau = \varepsilon e^{i\phi}$:

$$\int_\varepsilon d\tau \, \tau^\alpha = i \int_\pi^{-\pi} d\phi \varepsilon^{\alpha+1} e^{i(\alpha+1)\phi} = \frac{\varepsilon^{\alpha+1}}{\alpha + 1} \left\{ e^{-i(\alpha+1)\pi} - e^{i(\alpha+1)\pi} \right\} .$$

On the other hand, the contribution of the two contour lines from the circle to some point b above and below the cut contains a part

$$\int_{\varepsilon}^{b} d\tau\, \tau^{\alpha}\left(e^{-i\pi\alpha} - e^{i\pi\alpha}\right).$$

The integrand is the jump across the cut. This can be integrated to give

$$\frac{1}{\alpha+1}\left(b^{\alpha+1} - \varepsilon^{\alpha+1}\right)\left(e^{-i\pi\alpha} - e^{i\pi\alpha}\right).$$

Together with the contribution from the circle

$$\frac{1}{\alpha+1}\, b^{\alpha+1}\left(e^{-i\pi\alpha} - e^{i\pi\alpha}\right).$$

This is independent of ε, and the limit $\varepsilon \to 0$ can be taken. Note that the result is $-2\pi i$ in the limit $\alpha \to -1$, as should be for a clockwise contour.

In the expression for $f^{+}(\mu)$ (and $f^{-}(\mu)$) we have not bothered about the precise behaviour at the start of the cut. This is in principle accounted for by the θ function. This θ function gives only the contribution along the cut; the small circle has been ignored. This is allowed only if $n > 1$ (corresponding to $\alpha > -1$). Otherwise one must carefully specify what happens at threshold in f^{+}, f^{-}, and in the subsequent integrals over τ.

To see in detail how this goes consider a function of τ^2 having a cut from $\tau^2 = {}^2$ to $\tau^2 = \infty$, but otherwise analytic (and going sufficiently fast to zero at infinity). In the τ plane the function has cuts from $-a$ to $-\infty$ and $+a$ to $+\infty$. The dispersion relation leads in the τ plane to the following contour:

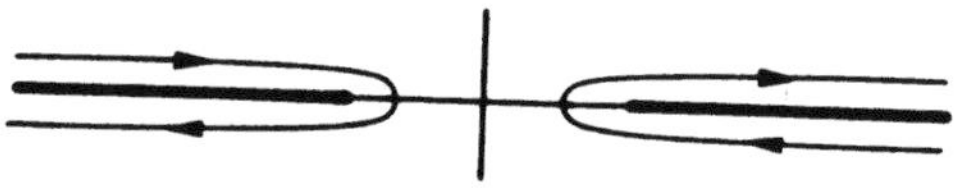

Consider the right-hand side contour. The cut starts at the point $\tau = a$. We may write

$$\int_{\circlearrowleft} \frac{1}{\tau - s - i\varepsilon}\, f(\tau^2) + \int_{a-\delta}^{\infty} \frac{1}{\tau - s - i\varepsilon}\, \bar{f}(\tau^2)\ ,$$

where $\circlearrowleft$ stands for the circle at a with radius δ, and $\bar{f}(\tau^2)$ is the jump over the cut, i.e.

$$\bar{f}(\tau^2) = \lim_{\delta'\to 0}\left\{ f(\tau^2 + i\delta') - f(\tau^2 - i\delta') \right\}.$$

This is the precise equivalent of our f^+. It is essential to first take the limit $\delta' = 0$ before the limit $\delta = 0$.

Let us now return to the question of subtractions. It is now possible to turn the reasoning around. We know that for I_n and its Fourier transforms the following properties hold:

i) $I_n(k)$ has poles for $n = 4, 6, \ldots$, and the $f_n^{\pm}(x)$ have no pole for $n = 4$.
ii) $I_n(x) = \theta(x_0)f_n^+(x) + \theta(-x_0)f_n^-(x)$ for $n < 4$.
iii) $I_n(x)$, $f_n^{\pm}(x)$ are Lorentz-invariant.

The functions f_n^+ and f_n^- are non-singular for $n = 4$; they are the cut diagrams and these are not divergent for $n = 4$. Concerning point (ii), the derivation of the cutting equation requires $n < 4$. Of course it is possible to define things by analytic continuation, but our dispersion-like relations as exhibited above will hold only for $n < 4$. From (ii) and (iii) it follows immediately that the pole in $n = 4$ of $I_n(x)$ can at most be a δ function. The precise reasoning is as follows. Applying a Lorentz transformation to (ii) and insisting on Lorentz invariance we find

$$I_n(x) = f_n^+(x) = f_n^-(x) \quad \text{for} \quad x_0 = 0\,, \quad \bar{x} \neq 0\,.$$

Consider now the analytic continuation of (ii) to larger n values. If f^+ and f^- have no pole for $n = 4$, then the only way that $I_n(x)$ can have a pole for $n = 4$ is in the point $x_0 = 0$, $\vec{x} = 0$. Remember now that we have shown that the Fourier transform of I_n is of the form

$$\frac{1}{n-4} \times \text{finite function for } n < 5\,,$$

i.e. the Fourier transform of I_n exists for $n < 5$. Since the Fourier transform of the pole part is a function which is non-zero only for $x_0 = \vec{x} = 0$, the only possibility is that this pole part is a polynomial in the external momenta (and the various masses in the problem), i.e. in coordinate space a δ function and derivatives of δ functions.

Thus we know now that the residue of I_n at the pole is a polynomial. Differentiating I_n with respect to the external momentum k, in the region $n < 4$ where we can use a well-defined representation, see Eqs. (13.4) and (13.5), we see that this derivative is finite for $n = 4$. It follows that the pole part is independent of k, which is indeed what is found before, Eq. (13.9). Also, this is all we need to know for renormalization purposes.

13.3 Multiloop Diagrams

We must now extend the method to diagrams containing arbitrarily many loops. This is quite straightforward. First note that external momenta (and sources) span at most a four-dimensional space. We split n-dimensional space into this four-dimensional space and the rest

$$\int d_np_1d_np_2 \cdots d_np_k = \int d_4\underline{p}_1d_4\underline{p}_2 \cdots d_4\underline{p}_k \times$$

$$\times \int d_{n-4}p_1 \cdots d_{n-4}p_k \ . \tag{13.13}$$

The $\underline{p}_i$ are the components of the p_i in four-dimensional space. The integrand will depend on the scalar products of the $\underline{p}_i$ with themselves ant the external vectors, and furthermore on the scalar products of the p_i with themselves. Again, the integration over those angles that do not appear in the integrand may be performed. This is done as follows. Consider the integral

$$\int d_{n-4}p_i \ .$$

The argument of this integral is already integrated over $p_{i+1} \ldots p_k$; therefore the integral depends only in the scalar product of p_i with $p_1 \ldots p_{i-1}$. These vectors span an $i-1$ dimensional space, and we may write

$$\int d_{n-4}p_i = \int d_{i-1}\bar{p}_i \int d_{n-3-i}\bar{\bar{p}}_i \ . \tag{13.14}$$

Now the integrand will no longer depend on the direction of $\bar{\bar{p}}$, and introducing polar coordinates in $\bar{\bar{p}}$ space

$$\int d_{n-3-i}\bar{\bar{p}}_i = \frac{2\pi^{(n-3-i)/2}}{\Gamma\left(\frac{n-3-i}{2}\right)} \int_0^\infty \omega_i^{n-4-i} d\omega_i \ . \tag{13.15}$$

The ω_i represent the lengths of the vectors $\bar{\bar{p}}_i$. Introduce in k-dimensional space the vectors

$$q_1 = \begin{pmatrix} \omega_1 \\ 0 \\ \vdots \\ 0 \end{pmatrix}, \quad q_2 = \begin{pmatrix} (P_2)_1 \\ \omega_2 \\ 0 \\ \vdots \\ 0 \end{pmatrix}, \quad \ldots, \quad q_k = \begin{pmatrix} (\bar{P}_k)_1 \\ \vdots \\ (\bar{P}_k)_{k-1} \\ \omega_n \end{pmatrix}$$

Obviously

$$(q_i q_j) = (P_i P_j) .$$

Also, for $i \neq k$,

$$\int d_k q_i = \int d_{i-1} \bar{q}_i \int d_{k-i+1} \bar{\bar{q}}_i =$$

$$= \int d_{i-1} \bar{P}_i \frac{2\pi^{(k-i+1)/2}}{\Gamma\left(\frac{k-i+1}{2}\right)} \int_0^\infty \omega_i^{k-i} \, d\omega_i , \qquad (13.16)$$

but note

$$\int d_k q_k = \int d_{k-1} \bar{P}_k \int_{-\infty}^{\infty} d\omega_k , \qquad (13.17)$$

with ω_k running from $-\infty$ to $+\infty$. We have therefore comparing Eqs. (13.14) and (13.15) with (13.16) and (13.17)

$$\int d_{n-4} P_i = \pi^{(n-k-4)/2} \frac{\Gamma\left(\frac{k-i+1}{2}\right)}{\Gamma\left(\frac{n-3-i}{2}\right)} \int d_k q_i \omega_i^{n-4-k} \quad \text{for } i \neq k ,$$

$$\int d_{n-4} P_k = 2\pi^{(n-k-4)/2} \frac{\Gamma\left(\frac{1}{2}\right)}{\Gamma\left(\frac{n-3-k}{2}\right)} \int d_k q_k \omega_k^{n-4-k} \theta(\omega_k) ,$$

where we used $\Gamma(1/2) = \sqrt{\pi}$.

Finally note

$$\omega_1 \omega_2 \cdots \omega_k = \varepsilon_{i_1 \ldots i_k} (q_1)_{i_1} \cdots (q_k)_{i_k} \equiv \det q .$$

Since ω_1, ω_2, etc., are positive we can write

$$\theta(\omega_k) = \theta(\det q) \ .$$

We arrive thus at the equation

$$\int d_{n-4}P_1 \ldots d_{n-4}P_k = 2 \prod_{i=1}^{k} \pi^{(n-k-4)/2} \frac{\Gamma\left(\frac{k-i+1}{2}\right)}{\Gamma\left(\frac{n-3-i}{2}\right)} \times$$

$$\times \int d_k q_1 \ldots d_k q_k (\det q)^{n-4-k} \theta(\det q) \ . \qquad (13.18)$$

Next we have

$$(\det q)^{\alpha} = \frac{1}{(\alpha+k) \ldots (\alpha+1)} \left(\det \frac{\partial}{\partial q}\right) (\det q)^{\alpha+1} \ .$$

This equation can be arrived at by tedious work. Just write out the determinants with the help of the ε symbol with k indices. Note that for $\alpha = 0$ the equation follows trivially because of

$$\varepsilon_{i_1 \ldots i_k} \varepsilon_{i_1 \ldots i_k} = k!$$

Also for $k = 1$ the equation is trivial.

Now the integrand is a function of the scalar products $a_{ij} = (q_i, q_j)$. Let us consider the operation on such a function of the operator

$$\det \frac{\partial}{\partial q} \ .$$

One has

$$\frac{\partial}{\partial (q_i)_j} = \sum_s (q_s)_j \left(\frac{\partial}{\partial a_{si}} + \frac{\partial}{\partial a_{is}} \right)$$

$$\det \left(\frac{\partial}{\partial q}\right) = (\det q) \det \left(\frac{\partial}{\partial a_{ij}} + \frac{\partial}{\partial a_{ji}} \right) =$$

$$2^k (\det q) \det \left(\frac{1}{2} \frac{\partial}{\partial a_{ij}} + \frac{1}{2} \frac{\partial}{\partial a_{ji}} \right).$$

With these equations one can construct the analytic continuation of the above equation to small values of n

$$\int dq_1 \ldots dq_k (\det q)^{n-4-k} = \int dq_1 \ldots dq_k \frac{1}{(n-4)(n-5)\ldots(n-3-k)} \times$$

$$\times \left(\det \frac{\partial}{\partial q} \right) (\det q)^{n-3-k}$$

$$= \frac{1}{(n-4)\ldots(n-3-k)} \int dq_1 \ldots dq_k (\det q)^{n-3-k} \left(\det - \frac{\partial}{\partial q} \right)$$

$$= \frac{2^k}{(n-4)\ldots(n-3-k)} \int dq_1 \ldots dq_k (\det q)^{n-2-k} \left(\det - \frac{1}{2} \frac{\partial}{a_{ij}} - \right.$$

$$\left. - \frac{1}{2} \frac{\partial}{a_{ji}} \right) .$$

Performing this operation λ times gives

$$\int d_{n-4}P_1 \ldots d_{n-4}P_k = 2 \prod_{i=1}^{k} \pi^{(n-k-4)/2} \frac{\Gamma\left(\frac{k-i-1}{2}\right)}{\Gamma\left(\frac{n-3-i}{2} + \lambda\right)} \times$$

$$\times \int d_k q_1 \ldots d_k q_k \theta(\det q)(\det q)^{n-4-k+2\lambda} \left(\det - \frac{1}{2} \frac{\partial}{\partial a_{ij}} - \right.$$

$$\left. - \frac{1}{2} \frac{\partial}{\partial a_{ji}} \quad ^{\lambda} \right) . \qquad (13.19)$$

Rather than worrying whether this derivation is correct or not we simply take this equation as the definition of the k-loop diagrams for non-integer n. It is not very difficult to verify that for finite diagrams the result coincides with the above equation for n = 4 (see below). In doing such work it is often advantageous to go back to the special coordinate system with which we started; the expression in terms of the q_i is mainly useful for invariance considerations. For instance shifts of integration variables such as

$$q_1 \longrightarrow q_1 + q_2$$

leave both d_kq_1 and d_kq_2 as well as det q unchanged as long as the integrals converge (which they do for sufficiently small n).

Equation (13.19) is much too complicated for practical work, and we will instead use the notation

$$\int d_np_1 \ldots \int d_np_n \ .$$

All the above work is to show how things can be defined for non-integer n.

The final step consists of showing that the algebraic properties necessary for diagrammatic analysis hold for this in the same way as they do for integer n. We have seen already that this is so for shifts of integration variables. Furthermore, clearly

$$\int d_np_1 \int d_np_2 = \int d_np_2 \int d_np_1$$

(always assuming that n is chosen such that the integrals converge).

Next there is the following problem. It seems that the definition depends on the number of loops. Then in considering cutting equations one has relations involving all kinds of possibilities on both sides of the cut, and it may seem hard to understand what happens. The answer to this is that the definition involves for k loops a k-dimensional q-space, but nothing prevents us from using a larger space. Thus in any equation one can use for k simply the largest number of closed loops that occur in any one term. And there is no difficulty in decreasing the k used if the number of loops is smaller than this k: simply perform the integrations over those directions that od not appear in the integrand.

Let us now consider the limit $n \rightarrow 4$ in the case where the original integral exists. This means that there are no difficulties with the ultraviolet behaviour. Using Eq. (13.19) with λ sufficiently large we may first do the q_1 integration. Using coordinates as described in the beginning one arrives at an integral of the type as encountered in the one-loop case

$$\frac{1}{\Gamma[\frac{1}{2}(n-4)+\lambda]}\int_0^\infty d\omega_1 \omega_1^{n-5+2\lambda}\left(-\frac{\partial}{\partial\omega^2}\right)^\lambda f(\omega_1^2)\ .$$

Keeping n in the neighbourhood of four it is possible to perform partial integration with respect to ω_1 only

$$\frac{1}{\Gamma[\frac{1}{2}(n-4)+1]}\int_0^\infty d\omega_1 \omega_1^{n-3}\left(-\frac{\partial}{\partial\omega_1^2}\right) f(\omega_1^2)\ .$$

Next we write $\omega_1^{n-3} = \omega_1 \cdot \omega_1^{\varepsilon} = \omega_1(1+\varepsilon \ln\omega_1 + O(\varepsilon^2))$, with $\varepsilon = n-4$. Since $\omega_1 \cdot d\omega_1 = 1/2 d\omega^2$, we find

$$\frac{1}{\Gamma[\frac{1}{2}(n-4)+1]}\left\{\frac{1}{2}f(0)+\frac{n-4}{2}\int_0^\infty d\omega_1^2 \ln\omega_1\left(-\frac{\partial}{\partial\omega^2}\right)f(\omega^2) + O[(n-4)^2]\right\}\ .$$

We see that the integral reduces to what one would have with $q_1 = 0$, plus a finite amount proportional to n - 4. Since q_1 was the "unphysical" part of the momentum p_1 we see that we recover in this way the expression that we started from. Moreover, if the original expression was finite, then the result for values of n close to four differs by finite terms proportional to n - 4.

If the integral does not exists for n = 4, then we are interested in constructing an analytic continuation to larger n values. This can be done as follows. Select a number of closed loops in the diagram. Call the loop-momenta associated with these loops p_1 ... p_s (s loops selected). Next insert

$$1 = \frac{1}{s\cdot n}\sum_{j=1}^{n}\sum_{i=1}^{s}\frac{d(p_i)_j}{d(p_i)_j}\ ,$$

which is symbolic for

$$\frac{1}{4k}\sum_{j=1}^{4}\sum_{i=1}^{k}\frac{dp_{ij}}{dp_{ij}}+\frac{1}{k^2}\sum_{j=1}^{k}\sum_{i=1}^{k}\frac{dq_{ij}}{dq_{ij}}$$

and perform partial integrations. The result is the original integral I plus an integral I′ that is better convergent with respect to the loops selected. Solving the equation with respect to I we get

$$I = \frac{1}{sn - \lambda} I'$$

in analogy with Eq. (13.6), obtained in the one-loop case. Here $(-\lambda)$ is the number obtained by counting the powers of the momenta $p_1 \ldots p_s$ in the integrand. That is the very nice thing about this method: there is a direct relation between power counting and the location of the poles in the complex n plane.

Here we find a pole for $n = \lambda/s$. If $4s - \lambda = 2, 1, 0$, we have quadratic, linear, or logarithmic divergencies with respect to the $p_1 \ldots p_s$ integrations.

It is clear that we now can have poles for $n = \lambda/s$, with λ and s integers. Diagrams with many closed loops give many poles in the complex n plane. The first pole would be at $n = 4 - 2/s$, $4 - 1/s$, 4 for quadratic, etc., divergent integrals, the next is one (generally however two) units 1/s further in the direction of increasing n, etc.

13.4 The Algebra of n-Dimensional Integrals

It is now very important to know how the previously discussed combinatorics survives all the definitions. In doing vector algebra one manipulates vectors and Kronecker δ symbols according to the rules

$$\begin{aligned}
\delta_{\mu\nu} p_\nu &= p_\mu , \\
p_\mu p_\mu &= p^2 , \\
\delta_{\mu\nu} \delta_{\nu\alpha} &= \delta_{\mu\alpha} , \\
\delta_{\mu\mu} &= n .
\end{aligned}$$

The only place where the dimension comes in is in the trace of the δ symbol. This δ symbol also appears naturally when performing integrals; for instance (see Appendix B)

$$\int d_n p \frac{p_\mu p_\nu}{(p^2 + 2pk = m^2)^\alpha} = \frac{i\pi^{n/2}}{(m^2 - k^2)^{\alpha - n/2}} \frac{1}{\Gamma(\alpha)} \times$$

$$\left\{\Gamma\left(\alpha-\frac{n}{2}\right)k_\mu k_\nu+\Gamma\left(\alpha-1-\frac{n}{2}\right)\delta_{\mu\nu}(m^2-k^2)\right\} .$$

The indices μ, ν are supposedly contracted with indices μ, ν of external quantities (such quantities are zero if the value of the index is larger than four) or other internal quantities. In doing combinatorics one will certainly meet identities of the form

$$\delta_{\mu\nu}p_\mu p_\nu = p^2 .$$

Now note

$$\int d_n p \frac{p^2}{(p^2+2pk+m^2)} = \frac{i\pi^{n/2}}{(m^2-k^2)^{\alpha-n/2}} \frac{1}{\Gamma(\alpha)} \times$$

$$\times\left\{\Gamma\left(\alpha-\frac{n}{2}\right)k^2+\Gamma\left(\alpha-1-\frac{n}{2}\right)\frac{n}{2}(m^2-k^2)\right\} .$$

All these integrals are computed according to the previously given recipes. The indices μ, ν simply specify two extra dimensions, the two-dimensional space spanned by the objects with which μ and ν specify contractions.

Clearly the two equations are consistent only if we use the rule $\delta_{\mu\nu}\delta_{\mu\nu} = n$.

All this can also be rephrased as follows. If we take the rule $\delta_{\mu\mu} = n$, then the algebra of the integrals is the same as that of the integrands.

The situation is somewhat more complicated if there are fermions and γ matrices. Now γ matrices never occur in final answers; only traces occur. The only relevant rules are

$$\{\gamma^\mu, \gamma^\nu\} = 2\delta_{\mu\nu}\,\mathbb{1} \quad (\mathbb{1} = \text{unit matrix}) ,$$

$$\mathrm{tr}\,(\gamma^\mu\gamma^\nu) = 4\delta_{\mu\nu} .$$

The numbers 2 and 4 are not directly related to the dimensionality of space-time, and they play no role in combinatorial relations. The $\delta_{\mu\nu}$ must of course be treated as indicated above.

Some considerations are in order now to establish that our regularized diagrams satisfy Ward identities for any value of the parameter n. The combinatorial proof of Ward identities involves (i) vector algebra and (ii) shifting of integration variables. We have already shown in Subsection 13.3 that shifting of integration variables is actually allowed because of the invariance of det q in Eq. (13.19). It is also easy to see now that, in the sense defined above, the vector algebra goes through unchanged for any n.

Difficulties arise as soon as in the Ward identities there appear quantities that have the desired properties only in four-dimensional space, like γ^5 or the completely antisymmetric tensor $\varepsilon_{\mu\nu\rho\sigma}$. Then the scheme breaks down, and there are violations of the Ward identities proportional to n - 4. If there are infinities (i.e. poles for n = 4) there may be finite violations of the Ward identities in the limit n = 4. This is what happens in the case of the famous Bell-Jackiw-Adler anomalies.

13.5 Renormalization

Ever since the invention of relativistic quantum electrodynamics, work has been devoted to the problem of renormalization. Mainly, there is the line of Bogoliubov-Parasiuk-Hepp-Zimmerman and the line of Stueckelberg-Petermann-Bogoliubov-Epstein-Glaser. Of course, both treatments have a lot in common; the BPHZ method however seems at a disadvantage in the sense that there seem to be unnecessary complications. Unfortunately these complications are such as to inhibit greatly the treatment of gauge theories.

The SPBEG method, on the other hand, can be taken over unchanged, and also accomodates nicely with the dimensional regularization scheme. The fundamental ingredients are unitarity and causality precisely in the form of the cutting equations. The only complication that remains in the case of gauge theories is the problem of dresses/bare propagators discussed in Section 9. It seems unlikely that this is a fundamental difficulty.

We will sketch in a rough way how renormalization proceeds in the SPBEG method. For more details we refer to the works of these authors.

In the foregoing a definition for diagrams also for non-integer n (= number of dimensions) has been given. This definition is such that:

i) cutting equations hold for all n;

ii) Ward identities hold for all n, provided the ε tensor and γ^5 do not play a role in the combinatorics of these Ward identities;

iii) divergences manifest themselves as poles for n = 4 in the complex n plane.

The problem is now to add counterterms to the Lagrangian such that the poles cancel out. The essential point is that this has to be done order by order in perturbation theory, and since this is precisely the origin of a lot of trouble in connection with gauge theories we will focus attention on that point.

Consider one-loop self-energy diagrams in quantum electromagnetics. They are:

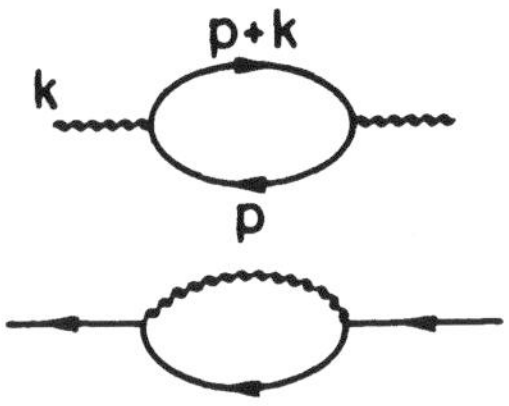

The Feynman rules are:

 $\dfrac{1}{(2\pi)^4 i}\,\dfrac{-i\gamma p + m}{p^2 + m^2 - i\varepsilon}$ 

$\dfrac{1}{(2\pi)^4 i}\,\dfrac{\delta_{\mu\nu}}{k^2 - i\varepsilon}$,

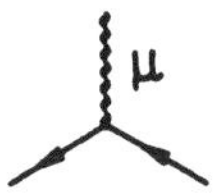

$(2\pi)^4 i \cdot ie\gamma^\mu$.

They follow from the Lagrangian

$$\mathcal{L} = -\frac{1}{4}(\partial_\mu A_\nu - \partial_\nu A_\mu)^2 - \bar\psi(\gamma\partial + m)\psi + ieA_\mu\bar\psi\gamma^\mu\psi - \frac{1}{2}(\partial_\mu A_\mu)^2 .$$

The first diagram gives

$$I_n^\gamma = -e^2 \int d_n p \,\frac{\mathrm{tr}\,\{\gamma^\mu[-i\gamma(p+k) + m]\gamma^\nu(-i\gamma p + m)\}}{(p^2 + m^2 - i\varepsilon)\,[(p+k)^2 + m^2 - i\varepsilon]} .$$

We must work out the trace using only $\{\gamma^\mu\gamma^\nu\} = 2\delta_{\mu\nu}\mathbb{1}$ and $\mathrm{tr}(\gamma^\mu\gamma^\nu) = 4\delta_{\mu\nu}$. The technique is to reduce a trace of λ matrices to traces of $\lambda - 2$ matrices. For four matrices

$$\mathrm{tr}\,(\gamma^\mu\gamma^\alpha\gamma^\nu\gamma^\beta) = -\mathrm{tr}\,(\gamma^\alpha\gamma^\mu\gamma^\nu\gamma^\beta) + 8\delta_{\mu\alpha}\delta_{\nu\beta} = \ldots = -\mathrm{tr}(\gamma^\alpha\gamma^\nu\gamma^\beta\gamma^\mu) +$$

$$+ 8(\delta_{\mu\alpha}\delta_{\nu\beta} - \delta_{\mu\nu}\delta_{\alpha\beta} + \delta_{\mu\beta}\delta_{\alpha\nu})$$

or

$$\mathrm{tr}\,(\gamma^\mu\gamma^\alpha\gamma^\nu\gamma^\beta) = 4\delta_{\mu\alpha}\delta_{\nu\beta} - 4\delta_{\mu\nu}\delta_{\alpha\beta} + 4\delta_{\mu\beta}\delta_{\alpha\nu} \,.$$

This gives

$$I_n^\gamma = -4e^2 \int d_n p \, \frac{m^2\delta_{\mu\nu} - (p+k)_\mu p_\nu - p_\mu(p+k)_\nu + \delta_{\mu\nu}(p^2 + pk)}{(p^2 + m^2 - i\varepsilon)[(p+k)^2 + m^2 - i\varepsilon]} \,.$$

Using Feynman parameters and the equations of Appendix B this integral can be computed giving

$$I_n = -\frac{4ie^2\pi^{n/2}}{\Gamma(2)} \int dx \, \frac{\Gamma\left(\frac{4-n}{2}\right) 2x(1-x)\,(k_\mu k_\nu - k^2\delta_{\mu\nu})}{[m^2 + k^2x(1-x)]^{2-n/2}} \,.$$

The quadratic divergence (pole at $n = 2$) has cancelled out through $(1 - n/2)\Gamma(1 - n/2) = \Gamma(2 - n/2)$.

The second diagram gives

$$I_n^e = -e^2 \int d_n p \, \frac{\gamma^\mu(-i\gamma p + m)\gamma^\mu}{(p^2 + m^2 - i\varepsilon)[(p+k)^2 - i\varepsilon]} \,.$$

From the anticommutation rules, and the rule $\delta_{\mu\mu} = n$

$$\gamma^\mu\gamma^\mu = n\mathbb{1} \,.$$

Further

$$\gamma^\mu\gamma p\gamma^\mu = -\gamma^\mu\gamma^\mu\gamma p + 2\gamma p = (2-n)\gamma p$$

(if $n = 4$ this reduces to the well-known Chisholm rule. Such rules are not valid in n-dimensional space). The integral may now be worked out

$$I_n^e = -\frac{ie^2\pi^{n/2}}{\Gamma(2)} \Gamma\left(\frac{4-n}{2}\right) \int_0^1 dx \, \frac{\mathbb{1}\, nm + (2-n)i\gamma kx}{[m^2(1-x) + k^2x(1-x)]^{2-n/2}} \,.$$

Both I_n^γ and I_n^e have a pole at $n = 4$. The residues are easily computed

$$PP(I_n^\gamma) = \frac{8}{3} i\pi^2 e^2 \frac{1}{n-4} (k_\mu k_\nu - k^2\delta_{\mu\nu}) ,$$

$$PP(I_n^e) = 2i\pi^2 e^2 \frac{1}{n-4} (4m\mathbb{1} - i\gamma k) .$$

If we introduce in the Lagrangian (remember that vertices and terms in the Lagrangian differ by a factor $(2\pi)^4 i$ the counter-terms

$$\frac{e^2}{24\pi^2(n-4)} (\partial_\mu A_\nu - \partial_\nu A_\mu)^2 + \frac{e^2}{8\pi^2(n-4)} \bar\psi(\gamma\partial - 4m)\psi ,$$

then the two-point functions up to order e^2 will be free of poles for $n = 4$, and one can take the limit $n = 4$. To have all one-loop diagrams finite a counterterm to cancel vertex divergencies must be introduced.

Performing a similar calculation (simplified very much from the beginning if only the pole part is needed, see Appendix B) for the vertex diagram leads to the counterterm

$$- \frac{ie^3}{8\pi^2(n-4)} \bar\psi\gamma^\nu\psi A_\nu .$$

The remarkable fact is that the counterterms are gauge invariant by themselves. This would not have been so if the coefficients of $\bar\psi\gamma\partial\psi$ and $-iA_\nu\bar\psi\gamma^\nu\psi$ had been different. It will become clear that this is special to this gauge; in the case of a gauge where the ghost loops are non-zero the result is different, and the counterterms are in general not gauge invariant by themselves.

There are now two separate questions to be discussed. Including the counterterms, all results up to a certain order are finite. This order is such that one has one closed loop but no "countervertex", or a tree diagram including at most one countervertex. An example is electron-electron scattering to order e^4 in amplitude. Some examples of contributing diagrams are:

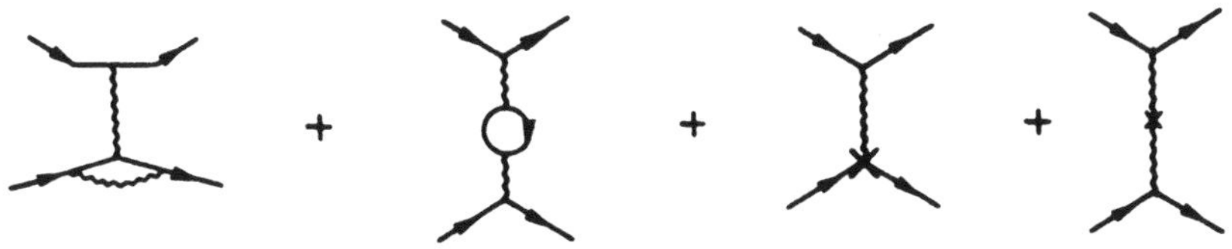

The crosses denote counterterms. The first and third diagrams (and the second and fourth) together are finite.

The question is now whether it is possible to make the theory finite up to the next order in e^2 by introducing further counterterms of the type shown above. Next one may ask how to understand in more detail the gauge structure of the renormalized theory, i.e. the Lagrangian including the counterterms. Both questions will be investigated now.

13.6 Overlapping Divergences

The problem is the following. Can the renormalization procedure be carried through order by order. First we must state more precisely what we mean, because in the Lagrangian we have counterterms of order e^2 and e^3. To this purpose we introduce a parameter η; and all counterterms found from the analysis of one-loop diagrams get this parameter as coefficient. Thus we have now

$$\mathcal{L}_{\text{counter}} = \eta \frac{1}{n-4} \left\{ \frac{e^2}{24\pi^2} (\partial_\mu A_\nu - \partial_\nu A_\mu)^2 + \right.$$
$$\left. + \frac{e^2}{8\pi^2} \bar{\psi}(\gamma\partial - 4m)\psi - \frac{ie^3}{8\pi^2} A_\mu \bar{\psi}\gamma^\mu\psi \right\} .$$

If now in a diagram the number of closed loops is i than we may associate with each diagram a factor L^i. The S matrix is finite up to first order in $(L + \eta)$, in the limit $\eta = 1$, $L = 1$. Thus at most one closed loop no η vertex, or no closed loop one η vertex.

It may perhaps be noted that if the number of ingoing and outgoing lines is given, then specifying $L + \eta$ is equivalent to specifying a certain order in e.

Let us now consider diagrams of order $(L + \eta)^2$, for instance photon self-energy diagrams. There are diagrams of order L^2, i.e. two closed loops, of order $L\eta$ and of order η^2:

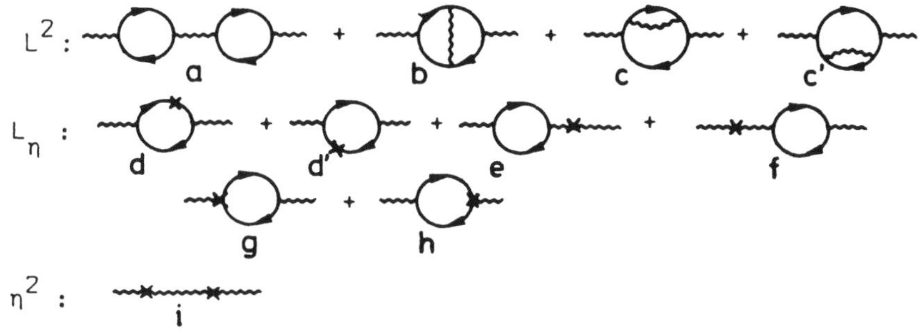

It is now necessary to introduce a classification. We divide all diagrams into two sets:

i) the set of all diagrams of order L^2, $L\eta$ or η^2 that can be disconnected by removing one propagaotr; here (a, e, f, and i);
ii) the rest, called the overlapping diagrams.

Set (i) are the non-overlapping diagrams. No new divergences occur, as can be verified readily. In fact, on both sides of the propagator in question one finds precisely what has been made previously:

a + e + f + i = 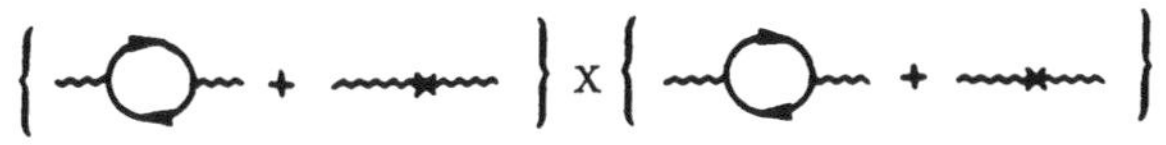

The overlapping diagrams may contain new divergencies, and we must try to prove that these new divergencies behave as local counterterms. In principle, the proof is very simple and based on the use of cutting equations. The first observation is that all cut diagrams (always taking together diagrams of a given order in $(L + \eta)$ on each side of the cut) are finite, that is there is no pole for $n = 4$. The proof of this is easy; these cut diagrams are of the structure of a product of diagrams of lower order in $(L + \eta)$, which are supposedly finite, and an integration over intermediate states. Since for given energy the available phase space is finite the result follows. This assumes that the diagrams have no-integrable singularities in phase space; the latter would correspond to infinite transition probabilities in lower order, which we take not to exist.

Now let us number the vertices to which the external lines are connected. Here there are only two such vertices, and we call them 1 and 2. According to our cutting relations we have, integrating over all x except over x_1 and x_2

$$f(x) = \theta(x_0)\Delta^+ + \theta(-x_0)\Delta^-$$

with $x = x_2 - x_1$. Here Δ^+ contains all cut diagrams with x_1 in the unshadowed region, and Δ^- all cut diagrams with x_2 in the unshadowed region. This is of precisely the same structure as discussed before; we know that if $\Delta^\pm$ are free of poles for $n = 4$, then $f(x)$ can only have a pole part that is a δ function or derivative of a δ function. In other words, the pole part in $n = 4$ must be a polynomial in the external momentum. For arbitrary diagrams this must be true for any combination of "external" vertices (= vertices that have at least one external line), and thus for any of the external momenta. Due to the fact that power counting and the existence of poles for $n = 4$ are in a one-one relationship it can easily be established that the polynomials are at most of a certain degree; here, photon self energies, at most of degree 2. To do all this properly it is necessary to distinguish between over-all divergences and subdivergences (the former disappear if any of the propagators is opened), and show that there are no subdivergences if all the lower order counterterms are included. Next it must be shown that after a certain number of differentiations with respect to the external momenta the over-all divergence disappears. All this is trivial: opening up a propagator in a self-energy diagram is equivalent to considering diagrams with four external lines of lower order in $(L + \eta)$, thus already finite. Over-all divergences correspond to over-all power counting (partial differentiation with respect to all loop momenta), and differentiation with respect to any external momentum lowers this power by one.

13.7 The Order of the Poles

Renormalization may now be performed order by order in $(L + \eta)$. One starts with diagrams with L^1 (one closed loop) and finds the necessary counterterms. They get a factor η in the Lagrangian. Next one considers diagrams of order L^2 (two closed loops) or $(L\eta)^1$ (one closed loop, one counter vertex) or η^2 (tree diagrams, two counter vertices). The total can be made finite by adding further counterterms. Such terms get a factor η^2 in the Lagrangian. In this way one can go on and obtain a counter Lagrangian in the form of a power series in . For $\eta = 1$ the complete theory is finite (there are no poles for $n = 4$).

We have already seen that the terms of order η contain simple poles only. We now want to indicate that the terms of order η^2 contain poles of the form $(n - 4)^{-2}$, and furthermore, the coefficient of such a pole is determined by the order η terms.

First of all, it is trivial to see that there are no quadratic poles at order η^2 if there are no poles at order η. If there are no poles at order η then there are no subdivergences in two-loop diagrams. But over-all divergences of diagrams of order L^2 are simple poles (this follows from partial integration), i.e. there are no quadratic poles.

To see that there are in general quadratic poles consider the sum of photon self-energy diagrams (c) and (d). The sub-integral in (c) may be done and gives rise to an expression of the form

$$\Gamma\left(2 - \frac{n}{2}\right)(p)^{n-3} ,$$

where p is the momentum going into the sub-loop. This equation is symbolic in so far as that we have indicated only the momentum dependence with respect to power counting. The above expression can then be obtained from dimensional arguments. The sum of diagrams (c) and (d) is of the form

$$\int d_n p(p)^{-3}\left\{\Gamma\left(2 - \frac{n}{2}\right)(p)^{n-3} - \frac{2}{4 - n}(p)^1\right\}.$$

Simplification can go too far; we write $p = (p^2 + m^2)^{1/2}$. Next

$$\int d_n p(p^2 + m^2)^\lambda = \Gamma\left(-\lambda - \frac{n}{2}\right)(m^2)^{n/2+\lambda}$$

and the integral becomes

$$\Gamma(-n + 3)\Gamma\left(2 - \frac{n}{2}\right)(m^2)^{n-3} - \Gamma\left(1 - \frac{n}{2}\right)\frac{2}{4 - n}(m^2)^{n/2-1} .$$

Since

$$\Gamma(3 - n) = \frac{1}{3 - n}\Gamma(4 - n) , \quad \Gamma\left(1 - \frac{n}{2}\right) = \frac{2}{2 - n}\Gamma\left(2 - \frac{n}{2}\right),$$

we find for the double pole

$$\frac{1}{3 - n}\,\frac{1}{4 - n}\,\frac{2}{4 - n}\,m^2 - \frac{2}{2 - n}\,\frac{2}{4 - n}\,\frac{2}{4 - n}\,m^2 \longrightarrow \left(\frac{1}{4 - n}\right)^2 2m^2 .$$

If we write

$$(m^2)^{n-3} = m^2 + m^2(n-4)\ln m^2 ,$$

$$(m^2)^{n/2-1} = m^2 + m^2 \frac{n-4}{2} \ln m^2 ,$$

we discover that the $\ln m^2$ term has no pole

$$\frac{m^2 \ln m^2}{(4-n)^2} \left\{ \frac{2}{3-n}(n-4) - \frac{8}{2-n} \frac{n-4}{2} \right\} =$$

$$= \frac{m^2 \ln m^2}{4-n} \left(\frac{2}{3-n} - \frac{4}{2-n} \right) .$$

The residue of the pole for $n = 4$ is zero. This is as should be because a term $\ln m^2$ is non-local, and not admissible as counterterm. Remember that m^2 stands for invariants made up from external momenta, masses, tec.

The above argument is very general and in fact based only on loop and power counting. It can be extended to arbitrary order, and a precise relation between the coefficients of the various higher order poles can be found. (G. 't Hooft, Nucl. Phys. B61 (1973) 455, see Chapter 3 of this book.)

13.8 Order-by-Order Renormalization

Consider the following two-loop diagram:

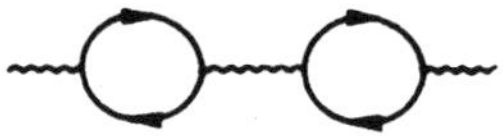

Suppose the self-energy bubble is a function of n and the momentum k of the form:

$$f(k^2) = \frac{1}{n-4} f_1(k^2) + f_2(k^2) + (n-4) f_3(k^2) .$$

We have omitted a factor $k^2\delta_{\mu\nu} - k_\mu k_\nu$. The two-loop diagram gives

$$[f(k^2)]^2 = \left(\frac{1}{n-4} \right)^2 f_1^2 + 2 \frac{f_1 f_2}{n-4} + f_2^2 + 2 f_1 f_3 + O(n-4) .$$

If we simply throw away the pole parts the result is for $n = 4$

$$f_2^2 + 2 f_1 f_3 .$$

However, this result is wrong because it violates unitarity. Suppose now we do order-by-order subtraction. Then one has after the treatment of one closed loop:

 $= f_2 + (n-4)f_3$.

The physical result is the limit n = 4 abd is equal to $f_2(k^2)$. Cutting two-loop diagrams should therefore not involve f_3^2. Next include the proper counterterms and consider:

$$= f_2^2 + O(n-4) \, .$$

Indeed one obtains the correct result consistent with unitarity. This demonstrates that order-by-order renormalization is <u>not</u> equivalent to throwing away poles and their residues of the unrenormalized S-matrix. In a way that is a pity, because the last method is so much easier. But there is no escape, one must do things step by step, that is order by order in the parameter η.

13.9 Renormalization and Slavnov-Taylor Identities

It has been shown that the one-loop counterterms are gauge invariant by themwelves, in the case of Lorentz gauge quantum electrodynamics. The question is whether we can understand this, and to what extent this is general phenomenon.

As a first step we note that combinatorial proofs can be read backwards, at least if they are of the local variety. In other words, Ward identities can be translated backwards into a symmetry if the Lagrangian. Therefore, if the Ward identities of the Lagrangian including counterterms are identical to those without counterterms, then both Lagrangians satisfy the same gauge invariance. Let us consider the S-T identities of the unrenormalized theory including electron sources:

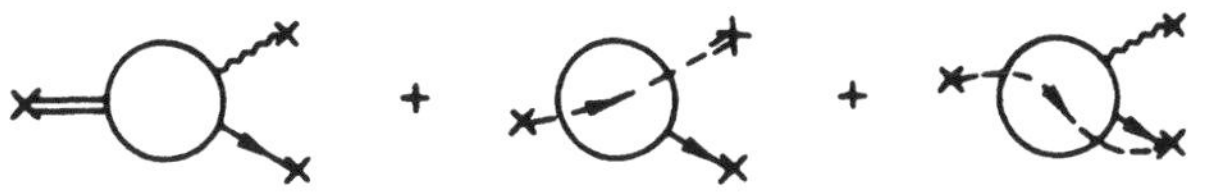 = 0

We have drawn one photon and one electron source explicitly, leaving other sources understood.

Since these identities are true for any value of the dimensional parameter n, they are in particular true for the pole parts.

Consider now one-loop diagrams. The pole parts in the above identity are:

i) the pole parts found in the Green's functions themselves;
ii) the pole part arising from a closed loop, involving the electron-ghost vertex shown in the last diagram.

Now this last vertex is defined by the behaviour of the Lagrangian under gauge transformations, but it is not present in the Lagrangian itself. In $\mathcal{L}$ one only finds $\bar{J}\psi$; if we introduce counterterms in the Lagrangian that make the Green's functions finite (including Green's functions with ingoing and outgoing ghost lines) then the above S-T identity can remain true only if we throw away the pole part of the type (ii). That is, for the renormalized Lagrangian, the S-T identities take the form:

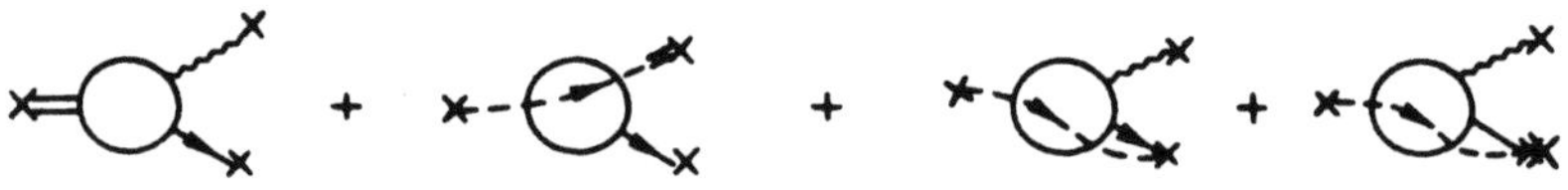

Here the double cross stands for minus the pole part of any diagram such as:

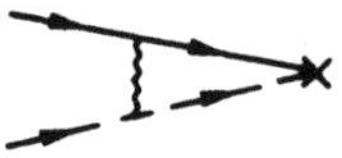

Clearly we can understand this identity as an S-T identity if we redefine the behaviour of the term $\bar{J}\psi$ under gauge transformations. If we say that under a renormalized gauge transformation ψ transforms as

$$\psi \to \psi + ie\Lambda\psi + i\,\frac{Z}{n-4}\,\Lambda\psi \;,$$

where Z is minus the residue of the pole part of the type (ii), then the above identity is again precisely an S-T identity. We conclude that the Lagrangian including counterterms is invariant for renormalized gauge transformations.

To make the statement more precise we must add on a factor η to the pole term in the renormalized gauge transformation. Since we allow as a first step only one closed loop the statement is only true up to first order in η. It is unfortunately somewhat complicated to extend this work to arbitrary order in η, but we will do it in the next section.

In quantum electrodynamics things are never really complicated, because in the usual gauges (Lorentz, Landau, etc.) there is no ghost vertex, ergo there are no pole parts of the type (ii). Then the renormalized Lagrangian is invariant with respect to gauge transformations that transform an electron source coupling as before, i.e.

$$\psi \to \psi + ie\Lambda\psi$$

with the same e as in the unrenormalized Lagrangian. The result is

$\mathcal{L}_{\text{unren}}$ is invariant under the transformation (Λ),

$\mathcal{L}_{\text{unren}} + \mathcal{L}_{\text{counter}}$ is invariant under the transformation (Λ),

therefore

$\mathcal{L}_{\text{counter}}$ alone is invariant.

In the general case this is not true.

13.10 Higher-Order Counterterms

Doing things order by order we suspect (and have shown this to be true up to first order in η) that $\mathcal{L}_{\text{ren}}$ is of the form

$$\mathcal{L}_{\text{ren}}(\eta) = \mathcal{L}_{\text{unren}} + \mathcal{L}_{\text{counter}} ,$$

$$\mathcal{L}_{\text{counter}} = \eta\mathcal{L}_1 + \eta^2\mathcal{L}_2 + \dots ,$$

where $\mathcal{L}_1$, $\mathcal{L}_2$, etc., contain factors $1/(n - 4)$. For $\eta = 1$ the theory is finite. The renormalized Lagrangian $\mathcal{L}(\eta)$ is invariant under gauge transformations of the form

$$A_\mu \to A_\mu + (t_o + t_1\eta + t_2\eta^2 + \dots)\Lambda ,$$

$$\psi \to \psi + (t_o' + t_1'\eta + t_2'\eta^2 + \dots)\Lambda\psi ,$$

where the t and t′ contain factors $1/(n - 4)$.

Let us suppose this latter statement to be true up to order k in η. Consider now:

i) diagrams containing only vertices of the unrenormalized Lagrangian with k + 1 closed loops; such diagrams exhibit poles up to degree k + 1 and satisfy the unrenormalized S-T identities;

ii) diagrams containing counterterms (corresponding to vertices of $\mathcal{L}_1 \ldots \mathcal{L}_k$, $t_0 \ldots t_k$, $t'_0 \ldots t'_k$) but with at least one closed loop;

iii) diagrams of order η^{k+1} (containing thus no closed loop).

Example for two closed loop photon self-energy diagrams:

Set (i) = diagrams (a), (b), (c).
Set (ii) = diagrams (d) to (h).
Set (iii) = diagram (i) plus a new diagram of the form 
that is needed to remove the divergencies of the overlapping diagrams. Now (i) + (ii) + (iii) give a finite theory as $n \to 4$ and $\eta \to 1$. That is simply how they were defined. Next (i) + (ii) satisfy Ward identities (the theory was assumed to be invariant up to order η^k). Therefore the residues of the pole of set (iii) satisfy the S-T identities. Because the diagrams (iii) are tree diagrams the dimension n appears nowhere except in the pole factors $1/(n - 4)$. Therefore the diagrams (iii) satisfy the S-T identities, which means invariance of the renormalized Lagrangian under the renormalized gauge transformation considering terms of order η^{k-1} only. This completes the proof by induction.

Acknowledgements

The authors are deeply indebted to Dr. R. Barbieri, who helped in writing a first approximation to this report. One of the authors (M.V.) wishes to express his gratitude for the hospitality and pleasant atmosphere encountered at the Scuola Normale at Pisa; the lectures given there have been a first test of the present material.

Finally, we would like to thank the CERN Document Reproduction Service for their excellent work on this report (in particular, Mme Paulette Estier for the speedy and accurate typing of the text), and Mme E. Ponssen of the CERN MSC Drawing Office for her efficient co-operation.

APPENDIX A

FEYNMAN RULES

The most frequently encountered Feynman rules will be summarized here. Also combinatorial factors will be discussed. The external sources to be employed for S-matrix definition must be normalized such that they emit or absorb one particle. The normalization formulae follow from the propagators; the "ingoing and outgoing line wave-functions" indicated below are the product of those sources, the associated propagator and the associated mass-shell factor $k^2 + m^2$.

Spin-0 particles

Propagator:

$$\frac{1}{(2\pi)^4 i} \frac{1}{k^2 + m^2 - i\varepsilon} \, . \qquad \text{(A.1)}$$

In shadowed region:

$$- \frac{1}{(2\pi)^4 i} \frac{1}{k^2 + m^2 + i\varepsilon} \, . \qquad \text{(A.2)}$$

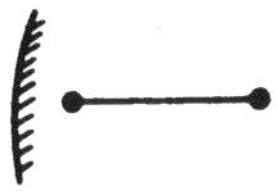

Cut propagator:

$$\frac{1}{(2\pi)^3} \theta(k_0) \delta(k^2 + m^2) \, . \qquad \text{(A.3)}$$

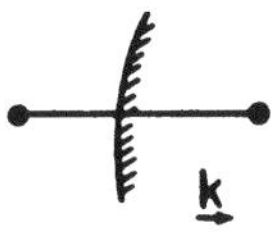

Wave-function:

$$1 \, . \qquad \text{(A.4)}$$

Spin-1/2 particles

Propagator:

$$\frac{1}{(2\pi)^4 i} \frac{-i\gamma k + m}{k^2 + m^2 - i\varepsilon} \, . \qquad \text{(A.5)}$$

In shadowed region:

$$- \frac{1}{(2\pi)^4 i} \frac{-i\gamma k + m}{k^2 + m^2 + i\varepsilon} \, . \qquad \text{(A.6)}$$

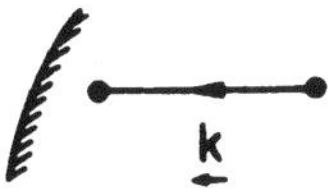

Cut propagator:

$$\frac{1}{(2\pi)^3}\,(-i\gamma k + m)\theta(-k_0)\delta(k^2 + m^2)\ . \qquad (A.7)$$

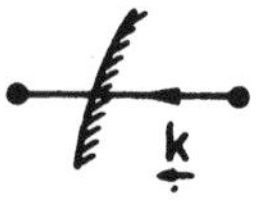

$$\frac{1}{(2\pi)^3}\,(-i\gamma k + m)\theta(k_0)\delta(k^2 + m^2)\ . \qquad (A.8)$$

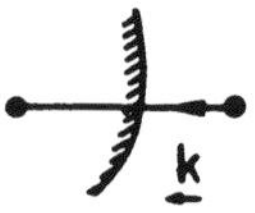

$$\left.\begin{array}{lll} \text{Ingoing particle wave-function :} & u^a(k)\sqrt{2k_0} & ;\ a = 1,\ 2 \\ \text{Ingoing antiparticle wave-function:} & -\bar{u}^a(k)\sqrt{2k_0} & ;\ a = 3,\ 4 \\ \text{Outgoing particle wave-function:} & \bar{u}^a(k)\sqrt{2k_0} & ;\ a = 1,\ 2 \\ \text{Outgoing antiparticle wave-function:} & u^a(k)\sqrt{2k_0} & ;\ a = 3,\ 4 \end{array}\right\} . \qquad (A.9)$$

Note the minus sign for the incoming antiparticle. The momentum k is directed inwards for incoming particles and outwards for outgoing particles. As usual $\bar{u} = u^*\gamma^4$.

The spinors are solutions of the Dirac equation (note that $k_0 = +\sqrt{\vec{k}^2 + m^2}$)

$$\begin{aligned} (i\gamma^\mu k_\mu + m)u^a(k) &= 0\ , \quad a = 1,\ 2\ , \\ (-i\gamma^\mu k_\mu + m)u^a(k) &= 0\ , \quad a = 3,\ 4\ . \end{aligned} \qquad (A.10)$$

In the 4 x 4 representation, with $\gamma^5 = \gamma^1\gamma^2\gamma^3\gamma^4$

$$\gamma^0 = \begin{pmatrix} 0 & -i\sigma_j \\ i\sigma_j & 0 \end{pmatrix},\quad \gamma^4 = \begin{pmatrix} 1 & 0 \\ 0 & -1 \end{pmatrix},\quad \gamma^5 = \begin{pmatrix} 0 & -1 \\ -1 & 0 \end{pmatrix}, \qquad (A.11)$$

$$\sigma_1 = \begin{pmatrix} 0 & 1 \\ 1 & 0 \end{pmatrix},\quad \sigma_2 = \begin{pmatrix} 0 & -i \\ i & 0 \end{pmatrix},\quad \sigma_3 = \begin{pmatrix} 1 & 0 \\ 0 & -1 \end{pmatrix},\quad 1 = \begin{pmatrix} 1 & 0 \\ 0 & 1 \end{pmatrix},$$

these solutions are given in the following table:

	$\uparrow u^1(k)$	$\downarrow u^2(k)$	$\downarrow u^3(k)$	$\uparrow u^4(k)$
$\sqrt{\dfrac{m+k_o}{2k_o}}\ \times$	1	0	$-\dfrac{k_3}{m+k_o}$	$\dfrac{k_1 - ik_2}{m+k_o}$
	0	1	$-\dfrac{k_1 + ik_2}{m+k_o}$	$-\dfrac{k_3}{m+k_o}$
	$\dfrac{k_3}{m+k_o}$	$\dfrac{k_1 - ik_2}{m+k_o}$	-1	0
	$\dfrac{k_1 + ik_2}{m+k_o}$	$\dfrac{-k_3}{m+k_o}$	0	1

The arrows denote spin up/down assignments in the k rest frame.
Normalization:

$$\sum_{\alpha=1}^{4} u_\alpha^{*i}(k) u_\alpha^j(k) = \delta_{ij} \ . \qquad \text{(A.12)}$$

Spin summations:

$$\sum_{i=1}^{2} u_\beta^i(k)\bar{u}_\alpha^i(k) = \frac{1}{2k_o}(-i\gamma k + m)_{\beta\alpha} \ ,$$

$$\sum_{i=3}^{4} u_\beta^i(k)\bar{u}_\alpha^i(k) = -\frac{1}{2k_o}(i\gamma k + m)_{\beta\alpha} \ . \qquad \text{(A.13)}$$

In connection with parity P, charge conjugation C, and time-reversal T, the following matrices and transformation properties are of relevance.

The matrice γ^4 is the transformation matrix connected with space reflection

$$u^{\alpha}(-\vec{k}, k_o) = -\gamma^4 u^{\alpha}(\vec{k}, k_o) , \quad \text{particle} \quad \alpha = 1, 2 ,$$

$$u^{\alpha}(-\vec{k}, k_o) = -\gamma^4 u^{\alpha}(\vec{k}, k) , \quad \text{antiparticle } \alpha = 3, 4 .$$

$$\gamma^4\gamma^{\mu}\gamma^4 = \begin{cases} -\gamma^{\mu} & \mu = 1, 2, 3 \\ \gamma^{\mu} & \mu = 4 . \end{cases} \tag{A.14}$$

The matrix C transforms an incoming particle into an incoming antiparticle, etc.

$$\begin{aligned} u^{\alpha}(k) &= -C\bar{u}^{\beta}(k) \quad \begin{cases} \alpha = 4, \ \beta = 1 \\ \alpha = 3, \ \beta = 2 \end{cases} \\ \bar{u}^{\alpha}(k) &= u^{\beta}(k)C^{-1} \quad \begin{cases} \alpha = 1, \ \beta = 4 \\ \alpha = 2, \ \beta = 3 \end{cases} \end{aligned} \tag{A.15}$$

$$C^{-1}\gamma^{\alpha}C = -\tilde{\gamma}^{\alpha} \qquad (\tilde{} = \text{transpose})$$

$$C^{-1}\gamma^5 C = \tilde{\gamma}^5 ,$$

where $C = \gamma^2\gamma^4$, $\tilde{C} = C^{-1} = -C$.

In connection with time-reversal we have the matrix D

$$\begin{aligned} u^{\alpha}(-\vec{k}, k_o) &= -D\tilde{\gamma}^4\bar{u}^{\beta}(\vec{k}, k_o) & \alpha = 2, \ \beta = 1 \\ &= D\tilde{\gamma}^4\bar{u}^{\beta}(\vec{k}, k_o) & \alpha = 1, \ \beta = 2 \\ \bar{u}^{\alpha}(-\vec{k}, k_o) &= -u^{\beta}(\vec{k}, k_o)\tilde{\gamma}^4 D^{-1} & \alpha = 3, \ \beta = 4 \\ &= u^{\beta}(\vec{k}, k_o)\tilde{\gamma}^4 D^{-1} & \alpha = 4, \ \beta = 3 . \end{aligned} \tag{A.16}$$

This changes spin, direction of three-momentum and furthermore exchanges in- and out-states;

$$D = -\gamma^5 C, \quad D^{-1}\gamma^{\mu}D = \tilde{\gamma}^{\mu} , \quad \mu = 1, 2, 3, 4, 5 .$$

Explicitly

$$C = \begin{pmatrix} 0 & 0 & 0 & 1 \\ 0 & 0 & -1 & 0 \\ 0 & 1 & 0 & 0 \\ -1 & 0 & 0 & 0 \end{pmatrix}, \quad D = \begin{pmatrix} 0 & 1 & 0 & 0 \\ -1 & 0 & 0 & 0 \\ 0 & 0 & 0 & 1 \\ 0 & 0 & -1 & 0 \end{pmatrix}. \tag{A.17}$$

Spin-1 particles

Propagator:

$$\frac{1}{(2\pi)^4 i} \frac{\delta_{\mu\nu} + k_\mu k_\nu/m^2}{k^2 + m^2 - i\varepsilon} .$$

In shadowed region:

$$- \frac{1}{(2\pi)^4 i} \frac{\delta_{\mu\nu} + k_\mu k_\nu/m^2}{k^2 + m^2 + i\varepsilon} .$$

Cut propagator:

$$\frac{1}{(2\pi)^3} (\delta_{\mu\nu} + k_\mu k_\nu/m^2)\theta(k_o)\delta(k^2 + m^2) .$$

Particle wave-functions:

$$e_\mu(k) \quad \text{with} \quad e^*_\mu(k)e_\mu(k) = 1 \quad \text{and} \quad k_\mu e_\mu = 0 .$$

There are only three wave-functions, because the propagator matrix has one eigenvalue zero. In the k rest system the various assignments are

	Ingoing	Outgoing
Spin-up:	$e_\mu = \frac{1}{\sqrt{2}}(1, i, 0, 0)$	$e_\mu = \frac{1}{\sqrt{2}}(1, -i, 0, 0)$
Spin-down:	$e_\mu = \frac{1}{\sqrt{2}}(1, -i, 0, 0)$	$e_\mu = \frac{1}{\sqrt{2}}(1, i, 0, 0)$
Spin-z component zero:	$e_\mu = (0, 0, 1, 0)$	$e_\mu = (0, 0, 1, 0)$.

For outgoing particles we must take these expressions to have correct phases. Indeed, the residue of the two-point spin-up/ spin-up amplitude is equal to one.

Combinatorial factors

These are best explained by considering a few examples. Let the interaction Lagrangian for a scalar field be

$$\mathcal{L}_I = \frac{\alpha}{3!}\phi^3 + \frac{\beta}{4!}\phi^4 .$$

The vertices are :

$(2\pi)^4 i\alpha$

$(2\pi)^4 i\beta$.

The lowest order self-energy diagram is:

Draw two points x_1 and x_2 and draw in each of these points the α vertex:

Now count in how many ways the lines can be connected with the same topological result. External line 1 can be attached in six, after that line 2 in three ways. After that there are two ways to connect the remaining lines such that the desired diagram results. Thus there are altogether 6 x 3 x 2 combinations. Now divide by the permutational factors of the vertices (here 3! for each α vertex). Finally divide by the number of permutations of the points x that have identical vertices. Here 2! The total result is

$$\frac{6 \times 3 \times 2}{3!\ 3!\ 2!} = \frac{1}{2} .$$

As another example consider the diagram:

There are three x points:

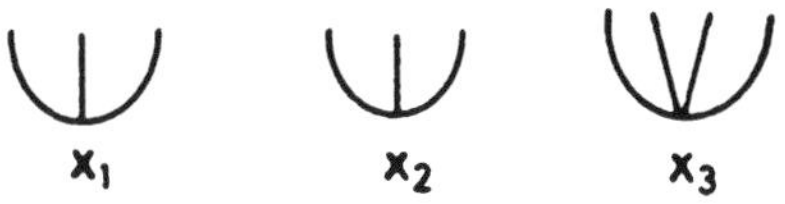

Line 1: six ways. Line 2: four ways. Then we have for instance:

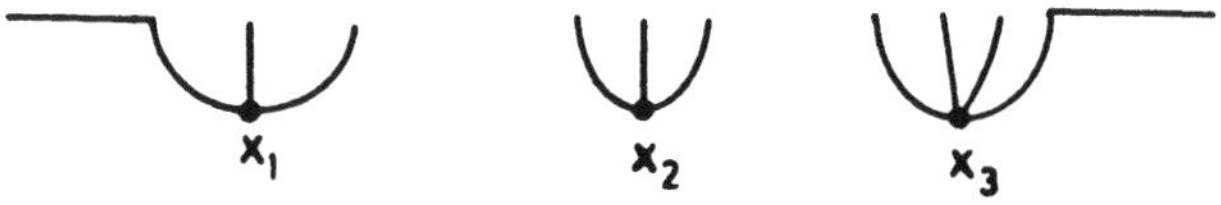

There are 6 x 3 x 2 ways to connect the rest such as to get the desired topology. We must divide by vertex factors (3! 3! 4!) and by 2! (permutation of the identical vertex points x_1 and x_2). The result is

$$\frac{6 \times 4 \times 6 \times 3 \times 2}{3! \;\; 3! \;\; 4! \;\; 2!} = \frac{1}{2} \; .$$

Final example: two identical sources connected by a scalar line:

Factor: $\frac{1}{2!}$.

For two non-dentical sources the factor is 1:

Factor: 1 .

Topology of quantum electrodynamics

We now show that the vertices of any diagram of quantum electrodynamics can be numbered in a unique way. Let the external momenta be $k_1 \; ... \; k_n$.

Step 1

Start at the electron line with the lowest momentum index. Follow the arrow of the line. Number the vertices consecutively.

Step 2

Go back to the first vertex (or lowest numbered vertex of which the photon line was not exploited). Follow the photon line. We arrive then either at a vertex that is already numbered, at an external photon line, or at another electron line. In the first two cases, take the next vertex along the electron line of step 1 and restart at step 1. When arriving at a new vertex, number again consecutively following the electron line along the arrow. When hitting the end, or an already numbered vertex, go back against the arrow and number all the vertices on the electron line before the vertex that was the entrance point of that line. After that restart 2.

Example:

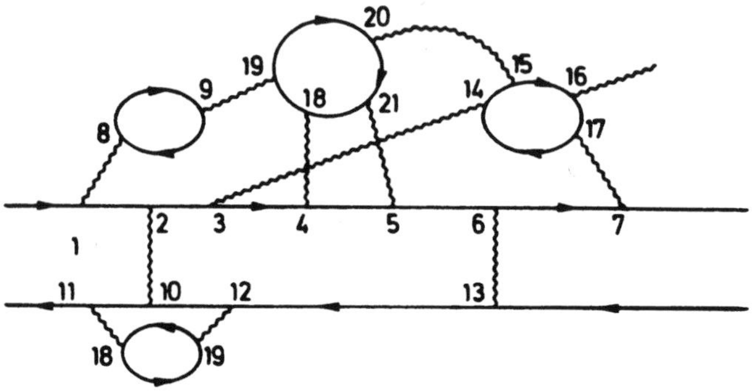

APPENDIX B

SOME USEFUL FORMULAE

$$\int d_n x f(x) = \int f(x) r^{n-1}\, dr\, \sin^{n-2}\theta_{n-1} d\theta_{n-1} \sin^{n-3} \times$$

$$\times\, \theta_{n-2}\, d\theta_{n-2} \cdots d\theta_1 \tag{B.1}$$

with $0 \leq \theta i \leq \pi$, except $0 \leq \theta_1 \leq 2\pi$. If $f(x)$ depends only on

$$r = \sqrt{x_1^2 + \ldots x_n^2}$$

one may perform the integration over angles using

$$\int_0^\pi \sin^m\theta\, d\theta = \sqrt{\pi}\, \frac{\Gamma\left(\frac{m+1}{2}\right)}{\Gamma\left(\frac{m+2}{2}\right)} \tag{B.2}$$

leading to

$$\int d_n x f(r) = \frac{2\pi^{n/2}}{\Gamma\left(\frac{n}{2}\right)} \int f(r) r^{n-1}\, dr\,, \tag{B.3}$$

$$\int_0^\infty dx\, \frac{x^\beta}{(x^2 + M^2)^\alpha} = \frac{1}{2}\, \frac{\Gamma\left(\frac{\beta+1}{2}\right) \Gamma\left(\alpha - \frac{\beta+1}{2}\right)}{\Gamma(\alpha)(M^2)^{\alpha-(\beta+1)/2}}\,. \tag{B.4}$$

Keeping the prescriptions and definitions of Section 13 in mind, the following equations hold for arbitrary n

$$\int d_n p\, \frac{1}{(p^2 + 2kp + m^2)^\alpha} = \frac{i\pi^{n/2}}{(m^2 - k^2)^{\alpha - n/2}}\, \frac{\Gamma\left(\alpha - \frac{n}{2}\right)}{\Gamma(\alpha)}\,, \tag{B.5}$$

$$\int d_n p\, \frac{p_\mu}{(p^2 + 2kp + m^2)^\alpha} = \frac{i\pi^{n/2}}{(m^2 - k^2)^{\alpha - n/2}}\, \frac{\Gamma\left(\alpha - \frac{n}{2}\right)}{\Gamma(\alpha)} (-k_\mu)\,, \tag{B.6}$$

$$\int d_n p\, \frac{p^2}{(p^2 + 2kp + m^2)^\alpha} = \frac{i\pi^{n/2}}{(m^2 - k^2)^{\alpha - n/2}}\, \frac{1}{\Gamma(\alpha)} \times$$

$$\times \left\{ \Gamma\left(\alpha - \frac{n}{2}\right) k^2 + \Gamma\left(\alpha - 1 - \frac{n}{2}\right) \frac{n}{2} (m^2 - k^2) \right\}\,, \tag{B.7}$$

$$\int d_n p \frac{p_\mu p_\nu}{(p^2 + 2kp + m^2)^\alpha} = \frac{i\pi^{n/2}}{(m^2 - k^2)^{\alpha - n/2}} \frac{1}{\Gamma(\alpha)} \times$$
$$\times \left\{ \Gamma\left(\alpha - \frac{n}{2}\right) k_\mu k_\nu + \Gamma\left(\alpha - 1 - \frac{n}{2}\right) \frac{1}{2} \delta_{\mu\nu} (m^2 - k^2) \right\} , \tag{B.8}$$

$$\int d_n p \frac{p_\mu p_\nu p_\lambda}{(p^2 + 2kp + m^2)^\alpha} = \frac{i\pi^{n/2}}{(m^2 - k^2)^{\alpha - n/2}} \frac{1}{\Gamma(\alpha)} \times$$
$$\times \left\{ -\Gamma\left(\alpha - \frac{n}{2}\right) k_\mu k_\nu k_\lambda - \Gamma\left(\alpha - 1 - \frac{n}{2}\right) \frac{1}{2} (\delta_{\mu\nu} k_\lambda + \delta_{\mu\lambda} k_\nu + \right.$$
$$\left. + \delta_{\nu\lambda} k_\mu) (m^2 - k^2) \right\} , \tag{B.9}$$

$$\int d_n p \frac{p^2 p_\mu}{(p^2 + 2kp + m^2)^\alpha} = \frac{i\pi^{n/2}}{(m^2 - k^2)^{\alpha - n/2}} \frac{1}{\Gamma(\alpha)} (-k_\mu) \times$$
$$\times \left\{ \Gamma\left(\alpha - \frac{n}{2}\right) k^2 + \Gamma\left(\alpha - \frac{n}{2} - 1\right) \frac{n+2}{2} (m^2 - k^2) \right\} . \tag{B.10}$$

The above equations contain indices μ, ν, λ. These indices are understood to be contracted with arbitrary n-vectors q_1, q_2, etc. In computing the integrals one first integrates over the part of n-space orthogonal to the vectors k, q_1, q_2, etc., using Eqs. (B.1) to (B.4). After that the expressions are meaningful also for non-integer n. Note that formally Eqs. (B.6) to (B.10) may be obtained from Eq. (B.5) by differentiation with respect to k, or by using $p^2 = (p^2 + 2pk + m^2) - 2pk - m^2$.

To show that integrals over polynomials give zero within the dimensional regularization scheme is very simple. Consider, for example,

$$I(\alpha) = \int d_n p (p^2)^\alpha ,$$

where α is some integer greater than or equal to zero. According to Eq. (13.19) in the case of only one loop, we have

$$I(\alpha) = \int d^4 \underline{p} \int_0^\infty d\omega\, \omega^{n-5} \frac{2\pi^{n/2-2}}{\Gamma\left(\frac{n-4}{2}\right)} (\underline{p}^2 + \omega^2)^\alpha .$$

By partial integrations (see Subsection 13.2) we get now

$$I(\alpha) = \frac{2\pi^{n/2-2}}{\Gamma\left(\frac{n-4}{2}+\lambda\right)} \int d^4\underline{p} \int_0^\infty d\omega\, \omega^{n-5+2\lambda} \left(-\frac{\partial}{\partial\omega^2}\right)^\lambda (\underline{p}^2+\omega^2)^\alpha ,$$

which gives zero for $\lambda > \alpha$.

A nice example, suggested by B. Lautrup is the following. Consider the following integral

$$I_\mu = \int d^4k \frac{k_\mu}{(k^2+m^2)^2} ,$$

which gives zero, because of symmetric integration, if one regularizes, for example, as follows

$$I_\mu \to \int d^4k k_\mu \left[\frac{1}{(k^2+m^2)^2} - \frac{1}{(k^2+\Lambda^2)^2} \right] .$$

It is also zero in the dimensional cut-off scheme according to Eq. (B.6).

Let us now shift the integration variable, forgetting about regulators

$$I_\mu = \int d^4k \frac{k_\mu + p_\mu}{[(k+p)^2+m^2]^2} . \tag{B.11}$$

Expanding the denominators, we get

$$I_\mu = \int \frac{d^4k}{(k^2+m^2)^2} \left[k_\mu + p_\mu - 4\frac{k_\mu (k\cdot p)}{k^2+m^2} \right] + O(p^2) , \tag{B.12}$$

which by symmetric integration ($k_\mu \to 0$, $k_\mu k_\nu \to \frac{1}{4}\delta_{\mu\nu}k^2$) gives

$$I_\mu = m^2 p_\mu \int \frac{d^4k}{(k^2+m^2)^3} + O(p^2) \neq 0 .$$

Using dimensional regularization, which means $k_\mu \to 0$ but $k_\mu k_\nu \to$ $\to (\delta_{\mu\nu}/n)k^2$, we get from Eq. (B.12)

$$I_\mu^n = p_\mu \int d^n k \frac{k^2\left(1 - \frac{4}{n}\right) + m^2}{(k^2 + m^2)^3} + O(p^2) .$$

From Eqs. (B.5) and (B.7)

$$\int d^n k \frac{1}{(k^2 + m^2)^3} = i\pi^{n/2}(m^2)^{n/2-3} \frac{\Gamma\left(3 - \frac{n}{2}\right)}{\Gamma(3)} ,$$

$$\int d^n k \frac{k^2}{(k^2 + m^2)^3} = i\pi^{n/2}(m^2)^{n/2-2} \frac{n}{2} \frac{\Gamma\left(2 - \frac{n}{2}\right)}{\Gamma(3)} .$$

Then

$$I_\mu^n = p_\mu i \frac{\pi^{n/2}}{2} (m^2)^{n/2-2} \left[\left(1 - \frac{4}{n}\right) \frac{n}{2} \Gamma\left(2 - \frac{n}{2}\right) + \right.$$
$$\left. + \Gamma\left(3 - \frac{n}{2}\right)\right] + O(p^2) .$$

In the limit $n \to 4$, remembering that

$$\Gamma(z) \overset{z\to -n}{\simeq} \frac{(-1)^n}{n!} \frac{1}{z + n} ,$$

the coefficient of the p_μ term turns out to be exactly zero. Of course, from Eqs. (B.5) to (B.7), I_μ in Eq. (B.11) gives zero to any order in p.

In computing pole parts it is very advantageous to develop denominators. Take Eq. (B.5). For $\alpha = 2$ we find the pole part

$$PP[(B.5), \alpha = 2] = \frac{-2i\pi^2}{n - 4} = Z_0 . \qquad (B.13)$$

This Z_0 is a basic factor. Every logarithmically divergent integral has this factor, and further vectors, δ functions, etc.

$$PP \int d_n p \frac{p_\alpha p_\beta}{(p^2 + m^2)^3} = Z_0 \frac{1}{4} \delta_{\alpha\beta} , \qquad (B.14)$$

$$PP \int d_n p \frac{p_\alpha p_\beta p_\mu p_\nu}{(p^2 + m^2)^4} = Z_o \frac{1}{24} (\delta_{\alpha\beta}\delta_{\mu\nu} + \delta_{\alpha\mu}\delta_{\beta\nu} + \delta_{\alpha\nu}\delta_{\beta\mu}) \,. \tag{B.15}$$

The four on the right-hand side follows from symmetry considerations; the coefficient follows because multiplication with $\delta_{\alpha\beta}$ gives the previous integral. We leave it to the reader to find the general equation.

For other than logarithmically divergent integrals the denominator must be developed. For instance

$$PP \int d_n p \frac{p_\alpha}{(p^2 + 2pk + m^2)^2} = -k_\alpha Z_o \,, \tag{B.16}$$

where we used Eq. (B.14) together with

$$\frac{1}{(p^2 + 2pk + m^2)^2} \simeq \frac{1}{p^4} \left\{ 1 - 4 \frac{pk}{p^2} + O(p^{-2}) \right\} \,.$$

Linearly, quadratically, etc., divergent integrals that have no dependence on masses or external momenta can be put equal to zero.

The result Eq. (B.16) coincides with what can be deduced from Eq. (B.6).

APPENDIX C

DEFINITION OF THE FIELDS FOR DRESSED PARTICLES

Also the matrix elements of fields and products of fields (such as encountered in currents) can be defined in terms of diagrams. It is then possible to derive, or rather verify, the equations of motion for the fields. This provides for the link between diagrams and the canonical operator formalism.

Since things tend to be technically complicated we will limit ourselves to a simple case, namely three real scalar fields interacting in the most simple way. The Lagrangian is taken to be

$$\mathcal{L} = \frac{1}{2} A(\partial^2 - m_A^2)A + \frac{1}{2} B(\partial^2 - m_B^2)B + \frac{1}{2} C(\partial^2 - m_C^2)C +$$

$$+ gABC + J_A A + J_B B + J_C C \,. \qquad \text{(C.1)}$$

The bare propagators will be denoted by the symbols Δ^b_{FA}, Δ^b_{FB} and Δ^b_{FC}, the dressed propagators by $\bar{\Delta}_{FA}$, $\bar{\Delta}_{FB}$ and $\bar{\Delta}_{FC}$. For example

$$\Delta^b_{FA} = \frac{1}{(2\pi)^4 i} \frac{1}{k^2 + m_A^2 - i\varepsilon} \,,$$

$$\bar{\Delta}_{FA} = \frac{1}{(2\pi)^4 i} \frac{1}{Z_A^2(k^2 + M_A^2) - \Gamma_{1A}(k^2) - i\varepsilon} \,. \qquad \text{(C.2)}$$

The pole part of the dressed propagator plays an important role and will be denoted by Δ_F

$$\Delta_{FA} = \frac{1}{(2\pi)^4 i} \frac{1}{Z_A^2(k^2 + M_A^2) - i\varepsilon} \,. \qquad \text{(C.3)}$$

The result (C.2) has been obtained as follows (see Section 9, in particular Eq. (9.3)). The function $\Gamma_A(k^2)$ is the sum of all irreducible self-energy diagrams for the A-field. The dressed propagator is of the form $(k^2 + m_A^2 - \Gamma_A)^{-1}$. This expression will have a pole for some value of k^2, say for $k^2 = -M_A^2$. Then we can expand Γ_A around the point $k^2 = -M_A^2$

$$\Gamma_A(k^2) = \delta m_A^2 + (k^2 + M_A^2)F_A + \Gamma_{1A}(k^2) ,$$

$$Z_A^2 \equiv 1 - F_A , \qquad \delta m_A^2 = m_A^2 - M_A^2 , \tag{C.4}$$

where Γ_{1A} is of order $(k^2 + M_A^2)^2$. Insertion of this expression leads to Eq. (C.2).

Next to the propagators we define <u>external line factors</u> $N_A(k^2)$, etc. They are the ratio of the dressed propagators and their pole parts

$$N_A(k^2) = \frac{\bar{\Delta}_{FA}}{Z_A \Delta_{FA}} . \tag{C.5}$$

In the limit $k^2 = -M_A^2$ this is precisely the factor occurring in external lines when passing from Green's function to S-matrix. Finally we have the important Δ^+ and Δ^- functions

$$\Delta_A^{\pm} = \frac{1}{(2\pi)^3} \frac{1}{Z_A^2} \theta(\pm k_0)\delta(k^2 + M_A^2) .$$

Consider now any Green's function involving at least one A-field source. For all except this one source we follow the procedure as used in obtaining the S-matrix, that is all dressed propagators and associated sources are replaced by factors N and the mass-shell limit is taken. For the singled out A-field source we replace the dressed propagator and source by $N_A(k^2)$, but do not take the limit $k^2 = -M^2$. The Fourier transform with respect to k of the function so obtained is defined to be the matrix element (for a given order in the coupling constant with the appropriate in- and out-states) of an operator denoted by

$$\frac{\delta S}{\delta A(x)} . \tag{C.6}$$

(The notation used here should not be confused with notations of the type used in Section 9.) It is, roughly speaking, obtained from the S-matrix by taking off one external A-line and replacing that line by the factor $N_A(k^2)$. Diagrammatically:

$$< \beta| \frac{\delta S}{\delta A(x)} |\alpha > = \quad \alpha \Big| \raisebox{-1ex}{\Large\bigcirc} \Big\} \beta \; ; \qquad \text{c}\!\!-\!\!-\!\!- \equiv N_A(k^2)$$

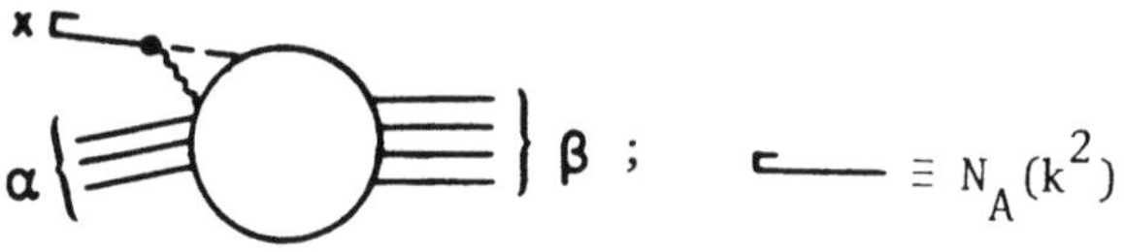

with the notation: (C.7)

$$\bullet\!\!-\!\!-\!\!\bullet \; , \; \bullet\!-\!-\!-\!\bullet \; , \; \bullet\!\!\sim\!\!\sim\!\!\bullet \qquad = \text{A-, B-, C-line.}$$

It is to be noted that the propagators used are completely dressed operators, and therefore self-energy insertions are not to be contained in Eq. (C.7). In particular there are no contributions of the type:

$$\text{(C.8)}$$

However, the factor $N_A(k^2)$ implies really the insertion of irreducible self-energy parts. Working out Eq. (C.5) we see (compare Eq. (C.4))

$$N_A(k^2) = \frac{1}{Z_A} \left[1 + \left\{ \Gamma_A(k^2) - i(2\pi)^4 (m_A^2 - M_A^2) - \right. \right.$$

$$\left. \left. - i(2\pi)^4(k^2 + M_A^2)F_A \right\} \bar{\Delta}_{FA}(k^2) \right] .$$

Diagrammatically:

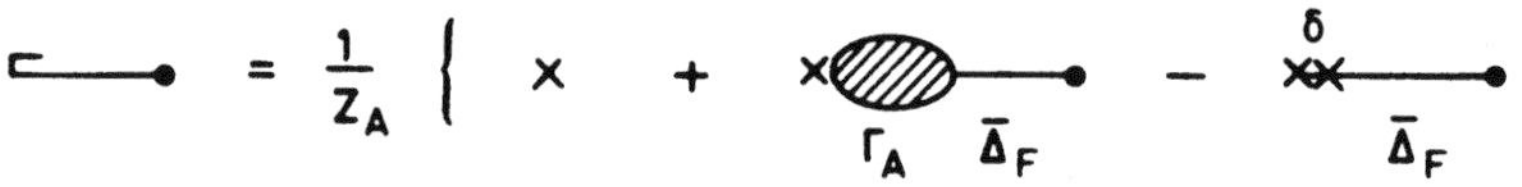

$$\text{c}\!\!-\!\!-\!\!\bullet \;\; = \frac{1}{Z_A} \left\{ \; \text{x} \;\; + \;\; \text{x}\underset{\Gamma_A \quad \bar{\Delta}_F}{\raisebox{-0.5ex}{\Large\oslash}}\!\!-\!\!\bullet \;\; - \;\; \underset{\bar{\Delta}_F}{\text{xx}}\!\!-\!\!\bullet \right\} \quad \text{(C.9)}$$

$$\delta = (2\pi)^4 i(m_A^2 - M_A^2) + (2\pi)^4 i(k^2 + M_A^2)F_A .$$

Remember that Γ_A starts with a B- and a C-line, and that N_A is attached to a B-C vertex (see Eq. (C.7)). We see that the right-hand side of Eq. (C.7) consists of <u>all</u> skeleton diagrams starting with an amputated B-C vertex, apart from the δ-correction.

We now define the product of this object and the matrix $S^\dagger$. It is obtained by connecting diagrams of $\delta S/\delta A$ to diagrams of $S^\dagger$

by means of Δ^+ functions:

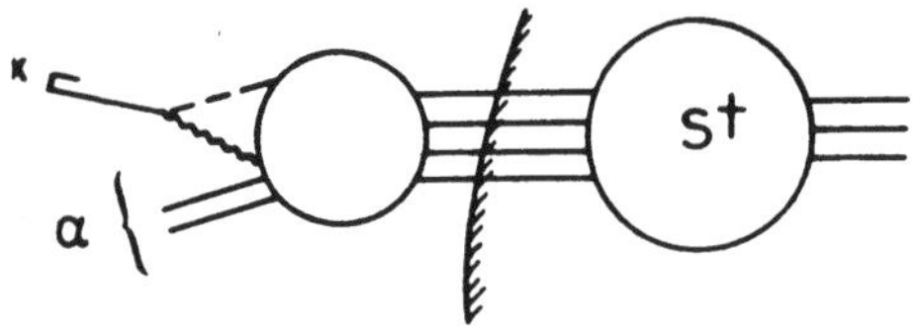

$$\Big\} \, \beta \equiv \langle \beta | S^\dagger \frac{\delta S}{\delta A(X)} | \alpha \rangle \qquad \text{(C.10)}$$

This is a collection of cut diagrams, with $S^\dagger$ corresponding to the part in the shadowed region. This definition of the product is the same as that encountered in the expression $S^\dagger S$.

It turns out that the differentiation symbol δ/δ in $\delta S/\delta A$ has more than formal meaning. With the help of the cutting equations it is easy to show that

$$S^\dagger \frac{\delta S}{\delta A(x)} + \frac{\delta S^\dagger}{\delta A(x)} S = 0 \, . \qquad \text{(C.11)}$$

The second term has the point x to the right of the cutting line. This can be expressed formally by writing $\delta(S^\dagger S)/\delta A(x) = 0$, which is what is expected if unitarity holds, $S^\dagger S = 1$. One may speak of generalized unitarity, because the A-line is off mass-shell.

The A-field current $j_A(x)$ is defined by

$$J_A(x) = i S^\dagger \frac{\delta S}{\delta A(x)} \, . \qquad \text{(C.12)}$$

By virtue of Eq. (C.11) it follows that $j_A(x)$ is Hermitian.

To define the matrix elements of the field A consider the equation of motion

$$(\partial^2 - M_A^2) A(x) = j_A(x) \, . \qquad \text{(C.13)}$$

This is not directly the equation of motion that one would write down given the Lagrangian (C.1), because we have the mass M_A^2 (defined by the location of the pole of the dressed propagator) instead of m_A^2.

The equation of motion (C.13) can be rewritten as an integral equation

$$\frac{1}{Z_A} A(x) = \frac{1}{Z_A} A_{in}(x) - iZ_A \int d_4x' \Delta_{RA}(x - x') J_A(x') \ . \qquad (C.14)$$

The retarded Δ function is

$$\Delta_{RA}(x) = \frac{1}{(2\pi)^4 i} \frac{1}{Z_A^2} \int d_4k e^{ikx} \frac{1}{k^2 + M_A^2 - i\epsilon k_o} \ . \qquad (C.15)$$

This function is zero unless $x_o > 0$. In fact

$$\Delta_R(x) = \theta(x_o)\left\{\Delta^+(x) - \Delta^-(x)\right\} \ . \qquad (C.16)$$

In passing, we note the identities

$$\begin{aligned} \Delta_R - \Delta^+ &= -\Delta_F^* \ , \\ \Delta_R + \Delta^- &= \Delta_F \ . \end{aligned} \qquad (C.17)$$

Equation (C.14) defines the A-field in terms of diagrams. It satisfies the weak or asymptotic definition

$$\lim_{x_o \to -\infty} < \alpha|A(x)|\beta> = \lim_{x_o \to -\infty} < \alpha|A_{in}(x)|\beta > .$$

The field $A_{in}(x)$ is a free field satisfying the equation of motion (C.13) with $j = 0$.

We will write Eq. (C.14) in terms of diagrams, and to that purpose we must introduce the "ordered product". We write

$$\frac{1}{Z_A} A_{in}(x) = \frac{1}{Z_A} A_{in}(x) S^\dagger S = \frac{1}{Z_A} :A_{in}(x) S^\dagger: S +$$

$$+ Z_A \int d_4x' \Delta_A^+(x - x') \frac{\delta S'}{\delta A(x')} S \ . \qquad (C.18)$$

The double dots imply that A_{in} is not to be connected by a Δ^+-line to $S^\dagger$. Below this will be shown diagrammatically. The function Δ_A^+ has been defined before. Due to the presence of this Δ^+ only the mass-shell value of the factor N is required in $\delta S^\dagger/\delta A$, and this is $1/Z_A$. Using Eq. (C.18) we can rewrite the integral equation of motion (C.14) in the form

$$\frac{1}{Z_A} A(x) = \left[\frac{1}{Z_A} :A_{in}(x)S^{\dagger}: - iZ_A \int d_4x' \left\{ \Delta_{RA}(x - x') - \right.\right.$$

$$\left.\left. - \Delta_A^{+}(x - x') \right\} (-i) \frac{\delta S^{\dagger}}{\delta A(x')} \right] S \; . \qquad \text{(C.19)}$$

Keeping in mind Eq. (C.17) as well as Eq. (C.5) we see that $A(x)/Z_A$ can be pictured as follows:

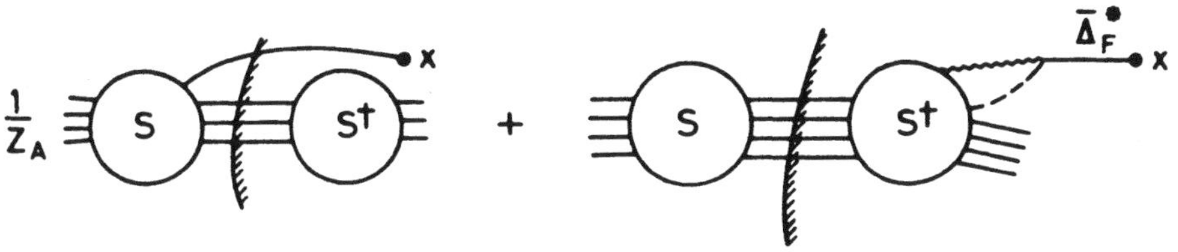

This then is the diagrammatic expression for the matrix elements of the field A. Similar expressions can be derived for the fields B and C. All kinds of relations from canonical field theory can be derived using these expressions. For instance

$$j_A(x) = - \frac{g}{Z_A Z_B Z_C} B(x)\dot{C}(x) - \frac{\delta m_A^2}{Z_A^2} A(x) - \frac{F_A(\partial^2 - M_A^2)}{Z_A^2} A(x)$$

(C.21)

with F_A and δm from Eq. (C.4).

CHAPTER 2.3

GAUGE THEORIES WITH UNIFIED WEAK, ELECTROMAGNETIC, AND STRONG INTERACTIONS*

by

G. 'T HOOFT
University of Utrecht

1. Introduction

Only half a decade ago, quantum field theory was considered as just one of the many different approaches to particle physics, and there were many reasons not to take it too seriously. In the first place the only possible "elementary" particles were spin zero bosons, spin $\frac{1}{2}$ fermions, and photons. All other particles, in particular the ρ, the N^*, and a possible intermediate vector boson, had to be composite. To make such particles we need strong couplings, and that would lead us immediately outside the region where renormalized perturbation series make sense. And if we wanted to mimic the observed weak interactions using scalar fields, then we would need an improbable type of conspiracy between the coupling constants to get the $V - A$ structure[1]. Finally, it seemed to be impossible to reproduce the observed simple behaviour of certain inclusive electron-scattering cross sections under scaling of the momenta involved, in terms of any of the existing renormalizable theories[2]. No wonder that people looked for different tools, like current algebra's, bootstrap theories and other nonperturbative approaches.

Theories with a non-Abelian, local gauge invariance, were known[3], and even considered interesting and suggestive as possible theories for weak

*Rapporteur's talk given at the E.P.S. International Conference on High Energy Physics, Palermo, Sicily, 23–28 June 1975.

interactions[4,5], but they made a very slow start in particle physics, because it seemed that they did not solve very much since unitarity and/or renormalizability were not understood and it remained impossible to do better than lowest order calculations.

When finally the Feynman rules for gauge theories were settled[6] and the renormalization procedure in the presence of spontaneous symmetry breakdown understood[7–19], it was immediately realized that there might exist a simple Gauge Model for all particles and all interactions in the world. The first who would find the Model would obtain a theory for all particles, and immortality. Thus the Great Model Rush began[29,35,36,44–53,56,58].

First, one looks at the leptons. The observed ones can easily be arranged in a symmetry pattern consistent with experiment[5]: SU(2) × U(1). But if we assume that other leptons exist which are so heavy that they have not yet been observed then there are many other possibilities. To settle the matter we have to look at the hadrons.

The observed hadron spectrum is so complicated with its octuplets, nonets and decuplets that it would have been a miracle if they would fit in a simple gauge theory like the leptons. They don't. To reproduce the nice SU(3) × SU(3) structure one is forced to take the "quarks", the building blocks of the hadrons, as elementary fields. The existing hadrons are then all assumed to be composite. To bind those quarks together we need strong forces and here we are, back at our starting point. What have we won?

We have won quite a lot, because the tools we can use, renormalized gauge theories, are much more powerful than the old renormalizable theories. Not only do we have indications that they exhaust all possible renormalizable interactions[20] but there is also a completely new property: the behaviour of some of these theories under scaling of all coordinates and momenta[21]. If you look at such a system of particles through a microscope, then what you see is a similar system of particles, but their interactions have reduced. The theory is "asymptotically free"[21–23]. The old theories always show a messy, strongly interacting soup when you look through the microscope[24]. If you assume that any theory should be defined by giving its behaviour at small distances, then the old theories would be very ill-defined, contrary to the gauge theories.

But when it comes to model building, then it is still awkward that the forces between the quarks are strong, because that makes gauge theories not very predictive and there are countless possibilities. I have seen theories with 3, 4, 6, 9, 12, 18 and more quarks. How should we choose among all these different group structures? Before answering the question let us first make up the balance. What we are certain of is:

1) Gauge theories are renormalizable, if firstly the local symmetry is broken spontaneously, and secondly the Adler–Bell–Jackiw anomalies are arranged to cancel.
2) Global symmetries may be broken explicitly, so we can always get rid of Goldstone bosons.
3) We can make asymptotically free theories for strong interactions.
   Then the following statements are not absolute but have been learnt from general experience:
4) The Higgs mechanism is an expensive luxury: each time we introduce a Higgs field we have to accept many new free parameters in the system.
5) There are always more free parameters than there are masses in the theory, so we can never obtain a reliable mass relation for the elementary constituent particles. Of course, masses of composite systems are not free but can be calculated.
6) To break large groups like SU(4) or SU(3) × SU(3) by means of the Higgs mechanism is hopelessly complicated. Theories with a small gauge group like SU(2) × U(1) or large but unbroken groups are in a much better shape.

Of course, these are practical arguments, that distinguish useful from useless theories. But do they also distinguish good from false theories? Personally I tend to believe this. I find it very difficult to believe that nature would have created as many Higgs fields as are necessary to break the big symmetry groups. It is more natural to suppose that just one or two Higgs fields are present and some remaining local symmetry groups are not broken at all.

2. Strong Interaction Theory

a. Towards permanent color binding

There is a general consensus on the idea that gauge vector particles corresponding to the color group $SU(3)^c$ can provide for the necessary binding force between quarks, which transform as triplet representations of this group. The states with lowest energy are all singlets. This theory explains the observed selection rules and the SU(6) properties of the hadrons. But now there are essentially two possibilities.

The first possibility is that $SU(3)^c$ is broken by the Higgs mechanism, so that the masses of all colored objects are large, but finite. The ψ particles can be incorporated in this scheme: they may be the first colored objects, as you heard in the sessions on color theories. Besides the disadvantages of such theories I mentioned before, it is also difficult to arrange suppression of higher order contributions to $K^0 - \bar{K}^0$ mixing and $K_L \to \mu^+\mu^-$ decay[25] and the theories are not asymptotically free.

The alternative possibility is that $SU(3)^c$ is not broken at all. All colored objects like the quarks and the color vector bosons have strictly infinite masses[26,27]. I suspect that this situation can be obtained from the former one through a phase transition. Let me explain this.

In the Higgs broken color theories there exist "solitons", objects closely related to the magnetic monopole solutions[28] in the Georgi–Glashow model[29]. Now let me continuously vary the parameter μ^2 in the Higgs potential

$$V = \frac{1}{2}\mu^2\phi^2 + \frac{1}{8}\lambda\phi^4 \tag{1}$$

from negative to positive values (Fig. 1).

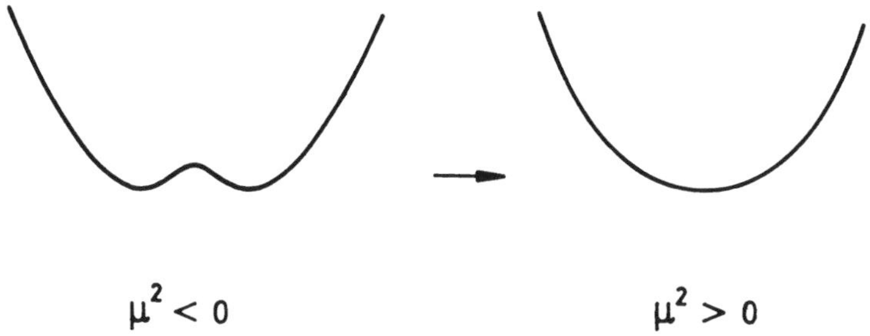

Fig. 1. The Higgs potential before and after the phase transition.

We keep λ fixed. Now the vacuum expectation value F_{Higgs} of the Higgs field ϕ is first roughly proportional to $|\mu|$ and so are the vector boson mass M_V and the soliton mass,

$$M_s \simeq \frac{4\pi M_V}{g^2}, \tag{2}$$

as indicated on the left-hand side of Fig. 2.

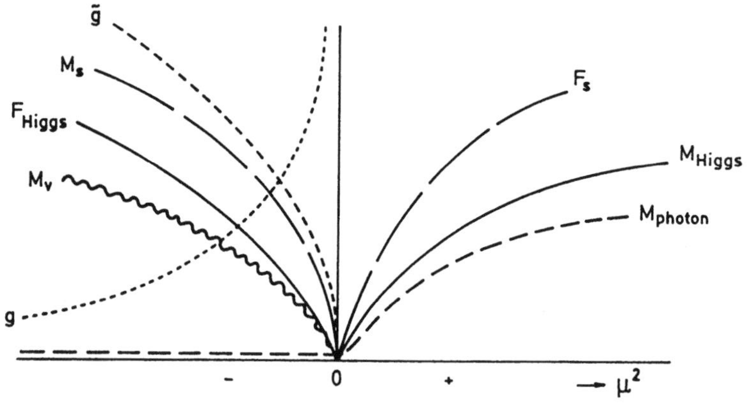

Fig. 2. The phase transition.

M_{Higgs} = mass of unbroken Higgs field
F_{Higgs} = vac. exp. value of Higgs field
M_s = mass of soliton
F_s = vac. exp. value of soliton field
g = color (electric) coupling constant
$\tilde{g}$ = soliton coupling constant ($g\tilde{g} = 2\pi n$)

What I assume is that when μ^2 becomes positive, it is the soliton's turn to develop a non-zero vacuum expectation value. Since it carries color-magnetic charges, the vacuum will behave like a superconductor for color-magnetic charges. What does that mean? Remember that in ordinary electric superconductors magnetic charges are confined by magnetic vortex lines; as described by Nielsen and Olesen[30]. We now have the opposite: it is the color charges that are confined by "electric" flux tubes. So we think that after the phase transition all color non-singlets will be tied together by "strings" into groups that are color singlets. In this phase the Higgs scalars play no physical role whatsoever and we may disregard them now.

The great shortcoming of this theory is that it is intuitive and as yet no mathematical framework exists. But there are various reasons to take it seriously.

Theoretically:

i) we see it happen if we replace continuous space-time by a sufficiently coarse lattice[27]: the action becomes that of the Nambu string.

ii) We see it happen in the only soluble asymptotically free gauge theories: Schwinger's model[31] and, even better, in the SU(∞) gauge theory in two space-time dimensions[32].

Experimentally:

a) the flux lines would behave like the dual string and thus explain the straight Regge-trajectories[33].

b) This is the simplest and probably the only asymptotically free theory that explains Bjorken behaviour[22].

c) The theory is closely related to the rather successful MIT bag model[34].

In principle there exists an intermediate possibility: we probably have several phase-transitions when we go from unbroken SU(3) to for instance SU(2), U(1) and finally complete breaking. We will not consider the possibility that we are in SU(2) or U(1), but we must remember that from the low lying states it is difficult to deduce in what phase we really are.

If we are in the unbroken phase then SU(3) color must commute with weak and electromagnetic SU(2) $\times$ U(1), and with SU(3) flavor*). Consequently, we are obliged to introduce charm[35], or even more new quarks perhaps[36]. Now let us consider the ψ particles.

b. Charmonium

The most celebrated theory for ψ is that it is a bound state of a charmed quark and its antiparticle[37]. Charmed quarks are assumed to be rather heavy. The size of the bound state wave function will therefore be small and we can look at the thing through a microscope (i.e. apply a scale transformation). Then we see rather small couplings, so we may use perturbation expansion to describe the system. The mathematics is do-able here! At first approximation the gluon gauge field behaves exactly like Maxwell fields, even the SU(3) structure constants can be absorbed in the coupling constant. We can calculate the

*) i.e. the familiar broken symmetry group that transforms p, n and λ into each other. The word "flavor" has been proposed by Gell-Mann to denote both isospin and strangeness.

Table 1.

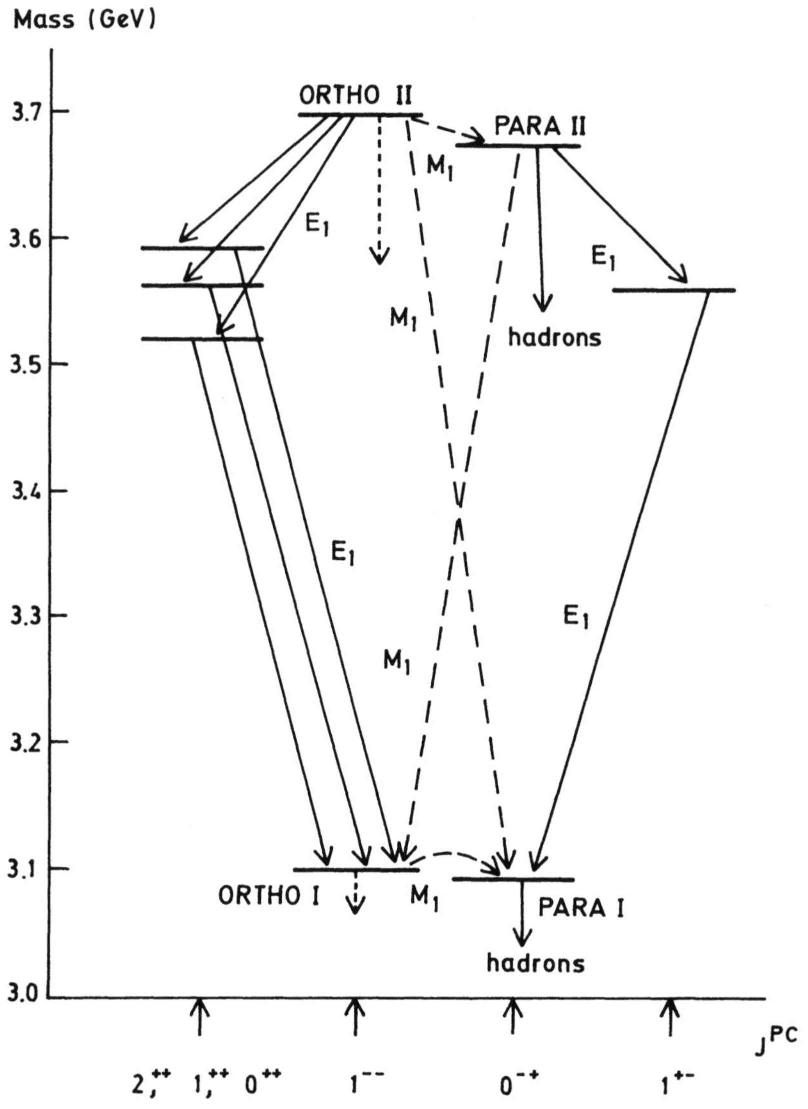

Mass (GeV)
3.7
3.6
3.5
3.4
3.3
3.2
3.1
3.0
ORTHO II
PARA II
M1
E1
E1
M1
hadrons
E1
E1
M1
ORTHO I
M1
PARA I
hadrons
J^PC
2,++ 1,++ 0++
1--
0-+
1+-

Table 2. SU(4) predictions

New Mesons (Y_Q ↑, → I_3)

charm	$JP = 0^-$	$JP = 1^-$	mass (GeV)
1	+ + 0	+ + 0	2
0	0	0	3.1
−1	0 − −	0 − −	2

New Baryons

charm	$JP = \frac{1}{2}^+$	$JP = \frac{3}{2}^+$	mass (GeV)
3		2+	5
2	2+ + +	2+ + +	3.7
	2+ + 0 +	2+ + 0 + 0	
1	+ 0 + 0 0	0	2.5

annihilation rate and level splittings exactly as in positronium. The annihilation rate of the positronium vector state is [38]

$$\Gamma = \frac{2(\pi^2 - 9)}{9\pi} m_e \alpha^6 (1 + 0(\alpha)) \tag{3}$$

For charmonium the formula would be*)

$$\Gamma = \frac{5(\pi^2 - 9)}{192\pi} m_{p'} \alpha_s^6 (1 + 0(\alpha_s)), \tag{4}$$

where $\alpha_s = g_s^2/3\pi$ (the subscript s standing for "strong").

It fits well in the renormalization group theory if α_s is around 1/3 at 2 GeV. That it is raised to the sixth power explains the stability of ψ. Note that this

*) The quark flavors in this paper are called p, n, p' and λ. More modern is the nomenclature u, d, c and s

is the rate with which the two quarks annihilate. It is independent (to first approximation) of the details of the hadronic final state[39].

Now this gauge theory for strong interactions gives precise predictions on the other charmonium states (Table 1) and the charmed hadrons (Table 2). They can be found in some nice papers by Appelquist, De Rújula, Politzer, Glashow and others[37].

It is very tempting to assume that this new quark is really the charmed one as predicted by weak interaction theories (see Sect. 3), but it could of course be an "unpredicted" quark.

If the predictions from this theory come out to be roughly correct then that would be a great success both for the asymptotically free gauge theory for strong interactions, and for the renormalization theories that predicted charm.

3. The Weak Interactions

a. SU(2) × U(1) theories

A simple SU(2) × U(1) pattern seems to be compatible with all experimental data on pure leptonic and semileptonic processes (neutral currents and charm), but with the pure hadronic weak interactions we still have a problem: there should be $\Delta I = 3/2$ and $\Delta I = 1/2$ transitions with similar strength and we only see $\Delta I = 1/2$. Also, these interactions seem to be somewhat stronger than other weak interaction processes. The traditional way to try to solve the problem is the possibility that $\Delta I = 1/2$ is "dynamically enhanced". Seen through our "microscope" at momenta of the order of the weak boson mass, the $\Delta I = 1/2$ and $\Delta I = 3/2$ parts of the weak interaction Hamiltonian may be equal in strength, but when we scale towards only 1 GeV, then $\Delta I = 1/2$ may be enhanced through its renormalization group equations. This mechanism really works, but can only give a factor between 6 and at most 14 in the amplitudes[40]. But there are many uncertainties, since we do not know exactly from where to where we should scale, and what the importance is of the higher order corrections. It has been argued that a similar mechanism might depress leptonic decay modes of charmed particles, thus giving them a better camouflage that prevents their detection[41] and the mechanism could influence parity and isospin violation[42].

An interesting alternative explanation of the $\Delta I = 1/2$ rule has recently been given by De Rújula, Georgi and Glashow[43]. In usual SU(2) × U(1) theories only the left-handed parts of the spinors may be SU(2) doublets, all right-handed parts are singlets. These authors however take the right-handed parts of p' and n to form a doublet. This adds to the hadronic current

$$\bar{p}'\gamma_\mu(1-\gamma_5)n\,, \tag{5}$$

which is only observable in purely hadronic events or charm decays. Note that there is no Cabibbo rotation. We get among others a term,

$$\cos\theta_c\,\bar{p}'\gamma_\mu(1+\gamma_5)\lambda \quad \bar{n}\gamma_\mu(1-\gamma_5)p'\,, \tag{6}$$

in the effective Hamiltonian. Note the absence of Cabibbo suppression by factors $\sin\theta_c$ and the $\Delta I = 1/2$ nature of this term.

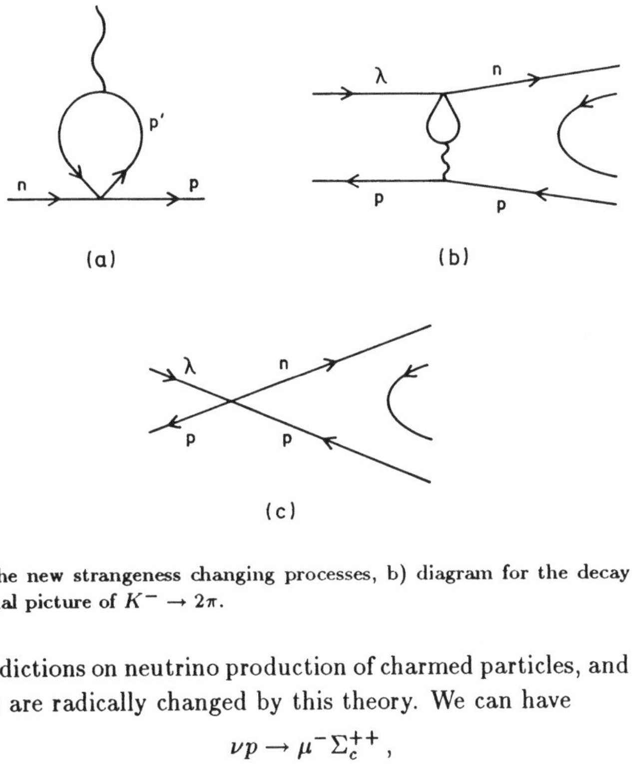

Fig. 3. a) The new strangeness changing processes, b) diagram for the decay $K^- \to 2\pi$, c) conventional picture of $K^- \to 2\pi$.

The predictions on neutrino production of charmed particles, and the charm decay rates are radically changed by this theory. We can have

$$\nu p \to \mu^-\Sigma_c^{++}\,, \tag{7}$$

etc. with no $\sin^2\theta_c$ suppression, and pseudo $\Delta S = 2$ processes can take place like

$$\begin{array}{l} \nu p \to \mu^- p K^+ D^0 \\ \qquad\qquad\quad \lfloor\!\!\to \mu^-\nu^- K^+ \end{array} \tag{8}$$

because of $D^0 - \bar{D}^0$ mixing (as in $K_0\bar{K}_0$ mixing). Experts however are skeptical about the idea. The charmed quark in this current has to violate Zweig's rule and the charmed quark loop must couple to gauge gluons before the current can contribute.

The idea is very recent and I have not yet seen any detailed calculations. We will soon know what the contribution of such diagrams can be. Furthermore, the $K_L - K_S$ mass difference, when naively calculated in this model seems to be too large compared with experiment.

b. Other and larger groups

SU(2) × U(1) theories do not really unify weak interactions and electromagnetism. The U(1) group may be considered as the fundamental electromagnetic group, and its photon is merely somewhat mixed with the neutral component of the weak SU(2) gauge field. True unification occurs only then when we have a single compact group.

1. *Heavy leptons*

The first example was originally invented to avoid neutral currents: the Georgi–Glashow O(3) model[29)]. Now we have a large class of models[44–52)] based on SU(2) × U(1), O(3), O(4), O(4) × U(1), SU(3), SU(3) × U(1), SU(3) × SU(3) etc., all predicting new leptons. An extensive discussion of these models is given by Albright, Jarlskog and Tjia[53)]. Table 3 gives the predicted leptons in these schemes. Albright, Jarlskog and Wolfenstein also analysed the possibilities to detect such objects by neutrino production[54)].

Of course, we can always extend the lepton spectrum in other ways, for instance by adding new representations[55)], so that we get more members in the series $\binom{e}{\nu_e}$, $\binom{\mu}{\nu_\mu}$, $\binom{x}{\nu_x}$, etc.

2. *The Pati-Salam Model*[56)]

Theoreticians are eagerly awaiting the discovery of the first heavy lepton. But that may take quite a while and in the meantime we search for more guidelines to disclose the symmetry structure of our world. One such guideline is that eventually we do not expect that baryons and leptons are essentially different. That is, they might belong to just one big multiplet. This, and other ideas of symmetry and simplicity led Pati and Salam to formulate their most recent "completely unified model". Leptons and quarks from just one representation of SU(4) × SU(4), of which color SU(3) and weak SU(2) × U(1) are subgroups. If, as argued before, color is unbroken then we are free to mix the photon (also unbroken) with colored bosons, so there is no physical difference between the charge assignments

Table 3. Heavy leptons

Gauge group	Muonic leptons	Refs.
SU(2) x U(1)	μ^- ν_μ	Weinberg, Salam[4,5,17,44]
	μ^- ν_μ M^0	Bjorken and Llewellyn Smith no. 3[51] Bég and Zee[50]
	μ^- ν_μ M^0 M^+	Bjorken and Llewellyn Smith no. 5[51] Prentki and Zumino[52]
	μ^- ν_μ M^+	Prentki and Zumino[52]
	μ^- ν_μ M_0 M_0' M^+	Bjorken and Llewellyn Smith no. 4[51]
$O(3) \simeq SU(2)$	μ^- ν_μ M_0 M^+	Georgi and Glashow[29] Bjorken and Llewellyn Smith no. 6[51]
O(4) = SU(2) x SU(2)	μ^- ν_μ M_0 M^+	Pais[45]
	μ^- ν_μ M^0 M_0' M^+	Cheng[45]
O(4) x U(1)	μ^- ν_μ M_0 M^+	Pais[46]
SU(3) x U(1)	μ^- ν_μ M^-	Schechter and Singer[47]
	M^0 μ^- ν_μ	Gupta and Mani[47]
	M'^0 M'^+ M^- M^0 M''^0 M^+ μ^- ν_μ	Albright, Jarlskog, Tjia[53]

Table 3. (*Continued*)

Gauge group	Leptons		Refs.
U(3)	left $E^0 \quad e^+$ ν_μ $\mu^- \quad M^0 \quad M^+$ $M^{0'}$ $E^- \quad E^{0'}$	right $M^0 \quad M^+$ $\bar{\nu}_c$ $E^- \quad E^0 \quad e^+$ $E^{0'}$ $\mu^- \quad M^{0'}$	Salam and Pati[48)]
	left $M^0 \quad \mu^+$ ν_e $e^- \quad E^0 \quad E^+$ $E^{0'}$ $M^- \quad M^{0'}$	right $E^0 \quad E^+$ $\bar{\nu}_\mu$ $M^- \quad M^0 \quad \mu^+$ $M^{0'}$ $E^- \quad E^{0'}$	Salam and Pati[48)]
SU(3) x SU(3)	$\nu_e, \bar{\nu}_\mu$ $e^- \quad \mu^+$ L^0		Achiman, Weinberg[49)]
	$\bar{\nu}_e \quad e^+$ L^0 $L^- \quad L^{0'} \quad L^+$ $\mu^- \quad \nu_\mu$		Achiman[49)]

$$\begin{pmatrix} -1/3 & -1/3 & -1/3 & -1 \\ 2/3 & 2/3 & 2/3 & 0 \\ -1/3 & -1/3 & -1/3 & -1 \\ 2/3 & 2/3 & 2/3 & 0 \end{pmatrix} \quad \text{and} \quad \begin{pmatrix} -1 & 0 & 0 & -1 \\ 0 & 1 & 1 & 0 \\ -1 & 0 & 0 & -1 \\ 0 & 1 & 1 & 0 \end{pmatrix}$$

because the photon may freely mix with components of the unbroken color vectors fields*).

*)If the integer-charge assignment is adopted however, then leptons couple directly to color gluons, and this would make interpretation of the experimentally observed ratio $R = \sigma(\text{hadrons})/\sigma(p^+p^-)$ in e^+e^- annihilation more complicated. So R must be computed from the non-integer charges. This 16-plet yields $R = 10/3$.

4. The Higgs Scalars

The ugly ducklings of all Unified Theories are the Higgs scalars. They usually bring along with them as many free parameters as there are masses in the theory or more. This makes the theories so flexible that tests become very difficult.

These scalars are needed for pure mathematical reasons: otherwise we cannot do perturbation expansions, and we have no other procedure at hand to do accurate calculations. But do we need them physically?[57)]

Various attempts have been made to answer this question negatively. First of all they are ugly, and physics must be clean. But that is purely emotional. Then: we do not observe scalars experimentally. But: there is so much that we do not observe: quarks, I.V.B.'s, etc.

Linde and Veltman raised the point that the scalars do something funny with gravity: their vacuum-expectation value gives the vacuum a very large energy-density. That should renormalize the cosmological constant. (Otherwise our universe would be as curved as the surface of an orange[58)]). On the other hand the net cosmological cosntant is very many orders of magnitude smaller than this. How will we ever be able to explain this miraculous cancellation? In this respect it is interesting to note an observation made by Zumino[59)]): in supersymmetric models there is no cosmological constant renormalization, even at the one-loop level.

Ross and Veltman then suggest that perhaps one should choose the scalars in such a way that the vacuum-energy density vanishes[60)]). In Weinberg's model that means that one should add an isospin 3/2 Higgs field. Such a Higgs field could reduce neutral current interactions, and that would be welcome to explain several experiments.

Personally I think that one has to consider the renormalization group to see what kind of scalars are possible. If we scale to small distances[106)] then the theory has nearly massless particles. Only a nearly exact symmetry principle can explain why their masses are so small at that scale. For fermions we have chiral symmetry for this. Scalar particles can be forced to be massless if they are the Goldstone bosons of some global symmetry. They then have the quantum numbers of the generators of that symmetry group. Since all global symmetry groups must commute with local gauge groups, it is difficult to get light scalar particles that are not gauge singlets.

According to this argument it is impossible to have scalar Higgs particles, except when they are strongly interacting, since in that case we cannot scale very far because of non-asymptotic-freedom.

A notable exception may be constructions using supersymmetries (see Sect. 5).

In connection with Higgs' scalars I want to clear up a generally believed misconception. It is not true that theories with a Higgs phenomenon in general cannot be asymptotically free. For many simple Higgs theories one can obtain asymptotic freedom, provided that certain very special relations are satisfied, between coupling constants that would otherwise be arbitrary[62]. These theories do not have a stable fixed point at the origin but I cannot think of any physical reason to require that. Examples of such theories are supersymmetric theories, with possible supersymmetry breaking masses. Quark theories of this nature do not exist.

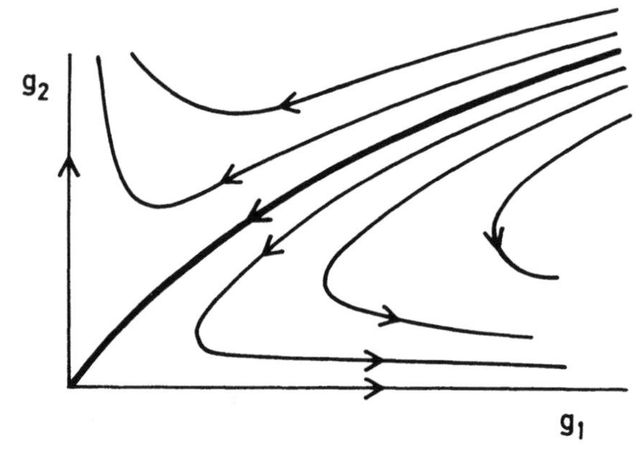

Fig. 4. Example of an unstable ultraviolet fixed point at the origin of parameter space. The arrows indicate the change of the coupling constants as momenta increase. The solid line is the collection of asymptotically free theories.

For weak interactions however I do not think that asymptotic freedom would be a good criterion, because any change of the theory beyond, say 10000 GeV would alter the relations between coupling constants completely.

5. Supersymmetric Models

Symmetry arguments can be deceptive. In the nineteenth century physicists argued that the Moon must be inhabited by animals, plants and people. This was based on theological and symmetry arguments (earth-moon-symmetry), similar to the ones we use today.

Keeping this warning in mind, let us now consider the supersymmetric models[63)]. For supersymmetry we have a special session. But an interesting attempt by Fayet[64)] must be mentioned here, who constructs a Weinberg-like model with supersymmetry. There, fermions and bosons sit in the same multiplet. Thus the photon joins the electro-neutrino. But which massless boson should join the muon-neutrino? This problem has not been solved, to my knowledge.

Even if supersymmetry would be ruled out by the overwhelming experimental evidence that fermion and boson masses are not the same, I think we do not have to drop the idea. Suppose that quantum field theory begins at the Planck length; we cannot go further than the Planck length as long as we do not know how to quantize General Relativity. And suppose that Quantum Field Theory is supersymmetric but the supersymmetry is very slightly broken (it is a global symmetry, so we may do that) and the breaking is described by a coefficient of the order of 10^{-40}. Suppose further that the representations happen to be such that there are no supersymmetric mass terms. Then at the mass scale of 1 GeV we would have all dimensionless couplings completely supersymmetric, but mass terms arise that break supersymmetry, because $1(\mathrm{GeV})^2$ is 10^{-40} in our natural units defined by the Planck length. I would like to call this relaxed supersymmetry and it is conceivable that many interesting models of this nature could be built[106)].

6. PC-Breaking

Theories without scalars are automatically PC invariant. To describe PC-breaking we can either introduce elementary scalar fields with PC $= -1$ and put PC-odd terms in the Lagrangian, or believe that all scalars are in principle composite. Then PC-breaking must be spontaneous as described by T.D. Lee[65)]. Usually however the scalars are not specified, but only the currents[66)].

De Rújula, Georgi and Glashow observe that their chiral current can easily be modified to incorporate PC-breaking, a scheme already proposed by Mohapatra and Pati[67)] in 1972:

$$\Delta J = \bar{p}'\gamma_\mu(1-\gamma_5)(n \cos\phi + i\lambda \sin\phi)\,, \qquad \phi \ll 1\,. \tag{9}$$

7. Quantum Gravity

Quantum gravity is still not understood, but an interesting formal interpretation is given by Christodoulou[68)]. He gives a completely new definition of time, in terms of the distance in "superspace" between two three-dimensional geometries. But his formalism is not yet in a shape that enables one to give interesting physical predictions, and he has not yet considered the problem of infinities.

Renormalizable interactions between gravity and matter have not yet been found[69,70] but Berends and Gastmans find suggestive cancellations in the gravitational corrections to anomalous magnetic moments of leptons[71].

Numerous authors tried to put terms like $\sqrt{g}\,R^2$ or/and $\sqrt{g}\,R_{\mu\nu}\,R^{\mu\nu}$ in the Lagrangian, and thus (re)obtain renormalizability. It is about as clever as jumping to the moon through a telescope. Personally, I am convinced that if you want a finite theory of gravity, you have to put new physics in[72]. Very interesting in that respect are the attempts by Scherk and Schwarz to start with dual models.

8. Dual Models

Though originally designed as models for strong interactions, the dual models have become interesting for other types of interactions as well. They are the only field theoretical scheme that start from an infinite mass spectrum, forming Regge-trajectories with slope α'. In the limit $\alpha' \to 0$ they can mimic not only many renormalizable theories but also gravitation (always in combination with matter fields). If the tachyon-problem and the 26-dimension problem can be overcome then one might end up with a big renormalizable theory that unifies everything[73].

Scherk and Schwarz point out that for $\alpha' \neq 0$ those models are equally or better convergent than renormalizable theories.

9. Two-Dimensional Field Theories

Field theories in one-space and one-time dimension are valuable playgrounds for testing certain mathematical theorems that are supposed to hold also in four dimensions. There is no time now to discuss all the interesting developments of the past years[74] but I do want to mention just three things.

In a recent beautiful paper S. Coleman[75] explains the complete equivalency between two seemingly different structures: the massive Thirring model on the one hand, and the Sine–Gordon model on the other. The solitons (extended particle solutions) in one theory correspond to the fermions of the other. Thus we get one of the very few theories that can be expanded both at small and at large values of the coupling constant.[*]

[*]Note added: this result would later be challenged, the equivalence of these two models is not complete.

Solitons, and their quantization procedure have been studied in two and four dimensions by many groups[28,76,77]. Even in theories with weak coupling constants, solitons interact strongly (they have very large cross sections!) so they could make interesting candidates for an alternative strong-interaction theory. No convincing soliton theory for strong interactions does yet exist, but several magnetic-monopole quark structures have been considered[30,77,78].

Thirdly we mention the gauge theories in two dimensions with gauge group U(N) or SU(N). They are exactly soluble for either $N = 1$, $m_{\text{fermion}} = 0$, or for: $N \to \infty$, m_{fermion} arbitrary, in which case the $1/N$ expansion is possible. These theories exhibit most clearly the quark confinement effect, when color is unbroken[31,32]. Quark confinement is almost trivial in two space-time dimensions because the Coulomb potential looks like Fig. 5.

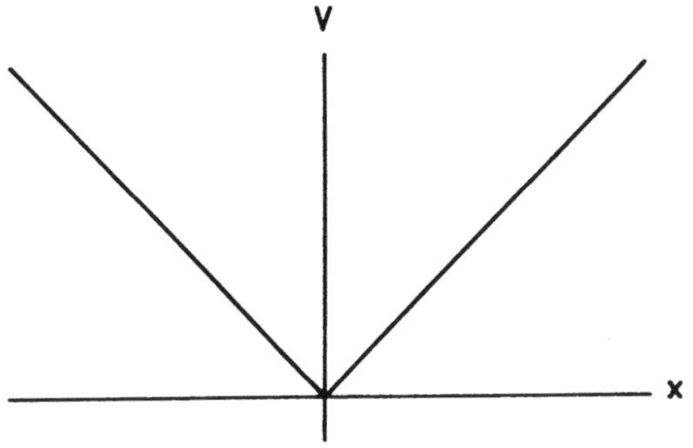

Fig. 5. The Coulomb potential in one space-like dimension.

QED in two dimensions (Schwinger's model) is not a good confinement theory because the electrons manage to screen electric charges completely, so the dressed electron is free, except for one bound state. At $N \to \infty$, probably at all $N > 1$, we get an infinity of bound states, "mesons" that interact with strength proportional to $1/N$. And they are on a nearly straight trajectory (actually a series of daughters because there is no angular momentum).

10. The Two Eta Problems

The latter one-space one-time dimensional model has lots of physically interesting properties. One is, that we can test ideas from current algebra. As in its four-dimensional analogue, we have in the limit where the quark masses m_p, $m_n \to 0$ an exact SU(2) × SU(2) symmetry, and the pion mass goes to zero in the sense[79,107]

$$m_\pi^2 \to C(m_p + m_n)\,. \qquad (10)$$

The proportionality constant contains the Regge slope. In two dimensions we have

$$C = g\sqrt{\pi} = 1/\sqrt{3\alpha'}\,. \qquad (11)$$

So if m_π is small, then m_p and m_n must be very small, in the order of 10 MeV. Now we can consider two old problems associated with the eta-meson. The first is that our theory seems to have U(2) × U(2) symmetry, not only SU(2) × SU(2). Thus, there should also be an isospin-zero particle, η, degenerate with π_0, whose mass should vanish. Experimentally however, $m_\eta^2 \gg m_\pi^2$. Also we do not understand why the $\pi_0 - \eta$ splitting goes according to $\pi_0 = \bar{p}p - \bar{n}n$; $\eta = \bar{p}p + \bar{n}n$ despite of possible mass differences of p and n. This problem reappears every now and then in the literature. Fritzsch, Gell-Mann and Leutwyler[80)] gave in 1972 a beautiful and simple solution: there is an Adler–Bell–Jackiw anomaly associated with the chiral U(1) subgroup[81)]:

$$\partial_\mu j_5^\mu = 2mj^5 + C\epsilon_{\mu\nu\alpha\beta}\,\mathrm{Tr}(G_{\mu\nu}G_{\alpha\beta})\,. \qquad (12)$$

The symmetry is broken and the eta-mass is raised. But then came the big confusing counter argument: we can redefine the axial current so that it is conserved, by writing

$$\tilde{j}_5^\mu = j_5^\mu - 2C\epsilon_{\mu\nu\alpha\beta}\,\mathrm{Tr}\,A_\nu\left(\partial_\alpha A_\beta - \partial_\beta A_\alpha + \frac{2}{3}g[A_\alpha, A_\beta]\right), \qquad (13)$$

so there should be a massless eta after all. Considerable effort has been made the past years to show that this counter argument is wrong. There are four counter counter arguments, all based on the fact that $\tilde{j}_5^\mu$ is not gauge-invariant:

i) $\tilde{j}_5^\mu$ is not gauge invariant, therefore the massless eta carries color, so it will be confined and thus removed from the physical spectrum[82)].
ii) The gluon field that occurs explicitly in the "corrected" equation for the axial current is a long-range field. The Goldstone theorem does not apply when long-range interactions are present.
iii) In the Lorentz-gauge there is no explicit long-range Coulomb force. But then there are negative metric states and this eta may be a ghost. It will be cancelled by other ghosts with wrong metric[83)].
iv) But the simplest argument is that we can calculate the eta mass exactly in two dimensions and see what happens. In two dimensions there is an anomaly only in the U(1) field[107)]:

$$\partial_\mu j_5^\mu = 2mj^5 + C\epsilon_{\mu\nu}F_{\mu\nu}\,. \qquad (14)$$

It is exactly this anomaly that raises the mass of the eta (it can be identified with Schwinger's photon), despite of the fact that we can find a new axial current:

$$\tilde{j}_5^\mu = j_5^\mu - 2C\epsilon_{\mu\nu}A_\nu \,, \tag{15}$$

which is conserved*).

The second eta problem is the decay

$$\eta \longrightarrow 3\pi \,.$$

It breaks G-parity and thus isospin. If we assume this decay to be electromagnetic then current algebra shows us that it should be suppressed by factors of m_π^2/m_η^2, which it does not seem to be[85]. I think there is a very simple explanation in terms of the present theory: the proton- and neutron-quark masses are free parameters in our theory, not "determined" by electromagnetism. Their difference follows from known hadron mass splittings and tends to unmix $\bar{p}p$ and $\bar{n}n$, that is, mix π^0 and η. We get

$$\eta \longrightarrow \pi^0 \longrightarrow 3\pi$$

with the correct order of magnitude[86]. The essential difficulty with the current algebra argument was that all SU(2) breaking effects were assumed to be electromagnetic. That would be beautiful, we could do current algebra by replacing ∂_μ by $\partial_\mu + iqA_\mu$ everywhere[87].

In a gauge theory this is wrong. As I emphasized in the beginning, mass differences cannot be explained by electromagnetism alone, they are arbitrary parameters and must be fixed by experiment. We'll have to live with that. Only for composite systems mass differences can be calculated. This suggests of course that we must make a theory with composite quarks. We leave that for a next generation of physicists.

Note that the breaking of SU(2) × SU(2) is governed by proton- and neutron quark masses alone. They are of order 10–20 MeV and differ by 5 MeV or so. So the breaking term of SU(2) × SU(2) also breaks SU(2) rather badly.

*)This argument is not quite correct, because in two space-time dimensions the Goldstone realization of a continuous symmetry is impossible[84]. However, if we let first $N = \infty$, then $m = 0$ then we do get nevertheless the Goldstone mode. This is possible because the meson-meson interactions decrease like $1/N$. Note that in any case there are no parity doublets for large N.

That is why you may not use PCAC and isospin together to get factors of m_π^2/m_η^2 in the decay $\eta \to 3\pi$*).

11. Misellaneous

a. Perturbation expansion

The renormalizability of the perturbation theory for gauge fields being well settled, there are still new developments. The dimensional renormalization procedure is the solution to all existence and uniqueness problems for the necessary gauge-invariant counterterms.

But if one wishes to circumvent the continuation to non-integer number of dimensions then the combinatorics is very hard. The Abelian Higgs–Kibble model can now be treated completely to all orders within the Zimmermann normal product formalism[88] if there are no massless particles. Slavnov identities can be satisfied to all orders in this procedure also in the non-Abelian Higgs–Kibble model if the group is semi-simple (no invariant U(1) group) and if no massless particles are present. Unitarity and gauge invariance have explicitly been proven this way in a particular SU(2) model.

Massless particles are very complicated this way, but advances have been made by Lowenstein and Becchi[89] in certain examples of massless Yang–Mills fields, and Clark and Rouet[90] for the Georgi–Glashow model.

It is noted by B. de Wit[91] that there is a technical restriction on the allowable form of the gauge-fixing term, relevant for supersymmetric models. The gauge-fixing term cannot have a non-vanishing vacuum expectation value in lowest order, otherwise contradictions arise. Of course, the Slavnov identities make the vacuum expectation value of this term vanish automatically in the usual formulation.

b. The background field method

The algebra in gauge theories is often quite involved. For certain calculations it would be of great help if gauge-invariance could be maintained throughout the calculation. On the other hand we must choose a gauge condition, which by definition spoils gauge-invariance right from the beginning. The trick is now to use the so-called background gauge[92]: the fields are split into a c-number, called background field, and a q-number, called quantum field. Only

*) As yet we have no satisfactory explanation for a possible $\Delta I = 3$ component in $\eta \to 3\pi$.

the quantum part must be fixed by a gauge condition, whereas gauge-invariance for the c-numbers can be maintained. The method was very successful in the case of gravity[70,72] and has also been applied to calculate anomalous dimensions of Wilson operators[93,94]. The method can be generalized for higher order irreducible graphs[105].

c. Two-loop-beta

The Callan–Symanzik beta function has now been calculated up to two loops for several gauge theories[95,96]. We have

$$\beta(g) = Ag^3 + Bg^5 + O(g^7)\,.$$

Now A can have either sign, and can also be very close to zero. In general, B will not vanish (it can be either positive or negative). We can then get either an UV or an IR stable fixed point close to zero. In certain supersymmetric models, A vanishes. The question is whether perhaps $\beta(g)$ vanishes identically for such a model. The answer appears to be: no, because for these models B has now been calculated also and it is non-zero[96].*)

d. The infrared problem

Massless Yang–Mills theories are very infrared divergent. As explained, we expect extremely complicated effects to occur, like flux tube formation and color confinement. A general argument is presented by Patrascioiu[97] and Swieca[98] that shows that if we have a local gauge-invariance and if we can have isolated regions in space (-particles) with non-vanishing total charge, then there must exist massless photons coupled to that charge. This is only proven for Abelian gauge symmetries, but if it would also hold for non-Abelian invariance, then the absence of massless colored photons must imply the absence of any colored particles (= color confinement).

e. Symmetry restoration at high temperatures

Just like a superconductor that becomes normal when the temperature is raised above a certain critical value, so can the vacuum of the Weinberg model become "normal" at a certain temperature[99]. The critical temperature is typically of the order of

$$kT \sim \frac{M_W}{e}\,.$$

*)Note added later: this is now known to be a mistake; β vanishes in all orders in these models.

This assumes that Hagedorn's limit on high temperatures[100] is invalid. Indeed it is invalid in the present quark theories, but the specific heat of the vacuum is very high because there are so many color components of fields. Observe that for SU(N) theories Hagedorn might be correct in the limit $N \to \infty$.

f. Symmetry restoration at high external fields

If we consider extremely strong magnetic fields then also the symmetry properties of the vacuum might change[101]. One can speculate on restoration of color symmetry, $e - \mu$ symmetry, parity or CP restoration, and vanishing of Cabibbo's angle. In still larger fields formation of magnetic monopoles[28] would make the vacuum unstable, as in strong electric fields.

g. Symmetry restoration at high densities

A very high fermion density means that $\bar{\psi}\psi$ and $\bar{\psi}\gamma_4\psi$ have a vacuum expectation value[102]. This also can have a symmetry restoring effect.

T.D. Lee, Margulies and Wick[103] argue that chiral SU(2) $\times$ SU(2) might be restored at very high nuclear densities. Thus the mass of one nucleon would go to zero and perhaps very heavy stable nuclei could be formed. The most recent calculations show a very remarkable phase transition at no more than twice the normal nuclear density. Although the result is of course model dependent, this work seems to predict stable large nuclei with binding energy of 150 MeV/nucleon.

At still higher densities we can speculate on more transition points. Again we think that the quark picture is more suitable than Hagedorn's picture[104].

12. Conclusions

a. Unifying everything

What I hoped to have made clear at this conference is that gauge fields are likely to describe all fundamental interactions including, in a sense, gravity. This is a breakthrough in particle physics and deserves to be called: "unification of all interactions".

b. Unifying nothing

But when we consider our present theory of strong interactions, the unbroken color version, then we see that it is unlikely to be really unified with weak and electromagnetic interactions, unless we go at ridiculously high energies, because the gauge coupling constants probably still differ considerably, and $SU(3)^{color}$ commutes with $SU(2) \times U(1)$. Also if we look at weak and electromagnetic interactions, we see that true unification has not yet been reached. At small distances strong interactions become weak, weak interactions become strong and electromagnetic ones stay electromagnetic, but no unification yet.

Perhaps our knowledge of the particle spectrum is still far too incomplete to enable us to unify their interactions.

I have given air to my own feeling that we are going in the wrong direction by choosing larger and larger gauge groups.

Perhaps we can use the confinement mechanism again to build quarks and leptons from still more elementary building blocks (chirps, growls, etc.).

Instead of "unifying" all particles and forces, it is much more important to unify knowledge.

Acknowledgement

I would like to thank D.A. Ross and M. Veltman for many interesting discussions.

References

1. C. Fronsdal, *Phys. Rev.* **B136** 1190 (1964). W. Kummer and G. Segrè, *Nucl. Phys.* **64** 585 (1965). G. Segrè, *Phys. Rev.* **181** 1996 (1969). L.F. Li and G. Segrè, *Phys. Rev.* **186** 1477 (1969). N. Christ, *Phys. Rev.* **176** 2086 (1968).
2. The literature on the renormalization group is vast. See for instance: M. Gell-Mann and F.E. Low, *Phys. Rev.* **95** 1300 (1954). C.G. Callan Jr., *Phys. Rev.* **D2** 1541 (1970). K. Symanzik, *Commun. Math. Phys.* **16** 48 (1970). K. Symanzik, *Commun. Math. Phys.* **18** 227 (1970). S. Coleman, Dilatations, lectures given at the International School "Ettore Majorana", Erice, 1971. S. Weinberg, *Phys. Rev.* **D8** 3497 (1973). B. Schroer, "A trip to Scalingland", to be publ. in the *Proceedings of the V Symposium on Theoretical Physics* in Rio de Janeiro, Jan. 1974. K. Wilson, *Phys. Rev.* **140B** 445 (1965); **179** 1499 (1969); **D2** 1438 (1970); **B4** 3174 (1971).

3. C.N. Yang and R. Mills, *Phys. Rev.* **96** 191 (1954). R. Shaw, Cambridge Ph.D. Thesis, unpubl.
4. J. Schwinger, *Ann. Phys.* (N.Y.) **2** 407 (1957). S.A. Bludman, Il *Nuovo Cimento* **9** 433 (1958). M. Gell–Mann, *Proc. 1960 Rochester Conf.*, p. 508. A. Salam and J.C. Ward, *Nuovo Cimento* **19** 165 (1961). A. Salam and J.C. Ward, *Phys. Lett.* **13** 168 (1964).
5. S. Weinberg, *Phys. Rev. Lett.* **19** 1264 (1967).
6. B.S. DeWitt, *Phys. Rev.* **162** 1195, 1239 (1967). L.D. Faddeev and V.N. Popov, *Phys. Lett.* **25B** 29 (1967). E.S. Fradkin and I.V. Tyutin, *Phys. Rev.* **D2** 2841 (1970). S. Mandelstam, *Phys. Rev.* **175** 1580, 1604 (1968). G. 't Hooft, *Nucl. Phys.* **B33** 173 (1971).
7. G. 't Hooft, *Nucl. Phys.* **B35** 167 (1971). G. 't Hooft and M. Veltman, *Nucl. Phys.* **B44** 189 (1972); *Nucl. Phys.* **B50** 318 (1972). B.W. Lee, *Phys. Rev.* **D5** 823 (1972); B.W. Lee and J. Zinn-Justin, *Phys. Rev.* **D5** 3121, 3137, 3155 (1972); **D7** 1049 (1973). For extensive reviews of the subject see refs. 8–19.
8. G. 't Hooft and M. Veltman, "DIAGRAMMAR", CERN Report 73/9 (1973), chapter 2.2 of this book.
9. B.W. Lee, in *Proceedings of the XVI International Conference on High Energy Physics*, ed. by J.D. Jackson and A. Roberts (NAI., Batavia, Ill. 1972) Vol IV, p. 249.
10. N. Veltman, "Gauge field theory", Invited talk presented at the *International Symposium on Electron and Photon Interactions at High Energies.* Bonn, 27–31 August 1973.
11. C.H. Llewellyn Smith, "Unified models of weak and electromagnetic interactions", Invited talk presented at the *International Symposium on Electron and Photon Interactions at High Energies*, Bonn, 27–31 August 1973.
12. S. Weinberg in *Proceedings of the 2^e^ Conference Internationale sur les Particules Elementaires*, Aix-en-Provence 1973. *Sup. au Journal de Physique*, Vol. 34, Fasc. 11–12, C1 1973, p. 45.
13. E.S. Abers and B.W. Lee, "Gauge theories", *Phys. Rep.* 9c, No. 1.
14. S. Coleman, "Secret symmetry. An introduction to spontaneous symmetry breakdown and gauge fields", lectures given at the 1973 Erice Summer School.
15. M.A. Beg and A. Sirlin, "Gauge theories of weak interactions", *Annual Rev. Nucl. Sci.* **24** 379 (1974).
16. J. Iliopoulos, "Progress in gauge theories", Rapporteur's talk given at the *17th International Conf. on High Energy Physics*, London 1974. See also the references therein.
17. S. Weinberg, *Sci. Am.* **231** 50 (1974).
18. J. Zinn-Justin, lectures given at the International Summer Inst. for Theoretical Physics, Bonn 1974.

19. S. Weinberg, *Rev. Mod. Phys.* **46** 255 (1974).
20. J.M. Cornwall, D.N. Levin and G. Tiktopoulos, *Phys. Rev. Lett.* **30** 1268 (1973); **31** 572 (1973); D.N. Levin and G. Tiktopoulos, preprint UCLA/75/TEP/7 (Los Angeles, March 1975). C.H. Llewellyn Smith, *Phys. Lett.* **B46** 233 (1973).
21. This property, and the magnitude of the one-loop beta function and the behaviour when fermions are present, were mentioned by me after Symanzik's talk at the Marseille Conference on Renormalization of Yang–Mills fields and Applications to Particle Physics, June 1972. At that time however, we did not believe that the small distance behaviour would be easy to study experimentally. The same things were rediscovered by H.D. Politzer, *Phys. Rev. Lett.* **30** 1346 (1973) who optimistically believed that this would make the perturbation expansion convergent, and by D.J. Gross and F. Wilczek, *Phys. Rev. Lett.* **30** 1343 (1973) who realized their importance in connection with Bjorken scaling.
22. D.J. Gross and F. Wilczek, *Phys. Rev.* **D8** 3633 (1973); **D9** 980 (1974). H. Georgi and H.D. Politzer, *Phys. Rev.* **D9** 416 (1974).
23. G. 't Hooft, *Nucl. Phys.* **B61** 455 (1973).
24. S. Coleman and D. Gross, *Phys. Rev. Lett.* **31** 851 (1973).
25. I. Bender, D. Gromes and J. Körner, *Nucl. Phys.* **B88** 525 (1975).
26. J. Kogut and L. Susskind, Tel Aviv preprint CLNS-263 (1974) *Phys. Rev.* **D9** 3501 (1974). G. 't Hooft, CERN preprint TH 1902 (July 1974), *Proceedings of the Colloquium on Recent Progress in Lagrangian Field Theory and Applications*, Marseille, June 24–28, 1974, p. 58. See further A. Neveu, this conference.
27. K.G. Wilson, *Phys. Rev.* **D10** 2445 (1974). R. Balian, J.M. Drouffe and C. Itzykson, *Phys. Rev.* **D10** 3376 (1974). C.P. Korthals Altes, *Proceedings of the Colloquium on Recent Progress in Lagrangian Field Theory and Applications*, Marseille, June 24–28, 1974, p. 102.
28. G. 't Hooft, *Nucl. Phys.* **B79** 276 (1974), chapter 4.2 of this book. B. Julia and A. Zee, *Phys. Rev.* **D11** 2227 (1975).
29. H. Georgi and S. L. Glashow, *Phys. Rev. Lett.* **28** 1494 (1972).
30. H.B. Nielsen and P. Olesen, *Nucl. Phys.* **B61** 45 (1973). Y. Nambu, *Phys. Rev.* **D10** 4262 (1974). S. Mandelstam, Berkeley preprint (Nov. 1974).
31. J. Schwinger, *Phys. Rev.* **128** 2425 (1962).
32. G. 't Hooft, *Nucl. Phys.* **B75** 461 (1974), chapter 6.2 of this book.
33. See P.H. Frampton's Lecture on dual theories at this conference.
34. A. Chodos, R.L. Jaffe, K. Johnson, C.B. Thorn and V.F. Weisskopf, *Phys. Rev.* **D9** 3471 (1974) and **D10** 2599 (1974).
35. S.L. Glashow, J. Iliopoulos and L. Maiani, *Phys. Rev.* **D2** 1285 (1970). See also: T. Cote and V.S. Mathur, Rochester preprint. UR-519-coo-3065-108 (1975).
36. R.M. Barnett, *Phys. Rev. Lett.* **34** 41 (1975). H. Fritzsch, M. Gell-Mann and P. Minkowski, to be published. H. Harari, *Phys. Lett.* **57B** 265 (1975).

37. A. De Rújula and S.L. Glashow, *Phys. Rev. Lett.* **34** 46 (1975). Th. Appelquist, A. De Rújula, H.D. Politzer and S.L. Glashow, Harvard preprint. B.J. Harrington, S.Y. Park and A. Yildiz, *Phys. Rev. Lett.* **34** 706 (1975). Th. Appelquist and H.D. Politzer, Inst. Adv. Stud. Princeton preprint coo 2220-43 (Feb. 1975). J. Ellis, *Acta Phys. Austriaca*, suppl 14 143 (1975).
38. A. Ore and J.L. Powell, *Phys. Rev.* **75** 1696 (1949).
39. Other explanations of the small decay width of ψ can be given if we assume more charmed quarks, see R.M. Barnett, ref. 36.
40. M.K. Gaillard and B.W. Lee, *Phys. Rev. Lett.* **33** 108 (1974). G. Altarelli and L. Maiani, *Phys. Lett.* **52B** 351 (1974). K. Wilson, *Phys. Rev.* **179** 1499 (1969). M.K. Gaillard, B.W. Lee and J. Rosner, Fermi Lab. Pub. 74:34 THY (1974).
41. G. Altarelli, N. Cabibbo and L. Maiani, *Nucl. Phys.* **B88** 285 (1975).
42. G. Altarelli, R.K. Ellis, L. Maiani and R. Petronzio, *Nucl. Phys.* **B88** 215 (1975).
43. A. De Rújula, H. Georgi and S.L. Glashow, *Nucl. Rev. Lett.* **35** 69 (1975).
44. J. Schechter and Y. Ueda, *Phys. Rev.* **D9** 736 (1970).
45. A. Pais, *Phys. Rev. Lett.* **29** 1712 (1972). A. Soni, *Phys. Rev.* **D9** 2092 (1974). T.P. Cheng, *Phys. Rev.* **D8** 496 (1973).
46. A. Pais, *Phys. Rev.* **D8** 625 (1973).
47. J. Schechter and M. Singer, *Phys. Rev.* **D9** 1769 (1974). V. Gupta and H.S. Mani, *Phys. Rev.* **D10** 1310 (1974).
48. J.C. Pati and A. Salam, *Phys. Rev.* **D8** 1240 (1973).
49. Y. Achiman, Heidelberg IHEP preprint (Dec. 1972). S. Weinberg, *Phys. Rev.* **D5** 1962 (1972).
50. M.A. Bég and A. Zee, *Phys. Rev. Lett.* **30** 675 (1973).
51. J.D. Bjorken and C.H. Llewellyn Smith, *Phys. Rev.* **D7** 887 (1973). C.H. Llewellyn Smith, *Nucl. Phys.* **B56** 325 (1973).
52. J. Prentki and B. Zumino, *Nucl. Phys.* **B47** 99 (1972). B.W. Lee, *Phys. Rev.* **D6** 1188 (1972).
53. C.H. Albright, C. Jarlskog and M.O. Tjia, *Nucl. Phys.* **B86** 535 (1975).
54. C.H. Albright and C. Jarlskog, *Nucl. Phys.* **B84** 467 (1975). C.H. Albright and C. Jarlskog and L. Wolfenstein, *Nucl. Phys.* **B84** 493 (1975).
55. For a review and 108 more references, see: M.L. Perl and P. Rapidis, SLAC-PUB-1496 preprint (Oct. 1974).
56. J.C. Pati and A. Salam, *Phys. Rev.* **D10** 275 (1974); **D11** 703 (1975); **D11** 1137 (1975); *Phys. Rev. Lett.* **31** 661 (1973), preprint IC/74/87 Trieste (Aug. 1974). J.C. Pati, "Particles, forces and the new mesons", Maryland tech. rep. 75–069 (1975).
57. Many attempts have been made to formulate "dynamical symmetry breakdown", i.e. without the simple Higgs scalar. See for instance: T.C. Cheng and E. Eichten, SLAC-PUB-1340 preprint (Nov. 1973); E.J. Eichten and F.L. Feinberg, *Phys. Rev.* **D10** 3254 (1974). F. Englert, J.M. Frère and P. Nicoletopoulos, Bruxelles preprint (1975); T. Goldman and P. Vinciarelli, *Phys. Rev.* **D10** 3431 (1974).

58. A.D. Linde, *Pis'ma Zh. Eksp. Teor. Fiz.* **19** (1974) 320 (*JETP Lett.* **19** (1974) 183). J. Dreitlein, *Phys. Rev. Lett.* **33** 1243 (1974). M. Veltman, *Phys. Rev. Lett.* **34** 777 (1975).
59. B. Zumino, *Nucl. Phys.* **B89** 535 (1975).
60. D.A. Ross and M. Veltman, Utrecht preprint (1 April 1975).
61. See also: H. Georgi, H.R. Quinn and S. Weinberg, Harvard preprint (1974).
62. M. Suzuki, *Nucl. Phys.* **B83** 269 (1974). E. Ma, *Phys. Rev.* **D11** 322 (1975). N.P. Chang, *Phys. Rev.* **D10** 2706 (1974).
63. See Abdus Salam's talk on Supersymmetries at this conference.
64. P. Fayet and J. Iliopoulos, *Phys. Lett.* **51B** 475 (1974). P. Fayet, *Nucl. Phys.* **B78** 14 (1974); preprint PTENS 74/7 (Nov. 1974, Ecole Normale Supérieure, Paris), *Nucl. Phys.* **B90** 104 (1975).
65. T.D. Lee, *Phys. Rev.* **D8** 1226 (1973); *Phys. Rep.* **9c** (Z).
66. See A. Pais and J. Primack, *Phys. Rev.* **D8** 3063 (1973) and references therein.
67. R.N. Mohapatra and J.C. Pati, *Phys. Rev.* **D11** 566 (1975).
68. D. Christodoulou, *Proceedings of the Academy of Athens*, **47** (Jan. 1972); CERN preprint TH 1894 (June 1974).
69. G. 't Hooft and M. Veltman, *Ann. Inst. H. Poincaré* **20** 69 (1974).
70. S. Deser and P. van Nieuwenhuizen, *Phys. Rev. Lett.* **32** 245 (1974), *Phys. Rev.* **D10** 401 (1974), *Lett. Nuovo Cimento* **11** 218 (1974), *Phys. Rev.* **D10** 411 (1974). S. Deser, H.S. Tsao and P. van Nieuwenhuizen, *Phys. Lett.* **50B** 491 (1974), *Phys. Rev.* **D10** 3337 (1974).
71. F.A. Berends and R. Gastmans, Leuven preprint (Oct. 1974). M.T. Grisaru, P. van Nieuwenhuizen and C.C. Wu, *Phys. Rev.* **D12** 3203 (1975).
72. See also: G. 't Hooft, in *Trends in Elementary Particle Theory, Proceedings of the International Summer Institute on Theoretical Physics* in Bonn 1974, ed. by H. Rollnik and K. Dietz, Springer-Verlag p. 92.
73. J. Scherk and J.H. Schwarz, *Nucl. Phys.* **B81** 118 (1974); *Phys. Lett.* **57B** 463 (1975).
74. D.J. Gross and A. Neveu, *Phys. Rev.* **D10** 3235 (1974). R.G. Root, *Phys. Rev.* **D11** 831 (1975). L.F. Li and J.F. Willemsen, *Phys. Rev.* **D10** 4087 (1974). J. Kogut and D.K. Sinclair, *Phys. Rev.* **D10** 4181 (1974). R.F. Dashen, S.K. Ma and R. Rajaraman, *Phys. Rev.* **D11** 1499 (1975). S.K. Ma and R. Rajaraman, *Phys. Rev.* **D11** 1701 (1975).
75. S. Coleman, Harvard preprint (1974). B. Schroer, Berlin preprint FUB HEP 5 (April 1975). J. Fröhlich, *Phys. Rev. Lett.* **34** 833 (1975).
76. R.F. Dashen, B. Hasslacher and A. Neveu, *Phys. Rev.* **D10** 4114, 4130, 4138 (1974). P. Vinciarelli, 1993-CERN preprint TH (March 1975).

M.B. Halpern, Berkeley preprint (April 1975). K. Cahill, Ecole Polytechnique A preprint 205-0375 (March 1975, Paris); *Phys. Lett.* **53B** 174 (1974). C.G. Callan and D.J. Gross, Princeton preprint (1975). L.D. Faddeev, Steklov Math. Inst. preprint, Leningrad 1975.

77. S. Mandelstam, Berkeley preprints (Nov. 1974; Feb. 1975); *Phys. Rev.* **D11** 3026 (1975).
78. A.P. Balachandran, H. Rupertsberger and J. Schechter, Syracuse preprint rep. no. SU-4205-41 (1974). G. Parisi, *Phys. Rev.* **D11** 970 (1975). A. Jevicki and P. Senjanovic, *Phys. Rev.* **D11** 860 (1975). See also interesting ideas by L.D. Faddeev, preprint MPI-PAE/Pth 16, Max-Planck Institute, München, June 1974.
79. This follows immediately if we assume that a term

$$c(\bar{p}p + \bar{n}n)$$

in the quark Lagrangian corresponds to a term

$$c'\sigma$$

in the symmetric Lagrangian for a sigma model of the pion, with c and c' of the same order of magnitude. See ref. 107.

80. H. Fritzsch, M. Gell–Mann and H. Leutwyler, *Phys. Lett.* **47B** 365 (1973); W. Bardeen, Stanford report (1974, unpublished).
81. J.S. Bell and R. Jackiw, *Nuovo Cimento* **60A** 47 (1969).
82. J. Kogut and L. Susskind, *Phys. Rev.* **D9** 3501 (1974); **D10** 3468 (1974); **D11** 3594 (1975).
83. S. Weinberg, *Phys. Rev.* **D11** 3583 (1975).
84. S. Coleman, *Commun. Math. Phys.* **31** 259 (1973).
85. D. Sutherland, *Phys. Lett.* **23** 384 (1966). J.S. Bell and D. Sutherland, *Nucl. Phys.* **B4** 315 (1968).
86. See for instance ref. 107, and W. Hudnall and J. Schlechter, *Phys. Rev.* **D9** 2111 (1974). I. Bars and M.B. Halpern, *Phys. Rev.* **D11** 956 (1975).
87. M. Veltman and J. Yellin, *Phys. Rev.* **154** 1469 (1967).
88. C. Becchi, A. Rouet and R. Stora, *Phys. Lett.* **52B** 344 (1974); lectures given at the *International School of Elementary Particle Physics*, Basko Polje, Yugoslavia Sept. 1974, and at the *Marseille Colloquium on Recent Progress in Lagrangian Field Theory and Applications*, June 1974; preprint 75, p. 723 (Marseille, April 1975).
89. J. H. Lowenstein and W. Zimmermann, preprints MPI-PAE/PTh 5 and 6/75 (München, New York, March 1975); *Commun. Math. Phys.* **44** 73 (1975). C. Becchi, to be published.
90. T. Clark and A. Rouet, to be published.
91. B. de Wit, *Phys. Rev.* **D12** 1843 (1975).
92. B.S. DeWitt, *Phys. Rev.* **162** 1195; 1239 (1967). J. Honerkamp, *Nucl. Phys.* **B18** 269 (1972); J. Honerkamp, *Proceedings Marseille Conf.*, 19–23 June 1972. G. 't Hooft, *Nucl. Phys.* **B62** 444 (1973). G. 't Hooft, lectures given at the *XIIth Winter School of Theoretical Physics* in Karpacz, Poland, (Feb. 1975). M.T. Grisaru, P. van Nieuwenhuizen and C.C. Wu, *Phys. Rev.* **D12** 3203 (1975).

93. S. Sarkar, *Nucl. Phys.* **B82** 447 (1974); **B83** 108 (1974). S. Sarkar and H. Strubbe, *Nucl. Phys.* **B90** 45 (1975).
94. H. Kluberg-Stern and J.B. Zuber, preprint DPh-T/75/28 (CEN-Saclay, March 1975); *Phys. Rev.* **D12** 467, 482 (1975).
95. D.R.T. Jones, *Nucl. Phys.* **B75** 531 (1974). A.A. Belavin and A.A. Migdal, Gorky State University, Gorky, USSR, preprint (Jan. 1974). W.E. Caswell, *Phys. Rev. Lett.* **33** 244 (1974).
96. D.R.T. Jones, *Nucl. Phys.* **B87** 127 (1975). S. Ferrara and B. Zumino, *Nucl. Phys.* **B79** 413 (1974). A. Salam and J. Strathdee, *Phys. Lett.* **51B** 353 (1974).
97. A. Patrascioiu, Inst. Adv. Study, Princeton preprint COO 2220-45 (April 1975).
98. J.A. Swieca, NYU/TR4/75 preprint (1975, New York).
99. C.W. Bernard, *Phys. Rev.* **D9** 3312 (1974). L. Dolan and R. Jackiw, *Phys. Rev.* **D9** 3320 (1974). S. Weinberg, *Phys. Rev.* **D9** 3357 (1974). B.J. Harrington and A. Yildiz, *Phys. Rev.* **D11** 779 (1975).
100. R. Hagedorn, *Nuovo Cimento Suppl.* **3** 147 (1965).
101. A. Salam and J. Strathdee, preprint IC/74/140, Trieste (Nov. 1974); IC/74/133 (Oct. 1974).
102. R.J. Harrington and A. Yildiz, *Phys. Rev.* **D11** 1705 (1975). A.D. Linde, Lebedev Phys. Inst. preprint no. 25 (1975).
103. T.D. Lee and G.C. Wick, *Phys. Rev.* **D9** 2291 (1974). T.D. Lee and M. Margulies, *Phys. Rev.* **D11** 1591 (1975).
104. J.C. Collins and M.J. Perry, *Phys. Rev. Lett.* **34** 1353 (1975).
105. G. 't Hooft, "The background field method in gauge field theories", lectures given at the *XIIth Winter School of Theoretical Physics* in Karpacz, Feb. 1975, and at the *Symposium on Interactions of Fundamental Particles*, Copenhagen, August 1975.
106. The general idea to use the Renormalization Group to relate the theory at our mass scales of order 1 GeV to theories at very small distances (Planck length), also occurs in: H. Georgi, H.R. Quinn and S. Weinberg, Harvard preprint (1974), and M.B. Halpern and W. Siegel, Berkeley preprint (Feb. 1975).
107. G. 't Hooft, lectures given at the *International School of Subnuclear Physics "Ettore Majorana"*, Erice 1975.

CHAPTER 3
THE RENORMALIZATION GROUP

CHAPTER 3
THE RENORMALIZATION GROUP

Introduction to the Renormalization Group in Quantum Field Theory [3.1]

My first calculations of renormalization coefficients for gauge theories date from 1970. But at that time the delicate gauge dependence of the counter terms were not yet completely understood and it was hard to avoid making errors.[13] By 1972 however I learned about a procedure that would simplify things considerably. While at CERN I met J. Honerkamp[14] who explained to me the so-called background field method, that had been applied by B. DeWitt in his early gravity calculations. The basic idea was due to Feynman. It simply amounts to writing all fields A^i (which may include the gauge fields as well as possible scalars) as

$$A^i = A^{i,\,cl} + A^{i,\,qu}\,,$$

where the "classical" fields $A^{i,\,cl}$ are required to obey the classical equations of motion, in the possible presence of sources, whereas $A^{i,\,qu}$ are now the "quantized variables", so that they serve as the integration variables in a functional integral.

My first reaction was that this could only be a book-keeping device that would neither change the physics nor the kind of calculations one will have to do. But then Honerkamp told me about the trick: the gauge condition needs to be imposed only on the quantum fields, while it may depend explicitly on the classical fields. While fixing the gauge degrees of freedom for the quantum field, one may keep the expressions gauge invariant with respect to background gauge transformations. If we now renormalize the one-loop diagrams the counterterms needed will automatically be gauge invariant.

[13]The *sign* of the β-function for gauge theories was clear to me, already then, but at that time I was unable to convince my advisor of the significance of this observation.

[14]J. Honerkamp, *Nucl. Phys.* **B48** (1972) 169.

The method enables one to produce a complete algebra for all one-loop β functions for all renormalizable theories in four space-time dimensions. We also use it in perturbative quantum gravity, and Chapter 8.1 gives another introductory text to this method. It does not work so nicely for diagrams with more loops because the overlapping divergences require counterterms also containing the quantum fields, so that the advantages disappear.

The paper of this chapter is an unpublished set of graduate lecture notes, which had to be polished a bit.

CHAPTER 3.1

THE RENORMALIZATION GROUP IN QUANTUM FIELD THEORY

Eight Graduate School Lectures

Doorwerth, 25-29 January 1988

by

G. 't Hooft

Institute for Theoretical Physics
Princetonplein 5, P.O. Box 80.006
3508 TA UTRECHT, The Netherlands

1 Introduction

The values one should assign to the free parameters of a quantum field theory, such as *masses* (m_i) and *coupling constants* of various types (λ_i, g_i), depend on the renormalization procedure adopted. Although the physical properties of these theories should not be scheme dependent, the calculational procedures indeed do, in a rather non-trivial way, depend on the subtraction scheme.

It was proposed by Stueckelberg and Peterman[1] in 1953 that one should require for any decent renormalization procedure that its results be invariant under the group of non-linear transformations among these free parameters, and they noted that this requirement would imply consistency conditions upon the renormalization constants. They called this group the "*Renormalization group*". One finds that modern renormalizable field theories meet these requirements by construction, and most of the requirements are rather trivial, with one exception.

Renormalization is always associated with a *regularization procedure*, which needs some sort of cut-off parameter Λ . One subgroup of the renormalization group coincides exactly with a redefinition of Λ:

$$\Lambda \to K\Lambda \ , \tag{1.1}$$

and one finds that this subgroup corresponds with the group of *scale transformations*: we are comparing a theory with itself with all masses and all momenta scaled by a common factor K.

Just because the renormalization procedure is rather delicate one finds that even those coupling parameters λ_i that one would expect to be scale-invariant since they are dimensionless, actually transform non-trivially. And so it turned out that the renormalization group has an important application in quantum field theory. Stueckelberg and Peterman had been thinking about the general diffeomorfism group in the multi-dimensional parameter space; but all that remains nowadays is the simple, one- dimensional group of scale transformations. Nevertheless, the name "Renormalization group" stuck.

The renormalization group becomes particularly non-trivial when the scale transformations that we consider are large. We get extra information on the behavior of amplitudes at extremely large or extremely small external momenta that would otherwise be difficult to obtain. But this not only holds for quantum field theory. The same or similar procedures can be applied to statistical models. Here one also uses the words "renormalization group"[2] although the rationale for that is somewhat more obscure: a model is scale transformed and then compared with the original. But it is sometimes observed that in a rigorous treatment the scale transformation can only be performed one-way, from smaller scales to larger scale theories, so that the group is actually a semi-group[3]!

In these lectures we briefly discuss renormalization. For the basic imput of our theory we do not need many details, such as the combinatoric proofs of its consistency[4], the uniqueness of the results as consequences of requirements such as causality and unitarity[5,6], and the independence of physically measurable quantities from freely chosen gauge fixing parameters[7]. All we want to know is the dependence of "counterterms" on cut-off parameters Λ, and the relations between "infinities" that follow.

We then discuss a simple theory with one coupling constant and one mass parameter such as $\lambda\phi^4$ and quantum electrodynamics. We note that there are three different kinds of theories: asymptotically free, infrared free, and scale-invariant.

Some calculations are done in the next chapters and then we discuss the more general theories, with an arbitrary number of masses and coupling constants.

Also for the more general case we want to compute the renormalization group coefficients, but now we need a more powerful method. We find that the various coefficients can be put in a simple algebra, as a consequence of various internal symmetries of our models. One then needs to do only a few calculations to obtain the most general sets of renormalization group coefficients, up to one loop.

Two-loop calculations are much too lengthy to discuss here in any detail, but we show that it is important to know the two-loop coefficients if we want to give a rigorous definition of certain (asymptotically free) quantum field theories. We touch upon the point raised in the beginning: maybe we *can* define models rigorously, but proofs that allow us to sum the perturbative expansion without being troubled by its possibly very divergent nature are still lacking.

Quantum chromodynamics cannot be understood without the renormalization group. Many attempts have been made to improve our formulation of this theory, for instance by "resumming" its perturbation expansion. Most of this subject goes beyond the scope of these lectures but we briefly indicate some sources of the bad behavior of the perturbative axpansion.

2. Renormalization

Even though a mathematically completely rigorous description of *quantum field theory* in 3+1 dimensions is not known, it is generally thought that many different quantum field systems exist: models that combine the requirements of *quantum mechanics* on the one hand, and *special relativity* on the other. In these lectures we will explain why, and which difficulties we encounter.

A central point in quantum field theory is that although we do not know how to construct rigorous models with *finite*, non-trivial coupling strengths, there are many *perturbative* models. These are all defined by postulating the (renormalized: see later) coupling strengths, λ_i, to be infinitesimal, and everything we want to know about the system is expressed as perturbative series in λ_i :

$$f\{\lambda_i\} = f_0 + f_{1i}\lambda_i + f_{2ij}\lambda_i\lambda_j + \dots , \tag{2.1}$$

in which all coefficients $f_{Nij..}$ (depending in some definite way On the masses m_j) are uniquely computable, but nothing is said about the convergence or divergence of the series. Indeed, quite generally one expects divergence of the form[6]

$$|f_{Nij..}| \simeq C_0 C^N N! \ , \tag{2.2}$$

so that for finite λ_i expressions (2.1), as they stand, are not meaningful.

We will not try to prove eq. (2.2) in these lectures. Here only we note that, if true, it would imply that (2.1) may define $f\{\lambda_i\}$ within a tiny "error bar": it makes a lot of sense in perturbative field theory to cut off the series (2.1) as soon as the N+1st term becomes larger than the Nth. This happens when

$$CN|\lambda_i| \cong 1 \ , \tag{2.3}$$

at which point#

$$|f_N(\lambda^N)| \simeq C_0 e^{-N} \simeq C_0 e^{-1/C|\lambda|} \ , \tag{2.4}$$

and if this is a good measure for the uncertainty in f then indeed such models may be quite acceptable physically as soon as the coupling strengths λ_i stay reasonably small.

A standard way to construct a perturbative quantum field theoretical model is to start off with a *Lagrange density* $\mathcal{L}$ and then construct functional integrals for the wanted amplitudes. One then discovers that the "bare" coupling constants $\lambda^B{}_i$ as they occur in the Lagrangian are not the appropriate parameters to use in the expansion series (2.1). The coefficients $f_{Nij..}$ would then be infinite.

We will now explain what this really means. Imagine that we introduced some sort of cut-off in our theory. This could be done in a number of ways. For one, we could simply postulate that no states in Hilbert space are considered in which any particle has spacelike momentum $\mathbf{k}$ with $|\mathbf{k}|>\Lambda_{max}$, for some large number Λ_{max} . In this case we require *unitarity* only to hold rigorously for S matrix elements ${}_{out}\langle\mathbf{k}_1,\mathbf{k}_2,\dots|\mathbf{p}_1,\mathbf{p}_2,\dots\rangle_{in}$ for which the total energy, $k_{10}+k_{20}+\dots = p_{10}+p_{20}+\dots$ does not exceed $2\Lambda_{max}$, because only then we are guaranteed that there are no intermediate states $|Q\rangle$ containing any of the forbidden particles in the unitarity condition

- - - - - -

#We use Stirling's formula: $N! \to \sqrt{2\pi N}\, N^N e^{-N}$.

$$\sum_Q S|Q\rangle\langle Q|S^\dagger = \mathbb{1} \; . \tag{2.5}$$

This would make the theory finite but no longer Lorentz-invariant. It would be unitary up till energies $2\Lambda_{max}$.

A better way to make the theory finite is to add "unphysical" particles. These are particles that contribute to the unitarity relation (2.5) intermediate states $|Q\rangle$ that may have "negative metric": the sign of $|Q\rangle\langle Q|$ is wrong, as if

$$\langle Q|Q\rangle < 0 \; , \tag{2.6}$$

(The normalization of a one-particle state is determined by the value, and sign, of the residue of the pole in the propagator).

Loop integrals can then be made to converge so fast that the finite-momentum cut-off Λ_{max} of the previous paragraph can safely go to infinity such that the limit exists, which now is Lorentz invariant by construction. If we choose the masses of all these unphysical particles to exceed a certain number Λ_{PV} then we also have unitarity up till energies Λ_{PV} or $2\Lambda_{PV}$ because energy conservation would forbid intermediate states $|Q\rangle$ containing these "sick" particles. The subscript PV stands for Pauli and Villars who first proposed this regularization method[8].

A third way to make amplitudes finite is much more subtle and will be used very often here. We imagine changing the theory a little bit by postulating[9] that the *number of space-time dimensions*, n, deviates slightly from 4:

$$n = 4-\varepsilon \; , \; \varepsilon \text{ infinitesimal.} \tag{2.7}$$

It is not possible to give a mathematically rigorous definition of what this means unless we formulate everything in perturbation expansions (or if we abandon translation invariance, see ref [10]).

In Feynman diagrams n only occurs in the closed loop integrals over space-time momenta k :

$$\int d^n k \to \int d^{4-\varepsilon} k \; , \tag{2.8}$$

and for large enough n such integrals may be well-defined, if the integrand contains only explicit external momenta p_i that span a $d<n$ dimensional subgroup of Minkowski space :

$$\int d^n k\ f\{k,p_i\} = \int d^{n-d}k_2 \int d^d k_1\ f\{k_1,p_i;k_2^2\} =$$

$$\frac{\pi^{(n-d)/2}}{\Gamma((n-d)/2)} \int d^d k_1 \int_0^\infty dr\ 2r^{n-d-1} f\{k_1,p_i;r^2\}\ . \tag{2.9}$$

The subtle point is that the integral (2.9) may diverge just about as badly as the original integral with n integer, but this time there is a unique way to define its "finite part" as long as n is non-integer. In most cases this finite part can simply be defined as the analytic continuation of the integral from n values where it converges, but sometimes it is also infra-red divergent because of poles in the integrand at finite or zero values of k_1 and/or r. In the latter case one can still define the finite part by splitting these divergent pieces apart, or by repeated partial integrations.

What one now observes is that these "dimensionally regularized" expressions, though finite at generic n, may still produce poles at $n \to 4$. They are of the form $1/\varepsilon$ (for one-loop diagrams), or $1/\varepsilon^2$, $1/\varepsilon^3$, ... at higher loops. The divergent term $1/\varepsilon$ plays a role very similar to $\log\Lambda_{max}$ or $\log\Lambda_{PV}$ in the other regularization schemes♩. Again, at sufficiently small ε unitarity is only very slightly violated, this time because the phase space integrals (which always converge, unless there is some infra-red divergence which must be handled separately) will approach the physical values.

Coming back to the coefficients $f_{Nij..}$ of eq. (2.1), one now observes that, quite generally, they diverge, either as $\Lambda_{PV} \to \infty$ or as $1/\varepsilon \to \infty$. So at first sight it seems that these theories are inconsistent.

However, the coefficients λ_i that we started off with are not directly physically observable. Throughout, we will adopt the philosophy (renormalization Ansatz) that one is allowed to choose these *arbitrary* numbers to vary with Λ_{max} or Λ_{PV} or ε : for instance, in the

- - - - - -

♩In those schemes one may also encounter higher divergences going as a power of Λ (linearly or quadratically). These have been eliminated in the dimensional scheme. One might say that our "finite part" procedure eliminated them.

dimensional procedure we substitute into the series (2.1):

$$\lambda_i \equiv \lambda^B{}_i = \lambda^R{}_i + a^{[1]}_i(\lambda^R)\,\varepsilon^{-1} + a^{[2]}_i(\lambda^R)\,\varepsilon^{-2} + \ldots \tag{2.10}$$

where now $\lambda^R{}_i$ are defined to be truly finite constants, independent of ε (or Λ if another regulator was chosen). The coefficients $a^{[1]}_i$ are of second or third order in λ^R and the higher order coefficients $a^{[\nu]}_i$ of successively higher order in λ^R.

At first sight it seems odd that $\lambda^B{}_i$ are both infinite and infinitesimal. We have to keep in mind that eq. (2.10) only makes sense in the limit $\lambda^R{}_i \to 0$ *first*, then $\varepsilon \to 0$. Indeed, the renormalization group will enable us to replace (2.10) by an expression in which the limits may be reversed.

Since both λ^R and λ^B are to be considered infinitesimal the perturbation expansion in λ^R is equivalent to the one in λ^B. But it is important to note that since the $a^{[\nu]}_i$ are higher order in λ the expansions are term by term different.

There is lots of arbitrariness in (2.10). All we want is that any finite value for $\lambda^R{}_i$ produces some finite values for the physically measured quantities $f\{\lambda^B{}_j\}$ in the limit $\varepsilon \to 0$. The series of functions f_{Nij} has poles at $\varepsilon=0$ and so does the series (2.10). We want to define the series (2.10) in such a way that, when substituted into (2.1) we get new expansions in $\lambda^R{}_i$ that are all free from divergences at $\varepsilon=0$. If this is possible the theory is called renormalizable.

To find out whether a theory is renormalizable is in principle very simple:

(*i*) Write down *all* interaction terms in the Lagrangian that agree with the chosen symmetries of your model.

(*ii*) Expect *all* possible terms to be required in the series (2.11) for the bare coupling constants, as long as these

(*a*) have the correct symmetry transformation rules under whatever symmetry that was imposed on the model, and

(*b*) have the correct *dimension*.

(*iii*) If, following the above rules, a certain subset of couplings does not get renormalized once they are put equal to zero then, and only then, we can leave such couplings out.

Requirement (*iib*) follows from the assumption that the infinite parts of $a_i^{[\nu]}$ cannot be chosen freely (two different choices cannot possibly both yield a finite f), so they should follow from the structure of the theory, but then they should not depend on our choice of units for lengths and masses.

In general, if space-time has 4 dimensions, only a finite number of coupling constants with positive or vanishing dimension can be constructed. But then, if we choose *all* coupling constants of positive or vanishing dimensions then step (*iii*) allows us to leave all the others out. Only a finite number of coupling constants are left. Eq. (2.10) then only contains a finite number of possible terms at each order. They should suffice to render everything finite. They do. One condition is that there are no "anomalies", but this is just the obvious (and very important) requirement that also our regulator procedure, whichever we chose, obeys the chosen symmetries.

The student should be cautioned that the above is not a *proof* of the renormalizability of field theories, but must be seen as a summary of the result of such a proof, which can be found in the literature[4-7].

3. $\lambda\phi^4$ theory

To illustrate our methods we could either take the simplest theory which is just a single self-interacting scalar, called $\lambda\phi^4$ theory, or we could take more complicated cases containing fermions and/or gauge fields. The disadvantage of the pure scalar theory is that there is no field renormalization at one loop, which makes it a lttle too simple. We will take this simplest case anyway, but we'll do as if there is a field renormalization.

The Lagrangian is

$$\mathcal{L} = -\tfrac{1}{2}\left((\partial\phi_B)^2+m_B{}^2\phi_B{}^2\right) - (\lambda^B/4!)\phi_B{}^4 + J(x)\phi_B \ . \tag{3.1}$$

Here $J(x)$ is a space-time dependent source insertion which we put in there for technical convenience. Later it will be put equal to zero. The sub- or superscript B stands for "bare" or unrenormalized. These quantities are not directly observed. We write

$$\lambda^B = \lambda^R + a^{[1]}(\lambda^R)\varepsilon^{-1} + a^{[2]}(\lambda^R)\varepsilon^{-2} + \dots , \tag{3.2}$$

$$m_B = m_R + b^{[1]}(\lambda^R, m_R)\varepsilon^{-1} + b^{[2]}(\lambda^R, m_R)\varepsilon^{-2} + \dots , \tag{3.3}$$

$$\phi_B = \left(1 + c^{[1]}(\lambda^R)\varepsilon^{-1} + c^{[2]}(\lambda^R)\varepsilon^{-2} + \dots\right)\phi_R , \tag{3.4}$$

where λ^R , m_R and ϕ_R are truly finite. The normalization of J is not yet fixed.

Physically observable features of the model are reveiled by computing vacuum expectation values $\langle 0|J(x)|0\rangle$, $\langle 0|J(x_1)J(x_2)|0\rangle$, etc. The first of these is zero. The second gives us the "physical" propagator. In diagrams Fig. 1.

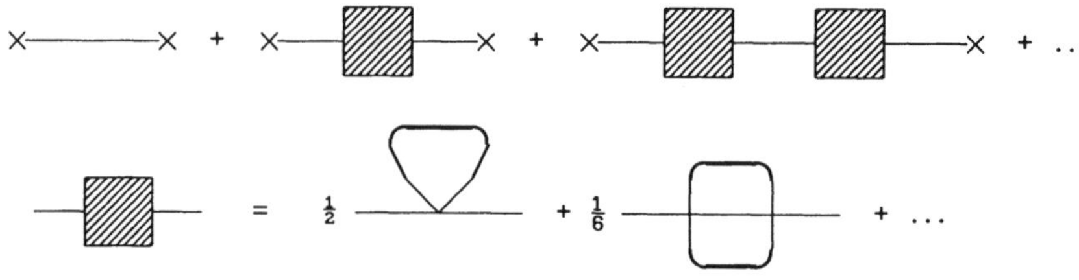

Fig. 1

The first diagram for the self energy insertion gives only a momentum-independent contribution. It can be removed by a mass renormalization. The second diagram has a non-trivial momentum dependence but is of too high order in λ_B for our considerations. In theories with fermions or gauge particles we have also diagrams of the kind given in Fig. 2

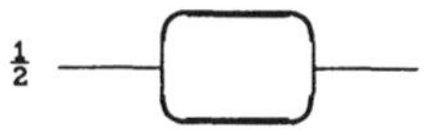

Fig. 2

We'll now pretend that there is a diagram of this kind in our example. In $4-\varepsilon$ dimensions the integral would typically give (the k^2 in the numerator has been put there just for dimensional reasons; in more elaborated theories the expressions are more complicated):

$$\Gamma_2(p) = (2\pi)^{\varepsilon-4} i(\lambda^B/2)\int d^{4-\varepsilon}k \frac{k^2}{((k+p)^2+m^2-i\varepsilon)(k^2+m^2-i\varepsilon)} =$$

$$= (2\pi)^{\varepsilon-4} i(\lambda^B/2)\int_0^1 dx \int d^{4-\varepsilon}k \frac{(k-xp)^2}{(k^2+m^2+x(1-x)p^2-i\varepsilon)^2} =$$

$$= \frac{-\lambda^B/2}{(4\pi)^{2-\varepsilon/2}} \int_0^1 dx \left(m^2+x(1-x)p^2\right)^{-\varepsilon/2}\Gamma(\varepsilon/2) \left[\frac{4-\varepsilon}{\varepsilon-2}\left(m^2+x(1-x)p^2\right)+x^2p^2\right] =$$

$$= \frac{\lambda^B/2}{(4\pi)^2} \int_0^1 dx \Big[\big((2x-3x^2)p^2+2m^2\big) \Big(\frac{2}{\varepsilon} - \gamma - \log\big(\frac{m^2+x(1-x)p^2}{4\pi}\big) \Big) +$$

$$+m^2 + x(1-x)p^2 \Big] + \mathcal{O}(\varepsilon) \quad . \tag{3.5}$$

Here, γ is Euler's constant: $\Gamma(\varepsilon/2) = 2/\varepsilon - \gamma + \mathcal{O}(\varepsilon)$. The insertion of the integral over the variable x is the famous "Feynman trick". Again, in our scalar theory the diagram of Fig. 2 actually does not occur. We have only the "seagull" diagram in Fig 1, which gives the p independent expression

$$\Gamma_2(p) = (2\pi)^{\varepsilon-4} i(\lambda^B/2) \int d^{4-\varepsilon}k \frac{1}{(k^2+m^2-i\varepsilon)} =$$

$$= \frac{-\lambda^B/2}{(4\pi)^{2-\varepsilon/2}} (m^2)^{-\varepsilon/2} \Gamma(\varepsilon/2) \frac{2}{\varepsilon-2} m^2 =$$

$$= \frac{\lambda^B/2}{(4\pi)^2} m^2 \Big[\frac{2}{\varepsilon} - 1 - \gamma - \log\big(\frac{m^2}{4\pi}\big) \Big] + \mathcal{O}(\varepsilon) \tag{3.6}$$

For pedagogical reasons we will pretend, temporarily, as if we are dealing with the more complicated expression (3.5) in our scalar field theory.

There are two things in eq. (3.5) that are of importance to us: the fact that there emerges a $1/\varepsilon$ term and the fact that we get a logarithmic function of p^2 . First look at the $1/\varepsilon$ term.

Let us substitute eqs. (3.2-4) into our Lagrangian. We get

$$\mathcal{L}(\phi_B) = \mathcal{L}(\phi_R) + (c^{[1]}/\varepsilon)\big(-(\partial\phi)^2 - m^2\phi^2 - \tfrac{1}{6}\lambda\phi^4\big) - b^{[1]} m\phi^2/\varepsilon - \tfrac{1}{24} a^{[1]}\phi^4/\varepsilon +$$

$$+ \text{higher orders.} \tag{3.7}$$

In the counter terms the distiction between B and R was neglected because that can be absorbed in the higher order corrections. The source term was not included because we are free to choose its normalization. We now add the correction terms in (3.7) to the diagrammatic rules treating them as if they were new interactions.

We see that extra terms should be added to the self energy expression of the form

$$-(2c^{[1]}/\varepsilon)(p^2+m^2) - 2b^{[1]}m/\varepsilon \ . \tag{3.8}$$

So if we choose

$$c^{[1]} = \frac{\lambda^B/2}{(4\pi)^2}\int_0^1 dx\ (2x-3x^2) \quad ; \tag{3.9}$$

$$b^{[1]} = \frac{\lambda^B/2}{(4\pi)^2}\int_0^1 dx\ (2-2x+3x^2)\ m_B \quad , \tag{3.10}$$

then the divergences of the form $1/\varepsilon^2$ in eq. (3.5) are cancelled for all momenta p. If we work with (3.6) we must choose

$$c^{[1]} = 0\ ; \quad b^{[1]} = \frac{\lambda^B/2}{(4\pi)^2}\, m_B . \tag{3.11}$$

Notice that the source is now coupled to ϕ^R, not ϕ^B. We discuss the choice for $a^{[1]}$ later.

Now consider the logarithm in (3.5). It seemed to come together with the $1/\varepsilon$ term, and, as we will see, it is indeed closely related. The important point is that, in dimensionally regularized expressions, the quantities inside the logarithm are *not dimensionless*! Clearly this is rather absurd, because if we would change our units of mass and momentum these expressions would change. However, such changes can easily be absorbed in *finite* renormalization terms in the Lagrangian, so indeed they will not be physically observable. In the other regularization schemes (momentum cut-off or Pauli-Villars) one always gets terms such as

$$\log\left(\frac{m^2 + x(1-x)p^2}{\Lambda^2}\right) \tag{3.12}$$

replacing the logarithm of (3.5). One may now understand why we compared $1/\varepsilon$ with $\log\Lambda$. The renormalization procedure in these older schemes is just as in the dimensional scheme: we put counter terms proportional to $\lambda\log\Lambda$ (and higher powers of this to remove the higher order divergences) in the Lagrangian.

In an "apparently scale invariant" theory, that is, if $m=0$, one would at first sight expect (3.5) to scale like p^2. But this is not so, because of the logarithm. It is unbounded above and below, so at a certain value of p^2 the self energy (3.5) will vanish, whereas for very large or very small p^2 the logarithm will dominate. But because the $\log\Lambda$ terms can be removed (or altered) by renormalizations, there is also the freedom to *choose* at will the value of p^2 where the self-energy vanishes. This is how one may first observe that a scale transformation of a theory maps the theory into itself only if this transformation is accompanied by (finite) renormalizations of the kind (3.3,4).

Also λ needs renormalization. This is seen when we consider the one-loop corrections to the scattering diagrams (Fig. 3)

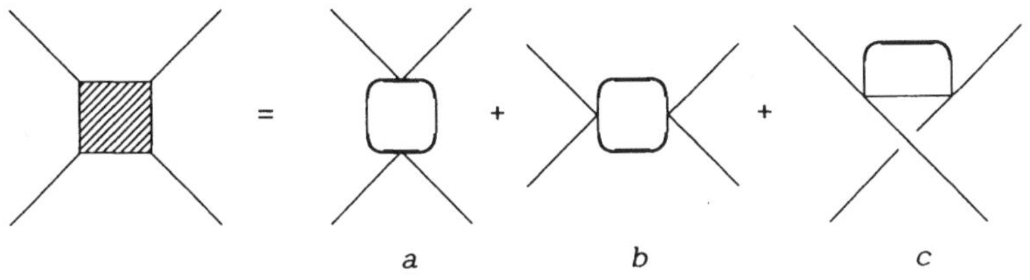

Fig. 3

Consider one of the terms in Fig. 3, for instance b. It is given by the integral

$$\Gamma_b(p) = -(2\pi)^{\varepsilon-4} i \, \frac{(\lambda^B)^2}{2} \int d^{4-\varepsilon}k \, \frac{1}{(k^2+m^2-i\varepsilon)((k+p)^2-i\varepsilon)} =$$

$$= -(2\pi)^{\varepsilon-4} i \, \frac{(\lambda^B)^2}{2} \int_0^1 dx \int d^{4-\varepsilon}k \, \frac{1}{(k^2+m^2+x(1-x)p^2-i\varepsilon)^2} =$$

$$= \frac{(\lambda^B)^2}{2(4\pi)^{2-\varepsilon/2}} \int_0^1 dx \left(m^2+x(1-x)p^2\right)^{-\varepsilon/2} \Gamma(\varepsilon/2) =$$

$$= \frac{(\lambda^B)^2}{2(4\pi)^2} \int_0^1 dx \left[\frac{2}{\varepsilon} - \gamma - \log\left(\frac{m^2+x(1-x)p^2}{4\pi}\right)\right] + \mathcal{O}(\varepsilon) \quad . \qquad (3.13)$$

Here p is the momentum through the horizontal channel in Fig. 3. The divergence at $\varepsilon=0$ can be removed, as before, by plugging in eq. (3.2), which would give in addition to (3.13):

$$- a^{[1]}(\lambda^R)/\varepsilon \ . \qquad (3.14)$$

Therefore we choose

$$a^{[1]} = 3\lambda^2/16\pi^2 \ , \tag{3.15}$$

where the factor 3 came from the fact that we want to cancel the divergences of the total of the three diagrams in Fig. 3. Again, the logarithm determines the scaling behavior of the theory when the mass effects become unimportant.

4. analysis of the scaling behaviour of dimensionally renormalized theories

When ε is an irrational number dimensionally regularized amplitudes can be uniquely defined. With this we mean that the mathematical procedure to extract the "finite part" of an integral in any Feynman diagram is unambiguous. The deeper reason for this is that the integrands, built out of propagators and vertices are allways rational expressions in the momenta to be integrated over. Integrals over *pure polynomials* and *pure powers* such as

$$\int d^{4-\varepsilon}k \ \mathrm{Pol}(k), \ \int dx \ x^{\alpha}, \ \int dx(x-a)^{\beta}, \tag{4.1}$$

can all be postulated to be zero. Convergent integrals are postulated to have their usual values. This way one succeeds in defining uinambiguous subtractions for all power-like divergences. Only the logarithmically divergent integral,

$$\int_0^{\infty} dx/x \ , \tag{4.2}$$

cannot be made finite because *infrared* divergences ($x \to 0$) are entangled with *ultraviolet* divergences ($x \to \infty$). If ε is irrational there are no logarithmic divergences (for half integer, third integer ε logarithmic divergences show up at two, three loops).

Consequently, all relations between diagrams that can be proved via partial integration, redefinitions of integration variables and combinatorics of diagrams, will hold automatically, and we need not worry about omitted boundary terms in integrals. This includes all Ward and Slavnov-Taylor identities, but it also includes *scaling* relations between diagrams.

Thus, at $\varepsilon \neq 0$ all "naive" scaling relations obtained by counting dimensions hold exactly.

What are the dimensions of the various parameters at $\varepsilon \neq 0$? In a functional integral we exponentiate the action S , so its dimension must be chosen to be zero. We have

$$S = \int d^{4-\varepsilon}x \ \mathcal{L}(x) \ , \tag{4.3}$$

so the Lagrange density $\mathcal{L}$ has dimension $\varepsilon-4$. Let us write

$$\mathcal{L} = \mu^{4-\varepsilon} \mathcal{L}_0 \ , \tag{4.4}$$

where μ is an arbitrary quantity with the dimension of a mass (unit of mass) and $\mathcal{L}_0$ is a dimensionless object.

The kinetic term in $\mathcal{L}$, being $-\frac{1}{2}(\partial\phi)^2$, carries the dimensionless unit $\frac{1}{2}$, therefore the dimensions must match. So we find the dimension of ϕ . We write

$$\phi^B = \mu^{1-\varepsilon/2} \phi_0^B \ . \tag{4.5}$$

Similarly,

$$m^B = \mu m_0^B \quad , \quad \lambda^B = \mu^{\varepsilon}\lambda_0^B \ . \tag{4.6}$$

Thus, although λ is scale-invariant in four dimensions, λ^B in $4-\varepsilon$ dimensions is not. This should be kept in mind when one uses eq. (3.2).

From now on we should insist that λ^R is kept dimensionless, but then eq. (3.2) can only be written down if we have some quantity μ to our disposal. μ will be chosen to be some arbitrary mass; in fact it will be related to that value of p^2 for which the self-energy or the vertex-correction vanishes. Thus we now write

$$\begin{aligned}
\lambda^B &= \mu^{\varepsilon}(\lambda^R + a^{[1]}(\lambda^R)\varepsilon^{-1} + a^{[2]}(\lambda^R)\varepsilon^{-2} + \dots) \ , \\
m^B &= \mu(m^R + b^{[1]}(\lambda^R)\varepsilon^{-1} + b^{[2]}(\lambda^R)\varepsilon^{-2} + \dots) \ , \\
\phi^B &= \mu^{1-\varepsilon/2} \phi^R(1 + c^{[1]}(\lambda^R)\varepsilon^{-1} + c^{[2]}(\lambda^R)\varepsilon^{-2} + \dots) \ .
\end{aligned} \tag{4.7}$$

It is important to know that the series (4.7) are constructed in such a way that *finite* expansion coefficients are obtained for *all* choices of the (infinitesimal) quantity λ^R . Therefore it should do no harm if we substitute in (4.7)

$$\lambda^R = \lambda^{R'} + e_1\varepsilon + e_2\varepsilon^2 + \dots \ ,$$

$$m^R = m^{R'} + f_1\varepsilon + f_2\varepsilon^2 + \ldots \,, \tag{4.8}$$

but this *does* give an entirely different series:

$$\lambda^B = \mu^\varepsilon\Big(\sum_{\nu=1}^{\infty} a^{[-\nu]'}\varepsilon^\nu + \lambda^{R''} + \sum_{\nu=1}^{\infty} a^{[\nu]'}\varepsilon^{-\nu} \Big) \,, \tag{4.9}$$

$$m^B = \mu \Big(\sum_{\nu=1}^{\infty} b^{[-\nu]'}\varepsilon^\nu + m^{R''} + \sum_{\nu=1}^{\infty} b^{[\nu]'}\varepsilon^{-\nu} \Big) \,, \tag{4.10}$$

and similar expressions for the bare fields ϕ^B, where the new coefficients $a^{[\nu]'}$ and $b^{[\nu]'}$ are functions of $\lambda^{R'}$ and $m^{R'}$, quite different from the original $a^{[\nu]}$ and $b^{[\nu]}$ as functions of λ^R and m^R ; also $\lambda^{R''}$ and $m^{R''}$ donot coincide with λ^R and m^R .

Clearly eqs (4.9) and (4.10) are a more general choice for the bare parameters than (4.7). The positive powers of ε are not necessary. In (4.7) the coefficients for the positive powers were chosen to vanish. They can allways be adjusted to zero by substituting (4.8) with judicious choices for e_ν and f_ν .

Since μ is essentially our unit of mass, all logarithms obtained in our integrations will have entries made dimensionless with powers of μ: $\log((p^2+m^2)/\mu^2)$. This is why μ will determine the order of magnitude of the momenta at which our loop corrections will (nearly) vanish. At those momenta perturbation expansion will converge relatively rapidly (at least the first few terms). If we want to understand how the system behaves at momenta very far away from these we have to make the transition to other μ values. Therefore, it is of importance to consider scale transformations:

$$\mu = \mu'(1+\sigma) \,, \quad \sigma \text{ infinitesimal.} \tag{4.11}$$

This turns expressions (4.7) into

$$\lambda^B = \mu'^{\varepsilon}(1+\sigma\varepsilon)\Big(\lambda^R + \sum_{\nu=1}^{\infty} a^{[\nu]}(\lambda^R,m^R)\varepsilon^{-\nu}\Big) =$$

$$= \mu'^{\varepsilon}\Big(\sigma\lambda^R\varepsilon + \lambda^R+\sigma a^{[1]} + \sum_{\nu=1}^{\infty} (a^{[\nu]}+\sigma a^{[\nu+1]})\varepsilon^{-\nu}\Big) \;; \tag{4.12}$$

$$m^R = \mu'(1+\sigma)\Big(m^R + \sum_{\nu=1}^{\infty} b^{[\nu]}(\lambda^R,m^R)\varepsilon^{-\nu}\Big) \,. \tag{4.13}$$

Notice that eq. (4.12) is of the form (4.9), not (4.7). In order to bring it into the form (4.7) we have to make a transformation (4.8). Let us try

$$\lambda^R = (1-\sigma\varepsilon)\ \tilde{\lambda}^R \ . \tag{4.14}$$

This removes the positive powers of ε from (4.12):

$$\lambda^B = \mu'\Big(\tilde{\lambda}^R + \sigma a^{[1]} - \sigma\tilde{\lambda}^R a^{[1]}_\lambda + \sum_{\nu=1}^{\infty} [a^{[\nu]}(\tilde{\lambda}^R, m^R) + \sigma a^{[\nu+1]} - \sigma\tilde{\lambda}^R a^{[\nu+1]}_\lambda]\varepsilon^{-\nu}\Big) \tag{4.15}$$

where $a^{[\nu]}_\lambda$ stands for $\partial a^{[\nu]}/\partial\lambda$, but also the expression for the bare mass changes:

$$m^B = \mu'\Big(m^R + \sigma m^R - \sigma\tilde{\lambda}^R b^{[1]}_\lambda + \sum_{\nu=1}^{\infty} [b^{[\nu]}(\tilde{\lambda}^R, m^R) + \sigma b^{[\nu]} - \sigma\lambda^R b^{[\nu+1]}_\lambda]\varepsilon^{-\nu}\Big) \ . \tag{4.16}$$

Note that we had to write the coefficients as functions of the newly chosen $\tilde{\lambda}^R$, which is now kept fixed as $\varepsilon \to 0$.

The series (4.15) and (4.16) and a similar one for the fields ϕ which we won't need for the moment, are now to substitute the old eqs (4.7). We see that the zeroth coefficients are

$$\lambda^{R'} = \tilde{\lambda}^R + \sigma a^{[1]} - \sigma\tilde{\lambda}^R a^{[1]}_\lambda \ ;$$

$$m^{R'} = m^R + \sigma m^R - \sigma\tilde{\lambda}^R b^{[1]}_\lambda \ . \tag{4.17}$$

It is essential that, with (4.14), we are free to keep either λ^R or $\tilde{\lambda}^R$ fixed as $\varepsilon \to 0$. They are both finite. When μ is chosen to be the unit of mass λ^R is fixed, and when μ' is chosen we'll decide to keep $\lambda^{R'}$ fixed, which implies that $\tilde{\lambda}^R$ is fixed. We wish to compare the two descriptions in the limit $\varepsilon=0$, so in (4.17) we may put $\tilde{\lambda}^R = \lambda^R$. The conclusion is that when we make the (infinitesimal) transition from μ to μ' with (4.11), we have to replace λ^R and m^R by:

$$\lambda^{R'} = \lambda^R + \sigma\left[1 - \lambda^R\frac{\partial}{\partial\lambda^R}\right] a^{[1]}(\lambda^R, m^R) \ ,$$

$$m^{R'} = m^R + \sigma\left[m^R - \lambda^R\frac{\partial}{\partial\lambda^R} b^{[1]}(\lambda^R, m^R)\right] \ . \tag{4.18}$$

This will often be written as[*]

$$\lambda^R = \lambda^{R'} + \beta(\lambda^R,m^R)\sigma \; ; \qquad m^R = m^{R'} + \alpha(\lambda^R,m^R)\sigma \; . \tag{4.19}$$

Here,

$$\beta(\lambda^R,m^R) = \left(\lambda^R\frac{\partial}{\partial\lambda^R} - 1\right) a^{[1]}(\lambda^R,m^R) \; ;$$

$$\alpha(\lambda^R,m^R) = \lambda^R\frac{\partial}{\partial\lambda^R} b^{[1]}(\lambda^R,m^R) - m^R \; . \tag{4.20}$$

The function $\beta(\lambda^R,m^R)$ describes the *anomalous dimension* of λ^R ; the function α also has a canonical part. From the above it follows that these functions can be computed from the counter terms needed in dimensional renormalization. At the one-loop level it is the only counter term; at two and more loops we also have the coefficients $a^{[2],[3],\ldots}$, $b^{[2],[3],\ldots}$, which govern higher poles ε^{-2} , ε^{-3} , etc. One will first have to substract these higher poles before being able to determine the coefficients of the lowest ones. Still, α and β are uniquely prescribed as perturbation expansions in λ^R.

In $\lambda\phi^4$ theory we have in lowest order in λ (see (3.13):

$$\beta(\lambda) = a^{[1]} = 3\lambda^2/16\pi^2 \; ; \tag{4.21}$$

and from the coefficient $b^{[1]}$ in (3.11),

$$\alpha(\lambda,m) = -m + \lambda m/32\pi^2 \; , \tag{4.22}$$

We have not yet written down the complete series that replaces (4.7) after the replacement $\mu \to \mu'$. at $\varepsilon \neq 0$ we substitute

$$\lambda^R = \lambda^{R'} + \beta\sigma - \varepsilon\lambda^R\sigma \; ;$$

$$m^R = m^{R'} + \alpha\sigma \; . \tag{4.23}$$

We get from (4.7):

$$\lambda^B = \mu'^{\varepsilon}\Big(\lambda^{R'} + \sum_{\nu=1}^{\infty} \varepsilon^{-\nu}[a^{[\nu]}(\lambda^{R'},m^{R'}) + \sigma a^{[\nu+1]} - \sigma\lambda^{R'} a^{[\nu+1]}_{\lambda} +$$

$$+ \beta\sigma\, a^{[\nu]}_{\lambda} + \alpha\sigma\, a^{[\nu]}_{m}]\Big) \; , \tag{4.24}$$

- - - - - -

[*]Our α differs by a factor $2m$ from Symanzik's α [13]. But in his notation usually he puts α=1, which we are here not allowed to do.

and

$$m^B = \mu' \left(m^{R'} + \sum_{\nu=1}^{\infty} \varepsilon^{-\nu} [b^{[\nu]}(\lambda^{R'}, m^{R'}) + \sigma b^{[\nu]} - \sigma\lambda^{R'} b_{\lambda}^{[\nu+1]} + \right.$$

$$\left. + \beta\sigma\, b_{\lambda}^{[\nu]} + \alpha\sigma\, b_{m}^{[\nu]}] \right) . \tag{4.25}$$

Note that, the way they are defined, all coefficients a , b , λ^R and m^R are dimensionless. Obviously, the functional dependence of $a^{[\nu]}$ and $b^{[\nu]}$ of $\lambda^{R'}$ and $m^{R'}$ should be independent of μ' . As argued before, the series (4.7) is the *only* choice of counter terms that should lead to a finite set of amplitudes in the limit $\varepsilon \to 0$. Therefore, all "correction terms" proportional to σ should cancel. Those terms proportional to σ that are of zeroth order in ε^{-1} were made to cancel by construction, but for the higher order ones this observation leads to important identities between the coefficients $a^{[\nu]}$ and $b^{[\nu]}$:

$$\beta\, a_{\lambda}^{[\nu]} + \alpha\, a_{m}^{[\nu]} = \lambda^R\, a_{\lambda}^{[\nu+1]} - a^{[\nu+1]} \qquad (\nu>0),$$

$$\beta\, b_{\lambda}^{[\nu]} + \alpha\, b_{m}^{[\nu]} + b^{[\nu]} = \lambda^R\, b_{\lambda}^{[\nu+1]} \qquad (\nu>0). \tag{4.26}$$

Since for $\nu>1$ the coefficients $a^{[\nu]}$ are at least of order $(\lambda^R)^2$ the equations (4.26) completely determine the higher order poles once we know the lowest order ones.

In ref[12] we stated that usually the coefficients $a^{[\nu]}$ are independent of the masses m , and the $b^{[\nu]}$ are proportional to m. This follows from simple power counting arguments when the divergent parts of integrals are considered as functions of the momenta *and* the masses m . However in some more complicated models not only the masses and coupling constants are renormalized, and not only are the fields ϕ multiplicatively renormalized, but one may also find renormalizations of the vacuum expectation values of the fields, so that one should substitute

$$\phi \to \phi + C , \tag{4.27}$$

where C contains poles in ε . For instance, if there is a ϕ^3 term next to the ϕ^4 term in the Lagrangian one might be inclined to adjust C so as to remove any terms in the Lagrangian that are *linear* in ϕ . But in this case the renormalized ϕ^3 and ϕ^4 couplings are defined in a more delicate way and then the above power counting argument breaks down, because a condition on C such as

$$\langle\phi\rangle = 0 \ , \tag{4.28}$$

introduces logarithmic dependence on masses in the theory.

Consider now the formal sum

$$\lambda^B = \lambda^R + \sum_{\nu=1}^{\infty} a^{[\nu]}(\lambda^R,m^R)\varepsilon^{-\nu} \ ; \tag{4.29}$$

$$m^B = m^R + \sum_{\nu=1}^{\infty} b^{[\nu]}(\lambda^R,m^R)\varepsilon^{-\nu} \ . \tag{4.30}$$

Putting

$$a^{[0]} = \lambda^R \ ; \quad b^{[0]} = m^R \ , \tag{4.31}$$

We find that (4.26) also holds for $\nu=0$, and (4.26) can be rewritten as

$$\left[\left(\beta(\lambda^R,m^R)-\varepsilon\lambda^R\right)\frac{\partial}{\partial\lambda^R} + \alpha(\lambda^R,m^R)\frac{\partial}{\partial m^R} + \varepsilon\right]\lambda^B = 0 \ ; \tag{4.32}$$

$$\left[\left(\beta(\lambda^R,m^R)-\varepsilon\lambda^R\right)\frac{\partial}{\partial\lambda^R} + \alpha(\lambda^R,m^R)\frac{\partial}{\partial m^R} + 1\right] m^B = 0 \ . \tag{4.33}$$

Equations of this sort are solved by considering trajectories in the space spanned by the parameters λ^R and m^R .

$$\frac{d\lambda^R(t)}{dt} = \beta(\lambda^R,m^R)-\varepsilon\lambda^R \ ; \tag{4.34}$$

$$\frac{dm^R(t)}{dt} = \alpha(\lambda^R,m^R) \ . \tag{4.35}$$

Then

$$\frac{d}{dt}\lambda^B(t) = -\varepsilon\lambda^B \ ; \ \frac{d}{dt}m^B(t) = -m^B \ . \tag{4.36}$$

Let's take the case (as in most two-parameter theories)

$$\beta = B_2\,(\lambda^R)^2 \ ; \quad \alpha = -1 + A_1\lambda^R. \tag{4.37}$$

Then the dependence between λ^B and λ^R is

$$\lambda^B = \frac{C\lambda^R}{\lambda^R - \varepsilon/B_2} \ , \tag{4.38}$$

but we also require that

$$\lambda^B = \lambda^R + \mathcal{O}(\lambda^R)^2 \ , \tag{4.39}$$

so

$$C = -\varepsilon/B_2 \ ; \quad \lambda^B = \frac{\lambda^R \varepsilon}{\varepsilon - B_2\lambda^R} \ . \tag{4.40}$$

This expression has been computed without special conditions on the order of the limits $\lambda^R \to 0$ and $\varepsilon \to 0$. So it replaces the series (3.2) or (4.7) when $\varepsilon \lesssim \lambda^R$. Note that in that case λ^R becomes negative unless $B_2<0$. In the limit $\varepsilon=0$ we have

$$\lambda^B = -\varepsilon/B_2 + \varepsilon^2/(B_2{}^2\lambda^R) + \dots \ , \tag{4.41}$$

so it starts out with a value independent of λ^R .

Actually, theories with negative values for the coupling constant are in trouble: the Hamiltonian will not be bounded from below. But because λ^B is so tiny these diseases (havivg to do with tunnelling out of the unstable vacuum) are not noticed at any finite order in perturbation expansion. Also we must stress that (4.40) is only valid as long as λ^R *and* λ^B stay small, because only then the perturbative expansions have some validity. So if $B_2>0$ the negative value of (4.41) cannot be trusted because we had to integrate the differential euations across the pole, where it cannot be correct. Now we could have taken the contour in t space to be *complex*, avoiding the pole, but this would require us to consider complex scale transformations and complex values of the coupling constant. where the Hamiltonian is also not well-behaved. Thus, eq. (4.41) has a formal validity only when both ε and λ^R are infinitesimal, in which case their relative sizes are not restricted.

We do see that the sign of B_2 is very important. If $B_2<0$, so that all bare coupling constants at $\varepsilon=0$ are positive, we call our theory *asymptotically free*. Let us consider this case for a moment, and assume that β depends only on λ^R , not on m^R (which is the case in most theories), such that

$$\beta(\lambda^R) = (\lambda^R)^2\tilde{B}(\lambda^R) \ , \tag{4.42}$$

with $\tilde{B}(0) = B_2 < 0$. One then proves from (4.32) and (3.39):

$$\lambda^B = \lambda^R \exp\Big(-\int_0^{\lambda^R} d\lambda \, \frac{1}{\lambda - \varepsilon/\tilde{B}(\lambda)}\Big) \ , \tag{4.43}$$

Let us take

$$\tilde{B}(\lambda) = B_2 + B_3\lambda + B_4\lambda^2 + \dots , \tag{4.44}$$

so that

$$\varepsilon/\tilde{B} = \varepsilon/B_2 - \varepsilon B_3\lambda/B_2^2 + \dots , \tag{4.45}$$

then this turns (4.43) into

$$\lambda^B = \lambda^R \exp\left[-\int_0^{\lambda^R} d\lambda \left(\frac{1}{\lambda-\varepsilon/B_2} - \frac{\varepsilon B_3\lambda/B_2^2}{(\lambda-\varepsilon/B_2)^2} + \dots\right)\right] =$$

$$= \lambda^R \exp\left[-\log\left(\frac{\lambda^R-\varepsilon/B_2}{-\varepsilon/B_2}\right) - \frac{\varepsilon B_3}{B_2^2}\left(\log\left(\frac{\lambda^R-\varepsilon/B_2}{-\varepsilon/B_2}\right) - \frac{\lambda^R}{\lambda^R-\varepsilon/B_2}\right) + \dots\right] , \tag{4.46}$$

which for small ε becomes

$$\lambda^B = -\varepsilon/B_2 + \frac{\varepsilon^2 B_3}{B_2^3}\log\varepsilon + \varepsilon^2\left(-\frac{1}{B_2^2\lambda^R} - \frac{B_3}{B_2^3}\log\lambda^R + O(1)\right) + O(\varepsilon^3\log\varepsilon) \tag{4.47}$$

The first two terms of this are independent of λ^R ; these are completely determined by the mathematics of the system: only with a λ^B that approaches zero this way when $\varepsilon\to 0$ the theory will give finite amplitudes. But the ε^2 term is arbitrary. The *higher* order terms, $\varepsilon^3\log\varepsilon$, etc. are therefore not important. They could be absorbed by λ^R.

We also note that eqs. (4.34) and (4.35) are is the *same* equations for λ^R and m^R as eqs (4.19) that describe the response of λ^R and m^R to an infinitesimal scale transformation $\mu \to \mu'=\mu(1+\sigma)$, when $\varepsilon=0$. For this reason the limit $\varepsilon \to 0$ in some respects corresponds to a limit towards short distance scales, just like the limit $\Lambda \to \infty$ for a more conventional regulator method.

The renormalization group equation

We have not yet discussed the renormalization of the fields $\phi(x)$ as it follows from eq. (4.7). It should be treated in exactly the same way as the mass renormalizations. Thus, in addition to eq. (4.33) we have

$$\left[(\beta-\varepsilon\lambda^R)\frac{\partial}{\partial\lambda^R} + \alpha\frac{\partial}{\partial m^R} + \gamma+1 - \varepsilon/2\right]\phi^B = 0 . \tag{5.1}$$

Here,

$$\gamma+1 = \left(\lambda^R\frac{\partial}{\partial\lambda^R} + \frac{1}{2}\right) c^{[1]} . \tag{5.2}$$

The differentiation $\partial/\partial\phi$ is lacking of course because the field ϕ was outside the brackets in (4.7).

Related to these equations is the change in the definition of the renormalized field ϕ^R when the unit of mass μ is changed. A transition (4.11) is associated with transformations (4.19) and:

$$\phi^R = \phi^{R'}\left(1+ \gamma\sigma\right) . \tag{5.3}$$

The combined transformations (4.11), (4.19) and (5.3) should leave the entire description of physical features invariant (renormalization group invariance):

$$\frac{\mathrm{d}}{\mathrm{d}\sigma}\,\Gamma^R(p_1, \;\ldots\; ,p_N) = 0 , \tag{5.4}$$

where $\Gamma^R = \langle\phi^R(p_1)\phi^R(p_2)\;\ldots\;\phi^R(p_N)\rangle$ is an "amputated" Green function. This can be rewritten as

$$\left[\frac{\mu\partial}{\partial\mu} + \alpha\frac{\partial}{\partial m^R} + \beta\frac{\partial}{\partial\lambda^R} + N\gamma\right]\Gamma^R(p_1, \;\ldots\; ,p_N) = 0 . \tag{5.5}$$

However, remember that m^R , and ϕ^R in (5.3), were defined to be dimensionless. But it is tempting to redefine them such that they have their canonical dimensions:

$$\mu m^R = m^{ren} \quad \text{(physical mass)} ;$$

$$\mu\phi^R = \phi^{ren} \quad \text{(physical field)} . \tag{5.6}$$

We then only keep the "anomalous parts" (coming from one loop and higher) of α and γ :

$$\alpha \to \mu(\alpha + m^R) \equiv \tilde{\alpha} \; ; \quad \gamma+1 \equiv \tilde{\gamma} \; . \tag{5.7}$$

$$\left(\frac{\mu\partial}{\partial\mu} + \tilde{\alpha}\,\frac{\partial}{\partial m^{ren}} + \beta\,\frac{\partial}{\partial\lambda^R} + N\tilde{\gamma}\right)\Gamma^{ren}(p_1, p_2, \ldots, p_N) = 0 \; . \tag{5.8}$$

This is *the renormalization group equation* as it follows from the rules of dimensional renormalization. The renormalization group equations used by Callan[11] and Symanzik[13] are equivalent but they look a little bit different. For Symanzik, μ is the mass parameter (hence his α is identical to one) and he does not have our $\tilde{\alpha}$ term, in stead of which his equations have a non-vanishing right hand side.

For our purposes we find the "renormalization group flow-equations" or Gell-mann Low equations[14], (4.11), (4.19) and (5.3) more informative: if

$$\mu \to \mu_0\, e^{\sigma} \; , \tag{5.9}$$

then

$$\frac{\partial\lambda^R}{\partial\sigma} = \beta(\lambda^R, m^{ren}) \; ; \quad \frac{\partial m^{ren}}{\partial\sigma} = \tilde{\alpha}(\lambda^R, m^{ren}) \; ; \quad \text{etc.} \tag{5.10}$$

(as stated before, in most cases β is independent of m^{ren} and $\tilde{\alpha}$ is linear in m^{ren}).

Three important cases must be distinguished:

(*i*) theories in which β (for at least one of its coupling constants) stays positive. This happens in most models; in *all* 4 dimensional theories that donot contain non-Abelian gauge fields (such as $\lambda\phi^4$, QED, and Yukawa theories). In this case the solution of (5.10) must show a strong, possibly divergent, effective coupling constant at high energies.

Suppose

$$\beta(\lambda) = b\lambda^2 \; . \tag{5.11}$$

Then we would get

$$\lambda(\sigma) = \frac{\lambda(\sigma_0)}{1-(\sigma-\sigma_0)b\lambda(\sigma_0)} \; , \tag{5.12}$$

which shows a *pole* at $\sigma = \sigma_0 + 1/b\lambda_0$, or

$$\mu = \mu_0 \, e^{1/b\lambda_0} \, . \qquad (5.13)$$

This is the so-called Landau singularity. The student is advized to study this singularity in the "chain of bubble approximation" for scalar field theories. The chain of bubbles represents faithfully what happens in an N component scalar field theory in the limit $N \to \infty$, λN fixed.

(*ii*) Theories in which all β coefficients start out negative. This may happen in non-Abelian gauge theories only (it is *guaranteed* only in a pure non-Abelian gauge theory). Then, according to (5.12) and (5.13) $\lambda(\mu)$ goes to zero logarithmically. This means that at high energy the perturbative expansion becomes better and better! We saw that in this case also the bare coupling constants formally tend to zero. There is a general consensus that theories of this type must have a rigorous mathematical footing (see introduction) but this could be proven only in very special cases[15]. The phenomenon is called *Asymptotic freedom* and was perhaps *the* most compelling reason for assuming a non-Abelian theory such as QCD for describing strong interactions. There is no Landau ghost in the far ultraviolet. There is one in the infrared (at very low values of μ), but one may well imagine that this region should be properly described by integrating out the field equations defined at high μ values, so that the Landau ghost may be just some sort of physical bound state.

(*iii*) In some very special cases one might find that $\beta(\lambda) = 0$ at all orders in perturbation theory. An example is $N=4$ supersymmetric Yang-Mills theory and superstrings in 2 dimensions Such a theory must be strictly scale-invariant. Quite generally one then also finds that such a theory must also be conformally invariant.

One must be careful to observe that the form taken by the renormalization group equations and the details of the expressions for the α and β functions depend a lot on details of the regularization scheme. In Pauli-Villars $\log\Lambda$ plays the role of ε and much of the rest goes the same way, except that many higher order coefficients will look different. A simple way to see this is as follows.

Suppose that two schemes choose a different convention for the *finite* part of the subtraction constants. Consequently,

$$\lambda^R = \lambda^{R'} + f(\lambda^{R'}) \ , \tag{5.14}$$

where $f(\lambda) = \mathcal{O}(\lambda^2)$, a given function. Now

$$\frac{\partial}{\partial\sigma}\lambda^R = \beta(\lambda^R) = b_2(\lambda^R)^2 + b_3(\lambda^R)^3 + \dots ;$$

$$\frac{\partial}{\partial\sigma}\lambda^{R'} = \beta'(\lambda^{R'}) = b_2'(\lambda^{R'})^2 + b_3'(\lambda^{R'})^3 + \dots . \tag{5.15}$$

Assume for simplicity that

$$f(\lambda) = c\lambda^2 \ . \tag{5.16}$$

It follows that (write $\lambda^{R'} = \lambda$):

$$\frac{\partial}{\partial\sigma}\lambda^R = b_2(\lambda^2+2c\lambda^3+c^2\lambda^4) + b_3(\lambda^3+3c\lambda^4+ \dots) + b_4\lambda^4 + \dots =$$

$$= (1+2c\lambda)\frac{\partial}{\partial\sigma}\lambda = (1+2c\lambda)\left(b_2'\lambda^2 + b_3'\lambda^3 + b_4'\lambda^4 + \dots\right) , \tag{5.17}$$

or

$$b_2' = b_2 \ ; \qquad b_3' = b_3 \ ; \qquad b_4' = b_4+cb_3+c^2b_2 \ ; \ \dots . \tag{5.18}$$

Observe that b_2 and b_3 are "universal": they are scheme independent; b_2 because it is the lowest order term, depending on the leading divergence at one loop, which is the same for all regulators; b_3 is universal because c happens to cancel out. This two-loop coefficient for the beta function ceases to be universal if we have more than one coupling constant[16]. The student is suggested to find a plausible substitution (5.14) in that case such that the two-loop coefficients change.

Having the freedom to redefine our coupling constants as suggested in (5.14) we could decide to choose c such that $b_3'=0$. It is not hard to convince oneself that if $b_2\neq 0$ a substitution of the form (5.14) with $f(\lambda)=\mathcal{O}(\lambda^3)$ can be found in such a way that

$$b_n' = 0 \ , \qquad n = 4,5, \dots \infty. \tag{5.19}$$

This $f(\lambda)$ is unique apart from scale transformations which obviously leave the functional form of $\beta(\lambda)$ invariant.

The words "minimal subtraction" have already been reserved for the dimensional renormalization counter terms (2.10) or (4.7) (which are unique in contrast with the more arbitrary choice (4.9)). We might call the choice that gives (5.19), implying the exact β function

$$\beta(\lambda') = b_2(\lambda')^2 + b_3(\lambda')^3 \quad \text{(no further terms)} , \tag{5.20}$$

an "ultra-minimal" subtraction.

A word of caution is of order. The original beta function $\beta(\lambda^R)$ is sometimes believed to have a fixed point:

$$\beta(\lambda_0) = 0 , \tag{5.21}$$

so that if $\lambda^R = \lambda_0$ the theory becomes scale-invariant. Of course this fixed point is totally unrelated to a possible fixed point of the truncated series (5.20). At such a point $\lambda^R = \lambda_0$ the realation (5.14) will tend to show singulariries. But we do learn that the question whether or not β has a fixed point may depend on how λ is defined when it becomes large. It is argued by many researchers that the question whether the theory is *scale invariant* should not depend on such definitions, and that therefore the existence of zeros of the β function should be scheme independent. But we stress that it is not known how to give any rigorous definition of 4 dimensional field theories beyond the perturbation expansion. Indeed there is reason to hope that a definition can be given precisely in the case that there is not such a zero, when the theory is asymptotically free $(\beta(\lambda)<0)$. In that case condition (5.20) determines λ uniquely apart from scale transformations.

6 The higher order terms in the many parameter case

The procedure of chapter 4 and the equations (4.26) can be generalized to an arbitrary number of parameters λ_k which may be found to have dimension

$$D_k = d_{(k)} + \varepsilon\rho_{(k)} , \tag{6.1}$$

in $n=4-\varepsilon$ dimensions. Some of these may be masses (for which ρ_k vanishes), others are three- or four-particle coupling constants (which have $d_k=0$) . We could even include fields ϕ, ψ, or gauge fixing parameters α . As before, we write

$$\lambda_k{}^B = \mu^{D_k}\left[\lambda_k{}^R + \sum_{\nu=1}^{\infty} a_k^{[\nu]}\{\lambda^R\}\varepsilon^{-\nu}\right] , \tag{6.2}$$

where in the brackets { } are those objects on which these quantities are allowed to depend. The renormalization of physically observable coupling strengths λ should not depend on gauge fixing parameters α or field strengths ϕ, ψ, so these form a set by themselves.

Consider the transformation

$$\mu = \mu'(1+\sigma) . \tag{6.3}$$

The equivalent of (4.12), (4.13) is

$$\lambda_k{}^B = (\mu')^{D_k}(1+d_{(k)}\sigma)\left[\lambda_k{}^R + \varepsilon\rho_{(k)}\lambda_k{}^R\sigma + \rho_{(k)}a_k^{[1]}\sigma + \right.$$
$$\left. + \sum_{\nu=1}^{\infty}(a_k^{[\nu]}+\rho_{(k)}a_k^{[\nu+1]}\sigma)\varepsilon^{-\nu}\right] . \tag{6.4}$$

Let us write

$$\lambda_k{}^R = \lambda_k{}^{R'} - \varepsilon\rho_{(k)}\lambda_k{}^{R'}\sigma + \beta_k\{\lambda^R\}\sigma , \tag{6.5}$$

where the functions β_k must be chosen such that the first term in (6.4) is just $\lambda_k{}^{R'}$. The ε term should remove the positive powers of ε in (6.4).

Eq. (6.4) turns into

$$\lambda_k{}^B = (\mu')^{D_k}(1+d_{(k)}\sigma)\left[\lambda_k{}^{R'} + \beta_k\sigma + \rho_{(k)}a_k^{[1]}\sigma - \sum_t a_{k,t}^{[1]}\rho_{(t)}\lambda_t{}^{R'}\sigma + \right.$$
$$\left. + \sum_{\nu=1}^{\infty}\left(a_k^{[\nu]}\{\lambda^{R'}\} + \rho_{(k)}a_k^{[\nu+1]}\sigma + \sum_t \beta_t a_{k,t}^{[\nu]}\sigma - \sum_t a_{k,t}^{[\nu+1]}\rho_{(t)}\lambda_t{}^{R'}\sigma\right)\varepsilon^{-\nu}\right] . \tag{6.6}$$

Here, $a_{k,t}^{[\nu]}$ stands for $\partial a_k^{[\nu]}/\partial\lambda_t{}^R$.

Thus we find how to express β_k in terms of the counter terms a_k :

$$\beta_k\{\lambda^R\} = -d_{(k)}\lambda_k^R + \sum_t a_{k,t}^{[1]}\rho_{(t)}\lambda_t{}^R - \rho_{(k)}a_k^{[1]} , \tag{6.7}$$

and since $(1+d_{(k)}\sigma)a_k^{[\nu]}\{\lambda^{R'}\}$ in (6.6) should be the same functions as $a_k^{[\nu]}\{\lambda\}$ in (6.2), we have the identities for $a_k^{[\nu]}$:

$$d_{(k)}a_k^{[\nu]} + \beta_t a_{k,t}^{[\nu]} = \sum_t a_{k,t}^{[\nu+1]}\rho_{(t)}\lambda_t^R - \rho_{(k)}a_k^{[\nu+1]} , \tag{6.8}$$

replacing our previous eqs. (4.26). In most cases the coefficients $a_k^{[\nu]}$ are polynomials of very low degrees in the masses and other dimensionful parameters, which simplifies the use of equs. (6.8).

7. An algebra for the one-loop counter terms

The first terms of the α, β and γ coefficients are by far the most important. They determine the leading scaling behaviour of the theory, and whether it is asymptotically free or not. The counter terms needed to make the one-loop Feynman diagrams finite are particularly simple to compute, although, if you start out doing first the complete one-loop calculation it can still be quite cumbersome. So it will be of help that a universal "master formula" can be written down that gives us all necessary one loop counter terms once and for all.

All 4 dimensional renormalizable field theories can be written in the form

$$\mathcal{L} = - G^a_{\mu\nu}G^a_{\mu\nu}/4 - (D_\mu\phi_i)^2/2 - V\{\phi_i\} - \bar\psi_i\gamma_\mu D_\mu\psi_i - $$

$$- \bar\psi_i\big(S_{ij}\{\phi\}+i\gamma_5 P_{ij}\{\phi\}\big)\psi_j , \tag{7.1}$$

where summation over repeated indices a, μ, ν, i, j, is implied. ϕ_i are a bunch of scalar fields and ψ_i are (possibly chiral) fermions. $G_{\mu\nu}$ are the gauge fields,

$$G^a_{\mu\nu} = \partial_\mu A^a_\nu - \partial_\nu A^a_\mu + f^{abc}A^b_\mu A^c_\nu , \tag{7.2}$$

where f^{abc} is any combination of coupling constants and structure functions that satisfies the Jacobi identity for some compact gauge Lie group. The function $V\{\phi\}$ is any quartic polynomial in the fields ϕ such as $\pm m^2\phi^2/2 + \lambda\phi^4/4!$ in a simple scalar field theory leaving open the question whether or not spontaneous symmetry breakdown occurs. The derivatives D_μ are *covariant* derivatives,

$$D_\mu\phi_i \equiv \partial_\mu\phi_i + T^a_{ij}A^a_\mu\phi_j \quad ; \quad D_\mu\psi_i \equiv \partial_\mu\psi_i + U^a_{ij}A^a_\mu\psi_j , \tag{7.3}$$

where again possible coupling constants were absorbed into the coefficients T and U . These must satisfy the usual commutation rules (and the U may be split into a left handed and a right handed part):

$$[T^a,T^b] = f^{abc}T^c \ ; \ U^a = U^a_L P_L + U^a_R P_R \ ; \ P_{\substack{L\\R}} = \frac{1\pm\gamma_5}{2} \ ; \quad [U^a_{\substack{L\\R}}, U^b_{\substack{L\\R}}] = f^{abc}U^c_{\substack{L\\R}} \ , \tag{7.4}$$

The functions S and P must be *linear* in ϕ , for instance, $S=m_\psi+g\phi$ in a Yukawa theory. And of course the entire Lagrangian must be invariant under the local gauge group.

We can now add to this Lagrangian a counter term $\Delta\mathcal{L}$ that can be chosen such that it absorbes all (one loop) infinities. The way to construct $\Delta\mathcal{L}$ is as follows.

We are interested in the complete set of one-loop diagrams. For instance we could insert (gauge-invariant) source terms in the Lagrangian and consider the dependence of the vacuum-vacuum transition amplitude on the strengths of these sources, expanded to one-loop order. A practical way to compute this amplitude is to shift the field variables:

$$A^a_\mu = A^{a\ cl}_\mu + A^{a\ qu}_\mu \ ; \quad \phi_i = \phi^{cl}_i + \phi^{qu}_i \ ; \quad \psi_i = \psi^{cl}_i + \psi^{qu}_i \tag{7.5}$$

where the "classical" fields are c numbers satisfying the classical equations of motion (with the source terms in), and the "quantum" fields are integrated over. Let us denote the bosonic quantum fields together as $\{\varphi_i\}$. The Lagrangian is now expanded in these:

$$\mathcal{L} = \mathcal{L}^{(classical)} + \mathcal{L}_1{}^i(x)\varphi_i(x) + \mathcal{L}_2{}^{ij}\varphi_i\varphi_j + \dots \ . \tag{7.6}$$

Now the fact that the classical fields satisfy the equations of motion implies that, when integrated over, the linear term $\mathcal{L}_1$ vanishes. The quadratic term generates propagators for the quantum fields and vertices with two quantum fields and other background fields. The higher order terms neglected in (7.5) generate vertices that cannot contribute to one-loop graphs, so we ignore them.

We rewrite the last term of (7.6) as

$$\mathcal{L}\{\varphi,x) = \partial_\mu\varphi_i W^{\mu\nu}_{ij}(x)\partial_\nu\varphi_j/2 + \varphi_i N^\mu_{ij}(x)\partial_\mu\varphi_j + \varphi_i M_{ij}(x)\varphi_j/2 \ . \tag{7.7}$$

Here, N and M depend on the classical background fields. We will see to it that W does not depend on the background fields.

In general, Lagrangian (7.7) is a gauge theory and should be quantized and renormalized according to the usual rules. A gauge-fixing term will be needed, and ghost fiels. As a gauge fixing term one could for instance take

$$\mathcal{L}^C = -\tfrac{1}{2}(D_\mu^{\ cl} A_\mu^{a\ qu})^2 \quad , \tag{7.8}$$

with

$$D_\mu^{\ cl} A_\mu^{a\ qu} = \partial_\mu A_\mu^{a\ qu} + f^{abc} A_\mu^{b\ cl} A_\mu^{c\ qu} \quad . \tag{7.9}$$

The reason why this choice is so clever is that we fixed the gauge, but still kept invariance under the transformation

$$\begin{aligned} A_\mu^{a\ cl'} &= A_\mu^{a\ cl} + f^{abc}\Lambda^b(x) A_\mu^{c\ cl} - g^{-1}\partial_\mu \Lambda^a(x) \quad , \\ A_\mu^{a\ qu'} &= A_\mu^{a\ qu} + f^{abc}\Lambda^b(x) A_\mu^{c\ qu} \quad . \end{aligned} \tag{7.10}$$

Consequently the required counter terms will also exhibit this gauge invariance. Since we will only need their dependence on A_μ^{cl}, the only allowed counter terms are gauge invariant ones. This would normally not have been the case, since field renormalizations (which are unobservable) may well depend on the gauge fixing.

We put (7.8) and the associated ghost (also invariant under (7.10)) in our Lagrangian (7.7). In general one can choose things such that

$$W_{ij}^{\mu\nu} = -\delta^{\mu\nu}\delta_{ij} \quad . \tag{7.11}$$

Now M will in general contain a mass term, but we will not include it in the propagator. Rather, we keep M as a vertex, realizing that from a certain order on (at sufficiently high powers of M) we will have only convergent diagrams, whereas we are really only interested in divergent renormalization counter terms. In other words, if there is a mass m we expand the propagator,

$$1/(k^2+m^2-i\varepsilon) = 1/(k^2-i\varepsilon) - m^2/(k^2-i\varepsilon)^2 + m^4/(k^2-i\varepsilon)^3 - \ldots \tag{7.12}$$

treating the correction terms as vertex insertions.

Consider now the one loop diagram of Fig. 4.

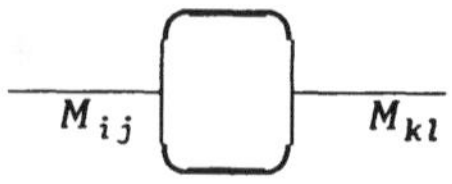

Fig. 4

The amplitude is

$$(\delta_{ik}\delta_{jl}+\delta_{il}\delta_{jk})\frac{1}{4(2\pi)^4 i}\int d^n k\,\frac{1}{(k^2-i\varepsilon)((k+p)^2-i\varepsilon)} \,. \tag{7.13}$$

If $n = 4-\varepsilon$ this is

$$(\delta_{ik}\delta_{jl}+\delta_{il}\delta_{jk})\frac{\Gamma(\varepsilon/2)}{4(4\pi)^{2+\varepsilon/2}}\int_0^1 dx[x(1-x)p^2]^{\varepsilon/2} \,, \tag{7.14}$$

which has a pole at $\varepsilon=0$:

$$(\delta_{ik}\delta_{jl}+\delta_{il}\delta_{jk})/(32\pi^2\varepsilon) + \text{finite} \,. \tag{7.15}$$

A counter term in $\Delta\mathcal{L}$ is needed:

$$\Delta\mathcal{L} = -(1/8\pi^2\varepsilon)\; M_{ij}M_{ji}/4 \,. \tag{7.16}$$

The complete $\Delta\mathcal{L}$ containing all possible combinations of M and N^μ can only have a relatively small number of terms. This is because simple power counting arguments tell us that they have to be polynomials of dimension 4: Now W is of dimension 0, N of dimension 1 and M of dimension 2. If W were non-trivial we could have an infinity of possible terms but now we must have

$$\Delta\mathcal{L} = (1/8\pi^2\varepsilon)\mathrm{Tr}(a_0M^2 + a_1(\partial_\mu N_\nu)^2 + a_2(\partial_\mu N_\mu)^2 + a_3MN^2 +$$

$$+ a_4N_\mu N_\nu\partial_\mu N_\nu + a_5(N^2)^2 + a_6(N_\mu N_\nu)^2 + a_7N^2(\partial_\mu N_\mu) \,. \tag{7.17}$$

Now it is easy to see that we can put some restrictions on M and N :

$$M_{ij}=M_{ji} \;;\quad N^\mu_{ij} = -N^\mu_{ji} \,, \tag{7.18}$$

by absorbing the symmetric part of $\partial_\mu N^\mu$ into M . However, one can do much more by observing a *new* "gauge" symmetry that is apparent in our system. Write

$$\mathcal{L} = -(\partial_\mu\varphi+N_\mu\varphi)^2/2 + \varphi X\varphi/2 , \tag{7.19}$$

with

$$X = M - N_\mu N^\mu . \tag{7.20}$$

We then have invariance under

$$\begin{aligned} \varphi'(x) &= \varphi(x) + \Lambda(x)\varphi(x) , \\ N_\mu' &= N_\mu - \partial_\mu\Lambda + [\Lambda,N_\mu] , \\ X' &= X + [\Lambda,X] . \end{aligned} \tag{7.21}$$

Here, Λ is infinitesimal.

Note that the dimension of this gauge group is as large as there are fields in the system, so it could be larger than the original gauge symmetry, and that the symmetry is exact also *after* a gauge fixing term is added (This was made possible by the fact that the background fields transform non-trivially; the gauge fixing term is allowed to depend on the background fields not in the same way as from the quantum fields).

A consequence is that $\Delta\mathcal{L}$ must have the same symmetry. The only combination of terms from (7.17) invariant under (7.21) is

$$\Delta\mathcal{L} = (1/8\pi^2\varepsilon)\mathrm{Tr}(aX^2+bY_{\mu\nu}Y_{\mu\nu}) , \tag{7.22}$$

where

$$Y_{\mu\nu} = \partial_\mu N_\nu - \partial_\nu N_\mu + N_\mu N_\nu - N_\nu N_\mu . \tag{7.23}$$

From (7.13) we find that

$$a = a_0 = -1/4 . \tag{7.24}$$

Similarly[17] one can find, by computing a few other diagrams,

$$b = (a_0-a_5)/2 = a_6/2 = -1/24 . \tag{7.25}$$

The excercise can be extended to include fermions[17]. The Lagrangian we then start off with is

$$\mathcal{L} = (7.7) - \bar{\psi}\gamma_\mu\partial_\mu\psi + \bar{\psi}F\psi + \bar{\psi}\alpha_i\varphi_i + \bar{\beta}_i\psi\varphi_i \ , \qquad (7.26)$$

plus higher orders in the quantum fields which are irrelevant for one loop calculations.

By making much use of our previous result, eq. (7.22), and taking the fermion minus sign into account, one finally obtains

$$\Delta\mathcal{L} = (1/8\pi^2\varepsilon)\mathrm{Tr}\Big(-X^2/4 - Y_{\mu\nu}Y_{\mu\nu}/24 + (1/4)\bar{\beta}\gamma_\mu(2\partial_\mu\alpha+F\gamma_\mu\alpha+2N_\mu\alpha)$$

$$+ H^2/8 + H_{\mu\nu}H_{\mu\nu}/48\Big) \ , \qquad (7.27)$$

where

$$H = \partial_\mu F\gamma_\mu - F\gamma_\mu F\gamma_\mu \ ;$$

$$H_{\mu\nu} = \partial_\mu F\gamma_\mu - \partial_\nu F\gamma_\mu + (-F\gamma_\mu F\gamma_\nu + F\gamma_\nu F\gamma_\mu)/2 \ . \qquad (7.28)$$

We see that this expression contains a few more simple numerical coefficients that had to be fixed by explicitly computing some diagrams.

The result obtained here, essentially algebraically, is not yet in the form we would like to see it, because the background fields are still hidden in the expressions X, Y, F, H, ... We now have to substitute V, S, P from (7.1) and the structure constants in here. The fields φ_i in (7.7) are in fact a combination of ϕ_i and $A_\mu^{a\ qu}$. This is a dull and lengthy process. But the result is worth while. One finally obtains[18]:

$$(8\pi^2\varepsilon)\Delta\mathcal{L} = \qquad (7.29)$$

$$G^a_{\mu\nu}G^b_{\mu\nu}[-(11/12)C_1^{ab} + (1/24)C_2^{ab} + (1/6)C_3^{ab}] - \Delta V - \bar{\psi}(\Delta S+i\gamma_5\Delta P)\psi \ ,$$

in which

$$C_1^{ab} = f^{apq}f^{bpq} \ , \qquad (7.30)$$

$$C_2^{ab} = -\mathrm{Tr}(T^aT^b) \ , \qquad (7.31)$$

$$C_3^{ab} = -\mathrm{Tr}(U^a_LU^b_L + U^a_RU^b_R) \ , \qquad (7.32)$$

$$\Delta V = (1/4)V_{ij}^2 + (3/2)V_i(T^2\phi)_i + (3/4)(\phi T^aT^b\phi)^2 \qquad (7.33)$$

$$+ \phi_iV_j\mathrm{Tr}(S_{,i}S_{,j}+P_{,i}P_{,j}) - \mathrm{Tr}(S^2+P^2)^2 + \mathrm{Tr}[S,T]^2 \ ,$$

$$V_i = \partial V/\partial\phi_i \ ; \ S_{,i} = \partial S/\partial\Phi_i \ , \text{ etc.} \tag{7.34}$$

And writing

$$W = S+iP \ ; \quad W^* = S-iP \ , \tag{7.35}$$

one gets

$$\Delta W = (1/4)W_{,i}W^*_{,i}W + (1/4)WW^*_{,i}W_{,i} + W_{,i}W^*W_{,i} +$$

$$+(3/2)U_R^2\, W + (3/2)WU_L^2 + W_{,i}\phi_j \mathrm{Tr}(S_{,i}S_{,j} + P_{,i}P_{,j}) \ . \tag{7.36}$$

One may wonder why there are no field renormalization terms such as $c(\partial_\mu\phi)^2$. The answer is that these can be rewritten as

$$-c\phi(\partial^2\phi) \tag{7.37}$$

which, via the equation of motion of the background fields, can be expressed in the other terms. It also can be noted that an infinitesimal renormalization of the field ϕ ,

$$\phi \to \quad \phi' = \phi + \delta\phi \tag{7.38}$$

adds to the Lagrangian terms proportional to the equation of motion:

$$\mathcal{L} \to \ \mathcal{L} + \delta\phi\ X \ , \tag{7.39}$$

where X vanishes because of the equation of motion. So field renormalizations, which by the way need not be gauge-invariant, are not seen if we use our background field method.

eqs (7.29) - (7.36) are our master formula for all one-loop counter terms. It is very instructive to play with a few examples to see what kind of renormalization group flow equations they produce.

8. Asymptotic freedom

Let us look at the one-loop results, eqs (7.29)-(7.36) more closely. How should we use them?

The term ΔV is easiest to understand. Take the case

$$V(\phi) = \lambda\phi^4/4! + \tfrac{1}{2}m^2\phi^2 \quad . \qquad (8.1)$$

There is only one field, so $i=j=1$;

$$V_{11} = \partial^2 V/\partial\phi^2 = \tfrac{1}{2}\lambda\phi^2 + m^2 \quad , \qquad (8.2)$$

and the first term in (7.33) is

$$\Delta V = \tfrac{1}{4}(\tfrac{1}{2}\lambda\phi^2 + m^2)^2 = (3/2)\lambda^2\phi^4/4! + \tfrac{1}{2}\lambda m^2\tfrac{1}{2}\phi^2 + \tfrac{1}{4}m^4 \quad , \qquad (8.3)$$

and including the factor $1/8\pi^2\varepsilon$ we get (note that $\Delta(m^2)=2m\Delta m$):

$$\Delta\lambda = (3/16\pi^2\varepsilon)\lambda^2 \ , \quad \Delta(m) = \lambda m/32\pi^2\varepsilon \quad ; \qquad (8.4)$$

compare eqs. (3.15) and (3.11). The constant $\tfrac{1}{4}m^4$ in (8.3) is immaterial.

If the field has N components, and $V=\lambda(\vec{\phi}^2)^2/8$, (a convenient choice of normalization) then

$$V_{ij} = \lambda(\phi_i\vec{\phi}^2/2)_{,j} = \lambda(\delta_{ij}\vec{\phi}^2/2 + \phi_i\phi_j) \quad ; \qquad (8.5)$$

$$\Delta V = \lambda^2(N/4 + 2)(\vec{\phi}^2)^2/4 \quad ; \qquad (8.6)$$

$$\Delta\lambda = (N+8)\lambda^2/16\pi^2\varepsilon \quad , \qquad (8.7)$$

which reproduces (8.4) for $N=1$ since we had a factor 3 difference in our definition of λ .

The next term in (7.33) contains T^2 . This represents the gauge coupling, i.e. which representation ϕ is of our gauge group. Suppose that in (7.3)

$$D_\mu\phi_a = \partial_\mu\phi_a + g\varepsilon^{abc}A_\mu^b\phi^c \quad , \qquad (8.8)$$

which would imply that ϕ is in the adjoint representation of the gauge group, and

$$T^a_{ij} = g\varepsilon^{iaj} \ . \tag{8.9}$$

Then

$$T^2_{ij} = -2g^2\delta_{ij} \ , \tag{8.10}$$

contributing to ΔV :

$$(3/2)(\lambda\vec{\phi}^2\phi_i/2)(-2g^2\phi_i) \to -12\lambda g^2\{(\vec{\phi}^2)^2/8\}/8\pi^2\varepsilon \ , \tag{8.11}$$

so that we get a *negative* counter term

$$\Delta\lambda = -12\lambda g^2/8\pi^2\varepsilon \ . \tag{8.12}$$

The third term in (7.33) gives again a positive contrubution:

$$(3/4)(\phi T^a T^b \phi)^2 = (3/4)g^4(\phi_a\phi_b - \vec{\phi}^2\delta_{ab})^2 \to 12g^4\{(\vec{\phi}^2)^2/8\}/8\pi^2\varepsilon \ . \tag{8.13}$$

It is easy to see now that if we insist on keeping the Hamilton density positive in a purely scalar theory most β functions will allways be positive (at least one of them), because of the definite sign of the first term in (7.29). But we also see that this is no longer obviously the case if gauge interactions are included. In practice however one can prove from all equations (7.29)-(7.36) (with much pain) that if we have scalars and only vectors (no spinors) then the middle term of (7.29) alone is not able to render the system asymptotically free (all β negative). There are too many terms with the wrong sign.

But now consider the *pure* gauge theory. We then only have:

$$\Delta\mathcal{L} = -(11/12)C_1^{ab}G^a_{\mu\nu}G^b_{\mu\nu}/8\pi^2\varepsilon \ , \tag{8.14}$$

and surely the Casimir operator C_1^{ab} is positive (or zero in the Abelian case). Mostly we diagonalize it:

$$C_1^{ab} = C_1(a)\delta^{ab} \ , \tag{8.15}$$

with $C_1 \geq 0$. Let us simply write $C_1(a) = C_1$. Since the coupling constant g was absorbed into the structure constants f we have that C_1 is proportional to g^2 . For an SU(2) gauge theory it is

$$C_1 = 2g^2 \ . \tag{8.16}$$

(8.14) renormalizes the fields A_μ . (One could have chosen to write (7.29) differently so as to avoid field renormalizations, but this notation came out to be the easiest.) Let us rewrite (absorbing temporarily the $1/8\pi^2\varepsilon$ into the coefficient C_1)

$$\mathcal{L} + \Delta\mathcal{L} = \mathcal{L}(A') \; ; \tag{8.17}$$

$$A' = [1+(11/6)C_1]A \; ; \tag{8.18}$$

but then

$$G^{a}_{\mu\nu}{}' = \partial_\mu A^{a}_\nu{}' - \partial_\nu A^{a}_\mu{}' + [1-(11/6)C_1]f^{abc}A^{b}_\mu{}'A^{c}_\nu{}' \; , \tag{8.19}$$

so that

$$\Delta g = -(11/6)C_1 g/8\pi^2\varepsilon \; , \tag{8.20}$$

which is $-(11/3)g^3/8\pi^2\varepsilon$ in SU(2).

Thus, the beta function for g is negative. Note that both fermions and bosons will tend to turn β in the other direction:

$$\Delta g = (-(11/6)C_1 + (1/12)C_2 + (1/3)C_3)g/8\pi^2\varepsilon \; . \tag{8.21}$$

Fields in the adjoint representation have

$$C_{2,3} = C_1 \; , \tag{8.22}$$

but in the elementary representation of SU(N) we have

$$C_{2,3} = C_1/2N \; , \tag{8.23}$$

(note that in (8.22) and (8.23) we consider chiral fermions. Massive, complex fermions have twice these values; furthermore, *scalar* fields were taken to be real. If they are complex we also need a factor two). If we have 11 massive fermions in the fundamental representation (for SU(2)) or $16\frac{1}{2}$ (for SU(3)) then β will switch sign.

Thus, QED, which has $C_1=0$, can never be asymptotically free. A purely scalar theory also can't, but also a scalar-fermion theory with Yukawa couplings cannot be asymptotically free. This is because in (7.32) onlu the terms with U^2W have a negative sign.

In gauge theories with fermions and scalars the situation is complicated. If one imposes the conditions that all beta functions come out negative one usually will find that all coupling constants should keep fixed ratios with each other under scale transformations:

$$\lambda_i = \lambda_0\tilde{\lambda}_i \ ; \ \beta_i\{\lambda\} = \lambda_i\beta_0(\lambda_0) \ . \qquad (8.24)$$

because only then they won't run away in different directions.

It is an interesting game to try to construct such models. There are plenty of them. The ratios λ_i often turn out to be rather ugly algebraical numbers except when one invokes symmetries such as supersymmetry. Supersymmetric Yang-Mills is asymptotically free ($N \le 2$) or scale-invariant ($N=4$) . In practice one finds that the fermions must couple sufficiently strongly to the gauge particles which happens typically if they are in the adjoint representation or higher. So in QCD it will be hard to add scalars while keeping asymptotic freedom because the quarks are all in the fundamental representation. Perhaps this is the reason why fundamental scalars that couple to the strong gauge group cannot exist.

Constructing *weakly* interacting systems while insisting on (8.24) , though an interesting game, is probably physically irrelevant because then the coupling constants run so slowly that the Landau ghost is extremely far away in the ultraviolet, where, certainly if it is beyond the point where gravitational forces are expected, none of our theories can be believed anyway.

Similarly to our equations (7.29)-(7.36) one can devise expressions that generate the two-loop contributions to the beta functions. But even in this compact algebraic notation the result is very lengthy[16].

9. Perturbation expansion at high orders and the renormalization group

The renormalization group gives us a good idea about the force law between quarks in QCD. The strength of the coupling constant increases with distance. So we get for the inter-quark force:

$$F(r) = \frac{g(r)^2}{r^2} , \qquad (9.1)$$

with

$$\partial g(r)/\partial r = -\beta(g) \ . \qquad (9.2)$$

and since $\beta<0$ we have a force law in which the strength of the force decreases not as quickly with the distance as $1/r^2$. At small distance scales quarks interact quite weakly, and indeed this fact, which was strongly indicated by experimental observations in the late 60's, has been a major reason to believe that a non-Abelian gauge theory must be responsible for the strong interactions.

But also for theoretical reasons an asymptotically free gauge theory (with fermions and possibly also scalars) is the only likely theory for any strongly interacting set of particles. One would believe that at least at very tiny distance scales the theory should be precisely defined. According to our own philosophy this is only the case if the coupling constant tends to zero. Now in any weakly interacting theory the coupling constants are small enough to make the perturbation expansion useful even if absolute formal convergence is lacking. So they need not become any smaller at small distance scales. But for strongly interacting theories this *is* necessary.

But even in asymptotically free theories the coupling constant tends to zero very slowly. Is this convergence fast enough to make the theory well-defined?

The answer to this question is not known. As should be evident from J. Smit's lectures at this graduate school it is often taken for granted that if, in a lattice gauge theory, the bare coupling constant is postulated to tend to zero according to

$$g^2(a) = 1/[A \log(1/a) + B \log\log(1/a)] , \tag{9.3}$$

as $a \to 0$, where A and B are determined by the renormalization group, then a sensible limit is obtained, independent of the details of the lattice.

There is no proof of this, apart from numerical evidence, which however at best can give only indications of how the system will tend to behave. Our problem is that, even at small distances, all features of our system are defined as perturbation series in g^2 , which diverge as $C^n g^{2n} n!$, as far as we know. The error, behaving as $\exp(-1/Cg^2)$ decreases rapidly as the scale becomes smaller. Does this uncertainty decrease fast enough to become irrelevant in the limit? Or could the error amplify as one integrates the equations back to large distances?

The question turns out to be not easy, particularly if we are dealing with models that are not exactly solvable.

One way to attack this problem is to search the source of the offending factors $n!$ in perturbation expansions. When one sums over large sets of Feynman diagrams one discovers that the *number* of diagrams with n loops increases like $n!$. Only in special cases they donot. One case is the so-called spherical model. This is $(\vec{\phi}^2)^2$ theory with N components, where we let N tend to infinity, such that $\lambda N = \tilde{\lambda}$ is kept fixed. This model is exactly solvable. The β function can be given in closed form, and indeed the limiting theory is well-defined. It is asymptotically free if $\tilde{\lambda}<0$, but in this case the negative sign of $\tilde{\lambda}$ does no harm. After all, λ itself tends to zero. The amount of action needed for a certain field configuration to tunnel into a mode that has negative energy is found to increase indefinitely with N , so that such a tunnelling is inhibited in the limit.

A less trivial model is *planar* field theory. Here, we have fields with *two* indices i and j which both can take N values. A typical example is SU(N) gauge theory. If one now allows N to tend to infinity, again keeping λN or g^2N fixed, one recovers all *planar* Feynman diagrams. There are infinitely many of these, and nobody knows how to sum these analytically. But their number now only grows geometrically with N (that is, without the $N!$).

But there is another source for factors $n!$ This can be understood in detail using the renormalization group, but here we give only an heuristic argument.

Consider a self-energy insertion diagram with one loop. It shows the following typical k dependence:

$$\Gamma(k^2) \cong \lambda k^2 \log k^2 \ , \tag{9.4}$$

as we have seen, and we know that the log is a crucial feature following from the renormalization group.

Now consider n of these diagrams connected in a series to the propagator. This diagram then behaves as

$$P(k^2) = (k^2-i\varepsilon)^{-1}\, \lambda^n (\log k^2)^n \ . \tag{9.5}$$

Suppose that, in turn, this diagram occurs inside a loop diagram, which we have to integrate:

$$\lambda^n \int d^4k\ F(k)(k^2-i\varepsilon)^{-1}(\log k^2)^n\ , \qquad (9.6)$$

and assume that the integral converges by power counting. If it didn't we make it converge by using subtractions.

Thus, at large k , the function $F(k)$ tends to k^{-4} . Assuming no infrared divergence the integral behaves as

$$\lambda^n \int_1^\infty d|k|(\log|k|)^n |k|^{-3}\ . \qquad (9.7)$$

Substituting $k^2 \to e^s$, we get

$$\int_0^\infty ds\ s^n e^{-s} \cong n!\ . \qquad (9.8)$$

This is how $n!$ appears that can ruin perturbation expansion.

It should be clear that the logarithmic behavior in eqs. (9.4)-(9.6) can be understood by studying scaling via the renormalization group. One then finds that the coefficients in front of the $n!$ can be obtained directly from the β function coefficients.

References

1. E.C.G. Stueckelberg and A. Peterman, Helv. Phys. Acta 26 (1953) 499; N.N. Bogoliubov and D.V. Shirkov, Introduction to the theory of quantized fields (Interscience, New York, 1959)
2. see for instance: M.N. Barber, Phys. Repts 29C (1977) 1
3. N.G. van Kampen, private communication
4. G. 't Hooft, Nucl. Phys. B33 (1971) 173; *ibid.* B35 (1971) 167;
5 M. Veltman, Physica 29 (1963) 186
6 G. 't Hooft and M. Veltman, DIAGRAMMAR, CERN report 73/9 (1973), repr. in "Particle interactions at Very High Energies", NATO Adv. Study Inst. Series, Sect B, Vol. 4b, p. 177.
7 G. 't Hooft and M. Veltman, Nucl. Phys. B50 (1972) 318
8 W. Pauli and F. Villars, Rev. Mod. Phys. 21 (1949) 434
9 G. 't Hooft and M. Veltman, Nucl. Phys. B44 (1972) 189; C.G. Bollini and J.J. Giambiagi, Phys. Letters 40B (1972) 566; J.F. Ashmore, Nouvo Cimento Letters 4 (1972) 289
10 G.L. Eyink, 1987 Ph.D. Thesis, Bruxelles
11 C.G. Callan, Phys. Rev. D2 (1970) 1541
12 G. 't Hooft, Nucl. Phys. B61 (1973) 455
13 K. Symanzik, Comm. Math. Phys. 18 (1970) 227; *ibid.* 23 (1971) 49; 45 (1975) 79
14 M. Gell-Mann and F. Low, Phys. Rev. 95 (1954) 1300
15 For instance, G. 't Hooft, Comm. Math. Phys. 86 (1982) 449; *ibid.* 88 (1983) 1
16 R. van Damme, Nucl. Phys. B227 (1983) 317
17 G. 't Hooft, Nucl. Phys. B62 (1973) 444
18 See ref. 16 and: G.'t Hooft, Nucl. Phys. B254 (1985) 11

CHAPTER 4
EXTENDED OBJECTS

CHAPTER 4
EXTENDED OBJECTS

Introduction to Extended Objects in Gauge Field Theories [4.1]

Up till now we maintained that quantum field theories (in four dimensions) can *only* be understood perturbatively. What we really meant however was that it is the infinite series of successive quantum corrections that must be addressed term by term. Even though the renormalization group allows us to partially resum this expansion, the manipulations are still permissible only when the coupling constant(s) is (are) sufficiently small.

Extended objects form another topic that suggests the possibility to transcend beyond the perturbation expansion. We get particles with masses inversely proportional to the coupling constants, and we get "instantons" that will produce tunneling effects with amplitudes depending exponentially on the inverse coupling constant. Indeed, phenomena described in this chapter may perhaps be called "nonperturbative", but we always have to remember that what we compute will require an infinite series of multiloop corrections, as usual. So, again, we must always assume the coupling constants to be sufficiently small.

The first paper in this chapter is an introduction to kinks, vortices, monopoles and instantons. This is a logical order. In its Section 4 it is explained how in certain gauge theories 2-dimensional objects (vortices) become unstable and how this leads to the existence of 3-dimensional things (monopoles). This is also the way I discovered them. Alexander Polyakov had followed a slightly different route. He had considered "hedgehog" configurations of scalar fields and asked whether one could construct particles this way. Following the arguments that I explain in Section 2, he found that gauge fields had to be added. That these things actually carry magnetic charge had been suggested to him by Lev Okun.

Note that this paper was written before the third generation of quarks and leptons had been firmly established. In a conservative "Standard Model" at that time therefore instantons gave rise to processes with $\Delta B = \Delta L = 2$ rather than 3.

Introduction to Magnetic Monopoles in Unified Gauge Theories [4.2]

The next paper is the one in which the discovery of magnetic monopoles as valid solutions in gauge theories was published. It gave rise to a renewed interest in monopoles by theoreticians, experimentalists and cosmologists alike. Experimental detection has never been confirmed, but, as I had pointed out in the Palermo Conference, Chapter 2.3, monopoles play a significant role in theory anyway. Most crucial is their part in the quark confinement mechanism, see in particular Chapter 7.

One later observation should be added to this paper. The coefficient $C(\beta)$ was computed by numerical variation techniques, and found to be close to one when β is small. It was an important result of Prasad and Sommerfield[15] that the equations simplify essentially when $\beta = 0$. The second order field equations then become the square of a first order equation (duality). This equation can be solved exactly and one then finds that $C(0) = 1$.

[15] M. K. Prasad and C. M. Sommerfield, *Phys. Rev. Lett.* **35** (1975) 760.

CHAPTER 4.1

EXTENDED OBJECTS IN GAUGE FIELD THEORIES

G. 't Hooft
University of Utrecht
The Netherlands

1. INTRODUCTION AND PROTOTYPES IN 1+1 DIMENSIONS

With the exception of the electron, the neutrino, and some other elementary particles, all objects in our physical world are known to be extended over some region in space. Thus, the subject "extended objects" is too large to be covered in just four lectures. However, most of these extended objects may be considered as bound states of more elementary constituents. If we exclude all those, then a very interesting small set of peculiar objects remains: objects that cannot be considered as just a bunch of particles, but as some smeared, but also more or less localized, configuration of fields. These fields must obey very special types of field equations to allow such smeared lumps of energy ("solitons") to be stable.

These may be two alternative reasons for a lump of field-energy to be stable against decay:

i) A conservation law might tell us that the total mass-energy of the decay products would be larger than our object itself. It is easy to prove that this cannot happen in a <u>linear</u> (i.e. non interacting) theory: If electric charge is given by

$$Q = ie \int [\phi^* \partial_0 \phi - (\partial_0 \phi^*)\phi] \, d^3x \, , \tag{1.1}$$

and the energy by

$$E = \int (\partial_0 \phi^* \partial_0 \phi + \vec{\partial}\phi^* \vec{\partial}\phi + m^2 \phi^* \phi) \, d^3x \quad , \tag{1.2}$$

Particles and Fields, Edited by D. H. Boal and A. N. Kamal (Plenum 1978).

then from

$$(\partial_o\phi^* \pm im\phi^*)(\partial_o\phi \mp im\phi) \geq 0 \ , \tag{1.3}$$

it follows that

$$E \geq \int [\partial_o\phi^*\partial_o\phi + m^2\phi^*\phi)\, d^3x \ \geq \ \frac{m|Q|}{e} \ , \tag{1.4}$$

whereas a bunch of charged particles at rest, with a total charge equal to Q, could have a total mass-energy not exceeding

$$m|\frac{Q}{e}| \ .$$

Consequently, a field configuration with $\vec{\partial}\phi \neq 0$ would not be stable by charge conservation alone. In interacting field theories however, one may construct many interesting stable lumps of fields simply because the above arguments do not apply when there are interactions. These have been considered by T.D. Lee and coworkers[1] and will not be treated in this series of talks.

ii) Topological stability. In these lectures we will only consider topologically stable structures. The statement that our objects are really field configurations implies that we consider field equations without bothering about the fact that small oscillations about the solutions ought to be quantized.

To explain topological stability we first turn our attention to models in one space - one time dimension[2]. Consider a simple scalar field ϕ satisfying

$$\partial_t^2\phi - \phi_x^2\phi + \frac{\partial}{\partial\phi} V(\phi) = 0 \ , \tag{1.5}$$

to which corresponds a Lagrangian:

$$\mathcal{L} = \frac{1}{2}(\partial_t\phi)^2 - \frac{1}{2}(\partial_x\phi)^2 - V(\phi) \ . \tag{1.6}$$

Eq. (1.5) corresponds to

$$\int \mathcal{L}[\phi+\eta]dxdt = \int \mathcal{L}[\phi]dxdt + O(\eta^2) \ . \tag{1.7}$$

The two cases we consider are

a) $$V(\phi) = \frac{\lambda}{4!}(\phi^2 - F^2)^2 \quad \text{(see Fig: 1)} \ . \tag{1.8}$$

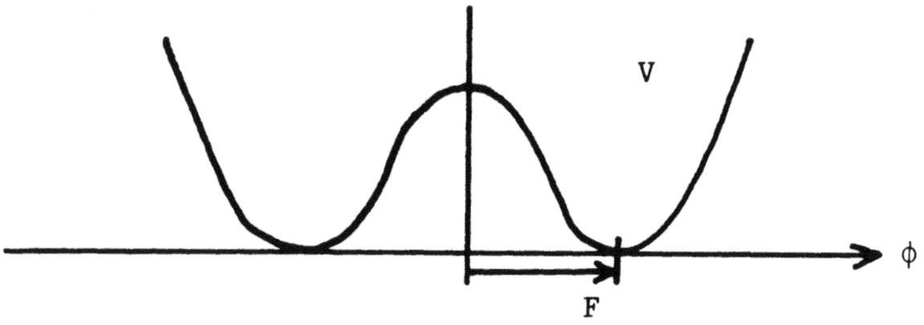

Fig. 1. Potential of the form $\lambda(\phi^2 - F^2)^2/4!$

It is customary then to write $\phi = F + \phi'$,

$$V = \frac{1}{2} m^2\phi'^2 + \frac{g}{3!} \phi'^3 + \frac{\lambda}{4!} \phi'^4 , \tag{1.9}$$

with

$$m^2 = \frac{1}{3} \lambda F^2 ; \qquad g = \lambda F . \tag{1.10}$$

b) $V(\phi) = A(1 - \cos \frac{2\pi\phi}{F})$ ("Sine-Gordon" model- see Fig.2). (1.11)

Here we write

$$A = m^4/\lambda , \qquad F = \frac{2\pi m}{\sqrt{\lambda}} , \tag{1.12}$$

so that

$$V = \frac{1}{2} m^2\phi^2 - \frac{\lambda}{4!} \phi^4 + \ldots \tag{1.13}$$

In both cases m stands for the mass of the physical particle, and λ is the usual coupling constant-expansion parameter.

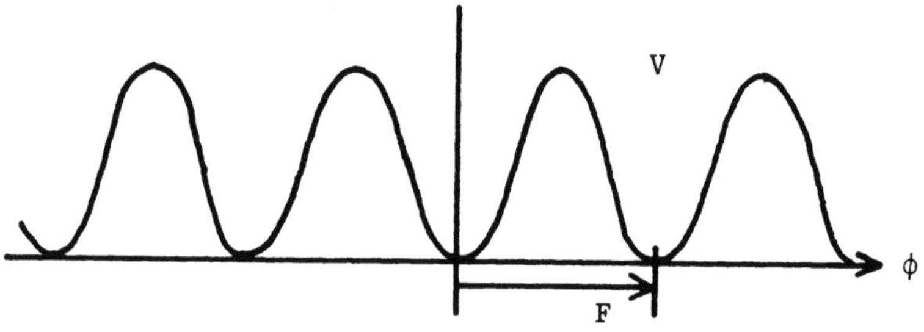

Fig. 2. Potential in "Sine-Gordon" model.

The energy of a field configuration is

$$H = \int \mathcal{H} d^3x \ ,$$
$$\mathcal{H} = \frac{1}{2}(\partial_t\phi)^2 + \frac{1}{2}(\partial_x\phi)^2 + V(\phi) \ . \tag{1.14}$$

The examples I gave here have the property that there is more than one way to make a zero-energy state[3] ("vacuum"):

$$\text{a)} \quad \phi = nF \ , \quad n = \pm 1 \ ,$$
$$\text{b)} \quad \phi = nF \ , \quad n = \ldots, -2, -1, 0, 1, 2, \ldots \tag{1.15}$$

The world in these models is assumed to be a one-dimensional line. We envisage the situation that at one side of the line ϕ has one value for n, at the other side a different value (see Fig. 3). We can have a transition region close to the origin which carries energy:

$$\phi \neq nF \ , \ n \text{ integer} \ ; \quad \partial_x\phi \neq 0 \ . \tag{1.16}$$

Assume that the situation is stationary:

$$\partial_t\phi = 0 \ . \tag{1.17}$$

We wish a finite total energy; $\mathcal{H}$ must approach zero at $x \to \pm\infty$. So ϕ must approach one if its possible vacuum values at $x \to \pm\infty$. Thus we find a discrete set of allowable boundary conditions.

In a stationary solution of the equation (1.6) also the energy is an extremum. The lowest-energy configuration under the given boundary conditions also satisfies this equation. It is obviously stable. In our examples the solutions are easily given:

$$\partial_x^2\phi = \partial V/\partial\phi \ , \tag{1.18}$$

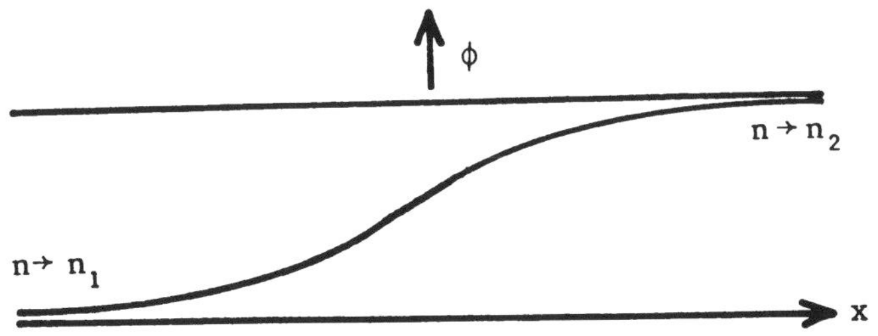

Fig. 3. A static form for ϕ with different boundary conditions at $x \to \pm\infty$.

$$\frac{d}{dx}\left[\frac{1}{2}(\partial_x\phi)^2 - V(\phi)\right] = 0 \ , \tag{1.19}$$

$$(\partial_x\phi)^2 = V(\phi) = 0 \quad \text{at} \quad x = \pm\infty, \tag{1.20}$$

therefore

$$\frac{dx}{d\phi} = [2V(\phi)]^{-\frac{1}{2}} \ . \tag{1.21}$$

Case a):

$$\phi(x) = F \tanh\frac{1}{2}mx \ , \tag{1.22}$$

Case b):

$$\phi(x) = \frac{2F}{\pi} \operatorname{arctg} e^{mx} \ . \tag{1.23}$$

Note that, as $x \to \infty$,

$$F \tanh\frac{1}{2}mx \to F(1 - 2e^{-mx}) \ , \tag{1.24}$$

$$\frac{2F}{\pi} \operatorname{arctg} e^{mx} \to F(1 - \frac{2}{\pi} e^{-mx}) \ , \tag{1.25}$$

thus the vacuum value is reached exponentially, with the mass of the physical particle in the exponent (see Fig. 3). The total energy of these objects ("solitons") is

$$E = \int\left[\frac{1}{2}(\partial_x\phi)^2 + V(\phi)\right]dx = 2\int V(\phi)dx \ , \tag{1.26}$$

Case a):

$$E = 2m^3/\lambda \ ,$$

Case b):

$$E = 8m^3/\lambda \ .$$

They behave as particles in every sense; for instance one can show that the relations between energy, momentum and velocity are as indicated by relativity theory.

We distinguish two types of stability requirements:

1) Topological stability. Since the boundary conditions in our case form a discrete set there can be no continuous transition

towards the vacuum boundary condition (which is the one for a bunch of ordinary elementary particles).

2) Stability against scaling. Let us return for a moment to n spacelike dimensions (instead of one). In the stationary case we always have

$$H = S + V \ , \quad S = \int \frac{1}{2}(\partial_x \phi)^2 d^n x \ , \quad V = \int V(\phi) d^n x \ . \tag{1.27}$$

The energy must be stationary under any infinitesimal variation. Let us try one special variation:

$$\phi(x) \to \phi(x/\Lambda) \ , \tag{1.28}$$

then

$$S \to \Lambda^{n-2} S \ ; \quad V \to \Lambda^n V \ . \tag{1.29}$$

We must have

$$\frac{\Lambda\partial}{\partial\Lambda}(S + V) = (n-2)\, S + nV = 0 \ . \tag{1.30}$$

But S and V are both positive. That is only compatible with (1.30) in the case $n = 1$. Since the decomposition (1.27) is possible in all scalar field theories, scalar theories have only stationary solitons in one space-dimension. In two or more dimensions it would always be energetically favorable for a system to shrink until it becomes pointlike. That is not an extended object by definition.

Finally, we emphasize that description of particles in terms of quantized fields in most cases is only possible if the coupling is not too strong:

$$\lambda/m^2 \ll 1 \tag{1.31}$$

(in 1+1 dimensions λ has the dimension of a mass-squared). Therefore, the mass of solitons, being proportional to m^3/λ, is always much greater than that of the original particles in the theory.

2. SOLITONS IN 2 SPACE DIMENSIONS AND STRINGS IN 3 SPACE DIMENSIONS

There is another reason why scalar theories have no topological solitons in 2 or more space dimensions. We can only have topological stability if the boundary condition differs from: $\phi \to$ constant at $|x| \to \infty$. Imagine that we have several field components $\vec{\phi}$ and assume that $V(\vec{\phi})$ has a whole continuum of minima. For instance we could take n field components (n= number of space-like dimensions) and have $V(\vec{\phi})$ = minimal if $|\vec{\phi}|$ = F. We could think of mapping

$$\vec{\phi}(\vec{x}) \to F \frac{\vec{x}}{|\vec{x}|} \quad \text{at} \quad |\vec{x}| \to \infty , \tag{2.1}$$

as a boundary condition. This would imply topological stability. But then

$$|\vec{\partial}\vec{\phi}| \to F/|\vec{x}| \quad , \tag{2.2}$$

and

$$2S = \int |\vec{\partial}\vec{\phi}|^2 \, d^n x \to \int F^2 d^n x/x^2 \;, \tag{2.3}$$

diverges at large x unless n = 1. This infrared divergence is closely related to the existence of massless Goldstone bosons in such a theory.

The only way known at present to re-obtain a topologically stable soliton in more than one spacelike dimension is to add a gauge field:

$$\partial_i \vec{\phi} \to D_i \vec{\phi} = (\partial_i + gA_i^a T^a)\vec{\phi} \,. \tag{2.4}$$

The space-components of the vector field A can be arranged such that $(D\vec{\phi})^2 \to 0$ more rapidly than $1/x^2$, and this way we can approach a physical vacuum so rapidly that the total energy converges. (Note also that by adding a gauge field the massless Goldstone boson disappears, so the theory has better infrared convergence. There may still be a massless vector boson but that is usually less harmful, as we will see.)

The simplest theory with a soliton in 2 space dimensions is scalar quantum-electrodynamics in the Higgs mode[4,5]:

$$\mathcal{L} = -\frac{1}{4} F_{\mu\nu}F_{\mu\nu} - (\partial_\mu - iqA_\mu)\psi^*(\partial_\mu + iqA_\mu)\psi - \frac{\lambda}{2}(\psi^*\psi - F^2)^2 . \tag{2.5}$$

Here q is the unit of charge. The 3+1 dimensional analogue of this model is well known in physics: it describes the superconductor. ψ is the simplified field ("order parameter") that essentially describes two-electron bound states with total spin zero (hence it is a scalar boson field). A_μ is the ordinary vector potential. $q = 2e$.

Inside the superconductor the photon behaves as a massive particle and electromagnetic fields are of short range. Long-range magnetic fields are forbidden because in order to create them we need potential differences (according to Maxwell's laws) and those do not occur inside a superconductor. But when placed in a strong magnetic field a superconductor will find an excited state in which it can allow magnetic flux to penetrate. This flux will go through the superconductor in narrow tubes. The flux in such a tube turns out to be quantized. If we consider a plane orthogonal to these tubes then we get equations for extended particle-like objects in this plane. They are the 2 dimensional solitons to be considered now. The flux tubes ("vortices", "strings"), in the three dimensional world are derived from them by adding a third coordinate (parallel to the axis) on which the fields do not depend.

The vacuum is described by

$$|\psi| = F . \tag{2.6}$$

We construct our soliton just as in the previous section; we choose to approach a vacuum rapidly at $|\vec{x}| \to \infty$, but in such a way that continuity requires a transition region at $x \to 0$. Write

$$x_1 = r \cos\theta ,$$

$$x_2 = r \sin\theta ,$$

then we choose as $r \to \infty$, $\psi \to Fe^{i\theta}$.

Continuity requires that ψ produces a zero somewhere near the origin of x-space. Since the vacuum value of $|\psi|$ is F, it costs energy to produce such a zero: we will obtain a topologically stable lump of energy near the origin of x-space. Let us look at the lowest-energy configuration. Assume it is time-independent and $A_0 = 0$. As a gauge condition we choose:

$$\partial_i A_i = 0 , \tag{2.7}$$

and, by symmetry arguments:

$$\psi = \chi(r)e^{i\theta} \quad , \qquad A_i = \varepsilon_{ij}x_j A(r) \quad . \tag{2.8}$$

One can easily check that the above is a self-consistent ansatz for a solution of the field equations.

To obtain the equation for χ and A it is easiest to reexpress the Lagrangian in terms of χ and A:

$$\int \mathcal{L}\, d^2\vec{x} = \int_0^\infty 2\pi r dr\left(-\frac{1}{2}r^2\left(\frac{dA}{dr}\right)^2 - \frac{1}{2}\left(\frac{d\chi}{dr}\right)^2 - \left(\frac{1}{r}+qAr\right)^2\chi^2 - \frac{\lambda}{2}(\chi^2-F^2)^2\right) - \pi r^2 A^2\Big|_0^\infty \tag{2.9}$$

Assuming $|A_i(\infty)| \to 0$ we may ignore the boundary term. As for boundary conditions at $r \to 0$ we must require

$$\begin{cases} \chi \to 0 & , \\ A \text{ stays finite} & . \end{cases} \tag{2.10}$$

At $r \to \infty$ the third and fourth terms in (2.9) converge only if

$$\begin{cases} \chi \to F & , \\ A \to -1/qr^2 & . \end{cases} \tag{2.11}$$

Since our system is assumed to be stationary ($\partial/\partial t = 0$), and the time-components of A_μ vanish, the energy will be just minus the Lagrangian.

$$E = \int_0^\infty \mathcal{E}(r)dr \quad , \tag{2.12}$$

$$\mathcal{E}(r) = 2\pi r\left(\frac{1}{2}r^2\left(\frac{dA}{dr}\right)^2 + \frac{1}{2}\left(\frac{d\chi}{dr}\right)^2 + \left(\frac{1}{r} + qAr\right)^2\chi^2 + \frac{1}{2}\lambda(\chi^2 - F^2)^2\right) \quad . \tag{2.13}$$

We must find the minimum of this energy under the above boundary conditions. The coupled differential equations,

$$\frac{d}{dr}\left(\frac{\partial \mathcal{E}}{\partial (dA/dr)}\right) = \frac{\partial \mathcal{E}}{\partial A} \quad ,$$
$$\frac{d}{dr}\left(\frac{\partial \mathcal{E}}{\partial (d\chi/dr)}\right) = \frac{\partial \mathcal{E}}{\partial \chi} \quad , \tag{2.14}$$

are not exactly soluble. The solution is sketched in Fig. 4. As in the 1+1 dimensional case, we expect that the fields at large distance from the origin deviate only exponentially from the vacuum values, with the masses of the two massive particles (vector boson and Higgs particle) in the exponents. Thus we obtain a well behaved, topologically stable soliton, particle-like in 2+1 dimensions, string- or vortex-like in 3+1 dimensions.

What will the mass of this particle be? Without solving the equations explicitly, we can make some scaling arguments. Put

$$F^2 = m_H^2/2\lambda = m_W^2/2q^2 \ , \tag{2.15}$$

where m_H and m_W are, respectively the Higgs and the vector boson masses, and

$$r = \bar{r}/m_W \ , \ \chi = \bar{\chi}\, m_W/q \ , \quad A = \bar{A}\, m_W^2/q \ , \quad \lambda/q^2 = \beta \ , \tag{2.16}$$

then

$$E = \frac{2\pi m_W^2}{q^2}\int \bar{r}d\bar{r}\left(\frac{1}{2}\bar{r}^2(\frac{d}{d\bar{r}}\bar{A})^2 + \frac{1}{2}(\frac{d}{d\bar{r}}\bar{\chi})^2 + (\frac{1}{\bar{r}}+\bar{A}\bar{r})^2\bar{\chi}^2 + \frac{1}{2}\beta(\bar{\chi}^2-\frac{1}{2})^2\right) \tag{2.17}$$

$$= \frac{2\pi m_W^2}{q^2}\ C_1(\beta)\ .$$

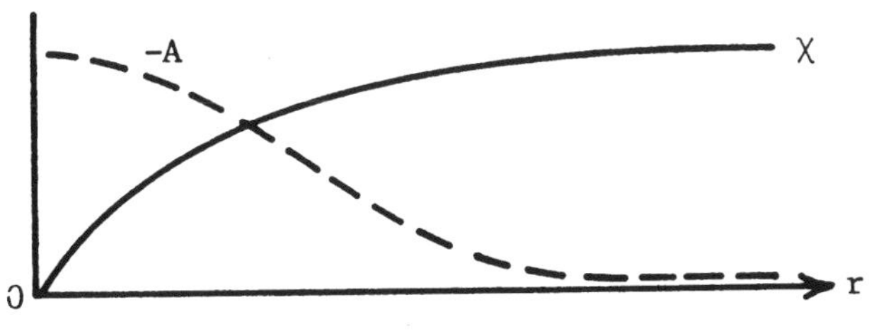

Fig. 4. Sketch of the solutions to Eq. (2.14).

Note the coupling constant downstairs. Of course, in 2+1 dimensions, q^2 has the dimension of a mass, hence m_W^2, not m_W. In 3+1 dimensions, the energy of a string is proportional to its length, and (2.17) gives the energy per unit of length, which is in units of mass-squared.

In the beginning of this section we anticipated that this vortex would be a magnetic flux tube. Let us now verify that by computing the magnetic field.

At spatial infinity the vector potential is found to be, from (2.11),

$$A_i \to -\varepsilon_{ij}x_j/qr^2 \quad . \tag{2.18}$$

Therefore

$$\oint A_i dx^i \to 2\pi/q \ . \tag{2.19}$$

So, indeed, we have a magnetic field going along the vortex with total flux $2\pi/q$, this in spite of the fact that the photon had become massive which would imply that electromagnetic fields are only short-range. There is no way of producing a broken fraction of that flux, therefore magnetic flux in a superconductor is quantized, by units

$$2\pi/q = \pi/e \ . \tag{2.20}$$

There is clearly also a physical reason for our vortex or string to be stable: the magnetic flux it contains cannot spread out in the superconductor and it is exactly conserved.

3. THE BOUNDARY CONDITION - HOMOTOPY CLASSES

The above was a pedestrian way to obtain the vortex in a superconductor. We will now reformulate the boundary condition more precisely in order to be able to produce more complicated objects later.

Remember we chose our boundary condition to be $\psi \to e^{i\theta_F}$ at $r \to \infty$. We could have chosen

$$\psi \to e^{in\theta_F} \quad ;$$

every choice of n would yield a particular configuration with lowest energy, dependent on n. If there would have been any way to make a continuous transition from one n to another, without violating $|\psi| = F$ at large distance from the origin, then that would cause an instability for whatever state had the higher energy. But there is no way to make this continuous transition. We say that the different choices of n are homotopically different[2]. The boundary has the topological shape of a circle. (In N dimensions it is the S_{N-1} sphere.) The vectors

$$\begin{pmatrix}\psi_1\\ \psi_2\end{pmatrix} \quad \text{with} \quad |\psi| = F$$

also form a circle. Two continuous mappings of a circle onto a circle are called "homotopic" if one can continuously be transformed into the other. All classes of mappings that are homotopic to each other form a "homotopy class". For the mappings of a circle onto a circle, or any S_{N-1} sphere onto an S_{N-1} sphere, these homotopy classes are labeled by an integer n running from $-\infty$ to ∞.

In order to formulate the boundary condition in the general case we introduce two types of vacuum: a <u>supervacuum</u> is a region of space or space-time where all vector fields vanish and all scalar fields have their vacuum value: $|\vec{\phi}| = F$ and are pointing in a preassigned direction in isospace, for instance the z-direction:

$$\begin{pmatrix}\psi_1\\ \psi_2\end{pmatrix} = \begin{pmatrix}F\\ 0\end{pmatrix} \quad \text{or} \quad \begin{pmatrix}\phi_1\\ \phi_2\\ \phi_3\end{pmatrix} = \begin{pmatrix}0\\ 0\\ F\end{pmatrix}, \text{ etc.}$$

Now in a gauge theory, like electromagnetism, we are allowed to make gauge rotations; for instance in the model of the previous section:

$$\begin{cases} A_\mu \to A_\mu - \frac{1}{q}\partial_\mu \Lambda(x) \,, \\ \psi \to e^{i\Lambda(x)}\psi \,. \end{cases} \tag{3.1}$$

After this, in general space-time-dependent, gauge rotation we obtain non-vanishing vector-potentials and the scalar fields will point in arbitrary direction in isospace. Physically, we still have a vacuum. This we will call a normal vacuum, but no longer supervacuum.

Any vacuum configuration is defined by specifying the gauge rotation $\Omega(x,t)$ that produces it from the supervacuum. Sometimes, if there is no complete spontaneous symmetry breaking, Ω is only defined up to a space-time independent constant rotation. We will ignore that for a moment.

Returning to the model of the previous section, based on the gauge group U(1), the gauge rotations Ω are determined by a point on the unit circle, and so the general vacuum is determined by an angle that may vary as a function of space-time. Now, infinity in two dimensions is characterized by an angle θ (the direction in which we go to infinity), so the boundary condition is defined by a mapping

$$\Omega(\theta)$$

of the unit circle onto an angle θ. As we saw before, this mapping must be in one of a discrete set of homotopy classes, labeled by $-\infty < n < +\infty$:

$$\Omega(\theta) = e^{in\theta} \quad . \tag{3.2}$$

Because n must stay integer there cannot be continuous transitions from a state containing n solitons of the same type into a state containing fewer solitons. Physically this is again the statement that magnetic flux going through a two-dimensional surface is conserved.

Now we are in a position that we can easily generalize our model into a similar model but with a non-Abelian gauge field instead of electromagnetism: the non-Abelian superconductor. We assume that the gauge rotations (3.1) and (3.2) are replaced by real, orthogonal, 3×3 matrix rotation $\Omega(x)$, with determinant one. They form the non-Abelian group SO(3). We assume, that, as in the U(1) superconductor, a Higgs field or set of Higgs fields with integer isospin break the symmetry spontaneously and completely. Does this model allow for stable flux tubes as its Abelian counterpart does? Do we have solitons in two-dimensional space?

To analyse this question we merely have to investigate the homotopy classes of the mappings of SO(3) matrices in the periodic space of angles θ. It is well known that the group SO(3) is locally isomorphic with SU(2), the group of 2×2 unitary matrices with determinant one. The correspondence is made by identifying 3-vectors with traceless, 2×2 tensors. An SU(2) matrix acting on the tensor corresponds to an SO(3) rotation of the 3-vector.

The SU(2) matrices are easier to handle. They can be decomposed as

$$\begin{aligned} \Omega &= a_o + i \sum_{\ell=1}^{3} a_\ell \sigma_\ell \quad , \\ \Omega^\dagger &= a_o^* - i \sum_{\ell=1}^{3} a_\ell^* \sigma_\ell \quad . \end{aligned} \tag{3.3}$$

The requirements $\Omega\Omega^\dagger = 1$ and $\det \Omega = 1$ correspond to:

a_o, a_ℓ are real and

$$\sum_{\ell=0}^{3} a_\ell^2 = 1 \; . \tag{3.4}$$

Clearly, the SU(2) matrices form an S_3 sphere (the surface of a sphere embedded in 4 dimensions).

The SO(3) matrices do not form an S_3 sphere, however. The reason is that an SU(2) rotation

$$\Omega = \begin{pmatrix} -1 & 0 \\ 0 & -1 \end{pmatrix}$$

on spinors leaves 2×2 tensors invariant. Consequently, two opposite points on the S_3-sphere correspond to the same SO(3) matrix (see Fig. 5). It is this feature that will allow a non-trivial homotopy class to emerge.

The homotopy classes here are entirely different from those in the U(1) model: any closed contour drawn on an S_3 sphere can be deformed continuously to a dot (see Fig. 6a), so the mappings of SU(2) onto the unit circle are characterized by only one homotopy class, namely that of the supervacuum. But if we are only concerned with SO(3) matrices one may construct one different homotopy class: all contours that start on one point of the S_3 sphere and close by ending up at the point opposite to A (see Fig. 6b). These are the only two homotopy classes in this case, this in contrast with the situation in the ordinary superconductor where we had an infinity of homotopy classes.

What is the consequence for our SO(3)-superconductor? The existence of one non-trivial homotopy class implies that we can

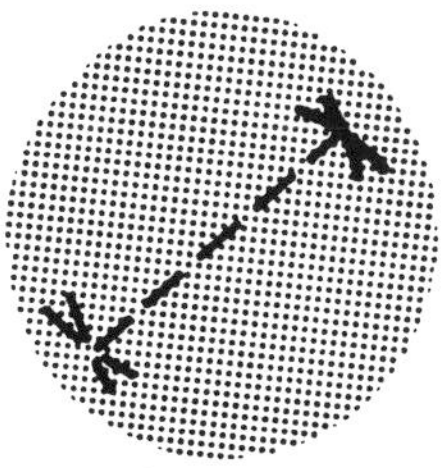

Fig. 5. The opposite points on S_3-sphere are identified.

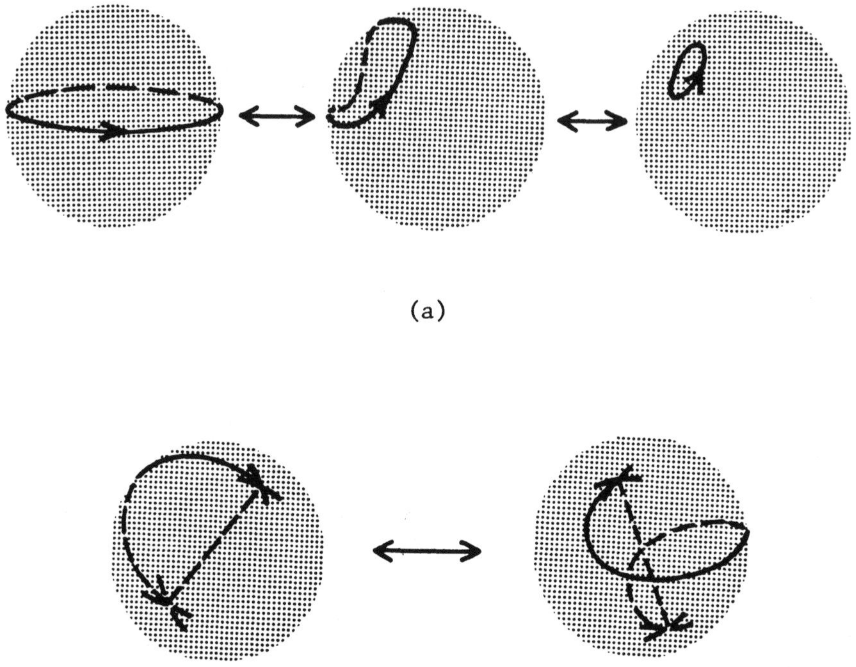

(a)

(b)

Fig 6. Two (and the only two) homotopy classes of closed paths for SO(3). Paths in (b) cannot be deformed into the paths in (a).

construct a non-trivial boundary condition for a stable soliton in 2 dimensions. So our model does admit a stable flux tube. However, now there is no longer an additive flux conservation law: if two of these solitons come together then the boundary of the total system is given by a contour in SO(3) space that goes through the S_3 sphere twice. That contour is in the trivial homotopy class, together with the boundary of a supervacuum. So the pair of solitons is not topologically stable which means they may annihilate each other. Thus, only single flux tubes are stable. Pairs of vortices, regardless their orientation, may annihilate each other, forming a shower of ordinary particles.

4. STRINGS AND MONOPOLES

Next, let us discuss a model very similar to the previous one. Again our gauge group is SO(3), but now we assume only one Higgs triplet, breaking it down to U(1). This is the so-called Georgi-Glashow model[6] introduced five years ago as a possible candidate for the weak interactions. It contains a massless U(1) photon, a massive charged "intermediate vector boson", and a neutral massive Higgs particle. Designed to avoid the necessity of neutral currents and charm, at the price of having to introduce heavy leptons, this model was doomed to become obsolete as a realistic description of observed weak interactions. However, being one of the very few truly unifying models it has some peculiar features which we now wish to focus on.

To create a situation exactly as in the previous model, we consider an ordinary superconductor[7] in a world whose high-energy behavior is described by the Georgi-Glashow model. The order parameter then provides for breaking of the residual U(1) symmetry.

According to the previous section this superconductor differs from the ordinary superconductor by the fact that two flux tubes pointing in the same direction may annihilate each other, because the gauge group is SO(3), not U(1). But physically this looks very odd, since one would expect magnetic flux to be conserved. The only possible explanation of the breaking up of (double) flux tubes is that a pair of magnetic monopoles is formed. They carry magnetic charge $\pm 4\pi/e$.

This way one is naturally led to accept the idea that magnetic monopoles occur in the Georgi-Glashow model. They are not confined to the superconductor, which we had only introduced for pedagogical reasons.

Let us now focus on the near vacuum surrounding this magnetic monopole. At one side, say the positive z-direction, we assume a double magnetic flux tube to emerge. Around this flux tube we do not have a supervacuum but a gauge rotation of that, which is in SU(2) notation:

$$\Omega(z \to +\infty, \phi) = \begin{pmatrix} e^{i\phi} & 0 \\ 0 & e^{-i\phi} \end{pmatrix} \tag{4.1}$$

where ϕ is the angle around the z-axis. At the other side is no magnetic flux coming in, so

$$\Omega(z \to -\infty, \phi) = \text{indep. } (\phi) \ . \tag{4.2}$$

According to homotopy theory no discontinuity in Ω is needed elsewhere (except at the location of the monopole, where no Ω is needed because it is not a vacuum configuration). Indeed we may construct Ω everywhere as follows:

$$\Omega(\theta,\phi) = \cos\frac{1}{2}\theta\begin{pmatrix} e^{i\phi} & 0 \\ 0 & e^{-i\phi}\end{pmatrix} + \sin\frac{1}{2}\theta\begin{pmatrix} 0 & i \\ i & 0\end{pmatrix} . \tag{4.3}$$

The Higgs field in this model is an isovector, which in a super-vacuum takes the form

$$\begin{pmatrix} 0 \\ 0 \\ F\end{pmatrix} . \tag{4.4}$$

In the surroundings of a single monopole it is

$$\Omega(\theta,\phi)\begin{pmatrix} 0 \\ 0 \\ F\end{pmatrix} = \frac{F}{|\vec{x}|}\begin{pmatrix} x \\ y \\ z\end{pmatrix} \tag{4.5}$$

where it is understood that the Ω of Eq. (4.3) has first been written in SO(3) notation.

It is clear that the boundary condition (4.5) for the Higgs field leaves a stable soliton at the center. It is also clear from the previous discussion that this soliton will carry a magnetic charge equal to $4\pi/e$.

We skip further discussion of the monopole in these notes because we have little to add to the existing literature[6,8,9]. Just note that, as was the case for other solitons, its mass is inversely proportional to the coupling parameter:

$$M = \frac{m_W C_2(\lambda/e^2)}{\alpha} , \qquad \alpha = e^2/4\pi \tag{4.6}$$

5. INSTANTONS

5.1. Construction

We have seen extended objects in one, two and three dimensions. None of the models of the previous sections, in which these objects occur are of direct interest in high energy physics. The models of interest are two different types of gauge theories that are most extensively studied these days: the $SU(2) \times U(1)$ gauge theory for the weak and electromagnetic interactions, "spontaneously broken" to U(1) by an isodoublet Higgs field[10,11], and a pure gauge theory based on SU(3) for the strong interactions[12].

The weak interaction gauge group has essentially a <u>four</u> component Higgs field because the Higgs doublet is complex. Therefore we expect a topologically stable object not in three but in four dimensions. Why? The boundary of a four dimensional world is topologically an S_3 sphere. Also the complex Higgs doublet

$$\begin{pmatrix} \phi_1 \\ \phi_2 \end{pmatrix} , \quad \text{with} \quad |\phi|^2 = F^2 ,$$

form an S_3 sphere. The mappings of S_3 spheres on S_3 spheres are again characterized by non-trivial homotopy classes, labeled by an integer n that can run from $-\infty$ to ∞. Thus we can construct boundary conditions for a four dimensional space under which configurations with n soliton-like objects are stable.

In contrast to the cases in lower dimensions, also pure gauge theories allow for topologically stable soliton-like objects in four dimensions. This is most easily understood when we consider the simplest non-Abelian group SU(2). Remember that one way to define a vacuum at the boundary of a region is to specify the gauge rotations $\Omega(\vec{x})$ that produce this vacuum out of the super-vacuum. The gauge rotations Ω themselves form an S_3 sphere (see section 3), so here too we have the non-trivial homotopy classes at the boundary. In fact, the S_3 sphere of the gauge rotations Ω themselves, is the same as the S_3 sphere of the complex doublets

$$\begin{pmatrix} \phi_1 \\ \phi_2 \end{pmatrix} = \Omega \begin{pmatrix} F \\ 0 \end{pmatrix} ,$$

so one can say that the 4 dimensional solitons in the pure gauge theory are essentially the same as those in the theories broken by Higgs isodoublets.

In order to distinguish our 4-dimensional objects from the 3-dimensional solitons and the strings and surfaces, we give them a new name: "instantons". The name emphasizes that if one of the four dimensions is time, then these objects occupy only a short time-interval (instant).

We will now argue that there is an important quantity that is sensitive to the presence of instantons:

$$G^a_{\mu\nu}\tilde{G}^a_{\mu\nu} \quad ,$$

with

$$\tilde{G}^a_{\mu\nu} = \frac{1}{2}\,\varepsilon_{\mu\nu\alpha\beta}G^a_{\alpha\beta} \quad , \tag{5.1}$$

where $G^a_{\mu\nu}$ is the usual covariant curl of the gauge vector field A^a_μ:

$$G^a_{\mu\nu} = \partial_\mu A^a_\nu - \partial_\nu A^a_\mu + g\varepsilon_{abc}A^b_\mu A^c_\nu \quad . \tag{5.2}$$

$G\tilde{G}$ is gauge invariant, and it is a pure divergence:

$$G\tilde{G} = \partial_\mu K_\mu \quad ,$$

$$K_\mu = 2\varepsilon_{\mu\nu\alpha\beta}A^a_\nu(\partial_\alpha A^a_\beta + \frac{g}{3}\,\varepsilon_{abc}A^b_\alpha A^c_\beta) \quad . \tag{5.3}$$

However, as one can verify easily, K_μ is not gauge invariant. Let us see what the consequence is of that.

Consider a region of space-time enclosed by a large S_3 sphere. It is irrelevant as yet whether the space-time is Euclidean or Minkowskian; we are only concerned about topological stability and not yet interested in any equation of motion for the fields, not even the space-time metric. Suppose we have a supervacuum at the surrounding sphere:

$$A^a_\mu = 0 \quad ,$$

therefore

$$K_\mu = 0 \quad . \tag{5.4}$$

Because of (5.4) at the boundary, and because of (5.3), we find by means of Gauss' law:

$$\int G^a_{\mu\nu}\tilde{G}^a_{\mu\nu}d^4x = 0 \tag{5.5}$$

regardless of what values the fields take inside the sphere.

Now we take a non-trivial homotopy class of gauge rotations to get a new vacuum boundary condition:

$$\Omega(x_1,\ldots,x_4) = \frac{x_4 - ix_i\sigma_i}{\sqrt{x_4^2 + (x_i)^2}} \quad . \tag{5.6}$$

If we write the vector field in an SU(2) notation,

$$A_\mu = -\frac{i}{2}\sigma_a A^a_\mu \,, \tag{5.7}$$

then the transformation law is

$$A'_\mu = \Omega A_\mu \Omega^{-1} - \frac{1}{g}\partial_\mu \Omega \cdot \Omega^{-1} \tag{5.8}$$

We find that the vacuum at the boundary has

$$A'^a_\mu = \frac{2}{g}\eta^a_{\mu\nu}\frac{x^\nu}{x^2} \,, \tag{5.9}$$

with

$$\begin{aligned} \eta^a_{\mu\nu} &= \varepsilon_{a\mu\nu} \quad \text{for} \quad \mu,\nu = 1,2,3 \,, \\ \eta^a_{\mu 4} &= -\delta_{a\mu} \quad \text{for} \quad \mu = 1,2,3 \,, \\ \eta^a_{4\nu} &= \delta_{a\nu} \quad \text{for} \quad \nu = 1,2,3 \,. \end{aligned} \tag{5.10}$$

Now although the field (5.9) is physically equivalent to the vacuum (for instance, $G^a_{\mu\nu}$ at the boundary vanishes), we nevertheless get a non-vanishing value for K_μ. This was possible because K_μ is not gauge-invariant:

$$K'_\mu = 16x_\mu/g^2|x|^4, \tag{5.11}$$

where

$$|x|^2 = x_4^2 + x_i^2 \quad .$$

If we now apply Gauss' law we find that inside the S_3 sphere $G_{\mu\nu}$ cannot vanish everywhere because

$$\int G\tilde{G}\, d^4x = \int \partial_\mu K_\mu d^4x = \int_{\text{boundary}} K_\perp\, d^3x = \frac{32\pi^2}{g^2} \quad . \tag{5.12}$$

Merely because the vacuum at our boundary is topologically distinct from the supervacuum we get a non-trivial physical field configuration inside.

Note that the result (5.12) holds true also if the SU(2) group considered would be embedded in a larger gauge group. This implies that instantons remain topologically stable even if the gauge group is enlarged. Remember the Abelian flux tubes that could become unstable if the U(1) group were part of a larger group; such a thing does not happen for instantons.

Now let us insert the field equations. It is easiest to consider first Euclidean space, where the boundary choice (5.6) has nice rotational properties. Is there a solution to the Euclidean field equations that approaches (5.9) at $x^2 \to \infty$ and stays regular at the origin? We try

$$A^a_\mu(x) = \frac{2}{g}\eta^a_{\mu\nu}x^\nu f(r) \ . \tag{5.13}$$

The field equation, $D_\mu G_{\mu\nu} = 0$, now reads

$$\frac{d^2}{(d\log r)^2}(r^2 f) = 4r^2 f - 12(r^2 f)^2 + 8(r^2 f)^3 \ . \tag{5.14}$$

This equation is scale-invariant, which makes it as easy to solve as the one-dimensional soliton, whose equation is translation-invariant. Besides the trivial solution, $f = 1/r^2$, we can have[13]

$$f(r) = \frac{1}{r^2 + \lambda} \ , \quad \lambda \text{ arbitrary} \ , \tag{5.15}$$

which has the required properties.

The total action for the solution is

$$S = -\frac{1}{4}\int G^a_{\mu\nu}G^a_{\mu\nu}d^4x \ . \tag{5.16}$$

Without actually computing S it is easy to derive an inequality for it. In Euclidean space $G^a_{\mu\nu}$ and $\tilde{G}^a_{\mu\nu}$ all have real components only. Therefore[13]

$$0 \le (G^a_{\mu\nu} \pm \tilde{G}^a_{\mu\nu})^2 = 2G^a_{\mu\nu}G^a_{\mu\nu} \pm 2G^a_{\mu\nu}\tilde{G}^a_{\mu\nu} \ ; \tag{5.17}$$

$$S \le \pm\frac{1}{4}\int G^a_{\mu\nu}\tilde{G}^a_{\mu\nu}d^4x \le -\frac{8\pi^2}{g^2} \ , \tag{5.18}$$

because of (5.12), and we can choose either sign. For our solution,

$$G^a_{\mu\nu} = -\frac{4\lambda}{g}\frac{\eta^a_{\mu\nu}}{(x^2+\lambda)^2} \ , \tag{5.19}$$

and we find

$$S = -8\pi^2/g^2 \ . \tag{5.20}$$

The inequality is saturated, because $G^a_{\mu\nu} = \tilde{G}^a_{\mu\nu}$ for our solution. Indeed, any solution of the (first order) equation $G_{\mu\nu} = \tilde{G}_{\mu\nu}$ automatically satisfies the second order equation $D_\mu G_{\mu\nu} = 0$ because $D_\mu \tilde{G}_{\mu\nu} = 0$ for any gauge field derived from a vector potential (5.2).

5.2. Euclidean Field Theory and Tunneling

A solution of the classical field equation in Euclidean space may seem to be of little physical relevance. What we really would like to consider are instanton events in Minkowski space. Consider again the S_3 boundary of a compact region in Minkowski space. The mapping $\Omega(x)$ at the boundary may be deformed a bit, as long as we stay in the same homotopy class. Thus, the instanton can be seen to correspond to a transition from a supervacuum at $t = -\infty$ (where $\Omega(\vec{x}) = 1$) to a gauge rotated vacuum at $t = +\infty$. There we can have

$$\Omega(\vec{x}) = \frac{\vec{x}^2 - 1 + ix^a\sigma^a}{\sqrt{(\vec{x}^2-1)^2 + \vec{x}^2}} \ . \tag{5.21}$$

At intermediate times there must be a region where $G_{\mu\nu} \neq 0$, because of Eq. (5.12). Because we have a vacuum at the beginning and at the end this field configuration cannot be a solution of the classical equations. Rather, it should be considered as a quantum-mechanical tunneling transition. How do we compute the amplitude of such a tunneling transition? It is instructive to compare this problem with the motion of a single particle through a one-dimensional potential well:

$$H = \frac{p^2}{2m} + V(x) \ , \tag{5.22}$$

in a region where $V(x) >> E$, assuming that Planck's constant is small. To first approximation

$$\frac{1}{|\psi|}\left|\frac{\partial\psi}{\partial x}\right| \approx \frac{1}{\hbar}\sqrt{2m(V-E)} \ . \tag{5.23}$$

The amplitude for a tunneling process is proportional to $e^{\tilde{S}}$ with

$$\tilde{S} = -\frac{1}{\hbar}\int_{x_a}^{x_b} \sqrt{2m(V-E)}\, dx \quad . \tag{5.24}$$

On the other hand, if a trajectory is in an energetically allowed region, E > V, then the wave function oscillates. The number of oscillations is given by

$$\frac{1}{2\pi\hbar}\int_{x_a}^{x_b} p dx = \frac{1}{2\pi\hbar}\int \sqrt{2m(E-V)}\, dx = \frac{1}{2\pi\hbar} R \quad . \tag{5.25}$$

R is related to the action integral:

$$R = \int pdx = \int p\dot{x}dt = \int (H+L)\, dt = \int (E+L)\, dt \quad .$$

If we normalize the energy to zero then

$$R = \int Ldt = S \ . \tag{5.26}$$

This is the total action of a motion from x_a to x_b with given energy. Note that the tunneling process is given by (5.24) which is the same expression except that the sign of V- E is flipped. The sign of the potential in:

$$m\frac{d^2x}{dt^2} = \frac{\partial V}{\partial x} \quad , \tag{5.27}$$

is also flipped if we replace t by it. If there is no classically allowed movement at zero energy from x_a to x_b at real times then there is an allowed transition for imaginary times. We find that the total action $\tilde{S}$ for this imaginary transition determines the tunneling amplitude[14]. In quantum field theory that corresponds to replacing Minkowski space by Euclidean space.

The physical tunneling amplitude from the supervacuum to a gauge rotated vacuum in a different homotopy class is governed to first approximation by the total action $\tilde{S}$ of a classical solution in Euclidean space. The instanton has, as we saw, $\tilde{S} = -8\pi^2/g^2$. So, we will find amplitudes proportion to

$$e^{-8\pi^2/g^2} \quad . \tag{5.28}$$

The superposition of the supervacuum and the gauge rotated vacua gives rise to different possibilities for the true ground state of Hilbert space, labeled by an arbitrary angle Θ. Discussion of this phenomenon is to be found in Refs. 15 and 16.

5.3. Symmetry Breaking through Instantons

The above holds for a gauge theory without fermions. Instantons there rearrange the ground state of the theory. They do not show up as special events[17].

A new phenomenon occurs if fermions are coupled to the theory. Consider some gauge-invariant fermion current J_μ. If there is a strong background gauge field $G^a_{\mu\nu}$, then there will be a vacuum polarization:

$$J_\mu \propto (G^a_{\mu\nu})^2 \quad . \tag{5.29}$$

That effect is computed in a triangular Feynman diagram (see Fig.7). One of the external legs denotes the space-time point x where $J_\mu(x)$ is measured. The other two legs are the two gauge photon external lines.

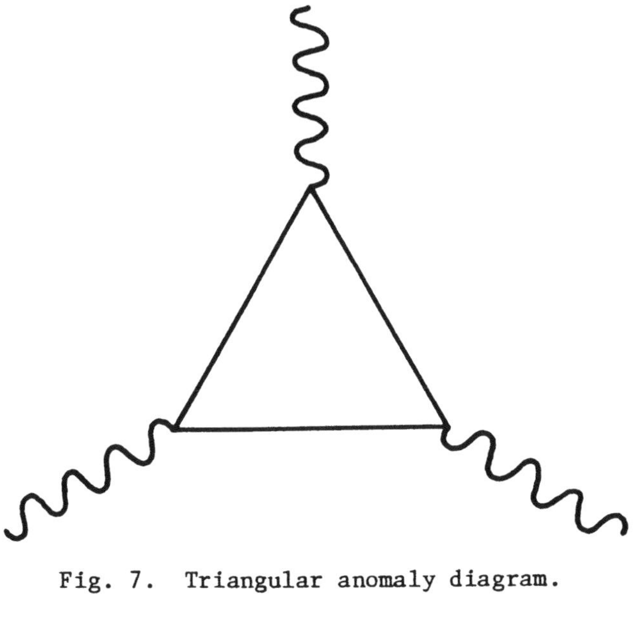

Fig. 7. Triangular anomaly diagram.

Let us assume, without yet specifying J, that it is conserved according to the Noether theorem. Then our diagram, which we could call

$$\Gamma_{\mu\alpha\beta}(q,k,k-q) \tag{5.30}$$

should satisfy

$$q_\mu \Gamma_{\mu\alpha\beta} = 0 \quad . \tag{5.31}$$

Gauge invariance further dictates that only transverse photons are coupled:

$$k_\alpha \Gamma_{\mu\alpha\beta} = 0 \quad ; \tag{5.32}$$

$$(k-q)_\beta \Gamma_{\mu\alpha\beta} = 0 \; . \tag{5.33}$$

Now the triangle diagram, when computed, shows an infinity that must be subtracted. It can happen that there are only two subtraction constants to be chosen. Eqs. (5.31)-(5.33) form three conditions for these two constants. They can be incompatible. In that case we have to keep gauge invariance in order to keep renormalizability. Consequently (5.31) is altered[18,19]. The result can be written as

$$\partial_\mu J_\mu = \frac{-ig^2 N}{16\pi^2} \cdot G^a_{\mu\nu}\tilde{G}^a_{\mu\nu} \quad . \tag{5.34}$$

Here N is an integer determined by the details which were left out in my discussion. This is called the Adler-Bell-Jackiw-anomaly.

Now consider Eq. (5.12)[20]. We find that an integer number of charge units Q belonging to the current J_μ are consumed by the instanton:

$$\Delta Q = \int d^4x \, \partial_\mu J_\mu = 2N \; . \tag{5.35}$$

In the strong-interaction gauge theory, a current J with this property is the chiral charge (total number of quarks with helicity + minus quarks with helicity -). Since the right hand side of (5.34) is a singlet under flavor SU(3) it is the ninth axial vector current which is nonconserved through the instanton. This probably explains why there is no SU(3) singlet (or SU(2) singlet) pseudoscalar particle as light as the pion[21,22].

Since the vector currents remain conserved, the instanton appears to flip the helicity of one of each type of fermion involved (see Fig. 8). This symmetry breaking effect of the instanton can be written in terms of an effective interaction Lagrangian. The original chiral symmetry was

$$U(N)_{left} \times U(N)_{right} \quad . \tag{5.36}$$

It is broken into

$$SU(N)_{left} \times SU(N)_{right} \times U(1)_{vector} \quad . \tag{5.37}$$

An effective Lagrangian that does this breaking is

$$\Delta\mathcal{L}^{eff} = C \exp(\frac{-8\pi^2}{g^2} + i\Theta) \det_{st} [\bar{\psi}_s(1+\gamma_5)\psi_t] + h.c. \tag{5.38}$$

Actually the effective interaction is more complicated when color indices are included[22], and C still contains powers of g. In the exponent, Θ is an arbitrary angle associated with the symmetry breaking.

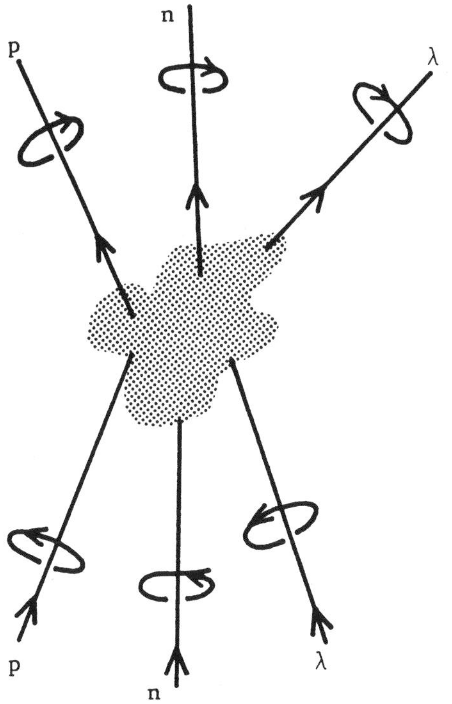

Fig. 8. The instanton appears to flip the helicity of one of each type of fermion involved.

Now let us turn to the instantons in the weak and electromagnetic guage theory SU(2) × U(1), with leptons, and non-strange, strange and charmed quarks which have color indices. This theory is anomaly-free, according to its commercials, as far as such is needed for renormalizability. Indeed, in triangle diagrams where all three external lines are gauge photons of whatever type, the anomalies (must) cancel. But if one of the external lines is one of the familiar conserved currents like baryon or lepton current (to which no known photons are coupled) then there are anomalies. We find, by inserting the anomaly equation (5.34) with the correct values for N, that one instanton gives

$\Delta Q_E = 1$ (electron-number nonconservation) ,

$\Delta Q_M = 1$ (muon-number nonconservation) ,

$\Delta Q_N = 3$ (non-strange quark nonconservation) ,

$\Delta Q_C = 3$ (strange/charm quark nonconservation) .

In total, two baryon and two lepton units are consumed by the instanton (strange and non-strange may mix through the Cabibbo angle). Thus we can get the decays

$PN \to e^{+-}\nu_\mu$ or $\mu^{+-}\nu_e$,

$NN \to \bar{\nu}_e\bar{\nu}_\mu$,

$PP \to e^+\mu^+$, etc.

The order of magnitude of these decays is given by

$$[e^{-8\pi^2/g^2}]^2 = e^{-16\pi^2/e^2\sin^{-2}\Theta_W} = e^{-4\pi\cdot 137\cdot\sin^2\Theta_W} . \qquad (5.39)$$

With a Weinberg angle $\sin^2\Theta_W = .35$, and assuming that the weak intermediate vector boson determines the scale, then we get lifetimes of the order of (give or take many orders of magnitude)

$$\tau \simeq 10^{225} \text{ sec} , \qquad (5.40)$$

corresponding to one deuteron decay in 10^{137} universes.

6. SOME REFLECTIONS ON CLASSICAL SOLUTIONS AND THE QUARK CONFINEMENT PROBLEM

Clearly, the study of non-trivial classical field configurations gives us a welcome extension of our knowledge and understanding of field theory beyond the usual perturbation expansion. One non-perturbative problem is particularly intriguing: quark confinement in QCD. There has been wide-spread speculation that the new classical solutions are somehow responsible for this phenomenon. Some authors believe that some plasma of instantons or instanton-like objects does the trick. We will not go that far. We will however exhibit some simple models with interesting properties. They will clearly show that a "phase transition" towards permanent confinement is not at all an absurd idea (it is actually much harder to make models with "nearly confinement" of quarks).

Our first model is based on an SU(3) gauge theory in 3+1 dimensions. The gauge group here is neither "color" nor "flavor", so let us call it "horror" (or: "terror"), for reasons that will become clear later (the model does not describe the observations on quarks very well).

Let us assume that the SU(3) symmetry is spontaneously completely broken by a conventional Higgs field. The Higgs field must be a "non-exotic" representation of horror (mathematically: a representation of SU(3)/Z(3)), e.g. an octet or decuplet representation.

In the following we will show that this model admits, like the U(1), and the SU(2) analogues, string-vortices, but these are again different. After that we will introduce quarks and show that they bind to the strings in the combination that we actually see in hadrons[23].

To see the topological properties of vortices in this model we consider, as we did in section 3, a two dimensional space, enclosed by a boundary which has the topology of a circle. The vacuum at this boundary is specified by a gauge rotation $\Omega(\theta)$, so the number of topologically stable structures is given by the number of homotopy classes of the mapping $\theta \to \Omega(\theta)$. The group SU(3) has an invariant subgroup Z(3) given by the three elements I, $e^{2\pi i/3}I$ and $e^{-2\pi i/3}I$. Our physical fields have been required to be invariant under Z(3). So, the mappings with $\Omega(2\pi) = e^{\pm 2\pi i/3}\Omega(0)$ are acceptable. They form two homotopy classes besides the trivial one: $\Omega(2\pi) = \Omega(0)$. See Fig. 9. We conclude that there are stable vortices. Oppositely oriented vortices are different: their boundaries are in different homotopy classes. But two vortices oriented in the same direction are in the same class as one vortex in the other direction (see Fig. 10), because

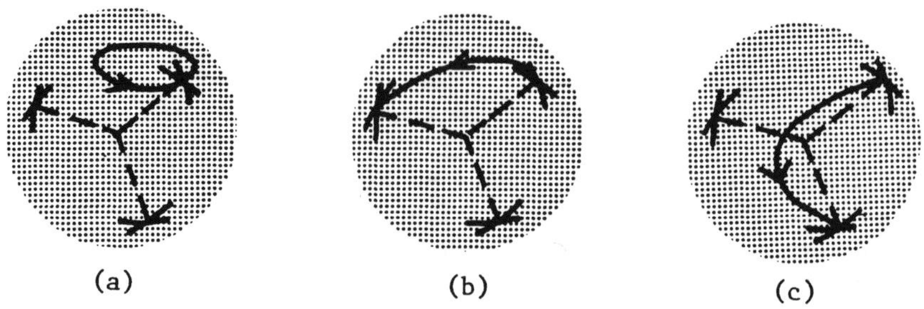

Fig. 9. Three homotopy classes of the mapping $\theta \to \Omega(\theta)$.

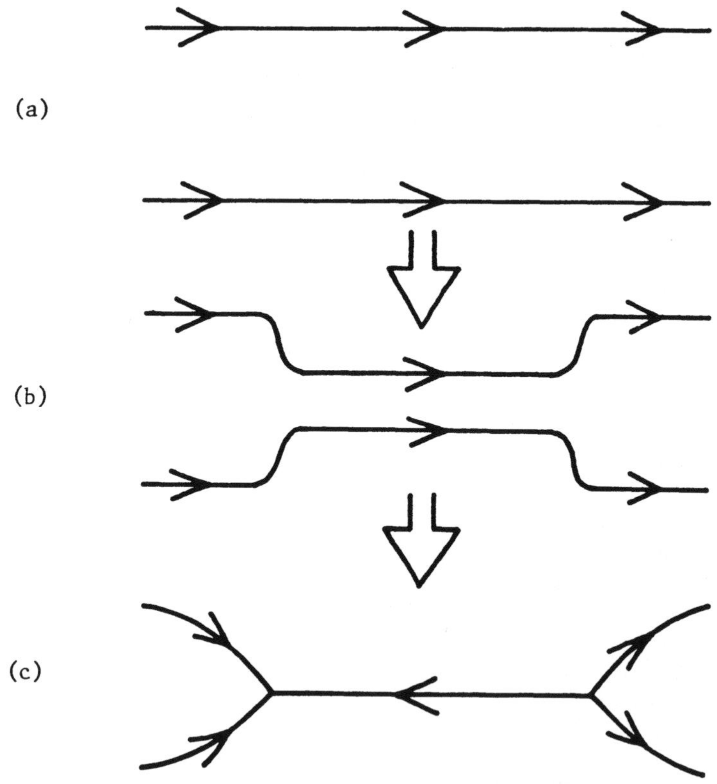

Fig. 10. Decay of two parallel vortices into a single vortex in opposite direction.

$e^{4\pi i/3} = e^{-2\pi i/3}$, so two parallel vortices will decay into an energetically more favorable single vortex in the other direction. Note that three-string connections are possible.

Quarks are now introduced as the end-points of strings. They could be compared with magnetic monopoles inside a superconductor. Note that quarks differ from antiquarks and because of the topological properties of these strings they only occur in the form of "hadronic" bound states (see Fig. 11).

We could go into lengthy details about actual construction of quark monopoles in the model, for instance by embedding SU(3)/Z(3) in a large gauge group that does not contain Z(3) itself. Then these monopoles themselves are allowed as classical solutions.

But the model is not very viable from a physical point of view because:

(i) The "horror" degrees of freedom are not known to exist; where are the "horror"-vector bosons etc.?

(ii) The Dirac monopoles will have monopole charges of the order of $2\pi/g_{horror}$. So, if quarks should behave as approximately free partons inside hadrons, g_{horror} should be so large that no perturbation expansion is likely to make sense, and

(iii) unlike monopoles (see section 4), quarks are light, relativistic, particles.

(iv) It isn't Q.C.D.

One thing on the other hand is very clear in this model: because the strings are topologically stable quarks are absolutely and permanently confined.

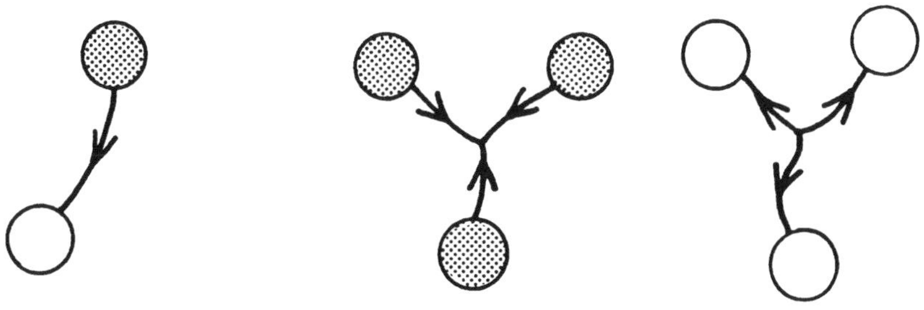

Fig. 11. Allowed quark and antiquark bound states.

The next model we consider is entirely different from the previous one. It is pure SU(N) color gauge theory (quantum-chromodynamics) but in 2 space - one time dimension. We claim that there is an elegant way to formulate the quark confinement mechanism here, although the dynamics is very complicated. For pedagogical reasons let us first add a set of Higgs fields to the system, which must be a representation of SU(N)/Z(N) as in the previous model. There is a complete spontaneous symmetry break-down and obviously no quark confinement. Let us also leave out the quarks now. Since we have only 2 space dimensions the model has no flux tubes but instead soliton particles. The boundary of this space can be in one of N homotopy classes, so if we have more than N/2 solitons they may decay: the number of solitons minus anti-solitons is only conserved modulo N. We now make a very important step[24]: define an operator field $\phi(\vec{x},t)$ that destroys one soliton (or creates one anti-soliton) at $(\vec{x},t)$. There is not enough time now to formulate a precise definition of $\phi(\vec{x},t)$. It suffices to state that in particular the Green's functions

$$<\phi(0)\phi^{\dagger}(x)>_0 \tag{6.1}$$

and

$$<\phi(x_1)\phi(x_2)\ \dots\ \phi(x_N)>_0 \tag{6.2}$$

can be defined accurately in terms of ordinary perturbation theory.[25] We can require that $\phi(\vec{x},t)$ produces or destroys only "bare" solitons: the vacuum (not supervacuum) remains at all $\vec{x}'$ with $|\vec{x}' - \vec{x}| \geq \varepsilon$ but a large field configuration is produced where $|\vec{x}' - \vec{x}| < \varepsilon$; then $\varepsilon \to 0$. Then (6.1) and (6.2) satisfy the Wightman axioms. We could assume that they may be more or less reproduced by an effective Lagrangian:

$$\mathcal{L}(\phi,\phi^*) = -\partial_\mu\phi^*\partial_\mu\phi - M^2\phi^*\phi - \frac{\lambda}{N!}\,(\phi^N + (\phi^*)^N)\ . \tag{6.3}$$

Here M is the calculable mass of the soliton. There will be many interaction terms; the one we wrote down is necessary to make (6.2) different from zero. Observe a Z(N) symmetry that leaves (6.1), (6.2) and (6.3) invariant:

$$\begin{aligned} \phi &\to e^{2\pi i/N}\phi\ , \\ \phi^* &\to e^{-2\pi i/N}\phi^*\ . \end{aligned} \tag{6.4}$$

Now what would happen if we vary the Higgs potential such that its second derivative at the origin turns from negative to positive? The field ϕ remains well defined. The soliton mass M^2 goes to zero if the Higgs mass goes to zero. It is therefore natural to assume that after the transition M^2 in (6.3) has changed sign. Our effective Lagrangian now could be

$$\mathcal{L}(\phi,\phi^*) = -\partial_\mu\phi^*\partial_\mu\phi - \frac{g}{2}(\phi^*\phi - F^2)^2 - \frac{\lambda}{N!}(\phi^N + (\phi^*)^N) . \qquad (6.5)$$

We imagine that then our Z(N) symmetry (6.4) will be spontaneously broken! We stress that we do not yet understand the details of the dynamics, but if this assumption is correct then the original model confines its quarks! Why? There are now N different vacuum states for the field ϕ:

$$\langle\phi\rangle_1 = e^{2\pi i/N}\langle\phi\rangle_2 = \dots \qquad (6.6)$$

If there are two different vacuum configurations in one plane then we obtain "Bloch walls" separating them. These Bloch walls can be considered to be closed strings. It is not hard to see that if we now add the quarks to the original model, then these will form end-points of the strings ("Dirac monopoles"). Since the strings carry energy per unit of length these quarks will be permanently confined by a linear potential.

We like to look at the transition from the SU(N) gauge theory to the Z(N) scalar theory as a dual transformation. The solitons in one theory are the elementary particles of the other. We note a peculiar antagonism: if the SU(N) symmetry is spontaneously broken, the Z(N) symmetry is intact and vice versa. After the dual transformation from the unbroken SU(N) theory to the broken Z(N) theory, quark confinement in the latter is obvious (once we accept the symmetry breaking antagonism).[26] What about 3+1 dimensional quantumchromodynamics? Our final speculation is that the "horror" model discussed in the beginning of this section is in a similar way the dual transform of quantumchromodynamics. The dual transformation in 3+1 dimensions is much more complicated than in 2+1 dimensions, but again, one can show that certain dual Green's functions corresponding to the dual gauge field may be constructed. The subject is obscured by the fact that we are not able to construct Lagrangians from the Green's functions, and that the horror coupling constant is inversely proportional to the color coupling constant so that it is impossible to compare the two perturbation series. However, if these dual transformations can be given a more sound footing then the quark confinement phenomenon will no longer look as mysterious as it does now.

REFERENCES

1) T.D. Lee and G.C. Wick, Phys. Rev. D9, 2291 (1974); T.D. Lee and M. Margulies, Phys. Rev. D11, 1591 (1975).

2) S. Coleman, in New Phenomena in Subnuclear Physics, International School of Subnuclear Physics "Ettore Majorana", Erice 1975, Part A, ed. A. Zichichi (Plenum Press, New York, 1977).

3) "State" here does not mean "state in Hilbert space" but rather a stationary classical field configuration.

4) H.B. Nielsen and P. Olesen, Nucl. Phys. B61, 45 (1973); H.B. Nielsen in Proceedings of the Adratic Summer Meeting on Particle Physics, Rovinj, Yugoslavia 1973, ed. M. Martinis et al. (North Holland, Amsterdam, 1974).

5) B. Zumino, in Renormalization and Invariance in Quantum Field Theory, NATO Adv. Summer Inst., Capri 1973, ed. E.R.Caianiello (Plenum Press, New York, 1974).

6) H. Georgi and S.L. Glashow, Phys. Rev. Lett. 32, 438 (1974).

7) For simplicity our order parameter is chosen to carry a single charge $q=e$, so that single flux tubes are quantized by units $2\pi/e$.

8) G. 't Hooft, Nucl. Phys. B79, 276 (1974); Nucl. Phys. B105, 538 (1976); A. Polyakov, JETP Lett. 20, 194 (1974).

9) B. Julia and A. Zee, Phys. Rev. D11, 2227 (1975).

10) S. Weinberg, Phys. Rev. Lett. 19, 1264 (1967).

11) G. 't Hooft, Nucl. Phys. B35, 167 (1971).

12) J. Kogut and L. Susskind, Phys. Rev. D9, 3501 (1974); K.G. Wilson, Phys. Rev. D10, 2445 (1974).

13) A.A. Belavin et al., Phys. Lett. 59B, 85 (1975).

14) S. Coleman, in The Why's of Subnuclear Physics, Erice lecture notes, Erice 1977.

15) R. Jackiw and C. Rebbi, Phys. Rev. Lett. 37, 172 (1976).

16) C. Callan, R. Dashen and D. Gross, Phys. Lett. 63B, 334 (1976).

17) It is to be remarked however that the instanton effects violate parity conservation if Θ is not a multiple of π. Parity violating events could then be searched for.

18) S.L. Adler, Phys. Rev. 177, 2426 (1969).

19) J.S. Bell and R. Jackiw, Il Nuovo Cimento A60, 47 (1969).

20) The factor i should be added in (5.12) since we are now dealing with a Minkowski metric where time components of vectors are taken to be imaginary.

21) H. Fritzsch, M. Gell-Mann and H. Leutwyler, Phys. Lett. 47B, 365 (1973).

22) G. 't Hooft, Phys. Rev. Lett. B37, 8 (1976), and G. 't Hooft, Phys. Rev. D14, 3432, (1976).

23) F. Englert, Lectures given at the Cargèse Summer School, July 1977.

24) S. Mandelstam, Phys. Rev. D11, 3026 (1975).

25) G. 't Hooft, to be published.

26) Compare the dual transformation in the 2 dimensional Ising model: the ordered phase is transformed into the disordered phase and vice versa.

CHAPTER 4.2

MAGNETIC MONOPOLES IN UNIFIED GAUGE THEORIES

G. 't HOOFT
CERN, Geneva

Received 31 May 1974

Abstract: It is shown that in all those gauge theories in which the electromagnetic group U(1) is taken to be a subgroup of a larger group with a compact covering group, like SU(2) or SU(3), genuine magnetic monopoles can be created as regular solutions of the field equations. Their mass is calculable and of order 137 M_W, where M_W is a typical vector boson mass.

1. Introduction

The present investigation is inspired by the work of Nielsen et al. [1], who found that quantized magnetic flux lines, in a superconductor, behave very much like the Nambu string [2]. Their solution consists of a kernel in the form of a thin tube which contains most of the flux lines and the energy; all physical fields decrease exponentially outside this kernel. Outside the kernel we do have a transverse vector potential A, but there it is rotation-free: if we put the kernel along the z axis, then

$$A(x) \propto (y, -x, 0)/(x^2 + y^2) . \tag{1.1}$$

$A(x)$ can be obtained by means of a gauge transformation $\Omega(\varphi)$ from the vacuum. Here φ is the angle about the z axis:

$$\Omega(0) = \Omega(2\pi) = 1 .$$

It is obvious that such a string cannot break since we cannot have an end point: it is impossible to replace a rotation over 2π continuously by $\Omega(\varphi) \to 1$. Or: magnetic monopoles do not occur in the system. Also it is easy to see that these strings are oriented: two strings with opposite direction can annihilate; if they have the same direction they may only join to form an even tighter string.

Now, let us suppose that the electromagnetism in the superconductor is in fact described by a unified gauge theory, in which the electromagnetic group U(1) is a subgroup of, say, SO(3). In such a non-Abelian theory one can only imagine non-oriented strings, because a rotation over 4π can be continuously shifted towards a fixed Ω. What happened with our original strings? The answer is simple: in an SO(3) gauge theory magnetic monopoles with twice the flux quantum (i.e., the

Nuclear Physics B79 (1974) 276–284. North-Holland Publishing Company

Schwinger [3,4] value), occur. Two of the original strings, oriented in the same direction, can now annihilate by formation of a monopole pair [5].

From now on we shall dispose of the original superconductor with its quantized flux lines. We consider free monopoles in the physical vacuum. That these monopoles are possible, as regular solutions of the field equations, can be understood in the following way. Imagine a sphere, with a magnetic flux Φ entering at one spot (see fig. 1). Immediately around that spot, on the contour C_0 in fig. 1, we must have a magnetic potential field $\boldsymbol{A}$, with $\oint(\boldsymbol{A}\cdot d\boldsymbol{x}) = \Phi$. It can be obtained from the vacuum by applying a gauge transformation Λ:

$$\boldsymbol{A} = \nabla \Lambda. \tag{1.2}$$

This Λ is multivalued. Now we require that all fields, which transform according to

$$\psi \to \psi\, e^{ni\Lambda} , \tag{1.3}$$

to remain single valued, so Φ must be an integer times 2π: we then have a complete gauge rotation along the contour in fig. 1.

In an Abelian gauge theory we must necessarily have some other spot on the sphere where the flux lines come out, because the rotation over $2k\pi$ cannot continuously change into a constant while we lower the contour C over the sphere. In a non-Abelian theory with compact covering group, however, for instance the group O(3), a rotation over 4π may be shifted towards a constant, without singularity: we may have a vacuum all around the sphere. In other theories, even rotations over 2π

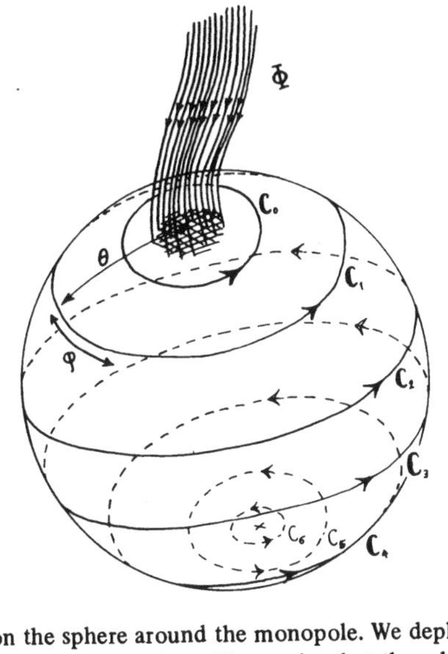

Fig. 1. The contour C on the sphere around the monopole. We deplace it from C_0 to C_1, etc., until it shrinks at the bottom of the sphere. We require that there be no singularity at that point.

may be shifted towards a constant. This is why a magnetic monopole with twice or sometimes once the flux quantum is allowed in a non-Abelian theory, if the electromagnetic group U(1) is a subgroup of a gauge group with compact covering group. There is no singularity anywhere in the sphere, nor is there the need for a Dirac string.

This is how we were led to consider solutions of the following type to the classical field equations in a non-Abelian Higgs–Kibble system: a small kernel occurs in the origin of three dimensional space. Outside that kernel a non-vanishing vector potential exists (and other non-physical fields) which can be obtained from the vacuum* by means of a gauge transformation $\Omega(\theta,\varphi)$. At one side of the sphere $(\cos\theta \to 1)$ we have a rotation over 4π, which goes to unity at the other side of the sphere $(\cos\theta \to -1)$. For such a rotation one can, for instance, take the following SU(2) matrix:

$$\Omega(\theta,\varphi) = \cos\tfrac{1}{2}\theta\begin{pmatrix} e^{i\varphi} & 0 \\ 0 & e^{-i\varphi}\end{pmatrix} + \sin\tfrac{1}{2}\theta\begin{pmatrix} 0 & i \\ i & 0\end{pmatrix}. \tag{1.4}$$

Now consider one rotation of the angle φ over 2π. At $\theta = 0$, this Ω rotates over 4π (the spinor rotates over 2π). At $\theta = \pi$, this Ω is a constant. One easily checks that

$$\Omega\Omega^\dagger = 1\ . \tag{1.5}$$

In the usual gauge theories one normally chooses the gauge in which the Higgs field is a vector in a fixed direction, say, along the positive z axis, in isospin space. Now, however, we take as a gauge condition that the Higgs field is $\Omega(\theta,\varphi)$ times this vector. As we shall see in sect. 2, this leads to a new boundary condition at infinity, to which corresponds a non-trivial solution of the field equations: a stable particle is sitting at the origin. It will be shown to be a magnetic monopole. If we want to be conservative and only permit the normal boundary condition at infinity, with Higgs fields pointing in the z direction, then still monopole-antimonopole pairs, arbitrarily far apart, are legitimate solutions of the field quations.

2. The model

We must have a model with a compact covering group. That, unfortunately, excludes the popular SU(2) × U(1) model of Weinberg and Salam [6]. There are two classes of possibilities.

(i) In models of the type described by Georgi and Glashow [7], based on SO(3), we can construct monopoles with a mass of the order of 137 M_W, where M_W is the

* As we shall see this vacuum will still contain a radial magnetic field. This is because the incoming field in fig. 1 will be spread over the whole sphere.

mass of the familiar intermediate vector boson. In the Georgi–Glashow model, $M_W < 53\ \text{GeV}/c^2$.

(ii) The Weinberg–Salam model can still be a good phenomenological description of processes with energies around hundreds of GeV, but may need extension to a larger gauge group at still higher energies. Weinberg [8] proposed SU(3) × SU(3) which would then be compact. Then the monopole mass would be 137 times the mass of one of the superheavy vector bosons.

We choose the first possibility for our sample calculations, because it is the simplest one. We take as our Lagrangian:

$$\mathcal{L} = -\tfrac{1}{4} G^a_{\mu\nu} G^a_{\mu\nu} - \tfrac{1}{2} D_\mu Q_a D_\mu Q_a - \tfrac{1}{2}\mu^2 Q_a^2 - \tfrac{1}{8}\lambda (Q_a^2)^2 , \tag{2.1}$$

where

$$G^a_{\mu\nu} = \partial_\mu W^a_\nu - \partial_\nu W^a_\mu + e\, \epsilon_{abc} W^b_\mu W^c_\nu ,$$

$$D_\mu Q_a = \partial_\mu Q_a + e\, \epsilon_{abc} W^b_\mu Q_c . \tag{2.2}$$

W^a_μ and Q_a are a triplet of vector fields and scalar fields, respectively.

We choose the parameter μ^2 to be negative so that the field Q gets a non-zero vacuum expectation value [6,7,9]:

$$\langle Q_a \rangle^2 = F^2 , \qquad \mu^2 = -\tfrac{1}{2}\lambda F^2 . \tag{2.3}$$

Two components of the vector field will acquire a mass:

$$M_{W_{1,2}} = eF , \tag{2.4}$$

whereas the third component describes the surviving Abelian electromagnetic interactions. The Higgs particle has a mass:

$$M_H = \sqrt{\lambda}\, F . \tag{2.5}$$

We are interested in a solution where the Higgs field is not rotated everywhere towards the positive z direction. If we apply the transformation Ω of eq. (1.5) to the isospin-one vector $F(0,0,1)$ we get

$$F(\sin\theta \cos\varphi,\ \sin\theta \sin\varphi,\ \cos\theta). \tag{2.6}$$

We shall take this isovector as our boundary condition for the Higgs field at spacelike infinity. As one can easily verify, it implies that the Higgs field must have at least one zero. This zero we take as the origin of our coordinate system.

We now ask for a solution of the field equations that is time-independent and spherically symmetric, apart from the obvious angle dependence. Introducing the vector

$$r_a = (x,y,z) , \qquad r_a^2 = r^2 , \tag{2.7}$$

we can write

$$Q_a(x,t) = r_a Q(r) , \qquad W^a_\mu(x,t) = \epsilon_{\mu ab} r_b W(r) , \tag{2.8}$$

where $\epsilon_{\mu ab}$ is the usual ϵ symbol if $\mu = 1, 2, 3$, and $\epsilon_{4ab} = 0$.

In terms of these variables the Lagrangian becomes

$$L = \int \mathcal{L}\, \mathrm{d}^3x = 4\pi \int_0^\infty r^2\, \mathrm{d}r \Big[-r^2 \left(\frac{\mathrm{d}W}{\mathrm{d}r}\right)^2 - 4rW\frac{\mathrm{d}W}{\mathrm{d}r} - 6W^2 - 2er^2W^3$$

$$- \tfrac{1}{2}e^2r^4W^4 - \tfrac{1}{2}r^2\left(\frac{\mathrm{d}Q}{\mathrm{d}r}\right)^2 - rQ\frac{\mathrm{d}Q}{\mathrm{d}r} - \tfrac{3}{2}Q^2 - 2er^2WQ^2 - e^2r^4W^2Q^2$$

$$+ \tfrac{1}{4}\lambda F^2r^2Q^2 - \tfrac{1}{8}\lambda r^4Q^4 - \tfrac{1}{8}\lambda F^4 \Big] , \tag{2.9}$$

where the constant has been added to give the vacuum a vanishing action integral. The field equations are obtained by requiring L to be stationary under small variations of the functions $W(r)$ and $Q(r)$. The energy of the system is then given by

$$E = -L , \tag{2.10}$$

since our system is stationary.

Before calculating this energy, let us concentrate on the boundary condition at $r \to \infty$. From the preceding arguments we already know that we must insist on

$$Q(r) \to F/r . \tag{2.11}$$

The field W must behave smoothly, as some negative power of r:

$$W(r) \to ar^{-n} . \tag{2.12}$$

From (2.9) we find the Lagrange equation

$$\frac{\mathrm{d}}{\mathrm{d}r}\left(2r^4\frac{\mathrm{d}W}{\mathrm{d}r} + 4r^3W\right)$$

$$= r^2\,[4r\frac{\mathrm{d}W}{\mathrm{d}r} + 12W + 6er^2W^2 + 2e^2r^4W^3 + 2er^2Q^2 + 2e^2r^4WQ^2] . \tag{2.13}$$

So, substituting (2.11) and (2.12),

$$(3-n)(4-2n)ar^{2-n} \underset{r\to\infty}{\longrightarrow} -4nar^{2-n} + 12ar^{2-n} + 6ea^2r^{4-2n} + 2e^2a^3r^{6-3n}$$

$$+ 2eF^2r^2 + 2e^2aF^2r^{4-n} . \tag{2.14}$$

The only solution is

$$n = 2 , \qquad a = -1/e . \tag{2.15}$$

So, far from the origin, the fields are

$$W^a_\mu(x,t) \to -\epsilon_{\mu ab} r_b/er^2\,, \qquad Q_a(x,t) \to Fr_a/r\,. \tag{2.16}$$

Now most of these fields are not physical. To find the physically observable fields, in particular the electromagnetic ones, $F_{\mu\nu}$, we must first give a gauge invariant definition, which will yield the usual definition in the gauge where the Higgs field lies along the z direction everywhere. We propose:

$$F_{\mu\nu} = \frac{1}{|Q|}\, Q_a G^a_{\mu\nu} - \frac{1}{e|Q|^3}\, \epsilon_{abc} Q_a (D_\mu Q_b)(D_\nu Q_c)\,, \tag{2.17}$$

because, if after a gauge rotation, $Q_a = |Q|(0,0,1)$ everywhere within some region, then we have there

$$F_{\mu\nu} = \partial_\mu W^3_\nu - \partial_\nu W^3_\mu\,,$$

as one can easily check. (Observe that the definition (2.17) satisfies the usual Maxwell equations, except where $Q^a = 0$; this is one other way of understanding the possibility of monopoles in this theory.) From (2.16), we get (see the definitions (2.2)):

$$Q_a G^a_{\mu\nu} = -\frac{F}{er^3}\, \epsilon_{\mu\nu a} r_a\,, \tag{2.18}$$

$$D_\mu Q_a = \partial_\mu Q_a + e\epsilon_{abc} W^b_\mu Q_c = 0\,. \tag{2.19}$$

Hence

$$F_{\mu\nu} = -\frac{1}{er^3}\, \epsilon_{\mu\nu a} r_a\,. \tag{2.20}$$

Again, the ϵ symbol has been defined to be zero as soon as one of its indices has the value 4. So, there is a radial magnetic field

$$B_a = r_a/er^3\,, \tag{2.21}$$

with a total flux

$$4\pi/e\,.$$

Hence, our solution is a magnetic monopole, as we expected. It satisfies Schwinger's condition

$$eg = 1 \tag{2.22}$$

(in units where $\hbar = 1$). In sect. 4, however, we show that in certain cases only Dirac's condition

$$eg = \tfrac{1}{2} n\,, \qquad n \text{ integer}\,,$$

is satisfied.

3. The mass of the monopole

Let us introduce dimensionless parameters:

$$w = W/F^2e\,, \qquad q = Q/F^2e\,,$$
$$x = eFr\,, \qquad \beta = \lambda/e^2 = M_{\mathrm{H}}^2/M_{\mathrm{W}}^2\,. \tag{3.1}$$

From (2.9) and (2.10) we find that the energy E of the system is the minimal value of

$$\frac{4\pi M_{\mathrm{W}}}{e^2}\int_0^\infty x^2\,\mathrm{d}x\left[x^2\left(\frac{\mathrm{d}w}{\mathrm{d}x}\right)^2 + 4xw\frac{\mathrm{d}w}{\mathrm{d}x} + 6w^2 + 2x^2w^3 + \tfrac{1}{2}x^4w^4 + \tfrac{1}{2}x^2\left(\frac{\mathrm{d}q}{\mathrm{d}x}\right)^2\right.$$
$$\left. + xq\frac{\mathrm{d}q}{\mathrm{d}x} + \tfrac{3}{2}q^2 + 2x^2wq^2 + x^4w^2q^2 - \tfrac{1}{4}\beta x^2q^2 + \tfrac{1}{8}\beta x^4q^4 + \tfrac{1}{8}\beta\right]. \tag{3.2}$$

The quantity between the brackets is dimensionless and the extremum can be found by inserting trial functions and adjusting their parameters.

We found that the mass of the monopole (which is equal to the energy E since the monopole is at rest) is

$$M_{\mathrm{m}} = \frac{4\pi}{e^2}M_{\mathrm{W}}C(\beta)\,, \tag{3.3}$$

where $C(\beta)$ is nearly independent of the parameter β. It varies from 1.1 for $\beta = 0.1$ to 1.44 for $\beta = 10$ *.

Only in the Georgi–Glashow model (for which we did this calculation) is the parameter M_{W} in eq. (3.3) really the mass of the conventional intermediate vector boson. In other models it will in general be the mass of that boson which corresponds to the gauge transformations of the compact covering group: some of the superheavies in Weinberg's SU(3) × SU(3) for instance.

4. Conclusions

The relation between charge quantization and the possible existence of magnetic monopoles has been speculated on for a long time [10] and it has been observed that the gauge theories with compact gauge groups provide for the necessary charge quantization [11]. On the other hand, solutions of the field equations with abnormally rotated boundary conditions for the Higgs fields have also been considered before [1,12]. Nevertheless, it had escaped to our notion until now that magnetic monopoles occur among the solutions in those theories, and that their properties are predictable and calculable.

* These values may be slightly too high, as a consequence of our approximation procedure.

Our way of formulating the theory of magnetic monopoles avoids the introduction of Dirac's string [3], We expect no fundamental problems in calculating quantum corrections to the solution although they might be complicated to carry out.

The prediction is the most striking for the Georgi–Glashow model, although even in that model the mass is so high that that might explain the negative experimental evidence so far. If Weinberg's SU(2) × U(1) model wins the race for the presently observed weak interactions, then we shall have to wait for its extension to a compact gauge model, and the predicted monopole mass will be again much higher. Finally, one important observation. In the Georgi–Glashow model, one may introduce isospin $\frac{1}{2}$ representations of the group SU(2) describing particles with charges $\pm\frac{1}{2}e$. In that case our monopoles do not obey Schwinger's condition, but only Dirac's condition

$$qg = \tfrac{1}{2} \,,$$

where q is the charge quantum and g the magnetic pole quantum, in spite of the fact that we have a completely quantized theory. Evidently, Schwinger's arguments do not hold for this theory [13]. We do have, in our model

$$\Delta qg = 1 \,,$$

where Δq is the charge-difference between members of a multiplet, but this is certainly not a general phenomenon. In Weinberg's SU(3) × SU(3) the monopole quantum is the Dirac one and in models where the leptons form an SU(3) × SU(3) octet [14] the monopole quantum is three times the Dirac value (note the possibility of fractionally charged quarks in that case).

We thank H. Strubbe for help with a computer calculation of the coefficient $C(\beta)$, and B. Zumino and D. Gross for interesting discussions.

References

[1] H.B. Nielsen and P. Olesen, Niels Bohr Institute preprint, Copenhagen (May 1973); B. Zumino, Lectures given at the 1973 Nato Summer Institute in Capri, CERN preprint TH. 1779 (1973).
[2] Y. Nambu, Proc. Int. Conf. on symmetries and quark models, Detroit, 1969 (Gordon and Breach, New York, 1970) p. 269;
L. Susskind, Nuovo Cimento 69A (1970) 457;
J.L. Gervais and B. Sakita, Phys. Rev. Letters 30 (1973) 716.
[3] P.A.M. Dirac, Proc. Roy. Soc. A133 (1934) 60; Phys. Rev. 74 (1948) 817.
[4] J. Schwinger, Phys. Rev. 144 (1966) 1087.
[5] G. Parisi, Columbia University preprint CO-2271-29.
[6] S. Weinberg, Phys. Rev. Letters 19 (1967) 1264.
[7] M. Georgi and S.L. Glashow, Phys. Rev. Letters 28 (1972) 1494.
[8] S. Weinberg, Phys. Rev. D5 (1972) 1962.

[9] F. Englert and R. Brout, Phys. Rev. Letters 13 (1964) 321;
P.W. Higgs, Phys. Letters 12 (1964) 132; Phys. Rev. Letters 13 (1964) 508; Phys. Rev. 145 (1966) 1156;
G.S. Guralnik, C.R. Hagen and T.W.B. Kibble, Phys. Rev. Letters 13 (1964) 585.
[10] D.M. Stevens, Magnetic monopoles: an updated bibliography, Virginia Poly. Inst. and State University preprint VPI-EPP-73-5 (October 1973).
[11] C.N. Yang, Phys. Rev. D1 (1970) 2360.
[12] A. Neveu and R. Dashen, Private communication.
[13] B. Zumino, Strong and weak interactions, 1966 Int. School of Physics, Erice, ed. A. Zichichi (Acad. Press, New York and London) p. 711.
[14] A. Salam and J.C. Pati, University of Maryland preprint (November 1972).

CHAPTER 5
INSTANTONS

CHAPTER 5
INSTANTONS

Introduction to Computation of the Quantum Effects Due to a Four-Dimensional Pseudoparticle [5.1]

An introduction to instantons was already given in the first paper of the previous chapter. They are topologically stable twists in four dimensions. The classical solution in Euclidean space describes tunneling phenomena. In contrast with all other amplitudes in quantum field theory the instanton related amplitudes are exponentially damped when the coupling constant is small.

Now this confronted me with a problem. As explained earlier, quantum field theory cannot be trusted with infinite accuracy, because the loop expansion has a built-in deficiency: the divergence at very high orders, irrespective of the value of the coupling parameters. The margin of error in any amplitude due to this divergence in general cannot be avoided (with the possible exception of asymptotically free theories), and it is of the same order of magnitude as the instanton effects. Therefore, are the instanton effects real?

Fortunately, the instanton effects violate a symmetry, so that they can clearly be distinguished from ordinary perturbative contributions which in the corresponding channels vanish strictly. The dynamical origin of the instanton effects due to tunneling seems to be straightforward. And there seems to be no obstacle against calculating them. It was not hard to estimate what the results of such calculations would be.

But are these exponentially damped phenomena really completely self-consistent? The true coupling constant of the theory must be the renormalized one. Is that also the parameter in the exponent? Are instanton effects really finite? This was my motivation for performing some very detailed calculations. I wanted to know exactly how the nuts and bolts fit together here, and what the possible surprises would be. I decided to calculate the prefactor in front of the exponent.

The surprise was that everything works out fine. There were quite a few hurdles to be taken. The prefactor can really be seen as a one-loop quantum correction in the exponent itself. To compute it we needed the renormalized determinants of infinite dimensional matrices $\mathcal{M}$, and several tricks were needed to obtain these. We first had to remove the zero modes, and for that we had to understand what their physical intepretation is. There are zero modes corresponding to translations and scale transformations, but also three due to global gauge rotations. The latter can only be understood in relation with the gauge fixing procedure. The zero modes, eight in total (if the gauge group is $SU(2)$), give rise to an extra factor g^{-8} in front of the exponent.

The original paper contained a few minor errors. Now all errors act multiplicatively on the final result, so as far as the final amplitude is concerned none of the errors was minor. But of course the real importance of our calculation was to see if it could be done at all, and if the answer would come out finite. The answer we found to this is of course not affected by small errors. Anyway, the version reproduced here has been re-edited to remove all errors known to us. We owe some of these refinements to meticulous calculations by Ore and Hasenfratz.[16]

The word "instanton" does not occur in the original publication. I had proposed this name in a previous paper, but the editors of *Phys. Rev. Lett.* and *Phys. Rev.* opposed the use of self-invented phrases. Their alternative ("Euclidean Gauge Field Pseudoparticle, EGFP for short") was so ugly that the name instanton caught on quickly.

Introduction to How Instantons Solve the $U(1)$ Problem [5.2]

Because of their "nonperturbative" nature, instantons give rise to a number of controverses. A very important recent issue is the question whether the symmetry breaking effects associated with instantons continue to be exponentially damped when collision processes at very high energies are considered. A number of authors have presented calculations suggesting sharply increasing amplitudes, and at energies above a few tens of TeV's the exponential could be completely cancelled so that the amplitudes could even hit the unitarity barrier. It is true that at such energies tunneling is formally no longer necessary. The transition could take place via so-called *sphalerons*. These are metastable field configurations resembling magnetic monopoles, but such that they can decay with equal probabilities into two channels with different winding numbers, hence different B and L.

But that the amplitudes should become strong is a highly dubious claim. Precisely because Feynman diagrams can no longer be trusted (they diverge) calculations based on diagrammatic techniques may well be incorrect. My own attempt at constructing a physically reasonable scenario for the efficiency of sphaleron production at high energies suggests that exponential suppression factors, comparable to

[16] F. R. Ore, *Phys. Rev.* **D16** (1977) 2577; A. Hasenfratz and P. Hasenfratz, *Nucl. Phys.* **B193** (1981) 210.

the one at low energies, $\exp(-8\pi^2/g^2)$, continue to be present at all energies. Only in high temperature plasmas such as the one that must have existed in the very early universe, the conditions for efficient sphaleron production seem to be realized.

So much for instantons in weak interaction theories. In the strong interactions they also occur, and here the exponential suppression is not important. The symmetry violated by instantons is chiral symmetry. This observation finally provided for a satisfactory answer to a famous problem in QCD: the $U(1)$ problem. According to its Lagrangian namely, QCD possesses a nearly perfect global (chiral) symmetry of the form $U(2) \otimes U(2)$, only broken by the light quark masses into the diagonal $U(2)$ subgroup of this group. This group has eight independent generators. Phenomenologically however, there is only a global $SU(2) \otimes SU(2) \otimes U(1)$ chiral symmetry, a group with seven generators, spontaneously broken into $SU(2) \otimes U(1)$. The three broken symmetry generators give rise to three Goldstone bosons, π^+ , π^0 and π^-. This means that one $U(1)$ subgroup of $U(2) \otimes U(2)$ is missing. The corresponding Goldstone boson would be one with the quantum numbers of the η particle. The mystery is that for this the η particle seems to be too heavy (550 MeV). Why is it so much heavier than the pions (140 MeV)? Remember that we have to compare the squares of the masses rather than the masses themselves. A spurious interaction, notably absent in the QCD Lagrangian, must be responsible for this explicit chiral symmetry breaking.

In my 1975 Palermo paper (Chapter 2.3) I already indicated what a number of authors had observed: chiral $U(1)$ is explicitly broken by an anomaly. The only remaining problem is: how does this anomaly raise the mass of the η particle? The answer is that the extra interaction responsible for this is precisely the one effectuated by the instanton. Indeed, the effective interaction vertex due to instantons has precisely the quantum numbers of an η mass term.

But things are not quite this simple. In a number of papers, Rodney Crewther criticised this point of view. He claimed that the theory, including all its instantons, still obeys what he called "anomalous Ward identities", and that these identities prohibit an η mass term.

These arguments seemed to be formal mathematics to me, not reflecting the true physical nature of the instanton-induced interactions. Then, in 1984, a pupil of Crewther's, G. A. Christos, published a review paper in *Physics Reports*[17] clarifying Crewther's arguments against the instanton solution of the $U(1)$ problem. This paper was written in a style that compelled me to react. The "anomalous Ward identities" do not refer to amplitudes relating to physically observable quantities, and do not have a proper large distance limit. The easiest way to see what happens is to write down effective Lagrangians that exhibit all symmetries and selection rules (including the spurious Ward identities) of the expected instanton interactions. This was published in another issue of Physics Reports, of which we include a reprint.

[17] G. A. Christos, *Phys. Rep.* **116** (1984) 251.

Introduction to Naturalness, Chiral Symmetry, and Spontaneous Chiral Symmetry Breaking [5.3]

The next paper is only indirectly related to instantons. It shows how topological winding numbers can be used to deduce information about the spectrum of light bound states in theories such as QCD. The paper, one of a series of lectures presented in a summer school in Cargèse in 1979, speculates on how the Standard Model may be extended beyond the magic threshold of 1 TeV. At present most particle physicists adhere to the theory that beyond that energy range particles form supersymmetric multiplets. This could be a natural way to protect the Higgs scalar from becoming too heavy. At the moment this is written there is however still no single direct evidence for supersymmetry, and certainly in 1979 there wasn't.

If there is no supersymmetry then there must be strong interactions in the TeV range. The symmetry pattern of these interactions must then be such that the presently observed particle spectrum is the spectrum of low-lying bound states. How do we compute this spectrum in a given theory? Take QCD as an example. At low energies we have a chiral σ model. Chiral symmetry is spontaneously broken, but in other versions of this dynamical system chiral symmetry may be realized in a different way. We could have massless nucleons instead of massless pions. How many massless bound states would be protected by the symmetries? What are the rules for more complicated variations of the theme "QCD"? This is the question the paper addresses. And an important theorem is discussed: all triangle anomalies for all fermionic currents (in particular the ones to which no gauge fields are coupled) must be the same for the bound state spectrum as they are for the original "bare" constituent particles. This leads to index theorems constraining the possibilities for the bound state spectrum.

The theorems were not powerful enough to constrain the spectrum completely, so I tried to deduce more theorems. They are discussed in Section 3.12.

The latter however are mere conjectures. It turns out that they cannot be completely true because the resulting spectrum would have fractional occupation numbers which is clearly absurd. My own conclusion was therefore that the various chiral symmetries *must* be spontaneously broken so that instead of massless fermions we get massless scalars. This would kill the bound state theories for quarks and leptons, and, disappointingly, such was my conclusion. Many authors since then argued that the conjectures of Section 3.12 are too stringent. But even the anomaly matching conditions alone (which surely must be true) are so restrictive that no really attractive bound state model for quarks and leptons could be produced. Does this imply that we have to accept the supersymmetry scenario? Or is Nature simply too complicated for us to guess how things go from what we know at present? Only the Future will tell how stupid we are now.

Computation of the quantum effects due to a four-dimensional pseudoparticle*

G. 't Hooft†
Physics Laboratories, Harvard University, Cambridge, Massachusetts 02138
(Received 28 June 1976)

A detailed quantitative calculation is carried out of the tunneling process described by the Belavin-Polyakov-Schwarz-Tyupkin field configuration. A certain chiral symmetry is violated as a consequence of the Adler-Bell-Jackiw anomaly. The collective motions of the pseudoparticle and all contributions from single loops of scalar, spinor, and vector fields are taken into account. The result is an effective interaction Lagrangian for the spinors.

I. INTRODUCTION

When one attempts to construct a realistic gauge theory for the observed weak, electromagnetic, and strong interactions, one is often confronted with the difficulty that most simple models have too much symmetry. In Nature, many symmetries are slightly broken, which leads to, for instance, the lepton masses, the quark masses, and *CP* violation. These symmetry violations, either explicit or spontaneous, have to be introduced artificially in the existing models.

There is one occasion where explicit symmetry violation is a necessary consequence of the laws of relativistic quantum theory: the Adler-Bell-Jackiw anomaly. The theory we consider is an SU(2) gauge theory with an arbitrary set of scalar fields and a number, N^f, of massless fermions. The apparent chiral symmetry of the form $U(N^f)\times U(N^f)$ is actually broken down to $SU(N^f)\times SU(N^f)\times U(1)$. This paper is devoted to a detailed computation of this effect.

The most essential ingredient in our theory is the localized classical solution of the field equations in Euclidean space-time, of the type found by Belavin *et al.*[1]

Although the main objective of this paper is the computation of the resulting effective symmetry-breaking Lagrangian in a weak-interaction theory, we present the calculations in such a way that they can also be used for possible color gauge theories of stong interactions based on the same classical field configurations. For such theories our intermediate expressions (12.5) and (12.8) will be applicable. Our final results are (15.1) together with the convergence factor (15.8).

Our general philosophy has been sketched in Ref. 2. We are dealing with amplitudes that depend on the coupling constant g in the following way:

$$g^{-C}\exp\left[-\frac{8\pi^2}{g^2}(1+a_1g^2+\cdots)\right]. \qquad (1.1)$$

The coefficient a_1 involves one-loop quantum corrections, and it determines the scale of the amplitude. Clearly, then, to understand the main features of such an amplitude, complete understanding of all one-loop quantum effects is desired. For instance, if one changes from one renormalization subtraction procedure to another, so that $g^2\to g^2+O(g^4)$, then this leads to a change in (1.1) by an overall multiplicative constant. Thus, the renormalization subtraction point μ may enter as a dimensional parameter in front of our expressions. This is just one of the reasons to suggest that our results will also have interesting applications in strong-interaction color gauge theories.

The underlying classical solutions only exist in Euclidean space, but they give rise to a particular symmetry-breaking amplitude that can easily be continued analytically to Minkowski space. We interpret this amplitude as the result of a certain tunneling effect from one vacuum to a gauge-rotated vacuum. We recall that, indeed, tunneling through a barrier can sometimes be described by means of a classical solution of the equation of motion in the imaginary time direction.[2,3]

We compute in Euclidean space the vacuum-to-vacuum amplitude in the presence of external sources, thus obtaining full Green's functions. Of course, we must limit ourselves to gauge-invariant sources only, but that will be no problem. It turns out to be trivial to amputate the obtained Green's function and get the effective vertex.

The various calculational steps are the following. We first give in Sec. II the functional integral expression for the amplitude, first in a conventional Feynman gauge: $C_1=\partial_\mu A_\mu$. Later, we go over to the so-called background gauge: $C_4=D_\mu A_\mu^{\rm qu}$. This is actually only correct up to an overall factor, as will be explained in Sec. XI. It is just for pedagogical reasons that we ignore this complication for a moment. It is in this gauge that the quantum excitations take a simple form: "Spin-orbit" couplings commute with the operator L^2

$= -\frac{1}{8} L_{\mu\nu} L_{\mu\nu}$, where

$$L_{\mu\nu} = x_\mu \frac{\partial}{\partial x_\nu} - x_\nu \frac{\partial}{\partial x_\mu}, \tag{1.2}$$

so that we can look at eigenstates of L^2. (In other gauges only total angular momentum $\vec{J} = \vec{L}$ + spin + isospin is conserved, not L^2.)

In Sec. III we consider the quantum fluctuations described by an eigenvalue equation,

$$\mathfrak{M}\psi = E\psi, \tag{1.3}$$

in order to compute det$\mathfrak{M}$. Now this looks like an ordinary scattering problem (in 4+1 dimensions), and, indeed, we show that the product of all nonzero eigenvalues E, in some large box, can be expressed in terms of the phase shift $\eta(k)$ as a function of the wave number k.

In Sec. IV we show that Eq. (1.3) is essentially the same for scalars, spinors, and vectors, from which we derive the important result that the product of all nonzero eigenvalues is the same for scalar fields as for each component of the spinor and vector fields. So, we turn to the (much easier) scalar case first.

But even for scalars Eq. (1.3) has no simple solutions in terms of well-known elementary functions. We decide not to compute det$\mathfrak{M}$ by solving (1.2), but we compute instead

$$\det[(1+x^2)\mathfrak{M}(1+x^2)] .$$

The factors $1+x^2$ drop out if we divide by the same determinant coming from the vacuum (i.e., the case $A^{\text{cl}} = 0$). The equation

$$\mathfrak{M}\psi = \frac{\lambda}{(1+x^2)^2}\psi \tag{1.4}$$

is a simple hypergeometric equation that can be solved under the given boundary condition (Sec. V).

Now we must find the product of all eigenvalues λ, but that diverges badly even if we divide by the values they take in the vacuum. We must find a gauge-invariant regulator, and the regulator determinant must be calculable. Dimensional regularization is not applicable here, but we can use background Pauli-Villars regulators. They give messy equations unless they have a space-time-dependent mass:

$$\frac{M_i^2}{(1+x^2)^2} . \tag{1.5}$$

The rules are formulated in Sec. V.

In Sec. VI we compute the product of the eigenvalues using this regulator. In Sec. VII we make the transition to regulators with fixed mass by observing that a change toward fixed regulator mass must correspond to a local, space-time-dependent counterterm in the Lagrangian. The effect of this counterterm is computed.

Using the result of Sec. IV we now find also the contributions of all nonvanishing eigenvalues for the vectors and spinors. But there are also vanishing eigenvalues. They are listed in Sec. VIII. For the vector fields, we have eight zero eigenvectors in addition to the ones computed via the theorem of Sec. IV. They are to be interpreted as translations (Sec. IX), dilatation (Sec. X), and isospin rotations (Sec. XI). The last need special care and can only be interpreted correctly when different gauge choices are compared. This leads to the factor mentioned in the beginning.

In Sec. XII we combine the results so far obtained and add the fermions. This intermediate result may be useful to strong-interaction theories. In Sec. XIII we reexpress the result in terms of the dimensionally renormalized coupling constant g^D, as opposed to the previous coupling constant which was renormalized in a Pauli-Villars manner. In Sec. XIV the external sources for the fermions are considered and the amputation operation for the Green's function is performed. We obtain the desired effective Lagrangian, but there is still one divergence. So far, we only had massless particles, and as a consequence of that there is still a scale parameter ρ over which we must integrate. Asymptotic freedom gives a natural cutoff for this integral in the ultraviolet direction, but there is still an infrared divergence. In weak-interaction theories the Higgs field is expected to provide for the infrared cutoff. Section XV shows how to compute this cutoff.

The Appendix lists the properties of the symbols η, $\bar{\eta}$ which are used many times throughout these calculations.

II. FORMULATION OF THE PROBLEM

Let a field theory in four space-time dimensions be given by the Lagrangian

$$\mathcal{L} = -\tfrac{1}{4} G^a_{\mu\nu} G^a_{\mu\nu} - D_\mu \Phi^* D_\mu \Phi - \bar{\psi}\gamma_\mu D_\mu \psi + \bar{\psi}_s \mathcal{G}_{st} \psi_t , \tag{2.1}$$

where the gauge group is SU(2):

$$G^a_{\mu\nu} = \partial_\mu A^a_\nu - \partial_\nu A^a_\mu + g\epsilon_{abc} A^b_\mu A^c_\nu . \tag{2.2}$$

The SU(2) indices will be called isospin indices. The scalars Φ, taken to be complex, may contain several multiplets of arbitrary isospin:

$$D_\mu \Phi = \partial_\mu \Phi - igT^a A^a_\mu \Phi , \quad [T^a, T^b] = i\epsilon_{abc} T^c . \tag{2.3}$$

The spinors ψ are taken to be isospin-$\frac{1}{2}$ doublets.

The total number of doublets is N^f. Mass terms and interaction terms between scalars and spinors are irrelevant for the time being.

We inserted a source term in a gauge-invariant way, with respect to which we will expand, in order to obtain Green's functions. The indices $s,t=1,\ldots,N^f$, called flavor indices, label the different isospin multiplets. Isospin and Dirac indices have been suppressed. $\mathcal{J}_{st}$ must be diagonal in the isospin indices but may contain Dirac γ matrices.

The system (2.1) seems to have a chiral $U(N^f)\times U(N^f)$ global symmetry, but actually has an Adler-Bell-Jackiw anomaly[4] associated with the chiral U(1) current, breaking the symmetry down to $SU(N^f)\times SU(N^f)\times U(1)$. The aim of this paper is to find that part of the amplitude that violates the chiral U(1) conservation.

The functional integral expression for the amplitude is

$$W={}_{\text{out}}\langle 0\,|\,0\rangle_{\text{in}}$$

$$=\int \mathfrak{D}A\,\mathfrak{D}\psi\,\mathfrak{D}\Phi\,\mathfrak{D}\phi\;\times\exp\left\{\int\left[\mathcal{L}-\tfrac{1}{2}C_1{}^2(A)+\mathcal{L}_1^{\text{ghost}}(\phi)\right]d^4x\right\},\qquad(2.4)$$

to be expanded with respect to $\mathcal{J}$. Here $C_1(A)$ is a gauge-fixing term, and $\mathcal{L}_1^{\text{ghost}}$ are the corresponding ghost terms.

As is argued in Ref. 2, the U(1)-breaking part of this amplitude comes from that region of superspace where the A field approaches the solutions described in Ref. 1:

$$A_\mu^a(x)^{\text{cl}}=\frac{2}{g}\,\frac{\eta_{a\mu\nu}(x-z)^\nu}{(x-z)^2+\rho^2}\,,\qquad(2.5)$$

where z_μ and ρ are five free parameters associated with translation invariance and scale invariance. The coefficients η are studied in the Appendix. Conjugate to (2.5) we have its mirror image, described by the coefficients $\bar\eta$ (see Appendix).

Now these solutions form a local extremum of our functional integrand, and therefore it makes sense to consider separately that contribution to W in (2.4) that is obtained through a new perturbation expansion around these new solutions, taking the integrand there to be approximately Gaussian. The fields Φ, ϕ, and ψ all remain infinitesimal so that their mutual interactions may be neglected in the first approximation. Of course, we must also integrate over the values of z_μ and ρ. This will be done by means of the collective-coordinate formalism.[5] One writes

$$A_\mu^a=A_\mu^{a\,\text{cl}}+A_\mu^{a\,\text{qu}}\,,\qquad(2.6)$$

and those values of A^{qu} that correspond to translations or dilatations are replaced by collective coordinates.

The integrand in (2.4) now becomes

$$\mathcal{L}(A^{\text{cl}})-\tfrac{1}{2}(D_\mu A_\nu^{\text{qu}})^2+\tfrac{1}{2}(D_\mu A_\mu^{\text{qu}})^2-gA_\nu^{a\,\text{qu}}\epsilon_{abc}G_{\mu\nu}^{b\,\text{cl}}A_\mu^{c\,\text{qu}}-D_\mu\Phi^*D_\mu\Phi-\bar\psi\gamma_\mu D_\mu\psi+\bar\psi\mathcal{J}\psi-\tfrac{1}{2}C_1{}^2+\mathcal{L}_1^{\text{ghost}}+\mathcal{O}(A^{\text{qu}},\Phi,\psi)^3\,,\qquad(2.7)$$

where

$$S^{\text{cl}}=\int\mathcal{L}(A^{\text{cl}})d^4x=-8\pi^2/g^2\,,\qquad(2.8)$$

and the "covariant derivative" D_μ only contains the background field A_μ^{cl}, for instance:

$$D_\mu A_\nu^{a\,\text{qu}}=\partial_\mu A_\nu^{a\,\text{qu}}+g\epsilon_{abc}A_\mu^{b\,\text{cl}}A_\nu^{c\,\text{qu}}\,,\qquad(2.9)$$

etc.

We abbreviate the integral over (2.7) by

$$S^{\text{cl}}-\tfrac{1}{2}A^{\text{qu}}\mathfrak{M}_A A^{\text{qu}}+\bar\psi\,\mathfrak{M}_\psi\psi-\Phi^*\mathfrak{M}_\Phi\Phi-\phi^*\mathfrak{M}_{\text{gh}}\phi\,,\qquad(2.10)$$

where the last term describes the Faddeev-Popov ghost. Thus, expression (2.4) is (ignoring temporarily the collective coordinates, and certain factors $\sqrt{\pi}$ from the Gaussian integration; see Sec. IX)

$$W=\exp(-8\pi^2/g^2)(\det\mathfrak{M}_A)^{-1/2}\det\mathfrak{M}_\psi(\det\mathfrak{M}_\Phi)^{-1}\times\det\mathfrak{M}_{\text{gh}}\,.\qquad(2.11)$$

The determinants will be computed by diagonalization:

$$\mathfrak{M}_i\psi=E_i\psi\,,\qquad(2.12)$$

after which we multiply all eigenvalues E. Since there are infinitely many very large eigenvalues, this infinite product diverges very badly. There are two procedures that will make it converge:

(i) The vacuum-to-vacuum amplitude in the absence of sources must be normalized to 1, so that the vacuum state has norm 1. This implies that W must be divided by the same expression with $A^{\text{cl}}=0$.

(ii) We must regularize and renormalize. The dimensional procedure is not available here because the four-dimensionality of the classical solution is crucial. We will use the so-called background Pauli-Villars regulators (Secs. IV and V).

Taking a closer look at the eigenvalue equations (2.12) as they follow from (2.7), we notice that the background field in there gives rise to couplings between spin, isospin, and (the four-dimensional equivalent of) orbital angular momentum, through the coefficients $\eta_{a\mu\nu}$ in (2.5). Now these couplings simplify enormously if we go over to a new gauge that explicitly depends on the background field[6]

$$C_4^a(A^{qu}) = D_\mu A_\mu^{a\,qu}$$
$$= \partial_\mu A_\mu^{a\,qu} + g\epsilon_{abc} A_\mu^{b\,cl} A_\mu^{c\,qu}. \tag{2.13}$$

Thus the third and eighth terms in (2.7) cancel.

This choice of gauge will lead to one complication, to be discussed in Sec. XI: The gauge for the vacuum-to-vacuum amplitude in the absence of sources, used for normalization, in the region $A_\mu \sim 0$, is usually invariant under global isospin rotations, but the classical solution (2.5) and the gauge (2.13) are not. Associated with this will be three spurious zero eigenvalues of $\mathfrak{M}_A$ that cannot be directly associated with global isospin rotations. The question is resolved in Sec. XI by careful comparison of the gauge C_1 with C_4 and some intermediate choices of gauge.

There will be five other zero eigenvalues of $\mathfrak{M}_A$ that of course must not be inserted in the product of eigenvalues directly, since they would render expression (2.11) infinite. They exactly correspond to the infinitesimal translations and dilatations of the classical solution, and, as discussed before, must be replaced by the corresponding collective coordinates (Secs. IX and X).

The matrices $\mathfrak{M}$ are now (ignoring temporarily the fermion source)

$$\mathfrak{M}_A A_\mu^{a\,qu} = -D^2 A_\mu^{a\,qu} - 2g\epsilon_{abc} G_{\mu\nu}^{b\,cl} A_\nu^{c\,qu},$$
$$-\mathfrak{M}_\psi{}^2 \psi = -D^2\psi + \tfrac{1}{4} i\tau^a G_{\mu\nu}^{a\,cl} \gamma_\mu \gamma_\nu \psi,$$
$$\mathfrak{M}_\Phi \Phi = -D^2 \Phi, \tag{2.14}$$
$$\mathfrak{M}_{gh} \phi = -D^2 \phi.$$

In order to substitute the classical solution (2.5) with $z=0$ and $\rho=1$ (generalization to other z and ρ will be straightforward), we introduce the space-time operators

$$L_1^a = -\tfrac{1}{2} i\eta_{a\mu\nu} x^\mu \frac{\partial}{\partial x^\nu},$$
$$L_2^a = -\tfrac{1}{2} i\bar{\eta}_{a\mu\nu} x^\mu \frac{\partial}{\partial x^\nu}, \tag{2.15}$$

with

$$[L_p^a, L_q^b] = i\delta_{pq}\epsilon_{abc} L_p^c,$$
$$L^2 = L_1{}^2 = L_2{}^2 = -\tfrac{1}{8}(x_\mu \partial_\nu - x_\nu \partial_\mu)^2. \tag{2.16}$$

They represent rotations in the two invariant SU(2) subgroups of the rotation group SO(4).

Isospin rotations will be generated by the operators T^a for the scalars, $T^a = \tfrac{1}{2}\tau^a$ for the spinors, and $T^b A_\mu^a = i\epsilon_{abc} A_\mu^c$ for the vectors. Then

$$D^2 = \left(\frac{\partial}{\partial r}\right)^2 + \frac{3}{r}\frac{\partial}{\partial r} - \frac{4}{r^2}L^2 - \frac{8}{r^2+1} T\cdot L_1 - \frac{4r^2}{(r^2+1)^2} T^2, \tag{2.17}$$

where $r^2 = (x-z)^2$. This clearly displays the isospin-orbit coupling.

The vector and spinor fields also have a spin-isospin coupling. For the spinors we define the spin operators

$$S_1^a \psi = -\tfrac{1}{8} i\eta_{a\mu\nu} \gamma_\mu \gamma_\nu \psi,$$
$$S_2^a \psi = -\tfrac{1}{8} i\bar{\eta}_{a\mu\nu} \gamma_\mu \gamma_\nu \psi, \tag{2.18}$$

satisfying

$$[S_p^a, S_q^b] = i\delta_{pq}\epsilon_{abc} S_p^c \tag{2.19}$$

and

$$S_1{}^2 = \frac{3}{4}\frac{1-\gamma_5}{2},$$
$$S_2{}^2 = \frac{3}{4}\frac{1+\gamma_5}{2}. \tag{2.20}$$

For the vector fields we define

$$S_1^a A_\mu^{qu} = -\tfrac{1}{2} i\eta_{a\mu\nu} A_\nu^{qu},$$
$$S_2^a A_\mu^{qu} = -\tfrac{1}{2} i\bar{\eta}_{a\mu\nu} A_\nu^{qu}, \tag{2.21}$$
$$S_1{}^2 = S_2{}^2 = \tfrac{3}{4}.$$

For the scalar fields $\vec{S}_1 = \vec{S}_2 = 0$. Thus right- and left-handed spinors are $(\tfrac{1}{2}, 0)$ and $(0, \tfrac{1}{2})$ representations of SO(4), and vectors are $(\tfrac{1}{2}, \tfrac{1}{2})$ representations. Scalars of course are $(0,0)$ representations.

In terms of the operators S and T, the spin-isospin couplings turn out to be universal for all particles. Substituting the classical value for $G_{\mu\nu}^{a\,cl}$ in (2.14) we find

$$\mathfrak{M} = -\left(\frac{\partial}{\partial r}\right)^2 - \frac{3}{r}\frac{\partial}{\partial r} + \frac{4}{r^2}L^2 + \frac{8}{1+r^2} T\cdot L_1$$
$$+ \frac{4r^2}{(1+r^2)^2} T^2 + \frac{16}{(1+r^2)^2} T\cdot S_1, \tag{2.22}$$

with $\mathfrak{M} = \mathfrak{M}_A$ or $-\mathfrak{M}_\psi{}^2$ or $\mathfrak{M}_\Phi$ or $\mathfrak{M}_\phi$.

Observe the absence of spin-orbit and isospin-orbit couplings that contain x_μ or $\partial/\partial x_\mu$ explicitly. It all goes via the orbital angular momentum operator L_1 and that implies that L^2 commutes with $\mathfrak{M}$. This would not be so in other gauges. Further, $\mathfrak{M}$ commutes with $\vec{J}_1 = \vec{L}_1 + \vec{S}_1 + \vec{T}$ and $\vec{L}_2$ and $\vec{S}_2$. Eigenvectors of $\mathfrak{M}$ can thus be characterized by the quantum numbers

s_1 and s_2 (both either 0 or $\frac{1}{2}$),

t (total isospin, arbitrary for the scalars, $\frac{1}{2}$ for the spinors, 1 for the vector and the ghost),

$l = 0, \frac{1}{2}, 1, \ldots,$

$$j_1 = l - s_1 - t, l - s_1 - t + 1, \ldots, l + s_1 + t, \quad \text{as long as } j_1 \geq 0, \tag{2.23}$$

$j_1^3 = -j_1, \ldots, +j_1,$

$s_2^3 = -s_2, \ldots, +s_2,$

$l_2^3 = -l, \ldots, +l.$

For normalization we need the corresponding operator $\mathfrak{M}$ for the case that the background field is zero:

$$\mathfrak{M}_0 = -\left(\frac{\partial}{\partial r}\right)^2 - \frac{3}{r}\frac{\partial}{\partial r} + \frac{4}{r^2}L^2. \tag{2.24}$$

III. DETERMINANTS AND PHASE SHIFTS

The eigenvalue equation (2.12) with $\mathfrak{M}$ as in (2.22) differs in no essential way from an ordinary Schrödinger scattering problem. In this section we show the relation between the corresponding scattering matrix and the desired determinant.

Temporarily, we put the system in a large spherical box with radius R. At the edge we have some boundary condition: either $\Psi(R) = 0$, or $\Psi'(R) = 0$ (or a linear combination thereof). Here Ψ stands for any of the scalar, spinor, or vector fields. In the case $\Psi'(R) = 0$ the vacuum operator $\mathfrak{M}_0$ has a zero eigenvalue corresponding to Ψ = constant, and also the lowest eigenvalue of $\mathfrak{M}$ may go to zero more rapidly than $1/R^2$ when $R \to \infty$. Such eigenvalues have to be considered separately (negative eigenvalues can be proved not to exist).

We here consider all other eigenvalues of $\mathfrak{M}$. They approach the ones of $\mathfrak{M}_0$ if $R \to \infty$. We wish to compute the product

$$\prod_{n=1}^{\infty} \frac{E(n)}{E_0(n)}. \tag{3.1}$$

The scattering matrix $S(k) = e^{2i\eta(k)}$ will be defined by comparing the solution of

$$\mathfrak{M}\Psi = k^2\Psi \tag{3.2}$$

with

$$\mathfrak{M}_0\Psi_0 = k^2\Psi_0 \tag{3.3}$$

both with boundary condition $\Psi \to Cr^{2l}$ at $r = 0$. Let

$$\Psi_0(r) \sim Cr^{-3/2}(e^{-ik(r+a)} + e^{+ik(r+a)}) = 2Cr^{-3/2}\cos k(r+a) \quad \text{for large } r \tag{3.4}$$

and

$$\Psi(r) \propto Cr^{-3/2}[e^{-ik(r+a)} + S(k)e^{ik(r+a)}] = 2C'r^{-3/2}\cos[k(r+a) + \eta(k)]. \tag{3.5}$$

If we require at $r = R$ the same boundary condition for Ψ_0 then we must solve

$$k(n)(R+a) + \eta(k(n)) = k_0(n)(R+a), \tag{3.6}$$

thus

$$\frac{k(n)}{k_0(n)} \to 1 - \frac{\eta(k(n))}{(R+a)k(n)}. \tag{3.7}$$

The level distance $\Delta k = k(n+1) - k(n)$ is in both cases, asymptotically for large R,

$$\Delta k = \frac{\pi}{R} + O\left(\frac{1}{R^2}\right). \tag{3.8}$$

We find that

$$\prod_{n=1}^{\infty} \frac{E(n)}{E_0(n)} = \exp\left\{2\sum_1^{\infty} \ln[k(n)/k_0(n)]\right\} = \exp\left\{2\frac{R}{\pi}\sum_1^{\infty} \Delta k\left[-\frac{\eta(k)}{Rk} + O\left(\frac{1}{R^2}\right)\right]\right\} \to \exp\left[-\frac{2}{\pi}\int_0^{\infty} \frac{\eta(k)}{k}\,dk\right], \tag{3.9}$$

provided that the integral converges at both ends.

At $k \to 0$ the integral (3.9) converges provided that the interaction potential decreases faster than $1/r^2$ as $r \to \infty$; at $k \to \infty$ the integral converges if the interaction potential is less singular than $1/r^2$ as $r \to 0$. The latter condition is satisfied if we compare $\mathfrak{M}$ and $\mathfrak{M}_0$ at the same values for the quantum number l; the first condition is satisfied if $(L_1 + T)^2$ for the interacting matrix is set equal to L^2 for the vacuum matrix. If we consider the combined effect of all values for L^2 and $(L_1 + T)^2$ both for the vacuum and for the interacting case then we can split the integral (3.9) somewhere in the middle, and combine the $k \to \infty$ parts, so that we get convergence everywhere.

An easier way to get convergence is to regularize:

$$\prod_{n=1}^{\infty} \frac{E(n)[E_0(n)+M^2]}{E_0(n)[E(n)+M^2]} \to \exp\left[-\frac{2}{\pi}\int_0^\infty \frac{\eta(k)}{k}\frac{M^2}{(k^2+M^2)}dk\right]. \tag{3.10}$$

Regulators will be introduced anyhow, so we will not encounter difficulties due to non-convergence of the integral in (3.9).

IV. ELIMINATION OF THE SPIN DEPENDENCE

In Eq. (2.22) the operators $T\cdot L_1$ and $T\cdot S_1$ do not commute. Only in the case that

$$|j_1 - l| = s+t$$

(as defined in 2.23) do they simultaneously diagonalize. If

$$|j_1 - l| < s+t$$

and

$$j_1 \neq 0, \quad l\neq 0, \quad s=\tfrac{1}{2}, \quad t\neq 0,$$

then we have a set of coupled differential equations for two dependent variables.

In any other case there would be no hope of solving this set of equations analytically, but here we can make use of a unique property of the equation

$$\mathfrak{M}\Psi = E\Psi\,, \tag{4.1}$$

which enables us to diagonalize it completely. If $s_1=0$ the equation could describe a left-handed fermion with isospin t:

$$\mathfrak{M}\psi = -(\gamma\cdot D)^2\psi = E\psi\,, \qquad \gamma_5\psi = +\psi\,. \tag{4.2}$$

But then we can define, if $E\neq 0$,

$$\psi' = \gamma\cdot D\psi \tag{4.3}$$

with

$$\mathfrak{M}\psi' = E\psi'\,, \qquad \gamma_5\psi' = -\psi' \tag{4.4}$$

Now ψ' has $s_1'=\frac{1}{2}$, and hence we found a solution for the set of coupled equations with $s_1'=\frac{1}{2}$ from a solution of the simpler equation with $s_1=0$. The operater $\gamma\cdot D$ in Eq. (4.3) does not commute with L^2, so if ψ has a given set of quantum numbers l, j_1, t then ψ' is a superposition of a state with $l' = l+\frac{1}{2}$ and one with $l'=l-\frac{1}{2}$. Now $\mathfrak{M}$ does commute with L^2, so if we project out the state with $l'=l+\frac{1}{2}$ or $l'=l-\frac{1}{2}$ then we get a new solution in both cases. Thus one solution with $s_1=0$ and quantum numbers l, j_1, t generates two solutions with $s_1'=\frac{1}{2}$, $l'=l\pm\frac{1}{2}$, $j_1'=j_1$, $t'=t$. In terms of the operators L, S, and T the new solutions to the coupled equations can be expressed in terms of the $S=0$ solutions as follows:

$$L_a'+S_a'=L_a\,, \quad l'=l\pm\tfrac{1}{2}\,, \qquad T'=T\,, \quad J'=J\,, \tag{4.5}$$

$$\Psi' = \left[\frac{1}{r}\,(2L^2-2L'^2+\tfrac{3}{2}) + \frac{4r}{1+r^2}\,S_1'\cdot T\right]\Psi + \frac{\partial}{\partial r}\Psi\,. \tag{4.6}$$

It is easy to check explicitly that if Ψ satisfies (2.22) with $S_1=0$, then the two wave functions Ψ' both satisfy (2.22) when the operators L, S, T are replaced by the primed ones.

Asymptotically, for large r, $\Psi'=(\partial/\partial r)\Psi$, and hence the phase shift $\eta(k)$ is the same for the primed case as for the original case. Consequently, the integral over the phase shifts as it occurs in (3.9) is the same for spinor and vector fields (with $s_1=\frac{1}{2}$) as it is for scalar fields (with $s_1=0$).

The above procedure becomes more delicate if $E=0$. Indeed, although scalar fields can easily be seen to have no zero-eigenvalue modes, spinor and vector fields do have them. In conclusion, the nonzero eigenvalues for the vector and spinor modes are the same as for the scalar modes, but the zero eigenvectors are different.

In the following sections we compute the universal value for the product. Note that also the regularized expressions (3.10) are equivalent because the $\eta(k)$ match for all k. The regulator of Eq. (3.10) corresponds to new fields with Lagrangians

$$\mathcal{L} = -\tfrac{1}{2}(D_\mu B_\nu)^2 - \tfrac{1}{2}M^2 B_\nu{}^2 - gB_\nu^a\epsilon_{abc}G_{\mu\nu}^b B_\mu^c \tag{4.7}$$

for vectors,

$$\mathcal{L} = \bar{\chi}[-(\gamma\cdot D)^2 - M^2]\chi \text{ for spinors}\,, \tag{4.8}$$

and

$$\mathcal{L} = -(D_\mu\xi)^* D_\mu\xi - M^2\xi^*\xi \text{ for scalars}\,. \tag{4.9}$$

Within the background field procedure it is obvious that such regulator fields make the one-loop amplitudes finite. Later (Sec. XIII) we will make the link with the more conventional dimensional regulators.

V. A NEW EIGENVALUE EQUATION AND NEW REGULATORS

As stated in the Introduction, the solutions to the equations $\mathfrak{M}\Psi = E\Psi$ even in the scalar case cannot be expressed in terms of simple elementary functions. But eventually we only need $\det\mathfrak{M}/\det\mathfrak{M}_0$, and this can be obtained in another way.

We write

$$V=\tfrac{1}{4}(1+r^2)\mathfrak{M}(1+r^2),\qquad V_0=\tfrac{1}{4}(1+r^2)\mathfrak{M}_0(1+r^2), \tag{5.1}$$

and, formally,

$$\det(\mathfrak{M}/\mathfrak{M}_0)=\det(V/V_0). \tag{5.2}$$

The equation

$$V\Psi=\lambda\Psi \tag{5.3}$$

corresponds to the expression

$$\left[\left(\frac{\partial}{\partial r}\right)^2+\frac{3}{r}\frac{\partial}{\partial r}-\frac{4}{r^2}L^2-\frac{4}{1+r^2}(J_1^2-L^2)+\frac{4(T^2+\lambda)}{(1+r^2)^2}\right]\Psi=0. \tag{5.4}$$

Write

$$x=\frac{1}{1+r^2},\qquad \Psi=r^{2l}(1+r^2)^{-l-j_1-1}\Phi(x), \tag{5.5}$$

then

$$\left\{\left(\frac{\partial}{\partial x}\right)^2+\left[\frac{2(j_1+1)}{x}-\frac{2(l+1)}{1-x}\right]\frac{\partial}{\partial x}+\frac{1}{x(1-x)}\left[T^2+\lambda-(l+j_1+1)(l+j_1+2)\right]\right\}\Phi=0. \tag{5.6}$$

This is a hypergeometric equation. The physical region is $1/(1+R^2)<x\leq 1$. In the Hilbert space of square-integrable wave functions the spectrum is now discrete, which implies that we can safely take the limit $R\to\infty$. The solutions for Φ are just polynomials:

$$\Phi(x)=\sum_{\nu=0}^{\infty}a_\nu x^\nu, \tag{5.7}$$

$$a_{\nu+1}=a_\nu\,\frac{(\nu-n)(\nu+n+2l+2j_1+3)}{(\nu+1)(\nu+2j_1+2)}, \tag{5.8}$$

where n is defined by

$$(n+l+j_1+1)(n+l+j_1+2)=T^2+\lambda. \tag{5.9}$$

If n = integer ≥ 0 then the series (5.7) breaks off. Otherwise Φ is not square-integrable. So we find the eigenvalues

$$\lambda_n=(n+l+j_1+1-t)(n+l+j_1+2+t),\qquad n=0,1,2,\ldots,\qquad T^2=t(t+1). \tag{5.10}$$

The vacuum case, $V_0\Psi=\lambda_0\Psi$, is solved by the same equation, but with $j_1=l$, $t=0$.

The product of these eigenvalues, even when divided by the vacuum values, still badly diverges so we must regularize. The regulators of Sec. IV are not very attractive here because they spoil the hypergeometric nature of the equations. More convenient here is a set of regulator fields with masses that all depend on space-time in a certain way. They are given by the Lagrangians (4.7)-(4.9) but with M^2 replaced by

$$\frac{4M^2}{(1+r^2)^2}. \tag{5.11}$$

We choose M^2 here so large that anywhere near the origin the regulator is heavy. Far from the origin the classical solution is expected to be close enough to the real vacuum, so that there the details of the regulators are irrelevant.

Of course the regulator procedure affects the definition of the subtracted coupling constant. In Sec. VII we link the regulator (5.11) with the more acceptable one of Sec. IV, and in Sec. XIII we make the link with the dimensional regulator.

The eigenvalues of the regulator are

$$\lambda_n^M=(n+l+j_1+1-t)(n+l+j_1+2+t)+M^2. \tag{5.12}$$

The regulators M_i with $i=1,\ldots,R$ are as usual of alternating metric $e_i=\pm 1$. Consequently, $\det\mathfrak{M}$ is replaced by

$$(\det\mathfrak{M})\prod_{i=1}^{R}(\det\mathfrak{M}_i)^{e_i}. \tag{5.13}$$

This converges rapidly if

$$\sum_1^R e_i=-1,\qquad \sum_1^R e_iM_i=0,\qquad \sum_1^R e_iM_i^2=0,\ldots, \tag{5.14}$$

and

$$\sum_1^R e_i\ln M_i\equiv-\ln M=\text{finite}. \tag{5.15}$$

Let $i=0$ denote the physical field, then

$$e_0=1,\quad M_0=0,\quad \sum_{i=0}^{R}e_i=0,\quad \text{etc.} \tag{5.16}$$

VI. THE REGULARIZED PRODUCT OF THE NONVANISHING EIGENVALUES

We now consider the logarithm of the regularized product of the nonvanishing eigenvalues, for a scalar field with total isospin t:

$$\ln\Pi(t) = \sum_{i=0}^{R} e_i \sum \ln \lambda^{M_i} \tag{6.1}$$

with

$$\lambda^{M_i} = (n+l+j_1+1-t)(n+l+j_1+2+t)+M_i^2 \tag{6.2}$$

(we imply that $e_0 = 1$ and $M_0 = 0$). The summation goes over the values of all quantum numbers. Now for given n, l, j_1 the degeneracy is $(2j_1+1)(2j_2+1) = (2j_1+1)(2l+1)$. The values of l, j_1, and n are restricted by

$$\begin{aligned}&\sigma \equiv l+j_1-t \geq 0\,,\\ &\tau \equiv j_1 - l + t \geq 0, \quad \tau \leq 2t\,, \quad n \geq 0\,.\end{aligned} \tag{6.3}$$

[Later we will divide $\Pi(t)$ by the vacuum value $\Pi_0(t)$, which is obtained by the same formulas as above and the following, but with t replaced by zero, and the degeneracy will be $(2t+1)(2l+1)^2$.]

We go over to the variables σ and τ as given by (6.3) and s with

$$s = n+l+j_1+\tfrac{3}{2}\,, \quad s \geq t+\sigma+\tfrac{3}{2}\,. \tag{6.4}$$

We find that

$$\ln\Pi(t) = \sum_{s=t+3/2}^{\infty} \sum_i e_i \sum_{\tau=0}^{2t} \sum_{\sigma=0}^{s-t-3/2} (\sigma+\tau+1)(2t+\sigma-\tau+1) \times \ln[s^2+M_i^2-(t+\tfrac{1}{2})^2]\,. \tag{6.5}$$

The summation over σ and τ gives

$$\ln\Pi(t) = \frac{2t+1}{3} \sum_{s=t+3/2}^{\infty} \sum_i e_i[s^3 - s(t+\tfrac{1}{2})^2] \times \ln[s^2+M_i^2-(t+\tfrac{1}{2})^2]\,. \tag{6.6}$$

The vacuum value $\Pi_0(t)$ is obtained from (6.6) by replacing t with zero and adding an additional multiplicity $2t+1$, thus

$$\ln\Pi_0(t) = \frac{2t+1}{3} \sum_{s=3/2}^{\infty} \sum_i e_i(s^3 - \tfrac{1}{4}s) \times \ln(s^2+M_i^2-\tfrac{1}{4})\,. \tag{6.7}$$

Now we interchange the summation over s and i, letting first s go from $t+\frac{3}{2}$ to Λ and taking $\Lambda \to \infty$ in the end. We get

$$\ln[\Pi(t)/\Pi_0(t)] = \frac{2t+1}{3} \sum_i e_i(A^{M_i}(t+\tfrac{1}{2}) - A^{M_i}(\tfrac{1}{2})) \tag{6.8}$$

with

$$A^{M_i}(\phi) = \sum_{s=\sigma+1}^{\Lambda} (s^3 - s\phi^2)\ln(s^2+M_i^2-\phi^2)\,. \tag{6.9}$$

Let us first consider the regulator contribution. Then M is large. We may consider the logarithm as a slowly varying function, and approximate the summation by means of the Euler-Maclaurin formula,

$$\sum_{s=p+1}^{\Lambda} f(s) = \int_p^{\Lambda} f(x)dx + [\tfrac{1}{2}f(x) + \tfrac{1}{12}f'(x) - \tfrac{1}{720}f'''(x)\cdots]\Big|_p^{\Lambda}\,, \tag{6.10}$$

and we obtain

$$A^M(\phi) = \text{indep}(\phi) + \phi^2(-\tfrac{1}{2}M^2 - \Lambda^2\ln\Lambda - \Lambda\ln\Lambda - \tfrac{1}{2}\Lambda - \tfrac{1}{6}\ln\Lambda - \tfrac{1}{4} - \tfrac{1}{6}\ln M^2) + \tfrac{1}{4}\phi^4(2\ln\Lambda+1) + O\left(\frac{1}{M^2}\right) + O\left(\frac{1}{\Lambda}\right). \tag{6.11}$$

The first term stands for an array of expressions, all independent of ϕ, and is not needed because it cancels out in Eq. (6.8).

For $A^0(\phi)$ the series (6.10) will not converge at $x = p$ so it cannot be used. After some purely algebraic manipulations we find

$$\begin{aligned}A^0(\phi) &= \sum_{s=\phi+1}^{\Lambda} s(s+\phi)(s-\phi)[\ln(s+\phi)+\ln(s-\phi)]\\ &= \text{indep}(\phi) + 4\phi^2\sum_{s=1}^{\Lambda} s\ln s + \sum_{s=1}^{2\phi} s(2\phi-s)(s-\phi)\ln s\\ &\quad + \phi^2(-3\Lambda^2-3\Lambda-\tfrac{1}{2})\ln\Lambda + \phi^2(\Lambda^2-\tfrac{1}{2}\Lambda-\tfrac{7}{12}) + \phi^4(\tfrac{1}{2}\ln\Lambda-\tfrac{1}{12}) + O\left(\frac{1}{\Lambda}\right).\end{aligned} \tag{6.12}$$

Now we insert (6.11) and (6.12) into (6.8):

$$\ln[\Pi(t)/\Pi_0(t)] = \frac{2t+1}{3}\left\{\sum_{s=1}^{2t+1} s(2t+1-s)(s-t-\tfrac{1}{2})\ln s + t(t+1)\left[4\sum_{1}^{\Lambda} s\ln s - 2\Lambda^2\ln\Lambda - 2\Lambda\ln\Lambda - \tfrac{1}{3}\ln\Lambda + \Lambda^2 + \tfrac{1}{3}\ln M - \tfrac{1}{3}t(t+1) - \tfrac{1}{2}\right]\right\}. \tag{6.13}$$

We made use of $\sum_0^R e_i = 0$, $\sum_0^R e_i M_i^2 = 0$, $\sum_1^R e_i \ln M_i^2 \equiv -\ln M$. The limit $\Lambda \to \infty$ exists. Defining

$$R = \lim_{\Lambda\to\infty}\left(\sum_{s=1}^{\Lambda} s\ln s - \tfrac{1}{2}\Lambda^2\ln\Lambda - \tfrac{1}{2}\Lambda\ln\Lambda - \tfrac{1}{12}\ln\Lambda + \tfrac{1}{4}\Lambda^2\right)$$

$$= 0.248\,754\,477\,, \tag{6.14}$$

we find that

$$\ln[\Pi(t)/\Pi_0(t)] = \frac{2t+1}{3}\left[t(t+1)\left(\frac{1}{3}\ln M + 4R - \frac{1}{3}t(t+1) - \frac{1}{2}\right) + \sum_{s=1}^{2t+1} s(2t+1-s)(s-t-\frac{1}{2})\ln s\right]. \tag{6.15}$$

R is related to the Riemann zeta function $\zeta(z)$ as follows:

$$R = \tfrac{1}{12} - \zeta'(-1)$$

$$= \frac{\ln 2\pi + \gamma}{12} - \frac{\zeta'(2)}{2\pi^2}$$

$$= \tfrac{1}{12}(\ln 2\pi + \gamma) + \frac{1}{2\pi^2}\sum_{s=1}^{\infty}\frac{\ln s}{s^2}\,,$$

$\gamma = 0.577\,215\,664\,9$ is Euler's constant,

and

$$-\zeta'(2) = \sum \frac{\ln s}{s^2}$$

$$= 0.937\,548\,254\,315\,844\,.$$

VII. THE FIXED MASS REGULATOR

Equation (6.15) gives the regularized product of all nonvanishing eigenvalues of $\mathfrak{M}$. But the regulator used was a very unsatisfactory one, from a physical point of view, because the regulator mass μ depends on space-time:

$$\mu^2 = \frac{4M^2}{(1+r^2)^2}\,. \tag{7.1}$$

This μ must be interpreted as the subtraction point of the coupling constant g. Now g does not occur in $\Pi(t)/\Pi_0(t)$, but it does occur in the expression for the total action for the classical solution, and as we emphasized in the Introduction, any change in the subtraction procedure is important. The problem here is that we wish to make a space-time-dependent change in the subtraction point, from μ to a fixed μ_0. We solve that in the following way.

The effect of a change in the regulator mass can be absorbed by a counterterm in the Lagrangian, and hence is local in space-time. So we expect that, if we make a space-time-dependent change in the regulator mass, then this change can be absorbed by a space-time-dependent counterterm. Moreover, since our regulators are both gauge-invariant, this counterterm is gauge invariant. For space-time-independent regulators, this counterterm can be computed by totally conventional methods:

$$\Delta\mathcal{L} = \frac{-g^2}{32\pi^2}G_{\mu\nu}G_{\mu\nu}\times\frac{1}{9}t(t+1)(2t+1)\ln(\mu/\mu_0)\,. \tag{7.2}$$

From locality we deduce that the same formula must also be true for space-time-dependent regulator mass $\mu(x)$, simply because no other gauge-invariant, local expressions of the same dimensionality exist. Inserting the classical value for $G_{\mu\nu}$,

$$G^{a\,\mathrm{cl}}_{\mu\nu} = -\frac{4}{g}\frac{\eta_{a\mu\nu}}{(1+r^2)^2}\,, \tag{7.3}$$

and expression (7.1) for μ, we get

$$\Delta S^{\mathrm{cl}} = \int \Delta\mathcal{L}\, d^4x$$

$$= \frac{16\times 12\pi^2}{32\pi^2\times 9}t(t+1)(2t+1)\times\int_0^\infty \frac{r^2 dr^2}{(1+r^2)^4}\ln\frac{\mu_0}{2M}(1+r^2)$$

$$= \tfrac{2}{3}t(t+1)(2t+1)\left(\tfrac{1}{6}\ln\frac{\mu_0}{2M} + \tfrac{5}{36}\right). \tag{7.4}$$

In the expression

$$\frac{\Pi_0(t)}{\Pi(t)} S^{\text{cl}} ,$$

with $\Pi(t)/\Pi_0(t)$ as computed in (6.15), we must correct S^{cl} with the above ΔS^{cl}, in order to get the corresponding expression with g subtracted with fixed mass regulators, as defined in (4.9). The regulator masses M_i in there must be such that

$$\sum_1^R e_i \ln M_i = -\ln \mu_0 . \tag{7.5}$$

Expression (7.4) must be added to (6.15). Thus we get

$$\ln[\Pi(t)/\Pi_0(t)] = \frac{t(t+1)(2t+1)}{3}\left[\tfrac{1}{3}\ln\frac{\mu_0}{2} + 4R + \sum_{s=1}^{2t+1} s(2t+1-s)(s-t-\tfrac{1}{2})\ln s - \tfrac{1}{3}t(t+1) - \tfrac{2}{9}\right] . \tag{7.6}$$

We note that the coefficient of the regulator term in (6.15) has the correct value. It matches the coefficient of (7.2) that has been computed independently. The regulator in this expression, (7.6), is the same as the one used in (3.10), and so we can use the result of Sec. IV to do the spinor and vector fields.

In Sec. IV we proved that the nonzero eigenvalues for vector and spinor fields are the same as for scalar fields, but we must take some multiplicity factors into account. Equation (7.6) holds for one complex scalar multiplet with isospin t. Fields with integer isospin may be real and then we have to multiply by $\frac{1}{2}$. The vector field has four components but is real, and hence its value for $\ln(\Pi/\Pi_0)$ is twice expression (7.6), with $t=1$. The complex Faddeev-Popov ghost has Fermi statistics and contributes with one unit, but opposite sign. Thus, altogether, the vector field contributes just like one complex scalar with $t=1$.

For fermions we must compute $\det \mathfrak{M}_\phi$, but the theorem of Sec. IV applies to $\mathfrak{M}_\phi^2$. The fermions have four Dirac components. So, altogether, fermions contribute just like two complex scalars, but the sign in $\ln(\Pi/\Pi_0)$ is opposite because of fermi statistics.

The above summarizes in words the complete contribution of all nonzero eigenstates to the functional determinants. But the spinor and vector fields have a few more modes, with $E=0$, and also the regulators have corresponding new modes, with $E=\mu_0^2$.

VIII. THE ZERO EIGENSTATES

First we consider the vector fields. We have $s_1=\frac{1}{2}$, $t=1$. Careful study of the operator $\mathfrak{M}$, Eq. (2.22), enables us to list the square-intergrable zero eigenstates as follows:

(i) $j_1=\frac{1}{2}$, $l=0$: $\Psi=(1+r^2)^{-2}$

$(j_2=s_2=\frac{1}{2})$, (8.1)

multiplicity $=(2j_1+1)(2j_2+1)=4$.

(ii) $j_1=0$, $l=\frac{1}{2}$: $\Psi=r(1+r^2)^{-2}$.

There are two possibilities for the other quantum numbers:

(a) $j_2=0$, multiplicity $=1$, (8.2)

(b) $j_2=1$, multiplicity $=3$. (8.3)

This completes the set of zero eigenstates. We interpret these as follows. States (i) have $j_1=j_2=\frac{1}{2}$, that is, the quantum numbers of an infinitesimal translation. The translations are considered in Sec. IX. State (iia) is the only singlet. It will correspond to the infinitesimal dilatation, Sec. X. State (iib) is just an anomaly. It will be discussed in Sec. XI. It is indirectly connected with infinitesimal global isospin rotations.

Spinors have similar sets of eigenstates, but their interpretation will be totally different. If $t=1$, then the eigenstates are essentially the same as the vector ones, but their multiplicity is half of that because $s_2=0$. In this paper we limit ourselves to $t=\frac{1}{2}$. Then there is just one zero eigenstate:

$$j_1=0,\quad l=j_2=0,\quad \Psi=(1+r^2)^{-3/2} . \tag{8.4}$$

Its multiplicity is of course N^f if there are N^f flavors. It leads to an N^f-fold zero in the amplitude [note that in (2.11) the amplitude is proportional to $\det\mathfrak{M}_\phi$ and thus is proportional to the product of the eigenvalues of $\mathfrak{M}_\phi$; if we have N^f zero eigenvalues then W has an N^f-fold zero]. But this zero will be removed if we switch on the fermion source $\mathcal{J}$ in the Lagrangian (2.1). In Sec. XIV we will construct the resulting N^f-point Green's function.

In strong-interaction theories the fermion mass will also remove this zero. The zero eigenstates must also be included in the regulator contributions. From Sec. VII on, our regulator mass is fixed and is essentially equal to μ_0. Every zero eigenvector of the operator $\mathfrak{M}$, Eq. (2.22), will be accompanied by a factor μ_0^{-2} for the regulator (a zero eigenvector of $\mathfrak{M}_\phi$ is accompanied by a factor μ_0^{-1}).

IX. COLLECTIVE COORDINATES: 1. TRANSLATIONS

Clearly, zero eigenvalues make no sense if they would be included in the products carefully computed in the previous sections. In the case of the vector fields, which we will now discuss, they would render the functional integral infinite because they are in the denominator. It merely means that the integration in those directions is not Gaussian.

Let us first consider the four modes (8.1). The angular dependence and index dependence can be read off from the quantum numbers. Written in full, the mode corresponds to the quantum field fluctuation (with arbitrarily chosen norm):

$$A_\mu^{a\,qu}(\nu) = 2\eta_{a\mu\nu}(1+r^2)^{-2}\,, \quad \nu = 1,\ldots,4. \tag{9.1}$$

This can be seen to be the space-time derivative of the classical solution up to a gauge transformation:

$$A_\mu^{a\,qu}(\nu) = -\frac{g}{2}\frac{\partial}{\partial z^\nu}\left\{\frac{2\eta_{a\mu\lambda}(x-z)^\lambda}{g[1+(x-z)^2]}\right\}\bigg|_{z=0} + D_\mu\Lambda^a(\nu)\,, \tag{9.2}$$

with

$$\Lambda^a(\nu) = -\eta_{a\nu\lambda}x^\lambda(1+x^2)^{-1}. \tag{9.3}$$

The gauge transformation is there because our gauge-fixing term depends on the background field.

If we want to replace the variable $\mathfrak{D}A^{qu}$ in this particular zero-mode direction by the collective variables dz^ν, then we must insert the corresponding Jacobian factor[5]

$$\int \mathfrak{D}A^{qu} \to \int \prod_\nu \left(\frac{2}{g}dz^\nu\right)\left\{\frac{1}{2\pi}\int [A_\mu^{a\,qu}(\nu)]^2 d^4x\right\}^{1/2}, \tag{9.4}$$

where $A_\mu^{a\,qu}(\nu)$ is the solution (9.1). The factor $2/g$ comes from the factor $g/2$ in (9.2). This way the result is independent of the normalization of $A_\mu^{a\,qu}(\nu)$ of (9.1). The factors $2\pi^{-1/2}$ arise from the fact that we compare this integral with Gaussian integrals of the form $\int dA \exp(-\frac{1}{2}A^2)$, and in these Gaussian integrals the factors $\sqrt{2\pi}$ that go with each eigenvalue had been suppressed previously. We could also have dragged along all factors $\sqrt{2\pi}$ at each of the eigenvalues of the matrices $\mathfrak{M}$, and then we would have noticed that the factors $\sqrt{2\pi}$ going with the corresponding modes of the regulators, which are still Gaussian, would have been left over. In (9.4) we just include these factors from the beginning.

The norm of the solution (9.1) is

$$\int A_\mu^{a\,qu}(\nu)A_\mu^{a\,qu}(\lambda)d^4x = 2\pi^2\delta_{\nu\lambda}\,, \tag{9.5}$$

so, together with the regulator, these four modes yield the factor

$$\left(\frac{2}{g}\right)^4\left(\frac{\mu_0^2}{2\pi}\right)^2(2\pi^2)^2 d^4z = 2^4\pi^2\mu_0^4 g^{-4}d^4z\,. \tag{9.6}$$

The integral over the collective coordinates z^μ will yield the total volume of space-time, if no massless fermions are present. If there are massless fermions, then we must include the sources $\mathcal{J}$, which break the translation invariance. In that case the z integration is rather like the integration over the location of an interaction vertex in a Feynman diagram in coordinate configuration, as we will see in Sec. XIV.

X. COLLECTIVE COORDINATES: 2. DILATATIONS

From the quantum numbers of the zero eigenstate (8.2) we deduce its angular and index dependence:

$$A_\mu^{a\,qu} = \eta_{a\mu\nu}x^\nu(1+x^2)^{-2}. \tag{10.1}$$

This is a pure infinitesimal dilatation of the classical solution:

$$A_\mu^{a\,qu} = \frac{g}{4}\frac{\partial}{\partial\rho}\left(\frac{2}{g}\frac{\eta_{a\mu\nu}x^\nu}{x^2+\rho^2}\right)\bigg|_{\rho=1}. \tag{10.2}$$

Thus, going from the integration variable A^{qu} in this direction to the collective variable ρ, we need a Jacobian factor:

$$\mathfrak{D}A_\mu^{a\,qu} \to \frac{4}{g}d\rho\left[\frac{1}{2\pi}\int (A_\mu^{a\,qu})^2 d^4x\right]^{1/2}. \tag{10.3}$$

The norm of the solution (10.1) is

$$\int (A_\mu^{a\,qu})^2 d^4x = \pi^2. \tag{10.4}$$

Thus, from this mode we obtain the factor

$$\frac{4\pi}{g}\left(\frac{\mu_0^2}{2\pi}\right)^{1/2} d\rho = 2^{3/2}\pi^{1/2}\mu_0 g^{-1}d\rho \tag{10.5}$$

at $\rho = 1$. Our system is not scale invariant because of the nontrivial renormalization-group behavior. The complete ρ dependence for $\rho \neq 1$ will be deduced from simple dimensional arguments (including renormalization group) in Sec. XII.

XI. GLOBAL GAUGE ROTATIONS AND THE GAUGE CONDITION

Discussion of the legitimacy of the background gauge-fixing term has been deliberately postponed to this section, because we wanted to derive first the existence of the three anomalous zero eigenstates (8.3). They have the explicit form

(arbitrary normalization)

$$\psi^a_\mu(b) = 2\eta_{ak\mu}\bar\eta_{bk\lambda}x^\lambda(1+x^2)^{-2}. \tag{11.1}$$

They are a pure gauge artifact

$$\psi^a_\mu(b) = D_\mu \psi^a(b)\,, \tag{11.2}$$

$$\psi^a(b) = \eta_{ak\nu}\bar\eta_{bk\lambda}x^\nu x^\lambda(1+x^2)^{-1}\,, \tag{11.3}$$

but $\psi^a(b)$ is not square-integrable. What is going on? Note that $\psi^a(b)$, since they are x dependent, do not generate a closed algebra of gauge rotations. They may not be replaced by a collective coordinate for global isospin rotations. To analyze this situation we first go back to a background-independent gauge-fixing term,

$$C^a_1(x) = \alpha\partial_\mu(A^{a\,\mathrm{cl}}_\mu + A^{a\,\mathrm{qu}}_\mu) = \alpha\partial_\mu A^{a\,\mathrm{qu}}_\mu\,, \tag{11.4}$$

where α is a free parameter. In this gauge we know exactly how to handle all zero eigenmodes: There are five for translations and dilatations and also three for global isospin rotations because global isospin is still an invariance in this gauge. To understand the latter we put the system in a large spherical box with volume V and assume that all (vector and ghost) fields vanish on the boundary. Let

$$\Lambda^a_1(b,x) = \delta^{ab} \tag{11.5}$$

generate an infinitesimal global isospin rotation. Then there is a zero eigenmode:

$$\psi^a_{1\mu}(b) = D_\mu\Lambda^a_1(b) = 2\epsilon_{acb}\eta_{c\mu\nu}x^\nu(1+x^2)^{-1}\,. \tag{11.6}$$

The subscript 1 is to remind us that this is a solution in the gauge C_1. Similarly as in the foregoing two sections, we can replace the integral over $\mathfrak{D}A^a_\mu$ by an integral over the collective coordinates $d\Lambda(b)$ by inserting the corresponding Jacobian factor

$$\mathfrak{D}A^a_\mu \to \int\prod_b d\Lambda(b)(\mathfrak{N}/2\pi)^{1/2} \tag{11.7}$$

with

$$\mathfrak{N} = \int[\psi^a_{1\mu}(b)]^2 d^4x\,, \tag{11.8}$$

which diverges as the volume V of space-time goes to infinity. The integral over the gauge rotation is just the volume of the group and yields

$$\int\prod_b d\Lambda^b = \frac{8\pi^2}{g^3}\,, \tag{11.9}$$

where the factor g^{-3} comes from our normalization of Λ_1 in (11.6). Thus, in this gauge, the zero eigenmode yields the factor

$$\frac{8\pi^2}{g^3}\left(\frac{\mathfrak{N}}{2\pi}\right)^{3/2}. \tag{11.10}$$

Now we will study different gauges, and for that we need to change the boundary condition (we know from Sec. III that that will not affect the finite eigenvalues, but the zero eigenmodes change) into a gauge-invariant one: The ghosts ϕ and gauge generators Λ must satisfy

$$\frac{\partial\phi}{\partial r}(R) = \frac{\partial\Lambda}{\partial r}(R) = 0\,, \tag{11.11}$$

where R is the radius of the box, and the vector fields

$$A_r(R) = 0\,,\quad \frac{\partial}{\partial r}A_\parallel(R) = 0\,, \tag{11.12}$$

where $A_\parallel$ is the vector component parallel to the boundary.

Now observe the following. The gauge term C_1, Eq. (11.4), does not fix the gauge completely which can be seen in two ways: (a) Global isospin rotations are still an invariance; (b) one component of the gauge term C_1 is identically zero:

$$\int C^a_1(x)d^4x = 0\,. \tag{11.13}$$

This leaves us the possibility of adding a constant to C_1, which is orthogonal to it, with which we fix the remaining global gauge

$$C^a_2(x) = \alpha\partial_\mu A^{a\,\mathrm{qu}}_\mu - \kappa\sum_b \delta^{ab}\int\psi^c_{1\mu}(b,y)A^{c\,\mathrm{qu}}_\mu(y)d^4y \tag{11.14}$$

with α,κ free parameters and $\psi^a_{1\mu}(b,y) = D_\mu\delta^{ab}$, as in (11.6). The integral over group space is now replaced by a Gaussian integral. The Gaussian volume is corrected for by the Faddeev-Popov ghost,

$$\mathcal{L}^{\mathrm{gh}}_2 = \phi^*_a(x)\left[\alpha\partial_\mu D_\mu\phi^a(x) - \kappa\sum_b\delta^{ab}\int\psi^c_{1\mu}(b,y)D_\mu\phi^c(y)d^4y\right], \tag{11.15}$$

so that the combined contribution of vector fields and ghosts is now independent of α and κ: The zero eigenvalues are replaced by

$$\lambda^{\mathrm{vector}}_2 = \kappa^2 V\mathfrak{N}$$

and $\qquad$ (11.16)

$$\lambda^{\mathrm{ghost}}_2 = \kappa\mathfrak{N}\,.$$

(Remember that the ghost, with the new boundary

condition, has now an eigenstate $\phi = $ constant.) Thus, instead of (11.10), this gauge gives

$$(\lambda_2^{\text{ghost}})^3(\lambda_2^{\text{vector}})^{-3/2} = (\mathfrak{N}/V)^{3/2}. \tag{11.17}$$

Conclusion: If the redundant eigenmodes are fixed by an additional component in the gauge-fixing term, then a correction factor is needed: Equation (11.10) divided by (11.17):

$$\frac{8\pi^2}{g^3}\left(\frac{V}{2\pi}\right)^{3/2}. \tag{11.18}$$

Here V is the volume of the spherical box.

The background gauge $C_4 = D_\mu A_\mu^{a\,\text{qu}}$ has a problem similar to the gauge C_1. Both the ghost (under the new boundary condition) and the vector field have one eigenvalue that vanishes like $1/V$ as $V \to \infty$ (not $1/R^2$, as the other eigenvalues). Let $\psi_4^a(b)$ be the three ghost eigenstates and $\psi_{4\mu}^a(b) = D_\mu \psi_4^a(b)$ be the vector ones. Let λ_4 be the ghost eigenvalue

$$D^2\psi_4^a(b) = -\lambda_4 \psi_4^a(b). \tag{11.19}$$

It is easy to see that

$$\lambda_4 = \frac{\int_V [\psi_{4\mu}^a(x)]^2 d^4x}{\int_V [\psi_4^a(x)]^2 d^4x}. \tag{11.20}$$

From (11.1) and (11.3) we see that this is $O(1/V)$. It is safer to have a gauge condition that fixes this gauge degree of freedom as $V \to \infty$,

$$C_3(x) = \alpha D_\mu A_\mu^{a\,\text{qu}}(x) - \kappa \sum_b \psi_4^a(b,x) \int \psi_{4\mu}^c(b,y) A_\mu^{c\,\text{qu}}(y) d^4y, \tag{11.21}$$

although the result, (11.25), will turn out to remain the same even if $\kappa = 0$, $\alpha = 1$. The ghost Lagrangian is

$$\mathcal{L}_3^{\text{gh}} = \phi_a^*(x)\Big[\alpha D^2\phi_a(x) - \kappa \sum_b \psi_4^a(b,x) \int \psi_{4\mu}^c(b,y) D_\mu \phi^c(y) d^4y\Big]. \tag{11.22}$$

In this gauge

$$\lambda_3^{\text{vector}} = \lambda_4\left(\alpha + \kappa \int [\psi_4^a(x)]^2 d^4x\right)^2 \tag{11.23}$$

and

$$\lambda_3^{\text{ghost}} = \lambda_4\left(\alpha + \kappa \int [\psi_4^a(x)]^2 d^4x\right), \tag{11.24}$$

where no summation over b is implied. In this gauge we find the contribution from the lowest eigenmodes:

$$(\lambda_3^{\text{ghost}})^3(\lambda_3^{\text{vector}})^{-3/2} = \lambda_4^{3/2}. \tag{11.25}$$

Using

$$\int [\psi^a(x)]^2 d^4x = V, \qquad \int [\psi_\mu^a(x)]^2 d^4x = 4\pi^2, \tag{11.26}$$

we find

$$\lambda_4 = 4\pi^2/V. \tag{11.27}$$

Thus, together with the correction factor (11.18), we find the correct contribution for the three eigenmodes (9.3), together with that of their regulator:

$$\frac{8\pi^2}{g^3}\left(\frac{V\mu_0^2}{2\pi}\right)^{3/2}\left(\frac{4\pi^2}{V}\right)^{3/2} = 2^{9/2}\pi^{7/2}\mu_0^3 g^{-3}. \tag{11.28}$$

XII. ASSEMBLING THE VECTOR, SCALAR, AND SPINOR TERMS

The eight zero-eigenvalue modes for the vector field give the factors (9.6), (10.5), and (11.28). Multiplying these gives

$$2^{10}\pi^6 g^{-8}\mu_0^8 d^4 z d\rho \tag{12.1}$$

for $\rho = 1$ (later we will find the ρ dependence).

The contributions from the nonvanishing eigenmodes both for the scalar and for the vector fields are essentially contained in formula (7.6). As we saw before, the vector fields, combined with the Faddeev-Popov ghost, together count as two real, or one complex, scalar with $t = 1$.

Let there be $N^s(t)$ scalar multiplets for each isospin t, where each complex scalar multiplet counts as one, and each real scalar multiplet counts as one-half. Then from (7.6) we obtain the total contribution from vector and scalar nonzero modes:

$$\frac{\Pi_0}{\Pi} = \exp\left\{-\left[\tfrac{2}{3} + \tfrac{1}{6}\sum_t N^s(t)C(t)\right]\ln\mu_0 - \alpha(1) - \sum_t N^s(t)\alpha(t)\right\}. \tag{12.2}$$

Here

$$C(t) = \tfrac{2}{3}t(t+1)(2t+1) \tag{12.3}$$

and

$$\alpha(t)=C(t)\left[2R-\tfrac{1}{6}\ln 2+\tfrac{1}{2}\sum_{s=1}^{2t+1} s(2t+1-s)\,(s-t-\tfrac{1}{2})\ln s-\tfrac{1}{6}t(t+1)-\tfrac{1}{9}\right]. \qquad (12.4)$$

The numerical values for $C(t)$ and $\alpha(t)$ are listed in Table I. Combining (12.1), (12.2), and the classical action (2.8) gives the total amplitude in the absence of fermions:

$$2^{10}\pi^6 g^{-8}\,\frac{d^4z\,d\rho}{\rho^5}\,\exp\left\{-\frac{8\pi^2}{g^2(\mu_0)}+\ln(\mu_0\rho)\left[\tfrac{22}{3}-\tfrac{1}{6}\sum_t N^s(t)C(t)\right]-\alpha(1)-\sum_t N^s(t)\alpha(t)\right\}. \qquad (12.5)$$

Note that the coefficient multiplying $\ln\mu_0$ coincides with the usual Callan-Symanzik β coefficient for $g^2(\mu_0)$ in such a way that (12.5) becomes independent of the subtraction point μ_0 if we choose $g^2(\mu_0)$ to obey the Gell-Mann-Low equation. We now also insert the ρ dependence if $\rho\neq 1$ by straightforward dimensional analysis.

The interpretation of (12.5) is best given in the language of path integrals: If $|0\rangle$ is the vacuum, and $|\bar{0}\rangle$ is the gauge-rotated vacuum, then (12.5) is the total contribution to $\langle\bar{0}|0\rangle$ from all paths in Euclidean space that have a pseudoparticle located at z within d^4z, having a scale between ρ and $\rho+d\rho$. The fermions can be introduced in two ways:

(i) If they have a mass $m \ll 1/\rho$ then only the lowest eigenvalue will depend critically on m.

(ii) If they are rigorously massless then the lowest eigenvalue depends critically on the external source $\mathcal{J}$.

In this paper we limit ourselves only to fermions with isospin $t=\frac{1}{2}$. In case (i) the contribution of the lowest modes will simply be

$$\left(\frac{m}{\mu_0}\right)^{N^f}. \qquad (12.6)$$

The nonvanishing eigenmodes are again obtained from (7.6), which represents the eigenvalues of $\mathfrak{M}_\psi{}^2$. Now we wish to compute $\det\mathfrak{M}_\psi$ and we take into account that the Dirac field has four components. Thus the fermion nonvanishing eigenmodes will give

$$\exp\left[\frac{N^f}{3}\,C(\tfrac{1}{2})\ln\mu_0+2N^f\alpha(\tfrac{1}{2})\right]. \qquad (12.7)$$

Together with (12.6) we find the total fermion factor that multiplies (12.5),

$$\rho^{N^f}m^{N^f}\exp[-\tfrac{2}{3}N^f\ln(\mu_0\rho)+2N^f\alpha(\tfrac{1}{2})], \qquad (12.8)$$

where we again inserted the factors ρ as they follow from dimensional arguments. Note that the well-known Callan-Symanzik β coefficient for $g^2(\mu_0)$ again matches the term in front of $\ln\mu_0$.

Equations (12.5) and (12.8) could be used as a starting point for a strong-interaction color theory where Euclidean pseudoparticles form a plasma-like statistical ensemble.[7] For that, one also needs to extend from SU(2) to SU(3). We will not do that in this paper.

In Sec. XIV we consider case (ii). In that case m must be replaced by the eigenvalue of the lowest mode as it is perturbed by the source insertion.

XIII. DIMENSIONAL RENORMALIZATION

The regulators used in Secs. VII–XII are what we call fixed mass Pauli-Villars regulators and they only make sense in the background-field formalism. They are given by the Lagrangians (4.7)–(4.9). In this section we wish to switch to another regulator scheme which is much more widely used in gauge theories: the dimensional method.[8] Let us emphasize again that if one switches to another regulator, then that affects the definition of $g(\mu)$ and that influences our calculation by an overall constant. We know that in the dimensional procedure the limit of large cutoff is replaced by a limit $n\to 4$, where n is the number of space-time dimensions, roughly in the following way:

$$\ln\Lambda \to \frac{1}{4-n} + \text{finite}. \qquad (13.1)$$

In (12.5) and (12.7) the regulator mass μ_0 plays the role of the cutoff Λ. Clearly then, the finite part in (13.1) will be relevant. In this section we derive that finite part, in ordinary perturbation

TABLE I. Numerical values of the coefficients $C(t)$ and $\alpha(t)$ as they occur in the text.

t	$C(t)$	$\alpha(t)$	
0	0	0	
$\frac{1}{2}$	1	$2R-\frac{1}{6}\ln 2-\frac{17}{72}$	$=0.145\,873$
1	4	$8R+\frac{1}{3}\ln 2-\frac{16}{9}$	$=0.443\,307$
$\frac{3}{2}$	10	$20R+4\ln 3-\frac{5}{3}\ln 2-\frac{265}{36}$	$=0.853\,182$

$$R=\tfrac{1}{12}(\ln 2\pi+\gamma)+\frac{1}{2\pi^2}\sum_{2}^{\infty}\frac{\ln s}{s^2}=0.248\,754\,477\,033\,784$$

theory. It corresponds to a finite counterterm in the Lagrangian. It is easy to compute this finite counterterm when, again, one makes use of the background fields. There are some diagrams to be computed and the rest is algebra. This algebra is identical to the algebra devised in Ref. 9. Symmetry arguments restrict the possible form of the finite counterterm in just two independent terms, X^2 and $Y_{\mu\nu}Y_{\mu\nu}$, in the language of Ref. 9. The first of these is obtained by comparing the integral

$$\frac{1}{(2\pi)^n}\int d^n k \frac{1}{(k^2+\mu_0{}^2)^2} \tag{13.2}$$

in the limits $\mu_0 \to \infty$ and $n \to 4$. For definiteness, we specify the theory at $n \neq 4$ dimensions: All trivial factors $(2\pi)^2$ must also be replaced by $(2\pi)^n$, which leads to the factor $(2\pi)^{-n}$ in (13.2).

The integral (13.2) is in this limit

$$\frac{1}{(4\pi)^2}\left[\frac{2}{4-n}-\gamma-2\ln\mu_0+\ln 4\pi+O(4-n)+O\left(\frac{1}{\mu_0{}^2}\right)\right], \tag{13.3}$$

where γ is again Euler's constant. So here

$$\ln\mu_0 \to \frac{1}{4-n}-\frac{1}{2}\gamma+\frac{1}{2}\ln 4\pi\,. \tag{13.4}$$

The coefficient in front of $Y_{\mu\nu}Y_{\mu\nu}$ is obtained in the same way by comparing the integral

$$\sum_i e_i \frac{1}{(2\pi)^2}\int d^n k \frac{k_\mu k_\nu k_\alpha k_\beta}{(k^2+\mu_i^2)^2} \tag{13.5}$$

in the two limits, where the signs e_i are defined as in Eqs. (5.15) and (5.16), replacing M by μ_0. This time we get

$$\sum_{i=0}^{R} e_i \frac{1}{(4\pi)^2 4!}\left[\frac{2}{4-n}-\gamma-2\ln\mu_i+\ln 4\pi+O(4-n)+O(\frac{1}{\mu_0^2})\right]\left(\delta_{\mu\nu}\delta_{\alpha\beta}+\delta_{\mu\alpha}\delta_{\nu\beta}+\delta_{\mu\beta}\delta_{\nu\alpha}\right). \tag{13.6}$$

So we see that for both the X^2 terms and the $Y_{\mu\nu}Y_{\mu\nu}$ terms the substitution (13.4) is to be made. However, we have to remember that in dimensional regularization the number of the fields A_m^{qu} is n rather than 4. For these fields one therefore must replace

$$\ln\mu_0 \to \frac{1}{4-n}-\frac{1}{2}\gamma+\frac{1}{2}\ln 4\pi-1\,. \tag{13.7}$$

In conclusion, (12.5) is to be replaced by

$$2^{10}\pi^6 g^{-8}\rho^{-5}d^4z\,d\rho\exp\left\{-\frac{8\pi^2}{g_B{}^2(n)}+\left(\ln\rho+\frac{1}{4-n}\right)\left[\tfrac{22}{3}-\tfrac{1}{6}\sum_t N^s(t)C(t)\right]+A-\sum_t N^s(t)A(t)\right\}, \tag{13.8}$$

with

$$A=-\alpha(1)+\frac{11}{3}(\ln 4\pi-\gamma)+\frac{1}{3}=7.05399103 \tag{13.9}$$

and

$$A(t)=-a(t)+\frac{1}{12}(\ln 4\pi-\gamma)C(t).$$

Numerically,

$$A(0)=0,\ A(\tfrac{1}{2})=0.30869069,\ A(1)=1.09457662,$$
$$A(\tfrac{3}{2})=2.48135610 \tag{13.10}$$

Similarly for the fermion factor

$$\rho^{N^f}m^{N^f}\exp\left[-\tfrac{2}{3}N^f\left(\ln\rho+\frac{1}{4-n}\right)-N^f B\right] \tag{13.11}$$

with

$$B=-2\alpha(\tfrac{1}{2})+\tfrac{1}{3}(\ln 4\pi-\gamma) = 0.35952290\,. \tag{13.12}$$

If we define the subtracted coupling constant as in Ref. 10,

$$g_B(n)=\mu^{4-n}\left(g_R^D(\mu)+\frac{a_1}{n-4}+\cdots\right), \tag{13.13}$$

with a_1 depending only on g_R but not on n or μ, then we can make the following replacements in (13.8) and (13.11):

$$g_B(n)\to g_R^D(\mu)\,,\qquad \ln\rho+\frac{1}{4-n}\to\ln(\rho\mu)\,. \tag{13.14}$$

Here the superscript D stands for the dimensional procedure which defines g_R^D. We see that the expression in terms of $g_R^D(\mu)$ differs slightly from the one in terms of $g(\mu_0)$.

XIV. THE FERMION SOURCE AND THE GREEN'S FUNCTION

We now consider the fermion zero eigenmode (9.4), and assume that the fermion mass (12.6) vanishes. In that case the source $\mathcal{J}_{st}$ in (2.1) must be taken into account, since the lowest eigenvalue will now mainly be determined by this source. We determine the lowest eigenvalues $E(i)$, $i=1,\ldots,N^f$ of the operator

$$\mathfrak{M}_\psi = -\gamma_\mu D_\mu \delta_{st} + \mathcal{J}_{st} \tag{14.1}$$

by perturbation theory, taking $\mathcal{J}$ as the small perturbation. The method is the standard one (the author thanks S. Coleman for an enlightening discussion on this point). The unperturbed, degenerate eigenmodes are (taking for simplicity $\rho=1$)

$$\psi_s^\alpha(t) = C(1+r^2)^{-3/2}u^\alpha \delta_{st}, \qquad \alpha=1,2, \quad s,t=1,\ldots,N^f. \tag{14.2}$$

The coefficients u^α contain besides the isospin index α a Dirac index. They satisfy

$$J_1^a u^\alpha = S_1^a u^\alpha + T_{\alpha\beta}^a u^\beta = 0,$$

or

$$(\eta_{a\mu\nu}\gamma_\mu\gamma_\nu + 4i\tau_a)u = 0 \tag{14.3}$$

and

$$\gamma_5 u = -u. \tag{14.4}$$

The coefficient C is determined by normalizing ψ,

$$C^{-2} = u^* u \int d^4x(1+x^2)^{-3} = \pi^2/2 \text{ if } u^*u=1. \tag{14.5}$$

Let

$$H_{st} = \langle \psi(s)|\mathfrak{M}_\psi|\psi(t)\rangle = \frac{2}{\pi^2}\int d^4x(1+x^2)^{-3}u_\alpha^* \mathcal{J}_{st}(x)u^\alpha, \tag{14.6}$$

then $E(i)$ are the N^f eigenvalues of H. We wish to compute

$$\prod_i E(i) = \det H = \left(\frac{2}{\pi^2}\right)^{N^f} \det_{st}\int d^4x(1+x^2)^{-3}u_\alpha^*\mathcal{J}_{st}(x)u^\alpha. \tag{14.7}$$

For large x^2, this amplitude has exactly the space-time structure of an N^f-point Green's function, where each source point is connected to the origin by two fermion lines. The integral over the collective coordinate z [which is at the origin in Eq. (14.7)] will correspond to the integration in coordinate configuration over the vertex variable.

Equation (14.7) should really be considered as our final result for the space-time dependence of the fermion Green's function. But it would be enlightening if we could represent it in terms of an effective Lagrangian.

We found that the effective Lagrangian can best first be written in the form

$$\mathcal{L}^{\text{eff}} = C\prod_{s=1}^{N^f}(\bar\psi_s\omega)(\bar\omega\psi_s), \tag{14.8}$$

where ω is some fixed Dirac spinor with isospin $\frac{1}{2}$. Owing to Fermi statistics, the various terms of the determinant in (14.7) will arise with the appropriate minus signs so that we may limit ourselves to sources $\mathcal{J}_{st}$ that are diagonal in s and t:

$$\mathcal{J}_{st} = \mathcal{J}(x)\delta_{st}. \tag{14.9}$$

Here $\mathcal{J}$ may still contain Dirac matrices.

In the presence of this source, the amplitude from the effective interaction (14.8) would be

$$C\prod_s[-\bar\omega S_F(x)\mathcal{J}(x)S_F(-x)\omega], \tag{14.10}$$

where the minus sign comes from the Fermi statistics and the $S_F(x)$ are the Dirac propagators for massless fermions in coordinate configuration,

$$S_F(x) = \frac{\gamma x}{2\pi^2(x^2)^2}. \tag{14.11}$$

Comparing this with (14.7), at large x^2, we find that we must require (leaving aside temporarily the other contributions to the overall constant)

$$(\gamma x\omega_\alpha)(\bar\omega_\alpha\gamma x) = x^2 u_\alpha\bar u_\alpha, \qquad C=(8\pi^2)^{N^f}, \tag{14.12}$$

where we may sum over the isospin index α but not over the Dirac components. Now from (14.3) one can derive

$$\sum_\alpha u_\alpha\bar u_\alpha = \tfrac{1}{2}(1-\gamma_5), \tag{14.13}$$

so we must require the ω_α to be such that

$$\sum_\alpha \omega_\alpha\bar\omega_\alpha = \tfrac{1}{2}(1+\gamma_5). \tag{14.14}$$

Thus, the ω_α are some parity reflection of u_α.

There is clearly no gauge-invariant solution to (14.14), so our effective Lagrangian (14.8) is apparently not gauge invariant. But note that we only wish to reproduce the amplitude (14.7) for gauge-invariant currents $\mathcal{J}_{st}$. Thus, any gauge rotation of (14.8) does the same job. We get a gauge-invariant $\mathcal{L}^{\text{eff}}$ if we average over the whole group of gauge rotations:

$$\mathcal{L}^{\text{eff}} = C\left\langle \prod_{s=1}^{N^f} (\bar{\psi}_s \omega)(\bar{\omega}\psi_s) \right\rangle , \tag{14.15}$$

where the brackets $\langle \ \rangle$ denote the average for all gauge rotations of ω. We can then derive

$$\langle \omega_\alpha \bar{\omega}_\beta \rangle = \tfrac{1}{4}\, \delta_{\alpha\beta}(1+\gamma_5) , \tag{14.16}$$

and, for instance,

$$\left\langle \prod_{s=1}^{2} (\bar{\psi}_s\omega)(\bar{\omega}\psi_s) \right\rangle = \tfrac{1}{24}\, (2\delta^{\beta_1}_{\alpha_1}\delta^{\beta_2}_{\alpha_2} - \delta^{\beta_2}_{\alpha_1}\delta^{\beta_1}_{\alpha_2})\epsilon^{st} \times \bar{\psi}_1^{\alpha_1}(1+\gamma_5)\psi_s^{\beta_1}\bar{\psi}_2^{\alpha_2}(1+\gamma_5)\psi_t^{\beta_2} . \tag{14.17}$$

The Lagrangian (14.15) only acts on the left-handed spinors. The parity-reflected Euclidean pseudoparticles will give a similar contribution acting on the right-handed spinors only. So in total we get $\mathcal{L}^{\text{eff}}$ of (14.15) plus its Hermitian conjugate.

Note that we obtain products of fermion fields, such as (14.17), that violate only chiral U(1) invariance. They have the chiral-symmetry properties of the determinant of an $N^f \times N^f$ matrix in flavor space and are therefore still invariant under chiral $SU(N^f)\times SU(N^f)$. The symmetry violation is associated with an arbitrary phase factor $e^{\pm i\omega}$ in front of the effective Lagrangians. If other mass terms or interaction terms occur in the Lagrangian that also violate chiral U(1), then they may have a phase factor different from these. We then find that our effective Lagrangian may violate P invariance, whereas C invariance is maintained. Thus we find that not only U(1) invariance but also PC invariance can be violated by our effect.

XV. CONVERGENCE OF THE ρ INTEGRATION

The entire expression that we now have for the effective Lagrangian is

$$\mathcal{L}^{\text{eff}}(z)d^4z = 2^{10+3N^f}\pi^{6+2N^f}g^{-8}d^4z \int \rho^{-5+3N^f}d\rho \exp\left\{ -\frac{8\pi^2}{[g_R^D(\mu)]^2} + \ln(\mu\rho)\left[\tfrac{22}{3} - \tfrac{1}{6}\sum_t N^s(t)C(t) - \tfrac{2}{3}N^f\right] + A - \sum_t N^s(t)A(t) - N^f B \right\} \left\langle \prod_{s=1}^{N^f} (\bar{\psi}_s\omega)(\bar{\omega}\psi_s) \right\rangle + \text{H.c.} , \tag{15.1}$$

with

$$\langle \omega_\alpha \bar{\omega}_\beta \rangle = \tfrac{1}{4}\, \delta_{\alpha\beta}(1+\gamma_5), \text{ etc.} ,$$

and the numbers $A, A(t), B, C(t)$ as defined before.

The ρ dependence has been changed because the effective Lagrangian (14.8) is not dimensionless.

We see that this integral converges as $\rho \to 0$ (except when there are very many scalars). But there is an infrared divergence as $\rho \to \infty$. In an unbroken color gauge theory for strong interactions this is just one of the various infrared disasters of the theory to which we have no answer. But in a weak-interaction theory it is expected that the Higgs field provides for the cutoff. Let there be a Higgs field with isospin q and vacuum expectation value F. Let its contribution to the original Lagrangian be

$$\mathcal{L}^H = -D_\mu\Phi^* D_\mu\Phi - V(\Phi) . \tag{15.2}$$

Formally, no classical solution exists now, because the Higgs Lagrangian tends to add to the total action of the pseudoparticle a contribution proportional to F^2, but this can always be reduced by scaling to smaller distances, until the action reaches the usual value $8\pi^2/g^2$ when the field configuration is singular.

On the other hand, it is clear that the quantum corrections, as can be seen in (15.1), act in the opposite way. There must be a region of values for ρ where the quantum effects compete with the effects due to the Higgs fields.

To handle this situation rigorously we alter slightly the philosophy of Sec. II. In Euclidean space it is not compulsory to consider only those classical fields for which the action is stationary. We will now look at approximate solutions of the classical equations, so that the total action is only a slowly varying function of one collective parameter, ρ.

We simply postulate the gauge field A to have the same configuration as before, with certain value for ρ, and now choose the Higgs field configuration in such a way that the total action is extreme. Only those infinitesimal variations that are pure scale transformations do not leave the action totally invariant, but nevertheless the parameter ρ gets the full treatment as a collective variable.

As will be verified explicitly, the dominant values for ρ will be those where the quantum effects and the Higgs contribution are equally important. Since the quantum effects are small we expect that there

$$\rho \ll 1/M_H, \tag{15.3}$$

which implies that the Higgs particle may be considered as approximately massless. Let us scale toward

$$\rho = 1, \quad |F| \sim 1, \quad M_H{}^2 \sim \lambda F^2 \ll 1.$$

The equation for this field will be approximately

$$D^2\Phi = 0, \qquad \Phi^2(r \to \infty) = F^2. \tag{15.4}$$

The solution to that is a zero-eigenvalue mode of the familiar operator (2.22):

$$j_1 = 0, \quad l = j_2 = q, \qquad |\Phi| = \left(\frac{x^2}{1+x^2}\right)^q |F|. \tag{15.5}$$

The contribution to the classical action is

$$S^H = \int [-D_\mu \Phi^* D_\mu \Phi - V(\Phi)] d^4x. \tag{15.6}$$

The first term is (observing that $x_\mu A_\mu^{\rm cl} = 0$)

$$-\int_V \partial_\mu(\Phi^* D_\mu \Phi) d^4x = -\int_S \Phi^* \partial_r \Phi \, d^3x = -4\pi^2 q F^2. \tag{15.7}$$

The second term in (15.6) is of order λF^4, where λ is a small coupling constant. If we scale back to arbitrary ρ, then the Higgs field factor in the total expression is

$$\exp S^H = \exp[-4\pi^2 q F^2 \rho^2 - O(\lambda F^4 \rho^4)]. \tag{15.8}$$

We see that the second term in the exponent may be neglected at first approximation.

Thus (15.8) multiplies the integrand in (15.1) and the ρ integration is now completely convergent. The integration over ρ yields a factor

$$\tfrac{1}{2}(4\pi^2 q F^2)^{2-(3/2)N^f - C}\,\Gamma(\tfrac{3}{2}N^f + C - 2), \tag{15.9}$$

where

$$C = \tfrac{11}{3} - \tfrac{1}{12}\sum_t N^s(t) C(t) - \tfrac{1}{3}N^f. \tag{15.10}$$

ACKNOWLEDGMENT

The author wishes to thank S. Coleman, R. Jackiw, C. Rebbi, and all other theorists at Harvard for their hospitality, encouragement, and discussions during the completion of this work.

APPENDIX: PROPERTIES OF THE η SYMBOLS

The group SO(4) is locally equivalent to SO(3) $\times$ SO(3). The antisymmetric tensors $A_{\mu\nu}$ in SO(4) having six components form a 3+3 representation of SO(3) $\times$ SO(3). The self-dual tensors

$$A_{\mu\nu} = \tfrac{1}{2}\epsilon_{\mu\nu\alpha\beta} A_{\alpha\beta} \tag{A1}$$

transform as 3-vectors of one SO(3) group. We now define the η symbols, in a way very similar to the Dirac γ matrices:

$$A_{\mu\nu} = \eta_{a\mu\nu} A_a, \qquad a = 1,2,3, \quad \mu,\nu = 1,\ldots,4 \tag{A2}$$

is a covariant mapping of SO(3) vectors on self-dual SO(4) tensors. A convenient representation is

$$\eta_{a\mu\nu} = \epsilon_{a\mu\nu}, \text{ if } \mu,\nu = 1,2,3, \quad \eta_{a4\nu} = -\delta_{a\nu}, \quad \eta_{a\mu4} = \delta_{a\mu}, \quad \eta_{a44} = 0. \tag{A3}$$

Let us also define

$$\bar\eta_{a\mu\nu} = (-1)^{\delta_{\mu4}+\delta_{\nu4}} \eta_{a\mu\nu}. \tag{A4}$$

The symbols $\bar\eta_{a\mu\nu}$ will then do the same with vectors of the other SO(3) group and tensors $B_{\mu\nu}$ that are minus their own dual.

We have the following identities:

$$\eta_{a\mu\nu} = \tfrac{1}{2}\epsilon_{\mu\nu\alpha\beta}\eta_{a\alpha\beta}, \quad \bar\eta_{a\mu\nu} = -\tfrac{1}{2}\epsilon_{\mu\nu\alpha\beta}\bar\eta_{a\alpha\beta}, \tag{A5}$$

$$\eta_{a\mu\nu} = -\eta_{a\nu\mu}, \tag{A6}$$

$$\eta_{a\mu\nu}\eta_{b\mu\nu} = 4\delta_{ab}, \tag{A7}$$

$$\eta_{a\mu\nu}\eta_{a\mu\lambda} = 3\delta_{\nu\lambda}, \tag{A8}$$

$$\eta_{a\mu\nu}\eta_{a\mu\nu} = 12, \tag{A9}$$

$$\eta_{a\mu\nu}\eta_{a k\lambda} = \delta_{\mu k}\delta_{\nu\lambda} - \delta_{\mu\lambda}\delta_{\nu k} + \epsilon_{\mu\nu k\lambda}, \tag{A10}$$

$$\delta_{k\lambda}\eta_{a\mu\nu} + \delta_{k\nu}\eta_{a\lambda\mu} + \delta_{k\mu}\eta_{a\nu\lambda} + \eta_{a\sigma k}\epsilon_{\lambda\mu\nu\sigma} = 0, \tag{A11}$$

$$\eta_{a\mu\nu}\eta_{b\mu\lambda} = \delta_{ab}\delta_{\nu\lambda} + \epsilon_{abc}\eta_{c\nu\lambda}, \tag{A12}$$

$$\epsilon_{abc}\eta_{b\mu\nu}\eta_{ck\lambda} = \delta_{\mu k}\eta_{a\nu\lambda} - \delta_{\mu\lambda}\eta_{a\nu k} - \delta_{\nu k}\eta_{a\mu\lambda} + \delta_{\nu\lambda}\eta_{a\mu k}, \tag{A13}$$

$$\eta_{a\mu\nu}\bar\eta_{b\mu\nu} = 0, \tag{A14}$$

$$\eta_{ak\mu}\bar\eta_{bk\lambda} = \eta_{ak\lambda}\bar\eta_{bk\mu}. \tag{A15}$$

*Work supported in part by the National Science Foundation under Grant No. MPS 75-20427.

†On leave from the University of Utrecht.

[1]A. A. Belavin *et al.*, Phys. Lett. 59B, 85 (1975).

[2]G. 't Hooft, Phys. Rev. Lett. 37, 8 (1976); R. Jackiw and C. Rebbi, *ibid.* 37, 172 (1976); Phys. Rev. D 14, 517 (1976); C. Callan, R. Dashen, and D. Gross, Phys. Lett. 63B, 334 (1976). See also F. R. Ore, Jr., Phys. Rev. D (to be published).

[3]R. Jackiw and S. Coleman (private communication).

[4]J. S. Bell and R. Jackiw, Nuovo Cimento 51, 47 (1969); S. L. Adler, Phys. Rev. 177, 2426 (1969).

[5]J. L. Gervais and B. Sakita, Phys. Rev. D 11, 2943 (1975); E. Tomboulis, *ibid.* 12, 1678 (1975).

[6]J. Honerkamp, Nucl. Phys. B48, 269 (1972); J. Honerkamp, in *Proceedings of the Colloquium on Renormalization of Yang-Mills Fields and Applications to Particle Physics, 1972,* edited by C. P. Korthals-Altes (C.N.R.S., Marseille, France, 1972).

[7]A. M. Polyakov, Phys. Lett. 59B, 82 (1975).

[8]G. 't Hooft and M. Veltman, Nucl. Phys. B44, 189 (1972); C. G. Bollini and J. J. Giambiagi, Phys. Lett. 40B, 566 (1972); J. F. Ashmore, Lett. Nuovo Cimento 4, 289 (1972); G. 't Hooft and M. Veltman, Report No. CERN 73-9, 1973 (unpublished).

[9]G. 't Hooft, Nucl. Phys. B62, 444 (1973).

[10]G. 't Hooft, Nucl. Phys. B61, 455 (1973).

CHAPTER 5.2

HOW INSTANTONS SOLVE THE U(1) PROBLEM

G. 't HOOFT

Institute for Theoretical Physics, Princetonplein 5, P.O. Box 80.006, 3508 TA Utrecht, The Netherlands

Abstract:

The gauge theory for strong interactions, QCD, has an apparent U(1) symmetry that is not realized in the real world. The violation of the U(1) symmetry can be attributed to a well-known anomaly in the regularization of the theory, which in field configurations called "instantons" can be seen to give rise to interactions that explicitly break the symmetry. A simple polynomial effective Lagrangian describes these effects qualitatively very well. In particular it is seen that no unwanted Goldstone bosons appear and the eta particle owes a large fraction of its mass to instantons. There is no need for field configurations with fractional winding numbers and it is explained how a spurious U(1) symmetry that remains in QCD even after introducing instantons, does not affect these results.

NORTH-HOLLAND - AMSTERDAM

Editorial Note

A review article should not lead to any major controversy even if it often implies that the author takes sides when conflicting results or approaches exist. Yet, when it covers a delicate and topical question, it may happen that some important controversy escapes the editor's attention. The very important and difficult "U(1) problem" was reviewed by G.A. Christos in a Physics Reports article published in 1984.

The present article by G. 't Hooft is meant to be a critical supplement to this former review. The editor is thankful to the present author to thus help to clarify this difficult question for the readers of Physics Reports.

M. Jacob

PHYSICS REPORTS (Review Section of Physics Letters) 142, No. 6 (1986) 357–387. North-Holland, Amsterdam

HOW INSTANTONS SOLVE THE U(1) PROBLEM

G. 't HOOFT

Institute for Theoretical Physics, Princetonplein 5, P.O. Box 80.006, 3508 TA Utrecht, The Netherlands

Received 18 April 1986

Contents:

1. Introduction

In addition to the usual hadronic symmetries the Lagrangian of the prevalent theory of the strong interactions, quantum chromodynamics (QCD) shows a chiral U(1) symmetry which is not realized, or at least badly broken, in the real world [1]. Now although it was soon established that the corresponding current conservation law is formally violated by quantum effects due to the Adler–Bell–Jackiw anomaly [2] it was for some time a mystery how effective U(1) violating interactions could take place to realize this violation, in particular because a less trivial variant of chiral U(1) symmetry still seemed to exist. Indeed, all perturbative calculations showed a persistence of the U(1) invariance.

With the discovery of instantons [3], and the form the Adler–Bell–Jackiw anomaly takes in these nonperturbative field configurations, this so-called U(1) problem was resolved [4, 5]. It was now clear how entire units of axial U(1) charge could appear or disappear into the vacuum without the need of (nearly) massless Goldstone bosons. In a world without instantons the η and η' particles would play the role of Goldstone bosons. Now the instantons provide them with an anomalous contribution to their masses.

In the view of most theorists the above arguments neatly explain why the η particle is considerably heavier than the pions (one must compare m_η^2 with m_π^2), and η' much heavier than the kaons.

Not everyone shares this opinion. In particular Crewther [6] argues that Ward identities can be written down whose solutions would still require either massless Goldstone bosons or gauge field configurations with fractional winding numbers, whereas experimental evidence denies the first and index theorems in QCD disfavor the second. If he were right then QCD would seem to be in serious trouble.

In a recent review article this dissident point of view was defended [7]. The way in which it refers to the present author's work calls for a reaction. At first sight the disagreement seems to be very deep. For instance the sign of the axial charge violation by the instanton is disputed; there are serious disagreements on the form of the effective interaction due to instantons, and the Crewther school insists that chiral U(1) is only spontaneously broken whereas we prefer to call this breaking an explicit one. Now as it turns out after closer study and discussions [8], much of the disagreement (but not all) can be traced back to linguistics and definitions. The aim of this paper is to demonstrate that using quite reasonable definitions of what a "symmetry" is supposed to mean for a theory, the "standard view" is absolutely correct; chiral U(1) is explicitly broken by instantons, and the sign of ΔQ_5 is as given by the anomaly equation; the effective Lagrangian due to instantons can be chosen to be local and polynomial in the mesonic fields, and the η and η' acquire masses due to instantons with integer winding numbers.

In order to make it clear to the reader what we are talking about we first consider a simple model (section 2) which we claim to be the relevant effective theory for the mesons in QCD (even though it is dismissed by ref. [7]). The model shows what the symmetry structure of the vacuum is and how the η particles obtain their masses. It also exhibits a curious periodicity structure with respect to the instanton θ angle, which was also noted in [7] but could not be cast in an easy language because of their refusal to consider models of this sort.

Now does the model of section 2 reflect the symmetry properties of QCD properly? Since refs. [6, 7, 8] express doubt in this respect we show in section 3 how it reflects the exactly defined operators and Green's functions of the exact theory. The calculation of the η-mass is redone, but now in terms of QCD parameters. We clearly do not pretend to "solve" QCD, so certain assumptions have to be made. The most important of these, quite consistent with all we know about QCD and the real world, is

$$\langle \bar{q}q \rangle = F \neq 0 . \tag{1.1}$$

Now refs. [6, 7] claim that in that case one needs gauge field configurations with fractional winding numbers. Our model calculations will clearly show that this is not the case.

How can it be then if our calculations appear to be theoretically sound and in close agreement with experiment, that they seem to contradict so-called "anomalous Ward identities"? To answer this question we were faced with unraveling some problems of communication. The current-algebraic methods of refs. [6, 7] were mainly developed before the rise of gauge theories but are subsequently applied to gauge-noninvariant sectors of Hilbert space. Their language is quite different from the one used in many papers on gauge theories [4] and due to incorrect "translations" several results of the present author were misquoted in [7]. After making the necessary corrections we try to analyze, in our own language, where the problem lies.

There are two classes of identities that one can write down for Green functions. One is the class of identities that follow from exactly preserved global or local symmetries. Local symmetries must always be exact symmetries. From those symmetries we get Ward identities [9] (in the Abelian case), or more generally Ward–Slavnov–Taylor [10] identities, which also follow from the (exact) Becchi–Rouet–Stora [11] global invariance.

On the other hand we have identities which follow from applying field transformations which may have the form of gauge transformations but which do not leave the Lagrangian (or, more precisely, the entire theory) invariant. These transformations are sometimes called Bell–Treiman transformations [12] in the literature, but "Veltman transformations" would be more appropriate [13]. The identities one gets reflect to some extent the dynamics of the theory and form a subclass of its Dyson–Schwinger equations. Thus when refs. [6, 7] perform chiral rotations of the fermionic fields, which do not leave θ invariant, they are performing a Veltman transformation, and their so-called "anomalous Ward identities" fall in this second class of equations among Green functions.

It is in the second derivative with respect to θ that refs. [6, 7] claim to get contradiction with our model calculations [8]: the second derivative of an insertion of the form

$$\mathrm{i}\theta F\tilde{F} , \tag{1.2}$$

in the Lagrangian vanishes, whereas the simple "effective field theory" requires an insertion of the form

$$\mathrm{e}^{\mathrm{i}\theta} \det(q_{\mathrm{L}}\bar{q}_{\mathrm{R}}) + \mathrm{h.c.} , \tag{1.3}$$

of which the second derivative produces the η mass. So they claim that in the real theory the η mass cannot be explained that simply.

In section 5 we explain why, in the modern formalism this problem does not arise at all. (1.2) should not be confused with ordinary Lagrange insertions and after resummation correctly reproduces (1.3). The canonical methods are not allowed if one tries first to quantize the gauge fields A_μ and only afterwards the fermion fields. The fermions have to be quantized and integrated out first. The phenomenon of "variable numbers of canonical fermionic variables" resolves the dilemma. We conclude, that the η mass is what it should be and there is no U(1) problem.

There is a further linguistic disagreement on whether the U(1) breaking of QCD should be called an explicit or a spontaneous symmetry breaking. In the effective Lagrangian model the symmetry is clearly broken explicitly. Several authors however refer to the θ angle as a property of the vacuum [5] in QCD. Of course as long as one agrees on the physical effects one is free to use whatever terminology seems appropriate. We merely point out that all physical consequences of the instanton effects in QCD (in particular the absence of a physical Goldstone boson) coincide with the ones of an explicit symmetry breaking. Our θ angle is as much a constant of Nature as any other physical parameter, to be compared for instance with the electron mass term which breaks the electron's chiral invariance.

Only if one adds nonphysical sectors to Hilbert space one may obtain an alternative description of the θ angle as a parameter induced by boundary effects producing a spontaneous symmetry breakdown. However, *any* explicit symmetry breaking can be turned into "spontaneous" symmetry breaking by artificially enlarging the Hilbert space. One gets no physically observable Goldstone boson however, so, the most convenient place to draw the dividing line between spontaneous and explicit global symmetry breaking is between the presence or absence of a Goldstone boson.

We explain this situation in section 6, where we show that any nonphysical symmetry can be forced upon a theory this way. In section 7 we show how the spurious U(1) symmetry that is used as a starting point in [7], actually belongs to this class. Section 8 shows that if one takes into account the enlarged Hilbert space with the variable θ angles then the effective theory of section 2 neatly obeys the so-called anomalous Ward identities. The effective model shows the vacuum structure so clearly that all problems with "fractional winding numbers" are removed.

The decay amplitude $\eta \to 3\pi$ posed problems similar to that of the η mass. As explained correctly in [7] this problem is resolved as soon as the U(1) breaking is understood so no further discussion of this decay is necessary. It fits well with theory.

Appendix A is a comment concerning the sign of axial charge violation under various boundary conditions.

Appendix B discusses the instanton-induced amplitude. The contribution from "small" instantons can be computed precisely but the infrared cutoff is uncertain. (Rough) Estimates of the amplitude in a simple color SU(2) theory give quite large values, which confirms that instantons may affect the symmetry structure of QCD sufficiently strongly such as to explain the known features of hadrons.

2. A simple model

Before really touching upon some of the more subtle aspects of the "U(1) problem" we first construct a simple "effective Lagrangian" model. Whether or not this model truly reflects the symmetry properties of QCD (which we do claim to be the case) is left to be discussed in the following sections. For simplicity the model of this section will be discussed only in the tree-approximation.

To be explicit we take the number of quark flavors to be two. Generalization towards any numbers of flavors (two and three are the most relevant numbers to be compared with the situation in the real world) will be completely straightforward at all stages in this section*. So we start with the "unbroken model" having global invariance of the form

* See however the remark following eq. (2.24).

$$\mathrm{U}(L)_{\mathrm{L}} \otimes \mathrm{U}(L)_{\mathrm{R}}, \qquad \text{with } L=2, \tag{2.1}$$

where the subscripts L and R refer to left and right, respectively. We consider complex meson fields with the quantum numbers of the quark–antiquark composite operator $\bar{q}_{\mathrm{R}j} q_{\mathrm{L}i}$. They transform under (2.1) as:

$$\phi'_{ij} = U^{\mathrm{L}}_{ik} \phi_{kl} U^{\mathrm{R}\dagger}_{lj}, \tag{2.2}$$

to be written simply as

$$\phi' = U^{\mathrm{L}} \phi U^{\mathrm{R}\dagger}. \tag{2.3}$$

Since we have no hermiticity condition on ϕ, there are eight physical particles σ, η, π_a and α_a ($a = 1,2,3$):

$$\phi = \tfrac{1}{2}(\sigma + \mathrm{i}\eta) + \tfrac{1}{2}(\boldsymbol{\alpha} + \mathrm{i}\boldsymbol{\pi}) \cdot \boldsymbol{\tau}, \tag{2.4}$$

where $\tau^{1,2,3}$ are the Pauli matrices. We take as our Lagrangian:

$$\mathscr{L} = -\mathrm{Tr}\, \partial_\mu \phi \partial_\mu \phi^\dagger - V(\phi). \tag{2.5}$$

A potential V_0 invariant under (2.1) is

$$V_0(\phi) = -\mu^2 \,\mathrm{Tr}\, \phi\phi^\dagger + \tfrac{1}{2}(\lambda_1 - \lambda_2)(\mathrm{Tr}\, \phi\phi^\dagger)^2 + \tfrac{1}{2}\lambda_2 \,\mathrm{Tr}(\phi\phi^\dagger)^2 \tag{2.6}$$

$$= -\frac{\mu^2}{2}(\sigma^2 + \eta^2 + \boldsymbol{\alpha}^2 + \boldsymbol{\pi}^2) + \frac{\lambda_1}{8}(\sigma^2 + \eta^2 + \boldsymbol{\alpha}^2 + \boldsymbol{\pi}^2)^2 + \frac{\lambda_2}{2}((\sigma\boldsymbol{\alpha} + \eta\boldsymbol{\pi})^2 + (\boldsymbol{\alpha} \wedge \boldsymbol{\pi})^2). \tag{2.7}$$

Assuming, as usual

$$\langle \sigma \rangle = f, \qquad \sigma = f + s, \tag{2.8}$$

we get, by taking the extremum of (2.7),

$$f^2 = 2\mu^2/\lambda_1; \tag{2.9}$$

$$V_0 = \frac{\lambda_1}{2}(fs + \tfrac{1}{2}(s^2 + \eta^2 + \boldsymbol{\alpha}^2 + \boldsymbol{\pi}^2))^2 + \frac{\lambda_2}{2}((f\boldsymbol{\alpha} + s\boldsymbol{\alpha} + \eta\boldsymbol{\pi})^2 + (\boldsymbol{\alpha} \wedge \boldsymbol{\pi})^2), \tag{2.10}$$

from which we read off:

$$m_s^2 = \lambda_1 f^2 = 2\mu^2, \qquad m_\eta^2 = 0, \qquad m_\alpha^2 = \lambda_2 f^2, \qquad m_\pi^2 = 0. \tag{2.11}$$

There are four Goldstone bosons, as expected from the U(2)⊗U(2) invariance, broken down to U(2) by (2.8).

We now consider two less symmetric additional terms in V:

$$V_m = U_m + U_m^* ;$$

$$U_m = \text{“} m\, e^{i\chi} \bar{q}_R q_L \text{”} = \tfrac{1}{4} m\, e^{i\chi}\, \mathrm{Tr}\, \phi = \tfrac{1}{2} m\, e^{i\chi} (\sigma + i\eta) , \tag{2.12}$$

and:

$$V_a = U_a + U_a^* ;$$

$$U_a = \text{“} \kappa\, e^{i\theta} \det(\bar{q}_R q_L) \text{”} = \kappa\, e^{i\theta} \det \phi$$

$$= \kappa\, e^{i\theta} ((\sigma + i\eta)^2 - (\boldsymbol{\alpha} + i\boldsymbol{\pi})^2) . \tag{2.13}$$

Here, m, χ, κ and θ are all free parameters. The terms between quotation marks are there just to show the algebraic structure, up to renormalization constants. Notice that V_0 still has the U(1) invariance

$$\phi \to e^{i\omega} \phi ,$$

or

$$(\sigma + i\eta) \to e^{i\omega} (\sigma + i\eta) , \qquad (\boldsymbol{\alpha} + i\boldsymbol{\pi}) \to e^{i\omega} (\boldsymbol{\alpha} + i\boldsymbol{\pi}) . \tag{2.14}$$

Therefore we are free to rotate

$$\chi \to \chi + \omega , \qquad \theta \to \theta + 2\omega . \tag{2.15}$$

(Here the 2 would be replaced by L in a theory with L flavors.)
Consider first the theory with $\kappa = 0$;

$$V = V_0 + V_m . \tag{2.16}$$

Then we can choose $\omega = \pi - \chi$, and

$$V_m = -m\sigma . \tag{2.17}$$

So χ is unphysical. Equation (2.9) is replaced by

$$f^2 = 2\mu^2/\lambda_1 + 2m/\lambda_1 f . \tag{2.18}$$

Consequently, in (2.11) we get

$$m_\pi^2 = m_\eta^2 = m/f . \tag{2.19}$$

Note that with the sign choice of (2.17), f must be the positive solution of (2.18).

Now take the other case, namely $m = 0$;

$$V = V_0 + V_a . \tag{2.20}$$

It is now convenient to choose

$$\omega = \tfrac{1}{2}(\pi - \theta) . \tag{2.21}$$

So here the angle θ is unphysical.

$$V_a = -2\kappa(\sigma^2 + \boldsymbol{\pi}^2 - \eta^2 - \boldsymbol{\alpha}^2) . \tag{2.22}$$

$$f^2 = 2\mu^2/\lambda_1 + 8\kappa/\lambda_1 . \tag{2.23}$$

The masses of the light particles become

$$m_\pi^2 = 0 ; \qquad m_\eta^2 = 8\kappa . \tag{2.24}$$

So the κ term contributes directly to the η mass and not to the pion mass. In case of more than two flavors the determinant in (2.13) will contain higher powers of ϕ, and extra factors f will occur in (2.24). But the mechanism generating a mass for the η' will not really be different from that of the η.

It is interesting to study the case when both m and κ are unequal to zero. We then cannot rotate both χ and θ independently and one of the two angles is physical. Since obviously the model Lagrangian is periodic with period 2π in θ, its physical consequences will be periodic in χ with period π, because of the invariance (2.15). In the general case we now also expect a nonvanishing value for

$$\langle \eta \rangle = g . \tag{2.25}$$

So we write $\eta = g + h$, where g is a c-number and h a field. The conditions for f and g are:

$$\begin{aligned} &\tfrac{1}{2}\lambda_1 f(f^2 + g^2) - \mu^2 f + m \cos\chi + 4\kappa f \cos\theta - 4\kappa g \sin\theta = 0 ; \\ &\tfrac{1}{2}\lambda_1 g(f^2 + g^2) - \mu^2 g - m \sin\chi - 4\kappa g \cos\theta - 4\kappa f \sin\theta = 0 . \end{aligned} \tag{2.26}$$

It will be convenient however to choose ω in (2.15) such that $g = 0$. According to (2.26) one then must have

$$m \sin\chi + 4\kappa f \sin\theta = 0 . \tag{2.27}$$

Now we find

$$m_\pi^2 = -\frac{m \cos\chi}{f} , \tag{2.28}$$

$$m_\eta^2 = -\frac{m \cos\chi}{f} - 8\kappa \cos\theta , \tag{2.29}$$

apart from an η–s mixing. The effect of this mixing however goes proportional to

$$16\kappa^2 \sin^2 \theta = m^2 \sin^2 \chi \tag{2.30}$$

and therefore is of higher order both in κ and m.

Consider now the function

$$F(\omega) = m \cos(\chi + \omega) + 2\kappa f \cos(\theta + 2\omega) . \tag{2.31}$$

Then (2.27) requires ω to be a solution of

$$F'(\omega) = 0 \tag{2.32}$$

(after which we insert the replacement (2.15)), and (2.29) implies:

$$fm_\eta^2 = F''(\omega) > 0 . \tag{2.33}$$

So, ω must be chosen such that (2.31) takes its minimum value. Furthermore, the replacement $\omega \rightarrow \omega + \pi$ switches the sign of the first but not of the second term in $F(\omega)$, so, if ω is chosen to be the *absolute* minimum of (2.31), then indeed also

$$m_\pi^2 > 0 , \tag{2.34}$$

except when there are two minima. In that case there is a phase transition with long range order (due to the massless pion) at the transition point.

This phase transition is at

$$\begin{aligned} &\chi = \pi/2 , \\ &\sin \theta = -m/4\kappa f , \\ &\cos \theta < 0 , \end{aligned} \tag{2.35}$$

or, if a rotation (2.15) is performed:

$$\begin{aligned} &\chi = 0 , \\ &\sin \theta = m/4\kappa f , \\ &0 < \theta < \pi/2 . \end{aligned} \tag{2.36}$$

A further critical point may occur at

$$\begin{aligned} &\chi = 0 \\ &\theta = \pi/2 \\ &m = 4\kappa f \end{aligned} \tag{2.37}$$

where also m_η vanishes. Of course these values of the parameters are not believed to be close to the ones describing real mesons. Phase transitions of this sort were indeed also described in ref. [7]. But because no specific model such as ours was considered, the periodicity structure in χ and θ was discussed in a somewhat untransparent way. Although obviously in our model we have a periodicity in θ with period 2π, ref. [7] suspected a period 4π. Indeed if ω in (2.15) shifts by an amount π, then $m\,e^{i\chi} \to -m\,e^{i\chi}$ and one might end up in an unstable analytic extension of the theory. Clearly the properties of the minimum of the potential $F(\omega)$, eq. (2.31), can only have period 2π in θ.

3. QCD

If Quantum Chromodynamics with L quark flavors, were to have an approximate $U(L)_L \times U(L)_R$ symmetry only broken by quark mass terms, the model of the previous section with $V = V_0 + V_m$ could then conveniently describe the qualitative features of mesons, but with η and π almost degenerate. (In the case $L > 2$ one merely has to substitute the 2×2 matrix ϕ by an $L\times L$ matrix.) That would be the case if somehow the effects of instantons could be suppressed.

Let us now consider instantons and write in a shorthand notation the functional integral I for a certain mesonic amplitude in QCD. For the ease of the discussion we assume all integrations to be in Euclidean space-time:

$$I = \int \mathrm{D}A \int \mathrm{D}\psi\, \mathrm{D}\bar{\psi}\, \exp[S_A + S_{A,\psi} + i\theta F\tilde{F} - J\bar{\psi}\psi]\,, \tag{3.1}$$

with

$$\mathrm{D}\psi\,\mathrm{D}\bar{\psi} = \prod \mathrm{d}^2\psi_L(x)\,\mathrm{d}^2\psi_R(x)\,\mathrm{d}^2\bar{\psi}_L(x)\,\mathrm{d}^2\bar{\psi}_R(x)\,, \tag{3.2}$$

$$\begin{aligned} S_A &= \int \mathrm{d}x(-\tfrac{1}{4}F^2_{\mu\nu}(A))\,, \\ S_{A,\psi} &= -\bar{\psi}_i(\gamma_\mu D^{(A)}_\mu + m_i)\psi_i\,. \end{aligned} \tag{3.3}$$

$$F\tilde{F} = \int \mathrm{d}x\,(g^2 F_{\mu\nu}\varepsilon_{\mu\nu\alpha\beta}F_{\alpha\beta}/64\pi^2)\,, \tag{3.4}$$

$$J\bar{\psi}\psi = J_\pi(\bar{\psi}_L\tau\psi_R - \bar{\psi}_R\tau\psi_L) + J_\eta(\bar{\psi}_L\psi_R - \bar{\psi}_R\psi_L) + \cdots \tag{3.5}$$

As we will see shortly, it is crucial that the integration over ψ is done first and the one over A afterwards.

Since we have no way of solving the theory exactly certain simplifying assumptions must be made. We now claim that the assumptions to be formulated next will in no way interfere with the known symmetry properties of the low-energy theory. Discussion of this claim will be postponed to sections 4–7.

An (anti-)instanton is a field configuration of the A fields with the property

$$\int_{\Delta V} F^a_{\mu\nu}F^a_{\alpha\beta}\varepsilon_{\mu\nu\alpha\beta}\,\mathrm{d}^4x = (-)64\pi^2/g^2\,, \tag{3.6}$$

where ΔV is the volume of a space-time region. Outside ΔV we have essentially $|F_{\mu\nu}| = 0$ but we cannot have $|A| = 0$ there because then (3.6) would vanish as the integrand is a total derivative.

The assumption we make is that the A integral can be split into an integral over instanton-locations and an integral over perturbative fluctuations around those instantons. We do this as follows. Let us divide space-time into four-dimensional boxes with volumes ΔV of the order of $1(\mathrm{fm})^4$. Each box may or may not contain one instanton or one anti-instanton. (There could be more than one instanton or anti-instanton in a single box, but we choose our boxes so small that such multi-instantons in one box become statistically insignificant.) The essential point is that since an instanton in a box ΔV will do nothing but gauge-rotate any of the fields outside ΔV, the instanton-numbers in each box are independent variables. Notice that at this point we do not require these twisted field configurations in the boxes to be exact solutions to the classical field configurations. This is why we have no difficulties confining each instanton to be completely inside one box, with only gauge rotations of the vacuum outside.

Let us then write

$$A = A_{\mathrm{inst}} + \delta A \tag{3.7}$$

where A_{inst} is due to the instantons only, then the integral over δA will essentially commute with the integral over the instantons. The δA integral is assumed to be responsible for the strong binding between the quarks. The confinement problem is *not* solved this way but is not relevant here since we decided to concentrate on low-energy phenomena only.

Note that the integration over A_{inst} is more than a summation over total winding number ν. Rather, if we write

$$\nu = \nu_+ - \nu_- \tag{3.8}$$

then the integral over A_{inst} closely corresponds to integration over the locations of the ν_+ instantons and the ν_- anti-instantons. It is important that we restrict ourselves to instantons with compact support (namely, limited to the confines of the box ΔV in which they belong). A larger instanton, if it occurs, should be represented as a small one in one of the boxes, with in addition a tail that is taken care of by integration over δA. "Very large" instantons are irrelevant because they would be superimposed by small ones. In short, in eq. (3.7), A_{inst} is defined to be a smooth field configuration that accounts for all winding numbers inside the boxes, and δA is defined to contribute to $\int F\tilde{F}$ by less than one unit in each box.

Now consider an isolated instanton located within one of our boxes, located at $x = x_1$. What is discussed at length in the literature is the fact that the ψ integration is now affected by the presence of a zero mode solution of the Dirac equation. If there were no other anti-instantons and no source term J then the fermionic integral, being proportional to the determinant of the operator $\gamma_\mu(\partial_\mu + \mathrm{i}gA_\mu)$, would vanish because of this one zero eigenvalue. If we do add the source term $J\bar{\psi}\psi$ the integral need not vanish. In ref. [4] it was derived that the instanton exactly acts as if it would contain a source for every fermionic flavor. Thus with one instanton located at $x = x_1$ the fermionic integral

$$\int \mathrm{D}\psi\,\mathrm{D}\bar{\psi}\,[\exp(S_{A,\psi} + J\bar{\psi}\psi)] \tag{3.9}$$

has the *same effect* as the integral

$$\kappa \int \mathrm{D}\psi\, \mathrm{D}\bar{\psi}\, [\exp(S_{0,\psi} + J\bar{\psi}\psi)] \cdot \det(\bar{\psi}_{\mathrm{R}}(x_1)\, \psi_{\mathrm{L}}(x_1))\,, \tag{3.10}$$

where κ may be computed from all one-loop corrections [4]. Indeed it was shown that the zero eigenmodes for all flavors which extend beyond the volume ΔV conveniently reproduce the fermionic propagators connecting x_1 with the sources J. The fact that (3.9) does have the same quantum selection properties as (3.10) can also be argued by realizing that a gauge-invariant regulator for the fermions had to be introduced, and instead of the *lowest* eigenmodes one could have concentrated on the much more localized *highest* fermionic states. The correctly regularized fermionic integral contains a mismatch by one unit for each flavor between the total number of left handed and right handed fermionic degrees of freedom. Since this happens both for the fermions and the antifermions $\bar{\psi}$ the determinant in (3.10) consists of products of L fermionic and L antifermionic fields.

Since we do require that the $\mathrm{SU}(L)_{\mathrm{L}} \otimes \mathrm{SU}(L)_{\mathrm{R}}$ is kept unharmed by the instantons, the determinant is at first sight the only allowed choice for (3.10) but, actually, if one does not suppress the color and spin indices, one can write down more expressions with the required symmetry properties.

Next consider ν_+ instantons, located at $x = x_i$. Following a declustering assumption which, at least to the present author's taste, is quite natural and does not require much discussion, we may assume these to act on the fermionic integrations as

$$\kappa^{\nu_+} \int \mathrm{D}\psi\, \mathrm{D}\bar{\psi}\, \mathrm{e}^{S_\psi} \prod_{i=1}^{\nu^+} \det(\psi_{\mathrm{L}}(x_i)\, \bar{\psi}_{\mathrm{R}}(x_i))\,. \tag{3.11}$$

Let us add the θ dependence and integrate over the instanton locations x_i:

$$\frac{1}{\nu_+!}\, \mathrm{e}^{\mathrm{i}\theta\nu_+} \prod_i \mathrm{d}^4 x_i\,. \tag{3.12}$$

The denominator $\nu_+!$ is due to exchange symmetry of the instantons.

We now extend our declustering assumption to the anti-instantons as well. This assumption was vigorously attacked in [6–8]. Indeed one might criticize it, for instance by suggesting that "merons" play a more crucial role [14]. We insist however that the assumption in no way interferes with the symmetry properties of our model. We will see in sections 7 and 8 that the anomalous Ward identities will be exactly satisfied by our model. To avoid confusion let us also stress that our declustering assumptions refer to the QCD part of the metric only, *not* to the contributions of the fermions which we denote explicitly. So there is no disagreement at all with the findings of ref. [15]. Indeed, our approach here is closely analogous to theirs.

Thus, consider ν_- anti-instantons. The complete instanton contribution to the functional integral is

$$\sum_{\nu_+=0}^{\infty} \sum_{\nu_-=0}^{\infty} \frac{\kappa^{\nu_+ + \nu_-}}{\nu_+!\,\nu_-!}\, \mathrm{e}^{\mathrm{i}\theta(\nu_+ - \nu_-)} \Big(\int \mathrm{d}^4 x\, \det(\psi_{\mathrm{L}}(x)\bar{\psi}_{\mathrm{L}}(x)) \Big)^{\nu_+} \Big(\int \mathrm{d}^4 x\, \det(\psi_{\mathrm{R}}(x)\bar{\psi}_{\mathrm{L}}(x)) \Big)^{\nu_-}\,. \tag{3.13}$$

The summations are now easy to carry out:

$$(3.13) = \exp \int \mathrm{d}^4 x [\kappa\, \mathrm{e}^{\mathrm{i}\theta} \det \psi_{\mathrm{L}}(x) \bar{\psi}_{\mathrm{R}}(x) + \kappa\, \mathrm{e}^{-\mathrm{i}\theta} \det \psi_{\mathrm{R}}(x) \bar{\psi}_{\mathrm{L}}(x)]\,, \tag{3.14}$$

which is precisely the effective interaction V_{a} of eq. (2.13). The remaining integrals over the fermionic fields ψ and the perturbative fields δA may well result in the effective Lagrangian model of section 2. Notice that, *before* we interchanged the $A_{\rm inst}$ and ψ, $\bar\psi$ integrations, we have made the substitution (3.10). This will be crucial for our later discussions. Once the substitution (3.10) has been made, the (A-field-dependent) extra fermionic degrees of freedom have been taken care of, and only then one is allowed to interchange the A and the ψ integrations. This is how (2.13) follows from (3.14).

4. Symmetries and currents

Let us split the generators $\Lambda_{\rm L}$ and $\Lambda_{\rm R}$ for the $\mathrm{U}(L)_{\rm L}\times\mathrm{U}(L)_{\rm R}$ transformations into scalar ones, Λ^a and Λ^0, and pseudoscalar ones, Λ_5^a and Λ_5^0. The infinitesimal transformation rules for the various fields considered thus far are:

$$\delta\psi_{\rm L} = -\tfrac{1}{2}\mathrm{i}\tau^a(\Lambda^a + \Lambda_5^a)\psi_{\rm L} + \mathrm{i}(\Lambda^0 + \Lambda_5^0)\psi_{\rm L}\,, \tag{4.1}$$

$$\delta\psi_{\rm R} = -\tfrac{1}{2}\mathrm{i}\tau^a(\Lambda^a - \Lambda_5^a)\psi_{\rm R} + \mathrm{i}(\Lambda^0 - \Lambda_5^0)\psi_{\rm R}\,, \tag{4.2}$$

$$\delta\phi = -\tfrac{1}{2}\mathrm{i}\Lambda^a[\tau^a,\phi] - \tfrac{1}{2}\mathrm{i}\Lambda_5^a\{\tau^a,\phi\}_+ + 2\mathrm{i}\Lambda_5^0\phi\,, \tag{4.3}$$

$$\delta\sigma = \Lambda_5^a\pi_a - 2\Lambda_5^0\eta\,, \tag{4.4}$$

$$\delta\eta = -\Lambda_5^a\alpha_a + 2\Lambda_5^0\sigma\,, \tag{4.5}$$

$$\delta\pi_a = \varepsilon_{abc}\Lambda^b\pi_c - \Lambda_5^a\sigma + 2\Lambda_5^0\alpha_a\,, \tag{4.6}$$

$$\delta\alpha_a = \varepsilon_{abc}\Lambda^b\alpha_c + \Lambda_5^a\eta - 2\Lambda_5^0\pi_a\,, \tag{4.7}$$

$$\delta\det\phi = 4\mathrm{i}\Lambda_5^0\det\phi\,. \tag{4.8}$$

In a theory with L flavors the factor 4 in eq. (4.8) must be replaced by $2L$. In a classical field theory the currents are most easily obtained by considering transformations (4.1)–(4.8) with space-time dependent $\Lambda_i(x)$. Their effect on the total action can be written as

$$\delta S = \int \mathrm{d}^4x(-F_i\Lambda_i(x) - J^i_\mu\partial_\mu\Lambda_i(x)) \tag{4.9}$$

(here $i = 1,\ldots,8$).
Since according to the equations of motion $\delta S = 0$ for all choices of $\Lambda_i(x)$, one has

$$\partial_\mu J^i_\mu(x) = F_i(x)\,. \tag{4.10}$$

A Lagrangian which gives invariance under the space-time independent Λ_i must have $F_i = 0$, so that the current J^i_μ is conserved.

We are now mainly concerned about the current $J_{\mu5}$ associated with Λ_5^0. The QCD Lagrangian (3.1) produces the current

$$J_{\mu 5} = \mathrm{i}\bar{\psi}\gamma_\mu \gamma_5 \psi\,, \tag{4.11}$$

and, prior to quantization:

$$\partial_\mu J_{\mu 5} = 2\mathrm{i}m\bar{\psi}\gamma_5\psi\,. \tag{4.12}$$

As is well known, however, eq. (4.12) does not survive renormalization. Renormalization cannot be performed in a chirally invariant way and therefore the symmetry cannot be maintained, unless we would be prepared to violate the local color gauge-invariance. But violation of color gauge invariance would cause violation of unitarity, so, in a correctly quantized theory, (4.12) breaks down. A diagrammatic analysis [2] shows that, at least to all orders of the perturbation expansion, one gets

$$\partial_\mu J_{\mu 5} = 2\mathrm{i}m\bar{\psi}\gamma_5\psi - \frac{\mathrm{i}Lg^2}{16\pi^2} F^a_{\mu\nu}\tilde{F}^a_{\mu\nu} \tag{4.13}$$

with

$$\tilde{F}^a_{\mu\nu} = \tfrac{1}{2}\varepsilon_{\mu\nu\alpha\beta}F^a_{\alpha\beta}\,. \tag{4.14}$$

We read off that, if we may ignore the mass term, then in a space-time volume V with ν_+ instantons and ν_- anti-instantons

$$\int_V \mathrm{d}^4x\, \partial_\mu J_{\mu 5} = -2\mathrm{i}L(\nu_+ - \nu_-)\,. \tag{4.15}$$

Here the factor i is an artefact of Euclidean space. Defining the charge Q_5 in a 3-volume V_3 by

$$Q_5 = \int_{V_3} J_0\, \mathrm{d}^3\boldsymbol{x} = \mathrm{i}\int_{V_3} J_4 \mathrm{d}^3\boldsymbol{x} = Q_\mathrm{R} - Q_\mathrm{L}\,, \tag{4.16}$$

each instanton causes a transition*

$$\Delta Q_5 = 2L\,. \tag{4.17}$$

This is called the "naive" equation in ref. [7]. Since we were working in a finite space-time volume V the nature of the "vacuum" has not yet entered into the discussion. Remarks on the language used here and in ref. [7] are postponed to appendix A.

Now let us write the corresponding equation in our effective Lagrangian model. Here,

$$J_{\mu 5} = 2\mathrm{i}\,\mathrm{Tr}\{(\partial_\mu \phi^*)\phi - \phi^*\partial_\mu\phi\} \tag{4.18}$$

and

* Apart from the disputed sign there are also differences in sign conventions with ref. [7].

$$\partial_\mu J_{\mu 5} = -2m(\eta \cos \chi + \sigma \sin \chi) + 16\kappa(\boldsymbol{\alpha} \cdot \boldsymbol{\pi} - \sigma\eta) \cos \theta + 8\kappa(\eta^2 + \boldsymbol{\alpha}^2 - \sigma^2 - \boldsymbol{\pi}^2) \sin \theta \,. \quad (4.19)$$

Before comparing this with eq. (4.13) of the QCD theory let us chirally rotate over an angle $\frac{1}{2}\chi$. The mass term in the original Lagrangian then becomes

$$-\bar{\psi} m \psi \cos \chi - \mathrm{i}\bar{\psi} m \gamma_5 \psi \sin \chi \,, \quad (4.20)$$

and then (4.13) becomes

$$\partial_\mu J_{\mu 5} = 2\mathrm{i} m \bar{\psi} \gamma_5 \psi \cos \chi - 2m \bar{\psi} \psi \sin \chi - \frac{\mathrm{i} L g^2}{16\pi^2} F^a_{\mu\nu} \tilde{F}^a_{\mu\nu} \,. \quad (4.21)$$

Therefore the first term in (4.19) can neatly be matched with the first terms of (4.21).

An issue raised in refs. [7, 8] is that there is an apparent discrepancy if we try to identify the last terms of (4.19) with the last term of (4.21). The last term of (4.21) contains the color fields only and there is absolutely no θ dependence here. But the last terms of (4.19) do show a crucial θ dependence. It is essentially

$$8\kappa \, \mathrm{Im}(\mathrm{e}^{\mathrm{i}\theta} \det \phi) \,. \quad (4.22)$$

Where did the θ dependence come from?

One way of arguing would be that the θ dependence of (4.22) is obvious. Chiral transformations are described by eq. (2.15) and any symmetry breaking term in a Lagrangian can obviously not be invariant at the same time. So the θ dependence of (4.22) is as it has to be. The symmetry breaking in QCD is not visible in its Lagrangian but is due to the θ dependence of the regularization procedure.

However, although this argument may explain why (4.19) shows a θ dependence and (4.21) does not, it does not explain why nevertheless these two theories can describe the same system. This is (partly) what the dispute is about. We claim that one can identify in the effective theory

$$\frac{128\pi^2\kappa}{Lg^2} \mathrm{Im}(\mathrm{e}^{\mathrm{i}\theta} \det \phi) = -\mathrm{i} F^a_{\mu\nu} \tilde{F}^a_{\mu\nu} \,, \quad (4.23)$$

so that, if $\theta \simeq 0$, one may identify $F\tilde{F}$ with the η field. At the same time we would also like to put

$$\bar{\phi} = \bar{q}_{\mathrm{R}} q_{\mathrm{L}} \,, \quad (4.24)$$

but this θ phase seems to be in disagreement with the canonical quantization procedure if A^a_μ, q_{R}, q_{L} and ϕ were to be considered as independent canonical variables. Another way of formulating this problem is that the right hand side of (4.23) seems to commute with the chiral charge operator Q_5 while the left hand side does not.

Notice that if we could somehow suppress instantons essentially $F\tilde{F}$ would vanish. The left hand side of (4.23) would vanish also, because $\kappa \to 0$. This suggests one simple answer to our problem: equation (4.23) violates axial charge conservation, but that is to be expected in a theory where axial charge is not conserved. Unfortunately some physicists insist in considering the U(1) violation by instantons as being "spontaneous" rather than explicit and therefore they rejected this simple answer. A rather curious attempt to bypass the problem was described in [7]. They first propose to replace our V_a of eq. (2.13) by

$$V'_a = \kappa\,\mathrm{Tr}(\log(\phi/\phi^\dagger))^2\,. \tag{4.25}$$

But this also does not commute with the axial charge operator and furthermore the logarithm is not single-valued so (4.25) makes no sense at all. So then they propose

$$\int_x V''_a = \kappa \int_x \left(\frac{-1}{\Box}\right)_x (\partial_\mu\,\mathrm{Tr}\log(\phi/\phi^\dagger))^2\,. \tag{4.26}$$

It is not obvious how this expression should be read such that it does make sense. If it is equivalent to (4.25) then clearly no improvement has been achieved. The problem of a multivalued logarithm has merely been substituted by the problem of an infrared divergent integral in x space. Equation (4.26) is then a clear example of linguistical gymnastics that should be avoided: formally it appears to be chirally invariant, yet it is equivalent to the local term (4.25), which is not.

We conclude in this section that the aforementioned problem is not solved by the logarithmic potentials V'_a of eqs. (4.25), (4.26). Let us call this problem the "U(1) dilemma". The correct resolution of the U(1) dilemma will be given in the next sections.

5. Solution of the U(1) dilemma

We must keep in mind how and why an effective Lagrangian is constructed. The word "effective" is meant to imply that such a model is not intended to describe the system in all circumstances. Rather, the model gives a simplified treatment of the system in a given range of energies and momenta. In this case we are interested in energies and momenta lower than, say, 1 GeV.

Now the complete theory contains variables at much higher frequencies. In as far as they play a role at lower energies, we must assume that they have been taken care of in the effective model. Consequently, the simple identification (4.24) is not correct as it stands. It should be read as

$$\phi \simeq (\bar{q}_R q_L)_{\text{low frequencies}}\,. \tag{5.1}$$

But what does "low frequency" mean? In a gauge theory the concept "frequency" need not be gauge-invariant. Therefore the splitting between high frequency and low frequency components of the quark fields must depend in general on the gluonic fields A_μ. This is why the contribution of the high frequency components of the quark fields to the axial current $J_{\mu 5}$ may depend explicitly on the A fields, a fact that is correctly expressed by the so-called "anomalous commutators" of [6, 7]. After integrating out the high frequency modes of the quark fields, but before integrating out the A fields, we have an expression for the axial current which has the following form:

$$J_{\mu 5} = 2\mathrm{i}\,\mathrm{Tr}\{(\partial_\mu \phi^*)\,\phi - \phi^*\partial_\mu\phi\} + J'_{\mu 5}(A)\,. \tag{5.2}$$

It is $J'_{\mu 5}(A)$ which is responsible for the nontrivial axial charge of the quantity $F\tilde{F}$ in (4.23). Let Q'_5 be the charge corresponding to $J'_{\mu 5}$. How does $F\tilde{F}$ commute with Q'_5?

Rather than $F\tilde{F}$ itself, it is the integral over some space-time volume ΔV,

$$\int_{\Delta V} F\tilde{F} = \nu^{+}_{\Delta V} - \nu^{-}_{\Delta V}\,, \tag{5.3}$$

that is relevant in (4.13). (We use the short hand notation of eq. (3.4).) Let us take ΔV so small that

$$\int_{\Delta V} F\tilde{F} = 0 \quad \text{or} \quad \pm 1\,. \tag{5.4}$$

(i) If $F\tilde{F}=0$ then we are not interested in its quantum numbers.
(ii) Whenever the right hand side of (5.3) is ± 1 we have an amplitude in which $\pm 2L$ units of axial charge are created.
(iii) The higher values of the right hand side are negligible.

The creation or annihilation of axial charges occurs because of the extra *high frequency* modes of $\bar{\psi}_L$, ψ_R or ψ_L and $\bar{\psi}_R$ that make the functional integral non-invariant. Let us call their contribution Z. If $\nu^{+}_{\Delta V}=1$ then

$$Z = \int \prod_{N} D\psi_L \prod_{N+1} D\psi_R \prod_{N} D\bar{\psi}_R \prod_{N+1} D\bar{\psi}_L (\exp S)\,, \tag{5.5}$$

where the subscripts under the multiplication symbols denote the numbers of variables to be integrated over. Then if

$$\begin{aligned} \psi_L &\to U_L \psi_L = e^{-i\omega}\psi_L\,, \\ \psi_R &\to U_R \psi_R = e^{i\omega}\psi_R\,, \end{aligned} \tag{5.6}$$

we have for all integrals over the anticommuting fields:

$$\begin{aligned} \int D\psi_L &\to e^{i\omega} \int D\psi_L\,, \\ \int D\psi_R &\to e^{-i\omega} \int D\psi_R\,, \end{aligned} \tag{5.7}$$

so that

$$Z \to e^{-2i\omega L} Z\,. \tag{5.8}$$

In this discussion we only include the high frequency components of the ψ fields. We see two things: the effective interaction Z due to an instanton transforms exactly as our insertion U_a of eq. (2.13), and secondly that, in a simplified picture where $F\tilde{F}$ takes integer values only (eq. (5.3)), the quantity $F\tilde{F}$, *after integration over the high frequency fermionic modes*, transforms with a factor

$$e^{\pm 2i\omega L}\,, \tag{5.9}$$

so that there is no longer any conflict* with (4.23). The transformation rules (4.1–4.8) hold for the effective fields. The terms containing Λ_5^0 tell us how the various fields commute with Q_5:

$$[Q_5, \Psi_L] = \Psi_L \tag{5.10}$$

$$[Q_5, \psi_R] = -\psi_R \tag{5.11}$$

$$[Q_5, \phi] = 2\phi \tag{5.12}$$

$$[Q_5, \sigma] = 2i\eta \tag{5.13}$$

etc.

6. Fictitious symmetry

The chiral U(1) symmetry breaking in QCD is an explicit one because the functional measure $\Pi\, D\psi$ fails to be chirally invariant when regularized in a gauge-invariant way [21]. This neatly explains why no massless Goldstone bosons are associated with this symmetry. Yet in several treatizes the words "spontaneous symmetry breaking" are used. How can this be?

Any broken global symmetry can formally be considered as a "spontaneously" broken one by a procedure consisting of two steps.

(i) Enlarge the physically accessible Hilbert space by adding all those Hilbert spaces of systems that would be obtained by applying the phoney symmetry transformation:

$$\mathfrak{H}' = \mathfrak{H} \times \mathrm{S} \tag{6.1}$$

where $\mathfrak{H}$ is the original Hilbert space and S the space of physical constants describing symmetry breaking.

(ii) Define the symmetry operator(s) as acting both in S and in $\mathfrak{H}$. We then obtain transformations in $\mathfrak{H}'$ that obviously leave the Hamiltonian H invariant. This procedure allows one to write down Ward identities for theories with symmetries broken explicitly by one or more terms in the Lagrangian. Since such identities were excessively used and advocated by Veltman in his early work on gauge theories with mass-insertions, we propose to refer to the above transformations as Veltman transformations [12].

Consider for example quantum electrodynamics with electron mass term

$$-m^*\bar{\psi}_L\psi_R - m\bar{\psi}_R\psi_L\,. \tag{6.2}$$

Then S is the space of complex numbers m. In this theory then, m is promoted to be an operator rather than a c-number. The chiral transformation

$$\begin{aligned} &\psi_{\substack{R\\L}} \to e^{\pm i\omega}\psi_{\substack{R\\L}} \\ &m \to e^{2i\omega} m \end{aligned} \tag{6.3}$$

* For the factor $e^{i\theta}$ see section 7.

is obviously an invariance of this theory. If in the "physical world"

$$\langle m \rangle = m = \text{real} \tag{6.4}$$

then one could argue that the symmetry (6.3) is "spontaneously broken".

The canonical charge operator $\tilde{Q}$ associated with (6.3) is now

$$\tilde{Q} = Q_{\mathrm{R}} - Q_{\mathrm{L}} + 2\left(m^* \frac{\partial}{\partial m^*} - m \frac{\partial}{\partial m}\right), \tag{6.5}$$

which commutes with (6.2). Thus, $\tilde{Q}$ is exactly conserved. But, since m is not a dynamical field, the new term cannot be written as an integral over 3-space, unless we enlarge the Hilbert space once again.

Let us now consider a Feynman diagram in which the mass term (6.2) occurs perturbatively as a two-prong vertex. Let there be a diagram with ν^+ insertions of the last term in (6.2), going with m, and ν^- of the first term, going with m^*. We have

$$\begin{aligned} \Delta Q_5 &= \Delta Q_{\mathrm{R}} - \Delta Q_{\mathrm{L}} = 2(\nu^+ - \nu^-) \\ \Delta \tilde{Q} &= 0\,. \end{aligned} \tag{6.6}$$

Only by brute force one could produce a current of which the fourth component would give a charge satisfying (6.6):

$$\begin{aligned} \tilde{J}_\mu &= J_{\mu 5} + K_\mu \\ \partial_\mu K_\mu(x) &= -2\mathrm{i}(\rho^+(x) - \rho^-(x)) \end{aligned} \tag{6.7}$$

where $\rho^{\pm}(x)$ is the density of the corresponding mass insertion vertices,

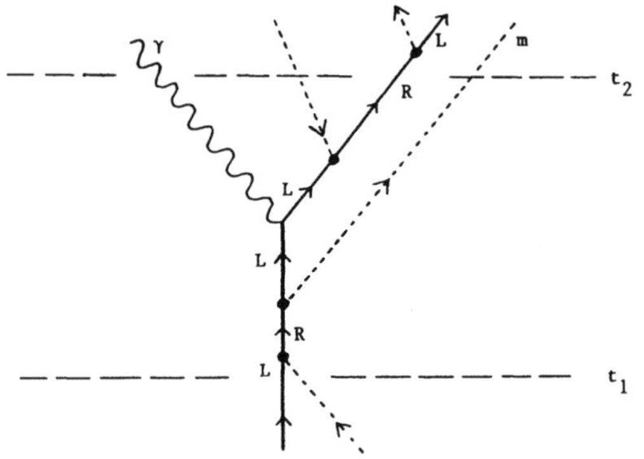

Fig. 1. Propagating electron (solid line) with mass terms. The propagators are expanded in m, yielding artificial particles (schizons, dotted lines) that carry away two units of axial charge, but no energy-momentum. Total chiral charge Q_5 is conserved. Here $Q_5(t_1) = Q_5(t_2) = 1$.

$$K_\mu(x) = -2\mathrm{i}\Box^{-1}\partial_\mu(\rho^+ - \rho^-)\,. \tag{6.8}$$

Clearly, K_μ is not locally observable. There is a nonobservable "Goldstone ghost" (the pole of $\Box^{-1}$). It goes without saying that $\bar{Q}$, although exactly conserved, and $\bar{J}_\mu$, are not very useful for canonical formalism. Yet the current $J^L_{\mu 5,\mathrm{sym}}$ and the charge Q^L_5 as used in ref. [7] are precisely of this form. This will be explained in the next section.

A neat way to implement the symmetry (6.3) is to treat the parameter m as a field: the "schizon", or "spurion", as those auxiliary objects are sometimes called to describe explicit symmetry breaking, such as isospin breaking by electromagnetism. The schizon field has a nonvanishing vacuum expectation value (6.4). Diagrammatically, a propagating electron could be represented by a diagram (fig. 1). Defining $\bar{Q} = \pm 2$ for the schizons we see that $\bar{Q}$ is absolutely conserved. Of course $\bar{Q}$ is also "spontaneously broken".

7. The "exactly conserved chiral charge" in a canonically quantized theory

The fictitious symmetry described in the previous section can be mimicked in a gauge theory in a way that looks very real. Consider instead of (4.11), the current

$$\begin{aligned} J_{\mu 5,\mathrm{sym}} &= J_{\mu 5} + K_\mu \\ K_\mu &= -\frac{g^2 \mathrm{i} L}{16\pi^2}\,\varepsilon_{\mu\nu\alpha\beta} A^a_\nu(\partial_\alpha A^a_\beta + \tfrac{1}{3} g f_{abc} A^b_\alpha A^c_\beta)\,. \end{aligned} \tag{7.1}$$

Then, in the limit $m \to 0$, one has

$$\partial_\mu J_{\mu 5,\mathrm{sym}} = 0\,. \tag{7.2}$$

The corresponding charge,

$$Q_{5,\mathrm{sym}} = \int J_{05,\mathrm{sym}}\,\mathrm{d}^3\boldsymbol{x} \tag{7.3}$$

generates "exact" chiral transformations. How does this operator act in Hilbert space?

To answer this question we must formulate the canonical quantization of the gluon field carefully. Conceptually the most transparent way is to first choose the temporal gauge:

$$A_0 = 0\,, \tag{7.4}$$

which leaves us formally the set of all states $|A(\boldsymbol{x}), \psi(\boldsymbol{x}), \bar{\psi}(\boldsymbol{x})\rangle$ at a given time t, where ψ and $\bar{\psi}$ should be seen as Grassmann numbers. Let us call the Hilbert space spanned by all these states the "huge" Hilbert space.

Then (7.4) leaves us invariance under all time-independent gauge transformations $\Omega = \Omega(\boldsymbol{x})$, so that the Hamiltonian in this space is invariant under a group G composed of gauge transformations $\Omega(\boldsymbol{x})$ that may vary from point to point. This generates an invariance at each $\boldsymbol{x}$, according to Noether's theorem. Writing

$$\Omega|A, \psi, \bar{\psi}\rangle = |A^{\Omega}\psi^{\Omega}\bar{\psi}^{\Omega}\rangle \tag{7.5}$$

where the subscript Ω indicates how the fields are gauge-transformed, we have

$$[H, \Omega] = 0 . \tag{7.6}$$

We can impose the gauge conditions of the second type:

$$\Omega|\Psi\rangle = |\Psi\rangle , \tag{7.7}$$

for all *infinitesimal* Ω, acting nontrivially only in a finite region of 3-space:

$$A_\mu^{\Omega}(x) = A(x) + D_\mu \Lambda(x) ,$$

Λ infinitesimal, and with compact support.

States $|\Psi\rangle$ satisfying (7.7) are said to be in the "large" Hilbert space (which is not as large as the "huge" one).

Finally, we consider all Ω with nontrivial winding number ν

$$\Omega_\nu|\Psi\rangle = e^{i\theta\nu}|\Psi\rangle . \tag{7.8}$$

These states $|\Psi\rangle$ are said to constitute the small, or physical Hilbert space at given θ.

Now notice that $J_{\mu 5,\mathrm{sym}}$ does not commute with Ω:

$$[J_{\mu 5,\mathrm{sym}}, \Omega] = \frac{iLg^2}{16\pi} D_\nu \Lambda \cdot \varepsilon_{\mu\nu\alpha\beta} F_{\alpha\beta} . \tag{7.9}$$

Therefore, $J_{\mu 5,\mathrm{sym}}$ cannot be considered to be an operator for states in the "large" Hilbert space. Acting on a state satisfying (7.7) it produces a state not satisfying (7.7).

Now the charge $Q_{5,\mathrm{sym}}$ of eq. (7.3) *does* commute with all Ω with $\nu = 0$, but not with the others:

$$[Q_{5,\mathrm{sym}}, \Omega_\nu] = 2iL\nu\Omega_\nu . \tag{7.10}$$

Therefore, $Q_{5,\mathrm{sym}}$ does act as an operator in the large Hilbert space, but not in the physical Hilbert space, because it mixes different θ values. We can write

$$\left[Q_{5,\mathrm{sym}} - 2iL \frac{\partial}{\partial\theta}, \Omega_\nu\right] = 0 . \tag{7.11}$$

We see that in every respect $Q_{5,\mathrm{sym}}$ behaves as $\tilde{Q}$ of the previous section, and $J_{\mu 5,\mathrm{sym}}$ as $\tilde{J}$.

A Goldstone boson would emerge in the theory if, besides the states satisfying (7.8), it could be possible to construct physical states in which θ would depend on space-time:

$$\theta \stackrel{?}{=} \theta(x, t) . \tag{7.12}$$

Here it is obvious that (7.12) would be in contradiction with (7.7) and (7.8): If we would compare different Ω_ν but with the same ν, such that the support of $\Omega^{(1)}$ would be in a region near $x^{(1)}$, and that of $\Omega^{(2)}$ near $x^{(2)}$, then the combination

$$\Omega^{(1)}\Omega^{(2)-1}$$

would have winding number zero. So the second gauge constraint would exclude any states for which $\theta(x^{(1)}) \neq \theta(x^{(2)})$. This is an important contrast with systems such as a ferromagnet, where local fluctuations are allowed, which, because of their large correlation lengths, correspond to massless excitations.

Because of the similarity between (7.11) and (6.5) we can consider $e^{i\theta}$ as a "schizon" field just as the electron mass term. Since θ cannot have any space-time dependence this schizon field cannot carry away any energy or momentum, just as m in the previous section.

Although $Q_{5,\mathrm{sym}}$ does not act in the "physical" Hilbert space, it is possible to write Ward identities [6, 7] due to its formal conservation,

$$\int d^4x\, \partial_\mu T\langle J^L_{\mu 5}(x), \mathrm{Op}\rangle = 2L \int d^4x\, \partial_\mu T\langle K_\mu(x), \mathrm{Op}\rangle + \int d^4x\, T\langle D_L(x), \mathrm{Op}\rangle + \langle [Q_{5,\mathrm{sym}}, \mathrm{Op}]\rangle\,, \tag{7.13}$$

where Op stands for any operator; and

$$D_L = 2im\bar{\psi}\gamma_5\psi\,, \tag{7.14}$$

which will vanish when $m \to 0$. K_μ satisfies

$$\partial_\mu K_\mu = -\frac{g^2 i}{32\pi^2} F^a_{\mu\nu}\tilde{F}^a_{\mu\nu}\,. \tag{7.15}$$

One can take

$$\mathrm{Op} = K_\nu(0)\,, \tag{7.16}$$

and assume

$$[Q_{5,\mathrm{sym}}, K_\nu] = 0 \tag{7.17}$$

while putting $D_L \to 0$. (7.18)

Now (7.17) is not obvious. Substituting (7.15) gives

$$[Q_{5,\mathrm{sym}}, F\tilde{F}] = 0\,. \tag{7.19}$$

On the other hand we showed in section 5 that $F\tilde{F}$ has nontrivial chiral transformation properties. This however corresponds to

$$[Q_5, F\tilde{F}] \neq 0\,. \tag{7.20}$$

Indeed,

$$\langle [Q_5, F\tilde{F}] \rangle \cong C \cdot f^2 \tag{7.21}$$

where the right hand side follows from the substitution (4.23) and the commutation rule (5.12). C is a constant and f the σ expectation value.

Equation (7.19) is a fundamental starting point of the discussions in refs. [6–8]. The difference between (7.19) and (7.20) must apparently be made up by the contribution of K_0 to $Q_{5,\mathrm{sym}}$. Now K_0 is not a physically observable field. Assigning to it the conventional commutation rules to be deduced from its composition in terms of color gauge fields is only allowed if one works in the "huge" Hilbert space including the gauge noninvariant states.

All we have to do to incorporate the fictitious symmetry generated by $Q_{5,\mathrm{sym}}$ into our model of effective fields described in section 2, is to add a schizon field, enlarging the Hilbert space. Let us call the schizon field

$$S = e^{i\theta} . \tag{7.22}$$

Our new identification is

$$F\tilde{F} = \frac{128\pi^2 i}{Lg^2} \operatorname{Im}(S \det \phi) , \tag{7.23}$$

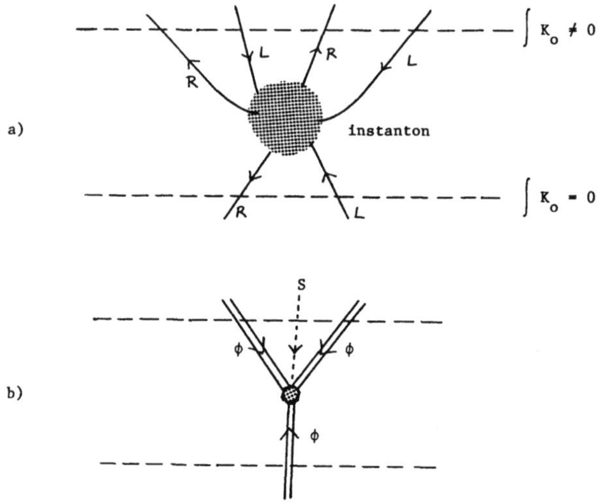

Fig. 2. Effective instanton action and its Q_5 symmetry properties. (a) Due to fermionic zero modes $2L$ units of Q_5 are absorbed at the site of the instanton. In the same time the charge generated by K_0 is not conserved. (b) In the effective theory the fermions are replaced by the ϕ field, and K_0 by a schizon S. The ϕ carry 2 units of Q_5 each, and S has $-2L$ units. S has a nonvanishing vacuum expectation value.

and if we postulate, in addition to (5.10)–(5.13) for $Q_{5,\mathrm{sym}}$ also (see 7.11):

$$[Q_{5,\mathrm{sym}}, S] = -2LS\,, \tag{7.24}$$

then with (5.12) we find that $F\tilde{F}$ commutes with $Q_{5,\mathrm{sym}}$. Substituting $\mathrm{e}^{\mathrm{i}\theta}\det\phi$ by $S\det\phi$ in (2.13) we see that indeed our effective field theory obeys the fictitious symmetry generated by $Q_{5,\mathrm{sym}}$. It must therefore also obey the so-called anomalous Ward identities. See fig. 2.

8. Diagrammatic interpretation of the anomalous Ward identities

The conclusion of the previous sections is that the model of section 2, with the substitution

$$\mathrm{e}^{\mathrm{i}\theta} \to S \tag{8.1}$$

obeys all anomalous Ward identities. It also exhibits in a very transparent way how the symmetries are now spontaneously broken. There are two vacuum expectation values:

$$\langle S \rangle = \mathrm{e}^{\mathrm{i}\theta} \tag{8.2}$$

$$\langle \sigma \rangle = f\,. \tag{8.3}$$

Both break $Q_{5,\mathrm{sym}}$ conservation. We can now draw Feynman diagrams in the Wigner representation by explicitly adding the vacuum bubbles due to (8.2) and (8.3). See fig. 3. By summing over the bubble insertions (geometric series which are trivial to sum), one reobtains the Goldstone representation of the particles. The σ bubbles tend to make the pions and eta massless, but the terms with κ (and the quark masses m) contribute linearly to m^2 for the various mesons. These diagrams clearly visualize where the masses come from and how the $Q_{5,\mathrm{sym}}$ charges are absorbed into the vacuum.

In ref. [7] an apparent problem was raised by their equations (4.27) and (4.28): they suggest the

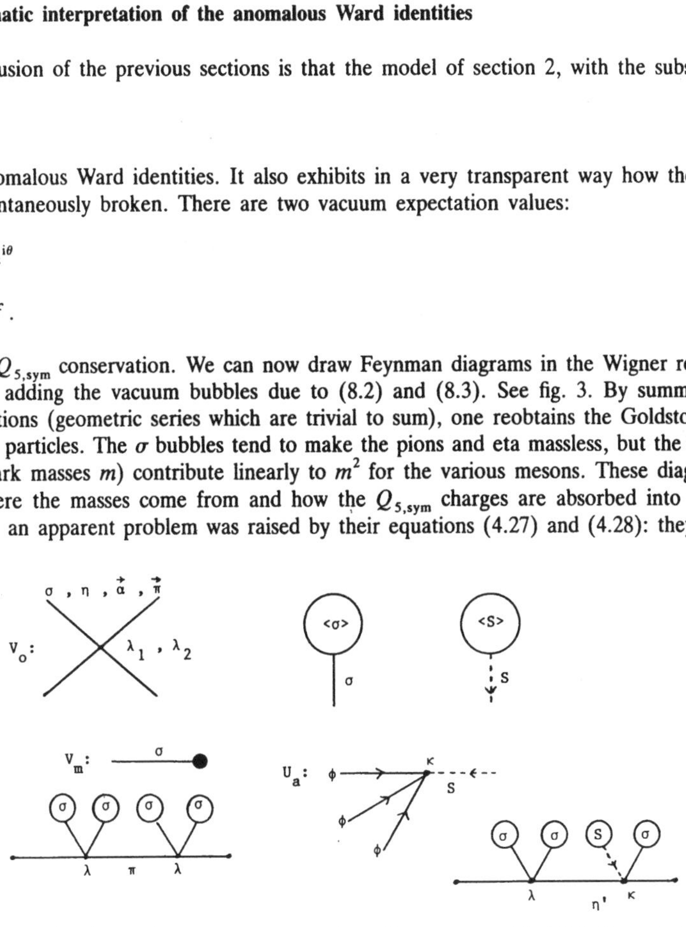

Fig. 3. Feynman rules in the Wigner mode. The σ blobs render the pion and the η massless. But the S blob gives a mass to η. Here we drew explicitly the η' propagator ($L = 3$). Its mass comes from the insertion at the right. Note that $m^2_{\eta'}$ is linear in $\langle S \rangle$ and in $\langle \sigma \rangle$: $m^2_{\eta'} \propto \kappa f$.

need for field configurations with fractional winding number ν, which would correspond with the breaking up of our S field into components with smaller $Q_{5,\mathrm{sym}}$ charges. In our diagrams we clearly see that there is no such need. If we have a Green function with an operator that creates only two chiral charges,

$$\chi_{\mathrm{Op}} = 2$$

then the sigma field can absorb these two, or add $2(L-1)$ more and have them absorbed by S. The vacuum simply isn't an eigenstate of Q_5 (nor $Q_{5,\mathrm{sym}}$) as it was assumed.

9. Conclusion

The disagreements between the approach of the Crewther's school to the U(1) problem, using anomalous Ward identities, and the more standard beliefs are not as wide as they appear. Their anomalous Ward identities, if applied with appropriate care, are perfectly valid for a simple effective field theory that clearly exhibits the most likely vacuum structure of QCD. It is important however to realize that the relevant exactly conserved chiral charge $Q_{5,\mathrm{sym}}$ is not physically observable, something which explains the need for the introduction of a spurion field S in the effective theory. It appears that the consequences of working with an unphysical symmetry were underestimated in ref. [7]. Some of the difficulties signalled in [7] were due to the too strong assumption that the vacuum is an eigenstate of Q_5. That such assumptions are unnecessary and probably wrong would have been realized if they had taken the effective theory more seriously.

The fact that the effective theory of section 2 displays the correct symmetry properties does not have to mean that it is accurate. Indeed it could be that instantons tend to split into "merons" [16], a dynamical property that might be a factor in the spontaneous chiral symmetry breaking mechanism [14]. But these aspects do not affect the symmetry transformation properties of the fields under consideration. More fields, describing higher resonances, could have to be added. The baryonic degrees of freedom are most likely to be considered as extended solutions of the effective field equations (skyrmions). That these skyrmions [17] indeed possess the relevant baryonic quantum numbers was discovered by Witten [18].

The author thanks R.J. Crewther for his patience in extensive discussions, even though no complete agreement was reached.

Appendix A. The sign of ΔQ_5

In ref. [7] the present author's work was claimed to be in error at various places. Although some minor technical corrections on the computed coefficients in the quantum corrections due to instantons were found (in the original publication see ref. [20]) and even some insignificant inaccuracies in the notation of a sign might occur, we stress here that none of those claims of ref. [7] were justified. In particular there are no fundamental discrepancies in the sign of ΔQ_5.

Let us here ignore the masses of the quarks. As formulated in section 4, an instanton in a finite space-time volume V causes a transition with

$$\Delta Q_5 = 2L \tag{A1}$$

where Q_5 is the gauge-invariant axial charge. Since $Q_{5,\text{sym}}$, as defined in section 7, is now strictly conserved one obviously has

$$\Delta Q_{5,\text{sym}} = 0\,, \tag{A2}$$

which is of course only defined in the "large" Hilbert space comprising all θ worlds.

Instead of (A1–A2), we read in ref. [7]:

$$\Delta Q_5 = 0\,; \qquad \Delta Q_{5,\text{sym}} = -2L\,. \tag{A3}$$

These are not the properties of a closed space-time volume such as we described, but represent the features of a Green's function where the asymptotic states are θ-vacua. Since $Q_{5,\text{sym}}$ contains explicitly the operator $\partial/\partial\theta$ in the "large" Hilbert space (cf. eq. (7.11)), the θ-vacuum is not invariant under $Q_{5,\text{sym}}$. This is why (A3) is not in conflict with $Q_{5,\text{sym}}$ conservation. But we also see that (A1–A2) and (A3) hold under different boundary conditions (the reason why $\Delta Q_5 = 0$ for Green's functions in a θ-vacuum is correctly explained in ref. [7]).

Appendix B. The integration over the instanton size ρ

To get even a rough estimate of the size of our instanton's contribution to an amplitude requires lengthy calculations. The most essential ingredients for such a calculation are provided in our previous paper [5.1] (see refs. [19] for the original publication, and the corrections given in [20]). Let us take for simplicity SU(2) as our color gauge group. An instanton with given size ρ then gives rise to an effective interaction of the form

$$Cg^{-8}\rho^{3N^{\rm f}-5}{\rm d}\rho\exp\Big\{-\frac{8\pi^2}{g_R^2(\mu)}+\log(\mu\rho)\Big[\frac{22}{3}-\frac{1}{6}\sum_t N^s(t)C(t)-\frac{2}{3}N^{\rm f}\Big]\Big\}$$
$$\Big\langle\prod_{s=1}^{N^{\rm f}}(\bar\psi_s\omega)(\bar\omega\psi_s)+\text{h.c.}\Big\rangle\,, \tag{B1}$$

where $N^{\rm f}$ is the number of fermions in the doublet representation, ρ is the size of the instanton, μ is an arbitrary mass unit enabling us to obtain a renormalization group invariant expression, and $N^s(t)$ is the number of scalar field representations with color t. The parameter C was computed in the previous paper, and it also depends on the scalars and fermions present:

$$C = 2^{10+3N^{\rm f}}\pi^{6+2N^{\rm f}}{\rm e}^{A-\sum_t N^s(t)A(t)-N^{\rm f}B}\,, \tag{B2}$$

where $A, A(t)$, and B are numbers computed in [5.1].

$$W = \exp(-8\pi^2/g^2)(\det \mathfrak{M}_A)^{-1/2} \det \mathfrak{M}_\psi \det \mathfrak{M}_{\mathrm{gh}} \tag{B3}$$

and the determinants can be computed by diagonalization:

$$\mathfrak{M}_i \psi = E_i \psi\,. \tag{B4}$$

It is clear that (B3) is highly divergent unless we formulate very precisely an appropriate subtraction procedure. A convenient method is to apply first a variety of Pauli–Villars regularization to the fields A^{qu} and ψ and then add correction terms to be obtained by comparing this regularization scheme to for instance dimensional regularization.

It is then found that (B3) is not the complete answer: there are zero eigenvalues of $\mathfrak{M}_A$, $\mathfrak{M}_\psi$ and $\mathfrak{M}_{\mathrm{gh}}$ which have to be considered separately. The corresponding eigenmodes must be replaced by collective coordinates including an appropriate Jacobian for this transformation. Since (B3) must be compared with the vacuum transition (in absence of instantons) the collective coordinate integration is to be divided by a norm factor determined by a Gaussian integral. The result is the effective Lagrangian (for the case that the color gauge group is SU(2)):

$$\begin{aligned}\mathfrak{L}^{\mathrm{eff}}(z) = {} & 2^{10+3N^{\mathrm{f}}}\pi^{6+2N^{\mathrm{f}}}g^{-8}\int \rho^{3N^{\mathrm{f}}-5}\,\mathrm{d}\rho \exp\Big\{-\frac{8\pi^2}{[g_{\mathrm{R}}^{\mathrm{D}}(\mu)]^2} \\ & +\log(\mu\rho)\Big[\frac{22}{3}-\frac{1}{6}\sum_t N^{\mathrm{s}}(t)C(t)-\tfrac{2}{3}N^{\mathrm{f}}\Big]+A-\sum_t N^{\mathrm{s}}(t)A(t)-N^{\mathrm{f}}B\Big\} \\ & \times\Big\langle\prod_{s=1}^{N^{\mathrm{f}}}(\bar\psi_s\omega)(\bar\omega\psi_5)\Big\rangle+\mathrm{h.c.}\,,\end{aligned} \tag{B5}$$

where N^{f} is the number of fermions in the doublet representation, ρ is a scalar parameter for the instanton, μ is an arbitrary mass unit enabling us to obtain a renormalization group invariant expression, and $N^{\mathrm{s}}(t)$ is the number of scalar field representations with color t.

Defining coefficients $\alpha(t)$ as in table 1, we have now:

$$\begin{aligned} A &= -\alpha(1)+\tfrac{11}{3}(\log 4\pi-\gamma)+\tfrac{1}{3}=7.053\,991\,03 \\ A(t) &= -\alpha(t)+\tfrac{1}{12}(\log 4\pi-\gamma)C(t) \\ A(1/2) &= 0.308\,690\,69 \\ A(1) &= 1.094\,576\,62 \\ A(3/2) &= 2.481\,356\,10 \\ B &= -2\alpha(1/2)+\tfrac{1}{3}(\log 4\pi-\gamma)=0.359\,522\,90\,. \end{aligned} \tag{B6}$$

Table 1

t	$C(t)$	$\alpha(t)$
0	0	0
1/2	1	$2R - \frac{1}{6}\log 2 - 17/72$
1	4	$8R + \frac{1}{3}\log 2 - 16/9$
3/2	10	$20R + 4\log 3 - \frac{5}{3}\log 2 - 265/36$

$$R = \frac{1}{12}(\log 2\pi + \gamma) + \frac{1}{2\pi^2}\sum_{2}^{\infty}\frac{\log s}{s^2} = 0.248\,754\,477.$$

Finally, the spinors ω are normalized by

$$\sum_{\alpha} \omega_\alpha \bar{\omega}_\alpha = \tfrac{1}{2}(1+\gamma_5) \tag{B7}$$

and required to be smeared in color space, such that for instance

$$\langle \omega_\alpha \bar{\omega}_\beta \rangle = \tfrac{1}{4}\delta_{\alpha\beta}(1+\gamma_5) \tag{B8}$$

and, in the case $L = N^{\mathrm{f}} = 2$:

$$\left\langle \prod_{s=1}^{2} (\bar{\psi}_s \omega)(\bar{\omega}\psi_s) \right\rangle = \frac{1}{24}\left(2\delta^{\beta_1}_{\alpha_1}\delta^{\beta_2}_{\alpha_2} - \delta^{\beta_2}_{\alpha_1}\delta^{\beta_1}_{\alpha_2}\right)\varepsilon^{st}\bar{\psi}^{\alpha_1}_1(1+\gamma_5)\psi^{\beta_1}_s\bar{\psi}^{\alpha_2}_2(1+\gamma_5)\psi^{\beta_2}_t \tag{B9}$$

where s and t are flavor indices and α_i, β_i color indices.

The ρ integral may seem to diverge in most interesting cases ($N^{\mathrm{f}} > 1$). Note however that it would be natural to choose

$$\mu = 1/\rho \tag{B10}$$

and substitute g^{-8} by the running value $g(\mu)^{-8}$. At large ρ one might take $g \propto \rho$ and thus improve the convergence. Of course the infrared end of the integral is quite uncertain because in our perturbative procedure the effects of confinement etc. have not been taken into account. This inhibits a precise evaluation of the amplitude. A rough estimate (for a color SU(2) theory) is obtained if we take at large ρ

$$g^2(1/\rho) \to 16\pi\rho^2\sigma \tag{B11}$$

where σ is the string constant. Then quarks with color charge 1/2 at a distance ρ from each other feel a force

$$\sigma = \tfrac{1}{4}g^2/4\pi\rho^2 . \tag{B12}$$

Our integral becomes, in the case $N^{\mathrm{f}} = 2$,

$$\mathfrak{L}^{\mathrm{eff}} = 2^{16}\pi^{10}\,\mathrm{e}^{A-2B}\mathfrak{L}_1 \int \frac{\rho\,\mathrm{d}\rho}{g^8(1/\rho)} \exp - 8\pi^2/g^2(1/\rho) . \tag{B13}$$

From (B11) we get

$$\rho\,\mathrm{d}\rho \cong g\,\mathrm{d}g/16\pi\sigma \tag{B14}$$

and using $x = 1/g^2$ the integral in (B13) is

$$\frac{1}{32\pi\sigma}\int_0^\infty \mathrm{d}x\,x^2\,\mathrm{e}^{-8\pi^2 x} = 2^{-13}\pi^{-7}\sigma^{-1} \tag{B15}$$

so that

$$\mathfrak{L}^{\mathrm{eff}} = 8\pi^3\,\mathrm{e}^{A-2B}\sigma^{-1}\mathfrak{L}_1 \tag{B16}$$

where $\mathfrak{L}_1$ is the Lagrangian (B9).

The result is uncomfortably large, but then the approximations used here (eq. B11) could at best only be expected to yield the order of magnitude of the expected interaction, which is clearly a strong one. Note that our coupling constant is subtraction scheme dependent. Since we did not yet include two loop order effects the subtraction scheme used strongly effects the final result. A factor 4π difference in the definition of μ in (B1) reduces the coefficient in front of the effective action (B16) by a factor $(4\pi)^3 = 2^6\pi^3$. This is just to illustrate how sensitively the amplitude obtained depends upon the assumptions.

References

[1] S.L. Glashow, in: Hadrons and their interactions, Erice 1967, ed. A. Zichichi (Acad. Press, New York, 1968) p. 83;
S.L. Glashow, R. Jackiw and S.-S. Shei, Phys. Rev. 187 (1969) 1916;
M. Gell-Mann, in: Proc. Third Topical Conf. on Particle Physics, Honolulu 1969, eds. W.A. Simonds and S.F. Tuan (Western Periodicals, Los Angeles, 1970) p. 1, and in: Elementary Particle Physics, Schladming 1972, ed. P. Urban (Springer-Verlag, 1972); Acta Physica Austriaca Suppl. IX (1972) 733;
H. Fritzsch and M. Gell-Mann, Proc. XVI Intern. Conf. on H.E.P., Chicago 1972, eds. J.D. Jackson and A. Roberts, Vol. 2, p. 135, and Phys. Lett. 47B (1973) 365.

[2] S.L. Adler, Phys. Rev. 177 (1969) 2426;
J.S. Bell and R. Jackiw, Nuovo Cim. 60A (1969) 47;
S.L. Adler and W.A. Bardeen, Phys. Rev. 182 (1969) 1517.

[3] A.A. Belavin, A.M. Polyakov, A.S. Schwartz and Yu.S. Tyupkin, Phys. Lett. 59B (1975) 85.

[4] G. 't Hooft, Phys. Rev. Lett. 37 (1976) 8; Phys. Rev. D14 (1976) 3432;
R. Jackiw and C. Rebbi, Phys. Rev. Lett. 37 (1976) 172;
C.G. Callan Jr., R.F. Dashen and D.J. Gross, Phys. Lett. 63B (1976) 334; Phys. Rev. D17 (1978) 2717.

[5] S. Coleman, in: The Whys of Subnuclear Physics, Erice 1977, ed. A. Zichichi (Plenum Press, New York, 1979) p. 805.

[6] R. Crewther, Phys. Lett. 70B (1977) 349; Riv. Nuovo Cim. 2 (1979) 63;
R. Crewther, in: Facts and Prospects of Gauge Theories, Schladming 1978, ed. P. Urban (Springer-Verlag, 1978); Acta Phys. Austriaca Suppl. XIX (1978) 47.

[7] G.A. Christos, Phys. Reports 116 (1984) 251.

[8] R. Crewther, private communication.

[9] J.C. Ward, Phys. Rev. 78 (1950) 1824;
Y. Takahashi, Nuovo Cim. 6 (1957) 370.

[10] A. Slavnov, Theor. and Math. Phys. 10 (1972) 153 (in Russian), transl. Theor. and Math. Phys. 10, p. 99;
J.C. Taylor, Nucl. Phys. B33 (1971) 436.

[11] C. Becchi, A. Rouet and R. Stora, Comm. Math. Phys. 42 (1975) 127; Ann. Phys. (N.Y.) 98 (1976) 287.
[12] M. Veltman, Nucl. Phys. B7 (1968) 637; Nucl. Phys. B21 (1970) 288.
[13] There is no paper by Bell or Treiman on this particular subject. See ref. [12].
[14] A.R. Zhitnitsky, The discrete chiral symmetry breaking in QCD as a manifestation of the Rubakov–Callan effect, Novosibirsk preprint 1986.
[15] C. Lee and W. Bardeen, Nucl. Phys. B153 (1979) 210.
[16] C.G. Callan Jr., R.F. Dashen and D.J. Gross, Phys. Lett. 66B (1977) 375; Phys. Rev. D17 (1978) 2717; Phys. Lett. 78B (1978) 307.
[17] T.H.R. Skyrme, Proc. Roy. Soc. A260 (1961) 127.
[18] E. Witten, Nucl. Phys. B223 (1983) 422; 433.
[19] G. 't Hooft, Phys. Rev. D14 (1976) 3432.
[20] F.R. Ore, Phys. Rev. D16 (1977) 2577;
G. 't Hooft, Phys. Rev. D18 (1978) 2199;
A. Hasenfratz and P. Hasenfratz, Nucl. Phys. B193 (1981) 210.
[21] K. Fujikawa, Phys. Rev. Lett. 42 (1979) 1195; Phys. Rev. D21 (1980) 2848.

CHAPTER 5.3

NATURALNESS, CHIRAL SYMMETRY, AND SPONTANEOUS CHIRAL SYMMETRY BREAKING

G. 't Hooft
Institute for Theoretical Physics
Utrecht, The Netherlands

ABSTRACT

A properly called "naturalness" is imposed on gauge theories. It is an order-of-magnitude restriction that must hold at all energy scales μ. To construct models with complete naturalness for elementary particles one needs more types of confining gauge theories besides quantum chromodynamics. We propose a search program for models with improved naturalness and concentrate on the possibility that presently elementary fermions can be considered as composite. Chiral symmetry must then be responsible for the masslessness of these fermions. Thus we search for QCD-like models where chiral symmetry is not or only partly broken spontaneously. They are restricted by index relations that often cannot be satisfied by other than unphysical fractional indices. This difficulty made the author's own search unsuccessful so far. As a by-product we find yet another reason why in ordinary QCD chiral symmetry must be broken spontaneously.

III1. INTRODUCTION

The concept of causality requires that macroscopic phenomena follow from microscopic equations. Thus the properties of liquids and solids follow from the microscopic properties of molecules and atoms. One may either consider these microscopic properties to have been chosen at random by Nature, or attempt to deduce these from even more fundamental equations at still smaller length and time scales. In either case, it is unlikely that the microscopic equations contain various free parameters that are carefully adjusted by Nature to give cancelling effects such that the macroscopic systems have some special properties. This is a

Dynamical Symmetry Breaking, A Collection of Reprints,
Edited by A. Farhi and R. Jackiw (World Scientific, 1982).

philosophy which we would like to apply to the unified gauge theories: the effective interactions at a large length scale, corresponding to a low energy scale μ_1, should follow from the properties at a much smaller length scale, or higher energy scale μ_2, without the requirement that various different parameters at the energy scale μ_2 match with an accuracy of the order of μ_1/μ_2. That would be unnatural. On the other hand, if at the energy scale μ_2 some parameters would be very small, say

$$\alpha(\mu_2) = \mathcal{O}(\mu_1/\mu_2) , \qquad \text{(III1)}$$

then this may still be natural, provided that this property would not be spoilt by any higher order effects. We now conjecture that the following dogma should be followed:
- at any energy scale μ, a physical parameter or set of physical parameters $\alpha_i(\mu)$ is allowed to be very small only if the replacement $\alpha_i(\mu) = 0$ would increase the symmetry of the system. -
In what follows this is what we mean by naturalness. It is clearly a weaker requirement than that of P. Dirac[1] who insists on having no small numbers at all. It is what one expects if at any mass scale $\mu > \mu_0$ some ununderstood theory with strong interactions determines a spectrum of particles with various good or bad symmetry properties. If at $\mu = \mu_0$ certain parameters come out to be small, say 10^{-5}, then that cannot be an accident; it must be the consequence of a near symmetry.

For instance, at a mass scale

$$\mu = 50 \text{ GeV},$$

the electron mass m_e is 10^{-5}. This is a small parameter. It is acceptable because $m_e = 0$ would imply an additional chiral symmetry corresponding to separate conservation of left handed and right handed electron-like leptons. This guarantees that all renormalizations of m_e are proportional to m_e itself. In sects. III2 and III3 we compare naturalness for quantum electrodynamics and ϕ^4 theory.

Gauge coupling constants and other (sets of) interaction constants may be small because putting them equal to zero would turn the gauge bosons or other particles into free particles so that they are separately conserved.

If within a set of small parameters one is several orders of magnitude smaller than another then the smallest must satisfy our "dogma" separately. As we will see, naturalness will put the severest restriction on the occurrence of scalar particles in renormalizable theories. In fact we conjecture that this is the reason why light, weakly interacting scalar particles are not seen.

CHIRAL SYMMETRY AND CHIRAL SYMMETRY BREAKING

It is our aim to use naturalness as a new guideline to construct models of elementary particles (sect. III4). In practice naturalness will be lost beyond a certain mass scale μ_0, to be referred to as "Naturalness Breakdown Mass Scale" (NBMS). This simply means that unknown particles with masses beyond that scale are ignored in our model. The NBMS is only defined as an order of magnitude and can be obtained for each renormalizable field theory. For present "unified theories", including the existing grand unified schemes, it is only about 1000 GeV. In sect. 5 we attempt to construct realistic models with an NBMS some orders of magnitude higher.

One parameter in our world is unnatural, according to our definition, already at a very low mass scale ($\mu_0 \sim 10^{-2}$ eV). This is the cosmological constant. Putting it equal to zero does not seem to increase the symmetry. Apparently gravitational effects do not obey naturalness in our formulation. We have nothing to say about this fundamental problem, accept to suggest that *only* gravitational effects violate naturalness. Quantum gravity is not understood anyhow so we exclude it from our naturalness requirements.

On the other hand it is quite remarkable that all other elementary particle interactions have a high degree of naturalness. No unnatural parameters occur in that energy range where our popular field theories could be checked experimentally. We consider this as important evidence in favor of the general hypothesis of naturalness. Pursuing naturalness beyond 1000 GeV will require theories that are immensely complex compared with some of the grand unified schemes.

A remarkable attempt towards a natural theory was made by Dimopoulos and Susskind [2]. These authors employ various kinds of confining gauge forces to obtain scalar bound states which may substitute the Higgs fields in the conventional schemes. In their model the observed fermions are still considered to be elementary.

Most likely a complete model of this kind has to be constructed step by step. One starts with the experimentally accessible aspects of the Glashow-Weinberg-Salam-Ward model. This model is natural if one restricts oneself to mass-energy scales below 1000 GeV. Beyond 1000 GeV one has to assume, as Dimopoulos and Susskind do, that the Higgs field is actually a fermion-antifermion composite field. Coupling this field to quarks and leptons in order to produce their mass, requires new scalar fields that cause naturalness to break down at 30 TeV or so. Dimopoulos and Susskind speculate further on how to remedy this. To supplement such ideas, we toyed with the idea that (some of) the presently "elementary" fermions may turn out to be bound states of an odd number of fermions when considered beyond 30 TeV. The binding mechanism would be similar to the one that keeps quarks inside the proton. However, the proton is not particularly light compared with the characteristic mass scale of quantum chromodynamics (QCD). Clearly our idea is only viable if something prevented our "baryons" from obtaining a mass (eventually a small mass may be due to some secondary perturbation).

The proton ows its mass to spontaneous breakdown of chiral symmetry, or so it seems according to a simple, fairly successful model of the mesonic and baryonic states in QCD: the Gell-Mann-Lévy sigma model[3]. Is it possible then that in some variant of QCD chiral symmetry is not spontaneously broken, or only partly, so that at least some chiral symmetry remains in the spectrum of fermionic bound states? In this article we will see that in general in SU(N) binding theories this is not allowed to happen, i.e. chiral symmetry must be broken spontaneously.

III2. NATURALNESS IN QUANTUM ELECTRODYNAMICS

Quantum Electrodynamics as a renormalizable model of electrons (and muons if desired) and photons is an example of a "natural" field theory. The parameters α, m_e (and m_μ) may be small independently. In particular m_e (and m_μ) are very small at large μ. The relevant symmetry here is chiral symmetry, for the electron and the muon separately. We need not be concerned about the Adler-Bell-Jackiw anomaly here because the photon field being Abelian cannot acquire non-trivial topological winding numbers[4].

There is a value of μ where Quantum Electrodynamics ceases to be useful, even as a model. The model is not asymptotically free, so there is an energy scale where all interactions become strong:

$$\mu_o \simeq m_e \exp(6\pi^2/e^2N_f) \ , \tag{III2}$$

where N_f is the number of light fermions. If some world would be described by such a theory at low energies, then a replacement of the theory would be necessary at or below energies of order μ_o.

III3. ϕ^4-THEORY

A renormalizable scalar field theory is described by the Lagrangian

$$\mathcal{L} = -\tfrac{1}{2}(\partial_\mu\phi)^2 - \tfrac{1}{2}m^2\phi^2 - \frac{1}{4!}\lambda\phi^4 \ . \tag{III3}$$

the interactions become strong at

$$\mu \simeq m \exp(16\pi^2/3\lambda) \ , \tag{III4}$$

but is it still natural there?

There are two parameters, λ and m. Of these, λ may be small because $\lambda = o$ would correspond to a non-interacting theory with total number of ϕ particles conserved. But is small m allowed? If we put m = o in the Lagrangian (III3) then the symmetry is not enhanced*). However we can take both m and λ to be small, because if $\lambda = m = o$ we have invariance under

$$\phi(x) \to \phi(x) + \Lambda \ . \tag{III5}$$

This would be an approximate symmetry of a new underlying theory at energies of order μ_o. Let the symmetry be broken by effects described by a dimensionless parameter ε. Both the mass term and the interaction term in the effective Lagrangian (III3) result from these symmetry breaking effects. Both are expected to be of order ε. Substituting the correct powers of μ_o to account for the dimensions of these parameters we have

$$\begin{aligned} \lambda &= \mathcal{O}(\varepsilon) \ , \\ m^2 &= \mathcal{O}(\varepsilon\mu_o^2) \ . \end{aligned} \tag{III6}$$

Therefore,

$$\mu_o = \mathcal{O}(m/\sqrt{\lambda}) \ . \tag{III7}$$

This value is much lower than eq. (III4). We now turn the argument around: if any "natural" underlying theory is to describe a scalar particle whose *effective* Lagrangian at low energies will be eq. (III3), then its energy scale cannot be given by (III4) but at best by (III7). We say that naturalness breaks down beyond $m/\sqrt{\lambda}$. It must be stressed that these are orders of magnitude. For instance one might prefer to consider λ/π^2 rather than λ to be the relevant parameter. μ_o then has to be multiplied by π. Furthermore, λ could be much smaller than ε because $\lambda = o$ separately also enhances the symmetry. Therefore, apart from factors π, eq. (III7) indicates a maximum value for μ_o.

Another way of looking at the problem of naturalness is by comparing field theory with statistical physics. The parameter m/μ would correspond to $(T-T_c)/T$ in a statistical ensemble. Why would the temperature T chosen by Nature to describe the elementary particles be so close to a critical temperature T_c? If $T_c \neq o$ then T may not be close to T_c just by accident.

III4. NATURALNESS IN THE WEINBERG-SALAM-GIM MODEL

The difficulties with the unnatural mass parameters only occur in theories with scalar fields. The only fundamental scalar

*) Conformal symmetry is violated at the quantum level.

field that occurs in the presently fashionable models is the Higgs field in the extended Weinberg-Salam model. The Higgs mass-squared, m_H^2, is up to a coefficient a fundamental parameter in the Lagrangian. It is small at energy scales $\mu \gg m_H$. Is there an approximate symmetry if $m_H \to o$? With some stretch of imagination we might consider a Goldstone-type symmetry:

$$\phi(x) \to \phi(x) + \text{const.} \tag{III8}$$

However we also had the local gauge transformations:

$$\phi(x) \to \Omega(x)\ \phi(x)\ . \tag{III9}$$

The transformations (III8) and (III9) only form a closed group if we also have invariance under

$$\phi(x) \to \phi(x) + C(x)\ . \tag{III10}$$

But then it becomes possible to transform ϕ away completely. The Higgs field would then become an unphysical field and that is not what we want. Alternatively, we could have that (III8) is an approximate symmetry only, and it is broken by all interactions that have to do with the symmetry (III9) which are the weak gauge field interactions. Their strength is $g^2/4\pi = \mathcal{O}(1/137)$. So at best we can have that the symmetry is broken by $\mathcal{O}(1/137)$ effects. Therefore

$$m_H^2/\mu^2 \gtrsim \mathcal{O}(1/137)\ .$$

Also the $\lambda\phi^4$ term in the Higgs field interactions breaks this symmetry. Therefore

$$m_H^2/\mu^2 \gtrsim \mathcal{O}(\lambda) \gtrsim \mathcal{O}(1/137)\ . \tag{III11}$$

Now

$$m_H^2 = \mathcal{O}(\lambda F_H^2)\ , \tag{III12}$$

where F_H is the vacuum expectation value of the Higgs field, known to be*)

$$F_H = (2G\sqrt{2})^{-1/2} = 174 \text{ GeV}\ . \tag{III13}$$

We now read off that

$$\mu \lesssim \mathcal{O}(F_H) = \mathcal{O}(174 \text{ GeV})\ . \tag{III14}$$

*) Some numerical values given during the lecture were incorrect. I here give corrected values.

This means that at energy scales much beyond F_H our model becomes more and more unnatural. Actually, factors of π have been omitted. In practice one factor of 5 or 10 is still not totally unacceptable. Notice that the actual value of m_H dropped out, except that

$$m_H = \mathcal{O}\left(\frac{\sqrt{\lambda}}{g} M_W\right) \gtrsim \mathcal{O}(M_W) \ . \qquad \text{(III15)}$$

Values for m_H of just a few GeV are unnatural.

III5. EXTENDING NATURALNESS

Equation (III14) tells us that at energy scales much beyond 174 GeV the standard model becomes unnatural. As long as the Higgs field H remains a fundamental scalar nothing much can be done about that. We therefore conclude, with Dimopoulos and Susskind[2)] that the "observed" Higgs field must be composite. A non-trivial strongly interacting field theory must be operative at 1000 GeV or so. An obvious and indeed likely possibility is that the Higgs field H can be written as

$$H = Z\bar{\psi}\psi \ , \qquad \text{(III16)}$$

where Z is a renormalization factor and ψ is a new quark-like object, a fermion with a new color-like interaction [2)]. We will refer to the object as meta-quark having meta-color. The theory will have all features of QCD so that we can copy the nomenclature of QCD with the prefix "meta-". The Higgs field is a meta-meson.

It is now tempting to assume that the meta-quarks transform the same way under weak SU(2) x U(1) as ordinary quarks. Take a doublet with left-handed components forming one gauge doublet and right handed components forming two gauge singlets. The meta-quarks are massless. Suppose that the meta-chiral symmetry is broken spontaneously just as in ordinary QCD. What would happen?

What happens is in ordinary QCD well described by the Gell-Mann-Lévy sigma model. The lightest mesons form a quartet of real fields, ϕ_{ij}, transforming as a

$$2^{\text{left}} \otimes 2^{\text{right}}$$

representation of

$$SU(2)^{\text{left}} \otimes SU(2)^{\text{right}}.$$

Since the weak interaction only deals with $SU(2)^{\text{left}}$ this quartet can also be considered as one complex doublet representation of weak SU(2). In ordinary QCD we have

$$\phi_{ij} = \sigma\delta_{ij} + i\tau^a_{ij}\pi^a \quad , \tag{III17}$$

and

$$<\sigma>_{vacuum} = \frac{1}{\sqrt{2}} f_\pi = 91 \text{ MeV} . \tag{III18}$$

The complex doublet is then

$$\phi_i = \frac{1}{\sqrt{2}} \begin{pmatrix} \sigma + i\pi^3 \\ \pi^2 + i\pi^1 \end{pmatrix} , \tag{III19}$$

and

$$<\phi_i>_{vacuum} = \begin{pmatrix} 1 \\ o \end{pmatrix} \times 64 \text{ MeV} . \tag{III20}$$

We conclude that if we transplant this theory to the TeV range then we get a scalar doublet field with a non-vanishing vacuum expectation value for free. All we have to do now is to match the numbers. If we scale all QCD masses by a scaling factor κ then we match

$$F_H = 174 \text{ GeV} = \kappa\ 64 \text{ MeV} ;$$

$$\kappa = 2700 . \tag{III21}$$

Now the mesonic sector of QCD is usually assumed to be reproduced in the 1/N expansion [5] where N is the number of colors (in QCD we have N = 3). The 4-meson coupling constant goes like 1/N. Then one would expect

$$f_\pi \propto \sqrt{N} . \tag{III22}$$

Therefore

$$\kappa = 2700 \sqrt{\frac{3}{N}} , \tag{III23}$$

if the metacolor group is SU(N).

Thus we obtain a model that reproduces the W-mass and predicts the Higgs mass. The Higgs is the meta-sigma particle. The ordinary sigma is a wide resonance at about 700 MeV[3], so that we predict

$$m_H = \kappa m_\sigma = 1900 \sqrt{\frac{3}{N}} \text{ GeV} , \tag{III24}$$

and it will be extremely difficult to detect among other strongly interacting objects.

III6. WHAT NEXT?

The model of the previous section is to our mind nearly inevitable, but there are problems. These have to do with the observed fermion masses. All leptons and quarks owe their masses to an interaction term of the form

$$g\bar{\psi}H\psi\ , \tag{III25}$$

where g is a coupling constant, ψ is the lepton or quark and H is the Higgs field. With (III16) this becomes a four-fermion interaction, a fundamental interaction in the new theory. Because it is non-renormalizable further structure is needed. In ref. 2 the obvious choice is made: a new "meta-weak interaction" gauge theory enters with new super-heavy intermediate vector bosons. But since H is a scalar this boson must be in the crossed channel, a rather awkward situation. (See option a in Figure 1.) A simpler theory is that a new scalar particle is exchanged in the direct channel. (See option b in Figure 1.)

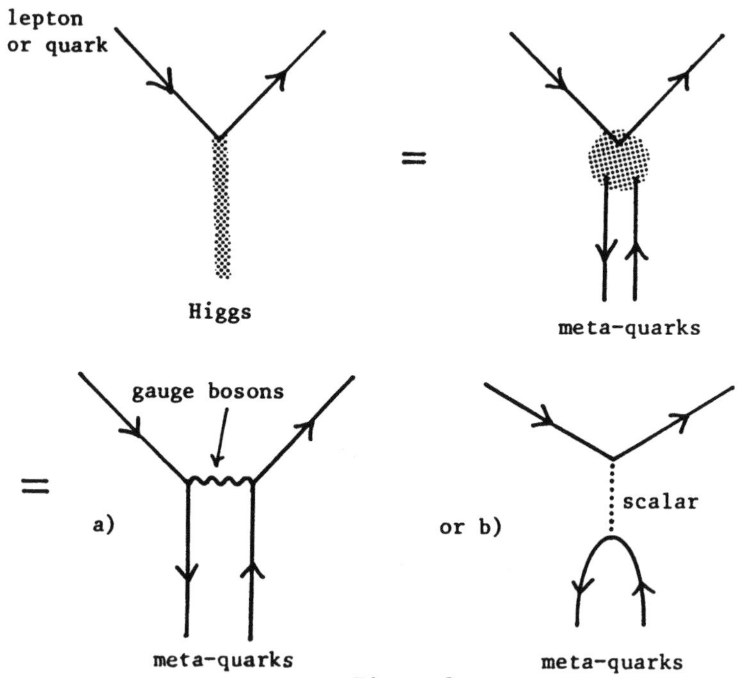

Figure 1.

Notice that in both cases new scalar fields are needed because in case a) something must cause the "spontaneous breakdown" of the new gauge symmetries. Therefore choice b) is simpler.
We removed a Higgs scalar and we get a scalar back. Does naturalness improve? The answer is yes. The coupling constant g in the interaction (III25) satisfies

$$g = g_1 g_2 / M_s^2 Z \ . \qquad \text{(III26)}$$

Here g_1 and g_2 are the couplings at the new vertices, M_s is the new scalar's mass, and Z is from (III16) and is of order

$$Z \sim \frac{1}{\sqrt{\frac{N}{3}}\,(\kappa\, m_\rho)^2} = \frac{\sqrt{N/3}}{(1800\ \text{GeV})^2} \ . \qquad \text{(III27)}$$

Suppose that the heaviest lepton or quark is about 10 GeV. For that fermion the coupling constant g is

$$g = \frac{m_f}{F} \simeq 1/20 \ .$$

We get

$$g_1 g_2 \simeq \left(\frac{M_s}{1800\ \text{GeV}}\right)^2 \sqrt{\frac{N}{3}} \cdot \frac{1}{20} \ .$$

Naturalness breaks down at

$$\mu = \mathcal{O}\left(\frac{M_s}{g_{1,2}}\right) = 8000 \ \sqrt[4]{\frac{3}{N}}\ \text{GeV} \ ,$$

an improvement of about a factor 50 compared with the situation in sect. III4. Presumably we are again allowed to multiply by factors like 5 or 10, before getting into real trouble.

Before speculating on how to go on from here to improve naturalness still further we must assure ourselves that all other alleys are blind ones. An intriguing possibility is that the presently observed fermions are composite. We would get option c), Figure 2.

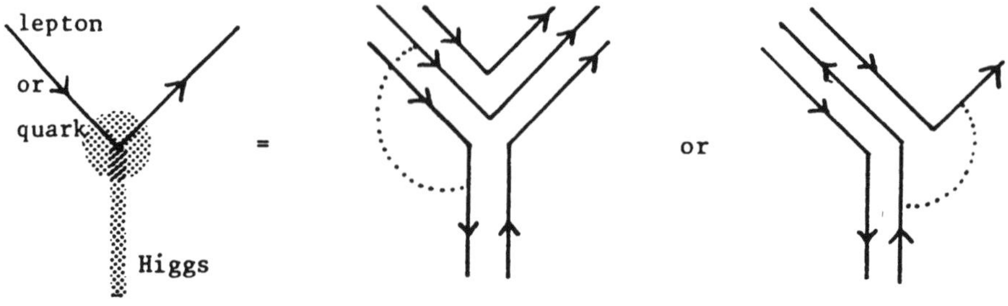

Fig. 2

The dotted line could be an ordinary weak interaction W or photon, that breaks an internal symmetry in the binding force for the new components. The new binding force could either act at the 1 TeV or at the 10-100 TeV range. It could either be an extension of metacolor or be a (color)" or paracolor force. Is such an idea viable?

Clearly, compared with the energy scale on which the binding forces take place, the composite fermions must be nearly massless. Again, this cannot be an accident. The chiral symmetry responsible for this must be present in the underlying theory. Apparently then, the underlying theory will possess a chiral symmetry which is not (or not completely) spontaneously broken, but reflected in the bound state spectrum in the Wigner mode: some massless chiral objects and parity doubled massive fermions. This possibility is most clearly described by the σ-model as a model for the lowest bound states occurring in ordinary quantum chromodynamics.

III7. THE σ MODEL

The fermion system in quantum chromodynamics shows an axial symmetry. To illuminate our problem let us consider the case of two flavors. The local color group is $SU(3)_c$. The subscript c here stands for color. The flavor symmetry group is $SU(2)_L \otimes SU(2)_R \otimes U(1)$ where the subscripts L and R stands for left and right and the group elements must be chosen to be space-time independent. We split the fermion fields ψ into left and right components:

$$\psi = \tfrac{1}{2}(1+\gamma_5)\psi_L + \tfrac{1}{2}(1-\gamma_5)\psi_R \ . \tag{III28}$$

$$\psi_L \text{ transforms as a } 3_c \otimes 2_L \otimes 1_R \otimes 2_{\mathcal{L}} \tag{III29}$$

$$\text{and } \psi_R \text{ transforms as a } 3_c \otimes 1_L \otimes 2_R \otimes \bar{2}_{\mathcal{L}} \tag{III30}$$

where the indices refer to the various groups. $\mathcal{L}$ stands for the Lorentzgroup SO(3,1), locally equivalent to SL(2,c) which has two different complex doublet representations $2_{\mathcal{L}}$ and $\bar{2}_{\mathcal{L}}$ (corresponding to the transformation law for the neutrino and antineutrino, respectively). The fields ψ_L and ψ_R have the same charge under U(1), whereas axial U(1) group (under which they would have opposite charges) is absent because of instanton effects[4].

The effect of the color gauge fields is to bind these fermions into mesons and baryons all of which must be color singlets. It would be nice if one could describe these hadronic fields as representations of $SU(2)_L \otimes SU(2)_R \otimes U(1)$ and the Lorentz group, and then cast their mutual interactions in the form of an effective Lagrangian, invariant under the flavor symmetry group. In the case at hand this is possible and the resulting construction is a successful and one-time popular model for pions and nucleons: the σ model[3]. We have a nucleon doublet

$$N = \tfrac{1}{2}(1+\gamma_5)N_L + \tfrac{1}{2}(1-\gamma_5)N_R \; , \tag{III31}$$

where

$$N_L \text{ transforms as a } 1_c \otimes 2_L \otimes 1_R \otimes 2 \quad , \tag{III32a}$$

$$\text{and } N_R \text{ transforms as a } 1_c \otimes 1_L \otimes 2_R \otimes \bar{2}_{\mathcal{L}} \quad . \tag{III32b}$$

Further we have a quartet of real scalar fields $(\sigma,\vec{\pi})$ which transform as a $1_c \otimes 2_L \otimes 2_{\mathcal{L}} \otimes 1_{\mathcal{L}}$. The Lagrangian is

$$\mathcal{L} = -\bar{N}[\gamma\partial+g_o(\sigma+i\vec{\tau}.\vec{\pi}\gamma_5)]N - \tfrac{1}{2}(\partial\pi)^2 - \tfrac{1}{2}(\partial\sigma)^2 - V(\sigma^2+\vec{\pi}^2) \; . \tag{III33}$$

Here V must be a rotationally invariant function.

Usually V is chosen such that its absolute minimum is away from the origin. Let V be minimal at $\sigma = v$ and $\vec{\pi} = o$. Here v is just a c-number. To obtain the physical particle spectrum we write

$$\sigma = v + s \tag{III34}$$

and we find

$$\mathcal{L} = -\tfrac{1}{2}\bar{N}(\gamma\partial+ g_o v)N - \tfrac{1}{2}(\partial\vec{\pi})^2 - \tfrac{1}{2}(\partial s)^2 - 2v^2V''(v^2)s^2 + \text{interaction terms} \; . \tag{III35}$$

Clearly, in this case the nucleons acquire a mass term $m_s = g_o v$ and the s particle has a mass $m_s^2 = 4v^2V''(v^2)$, whereas the pion remains strictly massless. The entire mass of the pion must be due to effects that explicitly break $SU(2)_L \times SU(2)_R$, such as a small mass term $m_q\bar{\psi}\psi$ for the quarks (III28). We say that in this case the flavor group $SU(2)_L \otimes SU(2)_R$ is spontaneously broken into the isospin group SU(2).

Another possibility however, apparently not realised in ordinary quantum chromodynamics, would be that $SU(2)_L \otimes SU(2)_R$ is *not* spontaneously broken. We would read off from the Lagrangian (III33) that the nucleons N would form a massless doublet and that the four fields $(\sigma,\vec{\pi})$ could be heavy. The dynamics of other confining gauge theories could differ sufficiently from ordinary QCD so that, rather than a spontaneous symmetry breakdown, massless "baryons" develop. The principle question we will concentrate on is why do these massless baryons form the representation (III32), and how does this generalize to other systems. We would let future generations worry about the question where exactly the absolute minimum of the effective potential V will appear.

III8. INDICES

We now consider any color group G_c. The fundamental fermions in our system must be non-trivial representation of G_c and we assume "confinement" to occur: all physical particles are bound states that are singlets under G_c. Assume that the fermions are all massless (later mass terms can be considered as a perturbation). We will have automatically some global symmetry which we call the flavor group G_F. (We only consider exact flavor symmetries, not spoilt by instanton effects.) Assume that G_F is not spontaneously broken. Which and how many representations of G_F will occur in the massless fermion spectrum of the baryonic bound states? We must formulate the problem more precisely. The massless nucleons in (III33) being bound states, may have many massive excitations. However, massive Fermion fields cannot transform as a $2_{\mathcal{L}}$ under Lorentz transformations; they must go as a $2_{\mathcal{L}} \oplus \bar{2}_{\mathcal{L}}$. That is because a mass term being a Lorentz invariant product of two fields at one point only links $2_{\mathcal{L}}$ representations with $\bar{2}_{\mathcal{L}}$ representations. Consider a given representation r of G_F. Let p be the number of field multiplets transforming as $r \otimes 2_{\mathcal{L}}$ and q be the number of field multiplets $r \otimes \bar{2}_{\mathcal{L}}$. Mass terms that link the $2_{\mathcal{L}}$ with $\bar{2}_{\mathcal{L}}$ fields are completely invariant and in general to be expected in the effective Lagrangian. But the absolute value of

$$\ell = p - q \qquad \text{(III36)}$$

is the minimal number of surviving massless chiral field multiplets. We will call ℓ the index corresponding to the representation r of G_F. By definition this index must be a (positive or negative) integer. In the sigma model it is postulated that

$$\text{index } (2_L \otimes 1_R) = 1 \tag{III37}$$

$$\text{index } (1_L \otimes 2_R) = -1$$

index (r) = o for all other representations r.

This tells us that if chiral symmetry is not broken spontaneously one massless nucleon doublet emerges. We wish to find out what massless fermionic bound states will come out in more general theories. Our problem is: how does (III37) generalize?

III9. ABSENCE OF MASSLESS BOUND STATES WITH SPIN 3/2 OR HIGHER

In the foregoing we only considered spin o and spin 1/2 bound states. Is it not possible that fundamentally massless bound states develop with higher spin? I believe to have strong arguments that this is indeed not possible. Let us consider the case of spin 3/2. Massive spin 3/2 fermions are described by a Lagrangian of the form

$$\mathcal{L} = \tfrac{1}{2}\bar{\psi}_\mu [\sigma_{\mu\nu}(\gamma\partial+m) + (\gamma\partial+m)\sigma_{\mu\nu}]\psi_\nu \ . \tag{III38}$$

Just like spin-one particles, this has a gauge-invariance if $m \to o$:

$$\psi_\mu \to \psi_\mu + \partial_\mu \eta(x) \ , \tag{III39}$$

where $\eta(x)$ is arbitrary. Indeed, massless spin 3/2 particles only occur in locally supersymmetric field theories. The field $\eta(x)$ is fundamentally unobservable.

Now in our model ψ_μ would be shorthand for some composite field: $\psi_\mu \to \psi\psi\psi$. However, then all components of this, including η, would be observables. If $m = o$ we would be forced to add a gauge fixing term that would turn η into an unacceptable ghost particle*).

We believe, therefore, that unitarity and locality forbid the occurrence of massless bound states with spin 3/2. The case for higher spin will not be any better. And so we concentrate on a bound state spectrum of spin 1/2 particles only.

*) Note added: during the lectures it was suggested by one attendant to consider only gauge-invariant fields as $\Psi_{\mu\nu} = \partial_\mu\psi_\nu - \partial_\nu\psi_\mu$. However, such fields must satisfy constraints: $\partial_{[\alpha}\Psi_{\mu\nu]} = o$. Composite field will never automatically satisfy such constraints.

III10. SPECTATOR GAUGE FIELDS AND -FERMIONS

So far, our model consisted of a strong interaction color gauge theory with gauge group G_C, coupled to chiral fermions in various representations r of G_c but of course in such a way that the anomalies cancel. The fermions are all massless and form multiplets of a global symmetry group, called G_F. For QCD this would be the flavor group. In the metacolor theory G_F would include all other fermion symmetries besides metacolor.

In order to study the mathematical problem raised above we will add another gauge connection field that turns G_F into a local symmetry group. The associated coupling constants may all be arbitrarily small, so that the dynamics of the strong color gauge interactions is not much affected. In particular the massless bound state spectrum should not change. One may either think of this new gauge field as a completely quantized field or simply as an artificial background field with possibly non-trivial topology. We will study the behavior of our system in the presence of this "spectator gauge field". As stated, its gauge group is G_F.

Note however, that some flavor transformations could be associated with anomalies. There are two types of anomalies:

i) those associated with $G_c \times G_F$, only occurring where the color field has a winding number. Only U(1) invariant subgroups of G_F contribute here. They simply correspond to small explicit violations of the G_F symmetry. From now on we will take as G_F only the anomaly-free part. Thus, for QCD with N flavors, G_F is not $U(N) \times U(N)$ but

$$G_F = SU(N) \otimes SU(N) \otimes U(1) .$$

ii) those associated with G_F alone. They only occur if the spectator gauge field is quantized. To remedy these we simply add "spectator fermions" coupled to G_F alone. Again, since these interactions are weak they should not influence the bound state spectrum.

Here, the spectator gauge fields and fermions are introduced as mathematical tools only. It just happens to be that they really do occur in Nature, for instance the weak and electromagnetic $SU(2) \times U(1)$ gauge fields coupled to quarks in QCD. The leptons then play the role of spectator fermions.

III11. ANOMALY CANCELLATION FOR THE BOUND STATE SPECTRUM

Let us now resume the particle content of our theory. At small distances we have a gauge group $G_c \otimes G_F$ with chiral fermions in several representations of this group. Those fermions which are trivial under G_C are only coupled weakly and are called "spectator fermions". All anomalies cancel, by construction.

At low energies, much lower than the mass scale where color binding occurs, we see only the G_F gauge group with its gauge fields. Coupled to these gauge fields are the massless bound states, forming new representations r of G_F, with either left- or right handed chirality. The numbers of left minus right handed fermion fields in the representations r are given by the as yet unknown indices $\ell(r)$. And finally we have the spectator fermions which are unchanged.

We now expect these very light objects to be described by a new local field theory, that is, a theory local with respect to the large distance scale that we now use. The central theme of our reasoning is now that this new theory must again be anomaly free. We simply cannot allow the contradictions that would arise if this were not so. Nature must arrange its new particle spectrum in such a way that unitarity is obeyed, and because of the large distance scale used the effective interactions are either vanishingly small or renormalizable. The requirement of anomaly cancellation in the new particle spectrum gives us equations for the indices $\ell(r)$, as we will see.

The reason why these equations are sometimes difficult or impossible to solve is that the new representations r must be different from the old ones; if $G_C = SU(N)$ then r must also be faithful representations of $G_F/Z(N)$. For instance in QCD we only allow for octet or decuplet representations of $(SU(3))_{flavor}$, whereas the original quarks were triplets.

However, the anomaly cancellation requirement, restrictive as it may be, does not fix the values of $\ell(r)$ completely. We must look for additional limitations.

III12 APPELQUIST-CARAZZONE DECOUPLING AND N-INDEPENDENCE

A further limitation is found by the following argument. Suppose we add a mass term for one of the colored fermions.

$$\Delta\mathcal{L} = m\,\bar{\psi}_{1L}\,\psi_{1R} + \text{h.c.}$$

Clearly this links one of the left handed fermions with one of the right handed ones and thus reduces the flavor group G_F into $G_F' \subset G_F$. Now let us gradually vary m from o to infinity. A famous theorem [5] tells us that in the limit $m \to \infty$ all effects due to this massive quark disappear. All bound states containing this quark should also disappear which they can only do by becoming very heavy. And they can only become heavy if they form representations r' of G_F' with total index $\ell'(r') = o$. Each representation r of G_F forms an array of representations r' of G_F'. Therefore

$$\ell'(r') = \sum_{r \text{ with } r' \subset r} \ell(r) \ . \tag{III40}$$

Apparently this expression must vanish.

Thus we found another requirement for the indices $\ell(r)$. The indices will be nearly but not quite uniquely determined now. Calculations show that this second requirement makes our indices $\ell(r)$ practically independent of the dimensions n_i of G_F. For instance, if G_c = SU(3) and if we have left- and righthanded quarks forming triplets and sextets then

$$G_F = SU(n_1)_L \otimes SU(n_2)_R \otimes SU(n_3)_L \otimes SU(n_4)_R \otimes U(1)^3 \tag{III41}$$

where $n_{1,2}$ refer to the triplets and $n_{3,4}$ to the sextets. G_c is anomaly-free if

$$n_1 - n_2 + 7(n_3 - n_4) = o \ . \tag{III42}$$

Here we have three independent numbers n_i.
If we write the representations r as Young tableaus then $\ell(r)$ could still depend explicitly on n_i.

However, suppose that someone would start as approximation of Bethe-Salpeter type to discover the zero mass bound state spectrum. He would study diagrams such as Fig. 3

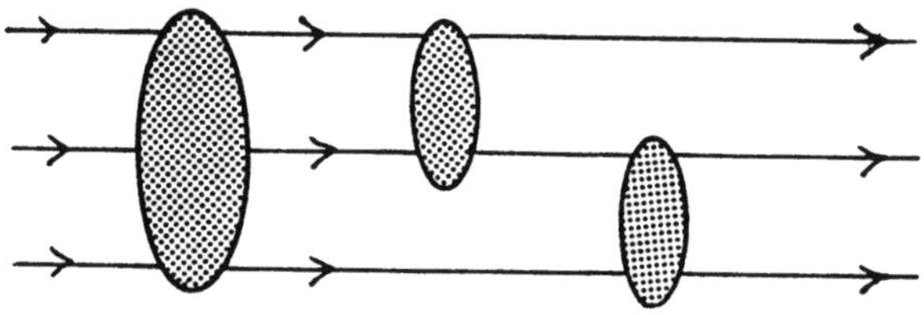

Fig. 3

The resulting indices $\ell(r)$ would follow from topological properties of the interactions represented by the blobs. It is unlikely that this topology would be seriously effected by details such as the contributions of diagrams containing additional closed fermion loops. However, that is the only way in which explicit n-dependence enters. It is therefore natural to assume $\ell(r)$ to be n-independent. This latter assumption fixes $\ell(r)$ completely. What is the result of these calculations?

III13. CALCULATIONS

Let G be any (reducible or irreducible) gauge group. Let chiral fermions in a representation r be coupled to the gauge fields by the covariant derivative

$$D_\mu = \partial_\mu + i\,\lambda^a(r)\,A^a_\mu \ , \qquad \text{(III43)}$$

where A^a_μ are the gauge fields and $\lambda^a(r)$ a set of matrices depending on the representation r. Let the left-handed fermions be in the representations r_L and the right-handed ones in r_R. Then the anomalies cancel if

$$\sum_L \mathrm{Tr}\{\lambda^a(r_L),\ \lambda^b(r_L)\}\ \lambda^c(r_L) =$$

$$\sum_R \mathrm{Tr}\{\lambda^a(r_R),\ {}^b(r_R)\}\ \lambda^c(r_R) \ . \qquad \text{(III44)}$$

The object $d^{abc}(r) = \mathrm{Tr}\{\lambda^a(r),\ \lambda^b(r)\}\ \lambda^c(r)$ can be computed for any r. In table 1 we give some examples. The fundamental representation r_o is represented by a Young tableau: □ . Let it have n components. We take the case that $\mathrm{Tr}\ \lambda(r_o) = o$. Write

$$\mathrm{Tr}\ I(r_o) = n \ , \qquad \mathrm{Tr}\ I(r) = N(r) \ ,$$

$$\mathrm{Tr}\ \lambda(r) = o \ ,$$

$$\mathrm{Tr}\ \lambda^a(r)\ \lambda^b(r) = C(r)\ \mathrm{Tr}\ \lambda^a(r_o)\ \lambda^b(r_o) \ ,$$

$$d^{abc}(r) = K(r)\ d^{abc}(r_o) \ . \qquad \text{(III45)}$$

We read off C and K from table 1.
Now III44 must hold both in the high energy region and in the low energy region. The contribution of the spectator fermions in both regions is the same. Thus we get for the bound states

$$\left(\sum_L - \sum_R\right) d^{abc}(r) = n_c\left(d^{abc}(r_{oL}) - d^{abc}(r_{oR})\right) \qquad \text{(III46)}$$

where a,b,c are indices of G_F and r_o is the fundamental representation of G_F. We have the factor n_c written explicitly, being the number of color components.

Let us now consider the case $G_C = SU(3)$; $G_F = SU_L(n) \otimes SU_R(n) \otimes U(1)$. We have n "quarks" in the fundamental representations. The representations r of the bound states must be in $G_F/Z(3)$. They are assumed to be built from three quarks, but we are free to choose their chirality. The expected representations

Table 1

r	N(r)	C(r)	K(r)
$\square$	n	1	1
$\square$	n	1	-1
$\square\square$, $\begin{smallmatrix}\square\\\square\end{smallmatrix}$	$\frac{n(n\pm1)}{2}$	$n\pm2$	$n\pm4$
$\square\square\square$, $\begin{smallmatrix}\square\\\square\\\square\end{smallmatrix}$	$\frac{n(n\pm1)(n\pm2)}{6}$	$\frac{(n\pm2)(n\pm3)}{2}$	$\frac{(n\pm3)(n\pm6)}{2}$
$\begin{smallmatrix}\square\square\\\square\end{smallmatrix}$	$\frac{n(n^2-1)}{3}$	n^2-3	n^2-9
$A\otimes B$	N(A)N(B)	C(A)N(B) + C(B)N(A)	K(A)N(B) + K(B)N(A)

are given in table 2, where also their indices are defined. Because of left-right symmetry these numbers change sign under interchange of left ↔ right.

Table 2

representation	index	representation	index
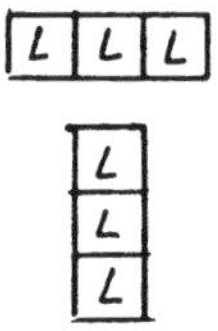	$\ell_{1\pm}$	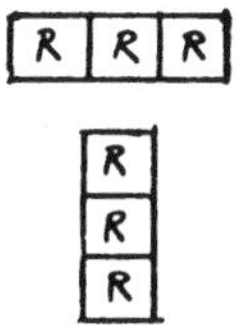	$-\ell_{1\pm}$
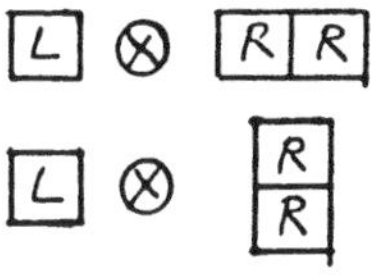	$\ell_{2\pm}$	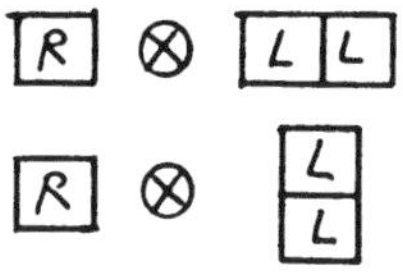	$-\ell_{2\pm}$
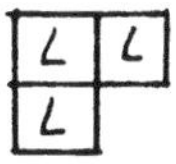	ℓ_3	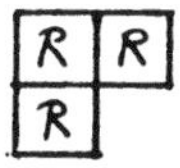	$-\ell_3$

For the time being we assume no other representations. In eq. III46 we may either choose a, b and c all to be $SU(n)_L$ indices, or choose a and b to be $SU(n)_L$ indices and c the U(1) index. We get two independent equations:

$$\sum_{\pm} \tfrac{1}{2}(n\pm3)(n\pm6)\ell_{1\pm} - \sum_{\pm} \tfrac{1}{2}n(n\pm7)\ell_{2\pm} + (n^2-9)\ell_3 = 3, \text{ if } n > 2 ,$$

and

$$\sum_{\pm} \tfrac{1}{2}(n\pm2)(n\pm3)\ell_{1\pm} - \sum_{\pm} \tfrac{1}{2}n(n\pm3)\ell_{2\pm} + (n^2-3)\ell_3 = 1, \text{ if } n > 1 . \tag{III47}$$

The Appelquist-Carazzone decoupling requirement, eq. (III40), gives us in addition two other equations:

$$\ell_{1+} - \ell_{2+} + \ell_3 = 0 ,$$

$$\ell_{1-} - \ell_{2-} + \ell_3 = 0 , \text{ both if } n > 2 . \tag{III48}$$

For n > 2 the general solution is

$$\ell_{1+} = \ell_{1-} = \ell ,$$
$$\ell_{2+} = \ell_{2-} = 3\ell - \frac{1}{3} ,$$
$$\ell_3 = 2\ell - \frac{1}{3} . \qquad \text{(III49)}$$

Here ℓ is still arbitrary. Clearly this result is unacceptable. We cannot allow any of the indices ℓ to be non-integer. Only for the case n = 2 (QCD with just two flavors) there is another solution. In that case ℓ_{2-} and ℓ_3 describe the same representation, and ℓ_{1-} an empty representation. We get

$$\ell_{2-} + \ell_3 = k = 1 - 10\,\ell_{1+} + 5\,\ell_{2+} . \qquad \text{(III50)}$$

According to the σ-model, $\ell_{1+} = \ell_{2+} = o$; $k = 1$. The σ-model is therefore a correct solution to our equations.

In the previous section we promised to determine the indices completely. This is done by imposing n-independence for the more general case including also other color representations such as sextets besides triplets. The resulting equations are not very illuminating, with rather ugly coefficients. One finds that in general no solution exists except when one assumes that all mixed representations have vanishing indices. With mixed representations we mean a product of two or more non-trivial representations of two or more non-Abelian invariant subgroups of G_F. If now we assume n-independence this must also hold if the number of sextets is zero. So ℓ_{2+} and ℓ_{2-} must vanish. We get

$$\ell_{1+} = \ell_{1-} = 1/9 ,$$
$$\ell_3 = -1/9 . \qquad \text{(III51)}$$

If all quarks were sextets, not triplets, we would get

$$\ell_{1+} = \ell_{1-} = 2/9 ,$$
$$\ell_3 = -2/9 . \qquad \text{(III52)}$$

In the case G_C = SU(5) the indices were also found. See table 3.

Table 3
indices for G_c = SU(5)

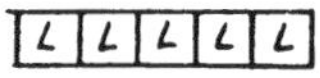

$\ell_{1+} = 1/25$

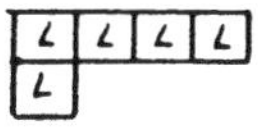

$\ell_{2+} = -1/25$

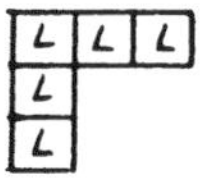

$\ell_3 = 1/25$

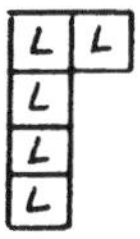

$\ell_{2-} = -1/25$

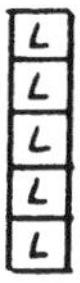

$\ell_{1-} = 1/25$

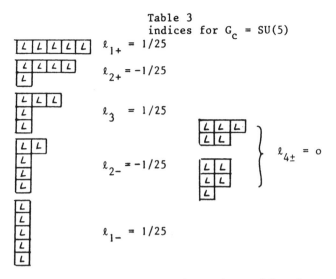

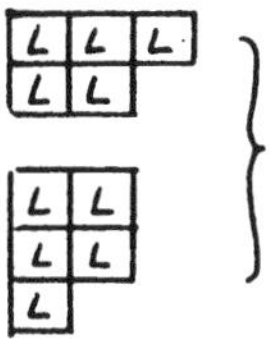

$\ell_{4\pm} = $ o

This clearly suggests a general tendency for SU(N) color groups to produce indices $\pm 1/N^2$ or o.

III14. CONCLUSIONS

Our result that the indices we searched for are fractional is clearly absurd. We nevertheless pursued this calculation in order to exhibit the general philosophy of this approach and to find out what a possible cure might be. Our starting point was that chiral symmetry is not broken spontaneously. Most likely this is untenable, as several authors have argued[6]. We find that explicit chiral symmetry in QCD leads to trouble in particular if the number of flavors is more than two. A daring conjecture is then that in QCD the strange quark, being rather light, is responsible for the spontaneous breakdown of chiral symmetry.
An interesting possibility is that in some generalized versions of QCD chiral symmetry is broken only partly, leaving a few massless chiral bound states. Indeed there are examples of models where our philosophy would then give integer indices, but since we must drop the requirement of n-dependence our result was not unique and it was always ugly. No such model seems to reproduce anything resembling the observed quark-lepton spectrum.

Finally there is the remote possibility that the paradoxes associated with higher spin massless bound states can be resolved. Perhaps the $\Delta(1236)$ plays a more subtle role in the σ-model than assumed so far (we took it to be a parity doublet).

We conclude that we are unable to construct a bound state theory for the presently fundamental fermions along the lines suggested above.

We thank R. van Damme for a calculation yielding the indices in the case G_c = SU(5).

REFERENCES

1. P.A.M. Dirac, Nature 139 (1937) 323, Proc. Roy. Soc. A165 (1938) 199, and in: Current Trends in the Theory of Fields, (Tallahassee 1978) AIP Conf. Proc. No 48, Particles and Fields Subseries No 15, ed. by Lannuti and Williams, p. 169.
2. S. Dimopoulos and L. Susskind, Nucl. Phys. B155 (1979) 237.
3. M. Gell-Mann and M. Lévy, Nuovo Cim. 16 (1960) 705.
   B.W. Lee, Chiral Dynamics, Gordon and Breach, New York, London, Paris 1972.
4. G. 't Hooft, Phys. Rev. Lett. 37 (1976) 8; Phys. Rev. D14 (1976) 3432.
   S. Coleman, "The Uses of Instantons", Erice Lectures 1977.
   R. Jackiw and C. Rebbi, Phys. Rev. Lett. 37 (1976) 172.
   C. Callan, R. Dashen and D. Gross, Phys. Lett. 63B (1976) 334.
5. G. 't Hooft, Nucl. Phys. B72 (1974) 461.
6. T. Appelquist and J. Carazzone, Phys. Rev. D11 (1975) 2856.
7. A. Casher, Chiral Symmetry Breaking in Quark Confining Theories, Tel Aviv preprint TAUP 734/79 (1979).

CHAPTER 6
PLANAR DIAGRAMS

CHAPTER 6
PLANAR DIAGRAMS

Introduction to Planar Diagram Field Theories [6.1]

Consider an $SU(N)$ gauge theory, possibly enlarged with scalars and/or fermions in the elementary or the adjoint representation. Take the limit $g \to 0$, $N \to \infty$, $g^2N = \tilde{g}^2$ fixed. In this limit only those Feynman diagrams survive that can be drawn on a plane without any crossings of the lines. We call these "planar diagrams". Planar diagrams resemble the world sheets of strings, and so one is led to hope that perhaps the stringlike behavior of the gluon field confining quarks can be understood in this limit. Unfortunately, planar diagrams form a set that is still too large to be summed analytically by any known method.

However, there is another important aspect to the planar diagrams. The perturbative expansion with respect to $\tilde{g}$ may converge much better than in theories with finite N. The reason for this is that the total number of planar diagrams is much smaller than the total of all diagrams. The only complication is that each individual diagram may give huge amplitudes at very high orders, because of the need for renormalization subtractions. The renormalized diagrams are *finite*, but huge.

The bulk of the first paper in this chapter is devoted to the question of whether perturbation expansion defines a theory with infinite accuracy in the infinite N limit. The answer is affirmative, but we could prove this only if the coupling constant is sufficiently small and all masses sufficiently large, a physically uninteresting case. So here we have a rigorously defined theory. Our construction is very technical. A casual reader may well be interested only in the first part where the planarity of the diagrams is proved, or possibly the less technical Section 16, where Borel summability is deduced. In Appendix A a curious model is discussed which is on the one hand an asymptotically free gauge theory and on the other hand there are a sufficient number of scalar fields to render all particles massive. To achieve this it was compulsory to also have fermions present.

Appendix B is the well-known "spherical model", a scalar N-component field in the $N \to \infty$ limit. This model is extremely instructive. In this limit one may allow the coupling constant to be negative (it *has* to become negative at high energies!); at sufficiently large N the vacuum will become sufficiently stable. The model shows that if the original renormalized mass parameter vanishes there will still be a "spontaneously generated" physical mass for the particle. There is a (negative) lower limit for the allowed values of the renormalized mass, at which the bound state mass tends to zero. The dotted region in Figure 21 is an unphysical solution for m and M for the one given (negative) value of m_R.

Introduction to A Two-Dimensional Model for Mesons [6.2]

Since our original motivation for studying the $N \to \infty$ limit was our hope to find solvable versions of QCD and to understand quark confinement we now present one successful attempt. In one-space, one-time dimension this limit can indeed be solved. This is because there a gauge choice can be made such that the gauge field self-interactions disappear. The surviving planar diagrams are then the so-called rainbow diagrams, see Figure 2. They obey simple closed Dyson–Schwinger equations which can be solved. The propagator can be found in closed form, the mesonic bound states can be expressed as Bethe–Salpeter like integral equations, which can easily be solved numerically.

At the end of this paper we add an epilogue.

CHAPTER 6.1

PLANAR DIAGRAM FIELD THEORIES

Gerard 't Hooft
Institute for Theoretical Physics
Princetonplein, 5 P.O. Box 80.006
3508 TA Utrecht, The Netherlands

ABSTRACT

In this compilation of lectures field theories are considered which consist of N component fields q_i interacting with N x N component matrix fields A_{ij} with internal (local or global) symmetry group SU(N) or SO(N). The double expansion in 1/N and $\tilde{g}^2 = Ng^2$ can be formulated in terms of Feynman diagrams with a planarity structure. If the mass is sufficiently large and $\tilde{g}^2$ sufficiently small then the(extremely non-trivial) expansion in $\tilde{g}^2$ at lowest order in 1/N is Borel summable. Exact limits on the behavior of the Borel integrand for the $\tilde{g}^2$ expansion are derived.

1. INTRODUCTION

In spite of considerable efforts it is still not known how to compute physical quantities reliably and accurately in any four-dimensional field theory with strong interactions. It seems quite likely that if any strong interaction field theory exists in which accurate calculations can be done, then that must be an asymptotically free non-Abelian gauge theory. In such theories the small-distance structure is completely described by solutions of the renormalization group equations[1]; and there are reasons to believe that the continuum theory can be uniquely defined as a limit of a lattice gauge theory[2], when the size of the meshes of the lattice tends to zero, together with the coupling constant, in a way prescribed by this renormalization group[3]. Indeed, one can prove using this formalism[4] that this limit exists up to any finite order in the perturbation expansion for small coupling.

However, this result has not been extended beyond pertur-

Progress in Gauge Field Theory, NATO Adv. Study Inst. Series
Edited by G 't Hooft *et al.* (Plenum, 1984).

bation expansion. It is important to realize that this might imply that theories such as "quantum chromodynamics" (QCD) are not based on solid mathematics, and indeed, it could be that physical numbers such as the ratio between the proton mass and the string constant do not follow unambiguously from QCD alone. In view of the qualitative successes of the recent Monte-Carlo computation techniques[5] the idea that hadronic properties could be shaped by forces other than QCD alone seems to be far-fetched, but it would be extremely important if this happened to be the case. More likely, we may simply have to improve our mathematics to show that QCD is indeed an unambiguous theory. Either way, it will be important to extend our understanding of the summability aspects of higher order perturbation theory as well as we can. The following constitutes just such an attempt.

There are two categories of divergences when one attempts to sum or resum perturbation expansion for a field theory in four space-time dimensions. One is simply the divergence due to the increasingly large numbers of Feynman diagrams to consider at higher orders. They grow roughly as $n!$ at order g^{2n}. This is a kind of divergence that already occurs if the functional integral is replaced by some ordinary finite-dimensional integral of similar type:

$$I(g^2) = \int d\vec{\phi}\ \ e^{-S(\vec{\phi})} \ ,$$

$$\begin{aligned} S(\vec{\phi}) &= \tfrac{1}{2}(\vec{\phi},M\vec{\phi}) + g \sum A_{ijk}\phi_i\phi_j\phi_k \\ &\quad + g^2 \sum B_{ijk\ell}\phi_i\phi_j\phi_k\phi_\ell \ . \end{aligned} \tag{1.1}$$

Here the diagrams themselves are bounded by geometric expressions but the numbers $K(n)$ of diagrams of order n are such that the expansion is only asymptotic:

$$\begin{aligned} I(g^2) &\to \sum_n K(n)\ C^n\ g^{2n} \ ; \\ K(n) &\to a^n\ n! \qquad , \end{aligned} \tag{1.2}$$

where a is determined by one of the stationary points of the action S, called "instantons".

However in four-dimensional field theories the diagrams themselves are not geometrically bounded. In some theories it has been shown[6] that diagrams of the nth order that required k ultraviolet subtractions (with essentially $k \lesssim n$) can be bounded at best by

$$b^n\, k!\, g^{2n} \, . \tag{1.3}$$

But since the total number of such diagrams grow at most as $n!/k!$ we still get bounds of the form (1.2) for the total amplitude, however with a replaced by a different coefficient. This is a different kind of divergence sometimes referred to as ultraviolet "renormalons"[4,7].

If a field theory is asymptotically free the corresponding coefficient b is negative and one might hope that the ultraviolet renormalons are relatively harmless. But in massless theories a similar kind of divergence will then be difficult to cope with: the infrared renormalons, which, as the word suggests, are due to a build-up of infrared divergences at very high orders: individual diagrams may still be convergent but their sum diverges again with n! The theories we will study more closely are governed by planar diagrams only. Their numbers grow only geometrically[8] so that divergences due to instantons are absent. These diagrams are akin to but more complicated than Bethe-Salpeter ladder diagrams and by trying to sum them we intend to learn much about the renormalon divergences.

Infinite color quantum chromodynamics is of course the most interesting example of a planar field theory but unfortunately our analysis cannot yet be carried out completely there. We do get bounds on the behavior of its Borel functions however (sect. 17).

Examples of large N field theories that we can handle our way are given in sect. 3 and appendix A.

2. FEYNMAN RULES FOR ARBITRARY N

In order to show that the set of planar Feynman diagrams becomes dominant at large N values we first formulate a generic theory at arbitrary N, with a coupling constant g as expansion parameter in the usual sense.

Let us express the fields as a finite number K of N-component vectors $\psi_i^a(x)$ ($a=1,\dots,K$; $i=1,\dots,N$), and a small number D of $N\times N$ matrices $A^b{}_i{}^j(x)$, where $b=1,\dots,D$ and $i,j=1,\dots,N$. Usually we will take ψ to be complex and $A_i{}^j$ to be Hermitean, in which case thesymmetry group will be U(N) or SU(N). The case that ψ_i are real and $A_i{}^j$ real and symmetric can easily be included after a few changes, in which case the symmetry group would be O(N) or SO(N), but the complex case seems to be more interesting from a physical point of view. In quantum chromodynamics K would be proportional to the number of flavors and D is the number of space-time dimensions plus two for the (non-Hermitean) ghost field.

In general then the Lagrangian has the form

$$\mathcal{L}(A,\psi,\psi^*) = - \sum_{a,\ldots} \psi^{*a}\left(M_0^{ab} + gM_1^{abc}A^c + \tfrac{1}{2}g^2M_2^{abcd}A^cA^d\right)\psi^b$$

$$- \mathrm{Tr}\left(\tfrac{1}{2}R_0^{ab}A^aA^b + \frac{1}{3}gR_1^{abc}A^aA^bA^c + \frac{1}{4}g^2R_2^{abcd}A^aA^bA^cA^d\right) , \qquad (2.1)$$

with

$$A^{a\ j}_{\ i} = \left(A^{a\ i}_{\ j}\right)^* . \qquad (2.2)$$

Here the usual matrix multiplication rule with respect to the indices i,j,... is implied, and Tr stands for trace with respect to these indices. The objects M and R carry no indices i,j but only "flavor" indices a,b,... . Furthermore $M_{0,1}$ and $R_{0,1}$ may contain the derivatives of $\psi(x)$ or $A(x)$. So the case that ψ are fermions is included: then a,b,... may include spinor indices. The coupling constant g has been put in (2.1) in such a way that it is a handy expansion parameter.

In order to keep track of the indices i,j,... it is convenient to split the fields $A^a{}_i{}^j$ into complex fields for i > j and real fields for i = j. One can then denote an upper index by an incoming arrow and a lower index by an outgoing arrow. The propagator is then denoted by a double line. In fig. 1, the A propagator stands for an $A_i{}^j$ propagator to the right if i > j; an $A_j{}^i$ propagator to the left if i < j and a real propagator if i = j.

It is crucial now that the coefficients M and P in the Lagrangian respect the U(N) (or O(N)) symmetry: they carry no indices i,j,... . Hence the vertices in the Feynman graphs only depend on these indices *via* Kronecker deltas. We indicate such a Kronecker delta in a vertex by connecting the corresponding index lines. Since we have a unitary invariance group these Kronecker deltas only connect upper indices with lower indices, therefore the index lines carry an *orientation* which is preserved at the vertices. This is where the unitary case differs from the real orthogonal groups: the restriction to real fields with O(N) symmetry corrèsponds to dropping the arrows in Fig. 1: the index lines then carry no orientation.

As for the rest the Feynman rules for computing a diagram are as usual. For instance, fermionic and ghost loops are associated with extra minus signs.

In some theories (such as SU(N) gauge theories) we have an extra constraint:

$$\mathrm{Tr}\ A^a = o . \qquad (2.3)$$

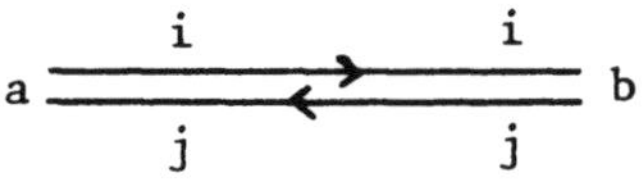

$P_{ab}(k) = \{R_o\}^{-1}_{ab}$ (N x N matrix propagator)

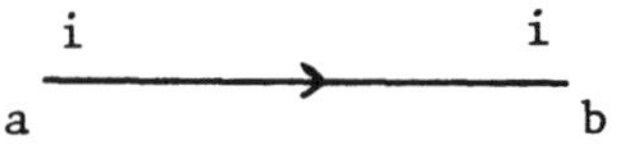

$P'_{ba}(k) = (M_o^{-1})_{ba}$ (N-vector propagator)

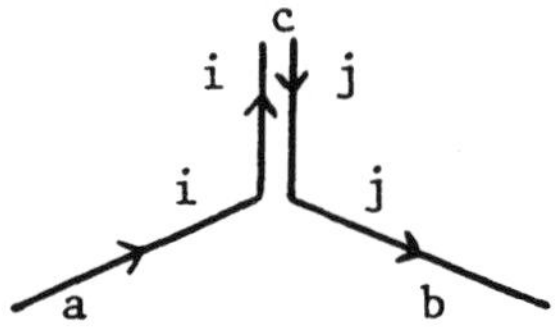

$-g\ M_1^{bac}$

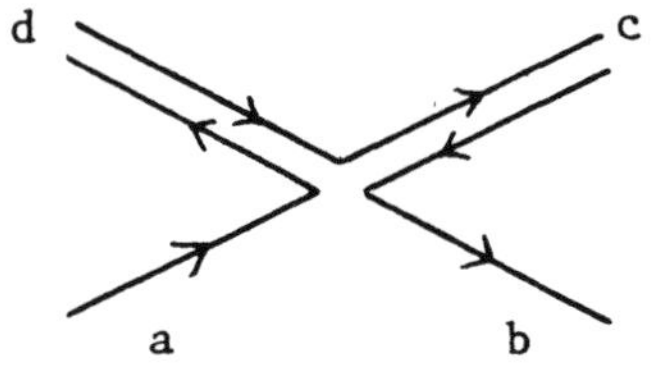

$-g^2\ M_2^{ba\ cd}$

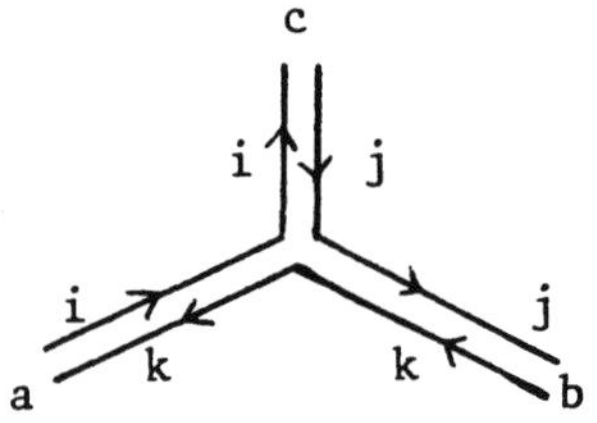

$-g\ R_1^{\{abc\}}$

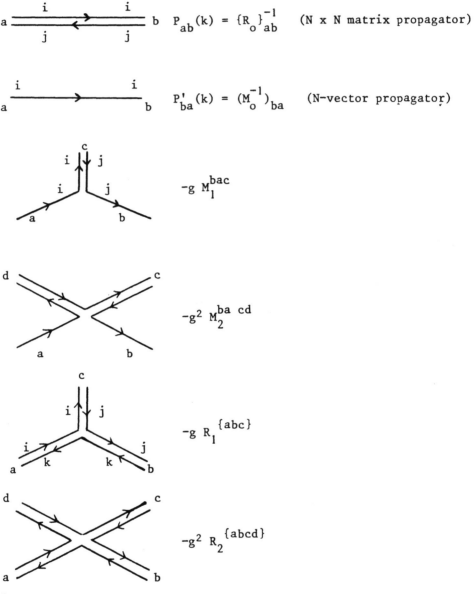

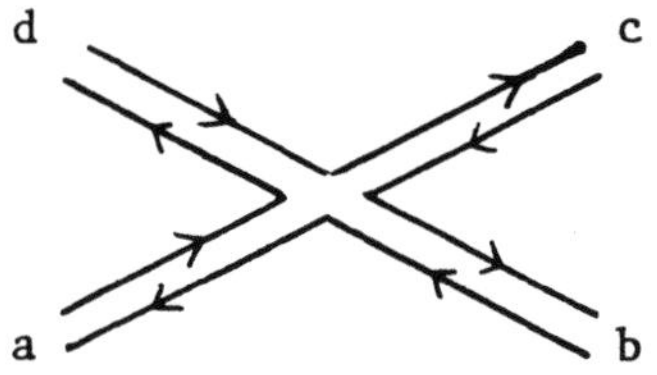

$-g^2\ R_2^{\{abcd\}}$

Fig. 1: Feynman Rules at arbitrary N. The accolades { } stand for cyclic symmetrization with respect to the indices a,b,...

In that case an extra projection operator is required in the propagator:

$$\delta_i^{\,k}\delta_\ell^{\,j} - \frac{1}{N}\,\delta_i^{\,j}\delta_\ell^{\,k} \ . \tag{2.4}$$

The second term in (2.4) corresponds to an extra piece in the propagator, as given in Fig. 2. We will see later that such terms are relatively unimportant as $N \to \infty$.

For defining amplitudes it is often useful to consider source terms that preserve the (global) symmetry:

$$\mathcal{L}^{\rm source} = J_\psi^{ab}(x)\psi^{*a}(x)\psi^b(x) + \tfrac{1}{2}J_A^{ab}(x)\,{\rm Tr}\,A^a(x)A^b(x) \ . \tag{2.5}$$

The corresponding notation in Feynman graphs is shown in Fig. 3.

3. THE $N \to \infty$ LIMIT AND PLANARITY

As usual, amplitudes and Green's functions are obtained by adding all possible (planar and non-planar) diagrams with their appropriate combinatorial factors. Note that, apart from the optional correction term in (2.4), the number N does not occur in Fig. 1. But, of course, the number N will enter into expressions for the amplitudes, and that is when an index-line closes. Such an index-loop gives rise to a factor

$$\sum_i \delta_i^{\,i} = N \ . \tag{3.1}$$

We are now in a position that we can classify the diagrams (with only gauge invariant sources as given by eq. (2.5)) according to their order in g and in N. Let there be given a connected diagram. First we consider the two-dimensional structure obtained by considering all closed index loops as the edges of little (simply connected) surface elements. All $N \times N$ matrix-propagators connect these surface elements into a bigger surface, whereas the N-vector-propagators form a natural boundary to the total surface. In the complex case the total surface is an oriented one; in the real case there is no orientation. In both cases the total surface may be multiply connected, containing "worm holes". For convenience we limit ourselves to the complex (oriented) case, and we close the surface by attaching extra surface elements to all N-vector-loops.

Let that surface have F faces (surface elements), P lines (propagators) and V vertices. We have $F = L + I + P_t$, where L is the number of N-vector-loops*) and I the number of index-loops; and we write (footnote: see next page)

$$a \;\; \overset{i}{\underset{i}{\longrightarrow}}\!\!\!\!\!\!\supset \quad \subset\!\!\!\overset{\ell}{\underset{\ell}{\longleftarrow}} \;\; b \qquad\qquad -\frac{1}{N}\{R_o\}^{-1}_{ab}$$

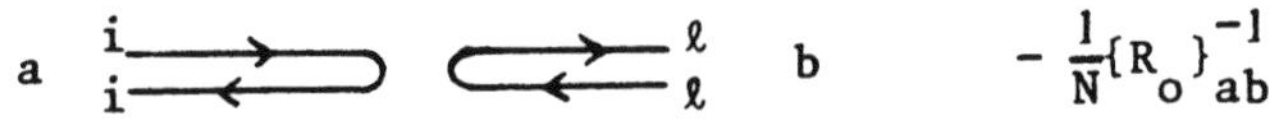

Fig. 2: Extra term in the propagator if Tr A is to be projected out.

J^{ab}_ψ (N-vector source)

J^{ab}_A (matrix source)

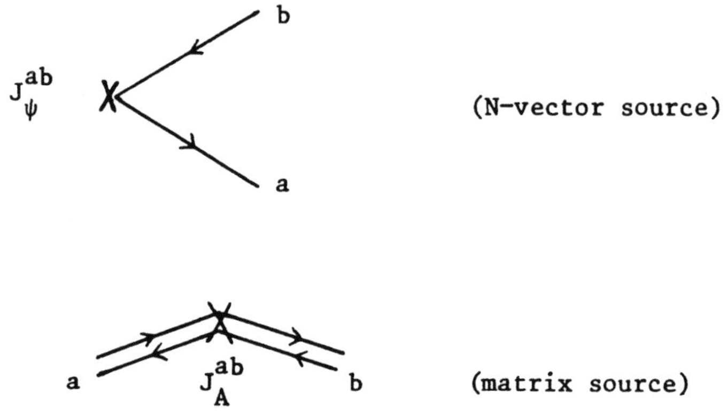

Fig. 3: Invariant source insertions.

$$V = \sum_n V_n ,$$

where V_n is the number of n-point vertices. (V_2 is the number of source insertions). The diagram is now associated with a factor

$$r = g^{V_3+2V_4} N^{I-P_t} . \tag{3.2}$$

Here P_t is the number of times the second term of (2.4) has been inserted to obtain traceless propagators. By drawing a dot at each end of each propagator we find that the total number of dots is

$$2P = \sum_n n V_n , \tag{3.3}$$

and eq. (3.2) can be written as

$$r = g^{2P-2V} N^{F-L-2P_t} . \tag{3.4}$$

Now we apply a well-known theorem of Euler:

$$F - P + V = 2 - 2H , \tag{3.5}$$

where H counts the number of "wormholes" in the surface and is therefore always positive (a sphere has H = o, a torus H = 1, etc.). And so,

$$r = (g^2N)^{\frac{1}{2}V_3+V_4} N^{2-2H-L-2P_t} . \tag{3.6}$$

Suppose we take the limit

$$N \to \infty , \ g \to o , \ g^2N = \tilde{g}^2 \ \text{(fixed)} . \tag{3.7}$$

If there are N-vector-sources then there must be at least one N-vector-loop:

$$L \geqslant 1 .$$

The leading diagrams in this limit have H = o, P_t = o and L = 1. They have one overall multiplicative factor N, and they are all planar: an open plane with the N-vector line at its edge (Fig. 4a).

*) "N-vector" here stands for N-component vector in U(N) space, so the N-vector-loops are the quark loops in quantum-chromodynamics.

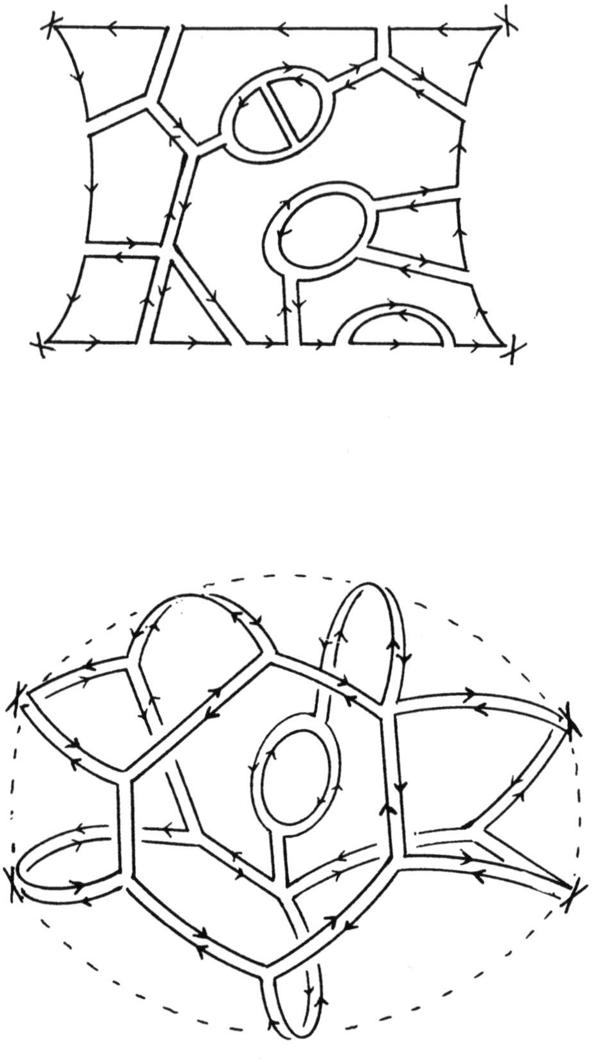

Fig. 4: Elements of the class of leading diagrams in the N → ∞ limit. a) If vector sources are present. b) In the absence of vector sources (e.g. pure gauge theory).

If there are only matrix sources then $L \geqslant o$. The leading diagrams all have the topology of a sphere and carry an overall factor N^2 (see Fig. 4b). We read off from eq. (3.6) that next to leading graphs are down by a factor $1/N$ for each additional N-vector-loop (= quark loop in quantum chromodynamics) and a factor $1/N^2$ for each "wormhole". Also the difference between U(N) and SU(N) theories disappears as $1/N^2$. It will be clear that this result depends only on the field variables being N-vectors and $N \times N$ matrices and the Lagrangian containing only single inner products or traces (not the *products* of inner products and/or traces). Diagrams with $L = 1$ and $H = o$ are the easiest to visualize. In the sequel we discuss convergence aspects of the summation of those diagrams in all orders of $\tilde{g}$. Our main examples are

1) U(N) (or SU(N)) gauge theories with fermions in the N representation;
2) purely Lorentz scalar fields, both in N and in $N \times N$ representations of U(N). That theory will be called $\mathrm{Tr}\lambda\phi^4$, or $-\mathrm{Tr}\lambda\phi^4$, if λ is given the unusual sign. Both SU(N) gauge theory and $-\mathrm{Tr}\lambda\phi^4$ are asymptotically free[9]. The latter has the advantage that one may add a mass term, so that it is also infrared convergent. However the fact that λ has the wrong sign implies that that theory only exists in the $N \to \infty$ limit, not for finite N. A model that combines all "good" features of the previous model is:
3) an SU(N) Higgs theory with N Higgs fields in the elementary representation, N fermions in the elementary representation and a fermion in the adjoint representation. A global SU(N) symmetry then survives. All vector, spinor and scalar particles are massive, and it is asymptotically free if $\tilde{h}^2/\tilde{g}^2 = 1$;

$$\tilde{\lambda}/\tilde{g}^2 = \frac{1}{8}(\sqrt{129}-5) \ , \tag{3.8}$$

where h is a Yukawa coupling constant and λ the Higgs self coupling. The reason for mentioning this model is that it is asymptotically free in the ultraviolet, and it is convergent in the infrared, so that our methods will enable us to construct it rigorously in the $N \to \infty$ limit (provided that masses are chosen sufficiently large and the coupling constant sufficiently small), and positivity of the Hamiltonian is guaranteed also for finite N so that there is every reason for hope that the theory makes sense also at finite N, contrary to the $-\lambda\mathrm{Tr}\phi^4$ theory. This model is described in appendix A.

4. THE SKELETON EXPANSION

From now on we consider diagrams of the type pictured in figure 4a ($H = o$, $L = 1$). They all have the same N dependence, so once we restricted ourselves to these planar diagrams only we may drop the indices i,j,... and replace the double-line propagators by single lines. Often we will forget the tilde (~) on g^2 because the factor N is always understood. Only the (few) indices a,b,... of eq. (2.1), as far as they do not refer to the SU(N) group(s), are kept. The details of this surviving index structure are not important for what follows, as long as the Feynman rules (Fig. 1) are of the general renormalizable type.

Our first concern will be the isolation of the ultraviolet divergent parts of the diagrams. For this we use an ancient device[10] called "skeleton-expansion" *). It can be applied to any graph, planar or not, but for the planar case it is particularly useful.

Consider a graph with at least five external lines. A one-particle irreducible subgraph is a subset of more than one vertices with the internal lines that connect these vertices, that is such that if one of the internal lines is cut through then the subgraph still remains connected. We now draw boxes around all one-particle irreducible subgraphs that have four of fewer external lines. In general one may get boxes that are partially overlapping. A box is *maximal* if it is not entirely contained inside a larger box.

Theorem: All maximal boxes are not-overlapping. This means that two different maximal boxes have no vertex in common.

Proof: If two maximal boxes A and B would overlap then at least one vertex x_1 would be both in A and B. There must be a vertex x_2 in A but not in B, otherwise A would not be maximal. Similarly there is an x_3 in B but not in A. Now A was irreducible, so that at least two lines connect x_1 with x_2. These are external lines of B but not of A U B. Now B may not have more than 4 external lines. So not more than two external lines of A U B are also external lines of B. The others may be external lines of A. But there can also be not more than two of those. So A U B has not more than four external lines and is also irreducible since A and B are, and they have a vertex in common. So we should draw a box around A U B. But then neither A nor B would be maximal, contrary to our assumption. No planarity was needed in this proof.

The skeleton graph of the diagram is now defined by replacing all maximal boxes by single "dressed" vertices. Any diagram can now be decomposed into its "skeleton" and the "meat", which is the collection of all vertex and self-energy insertions at every two-, three- and four-leg irreducible subgraph. In particular the self-

---

*) The method described here differs from Bjorken and Drell[10] in that we do not distinguish fermions from bosons, so that also subgraphs with four external lines are contracted.

energy insertions build up the so-called dressed propagator. We call the dressed three- and four-vertices and propagators the "basic Green functions" of the theory. They contain all ultraviolet divergences of the theory. The rest of the diagram, the "skeleton" built out of these basic Green functions is entirely void of ultraviolet divergencies because there are no further (sub)graphs with four of fewer external lines, which could be divergent.

The skeleton expansion is an important tool that will enable us to construct in a rigorous way the planar field theory. For, under fairly mild assumptions concerning the behavior of the basic Green functions we are able to prove that, given these basic Green functions, the sum of all skeleton graphs contributing to a certain amplitude *in Euclidean space* is absolutely convergent (not only Borel summable). This proof is produced in the next 6 sections. Clearly, this leaves us to construct the basic Green functions themselves. A recursive procedure for doing just that will be given in sects. 12-15. Indeed we will see that our original assumptions concerning these Green functions can be verified provided the masses are big and the coupling constants small, with one exception: in the scalar $-\lambda \mathrm{Tr}\phi^4$ theory the skeleton expansion always converges even if the bare (minimally subtracted) mass vanishes! (sect. 18)

5. TYPE IV PLANAR FEYNMAN RULES

We wish to prove the theorem mentioned in the previous section: given certain bounds for the basic Green functions, then the sum of all skeleton graphs containing these basic Green functions inside their "boxes" converges in the absolute. In fact we want a little more than that. In sects. 13-15 we will also require bounds on the total sum. Those in turn will give us the basic Green functions. We have to anticipate what bounds those will satisfy. In general one will find that the basic Green functions will behave much like the bare propagators and vertices, with deviations that are not worse than small powers of ratios of the various momenta. Note that all our amplitudes are *Euclidean*.

First we must know how the dressed propagators behave at high and low momenta. The following bounds are required:

$$|P_{ab}(k)| \leqslant \frac{Z(k)}{k^2+m^2}\,, \qquad \text{if } k^2 \geqslant o\,. \tag{5.1}$$

Here $P_{ab}(k)$ is the propagator. From now on we use the absolute value symbol for momenta to mean: $|p| = \sqrt{p^2+m^2}$. Then the field renormalization factor $Z(k)$ is approximately:

$$Z(k) \cong \left[\log\left(1+\frac{|k|}{m}\right)\right]^{\sigma} , \tag{5.2}$$

where σ is a coefficient that can be computed from perturbation expansion. The mass term in (5.1) is not crucial for our procedure but m in (5.2) can of course not be removed easily.

To write down the bounds on the three- and four-point Green functions in Euclidean space we introduce a convenient notation to indicate which external momenta are large and which are small.

For any planar Green function we label not the external momenta but the spaces in between two external lines by indices 1,2,3,... which have a cyclic ordering. An external line has momentum

$$p_{i,i+1} \underset{\text{def}}{=} p_i - p_{i+1} . \tag{5.3}$$

We have automatically momentum conservation,

$$\sum_i p_{i,i+1} = 0 , \tag{5.4}$$

and the p_i are defined up to an overall translation,

$$p_i \to p_i + q , \quad \text{all } i . \tag{5.5}$$

A channel (any in which possibly a resonance can occur) is given by a pair of indices, and the momentum through the channel is given by

$$p_{i,j} = p_i - p_j . \tag{5.6}$$

So we can look at the p_i as dots in Euclidean momentum space, and the distance between any pair of dots is the momentum through some channel. If we write

$$(((12)_{A_1}\ 3)_{A_2}\ 4)_{A_3} , \tag{5.7}$$

or simply $(((12)_1\ 3)_2\ 4)_3$, then this means:

$$|p_1-p_2| = A_1 ,$$

$$|p_1-p_3| = A_2 \gg A_1 ,$$

$$|p_1-p_4| = A_3 \gg A_2 . \tag{5.8}$$

So the brackets are around momenta that form close clusters.

Our bounds for the three- and four-point functions are now defined in table 1.

Table 1	
Bounds for the 3- and 4-point dressed Green functions. Z_{ij} stands for $Z(p_i-p_j)$. All other exceptional momentum configurations can be obtained by cyclic rotations and reflections of these. K_i are coefficients close to one.	
$((12)_1\ 3)_2$	$K_1(Z_{12}Z_{23}Z_{31})^{-\frac{1}{2}}\ A_2(A_2/A_1)^{\alpha}\ g(A_2)$
$(((12)_1\ 3)_2\ 4)_3$	$K_2^2(Z_{12}Z_{23}Z_{34}Z_{41})^{-\frac{1}{2}}\ (A_2/A_1)^{\alpha}\ (A_3/A_2)^{\beta}\ g^2(A_3)$
$((12)_1\ (34)_2)_3$	$K_3^2(Z_{12}Z_{23}Z_{34}Z_{41})^{-\frac{1}{2}}\ (A_3^2/A_1A_2)^{\alpha}\ g^2(A_3)$
$(((13)_1\ 2)_2\ 4)_3$	$K_4^2(Z_{12}Z_{23}Z_{34}Z_{41})^{-\frac{1}{2}}\ (A_2A_3/A_1^2)^{\beta}\ g^2(A_3)$
$((13)_1\ (24)_2)_3$	$K_5^2(Z_{12}Z_{23}Z_{34}Z_{41})^{-\frac{1}{2}}\ (A_3/A_1)^{2\beta}\ g^2(A_3)$
	if $A_1 \leqslant A_2$
$((123)_1\ 4)_2$	$K_6^2(Z_{12}Z_{23}Z_{34}Z_{41})^{-\frac{1}{2}}\ (A_2/A_1)^{\beta}\ g^2(A_2)$
$((12)_1\ 34)_2$	$K_7^2(Z_{12}Z_{23}Z_{34}Z_{41})^{-\frac{1}{2}}\ (A_2/A_1)^{\alpha}\ g^2(A_2)$

Here α and β are small positive coefficients. $g(x)$ is a slowly varying running coupling constant. For the time being all we need is some g with

$$\max_i K_i\ g(x) \leqslant g \qquad \text{for all } x\ , \tag{5.9}$$

where we also assume that possible summation over indices a,b,... is included in the K coefficients. Clearly the bare vertices would satisfy the bounds with $\alpha = \beta = 0$. Having positive α and β allows us to have any of the typical logarithmic expressions coming from the radiative corrections in these dressed Green functions. Indeed we will see later (sect. 13) that those logarithms will never surpass our power-laws.

Table 1 has been carefully designed such that it can be re-obtained in constructing the basic Green functions as we will see in sects. 12-14. First we notice that the field renormalization factors $Z(p_i-p_j)^{-\frac{1}{2}}$ cancel against corresponding factors in our bounds for the propagator (5.1). The power-laws of Table 1 can be conveniently expressed in terms of a revised set of Feynman rules. These are given in Fig. 5. We call them type IV Feynman rules after a fourth attempt to reformulate our bounds (types I, II and

$$\frac{1}{(k^2+m^2)^{1+\alpha}}$$ (dressed elementary propagator)

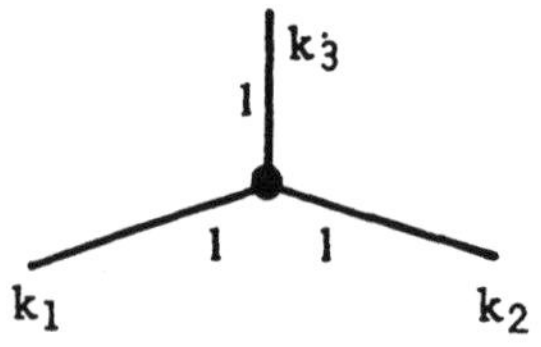

$$g[\max(|k_1|,|k_2|,|k_3|)]^{1+3\alpha}$$ (dressed 3-vertex)

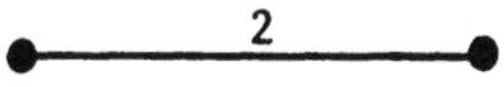

$$\frac{1}{|k|^{2\beta}}$$ (composite propagator)

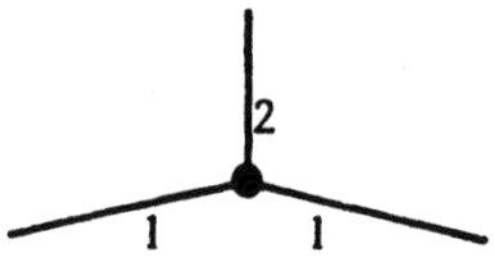

$$g[\max(|k_1|,|k_2|,|k_3|)]^{2\alpha+\beta}$$

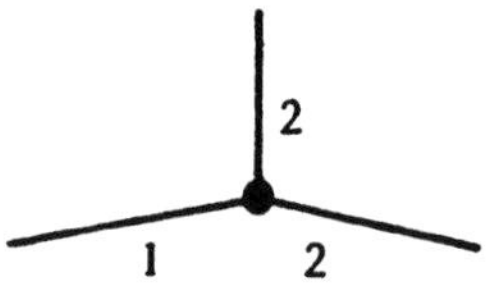

$$g[\max(|k_1|,|k_2|,|k_3|)]^{\alpha+2\beta-1}$$

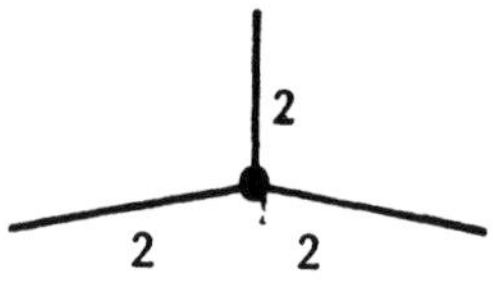

$$g[\max(|k_1|,|k_2|,|k_3|)]^{3\beta-2}$$

(generalized vertices)

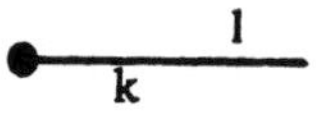

$$|k|^{-\alpha}$$ (external line)

Fig. 5: Type IV Feynman Rules. $|k|$ stands for $\sqrt{k^2+m^2}$.

III occur in refs. 11, 12 and are not needed here). The trick is to introduce a new kind of propagator, $\bullet\!\overset{2}{\text{———}}\!\bullet$, that represents an exchange of two or more of the original particles in the diagram we started off with.

The procedure adapted in these lectures deviates from earlier work[11] in particular by the introduction of the last two vertices in Fig. 5. Notice that they decrease whenever two of the three external momenta become large.

It is now a simple exercise to check that indeed any diagram built from basic Green functions that satisfy the bounds of Table 1 can also be bounded by corresponding diagram(s) built from type IV Feynman rules. The four-vertex is simply considered as a sum of two contributions both made by connecting two three-point vertices with a type 2 propagator, and the factors $|k|^{-\alpha}$ from the propagators in Fig. 5 are considered parts of the vertex functions (the mass term of the propagator may be left out; it is needed at a later stage). Type 2 propagators will also be referred to as "composite propagators".

Elementary power counting now tells us that the superficial degree of convergence, Z, of any (sub)graph with E_1 external single lines and E_2 external composite lines is given by

$$Z = (1-\alpha)E_1 + (2-\beta)E_2 - 4 \; . \tag{5.10}$$

Since we consider only skeleton graphs, all our graphs and subgraphs have

$$E_1 + 2E_2 \geqslant 5 \; . \tag{5.11}$$

Thus, Z is guaranteed to be positive if we restrict our coefficients by

$$o < \alpha < 1/5 \; ;$$

$$o < \beta < 2/5 \; . \tag{5.12}$$

(Infrared convergence would merely require $\alpha < 1$; $\beta < 2$, and is therefore guaranteed also.) So we know that with (5.12) all graphs and subgraphs are ultraviolet and infrared convergent. The theorem we now wish to prove is: the sum of all convergent type IV diagrams contributing to any given amplitude with 5 (or more) external lines converges in Euclidean space. It is bounded by the sum of all type IV tree graphs (graphs without closed loops) multiplied with a fixed finite coefficient.

A further restriction on the coefficients α and β will be necessary (eq. (8.15)).

6. NUMBER OF TYPE IV DIAGRAMS

The total number G(E,L) of connected or irreducible planar diagrams with E external lines and L closed loops in any finite set of Feynman rules, is bounded by a power law (in contrast with the non-planar diagrams that contribute for instance to the Lth order term in the expansion such as (1.1) for a simple functional integral):

$$G(E,L) \leqslant C_1^E C_2^L \, , \tag{6.1}$$

for some C_1 and C_2.

In some cases C_1 and C_2 can be computed exactly and even closed expressions for G(E,L) exist[8]. These mathematical exercises are beautiful but rather complicated and give us much more than we really need. In order to make these lectures reasonably self-sustained we will here derive a crude but simple derivation of ineq. (6.1) yielding C coefficients that can be much improved on, with a little more effort.

Let us ignore the distinction between the two types of propagators and just count the total number G(E,L) of connected planar ϕ^3 diagrams with a given configuration of E external lines and L closed loops. We have (see Fig. 6)

$$G(E+1,L) = G(E+2,L-1) + \sum_{n,L_1} G(n+1,L_1)G(E+1-n,L-L_1) \, . \tag{6.2}$$

$$G(E,L) = o \quad \text{if} \quad E < 2 \quad \text{or if} \quad L < o \, ;$$

$$G(2,o) = 1 \, . \tag{6.3}$$

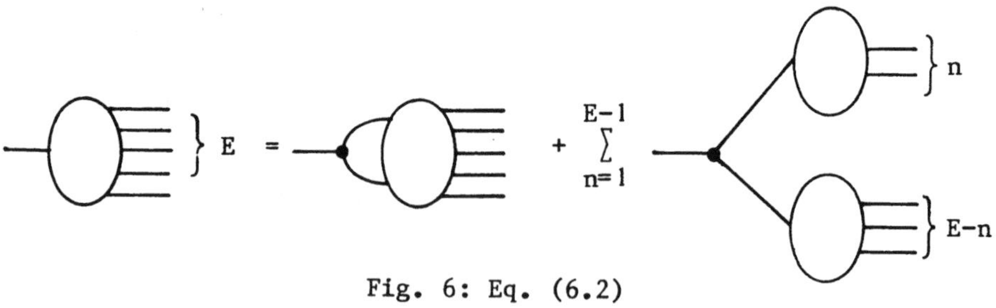

Fig. 6: Eq. (6.2)

We wish to solve, or at least find bounds for, G(E,L) from (6.2) with boundary condition (6.3). A good guess is to try

$$G(E,L) \leqslant \frac{C_0 C_1^E C_2^L}{(E-1)^2 (L+1)^2} \ , \ E \geqslant 2 \ , \ L \geqslant o \ , \tag{6.4}$$

which is compatible with (6.3) if

$$C_0 C_1^2 \geqslant 1 \ . \tag{6.5}$$

Using the inequality

$$\sum_{n=1}^{k} \frac{1}{n^2 (k-n)^2} \leqslant \frac{4}{k^2} \ , \tag{6.6}$$

we find that the r.h.s. of (6.2) will be bounded by

$$\frac{C_0 C_1^{E+2} C_2^{L-1}}{(E+1)^2 \ L^2} + \frac{16 C_0^2 C_1^{E+2} C_2^L}{(E-1)^2 (L+1)^2} \ , \tag{6.7}$$

which is smaller than

$$\frac{C_0 C_1^{E+1} C_2^L}{E^2 (L+1)^2} \ , \tag{6.8}$$

if

$$4C_1/C_2 + 64 C_0 C_1 \leqslant 1 \ . \tag{6.9}$$

This is not incompatible with (6.5) although the best "solution" to these two inequalities is a set of uncomfortably large values for C_0, C_1 and C_2. But we proved that they are finite.

The exact solution to eq. (6.2) is

$$G(E,L) = \frac{2^L (2E-2)!(2E+3L-4)!}{L!(E-1)!(E-2)!(2E+2L-2)!} \ , \tag{6.10}$$

which we will not derive here. Using

$$\frac{(A+B)!}{A!B!} \leqslant 2^{A+B-1} \ , \tag{6.11}$$

we find that in (6.1),

$$C_1 \leqslant 16 \ ; \ C_2 \leqslant 16 \ . \tag{6.12}$$

For fixed E, in the limit of large L,

$$C_2 \to 27/2 \ . \tag{6.13}$$

Similar expressions can be found for the set of irreducible diagrams. Since they are a subset of the connected diagrams we expect C coefficients equal to or smaller than the ones of eqs. (6.12) and (6.13). Limiting oneself to only convergent skeleton graphs will reduce these coefficients even further.

We have for the number of vertices V

$$V = E + 2L - 2 \ , \tag{6.14}$$

and the number of propagators P:

$$P = V + L - 1 \ . \tag{6.15}$$

So, if different kinds of vertices and propagators are counted separately then the number of diagrams is multiplied with

$$C_V^V C_P^P \ , \tag{6.16}$$

where C_V and C_P are some fixed coefficients. This does not alter our result qualitatively. Also if there are elementary 4-vertices then these can be considered as pairs of 3-vertices connected by a new kind of propagators, as we in fact did. So also in that case the numbers of diagrams are bounded by expressions in the form of eq. (6.1).

7. THE SMALLEST FACETS

We now wish to show that every planar type IV diagram with L loops is bounded by a coefficient C^L times a (set of) type IV tree graph(s), with the same momentum values at the E external lines. This will be done by complete induction. We will choose a closed loop somewhere in the diagram and bound it by a tree insertion. Now even in a planar diagram some closed loops can become quite large (i.e. have many vertices) and it will not be easy to write down general bounds for those. Can we always find a "small" loop somewhere?

We call the elementary loops of a planar diagram facets. Now Euler's theorem for planar graphs is:

$$V - P + L = 1 \ . \tag{7.1}$$

Take an irriducible diagram. Write

$$L = \sum_n F_n \ , \tag{7.2}$$

where F_n are the number of facets with exactly n vertices (or "corners"). Let

$$P = P_i + P_e \ , \tag{7.3}$$

where P_i is the number of internal propagators and P_e is the number of propagators at the edges of the diagram. Then, by putting a dot at every edge of each facet and counting the number of dots we get

$$\sum_n n\, F_n = 2P_i + P_e \ . \tag{7.4}$$

For the numbers V_n of n-point vertices we have similarly

$$\sum_n n\, V_n = 2P + E \ , \tag{7.5}$$

but in ourcase we only consider 3-point vertices (compare eqs. (6.14) and (6.15)):

$$3V = 2P + E \ . \tag{7.6}$$

Combining eqs. (7.1) - (7.6) we find

$$\sum_n (n-6)F_n = 2E - P_e - 6 \ . \tag{7.7}$$

This equation tells us that if a diagram has

$$L \geqslant 2E - 8 \tag{7.8}$$

then either it is a "seagull graph" ($P_e \leqslant 1$) which we usually are not interested in, or there must be at least one subloop with 6 or fewer external lines:

$$F_n > 0 \quad \text{for some} \quad n \leqslant 6 \ . \tag{7.9}$$

So diagrams with given E and large enough L must always contain facets that are either hexagons or even smaller.

In fact we can go further:
theorem: if a planar graph (with only 3-vertices) and all its irreducible subgraphs have $2E - P_e \geqslant 6$ then the entire graph obeys

$$L \leqslant \frac{E^2}{12} - \frac{E}{2} + 1 \ . \tag{7.10}$$

This simple theorem together with eq. (7.7) tells us that any diagram with a number of loops L exceeding the bound of (7.10) must have at least one elementary facet with 5 of fewer lines attached to it. Although we could do without it, it is a convenient theorem and now we devote the rest of this section to its proof (it could be skipped at first reading).

First we remark that if we have the theorem proven for all *irreducible* graphs up to a certain order, then it must also hold for reducible graphs up to the same order. This is because if we connect two graphs with one line we get a graph 3 with

$$L_3 = L_1 + L_2 \, ,$$

$$P_{e3} = P_{e1} + P_{e2} + 2 \, , \tag{7.11}$$

$$E_3 = E_1 + E_2 - 2 \, .$$

If L_1, E_1 and L_2, E_2 satisfy (7.10) then so do L_3 and E_3 (remember that E and L are integers and the smallest graph with $L > o$ has $E = 6$; propagators that form two edges of a diagram are counted twice in P_e).

Forthe irreducible graphs we prove (7.10) by a rather unusual induction procedure for planar graphs. We consider the outer rim of an irreducible graph and all the (in general not irreducible) graphs inside it (see Fig. 7). Let the entire graph have E external lines and P_e propagators at its sides. The subgraphs i inside the rim have e_i external lines and P_{ei} propagators at their sides. We count:

$$P_e = E + \sum_i e_i \, , \tag{7.12}$$

and the number of loops L of the entire diagram is

$$L = 1 + \sum_i (L_i + e_i - 1) \, . \tag{7.13}$$

Now each facet between the subgraphs and the rim must have at least 6 propagators as supposed, therefore

$$P_e + \sum P_{ei} + 2 \sum e_i \geqslant 6 \sum_i (e_i - 1) + 6 \, , \tag{7.14}$$

but if some of the subgraphs are single propagators we need to be more precise

$$P_e + \sum P_{ei} + 2 \sum e_i \geqslant 6 \sum_i (e_i - 1) + 6 + 2N_2 \, , \tag{7.15}$$

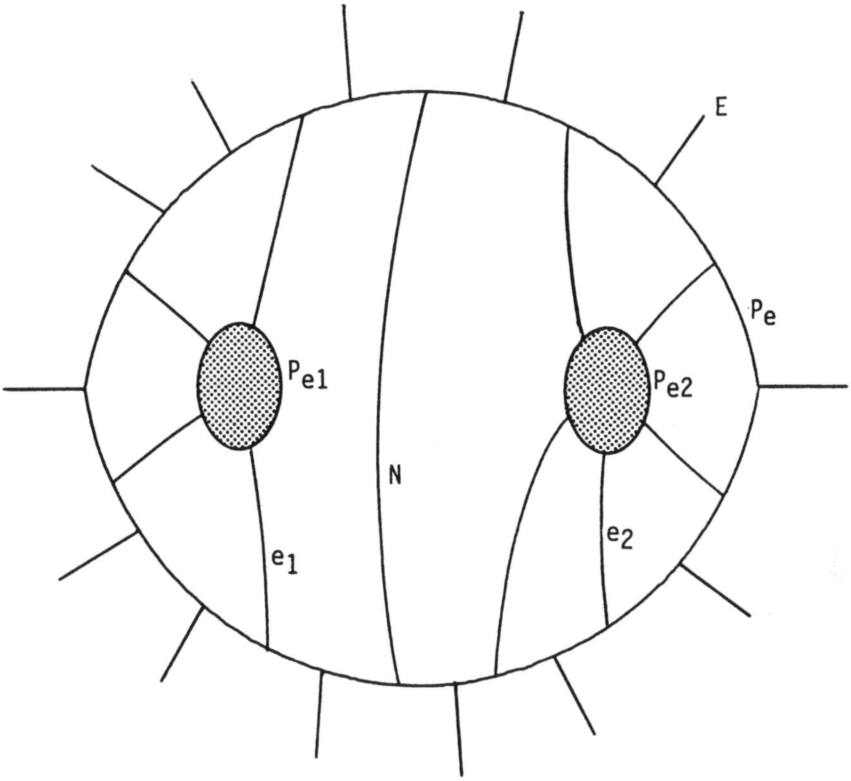

Fig. 7: Proving the theorem of sect. 7. The number E counts the external lines of the entire graph. P_e the number of sides and e_i and P_{ei} do the same for the subgraphs 1 and 2. N is the number of single propagators.

where N_2 is the number of single propagators, each of which contributes with e = 2 in eq. (7.13), and have P_{ei} = o. Now we use (7.12), and

$$P_{ei} \overset{!}{\leqslant} 2e_i - 6 \,, \tag{7.16}$$

as required, whereas $\Sigma(2e-6) + 2N_2 = o$ for the single propagators, to arrive at

$$E \geqslant \sum_i e_i + 6 \,. \tag{7.17}$$

From the assumption that all subraphs already satisfy (7.10) we get, writing $L_1 = \sum_i L_i$ and $E_1 = \sum_i e_i$:

$$L_1 \leqslant \frac{E_1^2}{12} - \frac{E_1}{2} + 1 \ , \qquad (7.18)$$

and from (7.13)

$$L \leqslant L_1 + E_1 \ . \qquad (7.19)$$

With (7.17) which reads $E \geqslant E_1 + 6$ we now see that (7.10) again holds for the entire diagram. The quadratic expression in (7.10) is the sharpest that can be derived from (7.17)-(7.19), and indeed large diagrams that saturate the inequality can be found, by joining hexagons into circular patterns.

Our conclusion is that if we wish to use an induction procedure to express a bound for diagrams with type IV Feynman rules and L loops in terms of one with a smaller number L' loops we can try to do that by replacing successively triangles, qudrangles and/or pentagons by type IV tree insertions, until the bound (7.10) is reached. In particular if E = 5 this leads us to a tree diagram. The next three sections show how this procedure works in detail.

8. TRIANGLES

Consider a (large) diagram with type IV Feynman rules. We had already decreed that it and all its subgraphs are ultraviolet and infrared convergent (divergent subgraphs had been absorbed into the vertices and propagators before). With eqs. (5.10)-(5.12) this means that each subgraph has

$$E_1 + 2E_2 \geqslant 5 \ , \qquad (8.1)$$

so, in particular, there are no self-energy blobs. First we use the inequality of Fig. 8 to replace composite propagators by ordinary dressed propagators one by one until ineq. (8.1) forbids any further such replacements. The inequality is readily proven: we write for the propagators with its vertices:

Fig. 8. A composite propagator is smaller than an elementary one.

$$|p_1|^{1+\alpha_1+\alpha_2+\gamma}\ |p_2|^{1+\alpha_3+\alpha_4+\gamma}\ |k^2|^{-1-\gamma}\ , \tag{8.2}$$

where k is the momentum through the propagator, $|p_1| \geqslant |k|$. $|p_2| \geqslant |k|$, and $\alpha_i = \alpha$ for a dressed propagator, and $\alpha_i = \beta-1$ for a composite propagator. At the left hand side $\gamma = \beta-1$, and at the right hand side $\gamma = \alpha$. Clearly we have

$$\left(\frac{|p_1||p_2|}{k^2}\right)^{\beta-1} \leqslant \left(\frac{|p_1||p_2|}{k^2}\right)^{\alpha}\ , \tag{8.3}$$

α and β being both close to zero due to (5.12).

Now consider all elementary triangle loops in our diagram. Under what conditions can we replace them by type IV 3-vertices (Fig. 9)? Due to (8.1) there can be at most one elementary (dressed external line, the others are composite: E_2 = 2 or 3;

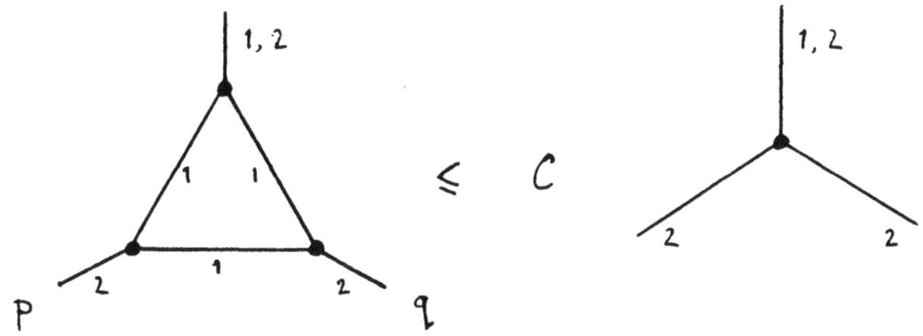

Fig. 9: Removal of elementary triangle facets.

E_1 = 1 or o. We write

$$\alpha_1 = \alpha \quad \text{or} \quad \beta - 1 \tag{8.4}$$

to cover both cases. Now let us replace the vertex functions by bounds that depend only on the momenta of the internal lines:

$$\Big(\max(|k|,|k+p|,|p|)\Big)^{\gamma} \leqslant R(\gamma)\Big(|k|^{\gamma} + |k+p|^{\gamma}\Big)\ , \tag{8.5}$$

with $\quad \gamma = 1+3\alpha\ ;\ R(\gamma) = 2^{3\alpha^*}$, or $\gamma = \beta+2\alpha \;;\; R(\gamma) = 1 \;.$ (8.6)

The integral over the loop momentum is then bounded by 8 terms all of the form

$$R \int d^4k \; (k^2)^{-\delta_1} (k+p)^{-2\delta_2} \; (k-q)^{-2\delta_3} \;, \tag{8.7}$$

where for a moment we ignored the mass term. It can be added easily later. We have convergence for all integrals:

$$Z = 2 \sum \delta - 4 = 1 - 2\beta - \alpha_1 > 0 \;. \tag{8.8}$$

and

$$\delta_1 \geqslant 1 + \alpha - \tfrac{1}{2}(2\alpha+\beta) - \tfrac{1}{2}(1+2\alpha+\alpha_1) = \tfrac{1}{2}(1-2\alpha-\alpha_1-\beta) \;; \tag{8.9}$$

The integral (8.7) can be done using Feynman multiplicators:

$$\frac{R\pi^2\Gamma(\Sigma\delta-2)}{\Gamma(\delta_1)\Gamma(\delta_2)\Gamma(\delta_3)} \int_0^1 dx_1dx_2dx_3 \; \frac{\delta(\Sigma x-1)\; x_1^{\delta_1-1}\; x_2^{\delta_2-1}\; x_3^{\delta_3-1}}{\left(p^2x_2x_3+q^2x_3x_1+(p+q)^2x_1x_2\right)^{\Sigma\delta_i-2}} \tag{8.10}$$

Now if $|p| \geqslant |q| \geqslant |p+q|$ (the other cases can be obtained by permutation) then $|q| \geqslant \frac{1}{2}|p|$, so our integrals are bounded by

$$C\Big[\max(|p|,|q|,|p+q|)\Big]^{-\frac{1}{2}Z} \tag{8.11}$$

where C is the sum of integrals of the type

$$\frac{\pi^2\Gamma(\Sigma\delta-2)}{\Gamma(\delta_1)\Gamma(\delta_2)\Gamma(\delta_3)} \int_0^1 dx_1dx_2dx_3 \; \frac{\delta(\Sigma x-1)\; x_1^{\delta_1-1}\; x_2^{\delta_2-1}\; x_3^{\delta_3-1}}{(x_1x_2+\frac{1}{2}x_1x_3)^{\frac{1}{2}Z}} \tag{8.12}$$

which can be further bounded (replacing x_1x_2 by $\frac{1}{2}x_1x_2$) by

$$C \leqslant \sum \frac{2^{\Sigma\delta-2}\; \pi^2\; \Gamma(\Sigma\delta-2)\Gamma(2-\delta_1)\Gamma(2-\delta_2-\delta_3)}{\Gamma(\delta_1)\; \Gamma(\delta_2+\delta_3)} \;, \tag{8.13}$$

if all integrals converge, of course. All entries in the Γ

functions must be positive. In particular, we must have

$$2 - \delta_2 - \delta_3 = \delta_1 - \tfrac{1}{2}Z > o \ . \qquad (8.14)$$

Now with (8.8) and (8.9) this corresponds to the condition:

$$\beta > 2\alpha \ , \qquad (8.15)$$

this is the extra restriction on the coefficients α and β to be combined with (5.12), and which we already alluded to in the end of sect. 5. A good choice may be

$$\alpha = 0.1 \ , \ \beta = 0.3 \ . \qquad (8.16)$$

We conclude that we proved the bound of Fig. 9, if α and β have values such as (8.16), and the number C in Fig. 9 is bounded by the sum of eight finite numbers in the form of eq. (8.13).

9. QUADRANGLES

We continue removing triangular facets from our diagram, replacing them by single 3-vertices, following the prescriptions of the previous sections. We get fewer and fewer loops, at the cost of at most a factor C for each loop. Either we end up with a tree diagram, in which case our argument is completed, or we may end up with a diagram that can still be arbitrarily large but only contains larger facets. According to sect. 7 there must be quadrangles and/or pentagons among these.

Before concentrating on the quadrangles we must realize that there still may be larger subgraphs with only three external lines. In that case we consider those first: a minimal triangular subgraph is a triangular subgraph that contains no further triangular subgraphs. If our diagram contains triangular subgraphs then we first consider a minimal triangular subgraph and attack quadrangles (later pentagons) in these. Otherwise we consider the quadrangles inside the entire diagram.

Let us again replace as many composite propagators ($\bullet\!\overset{2}{\text{———}}\!\bullet$) by single dressed propagators ($\bullet\!\overset{1}{\text{———}}\!\bullet$) as allowed by ineq. (8.1) for each subgraph. Then one can argue that as a result we must get at least one quadrangle somewhere whose own propagators are all of the elementary type ($\bullet\!\overset{1}{\text{———}}\!\bullet$), not composite ($\bullet\!\overset{2}{\text{———}}\!\bullet$). This is because facets with composite propagators now must be adjacent to 4-leg subgraphs (elementary facets or more complicated), and then these in turn must have facet(s) with elementary propagators. Also (although we will not really need this) one may argue that there will be quadrangles with not more than one external composite propagator, the others elementary (the one exception is the case when one of the adjacent quadranglular subgraphs has itself only

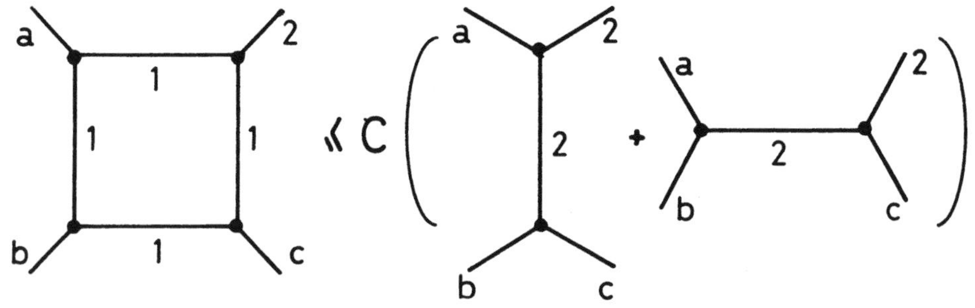

Fig. 10. Inequality for quadrangles. a, b and c may each be 1 or 2.

pentagons, but that case will be treated in the next section). As a result of these arguments, of all inequalities of the type given in Fig. 10 we only need to check the case that only one external propagator is composite, a = b = c = 1.

But in fact they hold quite generally, also in the other cases. This is essentially because of the careful construction of the effective Feynman rules of type IV in Fig. 5.

Rather than presenting the complete proof of the inequalities of Fig. 10 (5 different configurations) we will just present a simple algorithm that the reader can use to prove and understand these inequalities himself. In general we have integrals of the form

$$\int d^4k \prod_i \frac{1}{|k-p_i|^{2\delta_i}} \,. \tag{9.1}$$

We could write this as a diagram in Fig. 11, where the δ_i at the propagators now indicate their respective powers. The vertices are here ordinary point-vertices, not the type IV rules. Now write

$$\frac{1}{|k-p_1|^{\omega_2}|k-p_2|^{\omega_2}} \leqslant \frac{A(\omega_1,\omega_2)}{|p_1-p_2|^{\omega_2}|k-p_1|^{\omega_1}} + \frac{A(\omega_2,\omega_1)}{|p_1-p_2|^{\omega_1}|k-p_2|^{\omega_2}} \tag{9.2}$$

with

$$A(\omega_1,\omega_2) = \max\left[1,\left(\left(\frac{\omega_1}{\omega_1+\omega_2}\right)^{\omega_1} + \left(\frac{\omega_2}{\omega_1+\omega_2}\right)^{\omega_2}\right)^{-1}\right] . \tag{9.3}$$

Inserted in a diagram, this is the inequality pictured in Fig. 11.

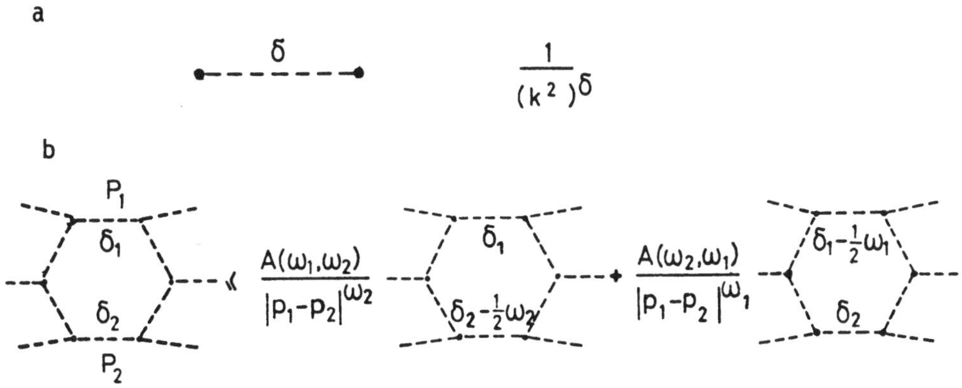

Fig. 11. a) Definition of a dotted propagator. Its vertices are just 1.
b) Extraction of a power of an external momentum.

We use it for instance when p_1-p_2 is the largest momentum of all channels, and if

$$\omega_1 \leqslant 2\delta_1 \ ; \ \omega_2 \leqslant 2\delta_2 \ ; \ \omega_i < Z \ , \tag{9.4}$$

where Z is the degree of convergence of the diagram: $Z = 2\Sigma\delta_i - 4$. We continue making such insertions, everytime reducing the diagrams to a convenient momentum dependent factor times a less convergent diagram. Finally we may have

$$Z < 2\delta_1 \ , \ Z < 2\delta_2 \ , \tag{9.5}$$

for two of its propagators. Then we use the inequality pictured in Fig. 12:

$$\int d^4k \prod_i (k-p_i)^{-2\delta_i} \leqslant C|p_1-p_2|^{-Z} \ . \tag{9.6}$$

Notice that in (9.5) we have a strictly unequal sign, contrary to ineq. (9.4). This C can be computed using Feynman multiplicators,

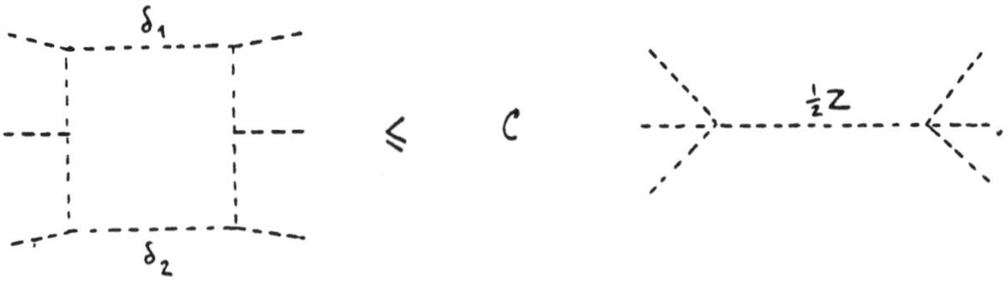

Fig. 12. Inequality holding if $\delta_{1,2}$ are strictly larger than $\frac{1}{2}Z$ (ineqs. (9.4), (9.5)).

much like in the previous section. We get

$$C \leqslant \frac{\pi^2\Gamma(\frac{1}{2}Z)}{\Gamma(2-\frac{1}{2}Z)} \prod_{i=1,2} \frac{\Gamma(\delta_i-\frac{1}{2}Z)}{\Gamma(\delta_i)} . \tag{9.7}$$

We see that ineqs. (9.2) and (9.6) have essentially the same effect: if two propagators have a power larger than a certain coefficient they allow us to obtain as a factor a corresponding "propagator" for the momentum in that particular channel. This is how proving the ineqs. of Fig. 10 can be reduced to purely algebraic manipulations. We discovered that ineq. (8.15) is again crucial. We notice that if the internal propagators of the quadrangle were elementary ones then the superficial degree of convergence of any of the other subgraphs of our diagram may change slightly, since in eq. (5.10) $\beta > 2\alpha$, but the left hand side of eq. (5.11) remains unchanged, because

$$\Delta E_1 = -2\Delta E_2 , \tag{9.8}$$

so our condition that all subgraphs be convergent remains fulfilled after the substitution of the inequalities of Fig. 10.

However, if one of the internal lines of the quadrangle had been a composite one ($\bullet\!\!\xrightarrow{\;2\;}\!\!\bullet$), then a subgraph would become more divergent, because we are unable to continue our scheme with something like a three-particle composite propagator ($\bullet\!\!\xrightarrow{\;3\;}\!\!\bullet$). A crucial point of our argument is that we will never really need such a thing, if we attack the quadrangle subgraphs in the right order.

10. PENTAGONS. CONVERGENCE OF THE SKELETON EXPANSION

As stated before, the order in which we reduce our diagram

into a tree diagram is:

1) remove triangular facets;
2) remove triangular subgraphs if any. By complete induction we prove this to be possible. Take a minimal triangular subgraph and go to 3;
3) remove quadrangular facets as far as possible. If any cannot be removed because of a crucial composite propagator in them, then
4) remove qudrangular subgraphs. After that we only have to
5) remove the pentagons.
6) If we happened to be dealing with a subgraph by branching at point 2 or 3, then by now that will have become a tree graph, because of the theorem in sect. 7. Go back to 1.

We still must verify point 5. If indeed our whole diagram contains pentagons then we can replace all propagators by elementary ones. But if we had branched at steps 2 or 3 then the subgraphs we are dealing with may still have composite external propagator(s). In that case it is easy to verify that there will be enough pentagons buried inside our subgraphs that do not need composite external lines. In that case we apply directly the inequality of Fig. 13. The procedure for proving Fig. 13 is

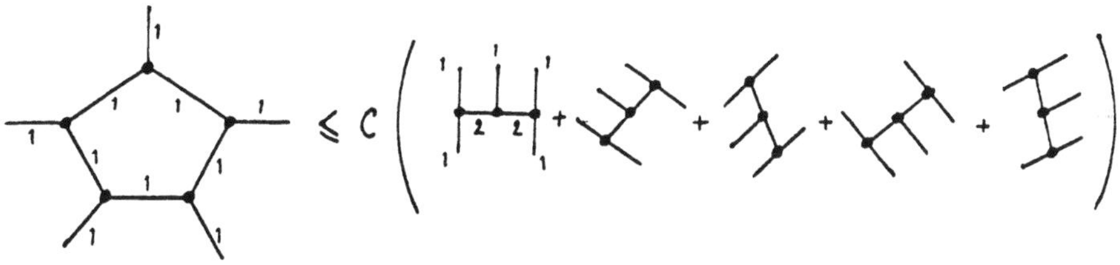

Fig. 13. Inequality for pentagons.

exactly as described for the quadrangles in the previous section. Again the degree of divergence of any of the adjacent subgraphs has not changed significantly. This now completes our proof by induction that any planar skeleton diagram with 5 external lines is equal to C^L times a diagram with type IV Feynman rules, where C is limited to fixed bounds. Since also the number of diagrams is an exponential function of L we see that for this set of graphs perturbation expansion in g has a finite radius of convergence. The proof given here is slightly more elegant than in Ref. 11, and also leads to tree graph expressions that are more useful for our manipulations.

If the diagram has 6 or more external lines then still a number of facets may be left, limited by ineq. (7.10), all having 6 or more propagators. If we wish we can still continue our procedure for these but that would be rather pointless: having a limited number of loops the diagram is finite anyhow. The difficulty would not so much be that no inequalities for hexagons etc. could be written down; they certainly exist, but our problem would be that the corresponding number C would not obviously be bounded by one universal constant. This is why our procedure would not work for nonplanar theories where ineq. (7.10) does not hold. In the non-planar case however similar theorems as ours have been derived[6].

11. BASIC GREEN FUNCTIONS

The conclusion of the previous section is that if we know the "basic Green functions", with which we mean the two- three- and four-point functions, and if these fall within the bounds given in Table 1, then all other Green functions are uniquely determined by a convergent sum. Clearly we take the value for the bound g^2 for the coupling constant (ineq. (5.9)) as determined by the inverse product of the coefficient C_2 found in sect. 6 and the maximum of the coefficients in the ineqs. pictured in Figs. 9, 10 and 13, times a combinatorial factor.

Now we wish not only to verify whether these bounds are indeed satisfied, but also we would like to have a convergent calculational scheme to obtain these basic Green functions. One way of doing this would be to use the Dyson-Schwinger equations. After all, the reason why those equations are usually unsoluble is that they contain all higher Green functions for which some rather unsatisfactory cut-off would be needed. Now here we are able to re-express these higher Green functions in terms of the basic ones and thus obtain a closed set of equations.

These Dyson-Schwinger equations however contain the bare coupling constants and therefore require subtractions. It is then hard to derive bounds for the results which depend on the difference between two (or more) divergent quantities. We decided to do these subtractions in a different way, such that only the finite, renormalized basic Green functions enter in our equations, not the bare coupling constants, in a way not unlike the old "bootstrap" models. Our equations, to be called "difference equations" will be solved iteratively and we will show that our iteration procedure converges. So we start with some *Ansatz* for the basic Green functions and derive from that an improved set of values using the difference equations. Actually this will be done in various steps. We start with assuming some function $g(x)$ for the *floating coupling constant*, where x is the momentum in the maximal channel (see Table 1):

$$x = \max_{i,j} |p_i - p_j| \ , \tag{11.1}$$

and a set of functions $g_{(i)}(x)$ with

$$g(x) \underset{\text{def}}{=} \max_{i} |g_{(i)}(x)| \ . \tag{11.2}$$

Here $g_{(i)}(x)$ is the set of independent numbers that determined the basic Green function at their "symmetry point":

$$|p_i - p_j| = x \quad \text{for all } i,j \ . \tag{11.3}$$

The index i in $g_{(i)}$ then simply counts all configurations in (11.3). With "independent" we mean that in some gauge theories we assume that the various Ward-Slavnov-Taylor[13] identities among the basic Green functions are fulfilled. This is not a very crucial point of our argument so we will skip any further discussion of these Ward or Slavnov-Taylor identities.

If the values of the basic n-point Green functions (n = 3 or 4) at their symmetry points are $A_i(x)$, then the relation between A_i and g_i is:

$$\begin{aligned} A_{3i}(x) &= \kappa^j_{i\mu}\, p_{j\mu}\, Z^{-3/2}(x)\, g_{ij}(x) \\ A_{4i}(x) &= Z^{-2}(x)\, g^2_{4i}(x) \ , \end{aligned} \tag{11.4}$$

where $\kappa^j_{i\mu}$ are coefficients of order one, and Z(x) is defined in (5.1) and (5.2). (We ignore for a moment the case of super-renormalizable couplings.) Our first Ansatz for $g_i(x)$ is a set of functions that is bounded by (11.2), with g(x) decreasing asymptotically to zero for large x as dictated by the lowest order term(s) of the renormalization group equations. We will find better equations for $g_i(x)$ as we go along. In any case we will require

$$\left|\frac{xd}{dx}\, g_i(x)\right| \leqslant \tilde{\beta}\, g^3(x) \tag{11.5}$$

for some finite coefficient $\tilde{\beta}$.

Our first *Ansatz* for the basic Green functions away from the symmetry points will be even more crude. All we know now is that they must satisfy the bounds of Table 1. In general one may start with choosing (11.4) to hold even away from the symmetry points, and

$$x = \max_{i,j} |p_i - p_j| \ . \tag{11.6}$$

After a few iterations we will get values still obeying the bounds of Table 1, and with *uncertainties* also given by Table 1 but K_i replaced by coefficients δK_i. Thus we start with

$$\delta K_i^{(0)} = K_i \ . \tag{11.7}$$

We will spiral towards improved *Ansätze* for the basic Green functions in two movements:

i) the "small spiral" is the use of difference equations to obtain improved values at exceptional momenta, *given the values $g_i(x)$ at the symmetry points*. These difference equations will be given in the next section.
ii) The "second spiral" is the use of a variant of the Gell-Mann-Low equation to obtain improved functions $g_i(x)$ from previous Ansätze for $g_i(x)$, making use of the convergent "small spiral" at every step. What is also needed at every step here is a set of integration constants determining the boundary condition of this Gell-Mann-Low equation. It must be ensured that these are always such that g(x) in ineq. (11.2) remains bounded:

$$g(x) < g_0 \ ,$$

where g_0 is limited by the coefficients K_i and the various coefficients C from sects. 6, 8, 9 and 10, as in ineq. (5.9).

12. DIFFERENCE EQUATIONS FOR BASIC GREEN FUNCTIONS

The Feynman rules of our set of theories must follow from a Lagrangian, as usual.
For brevity we ignore the Lorentz indices and such, because those details are not of much concern to us. Let the dressed propagator be

$$P(p) = -G_2^{-1}(p) \ , \tag{12.1}$$

and let the corresponding zeroth order, bare expressions be indicated by adding a superscript o. In massive theories:

$$P^0(p) = (p^2+m^2)^{-1} = -G_2^{0-1}(p) \ . \tag{12.2}$$

Define

$$G_2(p+k) - G_2(p) = G_{2\mu}(p|k)k_\mu \ , \tag{12.3}$$

so that

$$P(p+k) - P(k) = P(p+k)G_{2\mu}(p|k)k_\mu P(p) \ . \tag{12.4}$$

This gives us the "Feynman rule" for the difference of two dressed

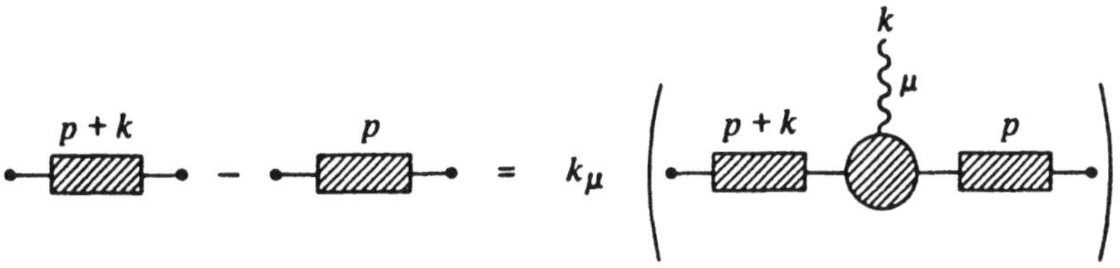

Fig. 14. Feynman rule for the difference of two dressed propagators. The 3-vertex at the right is the function $G_{2\mu}(p|k)$.

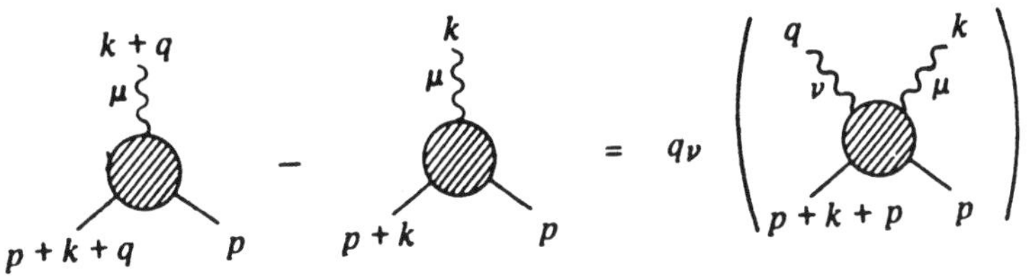

Fig. 15. Difference equation (12.6) for $G_{2\mu}$.

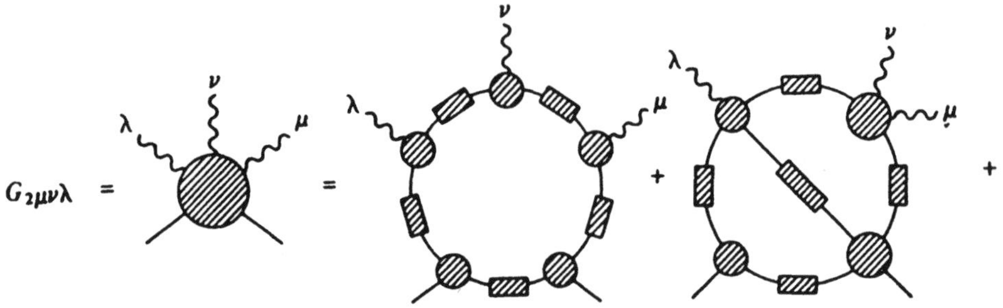

Fig. 16. Some arbitrarily chosen terms in the skeleton expansion for $G_{2\mu\nu\lambda}$.

propagators, depicted in Fig. 14. (Note that, in this section only, p and k denote external line momenta, not external loop momenta.)

We have also this Feynman rule for bare propagators. There $G^0_{2\mu}$ follows directly from the Lagrangian:

$$G^0_{2\mu} = -2p_\mu - k_\mu \ . \tag{12.5}$$

Continuing this way we define

$$G_{2\mu}(p|k+q) - G_{2\mu}(p|k) = G_{2\mu\nu}(p|k|q)q_\nu \ , \tag{12.6}$$

with

$$G^0_{2\mu\nu} = -\delta_{\mu\nu} \ . \tag{12.7}$$

In Feynman graphs this is sketched in Fig. 15. Differentiating once more we get

$$G_{2\mu\nu}(p|k|q+r) - G_{2\mu\nu}(p|k|q) = G_{2\mu\nu\lambda}(p|k|q|r)r_\lambda \ . \tag{12.8}$$

Of course $G_{2\mu\nu\lambda}$ can be computed formally in perturbation expansion. The rules for computing the new Green functions G_μ, $G_{\mu\nu}$, $G_{\mu\nu\lambda}$ are easy to establish. Let p_1 be one of the external loop momenta as defined in eq. (5.3). For a Green function $G(p_1)$ we have

$$G(p_1) = \int \dots \int dq_i \ f_1(p_1+q_1)f_2(p_1+q_2) \ \dots \ f_t(p_1+q_t).F \ , \tag{12.9}$$

where $f_i(q_i)$ are bare vertex and/or propagator functions adjacent to the external facet labeled by 1. The remainder F is independent of p_1. We write

$$G(p_1+k) - G(p_1) = \int \ldots \int F\,dq \left(\prod_1^t f(p_1+q_i+k) - \prod_1^t f(p_1+q_i) \right)$$

$$= \int \ldots \int F\,dq \sum_{s=1}^{t} \prod_1^{s-1} f(p_1+q_i+k) \left(f(p_1+q_s+k) - f(p_1+q_s) \right) \prod_{s+1}^{t} f(p_1+q_i), \tag{12.10}$$

which is just the rule for taking the difference of two products. We find the difference of two *dressed* Green functions G in terms of the difference of *bare* functions f. Therefore the "Feynman rules" for the diagrams at the right hand sides of Figs. 14, 15 and the l.h.s. of Fig. 16 consist of the usual combinatorial rules with new bare vertices given by the eqs. (12.5) and (12.7). These bare vertices occur only at the edge of the diagram.

We see that the power counting rules for divergences in $G_{2\mu\nu\lambda}$ are just as in 5-point functions in gauge theories. Since the global degree of divergence is negative we can expand in skeleton graphs. See Fig. 16, in which the blobs represent ordinary dressed propagators and dressed vertices or dressed functions G_μ and $G_{\mu\nu}$.

Notice that one might also need $G_{3\mu}(p_1,p_2|k)$ defined by

$$G_3(p_1,p_2+k) - G_3(p_1,p_2) = G_{3\mu}(p_1,p_2|k).k_\mu \ . \tag{12.11}$$

In short, the skeleton expansion expresses $G_{2\mu\nu\lambda}$ but also $G_{3\mu\nu}$ etc. in terms of the few basic functions $G_{2\mu}$, $G_{2\mu\nu}$, $G_{3\mu}$ and the basic Green functions $G_{2,3,4}$. Also the function $G_{4\mu}$, defined similarly, can thus be expressed. The corresponding Feynman rules should be clear and straightforward.

We conclude that the basic Green functions can in turn be expressed in terms of skeleton expansions, and, up to overall constants, these equations, if convergent, determine the Green functions completely. Notice that we never refer to the bare Lagrangian of the theory, so, perhaps surprisingly, these sets of equations are the same for all field theories. The difference between different field theories only comes about by choosing the integration constants differently.

Planarity however was crucial for this chapter, because only planar diagrams have well defined "edges": the new vertices only occur at the edge of a diagram.

13. FINDING THE BASIC GREEN FUNCTIONS AT EXCEPTIONAL MOMENTA (THE "SMALL SPIRAL")

In this section we regard the basic Green functions at their symmetry points as given, and use the difference equations of

Sect. 12 to express the values at exceptional momenta in terms of these. If $p_i - p_j$ is the momentum flowing through the planar channel ij, then in our difference equations we decide to keep

$$\mu = \max_{i,j} |p_i - p_j| \tag{13.1}$$

fixed. So the left hand side of our difference equations will show two Green functions with the same value for μ, one of which may be exceptional and the other at its symmetry point, and therefore known. (We use the concept of "exceptional momenta" as in ref. 14.)

Now the right hand side of these difference equations show a skeleton expansion of diagrams which of course again contain basic Green functions, also at exceptional momenta. But these only come in combinations of higher order, and the effect of exceptional momenta is relatively small, so at this point one might already suspect that when these equations are used recursively to determine the exceptional basic Green functions then this recursion might converge. This will indeed be the case under certain conditions as we will show in sect. 15.

Our iterative procedure must be such that after every step the bounds of Table 1 again be satisfied. This will be our guide to define the procedure. First we take the 4-point functions, and consider all cases of Table 1 separately.

The right hand side of our difference equations (Fig. 16) contains a skeleton expansion to which we apply the theorem mentioned in the end of sect. 5 and proven in sects. 5-10: the skeleton expansion for any 5-point Green function converges and is bound by tree diagrams constructed with type IV Feynman rules. Since the 5-point functions in Fig. 16 are irreducible, the internal lines in the resulting tree graph will always be composite propagators, as in the r.h.s. of Fig. 13. So we simply apply the type IV Feynman rules for 5 tree graphs to obtain bounds on the 5-point function in various exceptional regions of momentum space. Table 2 lists the results.

The power of $g^2(A_3)$ in the table applies where we consider the function $G_{4\mu}$. The other functions $G_{3\mu\nu}$ and $G_{2\mu\nu\lambda}$ have one and zero powers of $g(A_3)$, respectively. In front of all this comes a power series of the form

$$\sum_{n=1}^{\infty} C^n g_0^{2n} = Cg_0^2(1-Cg_0^2)^{-1} , \tag{13.2}$$

which converges provided that

Table 2

Bounds for the irreducible 5-point function at some exceptional momentum values.

$(((12)_1\ 53)_2\ 4)_3$	$\prod_i Z_{ii+1}^{-\frac{1}{2}}\ A_1^{-\alpha}\ A_2^{-1+\alpha-\beta}\ A_3^{\beta}\ g^2(A_3)$
$((12)_1\ 5\ (34)_2)_3$	$\prod_i Z_{ii+1}^{-\frac{1}{2}}\ A_1^{-\alpha}\ A_2^{-\alpha}\ A_3^{2\alpha-1}\ g^2(A_3)$
$(((513)_1\ 2)_2\ 4)_3$	$\prod_i Z_{ii+1}^{-\frac{1}{2}}\ A_1^{-1-2\beta}\ A_2^{\beta}\ A_3^{\beta}\ g^2(A_3)$
$((135)_1\ (24)_2)_3$	$\prod_i Z_{ii+1}^{-\frac{1}{2}}\ A_1^{-1-2\beta}\ A_3^{2\beta}\ g^2(A_3)$ if $A_2 > A_1$
$((1235)_1\ 4)_2$	$\prod_i Z_{ii+1}^{-\frac{1}{2}}\ A_1^{-1-\beta}\ A_2^{\beta}\ g^2(A_3)$

$$g_0 = \max_{\mu} |g(\mu)| < C^{-\frac{1}{2}} . \tag{13.3}$$

We now write an equation such as (12.8) as follows:

$$\tilde{G}_4\{(((12)_1\ 3)_2\ 4)_3\} = G_4\{((523)_2\ 4)_3\} + \\ + (p_1-p_5)_\mu\ G_{4\mu}\{(((12)_1\ 53)_2\ 4)_3\} , \tag{13.4}$$

where $r = p_1-p_5$. In this and following expressions the tilde (~) indicates which quantities are being replaced by new ones in the iteration procedure.

If the Ansatz holds for $G_4\{((523)_2\ 4)_3\}$ then the new exceptional function will obey

$$|\tilde{G}_4\{(((12)_1\ 3)_2\ 4)_3| \leqslant K_6^2(Z_{45}Z_{52}Z_{23}Z_{34})^{-\frac{1}{2}} \left(\frac{A_3}{A_2}\right)^{\beta} g^2(A_3) + \\ + (Z_{45}Z_{12}Z_{23}Z_{34})^{-\frac{1}{2}} \frac{Cg_0^2}{1-Cg_0^2}\ g^2(A_3) \left(\frac{A_2}{A_1}\right)^{\alpha} \left(\frac{A_3}{A_2}\right)^{\beta} . \tag{13.5}$$

Choosing
$$\frac{Cg_0^2}{1-Cg_0^2} \equiv \gamma \tag{13.6}$$

and considering that to a good approximation (since $|p_1-p_5| << |p_4-p_5|$):

$$Z_{45} = Z_{41} , \tag{13.7}$$

we find

$$|\tilde{G}_4\{(((12)_1\ 3)_2\ 4)_3\}| \leqslant (Z_{12}Z_{23}Z_{34}Z_{41})^{-\frac{1}{2}}\ g^2(A_3)\left(\frac{A_2}{A_1}\right)^\alpha \left(\frac{A_3}{A_2}\right)^\beta \cdot$$

$$\cdot \left(\gamma + \left(\frac{Z_{12}}{Z_{52}}\right)^{\frac{1}{2}} \left(\frac{A_1}{A_2}\right)^\alpha K_6^2\right) . \tag{13.8}$$

What is now needed is a bound for the last term in (13.8). Let

$$x_{12} = |p_{12}|/m \geqslant 1 \tag{13.9}$$

(remember that $|p|$ stands for $\sqrt{p^2+m^2}$),
and

$$f(x_{12}) = (\log(1+x_{12})^{\sigma/2} \cdot x_{12}^\alpha , \tag{13.10}$$

where σ is defined in (5.2).
When

$$x_{12} > \exp\ (-\sigma/2\alpha) = x_0 \tag{13.11}$$

this f is an increasing function, so that if

$$x_0 m \leqslant A_1 < A_2 \tag{13.12}$$

then

$$\frac{f(x_{12})}{f(x_{52})} < 1 . \tag{13.13}$$

The range $1 \leqslant x \leqslant x_0$ is compact, so there exists a finite number L such that

$$\frac{f(x_{12})}{f(x_{52})} \leqslant L \tag{13.14}$$

as soon as

$$x_{12} < x_{52} . \tag{13.15}$$

So we find that after one iteration given by (13.4), the new K_2 coefficient satisfies

$$K_2^2 \leqslant \gamma + K_6^2 L . \tag{13.16}$$

Similarly we derive

$$K_7^2 \leqslant \gamma + L \ , \tag{13.17}$$

when the difference equation is used to express $\tilde{G}\{((12)_1\ 34)_2\}$ in terms of $G\{(5234)_2\}$. Also we use

$$\tilde{G}_4\{((12)_1\ (34)_2)_3\} = G_4\{(52(34)_2)_3\} + $$
$$+ (p_1-p_5)_\mu\ G_{4\mu}\{((12)_1\ 5(34)_2)_3\} \ , \tag{13.18}$$

to find that after one step

$$K_3^2 \leqslant \gamma + K_7^2 L \leqslant \gamma + \gamma L + L^2 \ ; \tag{13.19}$$

and for the three-point function

$$K_1 \leqslant K_7^2 + L \leqslant \gamma + 2L \ . \tag{13.20}$$

The remaining coefficients K_{4-6} must be computed in a slightly different way. Consider K_4. We replace p_1 by p_5 now in such a way that

$$|p_5-p_3| \simeq 2|p_1-p_3| \ ; \qquad A_1 \to 2A_1 \tag{13.21}$$

and work with induction. Write

$$\tilde{\tilde{G}}_4\{(((13)_1\ 2)_2\ 4)_3\} = \tilde{G}_4\{(((53)_1\ 2)_2\ 4)_3\}$$
$$+ (p_1-p_5)_\mu\ G_{4\mu}\{(((513)_1\ 2)_2\ 4)_3\} \ . \tag{13.22}$$

Inspecting Tables 1 and 2 we find now

$$K_4^2 = \max\left(K_6^2 \ , \ \frac{\gamma}{1-2^{-2\beta}}\right) . \tag{13.23}$$

Applying the same technique we compute the fifth exceptional configuration of Table 1. We separate p_1 from p_3 until $A_1 \to A_2$. Then we separate alternatively p_2 from p_4 and p_1 from p_3 keeping $A_1 \simeq A_2$. This makes the rate of convergence slightly slower:

$$K_5^2 = \frac{2\gamma}{1-2^{-2\beta}} \ . \tag{13.24}$$

Finally K_6 is found by widening the separation between p_1, p_2 and p_3 in successive steps of factors of 2:

$$\tilde{\tilde{G}}_4\{((123)_1\ 4)_2\} = \tilde{G}_4\{((563)_1\ 4)_2\} \tag{13.25}$$
$$+ (p_5-p_1)_\mu\ G_{4\mu}\{((1235)_1\ 4)_2\} + (p_6-p_2)_\mu\ G_{4\mu}\{((2356)_1\ 4)_2\} \ ,$$

where

$$|p_5-p_6| = |p_3-p_6| = |p_5-p_3| \simeq 2|p_1-p_2| \,. \tag{13.26}$$

We find

$$|\tilde{G}_4\{((123)_1\ 4)_2|\} \leqslant \gamma Z(A_1)^{-1}\ Z(A_2)^{-1}\left(\frac{A_2}{A_1}\right)^{\beta} \,.$$

$$\sum_{n=1}^{^2\log\,(A_2/A_1)} \left(2^{-\beta n}\,.\,2\,\frac{Z(A_1)}{Z(2^n A_1)}\right) + |G_4\{(1234)_2\}| \,. \tag{13.27}$$

The sum can certainly be bounded:

$$\sum \leqslant L' \leqslant 2L(1-2^{\alpha-\beta})^{-1} \,. \tag{13.28}$$

Therefore

$$K_6^2 \leqslant \max(1,\gamma L') \,. \tag{13.29}$$

Thus all coefficients K_i have bounds that will be obeyed everywhere in the "small spiral" induction procedure. Note that these coefficients would blow up if $\alpha,\beta \to o$. In particular in (13.11) we need $\alpha > o$. Only if $g_0^2 \to o$ we can let $\alpha,\beta \to o$. It will be clear from the above arguments that our bounds are only very crude. Our present aim was to establish their existence and not to find optimal bounds.

In sect. 15 we show that the "small spiral" of iterations for the exceptional Green functions, given the non-exceptional ones in (13.27), actually converges geometrically.

14. NON-EXCEPTIONAL MOMENTA (THE "SECOND SPIRAL")

In order to formulate the complete recursion procedure for determining the basic Green functions we need relations that link these Green functions at different symmetry points. Again the difference equations are used:

$$G_4(p_1\ldots p_4) = G_4(2p_1,p_2p_3p_4) - p_\lambda G_{4\lambda}(p_1,2p_1,p_2p_3p_4)$$

$$= \ldots = G_4(2p_1,\ldots,2p_4) - \sum_{i=1}^{4} p_{i\lambda}G_{4\lambda}(p_1^{(i)},\ldots,p_5^{(i)}). \tag{14.1}$$

Here p_i and $p_j^{(i)}$ are external loop momenta. They are non-exceptional. We use a shorthand notation for (14.1). Writing $p_i^2 \simeq (p_i-p_j)^2 \simeq \mu^2$:

$$G_4(\mu) - G_4(2\mu) = \mu \sum_{i=1}^{4} G^{(i)}_{4\lambda}(2\mu,\mu) \ . \tag{14.2}$$

Similarly we have

$$G_{2,3}(\mu) - G_{2,3}(2\mu) = - \mu \sum_{i} G^{(i)}_{2,3\lambda}(2\mu,\mu) \ . \tag{14.3}$$

These are just discrete versions of the renormalization group equations. The right hand side of (14.2), [not (14.3)!] is to be expanded in a skeleton expansion which contains all basic Green functions at all μ, also away from their symmetry points. There we insert the values obtained after a previous iteration. It is our aim to derive from eq. (14.2) a Gell-Mann-Low equation[1] of the form

$$\frac{\mu\partial}{\partial\mu} g_i(\mu) = - \sum_{\ell=2}^{k} \beta^{(\ell)}_{ij_1\ldots j_\ell}\, g_{j_1}(\mu)\ldots g_{j_\ell}(\mu) + |g(\mu)|^N \rho_i(\mu) \ , \tag{14.4}$$

where $\beta^{(\ell)}$ are the first k coefficients of the β function, and they must coincide with the perturbatively computed β coefficients. Often (depending on the dimension of the coupling constant) only odd powers occur so that $k = N-2$ is odd. The rest function ρ must satisfy

$$|\rho_i(\mu)| \leqslant Q_N \tag{14.5}$$

for some constant $Q_N < \infty$. This inequality must hold in the sense that $|g(\mu)|^N \rho(\mu)$ must be a convergent expansion in the functions $g(\mu')$, with

$$\mu' \geqq m \tag{14.6}$$

(so that μ' may be smaller than μ), in such a way that the absolute value of each diagram contributes to Q_N and their total sum remains finite.

Now clearly eq. (14.2) is a difference equation, not a differential equation such as (14.4). Up till now differential equations were avoided because of infrared divergences. Just for ease of notation we have put (14.4) in differential form because the mathematical convergence questions that we are to consider now are insensitive to this simplification.

Consider the skeleton expansion of $G^{(i)}_{4\lambda}$ in (14.2). At each of the four external particle lines a factor $g(\mu_i)$ occurs with $\mu_i \geq \mu$, so it may seem easy to prove (14.4) from (14.2) with N = 3 or 4. However, we find it more convenient* to have an equation of the form (14.4) with $N \leq 7$, and our problem is that the internal vertices of the $G^{(i)}_{4\lambda}$ might have momenta which are less than μ. We will return to this question later in this section.

In proving the difference equation variant of (14.4) from (14.2) we have to make the transition from G_4 to g^2 and G_3 to g, and this involves the coefficients $Z(\mu)$, associated to the functions G_2, by equations of the form

$$G_2(\mu) = -\mu^2 Z^{-1}(\mu) \quad ;$$

$$G_3(\mu) = \mu Z^{-3/2}(\mu) g_3(\mu) \quad ; \tag{14.7}$$

$$G_4(\mu) = Z^{-2}(\mu) g_4^2(\mu) \quad .$$

where g_3, g_4 are just various components of the coupling constant g_i. In the following expressions we suppress these indices i when we are primarily interested in the dependence on μ (= $|p|$ at the symmetry point). Now from (14.2) and (14.3) we find not first order but third order differential equations for G_2, basically of the form

$$\frac{\partial^3}{\partial\mu^3} G_2 = G_{2,\lambda\lambda\lambda} = \mathcal{O}(g^2(\mu)Z^{-1}(\mu)/\mu) \ , \tag{14.8}$$

where $G_{2,\lambda\lambda\lambda}$ is just a shorthand notation for the combination of expandable functions $G_{2,\lambda\mu\nu}$ obtained after taking differences three times. Write

$$U_2(\mu) = -\frac{\partial^2}{\partial\mu^2} G_2(\mu) = -G_{2\lambda\lambda}(\mu) \ , \tag{14.9}$$

then

$$\frac{\mu\partial}{\partial\mu} U_2^{\prime}(\mu) = -\mu G_{2,\lambda\lambda\lambda}(\mu) \ , \tag{14.10}$$

and

$$\mu^2 Z^{-1}(\mu) = \int_m^n (\mu-\mu_1)U_2(\mu_1)d\mu_1 + A\mu + B \ . \tag{14.11}$$

* Closer analysis shows that actually N = 3 or 4 is sufficient to prove unique solubility. Only if we wish an exact, non-perturbative definition of the free parameters we need the higher N values. Note that not only Q_N but also g_0 may deteriorate as N increases.

Here A and B are free integration constants; A is usually determined by Lorentz invariance and B by the mass, fixed to be equal to m. In lowest order:

$$A = mU_2(m) \; ; \quad B = -\tfrac{1}{2}m^2U_2(m) \; . \tag{14.12}$$

This strange-looking form of the integration constants is an artifact coming from our substitution of difference equations by differential equations. Using difference equations we can impose Lorentz invariance by symmetrization in momentum space, so that only one (for each particle) integration constant is left: the mass term. We choose at all stages $\frac{1}{2}U_2(m) = Z(m) = 1$.

A convenient way to implement eq. (14.12) is to formally define $U_2(\mu) = 2$ if $o \leq \mu \leq m$, and replace the lower bound of the integral in (14.11) by zero. Then after symmetrization: $A = B = o$.

Equation (14.11) has a linearly convergent integral, whereas (14.10) is logarithmic. Together they determine the next iterative approximation to G . In fact we have

$$\mu G_{2,\lambda\lambda\lambda}(\mu) = Z^{-1}(\mu)f(\{g\}) \; , \tag{14.13}$$

and in $f(\{g\})$, Z occurs only indirectly. So the iteration converges fastest if we replace (14.10) by

$$\frac{\mu\partial}{\partial\mu}\tilde{U}_2(\mu) = -\frac{Z(\mu)}{\tilde{Z}(\mu)}G_{2,\lambda\lambda\lambda}(\mu) \; , \tag{14.14}$$

where the tilde denotes the new function $U_2(\mu)$.

One can however also use (14.9) with U_2 replaced by $\tilde{U}_2$.

We find

$$\frac{\mu\partial}{\partial\mu}Z^{-1} = -\int_{m/\mu}^{1} d\tau(1-\tau)\mu G_{2,\lambda\lambda\lambda}(\tau\mu) \; . \tag{14.15}$$

As stated before, the $\mathcal{O}\left(\frac{m}{\mu}\right)$ terms have been removed by symmetrization.

This equation allows us to remove the Z factors from the functions $G_{3,4}$ and arrive at first order renormalization group integrodifferential equations for $g_i(\mu)$.

For the 3-point functions we must write

$$U_3(\mu) = G_{3,\lambda}(\mu) = \frac{\partial G_3}{\partial\mu} \; ,$$

$$\frac{\mu\partial}{\partial\mu} U_3(\mu) = \mu G_{3,\lambda\lambda}(\mu) \ , \tag{14.16}$$

$$G_3(\mu) = \int_m^\mu \mu U_3(\mu) d\mu + C_3 \ , \tag{14.17}$$

$$\frac{\mu\partial G_3(\mu)}{\partial\mu} = \int_{m/\mu}^{1} d\tau \ \mu G_{3,\lambda\lambda}(\tau\mu) \ . \tag{14.18}$$

A potential difficulty in writing down the renormalization group equation even for N = 4 is the convolutions in (14.15) and (14.18) which contain Green functions at lower μ values, and so they depend on $g(\mu')$ with $\mu' < \mu$. So a further trick is needed to derive (14.4). This is accomplished by realizing that the integrals in (14.15) and (14.18) converge linearly in μ. Suppose we require at every iteration step (see eq. 11.5):

$$\left|\frac{\mu\partial}{\partial\mu} g(\mu)\right| \leqq \tilde{\beta}|g(\mu)|^3 \quad \text{and} \quad |g(\mu)| \leqq g_0 \tag{14.19}$$

for some $\beta < \infty$, $g_0 < \infty$. Then it is easy to show that if $\mu_1 \leqq \mu$, then

$$|g(\mu_1)| \leqq |g(\mu)| + C\left(\frac{\mu}{\mu_1}\right)^{\varepsilon} |g^3(\mu)| \ , \tag{14.20}$$

if

$$\varepsilon \geqq 3\tilde{\beta} g_0^2 + \tilde{\beta}/C \ . \tag{14.21}$$

So with C large enough and g_0 small enough we can make ε as small as we like. Inequality (14.20) is proven by differentiating with μ. This enables us to replace $g(\tau\mu)$ by $g(\mu)$ in (14.15) and (14.18) while the factor $\tau^{-\varepsilon}$ does no harm to our integrals. So we find bounds for $\frac{\mu\partial}{\partial\mu}\tilde{Z}^{-1}$ and $\frac{\mu\partial}{\partial\mu}\tilde{G}_3$ in terms of a power series of $g(\mu)$. We must terminate the series as soon as the factors $\tau^{-\varepsilon}$ accumulate to give τ^{-1}. This implies that N must be kept finite, otherwise $g_0 \to o$.

The same inequality (14.16) is used to go from N = 4 to N = 7 in these equations. If in a skeleton diagram a vertex is not associated with any external line, then it may be proportional to a factor $g(\mu')$ with $\mu' < \mu$. But using (14.16) we see that it may be replaced by $g(\mu)$ at the cost of a factor $(\mu/\mu_1)^{\varepsilon}$. At most three of these extra factors are needed. If the three corresponding vertices are chosen not to be too far away from one of the external vertices of the diagram (which we can always arrange), then this just corresponds to inserting an extra factor

$\left(\frac{p^{ext}}{p_1}\right)^{\varepsilon}$ at an external vertex. We now note that such factors still leave our integrals convergent. In the ultraviolet of course the diagrams converge even better than they already did, and in the infrared our degree of convergence was at least $1-\alpha$ (or $2-\beta$) as can easily be read off from eq. (5.10): adding the external propagators to any diagram one demands

$$Z + (2+2\alpha)E_1 + 2\beta E_2 < 4(E_1+E_2-1) \ . \tag{14.22}$$

Thus infrared convergence requires

$$1 - \alpha - T\varepsilon > o \tag{14.23}$$

where $T \lesssim 5$ is the number of times our inequality (14.20) was applied.

From the above considerations we conclude that an equation of the form (14.4) can be written down for any finite N, such that Q_N in inequality (14.5) remains finite. We do expect of course that Q_N might increase rapidly with N, but then we only want the equation for $N \leqq 7$. We are now in a position to formulate completely our recursive definition of the Green functions G_2, G_3, G_4 of the theory:

1) We start with a given set of trial functions $G_2(\mu)$, $G_3(\mu)$, $G_4(\mu)$ for the basic Green functions at their symmetry points. They determine our initial choice for the floating coupling constants $g_i(\mu)$ and the functions $Z_i(\mu)$. We require their asymptotic behaviour to satisfy (5.2), (5.9) and (11.5) (= eq. (14.19)).

2) We also start with an *Ansatz* for the exceptional Green functions that must obey the bounds of Table 1.

3) Use the difference equations of sect. 13 to improve the exceptional Green functions (the new values are indicated by a tilde ($\sim$))^l. These will again obey Table 1 as was shown in sect. 13. Repeat the procedure. It will converge towards fixed values for the exceptional Green functions (as we will argue in sect. 15). This we call the "small spiral".

4) With these values for the exceptional Green functions we are now able to compute the right hand side of the renormalization group equation for G_2, or rather Z^{-1}, from (14.15), using (14.20):

$$\frac{\mu\partial}{\partial\mu}\tilde{Z}_i^{-1}(\mu) = \tilde{Z}_i^{-1}(\mu)\left(\gamma_{ijk}g_j(\mu)g_k(\mu) + g^4(\mu)\sum(\mu)\right) , \tag{14.24}$$

where $\sum(\mu)$ is again bounded. Here γ_{ijk} are the one-loop γ coefficients[14] This gives us *improved* propagators. See sect. (15.b).

5) Now we can compute the right hand side of eq. (14.4). Before integrating eq. (14.4) it is advisable to apply Ward identities (if we were dealing with a gauge theory) in order to reduce the number of independent degrees of freedom at each μ. As is well known, in gauge theories one can determine all subtraction constants this way except those corresponding to the usual free coupling constants and gauge fixing parameters[15]. So the number of unknown functions $g_i(\mu)$ need not exceed the number of "independent" coupling constants of the theory*.

6) Eq. (14.4) is now integrated, giving improved expressions for $g_i(\mu)$. Now go back to 2. This is the "second spiral", which will be seen to converge towards fixed values of $g_i(\mu)$.

The question of convergence of these two spirals is now discussed in the following section.

15. CONVERGENCE OF THE PROCEDURE

a) Exceptional Momenta

In sect. 13 a procedure is outlined to obtain the Green functions at exceptional momenta, if the Green functions at the symmetry point are given. That procedure is recursive because eqs. (13.4), (13.18), (13.22) and (13.25) determine the Green functions $G_{2,3,4}$ in terms of the symmetry ones, and $G_{4\mu}$, $G_{3\mu\nu}$, $G_{2\mu\nu\lambda}$. But the latter still contain the previous ansatz for $G_{2,3,4}$. Fortunately it is easy to show that any error $\delta G_{2,3,4}$ will reduce in size, so that here the recursive procedure converges:

Let us indicate the bounds discussed in sects. 5 and 13 as

$$|G_n(p_1,\ldots,p_n)| \leqq B_n(p_1,\ldots,p_n) \,, \tag{15.1}$$

and assume that a first trial $G_n^{(1)}$ has an error

$$|\delta G_n^{(1)}| \leqq \varepsilon^{(1)} B_n \,, \tag{15.2}$$

with some $\varepsilon^{(1)} \leqq 2$.

Now $G_{4\mu}$, $G_{3\mu\nu}$, $G_{2\mu\nu\lambda}$ also satisfy inequalities of the form (15.1). Furthermore they were one order higher in g^2. So we have

* We put "independent" between quotation marks because our requirement of asymptotic freedom usually gives further relations among various running coupling constants, see appendix A

$$|\delta G_{4\mu}| \leqslant \varepsilon^{(1)} B_{4\mu} \sum_{n=4}^{\infty} n\, C^{n-2} g^{n-2} \,, \tag{5.3}$$

when the function $G_{4\mu}$ itself converges like

$$\sum C^n g^n \,,$$

and $B_{4\mu}$ is the bound for $G_{4\mu}$ itself, as given by Table 2.

The procedure of sect. 13 can be applied unaltered to the error δG_n in the Green functions. But there is a factor in front,

$$\varepsilon^{(1)} \sum_{n=2}^{\infty} (n+2)\, C^n g^n = \varepsilon^{(2)} \,. \tag{15.4}$$

This gives for the newly obtained exceptional Green functions an error

$$|\delta G_n^{(2)}| \leqslant \varepsilon^{(2)} B \tag{15.5}$$

and $\varepsilon^{(2)} < \varepsilon^{(1)}$ if we reduce the maximally allowed value for g, as given by (5.9), somewhat more:

$$g_{max} \to 0.6527\, g_{max} \,. \tag{15.6}$$

We stress that the above argument is only valid as long as the Green functions at their symmetry points were kept fixed and are determined by $g(\mu)$, bounded by (15.6).

b) The Z Factors

Knowing that at any stage $g(\mu)$ satisfies ineq. (14.9), we find that the solution of (14.24) is

$$\log \tilde{Z}_i(\mu) = \int^{\mu} d \log \mu_1 (\gamma_{ijk} g_j(\mu_1) g_k(\mu_1) + g^4(\mu_1) \textstyle\sum_i(\mu_1)) \,; \tag{15.7}$$

$$\tilde{Z}_i(\mu) = \left(\log \frac{\mu}{\Lambda}\right)^{\sigma_i} (1 + \mathcal{O}(g^2)) \,, \tag{15.8}$$

where the $\mathcal{O}(g^2)$ terms are again bounded by a coefficient times $g^2(\mu)$. These equations must be solved iteratively, because the right hand side of (15.7) contains skeleton expansions that again contain $Z(\mu)$, hidden in the function $\sum(\mu_1)$. It is not hard to convince oneself that such iterations converge. A change

$$|\delta Z| \leqq \varepsilon^{(1)} g^2 Z \tag{15.9}$$

yields a change in the function $\sum(\mu_1)$ bounded by

$$|\delta\sum| \leqq \varepsilon^{(1)} g_0^2 \sum , \tag{15.10}$$

so that

$$\frac{\delta\tilde{Z}(\mu)}{\tilde{Z}(\mu)} \leqq C\varepsilon^{(1)} g_0^2 g^2(\mu) \leqq \varepsilon^{(2)} g^2(\mu) \tag{15.11}$$

with $\varepsilon^{(2)} < \varepsilon^{(1)}$ if g_0 is small enough.

c) The coupling constants

We now consider the integro-differential equation (14.4). The solution is constructed iteratively by solving

$$\frac{\mu\partial}{\partial\mu}\tilde{g}_i(\mu) + \sum_{\ell=2}^{k} \beta^{(\ell)}_{ij_1\ldots j_\ell}\, \tilde{g}_{j_1}(\mu)\ldots\tilde{g}_{j_\ell}(\mu) = |g(\mu)|^N \rho_i(\mu) , \tag{15.12}$$

where the tilde denotes the next "improved" function $g_i(\mu)$. Our first Ansatz will be a solution of (15.12) with $\rho_i\{g(\mu),\mu\}$ replaced by zero. This certainly exists because the β coefficients are determined by perturbation expansion and therefore finite. The integration constants must be chosen such that for all $\mu \geqslant m$ we have

$$|g(\mu)| \leqslant \kappa\, g_{max} \tag{15.13}$$

where g_{max} is the previously determined maximally allowed value of $g(\mu)$ and κ is again a constant smaller than 1 to be determined later. In practice this requirement implies asymptotic freedom[3]:

$$\lim_{\mu\to\infty} g(\mu) = o . \tag{15.14}$$

(It is constructive to consider also complex solutions.)

If we now substitute this $g(\mu)$ in the right hand side of eq. (15.12) we may find a correction:

$$g(\mu) \to g(\mu) = g(\mu) + \delta g(\mu) , \tag{15.15}$$

for which we may require

$$|\delta g(\mu)| \leqslant \varepsilon\, g_{max} \quad \text{for all } \mu . \tag{15.16}$$

We must start with:

$$\varepsilon+\kappa < 1 \tag{15.17}$$

Will a recursive application of eq. (15.12) converge to a solution? Let the first Ansatz produce a change (15.16). The next correction is then, up to higher orders in δg, given by

$$\frac{\mu\partial}{\partial\mu}\,\delta\tilde{g}_i(\mu) + M_{ij}(\mu)\delta\tilde{g}_j(\mu) = \delta f_i(\mu) \tag{15.18}$$

(where M_{ij} is determined by differentiation of (15.12) with respect to $g_i(\mu)$.

To estimate $\delta f(\mu)$ we must find a limit for the change in ρ. Our argument that $|\rho| < Q_N$ came from adding the absolute values of all diagrams contributing to ρ, possibly after application of (14.20) several times. Replacing (14.20) then by

$$|\delta g(\mu_1)| \leqslant \delta g(\mu) + 3c\left(\frac{\mu}{\mu_1}\right)^{\varepsilon} g^2(\mu)\delta g(\mu) \tag{15.19}$$

which indeed is true if g satisfies (15.18), or

$$\left|\frac{\mu\partial}{\partial\mu}\,\delta g\right| \leqslant 3\tilde{\beta}\, g^2\, \delta g\ , \tag{15.20}$$

as can be derived from (14.20) and (14.21), we find that we can write

$$|\delta\rho| < \varepsilon C'\ , \tag{15.21}$$

with C' slightly larger than C, and

$$|\delta f(\mu)| \lesssim \varepsilon(N+1)C'\ g(\mu)^N\ . \tag{15.22}$$

Now asymptotically,

$$M_{ij}(\mu) \to M^0_{ij}/\log\mu\ , \tag{15.23}$$

where M^0_{ij} is determined by one-loop perturbation theory. If there is only one coupling constant it is the number 3/2. In the more general case we now assume it to be diagonalized:

$$M^0_{ij} = M(i)\delta_{ij}\ , \tag{15.24}$$

with one eigenvalue equal to 3/2. (Our arguments can easily be extended to the special situation when M^0_{ij} cannot be diagonalized, in which case the standard triangle form must be used.) The asymptotic form of the solution to (15.18) is

$$\delta g_i(\mu) = (\log\mu)^{-M(i)} \int_{\mu(i)}^{\mu} d\log\mu(\log\mu)^{M(i)}\,\delta f_i(\mu)\ , \qquad (15.25)$$

where $\mu(i)$ are integration constants. If $M(i) < \frac{N}{2} - 1$ then we choose $\mu(i) = \infty$. If $M(i) > \frac{N}{2} - 1$ we set $\mu(i) = m$. Then in both cases we get

$$|\delta\tilde{g}_i(\mu)| < \frac{\varepsilon(N+1)C''}{\left|\frac{N}{2} - M(i) - 1\right|}\ |g(\mu)|^{N-2}\ , \qquad (15.26)$$

where C" is related to C' and the first β coefficient. In a compact set of μ values where the deviation from (15.25) is appreciable we of course also have an inequality of the form (15.16).

If $M(i) = N/2 - 1$ then we simply pick another N value (which needs not be integer here), raising or lowering it by one unit. We see that we only need to consider $N \leqslant 4$. Comparing (15.26) with (15.16), noting that C" is independent of A, we see that if

$$\frac{(N+1)C''}{\left|\frac{N}{2} - M(i) - 1\right|}\ g_{max}^{N-3} < 1 \qquad (15.27)$$

then our procedure converges. Since C" stays constant or decreases with decreasing g_{max}, we find that a finite g_{max} will satisfy (15.27).

Also we should check whether $\tilde{g}(\mu)$ satisfies the *Ansatz* (14.19), with unchanged $\tilde{\beta}$. This however is obvious from the construction of $\tilde{g}$ through eq. (15.18).
Notice that the masses are adjusted in every step of the iteration for the Z functions, by choosing A and B in section 14. They are necessary now because we wish to confine the integrals (15.25) at $\mu \geqslant m$, limiting the solutions $g(\mu)$ to satisfy $|g(\mu)| \leqslant g$.

16. BOREL SUMMABILITY[16]

The fact that we obtained eq. (14.4) holding for $m \leqslant \mu < \infty$ is our central result. For simplicity of the following discussions we ignore the *next-to-leading* terms of the coefficients β. Let us take the case that the leading ones have $\ell = 3$. We find that

$$g_i^2(\mu) = \frac{b_i^2}{\log\mu^2 + C} + \mathcal{O}(g^4(\mu)) \qquad (16.1)$$

(the next-to-leading β coefficients give an unimportant correction in the denominator of the form $\log\log\mu^2$). The coefficients b_i are fixed by the leading renormalization group β coefficients. We must choose C such that

$$|g^2(\mu)| \leqslant g^2_{max} \quad \text{if} \quad \mu \geqslant m \; . \tag{16.2}$$

This is guaranteed if either

$$\mathrm{Re}\, C \gtrsim \frac{\max(b_i^2)}{g^2_{max}} - \log m^2 \equiv R - \log m^2 \tag{16.3a}$$

or

$$|\mathrm{Im}\, C| \gtrsim \frac{\max(b_i^2)}{g^2_{max}} \equiv R \; . \tag{16.3b}$$

Now the requirement of asymptotic freedom is usually so stringent that the constant C is the only free parameter besides the masses and possible dimension 1 coupling constants (a situation corresponding to the necessity of choosing all $\mu(i) = \infty$ in eq. (15.25)). But still we can do ordinary perturbation expansion, writing

$$g_i(m) = b_i g + \mathcal{O}(g^3) \; ; \tag{16.4}$$

$$C = \frac{1}{g^2} - \log m^2 \quad . \tag{16.5}$$

where g is now a regular expansion parameter. The $\mathcal{O}(g^3)$ terms are fixed by the asymptotic freedom requirement and can be computed perturbatively. We now claim that perturbation expansion in g is indeed Borel-summable:

$$G(g^2) = \int_0^\infty F(z) e^{-z/g^2} dz \; ; \tag{16.6a}$$

$$F(z) = \frac{1}{2\pi i} \int_C G(g^2) e^{+z/g} \, d(g^{-2}) \; , \tag{16.6b}$$

where the path C must be choosen to lie entirely in the region limited by eqs. (16.3) and (16.5) (see Fig. 17). Since in that region the Green functions $G(g^2)$ are approximately given by their perturbative values the integral (16.6b) will converge rapidly along this path, if

$$\mathrm{Re}\, z > 0 \; , \tag{16.7}$$

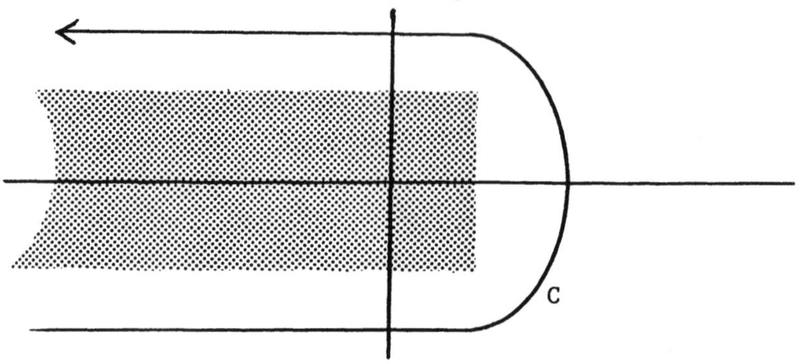

Fig. 17. The integration path C in eq. (16.7).

and F(z) will be bounded by

$$|F(z)| < A \exp(|z|/g^2_{max}) , \qquad (16.8)$$

where g_{max} is the allowed limit for g as derived in the previous sections.

However, eqs. (16.7) and (16.8) are not quite sufficient to prove Borel summability because we also want to have analyticity of F(z) for an open region around the origin. That this requirement is met can be seen as follows. Let us solve a variant of equation (14.4) of the following form:

$$\frac{\mu\partial}{\partial\mu} g_i(\mu,\Lambda) = \Big[- \sum_{\ell} \beta^{(\ell)}_{ij_1\ldots j_\ell}\, g_{j_1}(\mu,\Lambda)\ldots g_{j_\ell}(\mu,\Lambda) + |g(\mu,\Lambda)|^N \rho_i(\mu,\{g\}) \Big] \theta(\Lambda-\mu) , \qquad (16.9)$$

where $\theta(x)$ is the step function. The coefficients $\beta^{(\ell)}$ and the functional ρ are the same as before (constructed the same way via difference equations of sects. 12 and 13). Clearly we have, if $\mu > \Lambda$: $g(\mu,\Lambda) = g(\Lambda,\Lambda)$. And eq. (16.1) now reads

$$g_i(\mu) = \frac{b_i^2}{\log \mu^2 + C} + \mathcal{O}(g^4(\mu)) \; ; \; m \leqslant \mu \leqslant \Lambda . \qquad (16.10)$$

Our point is that this solution exists not only in the region (16.3), but also if

$$\mathrm{Re}\ C \leqslant -\frac{\max(b_i)}{g^2_{\max}} - \log \Lambda^2 \equiv -R-\log \Lambda^2 \ . \tag{16.11}$$

We can now close the contour C (see Fig. 18).

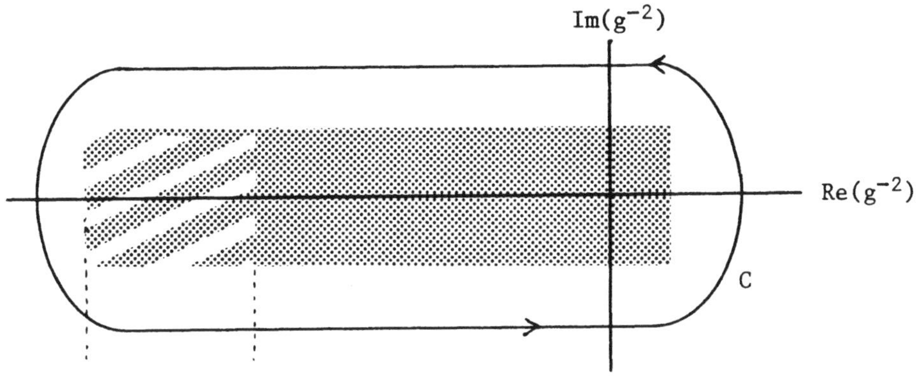

Fig. 18. The contour C of eq. (16.13); the forbidden regions of $G(g,\Lambda_i)$ are shaded.

$-R-\log\left(\frac{\Lambda_2^2}{m^2}\right)$ $\quad -R-\log\left(\frac{\Lambda_1^2}{m^2}\right)$

Now compare two different Λ values: Λ_1 and Λ_2, and compare the Green functions computed with these two Λ values, both as a function of g; taking

$$g_i(m,\Lambda) \stackrel{!}{=} b_i g + \mathcal{O}(g^3) \ . \tag{16.12}$$

We take these Green functions at (possibly exceptional) momentum values p, but always such that $\Lambda \gg |p|$. If they are computed directly following our algorithm then a slight Λ-dependence may still exist, coming from two sources: one is the fact that $g_i(\mu,\Lambda)$ depends on Λ because $\rho(\mu,\{g\})$ as a *functional* of g, may depend on $g(\mu')$ with $\mu' > \Lambda$. But clearly, since all integrals involved in the construction of ρ converge we expect this dependence to go like

$$|\delta G| \lesssim \Lambda^{\varepsilon(g)-1} \tag{6.13a}$$

or probably

$$|\delta G| \lesssim \Lambda^{\varepsilon(g)-2} \tag{16.13b}$$

(because linearly convergent equations can often be made quadratically convergent by symmetrization); here $\varepsilon(g) \downarrow o$ if $g \to o$. The second source of Λ-dependence comes from the application of difference equations to compute Green functions away from the symmetry point (sects. 12,13). Again, this error must behave like (16.13), because of convergence of the integrals involved.

The change in the Borel function F(z) can be read off from eq. (16.6b)

$$\left|F(z,\Lambda_2)-F(z,\Lambda_1)\right| \lesssim \frac{1}{2\pi i} \oint_C \Lambda_1^{\varepsilon(g)-2} e^{z/g^2}\, d(g^{-2}) \tag{16.14}$$

if $\Lambda_2 \gtrsim \Lambda_1$.
Under what conditions does this vanish in the limit $\Lambda_1 \to \infty$? In our integral g^{-2} ranges from $-R-\log\left(\frac{\Lambda_2^2}{m^2}\right)$ to R. So if Re $z \leqslant o$ then

$$\left|e^{z/g^2}\right| \lesssim \left(\frac{\Lambda_2^2}{m^2}\, e^R\right)^{-\mathrm{Re}z} . \tag{16.15}$$

If $|g^{-2}| \to \infty$ then $\varepsilon(g) \downarrow o$. This we may use on the far left of the curve C. Therefore ineq. (16.14) can be written as

$$|F(z,\Lambda_2) - F(z,\Lambda_1)| \lesssim e^R\left(\frac{\Lambda_2^2}{\Lambda_1^2}\right) \Lambda_2^{-2\,\mathrm{Re}z-2} . \tag{16.16}$$

By comparing a series of Λ values of the form $\Lambda_n = 2^n$, one easily sees that (16.16) guarantees convergence as soon as

$$\mathrm{Re}\; z > -1 , \tag{16.17}$$

which is a quite large region of analyticity of F(z). Note that $z = -1$ is the location of the first renormalon singularity[4,7]. Eq. (16.17) together with (16.8) implies complete Borel summability of this theory.

17. THE MASSLESS THEORY

In the sections 14-16 the mass m^2 had to be non-zero. This was the only way we could obtain the necessary ineq. (16.3a), allowing us to draw the contour C. What happens if we take the limit $m^2 \downarrow o$, considering not $g(m^2)$ but $g(\mu^2)$ at some fixed μ as expansion parameter? Our method to construct the entire theory then fails, but for some z values F(z) will still exist. The argument is analogous to the one of the previous section where we dealt with an ultraviolet problem. Now our difficulty is in the infrared. Comparing two different small values for m, we have

$$|\delta G| \lesssim (m^2)^{1-\varepsilon(g)} \; ; \tag{17.1}$$

$$|F(z,m_2) - F(z,m_1)| \lesssim \frac{1}{2\pi i} \oint_C (m_1^2)^{2-\varepsilon(g)} \, e^{z/g^2} \, d(g^{-2}) \; , \tag{17.2}$$

if $m_2 \lesssim m_1$.
Now if Re z > o then

$$\left| e^{z/g^2} \right| \lesssim \left(\frac{m^2}{\mu^2} \, e^R \right)^{\mathrm{Re}z} \; . \tag{17.3}$$

Thus

$$|F(z,m_2) - F(z,m_1)| \lesssim e^{R} \left(\frac{m_1^2}{m_2^2} \right) m_2^{2-2\mathrm{Re}z} \; . \tag{17.4}$$

Now we have convergence as $m^2 \downarrow o$ as long as

$$\mathrm{Re}\, z < 1 \; , \tag{17.5}$$

and, indeed, z = 1 is a point where an infrared renormalon singularity is to be expected. Thus, the Borel transform F(z) of the massless theory is analytic in the region

$$-1 < \mathrm{Re}z < 1 \; . \tag{17.6}$$

This result guarantees that perturbation expansion in g^2 diverges not worse than

$$g^{2n} \, n!$$

but is clearly not enough for Borel summability.

18. THE $- \lambda$ Tr ϕ^4 MODEL

A special case is the pure scalar planar field theory, with just one coupling constant λ and a mass m. If m is sufficiently large and λ sufficiently small then our analysis applies, and we find that all planar Green functions are uniquely determined. However, in this special case there is more: the Green functions can be uniquely determined as long as the masses in *all* channels are non-negative, and λ is negative. This was discovered by comparing with the much simpler "spherical model" (appendix B) which shows the same property.

The argument is fairly simple. Let us first take the theory at m^2 values which are so large that everything is well-defined. Now decrease the "bare mass" m_B^2 continuously.
The change in the dressed propagator

$$P(k,m^2) = -G_2^{-1}(k,m^2) \tag{18.1}$$

is determined by

$$\frac{\partial}{\partial m^2} G_2(k,m^2) \equiv G'(k,m^2) \tag{18.2}$$

which we take to start out at sufficiently large m^2.

Now the Feynman rules for

$$G''(k,m^2) \equiv \frac{\partial}{\partial m^2} G'(k,m^2) \ , \tag{18.3}$$

can easily be written down, just like those for

$$\frac{\partial}{\partial m^2} G_4(k_1,\ldots,k_4,m^2) \equiv G_4' \ . \tag{18.4}$$

Since G_2'' and G_4' are superficially convergent we can again express them in terms of a skeleton expansion containing only the functions G_4 and $-G_2'$ and the propagators P. *They are all positive* (remember that G_4 starts out as $-\lambda$, with $\lambda < o$), *and all integrals and summations converge*. Only G_4' has one surviving minus sign from differentiating one propagator with m . Thus:

$$G_2'' > o \ ; \tag{18.5}$$

$$G_4' < o \ . \tag{18.6}$$

If we let m^2 decrease then clearly G_2 will stay positive and G_2' negative. Their absolute values grow however, until a point is reached where either the sum of all diagrams will no longer converge, or the two-point function G_2 becomes zero. As soon as this happens the theory will be ill-defined. A tachyonic pole tends to develop, followed by catastrophes in all channels.
The point we wish to make in this section however is that as long as this does not happen, indeed all summations and integrals converge, so that our iterative procedure to produce the Green functions will also converge. For all those values of λ and m^2 this theory will be Borel summable.

This result only holds for the special case considered here, namely $-\lambda$ Tr ϕ^4 theory, because all skeleton diagrams that con-

tribute to some Green function, carry the same sign. They can never interfere destructively.
Notice that what also was needed here was convergence of the diagram expansion. Now we know that at finite N the non-planar graphs give a divergent contribution. Thus the "tachyons" will develop already at infinite m^2: the theory is fundamentally unstable. Of course we knew this already: λ after all has the wrong sign. The instantons that bring about the decay of our "false vacuum" carry on action S proportional to $-N/\tilde{\lambda}$ which is finite for finite N.

19. OUTLOOK

Apart from the model of sect. 18, the models we are able to construct explicitly now lack any appreciable structure, so they are physically not very interesting. Two (extremely difficult) things should clearly be tried to be done: one is the massless planar theories such as SU(∞) QCD. Clearly that theory should show an enormously intricate structure, including several possible phase-transitions. We still believe that more and better understanding of the infrared renormalons that limited analyticity of our borel functions in sect. 17 could help us to go beyond those singular points and may possibly "solve" that model (i.e. yield a demonstrably convergent calculational scheme).
Secondly one would try to use the same or similar skeleton techniques at finite N (non-planar diagrams). Of course now the skeleton expansion does not converge, but, in Borel-summing the skeleton expansion there should be no renormalons, and all divergences may be due entirely to instantonlike structures. More understanding of resummation techniques for these diagrams by saddle point methods could help us out. If such a program could work then that would enable us to write down SU(3) QCD in a finite (but small) box. QCD in the real world could then perhaps be obtained by gluing boxes together, as in lattice gauge theories.

Another thing yet to be done is to repeat our procedure now in Minkowski space instead of Euclidean space. Singling out the obvious singularities in Minkowski space may well be not so difficult, so perhaps this is a more reasonable challenge that we can leave for the interested student.

APPENDIX A. ASYMPTOTICALLY FREE INFRARED CONVERGENT PLANAR HIGGS MODEL

In discussing examples of planar field theories for which our analysis is applicable we found that pure SU(∞) gauge theory (with possibly a limited number of fermions) is asymptotically free as required, but unbounded in the infrared - so that even the ultraviolet limit cannot be treated exactly (see sects. 14 and 15).

$-\lambda\ \mathrm{Tr}\ \varphi^4$ is asymptotically free and can be given a mass term so that infrared convergence is also guaranteed. At $N \to \infty$ this is a fine planar theory, but at finite N the vacuum is unstable. The only theory that suffers from none of these defects is an SU(∞) gauge theory in which all bosons get a mass due to the Higgs mechanism. But then a new scalar self-coupling occurs that tends to be not asymptotically free. Asymptotic freedom is only secured if, curiously enough, several kinds of fermionic degrees of freedom are added. The following model is an example (similar examples can also be constructed at finite N, such as SU(2)).

In general a renormalizable model can be written as

$$\mathcal{L} = -\tfrac{1}{4}G^a_{\mu\nu}G^a_{\mu\nu} - \tfrac{1}{2}(D_\mu\phi_i)^2 - V(\phi_i) - \bar{\psi}(\gamma D + W(\phi))\psi\ , \tag{A.1}$$

where G is the covariant curl, ϕ_i is a set of scalar fields and ψ a set of spinors. V is a quartic and W a linear polynomial in ϕ. We write

$$\begin{aligned}
G^a_{\mu\nu} &= \partial_\mu A^a_\nu - \partial_\nu A^a_\mu + g^{abc}A^b_\mu A^c_\nu \\
D_\mu\phi_i &= \partial_\mu\phi_i + T^a_{ij}A^a_\mu\phi_j \\
D_\mu\psi_i &= \partial_\mu\psi_i + U^a_{ij}A^a_\mu\psi_j \\
W &= S + iP\gamma_5\ ;\ \hat{W} = S - iP\gamma_5 \\
U &= U_s + U_p\gamma_5\ ;\ \hat{U} = U_s - U_p\gamma_5 \\
C_1^{ab} &= g^{apq}g^{bpq}\ ;\ C_2^{ab} = -\mathrm{Tr}\,T^aT^b \\
C_3^{ab} &= -\mathrm{Tr}\left(U^a_LU^b_L + U^a_RU^b_R\right) = -2\mathrm{Tr}\left(U^a_sU^b_s + U^a_pU^b_p\right)\ .
\end{aligned} \tag{A.2}$$

The most compact way to write the complete set of one-loop β functions is to express them in terms of the one-loop counter-Lagrangian, $[8\pi^2(4-n)]^{-1}\Delta\mathcal{L}$, where $\Delta\mathcal{L}$ has been found to be[17], after performing the necessary field renormalizations,

$$\begin{aligned}
\Delta\mathcal{L} = {} & G^a_{\mu\nu}G^b_{\mu\nu}\left[\frac{11}{12}C_1^{ab} - \frac{1}{24}C_2^{ab} - \frac{1}{6}C_3^{ab}\right] \\
& + \frac{1}{4}V^2_{ij} + \frac{3}{2}V_i(T^2\phi)_i + \frac{3}{4}(\phi T^aT^b\phi)^2 \\
& + \bar{\psi}\left\{\frac{1}{4}W_i\hat{W}_iW + \frac{1}{4}W\hat{W}_iW_i + W_i\hat{W}W_i\right\}\psi \\
& + \frac{3}{2}\bar{\psi}(\hat{U}^2W + WU^2)\psi + \phi_i(V_j + \bar{\psi}W_j\psi)\mathrm{Tr}(S_iS_j + P_iP_j) \\
& - \mathrm{Tr}(S^2+P^2) + \mathrm{Tr}[S,P]^2\ .
\end{aligned} \tag{A.3}$$

Here V_i stands for $\partial V/\partial\varphi_i$, etc.
The scaling behavior of the coupling constants is then determined by

$$\frac{\mu\partial\mathcal{L}}{\partial\mu} = -\frac{\Delta\mathcal{L}}{8\pi^2} \; . \tag{A.4}$$

We now choose a model with $U(N)_{local} \times SU(N)_{global}$ symmetry. Besides the gauge field we have a scalar ϕ_i^s in the $N_{local} \times N_{global}$ representation, and two kinds of fermions:

$\psi_{(1)i}{}^{j}$ in $N_{local} \times N_{local}$ and $\psi_{(2)a}{}^{i}$ in $N_{local} \times N_{global}$. We choose

$$V(\phi) = \frac{\lambda}{2}\, \phi^{*s}_{i}\phi^{t}_{i}\; \phi^{*t}_{j}\phi^{s}_{j} \; , \tag{A.5}$$

and

$$\bar\psi W\psi = h\left[\bar\psi_{(2)}{}^{a}{}_{i}\,\phi^{*j}_{a}\;\psi_{(1)j}{}^{i} + \text{h.c.}\right] . \tag{A.6}$$

Writing

$$C_1^{ab} = C_2^{ab} = \tilde g^2\,\delta^{ab} \;\; ; \;\; C_3^{ab} = 3\tilde g^2\,\delta^{ab} \; , \tag{A.7}$$

we find:

$$-8\pi^2\,\frac{\mu\partial\tilde g^2}{\partial\mu} \equiv \Delta\tilde g^2 = \frac{3}{2}\,\tilde g^4 \;\; ; \tag{A.8}$$

and with $\tilde\lambda \equiv N\lambda$; $\tilde h^2 = Nh^2$, by substituting (A.5) and (A.6) in the expression (A.3) after some algebra, and in the limit $N \to \infty$:

$$\Delta\tilde h^2 = -3\tilde h^4 + \frac{9}{2}\,\tilde h^2\tilde g^2 \;\; ; \tag{A.9}$$

$$\Delta\tilde\lambda = -2\tilde\lambda + 3\tilde g^2\tilde\lambda - \frac{3}{4}\,\tilde g^4 - 4\tilde h^2\tilde\lambda + 4\tilde h^4 \; . \tag{A.10}$$

These equations (A.8)-(A.10) are ordinary differential equations whose solutions we can study. The signs in all terms are typical for any such models with three coupling constants g, λ and h. Only the relative magnitudes of the various terms differ from one model to another. For asymptotic freedom we need that the second and last terms of (A.10) and the last of (A.9) are sufficiently large. Usually this implies that the fermions must be in a sufficiently large representation of the gauge group, which explains our choice for the fermionic representations. Our model has an asymptotically free solution if all coupling constants stay in a fixed ratio with respect to each other:

$$\tilde\lambda = \hat\lambda\,\tilde g^2 \quad ; \quad \tilde h = \hat h\, g \; , \tag{A.11}$$

and then, from (A.8)-(A.10) we see:

$$\hat{h} = 1 ;$$

$$\hat{\lambda} = \frac{1}{8} (\sqrt{129}-5) . \qquad (A.12)$$

So indeed we have a solution with positive λ.

It now must be shown that in this model all particles can be made massive via the Higgs mechanism. We consider spontaneous breakdown of $SU(N)_{local} \times SU(N)_{global}$ into the diagonal $SU(N)_{global}$ subgroup. Take as a mass term

$$-\mu \, \phi_i^{*s} \, \phi_i^s . \qquad (A.13)$$

We can write V as

$$V = \frac{\lambda}{2} \left| \phi_i^{*s} \, \phi_i^t - F^2 \delta^{st} \right|^2 + \text{const.} \qquad (A.14)$$

Clearly this is minimal if

$$\phi_i^s = F \delta_i^s , \qquad (A.15)$$

or a gauge rotation thereof.
All vector bosons get an equal mass:

$$-D^*\phi \, D\phi \Rightarrow - g^2 F^2 A_\mu^2 ; \qquad (A.16)$$

$$M_A^2 = 2g^2 F^2 , \qquad (A.17)$$

and of course the scalars get a mass:

$$M_H^2 = \lambda F^2 . \qquad (A.18)$$

Thus the mass ratio is given by

$$\sqrt{\hat{\lambda}/2} = 0.6303 \; 6778 \ldots . \qquad (A.19)$$

This is a fixed number of this theory, but it will be affected by higher order corrections. The fermions can each be given a mass term:

$$- m_1 \bar{\psi}_{(1)} \psi_{(1)} - m_2 \bar{\psi}_{(2)} \psi_{(2)} , \qquad (A.20)$$

and the Yukawa force will give a mixing of a definite strength. The model described in this Appendix, is probably the simplest completely convergent planar field theory with absolutely stable vacuum. It is unlikely however that it would have a direct physical significance.

APPENDIX B. THE N-VECTOR MODEL IN THE N → ∞ LIMIT (SPHERICAL MODEL). SPONTANEOUS MASS GENERATION

When only N-vector fields are present (rather than N×N tensors) then the N → ∞ limit is easily obtained analytically. This is the quite illustrative spherical model. Let the bare Lagrangian be

$$\mathcal{L} = -\tfrac{1}{2}(\partial_\mu\vec{\phi})^2 - \tfrac{1}{2}m_B^2\vec{\phi}^2 - \frac{\lambda_B}{8}(\vec{\phi}^2)^2 \,. \tag{B.1}$$

The only diagrams that dominate in the N → ∞ limit are the chains of bubbles (Fig. 19).

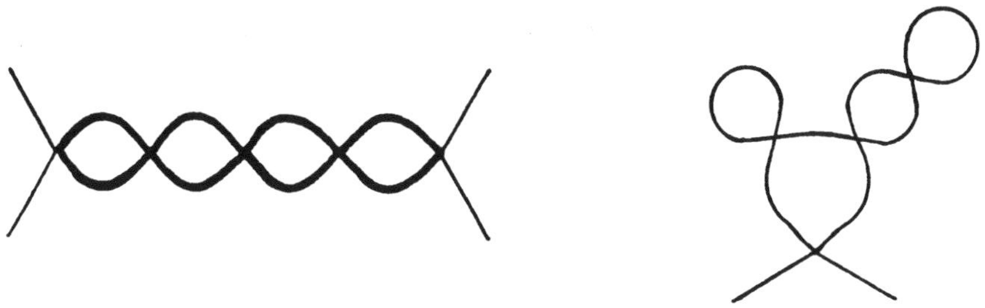

a b

Fig. 19. a) Dominating diagrams for the 4-point function.
b) Mass renormalization.

Let us remove some factors π^2 by defining

$$\tilde{\lambda}_B = N\lambda_B/16\pi^2 \,. \tag{B.2}$$

The diagrams of Fig. 19 are easily summed. Mass and coupling constant need to be renormalized. Dimensional renormalization is appropriate here. In terms of the finite constants $\tilde{\lambda}_R(\mu)$ and $m_R(\mu)$, chosen at some subtraction point μ, and the infinitesimal $\varepsilon = 4-n$, where n is the number of space-time dimensions, one finds:

$$\tilde{\lambda}_B = -\varepsilon\left(1 + \frac{\varepsilon}{\tilde{\lambda}_R(\mu)}\right)\mu^\varepsilon \,, \tag{B.3}$$

and

$$m_B = \mu m_R(\mu)\sqrt{\frac{-\varepsilon}{\tilde{\lambda}_R(\mu)}} \,. \tag{B.4}$$

The sum of all diagrams of type 19a gives an effective propagator of the form

$$F(q) = \frac{32\pi^2/N}{\gamma - \frac{2}{\tilde{\lambda}_R(\mu)} + \log\frac{\pi m^2}{\mu^2} + f(\frac{q^2}{m^2}) - i\varepsilon} , \tag{B.5}$$

where q is the exchanged momentum, γ is Euler's constant, and

$$f(\frac{q^2}{m^2}) = \int_0^1 dx \log\left| 1 + \frac{x(1-x)q^2}{m^2}\right| - \pi i \theta(-q^2-4m^2)\sqrt{1+\frac{4m^2}{q^2}} , \tag{B.6}$$

and m is the *physical* mass in the propagators of Fig. 19a; that is because these should include the renormalizations of the form of Fig. 19b.

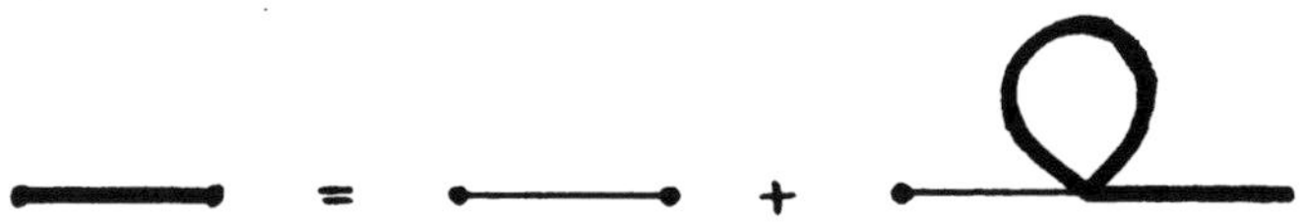

Fig. 20.

Fig. 20 shows how m follows from m_R:

$$2m_R^2\mu^2 = \tilde{\lambda}_R m^2 \log\left(\frac{\pi m^2}{\mu^2} + \gamma - 1 - \frac{2}{\tilde{\lambda}_R}\right) . \tag{B.7}$$

From (B.3) we see that the renormalization-group invariant combination is

$$\frac{1}{\tilde{\lambda}_R(\mu)} + \log\mu , \tag{B.8}$$

so that inevitably $\tilde{\lambda}_R < o$ at large μ. Indeed, in this model the vacuum would become unstable as soon as N is made finite. In the limit $N \to \infty$ however everything is still fine.

Now in Fig. 21 we plot both m_R^2 and the composite mass M^2, determined by the pole of F(q), as a function of the physical mass m^2. We see that at negative m_R^2 there are two solutions for m^2, but one should be rejected because M^2 would be negative, an indication for an unstable choice of vacuum.
The observation we wish to make in this appendix is that in the allowed region for m_R^2 we get an entirely positive 4-point function in Euclidean space ($\tilde{F}(q) > o$). If we chose m_R^2 to be fixed and vary $\tilde{\lambda}_R$ (or rather vary $\tilde{\lambda}_B$) then at $m_R^2 \geqslant o$ *all* $\tilde{\lambda}$ values are allowed, at negative m_R^2 only sufficiently small values. At $m_R^2 = o$ we see a

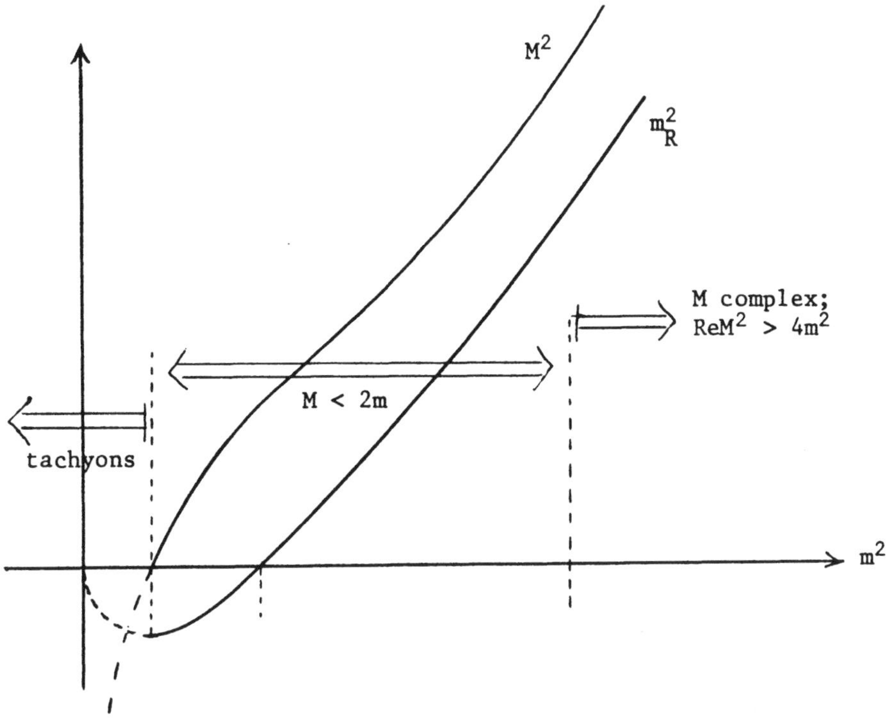

Fig. 21. Mass ratios at given value for $\tilde{\lambda}_R(\mu)$.

"spontaneous" generation of a finite value for m . Perturbation expansion in $\tilde{\lambda}_R$ would show the "infrared renormalon" difficulty. Apparently here the difficulty solves itself via this spontaneous mass generation.

REFERENCES

1. E.C.G. Stueckelberg and A. Peterman, Helv. Phys. Acta 26 (1953), 499; M. Gell-Mann and F. Low, Phys. Rev. 95 (1954) 1300.
2. K.G. Wilson, Phys. Rev. D10, 2445 (1974). K.G. Wilson, in "Recent Developments in Gauge Theories", ed. by G. 't Hooft et al., Plenum Press, New York and London, 1980, p. 363.
3. G. 't Hooft, Marseille Conference on Renormalization of Yang-Mills Fields and Applications to Particle Physics, June 1972, unpublished; H.D. Politzer, Phys. Rev. Lett. 30, 1346 (1973); D.J. Gross and F. Wilczek, Phys. Rev. Lett. 30, 1343 (1973).
4. G. 't Hooft, in "The Whys of Subnuclear Physics", Erice 1977, A. Zichichi ed. Plenum Press, New York and London 1979, p. 943.
5. M. Creutz, L. Jacobs and C. Rebbi, Phys. Rev. Lett. 42, 1390 (1979).
6. C. de Calan and V. Rivasseau, Comm. Math. Phys. 82, 69 (1981); 91, 265 (1983).

7) G. Parisi, Phys. Lett. 76B, 65 (1978) and Phys. Rep. 49, 215 (1979).
8) J. Koplik, A. Neveu and S. Nussinov, Nucl. Phys. B123, 109 (1977).
W.T. Tuttle, Can. J. Math. 14, 21 (1962).
9) See ref. 3.
10) J.D. Bjorken and S.D. Drell, Relativistic Quantum Mechanics (McGraw-Hill, New York, 1964).
11) G. 't Hooft, Commun. Math. Phys. 86, 449 (1982).
12) G. 't Hooft, Commun. Math. Phys. 88, 1 (1983).
13) J.C. Ward, Phys. Rev. 78 (1950) 1824.
Y. Takahashi, Nuovo Cimento 6 (1957) 370.
A. Slavnov, Theor. and Math. Phys. 10 (1972) 153 (in Russian).
English translation: Theor. and Math. Phys. 10, p. 99.
J.C. Taylor, Nucl. Phys. B33 (1971) 436.
14) K. Symanzik, Comm. Math. Physics 18 (1970) 227; 23 (1971) 49; Lett. Nuovo Cim. 6 (1973) 77; "Small-Distance Behaviour in Field Theory", Springer Tracts in Modern Physics vol. 57 (G. Höhler ed., 1971) p. 221.
15) G. 't Hooft, Nucl. Phys. B33 (1971) 173.
16) G. 't Hooft, Phys. Lett. 119B, 369 (1982).
17) G. 't Hooft, unpublished.
R. van Damme, Phys. Lett. 110B (1982) 239.

Nuclear Physics B75 (1974) 461–470. North-Holland Publishing Company

CHAPTER 6.2

A TWO-DIMENSIONAL MODEL FOR MESONS

G. 't HOOFT
CERN, Geneva

Received 21 February 1974

Abstract: A recently proposed gauge theory for strong interactions, in which the set of planar diagrams play a dominant role, is considered in one space and one time dimension. In this case, the planar diagrams can be reduced to self-energy and ladder diagrams, and they can be summed. The gauge field interactions resemble those of the quantized dual string, and the physical mass spectrum consists of a nearly straight "Regge trajectory".

It has widely been speculated that a quantized non-Abelian gauge field without Higgs fields, provides for the force that keeps the quarks inseparably together [1–4]. Due to the infra-red instability of the system, the gauge field flux lines should squeeze together to form a structure resembling the quantized dual string.

If all this is true, then the strong interactions will undoubtedly be by far the most complicated force in nature. It may therefore be of help that an amusingly simple model exists which exhibits the most remarkable feature of such a theory: the infinite potential well. In the model there is only one space, and one time dimension. There is a local gauge group U(N), of which the parameter N is so large that the perturbation expansion with respect to $1/N$ is reasonable.

Our Lagrangian is, like in ref. [4],

$$\mathcal{L} = \tfrac{1}{4} G_{\mu\nu i}{}^{j} G_{\mu\nu j}{}^{i} - \bar{q}^{ai}(\gamma_\mu D_\mu + m_{(a)})\, q^a{}_i\,, \tag{1}$$

where

$$G_{\mu\nu i}{}^{j} = \partial_\mu A_i{}^j{}_\nu - \partial_\nu A_i{}^j{}_\mu + g[A_\mu, A_\nu]_i{}^j\,; \tag{2.a}$$

$$D_\mu q^a{}_i = \partial_\mu q^a{}_i + g A_i{}^j{}_\mu q^a{}_j\,; \tag{2.b}$$

$$A_i{}^j{}_\mu(x) = -A^{*}_j{}^i{}_\mu(x)\,; \tag{2.c}$$

$$q^1 = p\,; \quad q^2 = n\,; \quad q^3 = \lambda\,. \tag{2.d}$$

The Lorentz indices μ, ν, can take the two values 0 and 1. It will be convenient to use light cone coordinates. For upper indices:

$$x^{\pm} = \frac{1}{\sqrt{2}}(x^1 \pm x^0) , \tag{3.a}$$

and for lower indices

$$p_{\pm} = \frac{1}{\sqrt{2}}(p_1 \pm p_0) ,$$

$$A_{\pm} = \frac{1}{\sqrt{2}}(A_1 \pm A_0) , \quad \text{etc.,} \tag{3.b}$$

where

$$p_1 = p^1 , \quad p_0 = -p^0 .$$

Our summation convention will then be as follows,

$$x_{\mu}p^{\mu} = x^{\mu}p^{\mu} = x_{\mu}p_{\mu} =$$

$$x_+p^+ + x_-p^- = x^+p^- + x^-p^+ = x_+p_- + x_-p_+ . \tag{4}$$

The model becomes particularly simple if we impose the light-cone gauge condition:

$$A_- = A^+ = 0 . \tag{5}$$

In that gauge we have

$$G_{+-} = -\partial_- A_+ , \tag{6}$$

and

$$\mathcal{L} = -\tfrac{1}{2}\mathrm{Tr}\,(\partial_- A_+)^2 - \bar{q}^a(\gamma\partial + m_{(a)} + g\gamma_- A_+)\,q^a . \tag{7}$$

There is no ghost in this gauge. If we take x^+ as our time direction, then we notice that the field A_+ is not an independent dynamical variable because it has no time derivative in the Lagrangian. But it does provide for a (non-local) Coulomb force between the Fermions.

The Feynman rules are given in fig. 1 (using the notation of ref. [4]).

The algebra for the γ matrices is

$$\gamma_-^2 = \gamma_+^2 = 0 , \tag{8.a}$$

$$\gamma_+\gamma_- + \gamma_-\gamma_+ = 2 . \tag{8.b}$$

Since the only vertex in the model is proportional to γ_- and $\gamma_-^2 = 0$, only that part of the quark propagator that is proportional to γ_+ will contribute. As a consequence

$$\frac{1}{k_-^2} \qquad \rightarrow \qquad \frac{1}{k_-^2}$$

$$\frac{m_0 - i\gamma_- k_+ - i\gamma_+ k_-}{m_a^2 + 2k_+ k_- - i\epsilon} \qquad \rightarrow \qquad \frac{-ik_-}{m_a^2 + 2k_+ k_- - i\epsilon}$$

$$-g\gamma_- \qquad \longrightarrow \qquad -2g$$

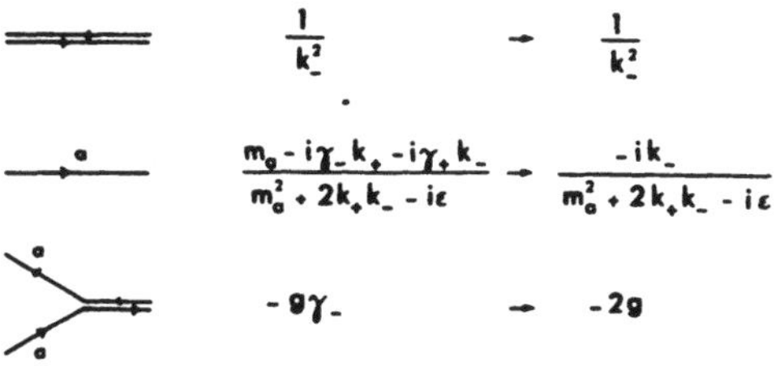

Fig. 1. Planar Feynman rules in the light cone gauge.

we can eliminate the γ matrices from our Feynman rules (see the right-hand side of fig. 1).

We now consider the limit $N \to \infty$; $g^2 N$ fixed, which corresponds to taking only the planar diagrams with no Fermion loops [4]. They are of the type of fig. 2. All gauge field lines must be between the Fermion lines and may not cross each other.

They are so much simpler than the diagrams of ref. [4], because the gauge fields do not interact with themselves. We have nothing but ladder diagrams with self-energy insertions for the Fermions. Let us first concentrate on these self-energy parts. Let $i\Gamma(k)$ stand for the sum of the irreducible self-energy parts (after having eliminated the γ matrices). The dressed propagator is

$$\frac{-ik_-}{m^2 + 2k_+ k_- - k_- \Gamma(k) - i\epsilon} . \tag{9}$$

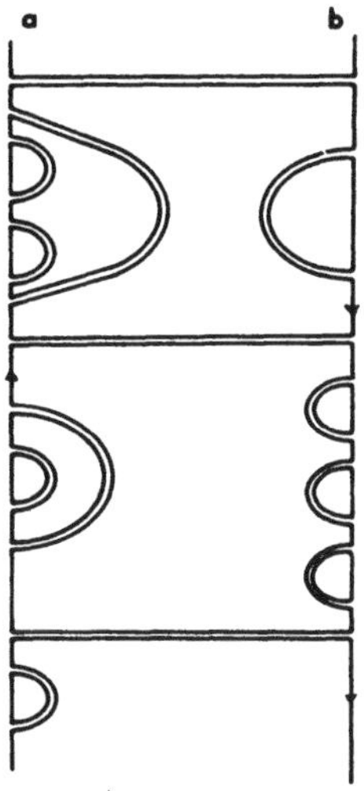

Fig. 2. Large diagram. a and b must have opposite $U(N)$ charge, but need not be each other's antiparticle.

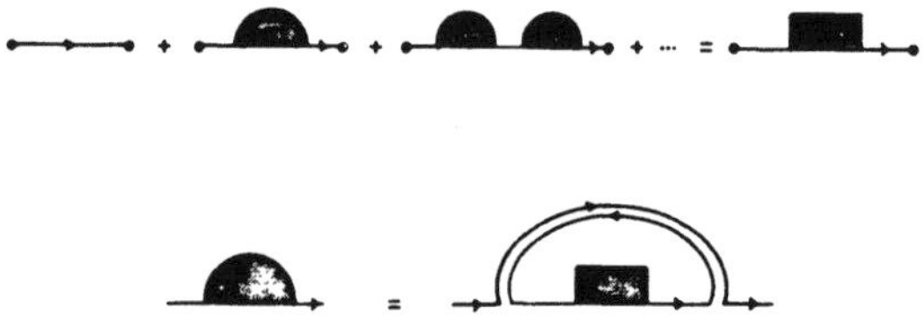

Fig. 3. Equations for the planar self-energy blob.

Since the gauge field lines must all be at one side of the Fermion line, we have a simple bootstrap equation (see fig. 3).

$$i\Gamma(p) = \frac{4g^2}{(2\pi)^2 i}\int \mathrm{d}k_+\,\mathrm{d}k_-\,\frac{1}{k_-^2}\,\frac{-i(k_- + p_-)}{(m^2 + [2(k_+ + p_+) - \Gamma(k+p)](k_- + p_-) - i\epsilon)}\,. \tag{10}$$

Observe that we can shift $k_+ + p_+ \to k_+$, so $\Gamma(p)$ must be independent of p_+, and

$$\Gamma(p_-) = \frac{ig^2}{\pi^2}\int \frac{\mathrm{d}k_-(k_- + p_-)}{k_-^2}\int \frac{\mathrm{d}k_+}{m^2 - (k_- + p_-)\Gamma(k_- + p_-) + 2(k_- + p_-)k_+ - i\epsilon}\,. \tag{11}$$

Let us consider the last integral in (11). It is ultra-violet divergent, but as it is well known, this is only a consequence of our rather singular gauge condition, eq. (5). Fortunately, the divergence is only logarithmic (we work in two dimensions), and a symmetric ultra-violet cut-off removes the infinity. But then the integral over k_+ is independent of Γ. It is

$$\frac{\pi i}{2|k_- + p_-|}\,,$$

so,

$$\Gamma(p_-) = -\frac{g^2}{2\pi}\int \frac{\mathrm{d}k_-}{k_-^2}\,\mathrm{sgn}(k_- + p_-)\,. \tag{12}$$

This integral is infra-red divergent. How should we make the infra-red cut-off? One can think of putting the system in a large but finite box, or turning off the interactions at large distances, or simply drill a hole in momentum space around $k\ = 0$. We shall take $\lambda < |k_-| < \infty$ as our integration region and postpone the limit $\lambda \to 0$ until it makes sense. We shall not try to justify this procedure here, except for the remark that our final result will be completely independent of λ, so even if a more thorough discussion would necessitate a more complicated momentum cut-off, this would in general make no difference for our final result.

We find from (12), that

$$\Gamma(p) = \Gamma(p_-) = -\frac{g^2}{\pi}\left(\frac{\operatorname{sgn}(p_-)}{\lambda} - \frac{1}{p_-}\right), \tag{13}$$

and the dressed propagator is

$$\frac{-ik_-}{m^2 - g^2/\pi + 2k_+k_- + g^2|k_-|/\pi\lambda - i\epsilon}. \tag{14}$$

Now because of the infra-red divergence, the pole of this propagator is shifted towards $k_+ \to \infty$ and we conclude that there is no physical single quark state. This will be confirmed by our study of the ladder diagrams, of which the spectrum has no continuum corresponding to a state with two free quarks.

The ladder diagrams satisfy a Bethe–Salpeter equation, depicted in fig. 4. Let $\psi(p, r)$ stand for an arbitrary blob out of which comes a quark with mass m_1 and

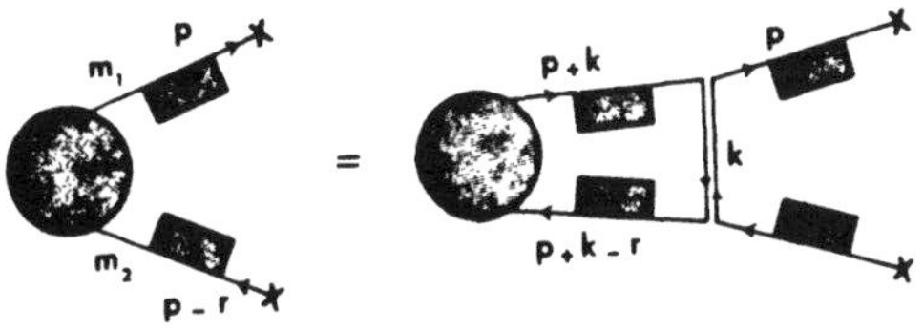

Fig. 4. Eq. (15).

momentum p, and an antiquark with mass m_2 and momentum $r - p$. Such a blob satisfies an inhomogeneous bootstrap equation. We are particularly interested in the homogeneous part of this equation, which governs the spectrum of two-particle states:

$$\psi(p,r) = -\frac{4g^2}{(2\pi)^2 i}(p_- - r_-)p_-\left[M_2^2 + 2(p_+ - r_+)(p_- - r_-) + \frac{g^2}{\pi\lambda}|p_- - r_-| - i\epsilon\right]^{-1}$$

$$\times\left[M_1^2 + 2p_+p_- + \frac{g^2}{\pi\lambda}|p_-| - i\epsilon\right]^{-1}\iint\frac{\psi(p+k,r)}{k_-^2}\,\mathrm{d}k_+\,\mathrm{d}k_-, \tag{15}$$

where

$$M_i^2 = m_i^2 - \frac{g^2}{\pi}. \tag{16}$$

writing

$$\varphi(p_-, r) = \int\psi(p_+, p_-, r)\,\mathrm{d}p_+, \tag{17}$$

we have for φ

$$\varphi(p_-,r) = -\frac{g^2}{(2\pi)^2 i}\left\{\int \mathrm{d}p_+\left[p_+ - r_+ + \frac{M_2^2}{2(p_- - r_-)} + \left(\frac{g^2}{2\pi\lambda} - i\epsilon\right)\mathrm{sgn}\,(p_- - r_-)\right]^{-1}\right.$$

$$\left.\times\left[p_+ + \frac{M_1^2}{2p_-} + \left(\frac{g^2}{2\pi\lambda} - i\epsilon\right)\mathrm{sgn}\,(p_-)\right]^{-1}\right\}\int\frac{\varphi(p_- + k_-, r)}{k_-^2}\,\mathrm{d}k_- \,. \tag{18}$$

One integral has been separated. This was possible because the Coulomb force is instantaneous. The p_+ integral is only non-zero if the integration path is *between* the poles, that is,

$$\mathrm{sgn}\,(p_- - r_-) = -\mathrm{sgn}\,(p_-)\,, \tag{19}$$

and can easily be performed. Thus, if we take $r_- > 0$, then

$$\varphi(p_-,r) = \frac{g^2}{2\pi}\,\theta(p_-)\,\theta(r_- - p_-)$$

$$\times\left[\frac{M_1^2}{2p_-} + \frac{M_2^2}{2(r_- - p_-)} + \frac{g^2}{\pi\lambda} + r_+\right]^{-1}\int\frac{\varphi(p_- + k_-, r)}{k_-^2}\,\mathrm{d}k_- \,. \tag{20}$$

The integral in eq. (20) is again infra-red divergent. Using the same cut-off as before, we find

$$\int\frac{\varphi(p_- + k_-, r)}{k_-^2}\,\mathrm{d}k_- = \frac{2}{\lambda}\,\varphi(p_-) + \mathrm{P}\int\frac{\varphi(p_- + k_-, r)}{k_-^2}\,\mathrm{d}k_- \,, \tag{21}$$

where the principal value integral is defined as

$$\mathrm{P}\int\frac{\varphi(k_-)\,\mathrm{d}k_-}{k_-^2} = \frac{1}{2}\int\frac{\varphi(k_- + i\epsilon)\,\mathrm{d}k_-}{(k_- + i\epsilon)^2} + \frac{1}{2}\int\frac{\varphi(k_- - i\epsilon)\,\mathrm{d}k_-}{(k_- - i\epsilon)^2}\,, \tag{22}$$

and is always finite.

Substituting (21) into (20) we find

$$-r_+\varphi(p_-,r) = \left(\frac{M_1^2}{2p_-} + \frac{M_2^2}{2(r_- - p_-)}\right)\varphi(p_-,r) - \frac{g^2}{2\pi}\,\mathrm{P}\int_{-p_-}^{r_- - p_-}\frac{\varphi(p_- + k_-, r)}{k_-^2}\,\mathrm{d}k_- \,. \tag{23}$$

The infra-red cut-off dependence has disappeared! In fact, we have here the exact form of the Hamiltonian discussed in ref. [4]. Let us introduce dimensionless units:

$$\alpha_{1,2} = \frac{\pi M_{1,2}^2}{g^2} = \frac{\pi m_{1,2}^2}{g^2} - 1\,, \quad -2r_+r_- = \frac{g^2}{\pi}\,\mu^2\,; \quad p_-/r_- = x\,; \tag{24}$$

μ is the mass of the two-particle state in units of $g/\sqrt{\pi}$.

Now we have the equation

$$\mu^2\varphi(x) = \left(\frac{\alpha_1}{x} + \frac{\alpha_2}{1-x}\right)\varphi(x) - \mathrm{P}\int_0^1 \frac{\varphi(y)}{(y-x)^2}\,\mathrm{d}y\,. \tag{25}$$

We were unable to solve this equation analytically. But much can be said, in particular about the spectrum. First, one must settle the boundary condition. At the boundary $x = 0$ the solutions $\varphi(x)$ may behave like $x^{\pm\beta_1}$, with

$$\pi\beta_1 \operatorname{cotg} \pi\beta_1 + \alpha_1 = 0\,, \tag{26}$$

but only in the Hilbert space of functions that vanish at the boundary the Hamiltonian (the right-hand side of (25)) is Hermitean:

$$(\psi, H\varphi) = \int\left(\frac{\alpha_1+1}{x} + \frac{\alpha_2+1}{1-x}\right)\varphi(x)\,\psi^*(x)\mathrm{d}x$$

$$+\frac{1}{2}\int_0^1\mathrm{d}x\int_0^1\mathrm{d}y\ \frac{(\varphi(x)-\varphi(y))(\psi^*(x)-\psi^*(y))}{(x-y)^2}\,. \tag{27}$$

In particular, the "eigenstate"

$$\varphi(x) = \left(\frac{x}{1-x}\right)^{\beta_1}$$

in the case $\alpha_1 = \alpha_2$, is not orthogonal to the ground state that does satisfy $\varphi(0) = \varphi(1) = 0$.

Also, from (27) it can be shown that the eigenstates φ^k with $\varphi^k(0) = \varphi^k(1) = 0$ form a complete set. We conclude that this is the correct boundary condition *. A rough approximation for the eigenstates φ^k is the following. The integral in (25) gives its main contribution if y is close to x. For a periodic function we have

$$\mathrm{P}\int_0^1 \frac{e^{i\omega y}}{(y-x)^2}\,\mathrm{d}y \simeq \mathrm{P}\int_{-\infty}^{\infty} \frac{e^{i\omega y}}{(y-x)^2}\,\mathrm{d}y = -\pi|\omega|\,e^{i\omega x}\,.$$

The boundary condition is $\varphi(0) = \varphi(1) = 0$. So if $\alpha_1, \alpha_2 \simeq 0$ then the eigenfunctions can be approximated by

$$\varphi^k(x) \simeq \sin k\pi x\,, \qquad k = 1, 2, \ldots\,, \tag{28}$$

with eigenvalues

$$\mu^2_{(k)} \simeq \pi^2 k\,. \tag{29}$$

This is a straight "Regge trajectory", and there is no continuum in the spectrum! The approximation is valid for large k, so (29) will determine the asymptotic form

* Footnote see next page.

of the trajectories whereas deviations from the straight line are expected near the origin as a consequence of the finiteness of the region of integration, and the contribution of the mass terms.

Further, one can easily deduce from (27), that the system has only positive eigenvalues if $\alpha_1, \alpha_2 > -1$. For $\alpha_1 = \alpha_2 = -1$ there is one eigenstate with eigenvalue zero ($\varphi = 1$). Evidently, tachyonic bound states only emerge if one or more of the original quarks were tachyons (see eq. (24)). A zero mass bound state occurs if both quarks have mass zero.

The physical interpretation is clear. The Coulomb force in a one-dimensional world has the form

$$V \propto |x_1 - x_2| \,,$$

which gives rise to an insurmountable potential well. Single quarks have no finite dressed propagators because they cannot be produced. Only colourless states can escape the Coulomb potential and are therefore free of infra-red ambiguities. Our result is completely different from the exact solution of two-dimensional massless quantum electrodynamics [5, 2], which should correspond to $N = 1$ in our case. The perturbation expansion with respect to $1/N$ is then evidently not a good approximation; in two dimensional massless Q.E.D. the spectrum consists of only one massive particle with the quantum numbers of the photon.

In order to check our ideas on the solutions of eq. (25), we devised a computer program that generates accurately the first 40 or so eigenvalues μ^2. We used a set of trial functions of the type $Ax^{\beta_1}(1-x)^{2-\beta_1} + B(1-x)^{\beta_2}x^{2-\beta_2} + \Sigma_{k=1}^{N} C_k \sin k\pi x$. The accuracy is typically of the order of 6 decimal places for the lowest eigenvalues, decreasing to 4 for the 40th eigenvalue, and less beyond the 40th.

A certain W.K.B. approximation that yields the formula

$$\mu^2_{(n)} \xrightarrow[n\to\infty]{} \pi^2 n + (\alpha_1 + \alpha_2) \log n + C^{\rm st}(\alpha_1, \alpha_2), \qquad n = 0, 1, \ldots \tag{30}$$

was confirmed qualitatively (the constant in front of the logarithm could not be checked accurately).

In fig. 5 we show the mass spectra for mesons built from equal mass quarks. In the case $m_q = m_{\bar{q}} = 1$ (or $\alpha_1 = \alpha_2 = 0$) the straight line is approached rapidly, and the constant in eq. (30) is likely to be exactly $\frac{3}{4}\pi^2$.

In fig. 6 we give some results for quarks with different masses. The mass difference for the nonets built from two triplets are shown in two cases:

(a) $m_1 = 0$; $\qquad m_2 = 0.200$; $\qquad m_3 = 0.400$,

(b) $m_1 = 0.80$; $\qquad m_2 = 1.00$; $\qquad m_3 = 1.20$,

in units of $g/\sqrt{\pi}$. The higher states seem to spread logarithmically, in accordance with

* This will certainly not be the last word on the boundary condition. For a more thorough study we would have to consider the unitarity condition for the interactions proportional to $1/N$. That is beyond the aim of this paper.

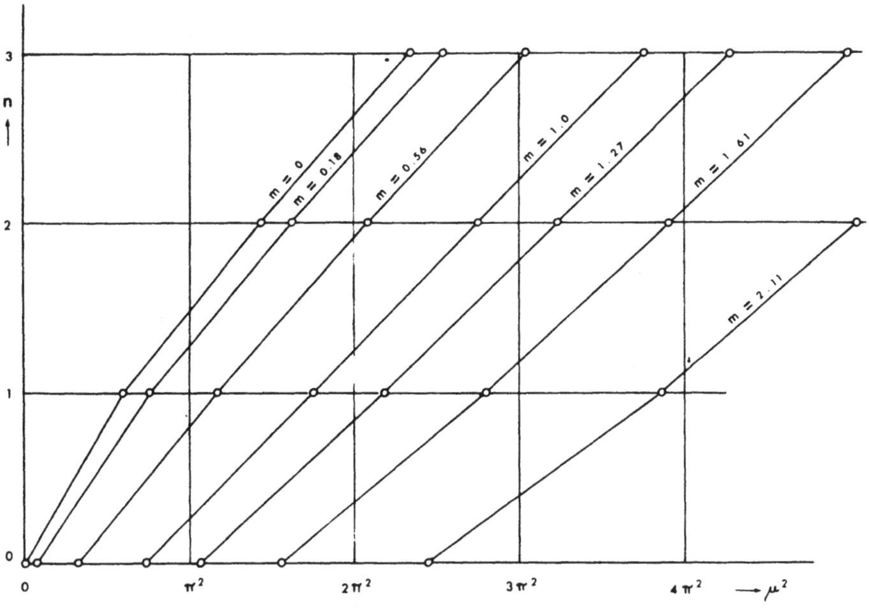

Fig. 5. "Regge trajectories" for mesons built from a quark-antiquark pair with equal mass, m, varying from 0 to 2.11 in units of $g/\sqrt{\pi}$. The squared mass of the bound states is in units g^2/π.

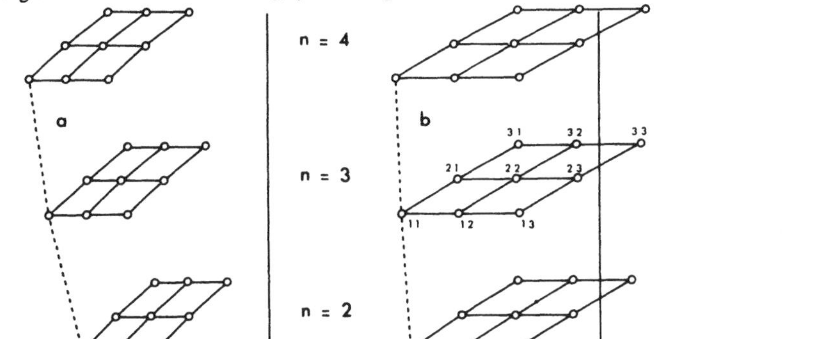

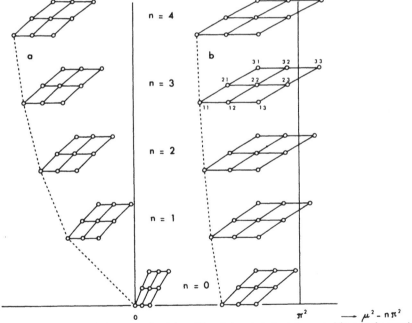

Fig. 6. Meson nonets built from quark triplets. The picture is to be interpreted just as the previous figure, but in order to get a better display of the mass differences the members of one nonet have been separated vertically, and the nth excited state has been shifted to the left by an amount $n\pi^2$. In case (a) the masses of the triplet are $m_1 = 0.00$; $m_2 = 0.20$; $m_3 = 0.40$ and in case (b) $m_1 = 0.80$; $m_2 = 1.00$; $m_3 = 1.20$. Again the unit of mass is $g/\sqrt{\pi}$.

eq. (30). But, contrary to eq. (30), it is rather the average mass, than the average squared mass of the quarks that determines the mass of the lower bound states.

Comparing our model with the real world we find two basic flaws. First there are no transverse motions, and hence there exists nothing like angular momentum, nor particles such as photons. Secondly, at $N = 3$ there exist also other colourless states: the baryons, built from three quarks or three antiquarks. In the $1/N$ expansion, they do not turn up. To determine their spectra one must use different approximation methods and we expect those calculations to become very tedious and the results difficult to interpret. The unitarity problem for finite N will also be tricky.

Details on our numbers and computer calculations can be obtained from the author or G. Komen, presently at CERN.

The author wishes to thank A. Neveu for discussions on the two dimensional gauge field model; N.G. van Kampen, M. Nauenberg and H. de Groot, for a contribution on the integral equation, and in particular G. Komen, who wrote the computer programs for the mass spectra.

References

[1] M.Y. Han and Y. Nambu, Phys. Rev. 139B (1965) 1006.
[2] A. Casher, J. Kogut and L. Susskind, Tel Aviv preprint TAUP-373-73 (June 1973).
[3] P. Olesen, A dual quark prison, Copenhagen, preprint NBI-HE-74-1 (1974).
[4] G. 't Hooft, CERN preprint TH 1786, Nucl. Phys. B, to be published.
[5] J. Schwinger, Phys. Rev. 128 (1962) 2425.

Apparently we find quite realistic meson spectra. The quark masses are free parameters, and in the limit $m_q \to 0$ the pions (and eta's!) are massless, which indicates that the chiral symmetry is spontaneously broken. This spontaneous symmetry breakdown may be surprising in view of the fact that there are theorems forbidding such a breakdown in two dimensions. The reason why these theorems are violated is that the massless mesons do not interact at all if $N = \infty$. One may expect that at finite N infrared divergences due to the massless pion and eta propagators will eventually restore chiral symmetry nonperturbatively.

The paper gave rise to another controversy. The only gauge choice in which the solutions can be given in elegant and practically closed form is the light cone gauge. However in this gauge the zero modes seem to be treated incorrectly, a problem related to the rather dangerous light cone boundary conditions. All we say about this in the paper is that the pole in the propagator is cut off by a principal value integration prescription. That this is actually correct is not so easy to see. As usual, one has to ask what the physical significance of the danger is. To understand the infrared structure of our model we have to take periodic boundary conditions, and these are only consistent with our procedure if the periodic variable is x^-. Now our "gauge condition" is $A_- = 0$, but actually this implies also a constraint: the total magnetic field generated by the closed loop formed by our periodic one-space is constrained to be zero. This is the total magnetic flux caught by our loop. Physically, we expect in the infrared limit that the loop should be so large that this single constraint should become unimportant. The principal value prescription can then be seen to be related to an extra gauge condition for the zero modes. The point here is that besides $A_- = 0$ we could fix the residual global gauge invariance by demanding $A_+(k_- = 0) = 0$. And finally, to judge whether the periodic boundary condition in the lightlike direction is legal, one has to check whether massless modes that survive can move with the speed of light so that acausal behavior might result. Well, we find that all observable objects, the physical mesons, come out to be massive (as long as $m_q \neq 0$), so there is no problem here. The final check comes when one notices that the light cone gauge in the left direction gives the same bound states as the light cone gauge in the right direction. Apparently our procedure was gauge independent.

An interesting problem was raised by D. Gross in a private discussion with the author.[18] Since we start off with a model with conservation of parity, all bound states should either be parity even or parity odd. The even states should be coupled to the current $\bar{\psi}_1\psi_2$ and the odd states to the current $\bar{\psi}_1\gamma_5\psi_2$. One expects the bound states to be alternatingly even and odd under parity. In terms of the solutions of the integral equation (25) one derives that the scalar current couples to a solution $\varphi(x)$ *via* the integral

$$\int_0^1 \left(\frac{m_1}{x} - \frac{m_2}{1-x}\right)\varphi(x)\mathrm{d}x\,,$$

[18] C. G. Callan, N. Coote, and D. J. Gross, *Phys. Rev.* **D13** (1976) 1649.

and the pseudoscalar current *via* the integral

$$\int_0^1 \left(\frac{m_1}{x} + \frac{m_2}{1-x}\right)\varphi(x)\mathrm{d}x\,,$$

where m_1 and m_2 are the masses of the quark and the antiquark inside the meson. If the masses are equal it is evident that for the even solutions the first integral vanishes and for the odd ones the second integral. But if the masses are unequal this is far from evident. The reason why it is so difficult to see is that our light cone gauge condition is not parity invariant.

Yet it is true. To see this, consider the product of the two integrals, and prove that it is equal to

$$\int_0^1 \mathrm{d}x \int_0^1 \mathrm{d}y \left(\frac{{m_1}^2}{xy} - \frac{{m_2}^2}{(1-x)(1-y)}\right)\varphi(x)\varphi(y) = \langle\varphi|[H,P]|\varphi\rangle\,,$$

where the operator H is, up to a constant, the right-hand side of Eq. (25), and P is defined by

$$P\varphi(x) = P\int_0^1 \frac{1}{y-x}\varphi(y)\mathrm{d}y\,.$$

This commutator expectation value must vanish for the eigenstates of H, hence one of the two integrals must vanish. All states are either even or odd under parity. This result confirms once again that our procedure is gauge independent.

CHAPTER 7
QUARK CONFINEMENT

CHAPTER 7

QUARK CONFINEMENT

Introduction to Confinement and Topology in Non-Abelian Gauge Theories [7.1]
Introduction to the Confinement Phenomenon in Quantum Field Theory [7.2]
Introduction to Can We Make Sense out of Quantum Chromodynamics? [7.3]

How can we understand the behavior of quarks in the theory of Quantum Chromodynamics? Is "ionization" of mesons and baryons into "free quarks" indeed absolutely forbidden? What is the mechanism for this absolute and permanent verdict of "confinement"? And finally, how do we calculate the details of the properties of mesons and baryons once the QCD Lagrangian is given?

The answers to these questions came very gradually. Our understanding became more and more complete. Precise calculations are still very difficult these days, but there seems to be no longer any fundamental obstacle. Confinement is a state of aggregation, comparable to the solid state, the liquid state, the gaseous state or the plasma state of ordinary materials. Other modes a system such as quantum chromodynamics could have condensed into, are various versions of Higgs phases or the Coulomb phase. Only in the Coulomb phase physical massless vector particles would persist in the physical spectrum, but, as will be demonstrated, these particles would have little left of what was once their non-Abelian character; they could only survive disguised as more or less ordinary photons (an exception to this would be QCD with such a large number of fermions, that the sign of the β function is reversed).

I wrote a number of papers on the subject, and a couple of summer school lecture notes. Three summer school lecture notes are reproduced here; for some of the technical details I refer the reader to the original papers. There is so much material to be discussed that the three papers reproduced here hardly overlap; they show different approaches to the confinement problem. In the first two papers the notion of "dual transformations", or the transition from order parameters to disorder parameters and back, plays a very important role.

It is instructive to consider these transformations in a box with periodic boundary conditions at finite temperature $T = 1/\beta$. This is done in the first paper, from Section VI onwards. Combining methods from field theory and statistical physics it is discovered there that powerful identities are obtained by rotating this box over 90° in Euclidean space. One finds distinct possibilities for the system to behave at large distance: the Higgs mode, the Coulomb mode and the confinement mode. The first and the last are each other's dual.

The last paper addresses the question of calculability. Its results are modest, if not disappointing. Even if our theory is asymptotically free, replacement of perturbation expansion by some guaranteedly convergent resummation procedure seems to be impossible. Only very modest formal improvement on the most naive perturbative expansions could be obtained. The much more pragmatic approach of Monte Carlo simulations on lattices seems to give much more satisfactory results in practice, but formal proofs that these methods converge to unique spectral data and transition amplitudes when the lattice becomes infinitely dense have never been given. I have nothing to add to this except the conjecture that indeed the Monte Carlo calculations define the theory rigorously seems to be quite plausible.

A small correction must be made in this last paper. As was correctly pointed out to me by Münster, the instanton singularities and the infrared renormalon singularities in QCD will not show up independently as suggested in Fig. 6, but they merge, due to a divergence of the integration over the instanton size parameters ρ.

Acta Physica Austriaca, Suppl. XXII, 531–586 (1980)

CHAPTER 7.1

CONFINEMENT AND TOPOLOGY IN NON-ABELIAN GAUGE THEORIES[+]

by

G. 't HOOFT
Institute for Theoretical Physics
Princetonplein 5, P.O.Box 80006
3508 TA Utrecht, The Netherlands

ABSTRACT

Pure non-Abelian theories with gauge group SU(N) are considered in 3 and 4 space-time dimensions. In 3 dimensions non-perturbative features invalidate ordinary coupling constant expansions. Disorder operators can be defined in 3 and 4 dimensions. Confinement is first explained in 3 dimensions as a spontaneous breakdown of a topologically defined Z(N) global symmetry of the theory. In 4 dimensions confinement can be seen as one of the various possible phases of the system by considering a box with periodic boundary conditions in the "thermodynamic limit". An exact duality equation allows either electric or magnetic flux tubes to be stable, but not both.

Special attention is given to the explicit oc-

[+]Lecture given at XIX. Internationale Universitätswochen für Kernphysik, Schladming, February 20 - 29, 1980.

currence of instanton configurations with an instanton angle θ in deriving duality.

I. INTRODUCTION

The first non-Abelian gauge theory that was recognized to describe interactions between elementary particles, to some extent, was the Weinberg-Salam-Ward-GIM model[1]. That model contains not only gauge fields but also a scalar field doublet H. Perturbation expansion was considered not about the point H = 0, but about the "vacuum value"

$$H = \binom{F}{0} .$$

Such a theory is usually called a theory with "spontaneous symmetry breakdown" [2]. In contrast one might consider "unbroken gauge theories" where perturbation expansion is only performed about a symmetric "vacuum". These theories are characterized by the absence of a mass term for the gauge vector bosons in the Lagrangian. The physical consequences of that are quite serious. The propagators now have their poles at $k^2=o$ and it will often happen that in the diagrams new divergences arise because such poles tend to coincide. These are fundamental infrared divergencies that imply a blow-up of the interactions at large distance scales. Often they make it nearly impossible to understand what the stable particle states are.

A particular example of such a system is "Quantum Chromodynamics", an unbroken gauge theory with gauge group SU(3), and in addition some fermions in the 3-representation of the group, called "quarks". We will investigate the possibility that these quarks are permanently confined inside bound structures that do not carry gauge quantum numbers. First of all this idea is not as absurd as it may seem. The converse would be equally difficult to understand. Gauge quantum numbers are a priori only defined up to local gauge transformations. The existence of *global* quantum numbers that would correspond to these local ones but would be detectable experimentally from a distance is not at all a prerequisite. We are nevertheless accustomed to attaching a global significance to local gauge transformation properties because we are familiar with the theories with spontaneous breakdown. The electron and its neutrino, for example, are usually said to form a gauge doublet, to be subjected to local gauge transformations. But actually these words are not properly used. Even the words "spontaneous breakdown" are formally not correct for local gauge theories (which is why I put them between quotation marks). The vacuum *never* breaks local gauge invariance because it itself is gauge invariant. All states in the physical Hilbert space are gauge-invariant. This may be confusing so let me illustrate what I mean by considering the familiar Weinberg et al model. The invariant Lagrangian is

$$L^{inv} = -\frac{1}{4}G^a_{\mu\nu}G^a_{\mu\nu}-\frac{1}{4}F_{\mu\nu}F_{\mu\nu}-D_\mu H^* D_\mu H-V(|H|)$$
$$- \bar{\psi}_L\gamma D\psi_L-\bar{e}_R\gamma De_R-k\bar{e}_R(H^*\psi_L)-k(\bar{\psi}_L H)e_R \; . \qquad (1.1)$$

Here H is the scalar Higgs doublet. The gauge group is SU(2)xU(1), to which correspond $A^a_\mu(G^a_{\mu\nu})$ and $A^o_\mu(F_{\mu\nu})$ The subscripts L and R denote left and right handed components of a Dirac field, obtained by the projection operators $\frac{1}{2}(1 \pm \gamma_5)$.

e_R is a singlet;

ψ_L is a doublet.

D_μ stands for covariant derivative.

The function $V(|H|)$ takes its minimum at $|H|=F$. Usually one takes

$$\langle H \rangle_{\text{vacuum}} = \binom{F}{0} , \qquad (1.2)$$

and perturbs around that value: $H = \binom{F + h_1}{h_2}$.

One identifies the components of ψ_L with neutrino and electron:

$$\psi_L = \binom{\nu_L}{e_L} . \qquad (1.3)$$

However, this model is *not* fundamentally different from a model with "permanent confinement". One could interpret the same physical particles as being all gauge singlets, bound states of the fundamental fields with extremely strong confining forces, due to the gauge fields A_μ^a of the group SU(2). We have scalar quarks (the Higgs field H) and fermionic quarks (the ψ_L field) both as fundamental doublets. Let us call them q. Then there are "mesons" ($q\bar{q}$) and "baryons" (qq). The neutrino is a "meson". Its field is the composite, SU(2)-invariant

$$H^* \psi_L = F \quad \nu_L + \text{negligible higher order terms.}$$

The e_L field is a "baryon", created by the SU(2)-invariant

$$\varepsilon_{ij} H_i \psi_{Lj} = F e_L + \ldots \quad , \tag{1.4}$$

the e_R field remains an SU(2) singlet.
Also bound states with angular momentum occur: The neutral intermediate vector boson is the "meson"

$$H^* D_\mu H = \frac{i}{2} g F^2 A_\mu^{(3)} + \text{total derivative} + \text{higher orders}, \tag{1.5}$$

if we split off the total derivative term (which corresponds to a spin-zero Higgs particle).
The $W^\pm$ are obtained from the "baryon" $\varepsilon_{ij} H_i D_\mu H_j$ and the Higgs particle can also be obtained from $H^* H$.
Apparently some mesonic and baryonic bound states survive perturbation expansion, most do not (only those containing a Higgs "quark" may survive).

Is there no fundamental difference then between a theory with spontaneous breakdown and a theory with confinement? Sometimes there is. In the above example the Higgs field was a faithful representation of SU(2). This is why the above procedure worked. But suppose that all scalar fields present were invariant under the center Z(N) of the gauge group SU(N), but some fermion fields were not. Then there are clearly two possibilites. The gauge symmetry is "broken" if physical objects exist that transform non-trivially under Z_N, such as the fundamental fermions. We call this the Higgs phase. If on the other hand all physical objects are invariant under Z_N, such as the mesons and the baryons, then we have permanent confinement.

Quantum Chromodynamics is such a theory where these distinct possibilities exist. It is unlikely that one will ever prove from first principles that permanent confinement takes place, simply because one can always imagine the Higgs mode to occur. If no fundamental scalar fields exist then one could introduce composite fields such as

$$H_{ab} = G^{a}_{\mu\nu} G^{b}_{\mu\nu} \quad ,$$

or

$$H^{j}_{i} = \bar{\psi}_{i} \psi^{j} \quad ,$$

and postulate nonvanishing vacuum expectation values for them:

$$\langle H_{ab} \rangle = F_{1} d_{ab8} + F_{2} d_{ab3}$$

or $$\langle H^{j}_{i} \rangle = F_{1} \lambda^{j}_{8i} + F_{2} \lambda^{j}_{3i} \quad .$$

In that case there would be no confinement. Whether or not $F_{1,2}$ are equal to zero will depend on details of the dynamics. Therefore, dynamics must be an ingredient of the confinement mechanism, not only topological arguments. What we will attempt in this lecture is to show that topological arguments imply for this theory the existence of phase regions, separated by sharp phase transition boundaries (usually of first order). One region corresponds to what is usually called "spontaneous breakdown", and will be referred to as Higgs phase. Another corresponds to absolute quark confinement. Still another phase exists which allows for long range Coulomb-like forces to occur. (Coulomb phase.)

It is illustrative to consider first pure gauge theories in 3 space-time dimensions. These differ in two important ways from their 4 dimensional counter parts. First, they are not scale-invariant in the classical limit. A consequence of that is that they do not have a computable small coupling constant expansion. Already at finite orders in the coupling constant g phenomena occur that are associated to a complex sort of vacuum instability. This is explained in Sects 2 and 3. Secondly, the topological properties are different. In three dimensions a "disorder parameter", being an operator-valued field $\Phi(\vec{x},t)$ can be defined. If

$$\langle\Phi(x,t)\rangle_{\text{vacuum}} \neq 0 \quad ,$$

then there is absolute confinement, as we explain in Sects. 4 and 5. The remaining sections are devoted to the four-dimensional case. They overlap to some extent lectures given at Cargèse[3] except for a more explicit consideration of effects due to instantons and their angle θ. Here it is convenient to introduce the "periodic box" (a cubic or rectangular box with periodic or pseudo-periodic boundary conditions). Again we have order and disorder operators but now they are defined not on space-time points (x,t) but on loops (C,t). In Sect. 7 it is explained how to interpret these operators in terms of magnetic and electric flux operators, and magnetic flux is defined in terms of the boundary conditions of the box. There are also other conserved quantities (Sect.8), to be interpreted as electric flux and instanton angle.

In Sect. 9 we consider the "hot box", a box at finite temperature $T = 1/k\beta$, and express the free energy F of such a box in terms of functional integrals with twisted periodic boundary conditions in 4-dimensional Euclidean space. In Sect. 10 we notice an exact duality relation fot the free energy of electric and magnetic fluxes. Section 11 shows that if electric confinement is assumed, magnetic confinement is excluded, and vice versa. The Coulomb phase realized for instance in the Georgi-Glashow model [4] is dually symmetric, as explained in sect. 12.

II. DIFFICULTIES IN THE PERTURBATION EXPANSION FOR QCD IN 2+1 DIMENSIONS

Consider pure gauge theory in 2+1 dimensions. The Lagrangian is

$$L = -\frac{1}{4} G^a_{\mu\nu} G^a_{\mu\nu} \quad , \tag{2.1}$$

with

$$G^a_{\mu\nu} = \partial_\mu A^a_\nu - \partial_\nu A^a_\mu + g f^{abc} A^b_\mu A^c_\nu \quad . \tag{2.2}$$

The functional integrals to be studied are

$$\int DA \exp(i\int L \, d^2\vec{x} \, dt + \text{source terms}) . \tag{2.3}$$

Clearly $\int L \; d^2x \, dt$ has to be dimensionless, therefore A^a_μ has dimension $(\text{mass})^{\frac{1}{2}}$ and g has dimension $(\text{mass})^{\frac{1}{2}}$.

Gauge-invariant quantities are ultra-violet convergent if they are regularized in a gauge invariant way (for instance by dimensional regularization). This is because the only possible counter terms would be

$$g^2 \, G_{\mu\nu} G_{\mu\nu} \; ; \; g^4 \, G_{\mu\nu} G_{\mu\nu} \; , \text{ etc.} \tag{2.4}$$

Which would all be of too high dimension. Conventionally this would imply that all Green's functions in Euclidean space would be well defined perturbatively. The physical theory (in Minkowsky space) would then be obtained by analytic continuation.

However, in our case we do have an infra-red problem, even in Euclidean space. Consider namely diagrams of the following type:

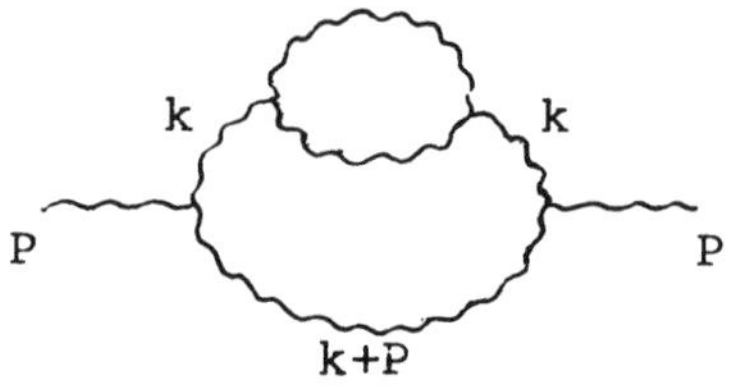

(2.5)

Simple power counting tells us that the small self-energy insertion is proportional to $g^2|k|$, where k is the momentum circulating in the large blob. Because of the two propagators the k integration has an infra-red divergent part:

$$\int d^3k \, \frac{|k|}{(k^2-i\varepsilon)^2} \text{ (remainder)} . \tag{2.6}$$

Here the divergence is only logarithmic, but it becomes worse if more self-energy insertions occur in the k-propagator. How should one cure such divergences?

To understand the physics of this infra-red divergence let us consider on easier case first: $g\phi^3$ theory without mass term:

$$L = -\frac{1}{2}(\partial_\mu\phi)^2 - \frac{g}{3!}\phi^3 , \tag{2.7}$$

this time in four space-time dimensions: Power counting tells us that ϕ has dimension (mass) [1] and g has dimension (mass)[1]. This theory is a little bit more artifical because an ultraviolet divergence has to be subtracted by a mass-counter term:

$$\Delta L \propto g^2 \phi^2 , \tag{2.8}$$

but still we could consider starting perturbation theory with no residual mass. If we take the same diagram (2.5) then now the self-energy insertion behaves as

$$g^2 \log |k| , \tag{2.9}$$

and the k integration is again infra-red divergent:

$$\int d^4k \frac{\log |k|}{(k^2-i\varepsilon)2} (\text{remainder}) . \tag{2.10}$$

In this case however the cure seems obvious: we were doing perturbation expansion at a very singular point. No problem arises if we first introduce a small mass term

$$L \to L - \frac{1}{2} \mu^2 \phi^2 , \tag{2.11}$$

and then let $\mu^2 \downarrow 0$ in the end. How are the infinities such as (2.10) "regularized" if we do that?

If k^2 in (2.10) is replaced by $k^2+\mu^2$ then the limit $\mu^2 \downarrow 0$ does not exist. However one may argue that this infinity is not very physical. Suppose we sum diagrams of the type:

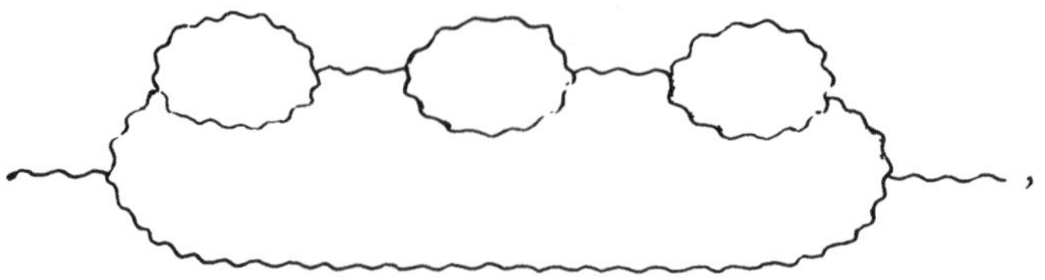

(2.12)

then the k integration is of the form

$$\int d^4k \, \frac{1}{k^2+a_1g^2\log|k|+\mu^2-i\varepsilon} \text{ (remainder)}. \qquad (2.13)$$

Clearly the limit $\mu^2 \downarrow 0$ exist here. It has a small imaginary part due to a tachyonic pole in the dressed propagator, which cannot be admitted in a real theory, but this is probably due to the fact that ϕ^3 theory is unstable and we ignore it. The point is that the "summed" theory is supposed to be free of infra-red divergences. In writing down (2.13) we replaced a sometimes divergent sum of bubles by an analytic and convergent expression. This expression could have been obtained directly if we wrote the Dyson-Schwinger equations for these diagrams:

(2.14)

If eq. (2.14) is used then the diagram

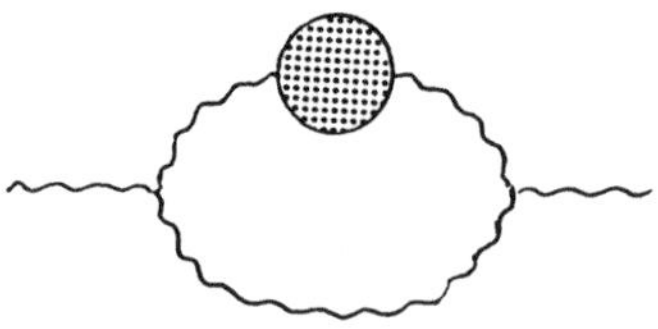

(2.15)

yields (2.13). Now (2,14) is only a truncated Dyson-Schwinger equation. If we would use the complete set of equations we would get the complete amplitude and this may be assumed to be free of divergences. (Also, in a good theory, free of tachyonic poles or cuts). A good theory should be a solution to its Dyson-Schwinger equations with dressed propagators not more singular than $1/k^2$.

Let us return to pure gauge theory in 2+1 dimensions. We can now roughly estimate how the integral (2.6) has to be cut-off, by replacing the bare propagators by dressed ones. This is not exactly according to Dyson-Schwinger equations but good enough for our purpose. The dressed propagator is

$$\frac{1}{k^2} f\left(g^2/|k|\right) \quad . \tag{2.16}$$

Let us assume that $f(z)$ has an expansion $1+a_1 z + a_2 z^2 + \ldots$ but is non singular at $z \to \infty$.
(This asymptotic expansion for f is not quite right, as as we will see later). Our integral is

$$I = \int d^3k \, \frac{f(g^2/\,|k|)}{k^2} \, R(k) \quad , \tag{2.17}$$

where $R(k)$ is the non-singular remainder. Let us expand R:

$$R(k) = R_o + R_\mu \, k_\mu + R_{\mu\nu} \, k_\mu \, k_\nu + \ldots , \tag{2.18}$$

and split the integral in two pieces:

$$\int d^3k = \int_{|k|<\varepsilon} d^3k + \int_{|k|>\varepsilon} d^3k \quad , \tag{2.19}$$

with $g^2 << \varepsilon << 1$. We write (2.19) as

$$I(g^2) = I_1(\varepsilon,g^2) + I_2(\varepsilon,g^2) \quad . \tag{2.20}$$

It is easy now to establish how $I_{1,2}$ behave for small ε and g^2:

$$\begin{aligned} I_2(\varepsilon,g^2) &= \\ &\int_{|k|>\varepsilon} \frac{d^3k}{k^2}\left(1+\frac{a_1g^2}{|k|}+\frac{a_2g^4}{|k|^2}+\ldots\right)(R_o+R_\mu k_\mu+R_{\mu\nu}k_\mu k_\nu\ldots) \\ &= \sum_n A_n(g^2)^n - 4\pi R_o\left(\varepsilon+g^2a_1\log\varepsilon-\frac{g^4a_2}{\varepsilon}\ldots\right) \\ &- \frac{4\pi}{3}R_{\mu\mu}\left(\frac{\varepsilon^3}{3}+\frac{g^2a_1\varepsilon^2}{2}+g^4a_2\varepsilon+g^6a_3\log\varepsilon\ldots\right) \end{aligned} \tag{2.21}$$

And if we write $k_\mu = g^2\omega_\mu$, then

$$\begin{aligned} I_1(\varepsilon,g^2) &= \\ g^2 &\int_{|\omega|<\varepsilon/g^2} \frac{d^3\omega}{\omega^2}\left(1+\frac{a_1}{\omega}+\frac{a_2}{\omega^2}+\ldots\right)(R_o+g^2R_\mu\omega_\mu+g^4R_{\mu\nu}\omega_\mu\omega_\nu\ldots) \\ &= \sum_n B_n(g^2)^n + 4\pi g^2R_o\left(\frac{\varepsilon}{g^2}+a_1\log\frac{\varepsilon}{g^2}+\ldots\right) \\ &+ \frac{4\pi}{3}g^6R_{\mu\mu}\left(\frac{\varepsilon^3}{3g^6}+a_1\frac{\varepsilon^2}{2g^4}+a_2\frac{\varepsilon}{g^2}+a_3\log\frac{\varepsilon}{g^2}+\ldots\right) \end{aligned} \tag{2.22}$$

Taking the two series together we find that, of course, the ε dependence disappears, but the asymptotic g^2-expansion does not only contain powers of g^2:

$$I(g^2) = \sum_n (A_n+B_n)(g^2)^n - 4\pi g^2 R_0 a_1 \log g^2$$
$$- \frac{4\pi}{3} g^6 R_{\mu\mu} a_3 \log g^2 + \dots \quad (2.23)$$

So now we found where the infra-red infinity of the integral (2.6) goes: $I(g^2)$ has no proper Taylor expansion in g^2. Eq. (2.23) shows terms with $\log g^2$. One consequence of that is that $f(g^2/_{|k|})$ in (2.16) will also develop logarithms further down in its expansion series. If we correct the previous analysis for this we find higher powers of logaritms further down the series.

Since a_1 and R_0 are known, the coefficient in front of the logarithm is well determined. On the other hand, the coefficient B_1 can only be determined if $f(z)$ is comletely known, because it must be integrated over. In general, at any fixed power of g^2, only the leading power of $\log g^2$ has a well determined coefficient. All other coefficients can only be evaluated if the complete non-perturbative solution of $f(z)$ is known. At higher orders also three- and more-point Green functions are required. Thus, the coefficients for the non-leading powers of $\log g^2$ can never be determined by perturbative means alone.

III. VACUUM STRUCTURE IN THE PERTURBATIVE 2+1 DIMENSIONAL THEORY. GAUGE INVARIANCE

Does the disease observed in the previous section have anything to do with confinement? One might argue that the dressed propagator does not exist in a theory with confinement. However, since it is not gauge-invariant, it might exist in certain convenient gauges, and we do not have to worry about spurious poles or cuts in this propagator because they need not correspond to physical states. In fact there are reasons to believe that in certain covariant gauges the propagator is actually very smooth and convergent.

Confinement is a property of the vacuum in the field theory and, as we will argue now, it is the vacuum that determines our unknown coefficients. The first unknown coefficient was B_1 in (2.22). It is

$$B_1 = \int_{|\omega|<1} \frac{d^3\omega}{\omega^2} f(1/|\omega|)\ R_o\ . \qquad (3.1)$$

Although unknown, its dependence on the remainder of the graph ,R, is very simple: it only depends on R (k=o). This implies that once it has been determined for one graph, it will be known for all other graphs as well. Let us take the diagram with R=1:

(3.2)

It corresponds to the computation of

$$<A^2(x)>_{vacuum}$$

Of course it is ultra-violet divergent, but that divergence can be assumed to be subtracted in the usual way. We conclude that if we assume some value for

$$C_o = <A^2(x)>^{subtr.}_{vacuum} \tag{3.3}$$

then all amplitudes can be expanded up to

$$\begin{aligned} \Gamma(k,\ldots,g^2) &= \Gamma_o(k,\ldots) + g^2\Gamma_1(k,\ldots) \\ &+ g^4\Gamma_2^o(k,\ldots) + g^4 \log g^2 \; \Gamma_2^1(k,\ldots) \\ &+ O(g^6) \quad . \end{aligned} \tag{3.4}$$

In particular, Γ_2^o depends on C_o. At higher orders one will need

$$\begin{aligned} C_2 &= <A^4(x)>^{subtr.}_{vacuum} \, , \\ C_3 &= <\partial A \; \partial A>^{subtr.}_{vacuum} \, , \end{aligned} \tag{3.5}$$

etc.

One may observe that A^2 in (3.3) is not gauge-invariant. Therefore, the actual value of C_o should not affect the Γ_2^o of gauge-invariant Green functions. Indeed, gauge-invariant Green functions have $R_o = o$, as a consequence of certain Ward-Takahashi identities.

But of course, the relevant quantities are

$$a_1 = \langle G_{\mu\nu}\, G_{\mu\nu} \rangle_{\text{vacuum}}^{\text{subtr.}}$$

$$b_1 = \langle \bar{\psi}\psi \rangle_{\text{vacuum}}^{\text{subtr.}}$$

$$a_2 = \langle (D_\alpha G_{\mu\nu})^2 \rangle_{\text{vacuum}}^{\text{subtr.}} \quad \text{etc.} \tag{3.6}$$

By power counting it is possible to tell where in the power series these coefficients center. a_1 has dimension 3; b_1 has dimension 2; a_2 has dimension 5. And g^2 has dimension 1. Therefore, a_1 will enter at order g^8; b_1 at order g^6 and a_2 at order g^{12}.

One may hope that iterative procedures exist to determine these coefficients semi-perturbatively: determine the known coefficients for $f(z)$; extrapolate to find a decent function for $z \to \infty$ (Padé?), then substitute this $f(z)$ to find the next coefficients and repeat. Whatever one does, the procedure will not be as straightforward as the determination of $\langle H \rangle$ from a Higgs Lagrangian. The unknowns this time form an infinite series. They are the vacuum expectation values of all composite operators.

It may seem that these vacuum experation values are only needed in the 2+1 dimensional theory, and that the problems discussed here are actually irrelevant for the real world which is in 3+1 dimensions. This, we emphasize, is not true. In 3+1 dimensions the same problem occurs, however not within the ordinary perturbation expansion. This perturbation expansion namely diverges at high orders. If one tries to rearrange the perturbation expansion to obtain convergent expressions, one finds that amplitudes can be conveniently expressed by a Borel formula:

$$\Gamma(g^2) = \int_0^\infty B(z)\, e^{-z/g^2}\, dz/g^2 \quad .$$

According to the renormalization group B(z) has dimension proportional to z. If this dimension coincides with that of $\langle G_{\mu\nu} G_{\mu\nu} \rangle$, etc. singularities arise and their treatment goes along the same lines as in the previous section. We refer to refs. [5,6]for further details.

IV. 2+1 DIMENSIONS NON PERTURBATIVELY: TOPOLOGICAL OPERATORS AND THEIR GREEN FUNCTIONS

In this section a new field is introduced that will enable us to get a better understanding of the possible vacuum structures in 2+1 dimensions. To set up our arguments step by step we begin with adding a Higgs field[7].The gauge group is SU(N) and the gauge symmetry is spontaneously and completely broken. The Higgs fields,H, must be a set of unique representations of SU(N)/Z(N) such as the octet and decuplet representations of SU(3)/Z(3). Thus all (vector and scalar) fields are invariant under the center Z(N) of the gauge group SU(N). (This is the subgroup of matrices $e^{2\pi in/N} I$, where n is integer.) Quarks, which are not invariant under Z(N), are not yet introduced at this stage.

Besides the massive photons and Higgs particle(s) this model contains one other class of particles:

extended soliton solutions that are stable because of a topological conservation law[+]. Consider, therefore, a region R in two-dimensional space surrounded by another region B where the energy density is zero (vacuum). In B, the Higgs field $H(\vec{x})$ satisfies

$$|\langle H(\vec{x})\rangle| = F , \tag{4.1}$$

where F is a fixed number. There must be a gauge rotation $\Omega(\vec{x})$ so that

$$\Omega(\vec{x})H(\vec{x}) = H_o , \tag{4.2}$$

where H_o is fixed. $\Omega(\vec{x})$ is determined up to elements of Z(N). We also require absence of singularities so $\Omega(\vec{x})$ is continuous. Consider a closed contour $C(\theta)$ in B parametrized by an angle θ with $0 \leq \theta \leq 2\pi$ and $C(0) = C(2\pi)$. Consider the case that C goes clockwise around R. Since B is not simply connected we may have

$$\Omega(2\pi) = e^{2\pi\, in/N}\,\Omega(0) , \tag{4.3}$$

with $0 \leq n \leq N$, n integer. Because of continuity, n is

---

[+]For a general introduction to solitons see e.g. Coleman [8]. For an introduction to this procedure see 't Hooft [9].

conserved. If $n \neq 0$ and if we require absence of singularities in R then the field configuration in R cannot be that of the vacuum or a gauge rotation thereof, so there must be some finite amount of energy in R. The field configuration with lowest energy E, in the case $n = 1$, describes a stable soliton with mass M=E. If $N > 2$ then solitons differ from antisolitons (which correspond to n=N-1) and the number of solitons minus antisolitons is conserved modulo N.

An alternative way to represent the fields corresponding to a soliton configuration is to extend $\Omega(\vec{x})$ to be also within R (which, however, may be possible only if we admit a singularity for Ω at some point $\vec{x}_o$ in R). We then apply Ω^{-1} to the above field configuration. This has the advantage that everywhere in B we keep $H=H_o$, regardless of the number n of solitons in R, but the price of that is to allow for a singularity $\vec{x}_o$ in R. We will refer to this as the "second representation" of the soliton, for later use.

A set of operators $\Phi(\vec{x})$ is now defined as follows. Let $|A_i(\vec{x}),H(\vec{x})\rangle$ be a state in Hilbert space which is an eigenstate of the space components of the vector fields and the Higgs fields, with $A_i(\vec{x})$ and $H(\vec{x})$ as given eigenvalues. Then

$$\Phi(\vec{x}_o)|A_i(\vec{x}),H(\vec{x})\rangle = |A_i^{\Omega[\vec{x}_o]}(\vec{x}),H^{\Omega[\vec{x}_o]}(\vec{x})\rangle , \tag{4.4}$$

where $\Omega^{[\vec{x}_o]}$ gauge rotation with the property that for every closed curve $C(\theta)$ that encloses $\vec{x}_o$ once we have

$$\Omega^{[\vec{x}_o]}(\theta=2\pi) = e^{\pm 2\pi i/N}\,\Omega^{[\vec{x}_o]}(\theta=0) , \tag{4.5}$$

where the minus sign holds for clockwise and the + sign

for anticlockwise C. When $\vec{x}_0$ is outside C, then

$$\Omega^{[\vec{x}_0]}(\theta=2\pi) = \Omega^{[\vec{x}_0]}(\theta=0) \quad . \tag{4.6}$$

The singularity of $\Omega^{[\vec{x}_0]}$ at $\vec{x}=\vec{x}_0$ must be smeared over an infinitesimal region around $\vec{x}_0$ but we will not consider this "renormalization" problem in this paper. In what sense is $\Phi(\vec{x})$ a *local* operator? The operator formalism in gauge theories is most conveniently formulated in the gauge $A_0(\vec{x},t)=0$. Then, time-independent continuous gauge rotations $\Omega(\vec{x})$ still form an invariance group. For physical states $|\psi\rangle$ however,

$$\langle AH|\psi\rangle = \langle A^{\Omega}\; H^{\Omega}|\psi\rangle \tag{4.7}$$

where Ω is any single-valued gauge rotation. So $\Phi(\vec{x})$ would have been trivial were it not that $\Omega^{[\vec{x}_0]}$ has a singularity at $\vec{x}_0$.

The operator $\Phi(\vec{x})$ leads physical states into physical states, and the details of Ω apart from (4.5) and (4.6) are irrelevant. It is now easy to verify that

$$\begin{aligned} \Phi(\vec{x})\Phi(\vec{y})\,|\psi\rangle &= \Phi(\vec{y})\Phi(\vec{x})\,|\psi\rangle \;, \\ \Phi^{+}(\vec{x})\Phi(\vec{y})\,|\psi\rangle &= \Phi(\vec{y})\Phi^{+}(\vec{x})\,|\psi\rangle, \end{aligned} \tag{4.8}$$

if $|\psi\rangle$ is physical state, because both the left- and the right-hand sides of (4.8) are completely defined by the singularities alone of the combined gauge rotations. Also, when $R(\vec{x})$ is a conventional gauge-invariant field operator composed of fields at $\vec{x}$ then obviously

$$[R(\vec{x}),\Phi(\vec{y})] = 0 \text{ for } \vec{x} \neq \vec{y}, \quad \text{but not necessarily for } \vec{x} = \vec{y} \tag{4.9}$$

Because of (4.8) and (4.9) Φ is considered to be a local field operator when it acts on physical states.

From its definition it must be clear that $\Phi(\vec{x})$ absorbs one topological unit, so we say that $\Phi(\vec{x})$ is the annihilation (creation) operator for one "bare" soliton (antisoliton) at $\vec{x}$ and $\Phi^+(x)$ is the creation (annihilation) operator for one "bare" soliton (antisoliton).

Let us now illustrate how one can compute Green functions involving $\Phi(x)$ by ordinary saddle-point techniques in a functional integral. Let us consider $<T(\Phi(0,t_1)\ \Phi^+(0,0))>= f(t_1)$ by computing the corresponding functional-integral expression:

$$f(t_1) = \frac{\int_C DADH \exp S(A,H)}{\int DADH \exp S(A,H)} , \tag{4.10}$$

where C is the set of field configurations where the fields make a sudden gauge jump at t=0 described by $\Omega^{[0]}$ (see (4.5) and (4.6)) and at $t=t_1$ they jump back by a transformation $\Omega^{[0]+}$ fields must be continuous everywhere else.

This was how $f(t_1)$ follows from the definitions but it is more elegant to transform a little further. By gauge transforming back in the region $0 < t < t_1$ we get that the fields are continuous everywhere except at $\vec{x}=0, 0 < t < t_1$. So there is a Dirac string [10] going from (0,0) to (0,t). At $0 < t < t_1$ we obtain in this way the soliton in its "second representation": a non-trivial field configuration with a singularity at the origin. In short: $\langle\Phi(x)\Phi^{+}(0)\rangle$ is obtained by integrating over field configurations with a Dirac string in space-time from 0 to x.

Let us compute $f(t_1)$ for Euclidean time $t_1 = i\tau$, τ real. Let the field theory be pure SU(N) without Higgs scalars. We must find the least negative action configuration with the given Dirac string. The conditions (4.5), (4.6) can be realized for an Abelian subgroup of gauge rotations Ω, so let us take

$$\Omega^{[0]^+}(\theta) = \begin{bmatrix} e^{-i\theta/N} & & & \emptyset \\ & \ddots & & \\ & & e^{-i\theta/N} & \\ \emptyset & & & e^{i\theta(1-1/N)} \end{bmatrix}, \qquad (4.11)$$

the last diagonal element being different from the others because we must have $\det\Omega = 1$ for all θ. Here θ is the angle around the time axis (remember space-time is here three dimensional). The singularity at the Dirac string must be the one obtained when $\Omega^{[0]^+}$ acts on the vacuum.

The transformations (4.11) form an Abelian subgroup, so, as an ansatz, the field configuration with this string singularity may be chosen to be an Abelian subset of fields corresponding to this subgroup:

$$\tfrac{1}{2}\lambda^a_{ij}\, A^a_\mu(x) = a_\mu(x)\, \lambda^{ij}(N) \ ;$$

$$\lambda^{ij}_{(N)} = \begin{bmatrix} 1 & & & & \emptyset \\ & 1 & & & \\ & & \ddots & & \\ & & & 1 & \\ \emptyset & & & & 1-N \end{bmatrix}, \qquad (4.12)$$

we have

$$L = \frac{1}{4} G^a_{\mu\nu} G^a_{\mu\nu} = -\frac{1}{2}N(N-1)F_{\mu\nu} F_{\mu\nu} ,$$

with

$$F_{\mu\nu} = \partial_\mu a_\nu - \partial_\nu a_\mu \quad . \tag{4.13}$$

Here λ^a_{ij} are the conventional Gell-Mann matrices extended to SU(N). Within this set of fields we just have the linear Maxwell equations, and the Dirac string is the one corresponding to two oppositely charged Dirac monopoles, one at 0 and one at t_1. Their magnetic charges are $\pm 2\pi/gN$.

The total action of this configuration is

$$\begin{aligned} S &= \frac{1}{2} N(N-1) \int F_{\mu\nu} F_{\mu\nu} d^3x = \\ &= -\frac{N-1}{2N} \left(\frac{4\pi^2}{g^2}\right)\left(Z_1 + Z_2 - \frac{1}{4\pi|\tau|}\right) , \end{aligned} \tag{4.14}$$

where Z_1 and Z_2 are the self-energies of the monopoles, which diverge but can be subtracted, leaving a space-time independent renormalization constant.

We now assume that (4.14) is indeed an absolute extremum for the total action of all field configurations with the given string singularity. We think that this assumption is plausible but present no proof. We thus obtain a first approximation to (4.10):

$$\begin{aligned} f(t_1) &= A \exp\left(\frac{(N-1)\pi}{2g^2|\tau|N} + O(\log(g^2\tau))\right) , \\ t_1 &= i\tau , \qquad \tau \ \text{real} \quad . \end{aligned} \tag{4.15}$$

Here A is a fixed constant obtained after subtraction of the infinite self-action at the sources. Note the plus sign in our exponent due to the attractive force between the monopole-antimonopole pair. Computation of the terms of higher order in $g^2\tau$ suffers from the obstacles mentioned in sect. 2. In any case, at large τ we expect no convergence. Of course, when the Higgs mechanism is turned on then the soliton acquires a finite mass, say M, and then at large τ we expect

$$f(t_1) \to A \exp(-M|\tau|) \ ,$$
$$\tau \to \infty \ ,$$
$$t_1 = i\tau,$$
$$\tau = \text{real} \ . \tag{4.16}$$

just as any ordinary (dressed) propagator.

However, there are also interactions, in particular the N-soliton processes, described essentially by

$$f(x_1,\ldots,x_N) = \langle T(\Phi(x_1)\Phi(x_2)\ldots\Phi(x_N))\rangle \tag{4.17}$$

where N is the group parameter. Again we choose $\{x_k\}$ to be Euclidean. There is some freedom in choosing the Dirac strings, for instance we can let one string leave at each point $x_1,\ldots,x_{N-1}$ and let these all assemble at x_N. Because of the modulo N conservation law, the N-1 quanta coming in at x_N are equivalent to one quantum leaving at x_N.

To find the field configuration with least negative action we could try again the ansatz (4.12) but then the result is that $x_1,\ldots,x_{N-1}$ repel each other and all attract x_N, an unsymmetric and therefore unlikely result. Obviously, any field configuration where the signs in the exponents such as (4.15) are positive, corresponding to attraction, will give much larger, therefore dominating, contributions to the amplitude. So let us try to produce such a field configuration now.

Observe that pure permutations of the N spinor components are good elements of SU(N), so by pure gauge rotations we are allowed to move the unequal diagonal element of (4.11) up or down the diagonal. So let us now again choose one Dirac string leaving at each of $x_1,\dots,x_{N-1}$ and all entering at x_N, but this time all these strings have their unequal diagonal element (see (4.11)) in a different position along the diagonal. At x_N the combined rotation is again of type (4.11) as one can easily verify, so this is a more symmetric configuration. Since the gauge transformations we performed so far are all diagonal elements of SU(N) they actually form the subgroup $[U(1)]^{N-1}$ of SU(N) which is Abelian still. Let us define

$$\lambda^{ij}(k) = \delta_{ij} - N\delta_{ik}\delta_{jk}, \quad k = 1,\dots N . \tag{4.18}$$

with

$$\sum_{k=1}^{N} \lambda_{(k)} = 0 .$$

so one of these λ matrices is actually redundant. Our present ansatz is

$$\frac{1}{2}\lambda^a_{ij} A^a_\mu(x) = \sum_{k=1}^{N} A^{(k)}_\mu(x)\lambda^{ij}_{(k)} , \tag{4.19}$$

without bothering about the invariance

$$A_\mu^{(k)}(x) \to A_\mu^{(k)} + B_\mu \ , \qquad \text{all } k \ . \tag{4.20}$$

The Lagrangian is

$$L = \frac{1}{4} G^a_{\mu\nu} G^a_{\mu\nu} = -\frac{1}{2} \sum_{k,\ell} \mathrm{Tr}\, \lambda_{(k)}\lambda_{(\ell)} F^{(k)}_{\mu\nu} F^{(\ell)}_{\mu\nu} =$$

$$= -\frac{1}{2} N(N-1) \sum_k F^{(k)}_{\mu\nu} F^{(k)}_{\mu\nu} +$$

$$+ \frac{1}{2} N \sum_{k\neq\ell} F^{(k)}_{\mu\nu} F^{(\ell)}_{\mu\nu} \ , \tag{4.21}$$

with

$$F^{(k)}_{\mu\nu} = \partial_\mu A^{(k)}_\nu - \partial_\nu A^{(k)}_\mu \ . \tag{4.22}$$

Notice the change of sign in (4.21) when fields of different index overlap. We now choose a Dirac monopole corresponding to the kth subgroup U(1) at x_k, so $A^{(k)}(x)$ is just the field of one monopole at $\mathbf{x}_k$ (remember that, because we stay in the Abelian subclass of fields, linear superpositions are allowed). The diagonal terms in (4.21) then only contribute to the monopole "self-energy" (more precisely: self-action) and only contribute to the already established Z factor that must be subtracted. Only the cross terms give non-trivial effects:

$$S = \frac{1}{2} \frac{4\pi^2}{g^2 N} \left[(N-1) \sum_{k=1}^{N} Z_k - \frac{1}{4\pi} \sum_{k>\ell} \frac{1}{|\mathbf{x}_k - \mathbf{x}_\ell|} \right] . \tag{4.23}$$

Again we assume that this represents the least negative action configuration without presenting the complete proof.

A good check of this equation is that if N-1 points come close and the Nth stays far away then we recover (4.14) apart from overall normalization, exactly as one should expect. So we conclude:

$$f(x_1,\ldots,x_N)=A'\exp\left[\frac{\pi}{2g^2N}\sum_{k>\ell}\frac{1}{|x_k-x_\ell|}+O(\log(g^2\tau))\right]. \tag{4.24}$$

for Euclidean $\{x_k\}$.

The above was mainly to illustrate how to compute in lowest-order perturbation expansion the Green functions that we will discuss further in sect.5. We did not prove that the field configurations we choose to expand about are really the minimal ones (which we do expect) but there is no need to elaborate on that point further here.

V. SPONTANEOUS SYMMETRY BREAKING AND CONFINEMENT

In sect.4 we found that some SU(N) gauge theories in 2+1 dimensions possess a topological quantum number, conserved modulo N, and that Green functions corresponding to exchange of this quantum may be computed. Our field behaves as a local complex scalar field (real for SU(2) in all respects. We expect that these Green functions satisfy all the usual Wightman axioms [11] except that they are more singular than usual at small distances. In fact, the Green functions as computed in sect.4 can be exactly reproduced in a theory with free scalar particles and non-polynomial sources

using superpropagator techniques [12]. We leave that as an exercise for the reader. In pure SU(N) the trouble is that only the quantum corrections give some interesting structure to these Green functions. The terms of higher orders in g^2 in (4.15) and (4.24) correspond to quantum loop corrections. They have not yet been computed, to the extent that they can be computed. However, let us assume that the resulting Green functions can be computed and are roughly generated by some effective Lagrangian with necessarily stong coupling, for instance:

$$L(\Phi,\Phi^*) = - \partial_\mu \Phi^* \partial_\mu \Phi - \mathbf{M}^2 \Phi^*\Phi - \frac{\lambda_1}{N!} (\Phi^N + (\Phi^*)^N) - \lambda_2 V(\Phi^*\Phi) \ . \qquad (5.1)$$

Here we have assumed that the Higgs mechanism (either by some explicit or by some dynamical Higgs field) makes all fields massive, generating also a large soliton mass M. Of course we expect many possible interaction terms but only the most important ones are added in (5.1): the λ_1 term produces the N-soliton interaction and is responsible for the non-vanishing result (4.24) for the expression (4.17), the N-point function. The λ_2 term is of course also expected and is included in (5.1) for reasons that will become clear later.

Observe now a Z(N) global symmetry that leaves the Lagrangian (5.1), and the Green functions considered in sect.4 invariant:

$$\Phi \rightarrow e^{2\pi i/N} \Phi,$$
$$\Phi^* \rightarrow e^{-2\pi i/N} \Phi^* \ . \qquad (5.2)$$

This is simply the symmetry associated with the topologically conserved soliton quantum number. Modulo N conservation laws correspond to Z(N) global symmetries.

As we saw, the mass term is essentially due to the Higgs mechanism. Roughly, $M^2 \propto \mu_H^2$ where μ_H^2 is the second derivative of the Higgs potential at the origin. We can now consider either switching off the Higgs field, or just changing the sign of μ_H^2 so that $\langle H \rangle \to 0$. What happens to M^2 (the soliton mass)? If it stays positive then the physical soliton has not gone away and the topological conservation law remains valid. The symmetry of the theory is as in the Higgs mode and we define this mode to be a "dynamical Higgs mode". However, a very good possibility is that M^2 also switches its sign, so that $\langle \Phi \rangle \to F \neq 0$. The topological global Z(N) symmetry may get spontaneously broken. This mode can be recognized directly from the Green functions of sect.4. The criterion is

$$|F|^2 = \lim_{|\tau| \to \infty} f(t_1), \quad t_1 = i\tau . \tag{5.3}$$

Just for amusement we might note that (4.15) indeed seems to give $F \neq 0$ but of course the quantum corrections may not be neglected and we do not know at present how to actually compute the limit (5.3).

Spontaneous breakdown of the topological Z(N) symmetry is a new phase the system may choose, depending on the dynamics. We may compare a bunch of molecules that chooses to be in a gaseous, liquid or solid phase depending on the dynamics and on the values of certain intensive parameters.

Let us study this new phase more closely. The vacuum will now have a Z(N) degeneracy, that is, an

N-fold degeneracy. Labeling these vacua by an index from 1 to N we have

$$\langle\Phi\rangle_1 = e^{2\pi i/N} \langle\Phi\rangle_2 = \ldots \quad . \tag{5.4}$$

Since the symmetry is discrete there are no Goldstone particles; all physical particles have some finite mass. Again we are able to construct a set of topologically stable objects: the Bloch walls that separate two different vacua. These vortex-like structures are stable because the vacua that surround them are stable (Bloch walls are vortex-like because our model is in 2 + 1 dimensions). The width of the Bloch wall or vortex is roughly proportional to the inverse of the lowest mass of all physical particles, and therefore finite. The Bloch wall carries a definite amount of energy per unit of length.

We will now show the relevance of the operator

$$A(C) = \mathrm{Tr}\, P \exp \oint_C ig\, A_k(\vec{x})\, dx^k \ , \tag{5.5}$$

for these Bloch walls. Here C is an arbitrary oriented contour in 2-dim. space and P stands for the path ordering of the integral. A_k^{ij} are the space components of the gauge vector field in the matrix notation. A(C) is a non-local gauge-invariant operator that does not commute with Φ . Let us explain that. As is well known, if C' is an open contour, then the operator

$$A(C',\vec{x},\vec{x}_2) = P \exp \int_{\vec{x}_1,C'}^{\vec{x}_2} ig\, A_k(\vec{x})dx^k \ ,$$

transforms under a gauge rotation Ω as

$$A^{\Omega}(C',\vec{x}_1,\vec{x}_2) = \Omega(\vec{x}_1)A(C',\vec{x}_1,\vec{x}_2)\Omega^{-1}(\vec{x}_2) \ . \qquad (5.6)$$

Now the operator $\Phi(\vec{x}_0)$ was defined by a gauge transformation Ω that is multivalued when followed over a contour that encloses $\vec{x}_0$. So when we close C' to obtain C, and C encloses $\vec{x}_0$ once,

$$A(C) = \mathrm{Tr}\, A(C,\vec{x}_1,\vec{x}_1) \ , \qquad (5.7)$$

then the value of A makes a jump by a factor exp $(\pm 2\pi i/N)$ when the operator $\Phi(\vec{x}_0)$ acts, so

$$A(C)\Phi(\vec{x}_0) = \Phi(\vec{x}_0)A(C)\exp(2\pi in/N) \ , \qquad (5.8)$$

where n counts the number of times that C winds around $\vec{x}_0$ in a clockwise fashion minus the number of times it winds around $\vec{x}_0$ anticlockwise. Eq. (5.8) is an extension of eq.(4.9) for non-local operators A(C). As we will see, (5.8) can be generalized to 3+1 dimensions. Now let us interpret (5.8) in a framework where $\Phi(\vec{x})$ is diagonalized. Then, as we see, A(C) is an operator that causes a jump by a factor $\exp(2\pi in/N)$ of $\Phi(\vec{x})$ for all $\vec{x}$ inside C. So A(C) causes a switch from one vacuum to another vacuum within C in the case that Z(N) is spontaneously broken. In other words A(C) creates a "bare" Bloch wall or vortex exactly at the curve C.

Our model does not yet include quarks. Quarks are not invariant under the center Z(N), so they do not admit a direct definition of $\Phi(\vec{x})$. This difficulty is to be expected when one considers the physics of the system. The vortices that were locally stable without quarks may now become locally unstable due to virtual quark-antiquark pair creation. Most authors therefore consider quark confinement to be a basic property of the glue surrounding the quarks, in wich quarks must be inserted perturbatively. Such a procedure is justified by the experimental evidence; all hadrons can be labeled according to the number and types of quarks they contain; none of them is said to be composed of an unspecifiable or infinite number of quarks. The number of gluons on the other hand can not easily be given. We shouldn't say it is zero for most hadrons, because we need the very soft gluons to provide the binding force.

How to introduce quarks at the perturbative level is further explained in ref. [13]. The outcome is that quarks are the end point of a vortex. The conventional operator

$$\bar{\psi}(\vec{x}_1)[P\,\exp(\int_{\vec{x}_1}^{\vec{x}_2} ig\, A_k(\vec{x})dx^k)]\psi(x_2) \,, \tag{5.9}$$

creates not only a quark pair but also a vortex in between them. This vortex is topologically stable if $\langle\Phi\rangle = F \neq 0$. If we have a configuration with N quarks then Φ makes a full rotation over 2π when it follows a closed contour around. This is why a "baryon" consisting of N quarks is not confined to anything else. Evidently, for real baryons N must be 3.

Our conclusion is as follows. In SU(N) gauge theories where all scalar fields are in representations that are invariant under the center Z(N) of SU(N) (such as octet or decuplet representations of SU(3)), there exists a non-trivial topological global

Z(N) invariance. If the Higgs mechanism breaks SU(N) completely then the vacuum is Z(N) invariant. However, we can also have spontaneous breakdown of Z(N) symmetry. If that breakdown is complete then we can have no Higgs mechanism for SU(N), because in that mode "colored" objects are permanently and completely confined by the infinitely rising linear potentials due to the Bloch-wall-vortices. We can also envisage the intermediate modes where a Higgs mechanism breaks SU(N) partly, and Z(N) is partly broken. Finally, if neither Higgs' effect, nor spontaneous breakdown of Z(N) take place, then there must be massless particles causing complicated long range interactions as we will show more explicitly for the 3+1 dimensional case. That may either correspond to a point where a higher order phase transition occurs, or to a new phase, e.g. the Coulomb or Georgi-Glashow phase, where an effective Abelian photon field survives at long distances, see sect. 12.

Eqs. (4.8) and 5.8) are the basic commutation relations satisfied by our topological fields Φ. They suggest a dual relationship between A and Φ. Indeed, one could start with a scalar theory exhibiting global Z(N) invariance and then define the topological operator A(C) through eq. (5.8), but it is impossible to see this way that A can be written as the ordered exponent of an integral of a vector potential, and also the gauge group SU(N) cannot be recovered. As we will explain later, the center Z(N) is more basic to this all then the complete group SU(N).

A good name for the field Φ is the "disorder parameter" [14] since it does not commute with the other, usual, fields wich have been called order parameters in solid-state physics. The fact that in the quark confinement phase the degenerate vacuum states are eigenstates of this disorder parameter shows a close analogy with the superconductor where the vacuum state is an eigenstate of the order parameter.

VI. SU(N) GAUGE THEORIES IN 3 + 1 DIMENSIONS

In the previous sections the construction of a scalar field and the successive formulation of the spontaneous breakdown of the topological Z(N) symmetry were only possible because the model was in 2 space, 1 time dimensions. Also the boundary between different but equivalent vacua can only serve as an vortex in 2+1 dimensions. It would have the topology of a sheet in 3+1 dimensions and therefore not be useful as a vortex of conserved electric flux. So in 3+1 dimensions the formulation of quark confinement must be considerably different from the 2+1 dimensional case. Nevertheless extension of our ideas to 3+1 dimensions is possible.

We concentrate on longe-range topological phenomena. One topological feature is the instanton, corresponding to a gauge field configuration with non-trivial Pontryagin or Second Chern Class number. This however has no direct implication for confinement. What is needed for confinement is something with the space-time structure of a string, i.e. a two dimensional manifold in 4 dim. space-time. Instantons are rather event-like, i.e. zero dimensional and can for instance give rise to new types of interactions that violate otherwise apparent symmetries.

As we will see, they do play a role, though be it a subtle one. A topological structure which is extended in two dimensional sheets exists in gauge theories, as has been first observed by Nielsen, Olesen [15] and Zumino [16]. They are crucial. We will exhibit them by compactifying space-time. For the instanton it had been convenient to compactify space-time to a sphere S_4. For our purposes a hypertorus

$$S_1 \times S_1 \times S_1 \times S_1$$

is more suitable [17]. One can also consider this to be a four dimensional cubic box with periodic boundary conditions. Inside, space-time is flat. The box may be arbitrarily large. To be explicit we put a pure SU(N) gauge theory in the box (no quarks yet). Now in the continuum theory the gauge fields themselves are representations of SU(N)/Z(N), where Z(N) is the center of the group SU(N):

$$Z(N) = \{e^{2\pi in/N}I;\ n = 0,\dots,N-1\} . \tag{6.1}$$

This is because any gauge transformation of the type (6.1) leaves $A_\mu(x)$ invariant. A consequence of this is the existence of another class of topological quantum numbers in this box besides the familiar Pontryagin number. Consider the most general possible periodic boundary condition for $A_\mu(x)$ in the box. Take first a plane $\{x_1,x_2\}$ in the 12 direction with fixed values of x_3 and x_4. One may have

$$\begin{aligned} A_\mu(a_1,x_2) &= \Omega_1(x_2)A_\mu(0,x_2) , \\ A_\mu(x_1,a_2) &= \Omega_2(x_1)A_\mu(x_1,0) . \end{aligned} \tag{6.2}$$

Here, a_1, a_2 are the periods.
ΩA_μ stands short for

$$\Omega A_\mu \ \Omega^{-1} + \frac{1}{gi} \partial_\mu \ \Omega^{-1} \quad . \tag{6.3}$$

The periodicity conditions for $\Omega_{1,2}(x)$ follow by considering (6.2) at the corners of the box:

$$\Omega_1(a_2)\Omega_2(0) = \Omega_2(a_1)\Omega_1(0)Z \ , \tag{6.4}$$

where Z is some element of Z(N).
One may now perform continuous gauge transformations on $A_\mu(x)$,

$$A_\mu(x_1,x_2) \to \Omega(x_1,x_2)A_\mu(x_1,x_2) \ , \tag{6.5}$$

where $\Omega(x_1,x_2)$ (non-periodic) can be arranged either such that $\Omega_2(x_1) \to I$ or such that $\Omega_1(x_2) \to I$, but not both, because Z in (6.4) remains invariant under (6.5) as one can easily verify. We call this element Z(1,2) because the 12 plane was chosen. By continuity Z(1,2) cannot depend on x_3 or x_4. For each $(\mu\nu)$ direction such a Z element exist, to be labeled by integers

$$n_{\mu\nu} = -n_{\nu\mu} \ , \tag{6.6}$$

defined modulo N. Clearly this gives

$$N^{\frac{d(d-1)}{2}} = N^6 \tag{6.7}$$

topological classes of gauge field configurations. Note that these classes disappear if a field in the fundamental representation of SU(N) is added to the system (these fields would make unacceptable jumps at the boundary). Indeed, to understand quark confinement it is necessary to understand pure gauge systems without quarks first.

As we shall see, the new topological classes will imply the existence of new vacuum parameters besides the well-known instanton[18] angle θ The latter still exists in our box, and will be associated with a topological quantum number ν , an arbitrary integer.

VII. ORDER AND DISORDER LOOP INTEGRALS

To elucidate the physical significance of the topological numbers $n_{\mu\nu}$ we first concentrate on gauge field theory in a three dimensional periodic box with time running from $-\infty$ to ∞. To be specific we will choose the temporal gauge,

$$A_4 = 0 \ . \tag{7.1}$$

(this is the gauge in wich rotation towards Euclidean space is particularly elegant). Space has the topology $(S_1)^3$. There is an infinite set of homotopy classes of closed oriented curves C in this space: C may wind any number of times in each of the three principal directions. For each curve C at each time t there is a quantum mechanical operator A(C,t) defined by

$$A(C,t) = \mathrm{Tr}\ P \exp \oint_C ig\ \vec{A}(\vec{x},t)\cdot d\vec{x}\ , \tag{7.2}$$

called Wilson loop or order parameter. Here P stands for path ordering of the factors $\vec{A}(\vec{x},t)$ when the exponents are expanded. The ordering is done with respect to the matrix indices. The $\vec{A}(\vec{x},t)$ are also operators in Hilbert space, but for different $\vec{x}$, same t, all $A(\vec{x},t)$ commute with each other. By analogy with ordinary electromagnetism we say that A(C) *measures* magnetic flux *through* C, and in the same time *creates* an electric flux line *along* C. Since A(C) is gauge-invariant under purely periodic gauge transformations, our versions of magnetic and electric flux are gauge-invariant. Therefore they are not directly linked to the gauge covariant curl $G^a_{\mu\nu}(\vec{x})$.

There exists a dual analogon of A(C) wich will be called B(C) or disorder loop operator [13]. C is again a closed oriented curve in $(S_1)^3$. A simple definition of B(C) could be made by postulating its equal-time commutation rules with A(C):

$$\begin{aligned} [A(C),\ A(C')] &= 0; \\ [B(C),\ B(C')] &= 0; \\ A(C)B(C') &= B(C')A(C)\ \exp 2\pi i n/N\ , \end{aligned} \tag{7.3}$$

where n is the number of times C' winds around C in a certain direction. Note that n is only well defined if either C or C' is in the trivial homotopy class (that is, can be shrunk to a point by continuous deformations). Therefore, if C' is in a nontrivial class we must choose C to be in a trivial class. Since these commutation rules (7.3) determine B(C) only up to factors that commute with A and B, we could make further requirements, for instance that B(C) be a unitary operator.

An explicit definition of B(C) can be given as follows. As in sect. 4, we go to the temporal gauge, $A_0 = 0$. We then must distinguish a "large Hilbert space" H of all field configurations A(x) from a "physical Hilbert space $\tilde{H} \subset H$. This $\tilde{H}$ is defined to be the subspace of H of all gauge invariant states:

$$\tilde{H} = \{|\psi>;\ <\vec{A}(\vec{x})|\psi> = <\Omega\vec{A}(\vec{x})|\psi>\} \quad , \tag{7.4}$$

where Ω is any infinitesimal gauge transformation in 3 dim. space. Often we will also write Ω for the corresponding rotation in **H**:

$$\tilde{H} = \{|\psi>;\ \Omega|\psi> = |\psi>,\ \Omega \text{ infinitesimal}\} \ . \tag{7.5}$$

Now consider a pseudo-gauge transformation $\Omega^{[C']}$ defined to be a genuine gauge transformation at all points $\vec{x} \notin C'$, but singular on C'. For any closed path $\mathbf{x}(\theta)$ with $0 < \theta < 2\pi$ twisting n times around C' we require

$$\Omega^{[C']}(x(2\pi)) = \Omega^{[C']}(x(0))e^{2\pi in/N} \ . \tag{7.6}$$

This discontinuity is not felt by the fields $A(\vec{x},t)$ wich are invariant under $Z(\mathbf{N})$. They do feel the singularity at C' however. We define B(C') as

$$\Omega^{[C']}$$

but with the singularity at C' smoothened; this corresponds to some form of regularization, and implies that the operator differs from an ordinary gauge transformation. Therefore, even for $|\psi> \varepsilon$ **H** we have

$$B(C')|\psi> \neq |\psi> \ . \tag{7.7}$$

For any regular gauge transformation Ω we have an Ω'

such that

$$\Omega\Omega^{[C']} = \Omega^{[C']}\,\Omega' \; . \tag{7.8}$$

Therefore, if $|\psi> \,\varepsilon\, \tilde{H}$ then $B(C')|\psi> \,\varepsilon\, \tilde{H}$, and $B(C')$ is gauge-invariant. We say that $B(C')$ *maesures* electric flux *through* C' and *creates* a magnetic flux line *along* C'.

We now want to find a conserved variety of Non-Abelian gauge-invariant magnetic flux in the 3-direction in the 3 dimensional periodic box. One might be temped to look for some curve C enclosing the box in the 12 direction so that A(C) maesures the flux through the box. That turns out not to work because such a flux is not guaranteed to be conserved. It is better to consider a curve C' in the 3-direction winding over the torus exactly once:

$$C' = \{\vec{x}(s), 0 \leq s \leq 1;\ \vec{x}(1) = \vec{x}(0)+(0,0,a_3)\}. \tag{7.9}$$

B(C') creates one magnetic flux line. But B(C') also changes the number n_{12} into $n_{12}+1$. This is because

$$\Omega^{[C']}$$

makes a Z(N) jump according to (7.6). If $\Omega_{1,2}(\vec{x})$ in (6.2) are still defined to be continuous then Z in (6.4) changes by one unit. Clearly, n_{12} measures the number of times an operator of the type B(C') has acted, i.e. the number of magnetic flux lines created. n_{12} is also conserved by continuity. We simply define

$$n_{ij} = \varepsilon_{ijk}\, m_k \quad , \tag{7.10}$$

with m_k the total magnetic flux in the k-direction. Note that $\vec{m}$ corresponds to the usual magnetic flux (apart from a numerical constant) in the Abelian case. Here, $\vec{m}$ is only defined as an integer modulo N.

VIII. NON-ABELIAN GAUGE-INVARIANT ELECTRIC FLUX IN THE BOX

As in the magnetic case, there exists no simple curve C such that the total electric flux through C, measured by B(C), corresponds to a conserved total flux through the box. We consider a curve C winding once over the torus in the 3-direction and consider the electric flux creation operator A(C). But first we must study some new conserved quantum numbers.

Let $|\psi\rangle$ be a state in the before mentioned little Hilbert space $\tilde{H}$. Then, according to eq. (7.5), $|\psi\rangle$ is invariant under *infinitesimal* gauge transformations Ω. But we also have some non-trivial homotopy classes of gauge transformations Ω. These are the pseudoperiodic ones:

$$\begin{aligned} \Omega(a_1,x_2,x_3) &= \Omega(0,x_2,x_3)Z_1 \;, \\ \Omega(x_1,a_2,x_3) &= \Omega(x_1,0,x_3)Z_2 \;, \\ \Omega(x_1,x_2,a_3) &= \Omega(x_1,x_2,0)Z_3 \;, \\ Z_{1,2,3} &\in \text{center } Z(N) \text{ of } SU(N) \;, \end{aligned} \tag{8.1}$$

and also those Ω which are periodic but do carry a non-trivial Pontryagin number ν. A little problem arises when we try to combine these two topological features. The $Z_{1,2,3}$ can be labeled by three integers $k_{1,2,3}$ between 0 and N:

$$Z_t = e^{2\pi i k_t/N} \;. \tag{8.2}$$

But how is ν defined? The best definition is obtained if we consider a field configuration in a *four* dimensional space, obtained by multiplying the box $(S_1)^3$ with a line segment:

$$0 \leq t \leq 1 \; .$$

Now choose a boundary condition: $A(t=1)=\Omega A(t=0)$. Then, if the fields in between are continuous,

$$P \equiv g^2 \int G_{\mu\nu} \tilde{G}_{\mu\nu} \, d^4x/32\pi^2 \tag{8.3}$$

is uniquely determined by Ω . On S_4 this would be the integer ν. Now however, it needs not to be integer anymore because of the twists in the periodic boundary conditions for $(S_1)^3$. We find

$$P = \frac{(\vec{m}\cdot\vec{k})}{N} + \nu \quad , \tag{8.4}$$

where ν is integer and $\vec{m}$ is the magnetization defined in the previous section. Notice that ν is only well defined if $\vec{m}$ and $\vec{k}$ are given as genuine integers, not modulo N. Taking this warning to heart, we write $\Omega[\vec{k},\nu]$ for any Ω in the homotopy class $[\vec{k},\nu]$.

Notice that not only do the $A_\mu(x)$ transform smoothly under $\Omega[\vec{k},\nu]$, since they are invariant under the Z(N) transformations of eq.(8.1), but also their boundary conditions do not change. These Ω commute therefore with the magnetic flux $\vec{m}$. If two Ω satisfy the same equation (8.1) and have the same ν, they may act differently on states of the big Hilbert space H, but since they differ only by regular gauge transformations they act identically on states in $\tilde{H}$, defined in (7.5). We may simultaneously diagonalize the Hamiltonian H, the magnetic flux $\vec{m}$, and $\Omega[k,\nu]$:

$$\Omega[\vec{k},\nu]|\psi> \quad = \quad e^{i\omega(\vec{k},\nu)}|\psi> \ , \tag{8.5}$$

where $\omega(\vec{k},\nu)$ are strictly conserved numbers. Now the Ω operators form a group. Defining for each Ω the number P as in (8.4) we have

$$\Omega[\vec{k}_1,p_1]\Omega[\vec{k}_2,p_2] = \Omega[\vec{k}_1+\vec{k}_2,p_1+p_2] \ , \tag{8.6}$$

so

$$\omega(\vec{k}_1,\nu_1) + \omega(k_2,\nu_2) = \omega(\vec{k}_1+\vec{k}_2,\nu_1+\nu_2) \ , \tag{8.7}$$

and

$$\omega(\vec{k} + N\vec{\ell},\nu) = \omega(\vec{k},\nu + (\vec{\ell}\cdot\vec{m})) \ , \tag{8.8}$$

if $\vec{\ell}$ is an integer. We find that ω must be linear in $\vec{k}$ and ν:

$$\omega(\vec{k},\nu) = \frac{2\pi}{N} (\vec{e}\cdot\vec{k}) + \frac{\theta}{N} (\vec{m}\cdot\vec{k}) + \theta\nu \ , \tag{8.9}$$

where e_i are integer numbers defined modulo N, and θ is the familiar instanton angle, defined to lie between 0 and 2π.

Now let us turn back to A(C) defined in eq.(7.2). If C is the curve considered in the beginning of this section, A(C) is not invariant under $\Omega[\vec{k},\nu]$ because

$$\begin{aligned} A(C) &\to \mathrm{Tr}\ \Omega(\vec{x}_1)[P \exp \int_C ig\ \vec{A}\ d\vec{x}]\ \Omega^{-1}(\vec{x}_1 + \vec{a}_3) \\ &= e^{-2\pi i k_3/N} A(C) \ . \end{aligned} \tag{8.10}$$

Therefore,

$$A(C)\Omega[\vec{k},\nu]|\psi> = \Omega[\vec{k},\nu]e^{-2\pi ik_3/N}A(C)|\psi> \qquad (8.11)$$

If $$\Omega[\vec{k},\nu]|\psi> = e^{i\omega(\vec{k},\nu)}|\psi> , \qquad (8.12)$$

and $$A(C)|\psi> = |\psi'> , \qquad (8.13)$$

then $$\Omega[\vec{k},\nu]\ |\psi'> = e^{i\omega(\vec{k},\nu)+2\pi ik_3/N}|\psi'> . \qquad (8.14)$$

Therefore $A(C)$ increases e_3 by one unit:

$$e_3A(C)|\psi> = A(C)(e_3+1)|\psi> \quad . \qquad (8.15)$$

e_3 is a good indicator for electric flux in the 3-direction, up to a constant. It is strictly conserved. However if we let θ run from 0 to 2π then $\vec{e}$ turns into $\vec{e} + \vec{m}$. It is therefore physically perhaps more appropriate to identify

$$\vec{e} + \frac{\theta}{2}\vec{m} \qquad (8.16)$$

as being the total electric flux in the three directions of the box.

IX. FREE ENERGY OF A GIVEN FLUX CONFIGURATION

Again we follow ref.[3] but for completeness we add the Pontryagin winding number ν.

Let us write down the free energy F of a given state $(\vec{e},\vec{m},\theta)$ at temperature $T = 1/k\beta$:

$$e^{-\beta F} = \underset{\tilde{H}}{\mathrm{Tr}}\ P_e(\vec{e})\ P_m(\vec{m})P_\theta(\theta)e^{-\beta H} \quad . \tag{9.1}$$

Here H is the Hamiltonian and $\tilde{H}$ the little Hilbert space. P are projection operators. $P_m(\vec{m})$ is simply defined to select a given set of $n_{ij} = \varepsilon_{ijk}\, m_k$, the three space-like indices of eq. (6.6). $P_e(\vec{e})P_\theta(\theta)$ is defined by selecting states $|\psi>$ with

$$\Omega[\vec{k},\nu]|\psi> = e^{\frac{2\pi i}{N}(\vec{k}\cdot\vec{e}) + \frac{\theta i}{N}(\vec{m}\cdot\vec{k}) + i\theta\nu}|\psi> \quad . \tag{9.2}$$

Therefore $P_e(\vec{e})P_\theta(\theta) =$

$$\frac{1}{N^3}\sum_{k,\nu} e^{-\frac{2\pi i}{N}(\vec{k}\cdot\vec{e}) - \frac{\theta i}{N}(\vec{m}\cdot\vec{k}) - i\theta\nu}\,\Omega[\vec{k},\nu] \quad . \tag{9.3}$$

Now $e^{-\beta H}$ is the evolution operator in imaginary time direction at interval β, expressed by a functional integral over a Euclidean box with sides (a_1,a_2,a_3,β):

$$<\vec{A}_{(1)}(\vec{x})|e^{-\beta H}|\vec{A}_{(2)}(\vec{x})> = \int DAe^{S(A)}\Big|_{\vec{A}(\vec{x},0)=\vec{A}_{(2)}(\vec{x})}^{\vec{A}(\vec{x},\beta)=\vec{A}_{(1)}(\vec{x})} \quad . \tag{9.4}$$

We may fix the gauge for $\vec{A}_{(2)}(\vec{x})$ for instance by choosing

$$\begin{aligned} A_{(2)3}(\vec{x}) &= 0 \ , \\ A_{(2)2}(x,y,0) &= 0 \ , \\ A_{(2)1}(x,0,0) &= 0 \ . \end{aligned} \tag{9.5}$$

We already had $A_4(\vec{x},t) = 0$. Since only states in $\tilde{H}$ are considered, we insert also a projection operator $\int_{\Omega\,\varepsilon\,I} D\Omega$ were I is the trivial homotopy class.

"Trace" means that we integrate over all $A_{(1)}=A_{(2)}$ therefore we get periodic boundary conditions in the 4-direction. Insertions of $\int_{\Omega\,\varepsilon\,I} D\Omega$ means that we have periodicity up to gauge transformations, in the completely unique gauge

$$A_4(\vec{x},\beta)=A_3(\vec{x},0)=A_2(x,y,0,0)=A_1(x,0,0,0)=0. \qquad (9.6)$$

Eq. (9.3) tells us that we have to consider twisted boundary conditions in the 41, 42, 43 directions and Fourier transform:

$$e^{-\beta F(\vec{e},\vec{m},\theta,\vec{a},\beta)} = \frac{1}{N^3}\sum_{\vec{k},\nu} e^{-\frac{2\pi i}{N}(\vec{k}\cdot\vec{e})-i\theta(\nu+\frac{\vec{m}\cdot\vec{k}}{N})} W\{\vec{k},\vec{m},\nu,a_\mu\}. \qquad (9.7)$$

Here $W\{\vec{k},\vec{m},\nu,a_\mu\}$ is the Euclidean functional integral with boundary conditions fixed by choosing $n_{ij} = \varepsilon_{ijk}\, m_k$; $n_{i4} = k_i$; $a_4 = \beta$, and a Pontryagin number ν. Because of the gauge choice (9.6) this functional integral must include integration over the Ω belonging to the given homotopy classes as they determine the boundary conditions such as (6.2). The definition of W is completely Euclidean symmetric. In the next chapter I show how to make use of this symmetry with respect to rotation over 90° in Euclidean space.

X. DUALITY

The Euclidean symmetry in eq. (9.7) suggests to consider the following SO(4) rotation:

$$\begin{bmatrix} 0 & -1 & & \\ 1 & 0 & & \\ & & 0 & 1 \\ & & -1 & 0 \end{bmatrix} \quad . \tag{10.1}$$

Let us introduce a notation for the first two components of a vector:

$$\begin{aligned} x_\mu &= (\vec{x}, x_4) \ , \\ \tilde{x} &= (x_1, x_2) \ , \\ \hat{x} &= (x_2, x_1) \ . \end{aligned} \tag{10.2}$$

We have, from eq. (9.7):

$$\exp\,[-\beta F(\tilde{e}, e_3, \tilde{m}, m_3, \theta, \tilde{a}, a_3, \beta)] \quad =$$

$$= \frac{1}{N^2} \sum_{\tilde{k}, \tilde{\ell}} \exp[\tfrac{2\pi i}{N}(-(\tilde{k}\cdot\tilde{e}) + (\tilde{\ell}\cdot\tilde{m})) - a_3 F(\tilde{\ell}, e_3, \tilde{k}, m_3, \theta, \hat{a}, \beta, a_3)] . \tag{10.3}$$

Notice that in this formula the transverse electric and magnetic fluxes are Fourier transformed and interchange positions. Notice also that, apart from a sign difference, there is a complete electric-magnetic symmetry in this expression, in spite of the fact that the definition of F in terms of W was not so symmetric. Eq. (10.3) is an exact property of our system. No approximation was made. We refer to it as "duality".

XI. LONG-DISTANCE BEHAVIOR COMPATIBLE WITH DUALITY

Eq. (10.3) shows that the instanton angle θ plays no role in duality. It does however affect the physical interpretation of $\vec{e}$ as electric flux, see (8.10). From now on we put $\theta = 0$ for simplicity, and omit it.

Let us now assume that the theory has a mass gap. No massless particles occur. Then asymptotic behavior at large distances will be approached exponentially. Then it is excluded that

$$F(\vec{e},\vec{m},\vec{a},\beta) \to 0, \quad \text{exponentially as} \quad \vec{a},\beta \to \infty \quad ,$$

for all $\vec{e}$ and $\vec{m}$, wich would clearly contradict (10.3). This means that at least some of the flux configurations must get a large energy content as $\vec{a}$, $\beta \to \infty$. These flux lines apparently cannot spread out and because they were created along curves C it is practically inescapable that they get a total energy wich will be proportional to their length:

$$E = \lim_{\beta\to\infty} F = \quad \rho a \quad . \tag{11.1}$$

However, duality will never enable us to determine whether it is the electric or the magnetic flux lines that behave this way. From the requirement that W in (9.7) is always positive one can deduce the impossibility of a third option, namely that only exotic combinations of electric and magnetic fluxes behave as strings (provided $\theta = 0$).

For further information we must make the physically quite plausible assumption of "factorizability":

$$F(\vec{e},\vec{m}) \to F_e(\vec{e}) + F_m(\vec{m}) \text{ if } \vec{a},\beta \to \infty \ . \tag{11.2}$$

Suppose that we have confinement in the electric domain:

$$F_e(0,0,1) \to \rho a_3 \tag{11.3}$$

where ρ is the fundamental string constant. Then we can derive from duality the behavior of $F_m(\vec{m})$.

First we improve (11.3) by applying statistical mechanics to obtain F_e for large but finite β. One obtains:

$$e^{-\beta F_e(e_1,e_2,0,\vec{a},\beta) + C(a,\beta)} =$$

$$= \sum_{n_1^\pm,n_2^\pm} \frac{1}{n_1^+!n_2^+!n_1^-!n_2^-!} \gamma_1^{n_1^++n_1^-} \gamma_2^{n_2^++n_2^-} \delta_N(n_1^+-n_1^--e_1)\delta_N(n_2^+-n_2^--e_2). \tag{11.4}$$

Here

$$\gamma_1 = \lambda a_2 a_3 \, e^{-\beta\rho a_1},$$

$$\gamma_2 = \lambda a_1 a_3 \, e^{-\beta\rho a_2},$$

$$\delta_N(x) = \frac{1}{N}\sum_{k=0}^{N-1} e^{2\pi ikx/N}$$

$$= \begin{cases} 1 & \text{if} \quad x = 0 \ (\text{mod } N) \\ 0 & \text{if} \quad x = \text{other integer number.} \end{cases} \tag{11.5}$$

The sum is over all nonnegative integer values of $n_i^\pm$ (the orientations $\pm$ are needed if $N \geq 3$). The γ's are Boltzmann factors associated with each string-like flux tube.

We now insert this, with (11.2), into (10.3) putting $e_3 = m_3 = 0$. One obtains

$$e^{-\beta F_m(m_1,m_2,0,\vec{a},\beta)} = C'e^{2\sum_a \gamma'_a \cos(2m_a\pi/N)}, \tag{11.6}$$

where C' is again a constant and

$$\gamma_1' = \lambda a_1 \beta \quad e^{-\rho a_2 a_3} ,$$
$$\gamma_2' = \lambda a_2 \beta \quad e^{-\rho a_1 a_3} . \qquad (11.7)$$

At $\beta \to \infty$ we get

$$F_m(\tilde{m},0,\vec{a},\beta) \to E_m(\tilde{m},0,\vec{a}) = \sum_i E_i(m_i,\vec{a}) ,$$

with

$$E_1(m_1,\vec{a}) = 2\lambda(1-\cos\frac{2\pi m_1}{N})a_1 e^{-\rho a_2 a_3} , \qquad (11.8)$$

and similarly for E_2 and E_3.

One reads off from eq. (11.8) that there will be no magnetic confinement, because if we let the box become wider the exponential factor

$$e^{-\rho a_2 a_3}$$

causes a rapid decrease of the energy of the magnetic flux. Notice the occurrence of the string constant ρ in there.

Of course we could equally well have started from the presumption that there were magnetic confinement. One then would conclude that there would be no electric confinement, because then the electric flux would have an energy given by (11.8).

XII. THE COULOMB PHASE

To see what might happen in the absence of a mass gap one could study the (first) Georgi-Glashow model [4] Here SU(2) is "broken spontaneously" into

U(1) by an isospin one Higgs field. Ordinary perturbation expansion tells us what happens in the infrared limit. There are electrically charged particles: $W^{\pm}$ (the charged vector particles). They carry *two* fundamental electric flux units ("quarks" with isospin ½ would have the fundamental flux unit $q_o = \pm \frac{1}{2} e$). There are also magnetically charged particles (monopoles, [19]. They also carry two fundamental magnetic flux units:

$$g = \frac{2\pi}{q_o} = \frac{4\pi}{e} \ . \qquad (12.1)$$

A given electric flux configuration of k flux units would have an energy

$$E = \frac{q_o^2 k^2 a_1}{2a_2 a_3} \ . \qquad (12.2)$$

At finite β however pair creation of $W^{\pm}$ takes place, so that we should take a statistical average over various values of the flux. Flux is only rigorously defined modulo $2q_o$. We have

$$e^{-\beta F_e(1,0,0)} = \frac{\sum_{k=-\infty}^{\infty} \exp[-\beta \frac{e^2 a_1}{2a_2 a_3}(k+\frac{1}{2})^2]}{\sum_{k=-\infty}^{\infty} \exp[-\beta \frac{e^2 a_1}{2a_2 a_3} k^2]} \ . \qquad (12.3)$$

Similarly, because of pair creation of magnetic monopoles

$$e^{-\beta F_m(1,0,0)} = \frac{\sum_{k=-\infty}^{\infty} \exp[-\beta \frac{8\pi^2 a_1}{e^2 a_2 a_3}(k+\frac{1}{2})^2]}{\sum_{k=-\infty}^{\infty} \exp[-\beta \frac{8\pi^2 a_1}{e^2 a_2 a_3} k^2]} \ . \qquad (12.4)$$

These expressions do satisfy duality, eq. (10.3). This is easily verified when one observes that

$$\sum_{k=-\infty}^{\infty} e^{-\lambda k^2} = \sqrt{\frac{\pi}{\lambda}} \sum_{k=-\infty}^{\infty} e^{-\pi^2 k^2/\lambda} \quad ,$$

and

$$\sum_{k=-\infty}^{\infty} (-1)^k e^{-\lambda k^2} = \sqrt{\frac{\pi}{\lambda}} \sum_{k=-\infty}^{\infty} e^{-\pi^2 (k+\frac{1}{2})^2/\lambda} \quad . \tag{12.5}$$

Notice now that this model realizes the dual formula in a symmetric way, contrary to the case that there is a mass gap. This dually symmetric mode will be referred to as the "Coulomb phase" or "Georgi-Glashow phase".

Suppose that Quantum Chromodynamics would be enriched with two free paramaters that would not destroy the basic topological features (for instance the mass of some heavy scalar fields in the adjoint representation). Then we would have a phase diagram as in the Figure below.

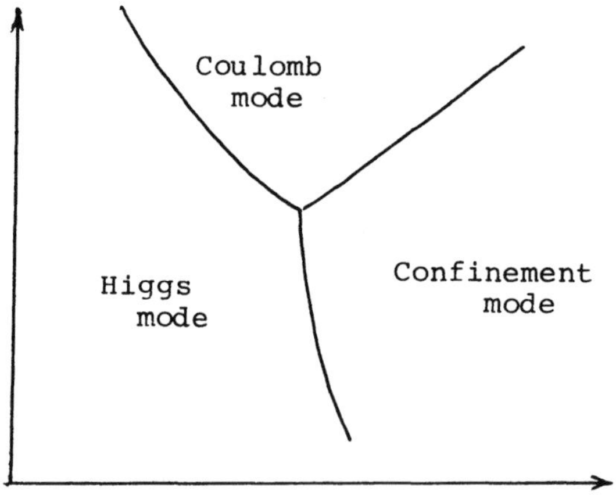

Fig. 1

Numerical calculations[20] suggest that the phase transition between the two confinement modes is a first order one. Real QCD is represented by one point in this diagram. Where will that point be? If it were in the Coulomb phase there would be long range, strongly interacting Abelian gluons contrary to experiment. In the Higgs mode quarks would have finite mass and escape easily. It could be still in the Higgs phase but very close to the border line with the confinement mode. If the phase transition were a second order one then that would imply long range correlation effects requiring light physical gluons. Again, they are not observed experimentally. If, wich is more likely, the phase transition is a first order one then even close to the border line not even approximate confinement would take place: quarks would be produced copiously. There is only one possibility: we are in the confinement mode. Electric flux lines cannot spread out. Quark confinement is absolute.

REFERENCES

1. S. Weinberg, Phys. Rev. Lett. 19 (1967) 1264.
   A. Salam and J. Ward, Phys. Lett. 13 (1964) 168.
   S.L. Glashow, J. Iliopulos and L. Maiani, Phys. Rev. D2 (1970) 1285.
2. E.S. Abers and B.W. Lee, Physics Reports 9C no. 1.
   S. Coleman in: Laws of Hadronic matter, Erice July 1973, ed. by A. Zichichi, Acad. Press, NY and London.
3. G. 't Hooft, lectures given at the Cargèse Summer Institute on "Recent Developments in Gauge Theories", 1979 (Plenum, New York, London) ed. G. 't Hooft et al. Lecture no II.
4. H. Georgi and S.L. Glashow, Phys. Rev. Lett. 28 (1972) 1494.
5. G. 't Hooft, in "The Whys of Subnuclear Physics", Erice 1977, ed. by A. Zichichi (Plenum, New York and London) p. 943.
6. G. Parisi, lectures given at the 1977 Cargèse Summer Institute.
7. P. Higgs, Phys. Rev. 145 (1966) 1156.
   T.W.B. Kibble, Phys. Rev. 155 (1967) 1554.
8. S. Coleman, in "New Phenomena in Subnuclear Physics", Erice 1975, Part A, ed. by A. Zichichi (Plenum press, New York and London).
9. G. 't Hooft, in "Particles and Fields", Banff 1977, ed. by D.H. Boal and A.N. Kamal, (Plenum Press, New York and London) p. 165.
10. P.A.M. Dirac, Proc. Roy. Soc. A133 (1931) 60; Phys. Rev. 74 (1948) 817.
11. R.F. Streater and A.S. Wightman, PCT, spin and statistics, and all that (Benjamin, New York and Amsterdam, 1964).
12. H. Lehmann and K. Pohlmeyer, Comm. Math. Phys. 20

(1971) 101;
A. Salam and J. Strathdee, Phys. Rev. D1 (1970) 3296.
13. G. 't Hooft, Nucl. Phys. B138 (1978) 1.
14. L.P. Kadanoff and H. Ceva, Phys. Rev. B3 (1971) 3918.
15. H.B. Nielsen and P. Olesen, Nucl. Phys. B61 (1973) 45; ibidem B160 (1979) 380.
16. B. Zumino, in "Renormalization and Invariance in Quantum Field Theory", ed. E.R. Caianiello, (Plenum Press, New York) p. 367.
17. G. 't Hooft, Nucl. Phys. B153 (1979) 141.
18. R. Jackiw and C. Rebbi, Phys. Rev. Lett. 37 (1976) 172.
C. Callan, R. Dashen and D. Gross, Phys. Lett. 63B (1976) 334 and Phys. Rev. D17 (1978) 2717.
19. G. 't Hooft, Nucl. Physics B79 (1974) 276
A.M. Polyakov, JETP Lett. 20 (1974) 194.
20. M. Creutz, L. Jacobs and C. Rebbi, Phys. Rev. D20 (1979) 1915.

CHAPTER 7.2

THE CONFINEMENT PHENOMENON IN QUANTUM FIELD THEORY

Gerard 't Hooft
Instituut voor Theoretische Fysica der Rijksuniversiteit
De Uithof
Utrecht, The Netherlands

0. ABSTRACT

In these written notes of four lectures it is explained how the phenomenon of permanent confinement of certain types of particles inside bound structures can be understood as a consequence of local gauge invariance and the topological properties of gauge field theories.

I. INTRODUCTION

The lagrangian of a quantum field theory describes the evolution of a certain number of degrees of freedom of the system, called *fields* as a function of space and time. In many cases this evolution is only straightforward in a small region of space-time and therefore these fields should be interpreted only as the *microscopic* variables of the theory. Only in the simplest cases these microscopic variables also correspond to actual physical particles (the *macroscopic* objects) but very often the connection is less straightforward for two reasons. One reason is that there may be a local gauge invariance. This is a class of transformations that transform a set of fields into another set of fields with the postulate that the new set describes the same physical situation as the old one. Therefore the physical fields only constitute some orthogonal subset of the original set of fields and one of the problems that we will study in these lectures is how to associate this subset to observable objects.

However even these observable objects ("transient particles") may in some cases not yet be the macroscopic physical particles. That is one second point : various kinds of Bose-condensation may

1981 Cargèse Summer School Lecture Notes on Fundamental Interactions, NATO Adv. Study Inst. Series B: Phys., Vol. 85, Edited by M. Lévy and J.-L. Basdevant (Plenum).

take place after which the spectrum of physical particles may look entirely different once again. We will study these condensation phenomena as we go along.

The most challenging application of our theoretical considerations is "quantum chromodynamics" or generalizations thereof. The "microscopic" lagrangian there contains only vector particles ("gluons") and spinors ("quarks") but neither of these are really physical. The spectrum of physical particles always consists of bound states of certain numbers of quarks and/or antiquarks with unspecified numbers of gluons. We will obtain a qualitative understanding of this transition from the microscopic to the macroscopic dynamical variables.

The first models that we will consider may seem to be a far way off from that desired goal but studying them will turn out to be crucial for obtaining a suitable frame and language in order to put the more advanced systems in a proper perspective.

II. SCALAR FIELD THEORY

Let us consider the lagrangian of a complex scalar field theory :

$$\mathcal{L}(\phi,\phi^*) = -\partial_\mu\phi^*\partial_\mu\phi - m^2\phi^*\phi - \frac{\lambda}{2}(\phi^*\phi)^2 \tag{2.1}$$

Equivalently we can use real variables :

$$\phi = \frac{1}{\sqrt{2}}(\phi_1 + i\phi_2)\ ; \qquad \phi^* = \frac{1}{\sqrt{2}}(\phi_1 - i\phi_2). \tag{2.2}$$

We write ϕ^* rather than $\phi^\dagger$ because here the fields are c-numbers, not operators, and the lagrangian must be seen in the context of a functional integral as described by B. de Wit in his lectures at this School.

The lagrangian (2.1) describes a system with two species of Bose-particles both with the same mass m but distinguishable in the quantum mechanical sense. One can either consider ϕ_1 and ϕ_2 as the two species, both equal to their own antiparticles, or consider ϕ as a particle and ϕ^* as its distinct antiparticle. At this level these descriptions are equivalent.

Obviously there is a *symmetry* under exchange of ϕ_1 and ϕ_2, or more generally

$$\begin{aligned}&\phi \rightarrow e^{i\Lambda}\phi\\ &\phi^* \rightarrow e^{-i\Lambda}\phi^*\end{aligned} \tag{2.3}$$

This is a group of rotations in the complex plane called U(1) and is only global, that is, Λ must be independent of space-time.

By Noether's theorem the model contains a conserved current,

$$j_\mu = \phi^* \partial_\mu \phi - (\partial_\mu \phi^*)\phi \tag{2.4}$$

with $$\partial_\mu j_\mu = 0 \tag{2.5}$$

Classically (2.5) is only true if ϕ, ϕ^* are required to obey the Euler-Lagrange equations generated by (2.1). Quantum mechanically (2.5) follows if we substitute for ϕ and ϕ^* the corresponding operators ϕ and $\phi^\dagger$.

Our model is trivial in the sense that, if $m^2 > 0$, the microscopic fields ϕ_1 and ϕ_2 directly correspond to the expected macroscopic physical particles (apart from possible stable bound states) and the symmetry (2.3) is also a symmetry between these particles.

III. BOSE CONDENSATION

Bose condensation is a well-known phenomenon in quantum statistical physics. Just in order to make the connection with our case of interest let us first consider an ideal non-relativistic Bose gas. The states are

$$|n(\vec{k}_1), n(\vec{k}_2), \ldots > \tag{3.1}$$

where

$$\vec{k}_i = \frac{\pi}{L}(\, l_1, l_2, l_3) \tag{3.2}$$

and l_1, l_2, l_3 are positive integers. L is the length of a side of the box in which the particles are contained. The energy of the states is

$$E = \sum_i n(\vec{k}_i) k^2/2M \tag{3.3}$$

where M is the mass of the (non-relativistic) particles. We take for the thermodynamic free energy F,

$$e^{-\beta F} = \sum_{\{n(\vec{k})\}} e^{-\beta[E(\{n(\vec{k})\}) - \mu N]} \tag{3.4}$$

where $N = \sum_i n(\vec{k}_i)$, $\beta = 1/T$

T is the temperature in natural units, and μ is the chemical potential.

We can easily solve (3.4) :

$$e^{-\beta F} = \prod_k \sum_{n=0}^{\infty} e^{\beta(\mu - k^2/2M)n} = \prod_k \frac{1}{1 - \exp \beta(\mu - k^2/2M)} \qquad (3.5)$$

and in the limit of infinite volume :

$$F = (2\pi)^{-3} TV \int d^3k \log(1 - \exp\beta(\mu - k^2/2M)) \qquad (3.6)$$

$$V = L^3$$

The particle number density is

$$\frac{N}{V} = -\frac{1}{V}\frac{\partial F}{\partial \mu} = \int d^3k \,(\exp \beta(-\mu + k^2/2M) - 1)^{-1} \qquad (3.7)$$

One easily notes that the formulas (3.5) - (3.7) explode if the chemical potential μ becomes larger than or equal to zero, a typical property of Bose gases.

Since this is a non-relativistic system it is convenient to introduce a field $\phi(\vec{x})$ in the following way :

$$\phi(\vec{x}) = 2\sqrt{\frac{2}{V}} \sum_{\vec{k}} a(\vec{k}) \prod_{i=1,2,3} \sin \frac{\pi(k_i x_i)}{L} \qquad (3.8)$$

where $a(\vec{k})$ is an operator that annihilates one particle with momentum $\vec{k}$ in the usual way. The hamiltonian is then

$$H = \sum_{\vec{k}} (\frac{k^2}{2M} - \mu)\, a^\dagger(\vec{k})\, a(\vec{k}) \qquad (3.9)$$

$$= \int d^3\vec{x}\, (\frac{1}{2M} \vec{\partial}\phi^\dagger(\vec{x})\vec{\partial}\phi(\vec{x}) - \mu\phi^\dagger(\vec{x})\phi(\vec{x})) \qquad (3.10)$$

where for convenience we included the chemical potential term so that this hamiltonian describes the complete system :

$$e^{-\beta F} = \mathrm{Tr}\, e^{-\beta H}$$

It is obvious here that μ should not be allowed to be positive.

Now however we can take into account the *repulsive* forces between the particles. When many particles are close together we expect an extra, positive contribution to H. A simple model for that is

$$H = \int d^3x \left(\frac{1}{2M}\,\partial\phi^\dagger\partial\phi - \mu\phi^\dagger\phi + \frac{\lambda}{2}\,\phi^{\dagger 2}\phi^2\right) \tag{3.11}$$

As long as μ is negative the λ term is just a small perturbation. But if μ is positive then the λ term is the only one that can stabilize the system.

Of course it is difficult to find the free energy of this revised system exactly, but an easy approximation, valid for λ not too large, is to substitute ϕ by a c-number (as defined in (3.8) it was an operator), and subsequently minimize H. We then find approximately the energy of the lowest eigenstate :

$$\phi^\dagger\phi \equiv \rho \tag{3.12}$$

$$\frac{\partial}{\partial\rho}\left(-\mu\rho + \frac{\lambda}{2}\,\rho^2\right) \simeq 0 \tag{3.13}$$

$$\rho \simeq \frac{\mu}{\lambda} \tag{3.14}$$

$$E_o \simeq -\frac{\mu^2}{2\lambda}\,V \tag{3.15}$$

$$e^{-\beta F} = \mathrm{Tr}\; e^{-\beta H} \simeq e^{-\beta E_o} \tag{3.16}$$

$$F \simeq E_o \tag{3.17}$$

$$\frac{N}{V} = \frac{1}{V}\int\phi^\dagger\phi\, d^3x = \rho \simeq \mu/\lambda\;, \quad (\mu > 0) \tag{3.18}$$

to be contrasted with (3.7), still approximately valid for $\mu < 0$. As μ turns from negative to positive a phase transition is said to take place; we suddenly get large values for the fields ϕ already in the lowest eigenstate of H. This is called Bose condensation. It takes place whenever the pressure is so high that the chemical potential becomes negative. Also the λ term must be sufficiently small.

The model described by our hamiltonian (3.11) resembles somewhat the model of the previous section. Note however that the mass

M comes in the derivative term and never vanishes. In the relativistic field theory M is replaced by 1/2, and $-\mu$ by the mass-squared, m^2. In that theory Bose-condensation can take place also : we must extend the allowed values for m^2 to negative values, clearly a more profound change in the particle properties. But what we get in return is that in the relativistic model, the vacuum itself, without any external pressure, can become a Bose-condensate.

IV. GOLDSTONE PARTICLES

In the previous section we discussed Bose-condensation in a statistical system only in a sketchy way because we will not need the details (for instance, ϕ and $\phi^\dagger$ were not canonical variables in the usual sense). We will be more precise for the case that is more relevant to us : the relativistic complex scalar case. We return to the lagrangian (2.1) and now assume m^2 to be negative. It is convenient to rewrite it as

$$\mathcal{L}(\phi,\phi^*) = -\partial_\mu\phi^*\partial_\mu\phi - \frac{\lambda}{2}(\phi^*\phi - F^2)^2 \tag{4.1}$$

where $m^2 = -\lambda F^2 < 0$ automatically, and an irrelevant constant, $-\lambda/2\ F^4$, has been added to the lagrangian.

The hamiltonian density of the system is

$$\mathcal{H} = \pi^\dagger\pi + \partial_i\phi^\dagger\partial_i\phi + \frac{\lambda}{2}(\phi^\dagger\phi - F^2)^2 \tag{4.2}$$

where $\pi,\pi^\dagger$ are the canonical momenta associated with $\phi,\phi^\dagger$ which are now operators rather than fields. The $\phi,\phi^\dagger$ dependent part has an extremum for

$$\phi^\dagger\phi = F^2 \tag{4.3}$$

from which

$$\phi = Fe^{i\omega}$$
$$\phi^\dagger = Fe^{-i\omega} \tag{4.4}$$

where ω is arbitrary but fixed to the same value everywhere in space.

Now if there were no $\pi^\dagger\pi$ term in $\mathcal{H}$ then (4.4) would be the exact solution to the Schrödinger equation for the lowest energy state. Every ω between 0 and 2π would describe a lowest-energy eigenstate of $H = \int \mathcal{H} d^3x$, each with eigenvalue $E_o = 0$. The question is whether the $\pi^\dagger\pi$ term causes sufficient fluctuations to lift this degeneracy. The answer is not so simple : in 1 space - 1 time dimension, yes; in 2 or more space dimensions, no. So let us limit ourselves to 3 space + 1 time dimensions. Then the vacuum (= lowest energy state) is degenerate and characterized by a phase angle ω. But (4.4) is not exactly valid due to the $\pi^\dagger\pi$ term. We replace it by

$$< 0|\phi|0 > = F'e^{i\omega} \tag{4.5}$$

where $|0>$ is the vacuum state and ω is the vacuum angle. From now on we will consider only the world surrounded by a vacuum with $\omega = 0$. Further, F' is close to F. In fact, because of ultraviolet divergences, subtractions must be made in F and ϕ, and we could choose these such that F' = F.

Actually our theory only makes sense if either λ is chosen to be rather small and F of order $1/\sqrt{\lambda}$, or if an adequate ultraviolet cutoff has been introduced. The reasons for this are deeper field theoretic arguments connected with the renormalization group that I will not go into. Let us assume that λ is rather small. Then the fluctuations of ϕ around F are also relatively small and it makes sense to split

$$\phi = F + \eta \tag{4.6}$$

and the lagrangian becomes

$$\mathcal{L} = - \partial_\mu\eta^*\partial_\mu\eta - \frac{\lambda}{2}(F(\eta+\eta^*) + \eta^*\eta)^2 \tag{4.7}$$

Writing

$$\eta = \frac{1}{\sqrt{2}}(\eta_1 + i\eta_2)$$

This becomes

$$\mathcal{L} = - \frac{1}{2}(\partial_\mu\eta_1^2 + \partial_\mu\eta_2^2) - \lambda F^2\eta_1^2 + \text{int} \tag{4.8}$$

where "int" stands for higher order terms in η_1, η_2. Notice now that one particle, η_1, obtained a mass

$$M_\eta = F\sqrt{2\lambda} \tag{4.9}$$

But its companion η_2 became massless.

The occurrence of a massless particle as soon as the vacuum expectation value of a field ϕ is not invariant under the continuous symmetry (2.3) has been first observed by J. Goldstone[1], and it is an exact property of the system, not related to our perturbative approximation (no higher order mass corrections). We conclude that after the phase-transition caused by Bose-condensation, the symmetry (2.3) is *spontaneously broken* (the degeneracy of ϕ_1 and ϕ_2 is not reproduced in η_1, η_2) and at the same time a massless particle appears : the Goldstone particle. Here we see the first example where the microscopic fields ϕ, ϕ^* in the lagrangian do not reflect accurately the physical spectrum, but the transition towards the η fields was still very simple. It is correct to characterize the vacuum by

$$< 0|\phi(\vec{x})|0 > = F \neq 0$$

and the vacuum is infinitely degenerate. Characterization of the *Higgs* mode, next section (V) will be very different!

V. THE HIGGS MECHANISM

We now switch on electromagnetic interactions*) simply by adding the Maxwell term to the lagrangian and replacing derivatives by covariant derivatives :

$$\mathcal{L}(\phi,\phi^*,A_\mu) = -\frac{1}{4}F_{\mu\nu}F_{\mu\nu} - (D_\mu\phi)^* D_\mu\phi - \frac{\lambda}{2}(\phi^*\phi - F^2)^2 \qquad (5.1)$$

where

$$F_{\mu\nu} = \partial_\mu A_\nu - \partial_\nu A_\mu$$

$$D_\mu\phi = (\partial_\mu + iqA_\mu)\phi \qquad (5.2)$$

q is the electric charge of the particle ϕ.

Indeed the previously introduced current j_μ, eq (2.4), is now the conserved electromagnetic current. But the invariance (2.3) can now be replaced by

$$\phi \to e^{i\Lambda(x)}\phi\ , \quad \phi^* \to e^{-i\Lambda(x)}\phi^*$$

$$A_\mu \to A_\mu - 1/q\ \partial_\mu\Lambda(x) \qquad (5.3)$$

---

*) It may seem to a superficial reader that these notes are just repeating the story[2] of the early 70's. However we are now not primarily interested in perturbative quantization but rather *non-perturbative characterization* of what happens.

which is a *local* invariance

As already mentioned in the Introduction, the consequence of this local invariance is that only a subspace of all (ϕ, ϕ^*, A_μ), namely the gauge-nonequivalent values, correspond to physical observables.

Traditionally, one now proceeds by choosing a gauge fixing procedure so that most, if not all, degeneracy is removed. One then performs perturbation expansions in λ and q^2. At zero λ and q^2 one can again ask whether or not

$$< \phi >_o = F \neq 0 \ ? \qquad (5.4)$$

and since the higher order corrections to $< \phi >_o$ are of higher order in λ and q^2 the qualitative distinction whether or not $< \phi_o > = 0$ remains valid at every order. And so we get the local variant on the Goldstone mechanism : after Bose-condensation of charged particles we get the Higgs mechanism [3]. However, there is a difficulty with (5.4), because ϕ is not gauge invariant. Smearing (5.4) over all of space-time may yield zero or not, depending on the gauge chosen. In a trivial gauge

$$\begin{aligned} \mathrm{Re}(\phi) &> 0 \\ \mathrm{Im}(\phi) &= 0 \end{aligned} \qquad (5.5)$$

we have

$$< \phi >_o > 0$$

always. We therefore propose to use criterion (5.4) *only* in perturbative considerations, where it is correct (as good as it can be) but *not* as an absolute non-perturbative criterion for the Higgs mode. Another criterion that cannot be used is whether or not the vacuum is degenerate. The problem there is that transformation (5.3) yields physically equivalent states, contrary to its global equivalent (2.3)!! Therefore all those vacuum states corresponding to different ω angles are now one and the same state. The vacuum is never degenerate if the symmetry is local. Local symmetries are *never* "spontaneously broken". Then why is this phrase so often used in connection with gauge theories ? Because, as I will show now, there certainly is such a thing as a Higgs mode and it usually can be described in some or other reasonable perturbation expansion around a Goldstone (= global) field theory.

Let us return to perturbation theory momentarily. We then write as usual

$$\phi(x) = F + \eta(x) \qquad (5.6)$$

$$\mathcal{L} = -\frac{1}{4}F_{\mu\nu}F_{\mu\nu} - q^2F^2A_\mu^2 - \sqrt{2}\,qFA_\mu D_\mu\eta_2 - (D_\mu\eta)^* D_\mu\eta - \lambda F^2\eta_1^2 + \text{int} \tag{5.7}$$

A convenient renormalizable gauge is obtained by adding the gauge fixing term

$$\mathcal{L}^c = -\frac{1}{2}(\partial_\mu A_\mu + \sqrt{2}\,qF\eta_2)^2 \tag{5.8}$$

so that

$$\mathcal{L} + \mathcal{L}^c = -\frac{1}{2}(\partial_\mu A_\nu)^2 - \frac{1}{2}M_A^2A^2 - D_\mu\eta^* D_\mu\eta - \frac{1}{2}M_\eta^2\eta_1^2 - \frac{1}{2}M_A^2\eta_2^2 + \text{int} \tag{5.9}$$

with

$$M_A = \sqrt{2}\,qF \qquad\qquad M_\eta = \sqrt{2\lambda}\,F \tag{5.10}$$

It is easily read off from this lagrangian that the vector particle A has a mass M_A and η_1 a mass M_η. The longitudinal component of the vector field A_μ and η_2 are ghosts, which both cancel against the Faddeev-Popov-DeWitt ghost [4,5] all having the same mass M_A.

So perturbation theory suggests that the Higgs theory behaves in a way very different from the symmetric or Coulomb theory : one of the two scalar fields ϕ disappears and the vector field obtains a mass so that the photon field is short-range only. This *is* a distinction that should survive beyond perturbation theory. Thus the criterion that electromagnetic forces become short-range is much more fundamental than either the vacuum value of the scalar field (5.4) or the "degeneracy of the vacuum". But, there is yet another new phenomenon in the Higgs mode contrary to the "unbroken" or Coulomb mode. This is important because the above does not yet distinguish a Higgs theory from just any non-gauge theory with massive vector particles.

VI. VORTEX TUBES

The non-relativistic version of the theory of the previous section is the superconductor : if electrically charged bosons (Cooper's bound state of an electron pair) Bose-condense then there the electric fields become short-range. Also magnetic fields are repelled completely (Meissner effect). Except when they become too strong. Then, because of magnetic flux conservation, they have to be allowed in. What happens is that a penetrating magnetic field forms narrow flux tubes. These flux tubes carry a multiple of a precisely defined quantum of magnetic flux. A little amendement to the perturbative Higgs theory can explain this.

By gauge transformations the "vacuum value" of the Higgs field ϕ can be changed into

$$< \phi(x) >_o = F e^{i\omega(x)} \tag{6.1}$$

if $\omega(x)$ is a continuous, differentiable function of space and time. However we can also consider perturbation theory around field configurations

$$\phi(x) = \rho(x)\, e^{i\omega(x)} \tag{6.2}$$

where $\rho(x) \simeq F$ nearly everywhere, but at some points ρ may be zero. At such points ω needs not be well defined and therefore in all the rest of space ω could be multivalued. For instance, if we take a closed contour C around a zero of $\rho(x)$ then following ω around C could give values that run from 0 to 2π instead of back to zero. The energy of such a field configuration is only finite provided that $D_i\phi(x)$ goes to zero sufficiently rapidly at ∞. Since $\partial_i\phi$ does not go to zero fast enough there must be a supplementary vector potential $A_i(x)$.

The easiest way to find that is by taking a gauge transformation that is regular at ∞ but singular at the origin.

$$e^{-i\Lambda} = \phi(x)/\rho(x) \tag{6.3}$$

In the new gauge $\partial_i\phi$ may vanish rapidly at ∞ and therefore $A_i(x)$ also. So in the old gauge

$$\oint A_i dx^i = \oint \frac{d\lambda}{q} = \frac{2\pi}{q} \tag{6.4}$$

which is the magnetic flux. One can compute the energy of the flux tube by assuming cylindrical symmetry and substituting (6.2) into (5.1). One then varies A_μ and ρ with the boundary condition (6.4) fixed, and minimizes the hamiltonian derived from (5.1).[6] One typically finds that the energy of a flux with length ℓ is

$$E = \alpha\ell \tag{6.4}$$

$$\text{with } \alpha = O(F^2) = O(M_A^2/q^2) \tag{6.5}$$

$$\text{if } q^2 = O(\lambda) \tag{6.6}$$

If finally a magnetic field is admitted inside a superconductor it can only come in some multiple of these vortices, never spread out because of the Meissner effect. Classically one may consider this as an aspect of the infinite conductivity of the material that is only broken down in sufficiently strong magnetic fields. The stable vortex configurations that we discussed here were first derived in the relativistic theory by Nielsen, Olesen and Zumino[6]. The existence of these macroscopic stable objects can be used as another characterization of the Higgs mechanism. They should also survive beyond perturbation expansion.

VII. DIRAC'S MAGNETIC MONOPOLES

At this stage it is useful to introduce the notion of a single magnetic charge à la Dirac[7]. It is not (yet) a dynamic particle but just a source or sink of magnetic flux, a spectator particle not dynamically involved in the lagrangian of the theory. A Dirac monopole can be visualized as the end point of an infinitely thin coil carrying a large electric current. The vector potential $\vec{A}$ is very large close to the coil, because of this electric current :

$$\oint \vec{A} d\vec{x} = \Phi \tag{7.1}$$

where Φ is the magnetic flux of the coil and the integral is over any contour going closely around it. Close to the coil dx is small, therefore $\vec{A}$ becomes large.

Nevertheless the effect of the coil on its surroundings comes only through the end points, if a gauge transformation exists that removes this large vector potential :

$$\oint \frac{d\Lambda}{q} = \Phi \tag{7.2}$$

Such gauge transformations Λ would be multivalued, but we require that $e^{i\Lambda}$ in (5.3) remains single-valued. So the jumps that Λ is allowed to make are multiples of 2π. Therefore the gauge transformation (7.2) turns single-valued field configurations into single-valued field configurations if

$$\Phi = 2\pi n/q \tag{7.3}$$

This is how Dirac found that the total amount of magnetic flux carried by a magnetic monopole must be quantized in units $2\pi/q$ where q is the smallest possible electric charge in the universe. This condition must be satisfied whenever we want a rotationally invariant quantized theory with magnetic monopoles and single valued fields. It is illustrative now to see what would happen with such a spectator particle inside a Higgs theory (or superconductor).

It is not accidental that the monopole quantum $2\pi/q$ coincides with the Nielsen-Olesen-Zumino vortex quantum. This implies that the monopole will be sitting at the end of an integer number of such vortices. Antimonopoles may be sitting at the other ends. Now the energy of such configurations is approximated by eq. (6.4). It is proportional to their separation distance. And so we notice that the monopoles inside a superconductor are kept together by an infinite potential well, the potential being simply linearly proportional to their separation. This is the first observation of a confinement feature in quantum field theory, although the confined objects were as yet spectators, not any of the participants of the field equations. That will come later (sect. XIII).

VIII. THE UNITARY GAUGE

So far we limited ourselves strictly to Abelian gauge theories. We knew what the microscopic field variables are, and now we know what particles and vortices survive at macroscopic distance scales. The latter depend critically on what kind, if any, of Bose condensation took place. Now in our introduction we also mentioned the *microscopic physical* variables. Formally the space H of these variables is given by

$$H = R/G \tag{8.1}$$

where R is the space of field variables and G is the (local) gauge group. How do we enumerate the variables in H ?

Traditionally one imposes a gauge condition on the fields in R, thus obtaining a subspace in R which could be representative for H. In section V we used the gauge fixing term (5.8). This is good enough if one intends to do perturbation expansion[2,5]. The ghost particles one obtains cancel each other and can be dealt with. However we claim that if a non-perturbative characterization of the physical variables is required then this is not good enough. Imagine that one tries to solve the Dyson-Schwinger equations of the theory in some nonperturbative way. Whenever a computed S-matrix element shows a pole one can never be sure whether or not this is due to a ghost or whether it is physical. Furthermore as we will see the ghosts will produce their own topological features called "phantom solitons" which are entirely non-physical. Therefore if we want some understanding of the physical variables we must go to a "unitary gauge" (a gauge with no ghosts)

Often the axial gauge

$$A_0 = 0 \tag{8.2}$$

is used in order to understand the physical Hilbert space. However, this leaves invariance with respect to time-independent gauge transformations :

$$\Lambda(\vec{x},t) = \Lambda(\vec{x}) \tag{8.3}$$

and so there is still a redundancy in our set of variables. It is not suitable for our purposes.

A completely ghost-free gauge can be formulated if we have a charged scalar field ϕ (if no such field is present one may consider building such a field by composing, say, two fermion fields). We do *not* require the Higgs phenomenon to take place. Regardless what condensation takes place at large distance scales one can look at the gauge

$$\begin{aligned} \mathrm{Re}(\phi) &= \rho > 0 \\ \mathrm{Im}(\phi) &= 0 \end{aligned} \tag{8.4}$$

This fixes the gauge function Λ locally, point by point in space-time, contrary to gauges such as eq. (5.8) where the condition on Λ requires solving a second order partial differential equation (the cause of the ghosts).

Within the unitary gauge (8.4) all components of the vector field A_μ are entirely observable. The complex scalar field ϕ is reduced to a real field ρ that can only take positive values. This would be a convenient description of the space of microscopic physical field variables were it not for one deficiency in the condition (8.4) : the original space of variables R certainly allows the scalar field ϕ to vanish at certain points in space-time. These points, defined by (for any $\phi \in R$)

$$\begin{aligned} \mathrm{Re}(\phi) &= 0 \\ \mathrm{Im}(\phi) &= 0 \end{aligned} \tag{8.5}$$

have the topological structure of a set of closed curves in 3-space, or closed surfaces in 3 + 1 dimensional space-time. At these points the condition (8.4) becomes singular : if $\phi = \rho e^{i\theta}$ then we must choose

$$\Lambda = -\theta \tag{8.6}$$

but the gradient of θ is easily seen to explode close to a zero of ϕ and therefore the vector potential $\vec{A}$, transforming as the gradient of Λ, will grow as the inverse power of the distance to this zero. Thus we find that the string-like structures, defined by (8.5) are separate degrees of freedom, giving a boundary condition on ρ ($\rho = 0$) and a prescribed singular boundary behavior of A_μ. This completes our discussion of the microscopic physical degrees of freedom for any Abelian gauge theory : we have observable vector fields A_μ, a truncated scalar field ρ ($\rho > 0$) and all possible closed strings *), on which there is a boundary condition for both ρ and A_μ

Note that only in the Higgs theory these physical variables are in the same time the *macroscopic* physical variables, although of course the macroscopic variables will be "dressed with a cloud of virtual particles" (the string becomes a vortex with finite thickness). In the "unbroken" Abelian theory of electromagnetism the macroscopic variables are harder to discuss. It appears that our vortices "Bose-condense" to form long range, non-energetic magnetic field lines : the ordinary magnetic field $\vec{B}$.

IX. PHANTOM SOLITONS

The gauge (8.5) is called "unitary gauge" because in that gauge all surviving fields will be physically observable. Their quanta will all contribute in the unitarity relation

$$\sum_n < a|S^\dagger|n >< n|S|b > = < a|b > \qquad (9.1)$$

However as soon as practical calculations are considered smoother gauge conditions are required. (8.5) is hard to implement if ϕ oscillates wildly at small distances. We will now argue that after a transition towards "smoother" gauge conditions not only ghost particles arise but also what we call "phantom solitons" : extended structures which are stable for topological reasons but nevertheless unphysical gauge artifacts [8].

Intermediate between the "renormalizable gauge" (5.8) and the unitary gauge we could choose

$$\arg(\phi) + \kappa\partial_\mu A_\mu = 0 \qquad (9.2)$$

where $\arg(\phi) = \mathrm{Im}(\log \phi)$ and κ is an arbitrary gauge parameter. The gauge condition (9.2)is smoother than the unitary gauge because at small distances, by power counting, the second term dominates and we come close to the renormalizable Lorentz gauge. One finds ghosts in this gauge which propagate with a mass

$$m_{gh} = (q/\kappa)^{1/2} \qquad (9.3)$$

so for small κ they become unimportant.

---

*) That is, strings without ends; they could run from ∞ to ∞.

Now imagine a Nielsen-Olesen-Zumino vortex tube in the form of a closed curve. What do the field configurations in the gauge (9.2) look like ? The gauge(9.2) is that particular gauge for which

$$W = \int d^4x(q^{-1}(\arg(\phi))^2 + \kappa A_\mu^2) \qquad (9.4)$$

has an extremum. Let us assume this is a minimum. The system then likes to arrange $\arg(\phi)$ to be zero as much as possible but not with too large vector potentials A_μ.

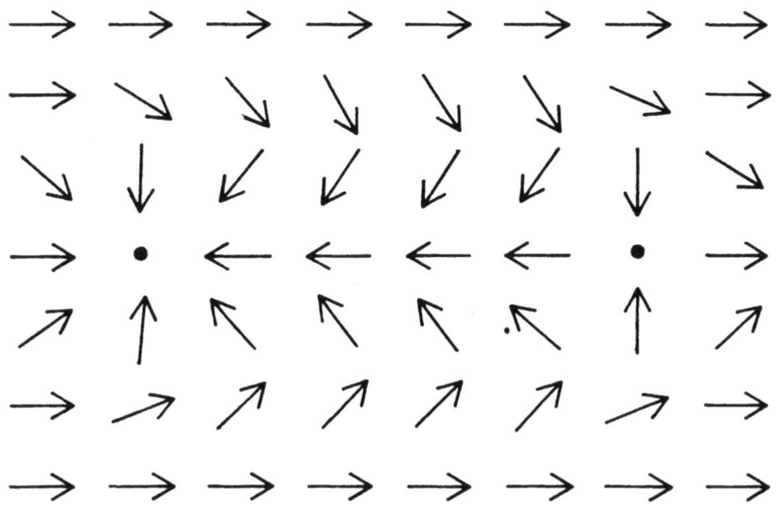

Fig 1

In fig 1 we pictured a cross section of the vortex. The plane is intersected twice, in opposite directions. At the intersection points the scalar field ϕ makes one complete rotation, again in opposite directions. The configuration in the figure is close to the optimal gauge (9.2), keeping ϕ as much as possible oriented towards the positive real axis. We see that the topology of the complete "twist" from top to bottom had to be preserved. This twist will cover an entire sheet spanned by the vortex. Certainly (9.2) will be obtained (i.e. (9.4)will be minimal) if that sheet has minimal surface. The equations for the field configurations inside the sheet are easy to solve if the sheet is considered to be locally sufficiently flat. Clearly the sheet is a gauge artifact. We think that structures of this sort will further obscure the physical interpretation of whatever solutions will be found to the Dyson-Schwinger equations in the gauge (9.2) or completely renormalizable gauges. For instance, bubbles made out of these sheets will form a whole Regge-like family of phantom particles.

X. NON-ABELIAN GAUGE THEORY

We now intend to perform the same procedure in non-Abelian gauge theories. For simplicity we restrict ourselves to the case that the gauge group is SU(N), with arbitrary N. The *microscopic* field variables are : a matrix vector field $A_\mu^{ij}(x)$ where μ is a Lorentz index and i, j run from 1 to N; and there may be Dirac spinor fields ψ_{iF}, $\bar{\psi}_{iF}$ where F is a flavor index. Scalar fields are not assumed to play a significant role but may be present too.

An element of the gauge group G (here SU(N)) is a space-time dependent unitary matrix $\Omega(x)$. Any transformation of the form

$$A'_\mu = \Omega(x)(A_\mu + \frac{i}{g}\partial_\mu)\Omega^{-1}(x)$$

$$\psi'_F = \Omega(x)\psi_F$$

$$\bar{\psi}'_F = \bar{\psi}_F\Omega^{-1}(x) \tag{10.1}$$

is postulated to describe the same physical situation as before, but of course gives a different set of values to the microscopic variables. If R is the space of microscopic variables, then R/G is the space of microscopic *physical* variables. Again we ask the question how to categorize or enumerate these physical variables.

In perturbation theory it is customary to impose a gauge condition, which implies that we find a subspace of R (the set of fields in R that satisfy the gauge condition) representative for R/G. A renormalizable gauge condition is

$$\partial_\mu A_\mu = 0 \tag{10.2}$$

but it is easy to see that this subspace of R does not accurately describe the physical degrees of freedom in R/G (even though it is accurate enough for perturbation theory). To see this, consider an infinitesimal perturbation in R

$$A_\mu \to A_\mu + \delta A_\mu \tag{10.3}$$

where δA_μ is entirely localized in a particular region δV of space-time. We may however have

$$\partial_\mu \delta A_\mu = f(x) \neq 0 \tag{10.4}$$

In order to impose the gauge condition (10.2) we now find an infinitesimal gauge transformation $\Omega = e^{ig\Lambda}$ that restores (10.2). It must satisfy

$$\partial_\mu D_\mu \Lambda = f(x) \tag{10.5}$$

but the inverse of the operator $\partial_\mu D_\mu$ is non-local. So in our subspace of R satisfying the gauge condition we find a perturbation which is spread out all over space-time, well outside δV. Clearly this perturbation outside δV is unphysical and in perturbation theory we learnt how to deal with this : the theory has "ghosts".

We know claim that beyond perturbation theory these ghosts obscure the physical contents of our theory. Therefore we shall look for a "unitary gauge". Our strategy is to determine this gauge in two steps. Let L be one of the largest Abelian subgroups of G, in our case $L = U(1)^{N-1}$.

$$L = U(1)^{N-1} \tag{10.6}$$

We call G/L the "non-Abelian part" of the gauge group and our first step will be to fix the "non-Abelian part" of the gauge redundancy. We choose a gauge condition C that reduces the space R into a subspace that could be called

$$H_1 = R/(G/L) \tag{10.7}$$

If the gauge group G has N^2-1 generators and L has N-1 generators, then we choose N^2-N real components for the gauge condition C, all invariant under $L \subset G$ but not G itself. The second step is the choice of an N-1 component gauge condition that fixes the remaining invariance under L :

$$H = H_1/L = R/G \tag{10.8}$$

But this second step is precisely the same as fixing the gauge in an ordinary Abelian gauge theory such as electromagnetism and is therefore much more trivial. To understand the physical contents of the theory one could just as well stop after obtaining (10.7) which is expected to describe Abelian charged particles and photons.

A variant on the Lorentz condition that reduces R to R/(G/L) is easy to find :

$$D^o_\mu A^{ch}_\mu = 0 \tag{10.9}$$

there D^o_μ is the L-covariant derivative, containing the diagonal part of the vector field A_μ only. A^{ch}_μ is the set of off-diagonal elements of A_μ only. Eq. (10.9) has indeed N^2-N components.

This gauge suffers from the ghost problem as much as the ordinary Lorentz gauge, and is therefore not suitable for understanding all physical degrees of freedom.

XI. UNITARY GAUGE

A unitary gauge must be picked in a way similar to the Abelian case. We need a field that transforms without derivatives under gauge transformations. We will limit ourselves to the case that this field, call it X, transforms as the adjoint representation under G.

$$X \rightarrow \Omega X \Omega^{-1} \tag{11.1}$$

Such a field namely can always be found. The simplest choice would be

$$X^{ij} = G^{ij}_{12} \tag{11.2}$$

which is one of the components of the covariant curl $G_{\mu\nu}$. This choice has the disadvantage of not being Lorentz-invariant. One may choose a composite field :

$$X^{ij} = G^{ik}_{\mu\nu} G^{kj}_{\mu\nu} \tag{11.3}$$

This however does not work if G = SU(2) because then X would be proportional to the identity matrix. We need a non-vanishing isovector part. We could choose

$$X^{ij} = G^{ik}_{\mu\nu} D^2 G^{jk}_{\mu\nu} \tag{11.4}$$

but this choice looks rather complicated. Perhaps the most practical choice would be to take ones refuge to an extra scalar field in the theory, giving it a sufficiently high mass value so that the theory is not changed perceptibly at low energies.

Our gauge condition will be that X is diagonal :

$$X = \begin{pmatrix} \lambda_1 & & & & 0 \\ & \lambda_2 & & & \\ & & \cdot & & \\ & & & \cdot & \\ & & & & \cdot \\ 0 & & & & \lambda_N \end{pmatrix} \tag{11.5}$$

where the eigenvalues λ_i may be ordered :

$$\lambda_1 > \lambda_2 > \ldots . > \lambda_N \tag{11.6}$$

What is the subgroup of the gauge transformations Ω under which (11.5) is invariant ? If we require

$$X' = \Omega X \Omega^{-1} = X$$

then

$$[X,\Omega] = 0 \tag{11.7}$$

therefore, Ω is also diagonal :

$$\Omega = \begin{pmatrix} e^{i\omega_1} & & & 0 \\ & \cdot & & \\ & & \cdot & \\ & & & \cdot \\ 0 & & & e^{i\omega_N} \end{pmatrix} \tag{11.8}$$

and since $\det\Omega = 1$, we have

$$\sum_{i=1}^{N} \omega_i = 0 \tag{11.9}$$

Indeed, this is the largest Abelian subgroup L of G. If we write

$$\Omega = e^{i\omega} \tag{11.10}$$

then the diagonal part A^o_μ of A_μ transforms as

$$A^o_\mu \to A^o_\mu - \frac{1}{g}\partial_\mu \omega \tag{11.11}$$

and the off-diagonal part A^{ch}_μ as

$$A^{ch\,ij}_\mu \to e^{i(\omega_i - \omega_j)} A^{ch\,ij}_\mu \tag{11.12}$$

Apart from the gauge transformations (11.11) and (11.12) all our fields are physically observable. So our physical degrees of freedom are

- N-1 "massless" photons
- 1/2 N(N-1) "massive" charged vector fields
- N scalar fields λ_i with the restriction : $\lambda_1 > \lambda_2 > \dots > \lambda_N$.

There is of course another constraint : depending on our choice for

X, we have that

$$\begin{pmatrix} \lambda_1 & & & \\ & \cdot & & \\ & & \cdot & \\ & & & \cdot \\ & & & & \lambda_N \end{pmatrix}$$

satisfies (11.2), (11.3), or (11.4). Of course we still have the local $U(1)^{N-1}$ symmetry to be removed by either conventional gauge fixing, or the procedure described in the previous chapters.

XII. A TOPOLOGICAL OBJECT

So far we assumed that the eigenvalues $\lambda_1, \ldots, \lambda_N$ coincide nowhere. What if they do, at some set of points in space-time ? At those points, the invariance group is larger and a problem emerges with our enumeration procedure. We now argue that these exceptional points : a) are pointlike in 3-space, describing particle-like trajectories in space-time, and b) correspond to singularities in the fields A_μ and ψ. The argument for the singularity is similar to the Abelian case, but the fact that these things are particle-like differs from the Abelian case, where the singular points were string-like.

The reason why the singularities in the generic case are pointlike is that the dimensionality of the space of field variables with two coinciding eigenvalues λ is three less than the space with non-coinciding eigenvalues. Therefore the dimensionality of the points in space or space-time where two eigenvalues coincide is three less than that of space-time itself. This statement holds for any generic N x N hermitian matrix field X(x).

What is the physical nature of such a particle-like singularity ? Since the eigenvalues λ_i were ordered, we only need to consider the case that two successive λ's coincide :

$$\lambda_j = \lambda_{j+1} \tag{12.1}$$

for certain j. Let us consider a close neighbourhood of such a point. Prior to the gauge-fixing we may take X to be (12.2), where D_1 and D_2 may safely be considered to be diagonalized because the other eigenvalues did not coincide. The three fields $\varepsilon_a(x)$ are small because we are close to the point where they vanish. With respect to that SU(2) subgroup of SU(N) that corresponds to rotations among the jth and $j + 1^{st}$ components, the fields ε_a form an isovector.

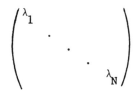

$$X = \begin{array}{c} \\ j \\ \\ j+1 \\ \\ \end{array} \left[\begin{array}{c|cc|c} D_1 & 0 & & 0 \\ \hline & \lambda + \varepsilon_3 & \varepsilon_1 - i\varepsilon_2 & \\ 0 & & & 0 \\ & \varepsilon_1 + i\varepsilon_2 & \lambda - \varepsilon_3 & \\ \hline 0 & 0 & & D_2 \end{array}\right] \tag{12.2}$$

(column labels: j, $j+1$)

One may write the center block as

$$X = \lambda I + \vec{\varepsilon}.\vec{\sigma} \tag{12.3}$$

where σ_a are the Pauli spin matrices. Close to a zero point of this $\vec{\varepsilon}$ field, the field $\vec{\varepsilon}$ has a hedgehog configuration. But gauge fixing i.e. diagonalization of X, corresponds to rotating ε_1 and ε_2 away such that ε_3 is positive ($\lambda_j > \lambda_{j+1}$). Thus is our unitary gauge,

$$\vec{\varepsilon} = \begin{pmatrix} 0 \\ 0 \\ +|\varepsilon_3| \end{pmatrix} \tag{12.4}$$

By now the reader may recognize this field configuration as the one for a magnetic monopole[9]. Indeed, fixing the L-gauge as well cannot be done without accepting a string-like singularity connecting zeros of opposite signature : the Dirac string.

The magnetic charges of the monopole can most easily be characterized with respect to the $U(1)^N$ subgroup of the extended gauge group U(N) :

$$\vec{m} = (\,0,\ \ldots,\ 0,\ \frac{2\pi}{g},\ -\frac{2\pi}{g},\ 0,\ \ldots,\ 0) \tag{12.5}$$

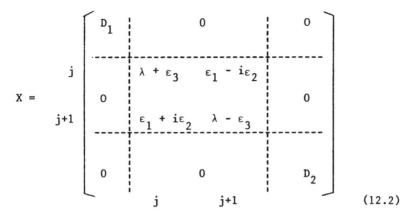

where the $\pm\, 2\pi/g$ are at the jth and $j+1^{st}$ position. g is here the fundamental electric charge of the elementary representation. We then see that $\vec{m}$ actually only acts in the subgroup $U(1)^N/U(1)$ of SU(N) because the sum of all its charges vanishes. It is constructive to notice a subtile difference between this magnetic charge spectrum and the spectrum of the electrically charged gauge particles A^{ij}_μ : from (11.12) we read off that these have electric charges

$$\vec{Q} = (0, \ldots, 0, +g, 0, \ldots, 0, -g, 0, \ldots, 0) \qquad (12.6)$$

where $\pm$ g occur at arbitrary, not necessarily adjacent, positions i and j. Again however the sum of the charges vanishes, so that we are really working in the Cartan group $U(1)^{N-1}$, not $U(1)^N$. Magnetic monopoles with

$$\vec{m} = (0, \ldots, 0, \frac{2\pi}{g}, 0, \ldots, 0, -\frac{2\pi}{g}, 0, \ldots, 0) \qquad (12.7)$$

are possible only if three or more eigenvalues coincide. The dimensionality of such points is at least 8 less than space-time so in general they do not occur at single points. Rather, they should be considered as bound states of "elementary" monopoles (12.5).

We conclude that we arrive at a picture where an Abelian gauge theory is enriched with magnetic monopoles, but because of the slightly different charge spectrum this picture is in general not "self dual". Contrary to the case discussed in section VII, the monopoles we have here are "dynamical", that is, they will inevitably take part in the dynamics of the system.

XIII. THE MACROSCOPIC VARIABLES

We have now arrived at a point where we could sketch a possible strategy for precise calculations for the dynamics of the system :

1) Consider the physical degrees of freedom in the space $H_1 = R/(G/L)$. We find N-1 sets of Maxwell fields, electrically charged fields (among which vector fields), and magnetically charged particles. The particular case of interest is now the possibility that magnetically charged particles "Bose-condense". If we ever are to understand such a mechanism in detail, the following step is probably necessary :

2) Eliminate the electric charges. With as much precision as possible we must compute all light-by-light scattering amplitudes and express them in term of an effective interaction lagrangian for the photon fields :

$\Gamma(A_\mu) = \sum$ 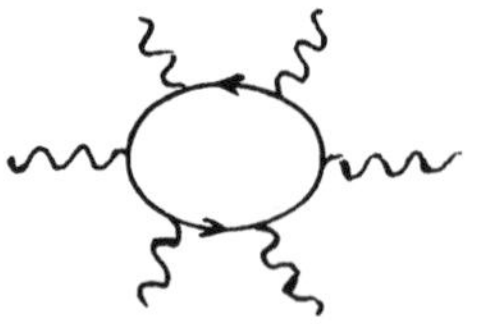 + higher orders (13.1)

3) Now perform the "dual transformation". Since we have only Maxwell fields and magnetic charges interacting with them, we could replace $\vec{B}$ by $\vec{E}$ and $\vec{E}$ by $-\vec{B}$, then introduce operator fields in the usual way for the monopole particles, which now look like ordinary electrically charged objects.

4) Work out the self-interactions among these magnetic monopoles. Set up a perturbation theory now in terms of $2\pi/g$. Then the question is :

5) Does, in terms of this perturbation theory, Bose condensation occur among these monopoles ? Is it reasonable to start with

$$< 0|\phi_{mon}|0 > \neq 0 \quad ? \tag{13.2}$$

If so, then the vacuum is a magnetic superconductor. The monopoles formally have a negative mass-squared. In this magnetic superconductor electric charges are confined. The descriptions of section VII apply qualitatively, after the interchange electric $\leftrightarrow$ magnetic.

XIV. THE DIRAC CONDITION IN THE ELECTRIC-MAGNETIC CHARGE SPECTRUM

Fig 2 represents the spectrum of possible charges in the case that the gauge group G is SU(2).

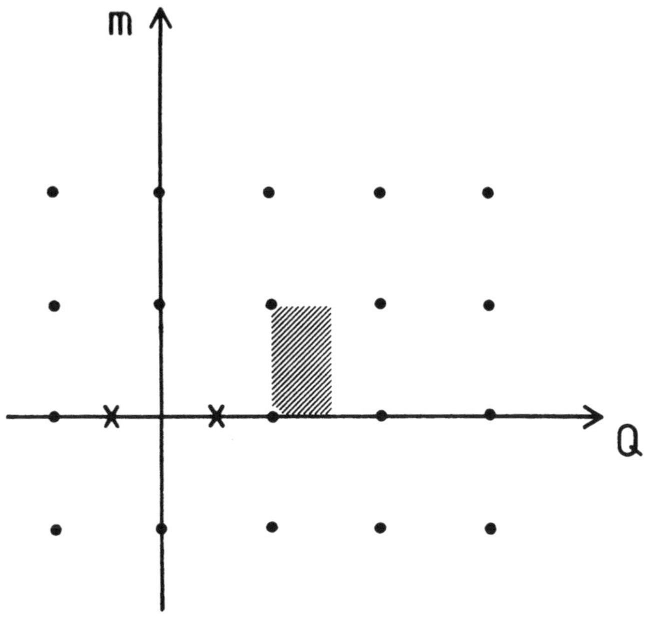

Fig 2

Horizontally is plotted the electric charge Q, vertically the magnetic charge $\vec{m}$. Elementary and bound state charges are indicated. The crosses represent a fundamental SU(2) doublet which may or may not have been added to the theory. The pure Maxwell field equations are now invariant under rotations of the figure about the origin. This is why one is always free to postulate that the fundamental fields carry no magnetic monopole charge. But does the lattice obtained in fig 2 have to be rectangular ? We will now argue that Dirac's quantization condition allows more kinds of lattices and "oblique" lattices indeed may result in a non-Abelian gauge theory.

The Dirac condition for a magnetic charge quantum m and the electric charge quantum q was

$$qm = 2\pi n \tag{14.1}$$

where n is integer. To be precise this corresponds to a quantization condition for the Lorentz force that the magnetically charged particle exerts on the electrically charged object. But now consider two particles 1 and 2 both with various kinds of magnetic and electric charges. Then the Lorentz force quantization corresponds to

$$\sum_i (q_i^{(1)} m_i^{(2)} - q_i^{(2)} m_i^{(1)}) = 2\pi n_{12} \tag{14.2}$$

where the index i refers to the label of the species of photons and n_{12} is an integer relevant for particles 1 and 2, to be referred to as the Dirac quantum of particle (1) with respect to particle (2).

In our specific case of SU(N) broken down to $U(1)^{N-1}$ we usually require

$$\sum_{i=1}^{N} q_i = 0 \qquad\qquad \sum_{i=1}^{N} m_i = 0 \tag{14.3}$$

Our charge lattice in this case will be spanned by 2(N-1) basic charges, to be labelled by an index A = 1, ..., 2N-2. Because of the invariance

$$\begin{pmatrix} m_i \\ \\ q_i \end{pmatrix} \rightarrow \begin{pmatrix} \cos\phi_i & \sin\phi_i \\ \\ -\sin\phi_i & \cos\phi_i \end{pmatrix} \begin{pmatrix} m_i \\ \\ q_i \end{pmatrix} \tag{14.4}$$

we may always take

$$m_i^{(A)} = 0 \qquad \text{for } A = 1, \ldots, N-1 \tag{14.5}$$

The gluons will provide us with a basis of electric charges:

$$q_i^{(A)} = g\delta_i^A - g\delta_i^{A+1} \quad \text{for } A = 1, \ldots, N-1 \tag{14.6}$$

(The fundamental representation, if it occurs, could have

$$q_i^{(k)} = g\delta_{ik} - g/N \text{).} \tag{14.7}$$

The magnetic monopoles have the remaining basic charges:

$$m_i^{(A)} = \frac{2\pi}{g}\delta_i^{A+1-N} - \frac{2\pi}{g}\delta_i^{A+2-N} \quad \text{for } A = N, \ldots, 2N-2 \tag{14.8}$$

It was Witten[10] who observed that monopoles may also carry electric charges. He found

$$q_i^{(A)} = \frac{\theta g^2}{4\pi^2} m_i^{(A)}, \quad \text{for } A = N, \ldots, 2N-2 \tag{14.9}$$

where θ is the instanton angle of the theory, $0 \leqslant \theta < 2\pi$. Notice that for any value of θ the Dirac condition (14.2) is fulfilled.

It can be seen that this phenomenon, eq. (14.9), follows from the lagrangian

$$\mathcal{L} = -\frac{1}{4}G^a_{\mu\nu}G^a_{\mu\nu} + \frac{\theta i g^2}{32\pi^2} G^a_{\mu\nu}\tilde{G}^a_{\mu\nu} + A^a_\mu J^a_\mu \tag{14.10}$$

there

$$\tilde{G}^a_{\mu\nu} = \frac{1}{2}\varepsilon_{\mu\nu\alpha\beta} G^a_{\alpha\beta} \tag{14.11}$$

and $G^a_{\mu\nu}\tilde{G}^a_{\mu\nu}$ corresponds to

$$4\sum_a \vec{E}^a.\vec{B}^a \tag{14.12}$$

The canonical argument can be found in refs 10), 11), 8).

In the case of SU(2) the charge lattice indeed becomes tilted now (Fig 3). It is remarkable that if θ runs from 0 to 2π then the charge lattice indeed turns back into itself, but the "elementary" monopole labelled by (2) in fig 3 is replaced by the "monopole gluon bound state" labelled (3). It seems that no fundamental distinction

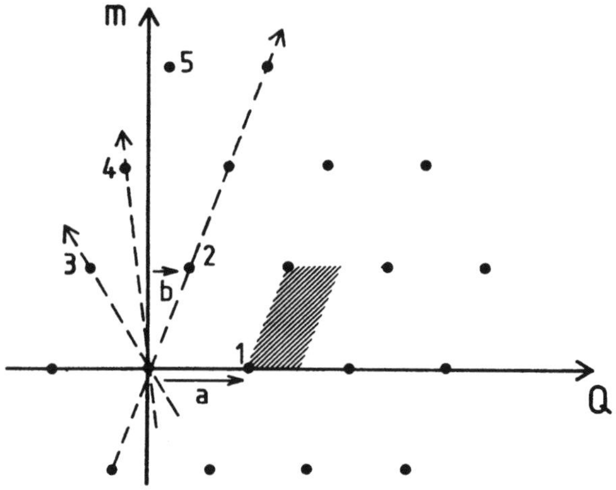

Fig 3

will be possible between monopoles and monopole-gluon bound states.

XV. OBLIQUE CONFINEMENT

We can now ask which of the objects in the lattice of fig 3 will form a Bose-condensate. If it is a purely electrically charged object, (1) in fig 3, then we get the familiar Higgs theory or electric superconductor. All charges that are not on the horizontal axis will then be confined by linear potentials, because of the arguments presented in section VII. By duality we now expect that also monopoles may conceivably undergo Bose condensation, for instance charge #(2) in fig 3. One cannot however have both electric and magnetic charges Bose-condense because if the electric ones condense then the magnetic ones have $m^2 \to +\infty$, not negative, and vice versa. In the case of larger N, Bose condensation can only take place among charges forming any linear sublattice of the original charge lattice, as long as its members all have vanishing Dirac quanta with respect to each other.

Now if θ is switched on, then point (3) gradually takes the place of (2). (2) and (3) cannot both Bose condense because they have a non-vanishing relative Dirac quantum. So it is either (2) or (3). It is likely that Bose condensation in the (2)-direction is replaced by condensation in the (3)-direction at $\pi < \theta < 3\pi$. This would then be a *phase transition* in θ, possibly of first order, just like the transition between Higgs and confinement.

Various attempts at dynamical calculations however indicate that at $\theta \simeq \pi$ the confinement mechanism is not strong. An explanation could simply be that the monopoles then carry large electric charges and therefore may have larger self-energies contributing positively to their mass-squared. Suggestions have been made that at $\theta \simeq \pi$ the Higgs mode reappears [12] or a Coulomb mode (no Bose condensation at all).

I suggest yet a different condensation mode that could possibly occur in theories with θ close to π. If neither (2) nor (3) condense because they carry large (but opposite) electric charges, then perhaps (4) which is a bound state of these two with much smaller electric charge condenses. This would only be possible if the lattice is oblique ($\theta \neq 0$) so this mode is referred to as "oblique confinement". A theory with oblique confinement shows some peculiar features. We stress that these will not occur in ordinary QCD because there we know that $\theta \simeq 0$. Our observations may be relevant for certain models with "technicolor" as we will show shortly.

Returning to the case that our gauge group was SU(2), we first argue that the "quarks" (or "preons") in this oblique confinement mode are not confined in the usual sense. That is, if we attach a flavor quantum number to every type of preon, then physical particles transforming as the fundamental representation of the flavor group do occur. The preons are the crosses in fig 4.

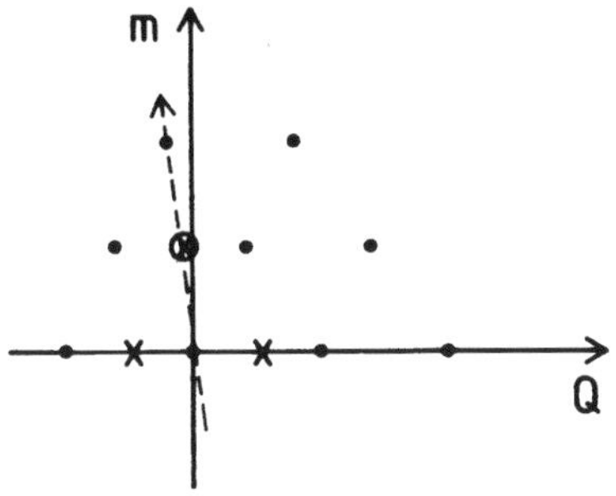

Fig 4.

Since they are not on the line connecting the origin to the condensed object (arrow) they are confined. However, a bound state of a monopole (without flavor quantum numbers) and one preon does occur on that line (⊗), and hence is physical, and has the same flavor as the preon itself. But a price had to be paid : spin and statistics-properties of the physical object are opposite to that of the original preon! If the preon is a fermion, the liberated object is a boson and vice-versa. This is a consequence of a general rule : if two particles have an odd relative Dirac quantum, then the *orbital* angular momentum in any bound state of the two is half-odd-integer, a well-known property of the Schrödinger equation of an electrically charged particle in the field of a magnetic point-source[13)]. That also the *statistics* of the bound state gets an extra fermionic contribution has been shown by Goldhaber[14)]. Let me give an outline of the argument.

If we wish to consider the statistics of two identical composite states A and B, both composed of a magnetic monopole M and an electric charge Q, (fig 5) then we would like to separate the center-of-mass motions of A and B from the orbital motions of M and Q inside A and B.

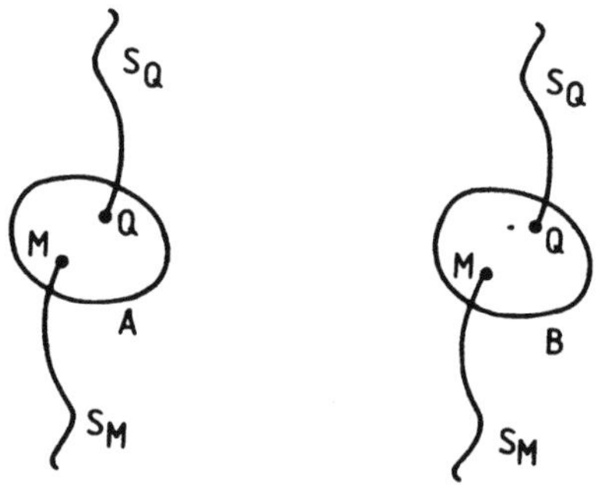

Fig 5

The particles Q see two Dirac strings, ending at both M's. The M's see two Dirac strings ending at both Q's. Now the center-of-mass motion of A can only be split from the orbital motion of M and Q if the magnetic strings run in a direction opposite to that of the electric strings (see fig 5), so that if M hits an electric string, in the same time Q hits a magnetic string. Now let us ignore the motion of M and Q inside A and B, but only consider A and B as a whole. Then A feels both strings at B (and vice versa), in fact, these two are connected in such a way that one string results, running from infinity to infinity. This could be expected because A and B have the same electromagnetic charge combination (they were identical) so *their* relative Dirac quantum vanishes. Their relative motion (A against B) is as if they only were electrically charged. Therefore we obtain the familiar Coulomb Schrödinger equation by removing this complete string by a single gauge transformation :

$$\psi_{AB} \to e^{i\phi_{AB}} \psi_{AB} \tag{15.1}$$

where ϕ_{AB} is the angle by which A rotates around the B-string.

Now notice that if we interchange A and B this angle is 180°, so that

$$e^{i\phi_{AB}} = -1$$

This is how one can see an extra minus sign appearing in the commutation properties of the particles A and B.

XVI. FERMIONS OUT OF BOSONS AND VICE-VERSA

Some exotic models can be constructed if we use oblique confinement as a starting point. Consider for instance an SU(3) gauge theory without fermions but with a scalar field ϕ in the fundamental representation. Let θ be close to π, and let us assume that the familiar Higgs mechanism takes place, described by

$$< \phi > = \begin{pmatrix} 0 \\ 0 \\ F \end{pmatrix} \tag{16.1}$$

(though formally incorrect, as explained in section V, the notation is useful and adequate for our purpose). SU(3) then "breaks spontaneously" into SU(2). One heavy neutral Higgs particle arises. The original octet of gauge fields splits into one neutral heavy vector particle and one SU(2) complex doublet of heavy vector particles.

The SU(2) triplet remains massless, and now we will assume that it produces oblique confinement in the SU(2) sector. The SU(2) doublet of heavy vector bosons will appear as physical particles but disguised as fermions! Here we have an example of fermionic gauge "bosons" without invoking anything resembling supersymmetry.

Another model worth considering is a simplistic weak interaction theory, based upon $SU(2)^{tc} \times (SU(2) \times U(1))^{ew}$. There "tc" stands for "technicolor" and ew for "electro-weak". Quarks and fermions are all in the usual representation of $(SU(2) \times U(1))^{ew}$ and singlets under $SU(2)^{tc}$. Now add one fermion multiplet transforming just like all other fermions under $(SU(2) \times U(1))^{ew}$ but also as a 2 under tc. Assume oblique confinement. Then this fermion will be liberated, but disguised as a boson. Probably it will have a spin-zero component. Since it is in line with the condensing monopole bound states it may Bose-condense itself. So we obtain a scalar field transforming just as the fermions under $(SU(2) \times U(1))^{ew}$ and a non-vanishing vacuum expectation value : a model for the Higgs particle. Indeed, its Bose condensation could well be responsible for the oblique confinement mode in the first place, so here it was not even necessary to consider the monopole-dyon bound state.

Unfortunately, elegant as it may be, this model seems to suffer from the same shortcomings as the more conventional technicolor ideas : it is hard to reproduce the required Yukawa couplings between this scalar field and the other fermions.

XVII. OTHER CONDENSATION MODES

It will be clear from the previous sections, by looking at the electric-magnetic charge lattice, that even more exotic forms of oblique confinement can be imagined. Just assume condensation of bound states with three or more monopoles. Such a condensation mode would be required for instance if we would wish to liberate the fundamental triplets in an SU(3) gauge theory. A fundamental exercise tells us then that these triplets do not switch their spin-statistics properties [8]. In any case, all these different confinement modes will be separated from each other by sharp phase transition boundaries, which should show up in the solutions of the theory when the parameter θ is varied.

In principle each point on the electric-magnetic charge lattice may correspond to a possible phase of the system. There may however be features which cannot easily be understood in terms of our intermediate physical degrees of freedom with Abelian electric and magnetic charges. We have in mind a condensation mode studied in more detail by Bais[15]. The simplest example is an SU(2) gauge theory with an isospin-two Higgs field, ϕ^{ab} (a, b = 1, 2, 3).

Let us assume

$$< \phi^{ab} > = F^{(a)}\delta^{ab} \qquad (17.1)$$

If we were just dealing with a global symmetry, we would say that SU(2) is spontaneously broken into a subgroup D (the invariance group of eq. (17.1)). Now D is the discrete subgroup of SU(2) corresponding to rotations of spinors over 90° :

$$D = (\pm I, \pm i\sigma^1, \pm i\sigma^2, \pm i\sigma^3) \qquad (17.2)$$

If the symmetry is local then D is not really a global invariance of the vacuum. What we do see is that magnetic vortex tubes can be constructed which are characterized by the following boundary condition at infinity :

$$A_\mu \to \frac{i}{g}\Omega\partial_\mu\Omega^{-1} \qquad (17.3)$$

where Ω is multivalued. If we go around the vortex once, then Ω turns into itself multiplied with an element D_1 of D. These vortices are non-commuting. Physically this means the following (fig 6)

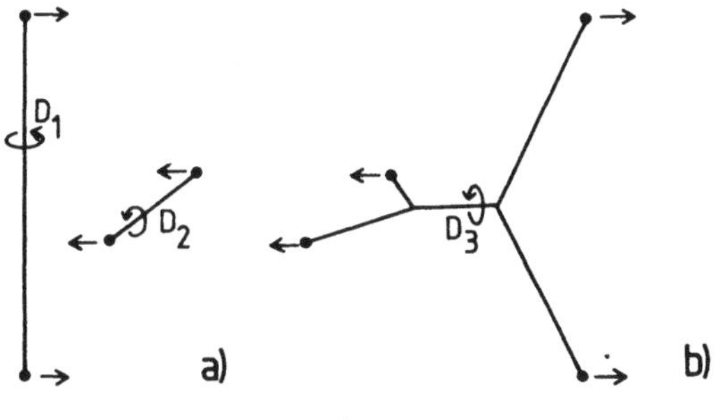

Fig 6

If two strings, characterized by different, non commuting elements D_1 and D_2, approach each other at right angles (fig 6a), then they cannot pass each other without leaving a connecting string D_3 (fig 6b). The element D_3 is given by

$$D_3 = D_1D_2D_1^{-1}D_2^{-1} \qquad (17.4)$$

Clearly the non-commuting properties of the original gauge group were crucial for understanding this phenomenon, so that our Abelian physical variables are not useful here. Indeed, we could ask the question whether the dual of this "Bais mode" exists, with electric strings having similar properties ? As yet, the answer to that question is unknown.

REFERENCES

1. J. Goldstone, Nuovo Cim. 19 154 (1961)
2. F.S. Abers and B.W. Lee, Phys Reports 9C, 1 (1973) and references therein.
3. P.W. Higgs, Phys. Lett. 12, 132 (1969), Phys. Lett. 13, 508 (1964), Phys. Rev. 145, 1156 (1966)
   F. Englert, R. Brout, Phys. Rev. Lett. 13, 321 (1964)
   G.S. Guralnik, C.R. Hagen and T.W.B. Kibble, Phys. Rev. Lett. 13 585 (1964)
   T.W.B. Kibble, Phys. Rev. 155, 627 (1967)
4. B.S. DeWitt, Phys. Rev. 162, (1967), 1195, 1239
5. G. 't Hooft and M. Veltman, "DIAGRAMMAR", CERN report 73/9 (1973)
6. H.B. Nielsen, P. Olesen, Nucl. Phys. B61, 45 (1973)
   B. Zumino, in "Renormalization and Invariance in Quantum Field Theory ed. E.R. Caianiello Plenum Press New York (1974) p. 367
7. P.A.M. Dirac, Proc. Roy. Soc. A133, 60 (1931)
   Phys. Rev. 74, 817 (1948)
8. G. 't Hooft, Nucl. Phys. B, to be published
9. A.M. Polyakov JETP Lett. 20, 194 (1974).
   G. 't Hooft, Nucl. Phys. B79, 276 (1974)
10. E. Witten, Phys. Lett. 86B, 283 (1979)
11. A. Salam and J. Strathdee, Lett. in Mathematical Physics 4, 505 (1980)
12. C. Callan, private communication (1979)
13. R. Jackiw and C. Rebbi, Phys. Rev. Lett. 36, 1116 (1976)
    G. 't Hooft and P. Hasenfratz, Phys. Rev. Lett 36, 1119 (1976)
14. A.F. Goldhaber, Phys. Rev. Lett. 36, 1122 (1976)
15. F.A. Bais, private communication.

CHAPTER 7.3

CAN WE MAKE SENSE OUT OF "QUANTUM CHROMODYNAMICS"?

G. 't Hooft
Institute for Theoretical Physics
University of Utrecht, Netherlands

1. INTRODUCTION

"Quantum Chromodynamics" is a pure gauge theory of fermions and vector bosons that is assumed to describe the observed strong interactions. To get an accurate theory it is mandatory to go beyond the usual perturbation expansion. Not only must we explore the mathematics of solving the field equations non-perturbatively; it is more important and more urgent first to find a decent formulation of these equations themselves, in such a way that it can be shown that the solution is uniquely determined by these equations. We understand how to renormalize the theory to any finite order in the perturbation expansion, but it is expected[1-5] that this expansion will diverge badly, for any value of the coupling constant. Thus the expansion itself does not yet define the theory. But the renormalization procedure is not known to work beyond the perturbation expansion. A clear example of the possible consequences of such an unsatisfactory situation was the recent suprising demonstration[6-10] that all non-Abelian gauge theories have parameters θ, in the form of an angle, that describe certain symmetry breaking phenomena in the theory, but never show up within the usual perturbation expansion because they occur in the

The Whys of Subnuclear Physics, Edited by A. Zichichi (Plenum).

combination

$$\exp[i\theta - \frac{8\pi^2}{g^2}]$$

where g is the gauge coupling constant. This discovery marks an increase in our understanding of non-perturbative field theory but this understanding is not yet complete and, in principle, more of such surprises could await us.

The situation can be compared with the infinity problems in the older theories of weak interactions. Those problems were solved by the gauge theories for which an acceptable and unique regularization and renormalization scheme was found. For our present strong interaction theory, again a "regularization scheme" must be found, this time for "regularizing" the infinities encountered in summing the perturbation expansion.

An interesting attempt to give a non-perturbative formulation of the (renormalized) theory is the introduction of a space-time lattice in various ways[11-13]. But, here also, a proof of uniqueness could not be given (will the continuum-limit yield one and only one theory?) and of course the θ phenomenon mentioned before was not observed in the lattice scheme. So, the lattice theories are still a long way off from answering our fundamental questions.

It is more important for us to make use as much as possible of the important pieces of information contained in the coupling constant expansion. Because of asymptotic freedom[14-16] this expansion tells us precisely what happens at asymptotically large external momenta and it would be a waste to throw that information away.

Which tools could we use to extend our definitions? One is a study of the theory at complex values of the coupling constant. That this is possible in some particular cases is explained in Section 4. We may find that Green's functions must become singular at certain points in the complex g^2 plane and stay regular at others.

Suppose we would discover that there are only singularities on the real axis for $-\infty \leq g^2 < -a$ (unfortunately the true situation is much less favourable). Then we could make a mapping:

$$g^2 \to u = 1 + 2g^{-2}a[1 - \sqrt{1+a^{-1}g^2}]$$

with the inverse

$$g^2 = 4au(1-u)^{-2}$$

The whole complex plane is now mapped onto the interior of the unit circle, with the singularities at the edge. If we rewrite the perturbation expansion in terms of u instead of g then we have convergence everywhere inside the circle*), which implies convergence for all g^2 not too close to the negative real axis. A definite improvement. Unfortunately, the structure for complex g^2 we find in Section 4 is too complicated for this method to work: the origin turns out to be an essential singularity. Still, I shall show how this knowledge of the complex structure may be used to give a very slight improvement in the perturbation expansion (Section 9).

The other tool to be used is the Borel resummation procedure, to be explained in Section 5. A new set of Green's functions" is considered whose perturbation expansion terms are defined to be the previous ones divided by (n-1)! where n is the order of the expansion terms. The singularities of these new functions can be found and the analytic continuation procedure as sketched above can be applied to obtain better convergence. There seem to emerge two types of singularities: one due to the instantons in the theory, the other due to renormalization phenomena. The latter are slightly controversial, they are the only ones that should occur in

---

*) We make use of a well-known theorem for analytic functions that says that the rate of convergence of an expansion around the origin is dictated by the singularity closest to the origin.

quantum electrodynamics, giving that theory of n! type divergence. The first few terms for the electron g-2 do not seem to diverge the way suggested by this singularity. Formally we can improve the perturbation expansion for QED but in practice the "improvement" seems to be bad.

Our work is not finished. What remains to be done is to prove that all singularities in the Borel variable have been found and to find a prescription how to deal with those singularities that are on the positive real axis. Finally, it must be shown that the integrals that link the Borel Green's functions with the original Green's functions make sense and a good theory is found (see Section 9).

The problem of quark confinement can probably be related to certain singularities on the positive real axis, because these singularities arise from infra-red divergences (Section 8).

2. DEFINITION OF THE COUPLING CONSTANT AND MASS PARAMETERS IN TERMS OF LARGE MOMENTUM LIMITS

This section contains the mathematical definitions of the parameters in the theory, so that the statements in the other sections can be made rigorous and free of unnecessary assumptions. It could be skipped at first reading.

The dimensional renormalization scheme is a convenient way of defining a perturbation series of off-mass shell Green's functions with some coupling constant $g_D(\mu)$ as an expansion parameter. The subscript D stands for dimensionally renormalized[17-20]. Each term of the perturbation series is finite. There is an arbitrariness in the choice of the subtraction point μ (which has the dimension of a mass). The theory is invariant under a simultaneous change in g_D and μ provided that

$$\frac{\mu\, d g_D^2}{d\mu} = \beta(g_D^2) = -\beta_1 g_D^4 + \beta_2 g_D^6 + \beta_3^D g_D^8 + \cdots \tag{2.1}$$

Here we show explicitly the minus sign for the first coefficient. This minus sign is a unique property of non-Abelian gauge fields and is responsible for "asymptotic freedom" (at increasing μ we get decreasing g^2, see Refs. 14-16).

The coefficients β_1 and β_2 are known[21-23]. Since we shall try to go beyond perturbation expansion we must be aware of two facts: First, the perturbation expansion is expected to diverge for all g^2 and is, therefore, at this stage, meaningless as soon as we substitute some finite value for g^2. Second, the dimensional procedure has only been defined in terms of the perturbation expansion (Feynman diagrams). Consequently, g_D^2 may not have any meaning at all as a finite number. The correct interpretation of these series is that they are asymptotic series valid for infinitesimal g only, or, equivalently, valid only for asymptotically large momentum: $p^2 = O(\mu^2) \to \infty$. Thus, g_D^2 may not be so good to use as a variable for a study of analytic structures at finite complex values.

We shall now introduce another parameter g_R^2, that may just as well be used instead of g_D^2. It is defined by the following requirements:

When $\mu \to \infty$, then

$$g_R^2(\mu) = g_D^2(\mu) + O(g_D^6(\mu)) \tag{2.2a}$$

$$\frac{\mu\, d}{d\mu} g_R^2(\mu) \equiv -\beta_1 g_R^4 + \beta_2 g_R^6 \tag{2.2b}$$

The series in (2.2b) must stop after the second term. In perturbation theory these requirements have a unique solution for g_R^2. For instance, we get that the rest term in (2.2a) is

$$\frac{\beta_3^D}{\beta_1} g_D^6 + \ldots \tag{2.3}$$

In the dimensional renormalization scheme also the mass parameter was cut-off-dependent (of course, one cannot define such a thing as a physical quark mass parameter, which would have been cut-off independent):

$$\frac{\mu d}{d\mu} m_D^2(\mu) = m_D^2(\mu)[-1 + \alpha_1 g_D^2 + \ldots] \tag{2.4}$$

Again we define $m_R^2(\mu)$ by

$$m_R^2(\mu) = m_D^2(\mu)(1 + O(g^2(\mu)) \text{ for } \mu \to \infty \tag{2.5a}$$

and

$$\frac{\mu d}{d\mu} m_R^2(\mu) = m_R^2(\mu)[-1 + \alpha_1 g_R^2] \tag{2.5b}$$

where the series in (2.5b) stops after the α_1 term.

In QCD the parameters α_1, β_1, β_2 are known[21-23])

$$\begin{aligned} \beta_1 &= (8\pi^2)^{-1}(11 - 2N_f/3) \\ \beta_2 &= (8\pi^2)^{-2}(19N_f/3 - 51) \\ \alpha_1 &= -(2\pi^2)^{-1} \end{aligned} \tag{2.6}$$

We emphasize that the new parameters g_R and m_R are better than the previous ones, because for any theory for which the perturbation expansion is indeed an asymptotic expansion, they are completely finite and non-trivial. Of course, they still depend on the subtraction point μ. At infinite μ they coincide with other definitions; at finite μ they are finite because we can solve Eqs. (2.2b) and (2.5b):

$$\frac{1}{\beta_1 g_R^2(\mu)} + \frac{\beta_2}{\beta_1^2} \log\left(\frac{\beta_1}{g^2(\mu)} - \beta_2\right) = \log(\mu/\mu_0) \tag{2.7a}$$

$$\frac{m_R(\mu)}{m_0} = \left(\frac{\beta_1}{g^2(\mu)} - \beta_2\right)^{\alpha_1/\beta_1} \tag{2.7b}$$

Here μ_0 and m_0 are integration constants. They are invariant under the renormalization group. Thus, μ_0 is a true parameter that fixes the gluon couplings, and has the dimensions of a mass. For each quark in the system we have a mass parameter m_0. As must be clear from the derivations, μ_0 and m_0 actually tell us how the theory behaves at asymptotically large energies and momenta.

The value of μ_0 for QCD is presumably of the order of the ρ mass, and m_0 will be a few MeV for the up and down quarks, 100 MeV or so for the strange quark, etc. For simplicity, we will often drop the terms containing β_2, and the quark masses will be put equal to zero.

3. THE RENORMALIZATION GROUP EQUATION

Now let us consider the Green's functions of the theory. For definiteness, take only the two-point functions (dimensionally renormalized)

$$G^D(K^2,\mu,g_R^2)=a_0(K^2)+g_R^2\,a_1^R(K^2,\mu)+\dots \qquad (3.1)$$

They satisfy a renormalization group equation[24]:

$$\left[\frac{\mu\partial}{\partial\mu}+\beta^R(g^2)\frac{\partial}{\partial g_R^2}+\gamma(g_R^2)\right]G^D(K^2,\mu,g_R^2)=0 \qquad (3.2)$$

Here β^R is the truncated, finite β function for the constant g_R^2 as it occurs in Eq. (2.2b), but $\gamma(g_R^2)$ is still an infinite series. We wish to do something about that also. Let us first make clear how to interpret Eq. (3.2). Consider the μ versus g^2 plane. Suppose we choose a special curve in that plane, where g_R^2 depends on μ such that

$$\frac{\mu d}{d\mu}g_R^2=\beta^R(g_R^2) \qquad (3.3)$$

then it follows from (3.2) that

$$\frac{\mu d}{d\mu}G^D(K^2,\mu,g_R^2(\mu))=-\gamma(g_R^2(\mu))G^D \qquad (3.4)$$

or

$$\frac{d}{dg_R^2(\mu)}\log G^D=-\frac{\gamma(g_R^2)}{\beta(g_R^2)}=\frac{z_0}{g_R^4}+\frac{z_1}{g_R^2}+z_2+\dots \qquad (3.5)$$

That implies that, if we stay on one of the curves (3.3), then (3.2) reduces to (3.5) which can easily be integrated. But the integration constant will still depend on k^2 and on the curve chosen, that is, on the constant μ_0, which we get in solving (3.3), see Eq. (2.7a). Thus we get

$$G^D = Z(g_R^2)\, G(K^2, \mu_o) \tag{3.6}$$

with

$$Z(g_R^2) = \exp -\left(\frac{z_o}{g_R^2} + z_1 \log g_R^2 + z_2\, g_R^2 + \ldots\right) \tag{3.7}$$

and for dimensional reasons, $G(k^2, \mu_0)$ can only depend on the ratio k^2/μ_0^2.

For our purposes it is now important to observe the following. The coefficients z_0 and z_1 are clearly very important as $g^2 \to 0$, but the terms $z_2 g_R^2 + \ldots$ in (3.7) can simply be absorbed in a redefinition of the coefficients $a_1, \ldots$ in (3.1). That way we get new, improved functions G^R that can be written as

$$\begin{aligned} G^R(K^2, \mu, g_R^2) &= a_o(K^2) + g_R^2\, a_1^R(K^2, \mu) + \ldots \\ &= (g_R^2)^{z_1} \exp\left(\frac{-z_o}{g_R^2}\right) G(K^2/\mu_o^2) \end{aligned} \tag{3.8}$$

In a typical example where we study the time ordered product of two operators

$$\bar{\psi}(o)\psi(o) \quad \text{and} \quad \bar{\psi}(x)\psi(x)$$

corresponding to the σ channel, we have

$$\begin{aligned} \gamma &= 2 - \pi^{-2} g_R^2 + \ldots \\ z_0 &= 2/\beta_1 \, . \\ z_1 &= -\pi^{-2}\beta_1^{-1} + 2\beta_2\beta_1^{-2} \end{aligned} \tag{3.9}$$

4. ANALYTIC STRUCTURE FOR COMPLEX g^2

In Eq. (3.8) we can write (neglecting for simplicity the β_2 terms)

$$G(K^2/\mu_0^2) = \mathcal{G}\left(\frac{1}{g^2} + \frac{1}{2}\beta_1 \log K^2/\mu^2\right) \tag{4.1}$$

This is a function of one single parameter

$$x = \frac{1}{g^2} + \frac{1}{2}\beta_1 \log K^2/\mu^2 \tag{4.2}$$

Complex x corresponds to either complex k^2, real g^2 or complex g^2, real k^2. Now, on physical grounds, we know what we should expect at real g^2, complex k^2 (Fig. 1). The singularities are at k^2 real and negative (i.e. Minkowskian). That is when $x = \text{real} + (\beta_1/2)\cdot(2n + 1)\pi i$, n integer. Choosing now k^2 real and positive we find the same singularities at

$$\frac{1}{g^2} = \text{real} + \frac{\beta_1}{2}(2n+1)\pi i \tag{4.3}$$

They are sketched in Fig. 2.

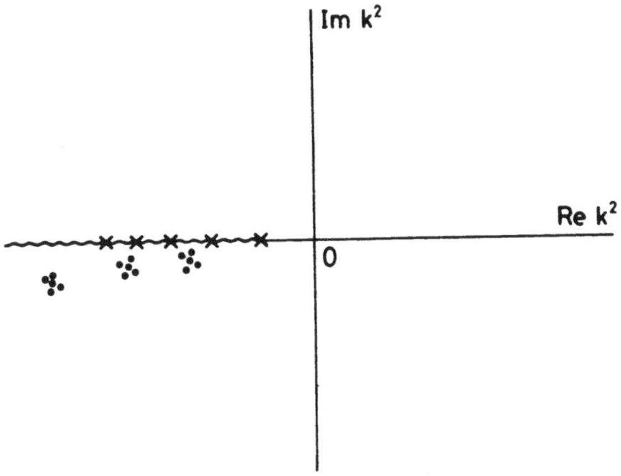

Fig. 1 Expected analytic structure of G^R for complex k^2. The wavy line is a cut (in Baryonic channels this cut starts away from the origin, in mesonic channels, since we have put $m_f = 0$, the cut starts at zero because the pion is massless). The dotted crosses are singularities to be expected in the second Riemann sheet (resonances).

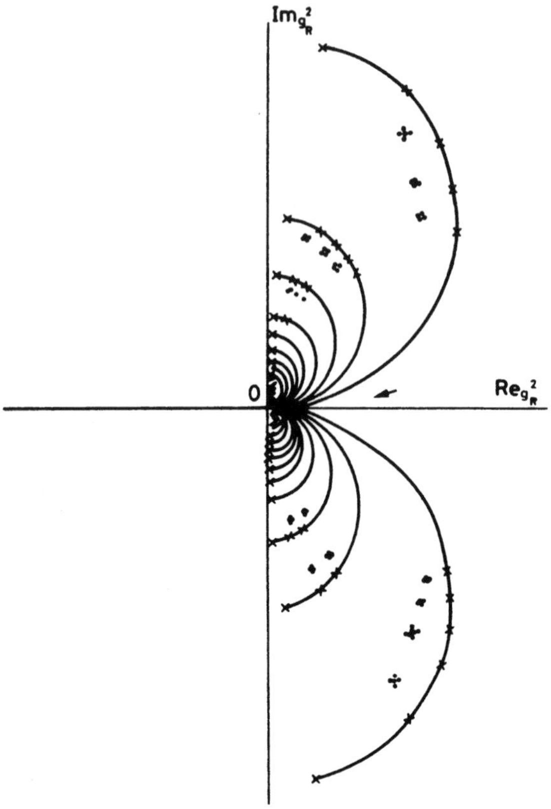

Fig. 2 Resulting analytic structure for complex g_R^2. The single cut of Fig. 1 now reproduces many times on semi circles. These semi-circles are only slightly distorted due to the β_2 term in Eq. (2.7a). The arrow shows the region where perturbation expansion is done. The cut on the left is due to the Z-factor in (3.8).

The conclusion of this section is obvious: we find such a bad accumulation of singularities at the origin that the analytic continuation procedure given in the introduction will never work. We must look for a more powerful technique.

5. BOREL RESUMMATION

We now assume that our Green's functions can be written as a Laplace transform of a special type

$$G^R(g^2) = \int_0^\infty F(z)\, e^{-z/g^2}\, dz \tag{5.1}$$

F(z) can be found perturbatively: If

$$G^R(g^2) = a_0 + a_1 g^2 + a_2 g^4 + a_3 g^6 + \dots \tag{5.2}$$

then

$$F(z) = a_0 \delta(z) + \frac{a_1}{0!} + \frac{a_2 z}{1!} + \frac{a_3 z^2}{2!} + \dots \tag{5.3}$$

where the δ function is understood to be included in the integral (5.1). One can check this trivially by inspection. The importance of this is that the series (5.3) converges much faster than (5.2). Contrary to (5.2) it may very well have a finite radius of convergence (this is at present believed to be the case for all renormalizable field theories). If F(z) can now be analytically continued to all real positive z, and if, for some g^2, the integral (5.1) converges, then the series (5.2) is called Borel summable. We will call F(z) the Borel function corresponding to the Green's function $G^R(g^2)$.

First, we wish to find out where we can expect singularities in F(z). Let us illustrate an interesting feature in the case of an over-simplistic "field theory", namely field theory at one space-time point. Remember that field theoretical amplitudes can be written as functional integrals, with a certain number of integration variables at each space-time point[10]. If we have one space-time point and one field, then there is just one integration to do

$$G = g\int dA\ \exp\left[-\frac{1}{2}A^2 - \frac{1}{g^2}V(gA)\right] \tag{5.4}$$

where $V(x) = V_3x^3 + V_4x^4 + \ldots$ and the factor g in front is just for convenience. g^2 comes out this way as the usual perturbation parameter. Note the minus signs in the integrand. This anticipates that we shall always consider Field theory in Euclidean space-time. Let us rescale the fields, and the action,

$$\begin{aligned} A' &= gA \\ -\frac{1}{2}A^2 - \frac{1}{g^2}V(gA) &= -\frac{1}{g^2}S'(A') \\ S'(A') &= \frac{1}{2}(A')^2 + V(A') \end{aligned} \tag{5.5}$$

Our integral becomes

$$G = \int dA'\ \exp\left[-\frac{1}{g^2}S'(A')\right] \tag{5.6}$$

Comparing this with the Borel expression (5.1), we immediately find F(z):

$$F(z) = \int dA'\,\delta(z - S'(A')) \tag{5.7}$$

Thus at given z we must find all solutions of $S'(A) = z$, which we can call $A_i(z)$. The result of the integral is

$$\sum_i \left[\frac{\partial S'(A)}{\partial A}\right]^{-1}_{A=A_i(z)} \tag{5.8}$$

This outcome reveals the singularities in the z plane: we must find the solutions of

$$\frac{\partial S'(A)}{\partial A} = 0 \tag{5.9}$$

to be recognized as the classical field equation of this system. At a solution $\bar{A}$ of (5.9) we have

$$\begin{aligned} &\bar{z} = S'(\bar{A}) \\ &\left[\frac{\partial S'}{\partial A}\right]^{-1} \to \left[2\,\frac{\partial^2 S'}{\partial A^2}\,(z-\bar{z})\right]^{-1/2} \\ &\text{as } z \to \bar{z} \end{aligned} \tag{5.10}$$

and so we find a square root branch point at $z = \bar{z}$.

Consider now multidimensional integrals of the same type. The integral (5.7) then corresponds to an integral over the contours in $\vec{A}$ space defined by the equation $S'(\vec{A}) = z$. You get singularities only at those values of z that are equal to the total action S' of a solution of the equation

$$\frac{\partial S'}{\partial \vec{A}} = 0 \tag{5.11}$$

because those are the contours that shrink to a point (in the case of a local extremum) or have crossing points (saddle points). Again, Eq. (5.11) is nothing but the classical Lagrange equation for the fields $\vec{A}$. Conclusion: to find singularities in F we have to search for finite solutions of the classical field equations in Euclidean space-time. Their rescaled action S' corresponds to singularity points $\bar{z}$ in the z place for the function F. In general, these singularities are branch points. In our actual four-dimensional field theories that are supposed to describe strong (or weak and electromagnetic) interactions; such solutions indeed occur, and are called "instantons", because they are more or less instantaneous and local in the Euclidean sense[25,26,6,10]. Their action (in the case of QCD) is $S' = 8\pi^2 n$, where n counts the "winding number"[8-10] and so we may expect singularities in the complex z plane at $z = 8\pi^2 n$.

6. UNIVERSALITY OF THE BOREL SINGULARITIES

The student might wonder whether the conclusions of the previous sections were not jumped to a little too easily. The connected Green's functions in field theories are not just multi(infinite) dimensional integrals but rather the ratio of such integrals with some source insertion and an integral for the vacuum, and then often differentiated with respect to those source insertions. Do all these additional manipulations not alter or replace these singularities and/or create new ones? Do different Green's functions perhaps not have their own singular points?

Let us for a moment forget the renormalization infinities, to which we devote a special section. Then the answer to these equations is reassuring. Multiplications, divisions, exponentiations can be carried out, after which we shall always find the singularities back in the same place as they were before, possibly with a different power behaviour. To understand this general property of Borel transforms, let us formulate some simple properties.

Let

$$G_i(g^2) = \int_0^\infty F_i(x)\, e^{-x/g^2}\, dx \tag{6.1}$$

Then, if

$$G_3(g^2) = G_1(g^2)\, G_2(g^2)$$

then

$$F_3(z) = \int_0^z F_1(z_1)\, F_2(z - z_1)\, dz_1 \qquad (6.2)$$

And let

$$G_2(g^2) = [G_1(g^2)]^{-1}, \quad G_1(0) = 1$$

then

$$F_2(z) = 2\delta(z) - F_1(z) - \int_\varepsilon^{z-\varepsilon} F_2(z')\, F_1(z - z')\, dz' \qquad (6.3)$$

Here the ε symbols are there just to tell us to leave the δ symbols out of the integration. It is easy to show that if (6.3) is solved iteratively, then the series converges for all z, as long as F_1 stays finite between ε and z.

Now note that we may choose the contours (0,z) so that they avoid singularities. Only if z is a singularity of either F_1 or F_2, or both, then $F_3(z)$ in Eq. (6.2) will be singular. Also $F_2(z)$ in Eq. (6.3) is only singular if F (z) is singular. Note, however, that if a singularity lies between 0 and z then the contour can be chosen in two (or more) ways, and, in general, we expect the outcome to depend on that. Thus if we start with pure pole singularities, they will propagate as branch points in the other Borel functions.

In quantum field theories, the Green's functions are related through many Schwinger-Dyson equations and Ward-Slavnov-Taylor identities. Since singularities survive the multiplications and divisions in these equations without displacement, they must occur in all Borel-Green's functions at the same universal values of z (unless miraculous cancellations occur; I think one can safely exclude that possibility). These singularities will, in general, be of the branch-point type.

In particular, those singularities that we obtained through the solutions of the classical equations, will stay at the same position for all Green's functions.

7. SINGULARITIES IN F(z) DUE TO INSTANTONS

Let us consider these classical equations in Euclidean space-time for the various field theories. First take $\lambda\phi^4$ theory, for simplicity, without mass term. Rescaling the fields and action the usual way

$$\varphi' = \sqrt{\lambda}\,\varphi \qquad S = -\frac{1}{\lambda} S' \tag{7.1}$$

we have

$$S' = \int d^4x \left[\frac{1}{2} (\partial\varphi')^2 + \frac{1}{4!} (\varphi')^4 \right] \tag{7.2}$$

It turns out[25] that a purely imaginary solution exists for the equations $\delta S/\delta\phi = 0$ (here, δ indicates derivative in the Euler sense), namely

$$\varphi = \frac{\rho i \sqrt{48}}{x^2 + \rho^2} \tag{7.3}$$

Here ρ is an arbitrary scale parameter (after all, our classical action is scale invariant).

In spite of this solution being purely imaginary, it is important to us because it indicates a singularity in F(z) away from the positive real axis. The corresponding value for S' is

$$S' = -16\pi^2 \tag{7.4}$$

So the singularity occurs at $z = -16\pi^2$, indeed away from the positive real axis. Such singularities are relatively harmless, since F is only needed for positive z. We may invoke the analytic continuation procedure sketched in the Introduction to improve convergence for the series in z.

Now, let us turn our attention to Quantum Chromodynamics. Here we have a real solution in Euclidean space:

$$A_\mu^{a'} = g A_\mu^a = \frac{2\eta_{a\mu\nu} x^\nu}{x^2 + \rho^2} \tag{7.5}$$

where $\eta_{a\mu\nu}$ are certain real coefficients[6,27] and ρ is again a free scale parameter. One finds for the action

$$S = -\frac{1}{g^2} S', \quad S' = 8\pi^2 \tag{7.6}$$

Thus $z = 8\pi^2$ is a singularity on the positive real axis. In fact, we can also have n instantons far apart from each other, so we also expect singularities*) at $z = 8\pi^2 n$. Now, Green's functions are obtained from F(z) by integrating from zero to infinity, over the positive real axis. Do the singularities on the real axis give unsurmountable problems? I think not, although the correct prescription will be complicated. A clue is the following. The single-instanton contribution to the amplitudes has been computed directly

---

*) A more precise analysis suggests that only those multi-instantons with zero total winding number (that is, as many instantons as anti-instantons) will give rise to ordinary singularities that limit the radius of convergence of F(z). The others give discontinuities rather than singularities.

in the small coupling constant limit. A typical result goes like[27-30)]

$$\Delta G(g^2) \rightsquigarrow C g^{-12} e^{-8\pi^2/g^2}(1+O(g^2)) \tag{7.7}$$

That is already a Green's function, the one we would like to obtain after integrating

$$\Delta G(g^2) = \int_0^\infty \Delta F(z)\, e^{-z/g^2}\, dz \tag{7.8}$$

A function F(z) that yields (7.7) exists

$$F(z) \to C \left(\frac{d}{dz}\right)^6 \delta(z-8\pi^2) \tag{7.9}$$

Indeed, a "singularity" at $z = 8\pi^2$. We see that, since all Green's functions will show the same exponential in their g dependence, the universality theorem of the previous section is obeyed. What is important is that by first computing (7.7) one can short-circuit the problem of defining an integration over such singular points. Thus the instanton singularities at the right-hand side on the real axis will not destroy our hopes of obtaining a convergent theory. The reason is that the physics of the instanton is understood. The situation is less clear for the other type of singularities that we discuss in the next section.

8. OTHER SINGULARITIES IN F

In principle, the instanton-singularities in F can also be understood within the context of ordinary perturbation expansion, by a statistical treatment of Feynman diagrams[31)]. We do not show the derivation here, but the following argument has been given. In the previous section, we have never bothered about the renormalization procedure that is supposed to make all diagrams finite. Suppose we had a strictly finite theory, with bounded propagators, bounded integrals and all that. Individual diagrams in such a theory are then bounded by a pure power law as a function of their order n. The only way that factors n! can arise is because there are n! diagrams at n^{th} order and they may not cancel each other very well. This is how in the statistical treatment the instanton singularity occurs. But in realistic four-dimensional renormalizable field theories, the power law for individual Feynman diagrams no longer holds. A simple example is quantum-electrodynamics. We consider the diagrams of the type shown in Fig. 3.

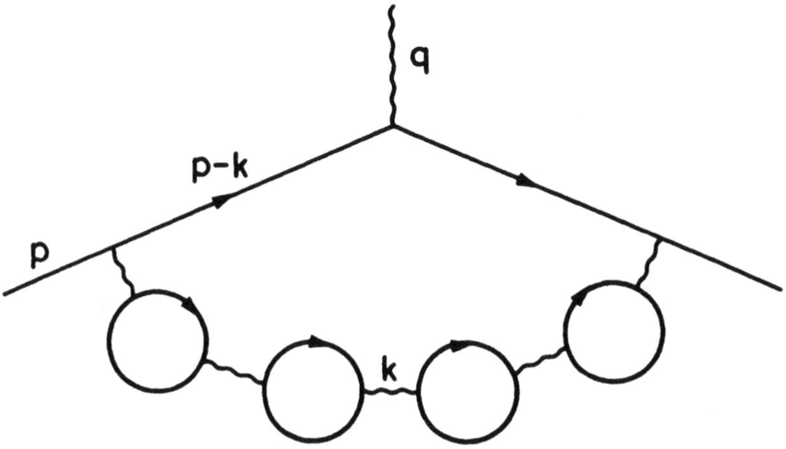

Fig. 3 Fourth member of a subclass of diagrams discussed in this section.

It is the class of diagrams with n electron bubbles in a row, which in itself closes again a loop. It is well known that each electron bubble separately behaves for large k^2 as

$$C\, K^2 \log K^2 \qquad (8.1)$$

and each propagator as $(k^2)^{-1}$. Thus, for large k^2 the integrand in the k variable behaves as

$$\frac{d^4k}{(k^2)^\alpha}(\log k^2)^n C^n \tag{8.2}$$

where α is some fixed power. After having made the necessary subtractions to make the integral converge, and in order to obtain physically relevant quantities, such as a magnetic moment, the leading coefficient α becomes 3 or larger. Let us replace $\log k^2$ by a new variable x, then (8.2) becomes proportional to

$$dx \cdot x^n e^{-(\alpha-2)x} C^n \tag{8.3}$$

Thus the integral over x will grow as $n \to \infty$ like

$$n!\,(\alpha-2)^{-n} C^n \tag{8.4}$$

A more precise analysis shows that C should be proportional to the first non-trivial β coefficient

$$C = -\frac{\beta_1}{2} \tag{8.5}$$

In the expansion for F(z) the factor n! is removed, as usual. It is clear that a new singularity develops at

$$\bar{z} = (\alpha-2)\left(-\frac{2}{\beta_1}\right) \tag{8.6}$$

It seems to be a universal phenomenon for all field theories, and not related to any instanton solution. Our definition for β_1 was positive for asymptotically free theories and negative otherwise. So, the singularity is at negative real z and therefore harmless if our theory is asymptotically free, but for non-asymptotically free theories such as QED and $\lambda\phi^4$, we have singularities on the positive real axis. Since a detailed understanding of the ultraviolet behaviour of non-asymptotically free theories is lacking, there may exist no cure for these singularities then. This is in contrast with the instanton singularities.

An important observation has been made by G. Parisi[32]. The ultraviolet behaviour of $\lambda\phi^4$ and QED are well understood in the limit $N \to \infty$, where N is the number of field components. A systematic study of the singular point (8.6) is then possible. Parisi found in $\lambda\phi^4$ theory a conspiracy between diagrams such that the first singularity at $\alpha = 3$ cancels. In total the integrals do behave as (8.2) but with $\alpha > 3$, after all necessary subtractions. At present, it is not understood whether this conspiracy is accidental for $\lambda\phi^4$ theory with N components, or whether it is a more general phenomenon. It does seem that only the first singularity may be subject to such cancellations.

In Figs. 4 and 5 we show the complex planes for the Borel variables z in $\lambda\phi^4$ theory and in QED, respectively. The singularities discussed in this section are called "renormalons" for short.

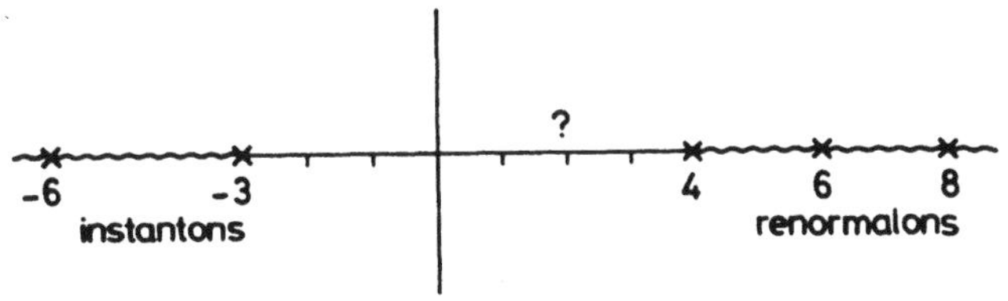

Fig. 4 Singularities in the Borel z variable for $\lambda\phi^4$. The units are $16\pi^2/3$. The question mark denotes the singularity that may be cancelled according to Parisi's mechanism.

The situation for QCD is more complex. Not only do we have the renormalons at points on the negative real axis but also there are such singularities on the positive real axis. They are due to the infra-red divergence of the theory. The mechanism is otherwise the

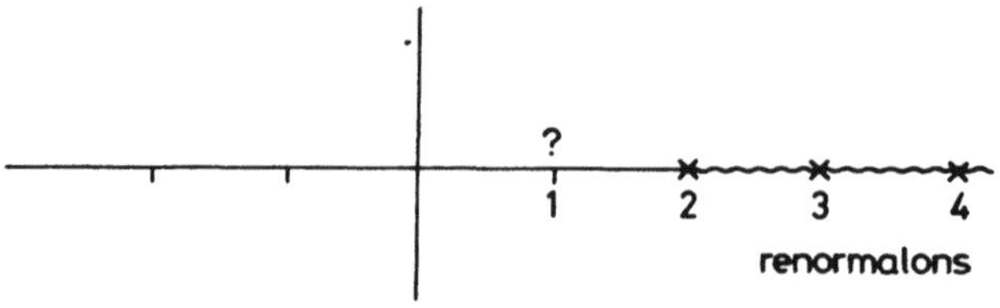

Fig. 5 Singularities for QED. Here the units are 3π, if α is the original expansion parameter.

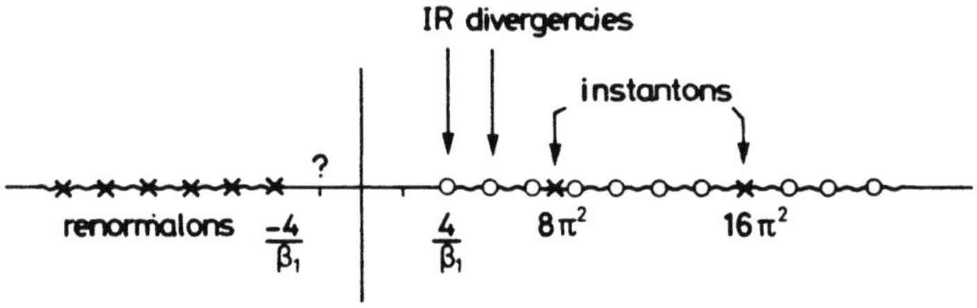

Fig. 6 Borel z plane for QCD. The circles denote IR divergences that might vanish or become unimportant in colour-free channels.

the same as discussed for the ultraviolet singularities (Fig. 6).

An interesting speculation is that these infra-red singularities are only surmountable in colourless channels, but the integration over these singularities becomes impossible in single quark- or gluon- channels. It is likely that these singularities are related to the quark confinement mechanism.

9. SECOND BOREL PROCEDURE

Many features of the singularities in the complex z plane of the Borel functions F(z) are still uncertain and ill-understood. But from the foregoing we derive some hopes that it will be possible to obtain F(z) for $0 \leq z < \infty$ for asymptotically free theories, such as QCD. The only thing to be investigated then is how the integral in

$$G(g^2) = \int_0^\infty F(z)\, e^{-z/g^2}\, dz \tag{9.1}$$

behaves at ∞. Does the integral converge? The answer to this is almost certainly: no. Consider massless QCD and its singularities in complex g_R^2 plane as derived in Section 3. According to Eq. (3.3) there are singularities when

$$1/g^2 = \text{real} + \frac{\beta_1}{2}(2n+1)\pi i \tag{9.2}$$

where the real number may be arbitrarily large. Substituting that in Eq. (9.1) we find that

$$\int_0^\infty F(z)\, e^{-z(\text{real} + \frac{\beta_1 \pi i}{2}(2n+1))}\, dz \tag{9.3}$$

must diverge. We had assumed that the singularities at finite z did not give rise to divergences. So F(z) must diverge at large z worse than any exponential of z. Note that (9.3) contains an oscillating term. It is likely then, that at $z \to \infty$, F(z) does not only grow very fast, but also oscillates with periods $4/\beta_1$ or fractions thereof.

Can we cure this disease? We have no further clue at hand which could provide us with any limit on the large z behaviour of F. But there is a way to express the unknown Green's functions in terms of a more convergent integral than (9.3). Let us treat the divergent integral (9.1) on the same footing as the divergent perturbation expansions which we had before. We consider a new, better converging integral

$$W(s) = \int_0^\infty \frac{F(x)\, dx\, s^{\beta_1 x/2}}{2\Gamma(\beta_1 x + 2)} \tag{9.4}$$

We may hope that this has a finite region of convergence, from which we can analytically continue. Note the analogy between (9.1) and (9.4) on the one hand, and (5.2) and (5.3) on the other. The integral relation between W and G, analogous to (5.1), is

$$G(g^2) = \int_0^\infty ds\, W(s) \exp[-\sqrt{s}\, e^{1/\beta_1 g^2} + 2/\beta_1 g^2] \tag{9.5}$$

Now, remembering that instead of varying g^2 we could vary k^2, replacing

$$2/\beta_1 g^2 \to \log K^2 \tag{9.6}$$

So that, ignoring the Z factor that distinguishes G from G^R (see Eq. (3.8)), one gets

$$G(K^2) \to K^2 \int_0^\infty ds\, W(s)\, e^{-\sqrt{s} K^2} \tag{9.7}$$

Now we can easily prove that, if our theory makes any sense at all, there may be no singularities in W(s) on the positive real axis, and the integral (9.7) must converge rapidly. Thus, if the integral (9.4) makes sense, then our problems are solved. The proof goes as follows:

$G(k^2)$ satisfies a dispersion relation: it is determined by its imaginary part. We have, at $k^2 = -a^2$, $a > 0$,

$$G(K^2 = a^2 + i\varepsilon) - G(K^2 = -a^2 - i\varepsilon) = -2\pi i \rho(a) \tag{9.8}$$

where ρ is usually a positive spectral function. Substituting (9.8) into (9.7) we get

$$\begin{aligned} \frac{-2\pi i \rho(a^2)}{a^2} &= \int_0^\infty 2\omega\, d\omega\, W(\omega^2)[e^{-i\omega a} - e^{i\omega a}] \\ &= -2\int_{-\infty}^{\infty} e^{i\omega a}\, \omega\, W(\omega^2)\, d\omega \end{aligned} \tag{9.9}$$

Thus, $\rho(a)$ and $W(\omega^2)$ are each other's Fourier transform. The inverse of (9.9) is

$$W(s) = \frac{1}{\sqrt{s}} \int_0^\infty \frac{\rho(a^2)}{a^2} \sin a\sqrt{s}\, da \tag{9.10}$$

A possible singularity at $a^2 = 0$ is an artifact of our simplifications and can be removed. It is important to observe that (9.10) severely limits the growth of $W(s)$ at large s so that (9.7) is always convergent.

Conclusion: this section results in an improvement on perturbation theory. The physically relevant quantities can be expressed in terms of integrals of the type (9.4), which converge better than the original ones of type (9.1). It is not known whether this improvement is sufficient, i.e., whether (9.4) actually converges in some neighbourhood of the origin.

Even if the important open questions mentioned in these lectures cannot be answered we think that refinement of these techniques will lead to an improved treatment of strong coupling theories.

REFERENCES

1) F.J. Dyson, Phys. Rev. 85 (1952) 861.

2) L.N. Lipatov, Leningrad Nucl. Phys. Inst. report (1976) (unpublished).

3) E. Brézin, J.C. Le Guillou and J. Zinn-Justin, Phys. Rev. D15 (1977) 1544 and (1977) 1558.

4) G. Parisi, Phys. Letters 66B (1977) 167.

5) C. Itzykson, G. Parisi and J.B. Zuber, Asymptotic estimates in Quantum Electrodynamics, CEN Saclay preprint.
The singularities I discuss in Section 8 of my lectures were assumed to be absent in this paper.

6) G. 't Hooft, Phys. Rev. Letters 37 (1976) 8.

7) A.M. Polyakov, Phys. Letters 59B (1975) 82 and unpublished work.

8) C. Callan, R. Dashen and D. Gross, Phys. Letters 63B (1976) 334.

9) R. Jackiw and C. Rebbi, Phys. Rev. Letters 37 (1976) 172.

10) See S. Coleman's lectures at this School.

11) K.G. Wilson, Phys. Rev. D10 (1974) 2445.

12) J. Kogut and L. Susskind, Phys. Rev. D11, 395 (1975).
L. Susskind, lectures at the Bonn Summer School (1974).

13) S.D. Drell, M. Weinstein, S. Yankielowicz, Phys. Rev. D14 (1976) 487 and DL4 (1976) 1627.

14) G. 't Hooft, Marseille Conf. on Renormalization of Yang-Mills fields and applications to particle physics, June (1972) (unpublished).

15) H.D. Politzer, Phys. Rev. Letters 30 (1973) 1346.

16) D.J. Gross and F. Wilczek, Phys. Rev. Letters 30 (1973) 1343.

17) G. 't Hooft and M. Veltman, Nuclear Phys. B44 (1972) 189.

18) C.G. Bollini and J.J. Giambiagi, Phys. Letters 40B (1972) 566.

19) J.F. Ashmore, Lettere al Nuovo Cimento 4 (1972) 289.

20) G. 't Hooft, Nuclear Phys. B61 (1973) 455.

21) D.R.T. Jones, Nuclear Phys. B75 (1974) 531.

22) A.A. Belavin and A.A. Migdal, Gorky State University pre-preprint (January 1974).

23) W.E. Caswell, Phys. Rev. Letters 33, (1974) 244.

24) S. Coleman, lectures given at the "Ettore Majorana" Int. School of Subnuclear Physics, Erice, Sicily (1971).
Note: we drop the inhomogeneous parts of the renormalization group equation, which can be avoided according to later formulations on the renormalization group (Ref. 20).

25) S. Fubini, Nuovo Cimento 34A (1976) 521.

26) A.A. Belavin et al., Phys. Letters 59B (1975) 85.

27) G. 't Hooft, Phys. Rev. D14 (1976) 3432.

28) F.R. Ore, "How to compute determinants compactly", MIT preprint (July 1977).

29) A.A. Belavin and A.M. Polyakov, Nordita preprint 77/1.

30) A.M. Polyakov, Nordita preprint 76/33 (Nuclear Phys. in press).

31) C.M. Bender and T.T. Wu, Phys. Rev. Letters 27 (1971) 461; Phys. Rev. D7 (1972) 1620.

32) G. Parisi, private communication.

CHAPTER 8
QUANTUM GRAVITY AND BLACK HOLES

CHAPTER 8
QUANTUM GRAVITY AND BLACK HOLES

Introduction to Quantum Gravity [8.1]

The years 1970–1975 gave us a remarkable insight in the general features common to all elementary particle theories. They must contain vector fields in the form of gauge fields, and spinors and scalars that must form representations of the vector field gauge group. The spectrum of all possible particle types that populate our universe will be represented in the most economical way possible, and if more types exist that are as yet unidentified then there are two corners to search for them: either they are very weakly interacting or they have very high masses.

The strength of these results is their completeness: we know exactly how to enumerate all possibilities. There are two weaknesses however. One is that there will be a built-in imprecision as explained earlier, because of the divergence of the loop expansion. We might try to avoid this by postulating asymptotic freedom, but this gives too stringent a restriction that can easily be made undone by suspecting as yet unknown particles in the hidden corners as just mentioned. The other weakness is that the number of different viable theories is still too large for comfort. Why did Nature make the choice she did?

On first sight the gravitational force makes things much worse. Including gravity makes our theory hopelessly nonrenormalizable; infinite series of uncontrollable counter terms undermine our ability to do meaningful calculations. But worse still, we do not at all understand how to formulate precisely the first principles for such theories. We can write down mathematical equations, but upon closer examination these turn out to be utterly meaningless. This problem persists up to this day, in spite of the commercials from the superstring adherents.

It is not hard to speculate on radically new theoretical ideas. I have toyed with discrete theories for space-time myself. But then one opens a Pandora's box full of schemes and formalisms that are completely void when it comes to making any firm predictions. In my opinion the art is not to obscure our view by clogging the scientific journals with speculations. The problem is to deduce from hard evidence features that *have* to be true for the world at Planckian time and distance scales.

What do we know about the perturbative expansion with respect to Newton's gravitational constant κ ? In the first paper of this chapter I address this question. It is shown that pure gravity, without any matter coupled to it, is one-loop renormalizable. This means that uncertainties due to nonrenormalizability and arbitrariness of counter terms are down by terms of order κ^2. An important lesson learnt here is that the "field variable" $g_{\mu\nu}(\mathbf{x},t)$ is not quite observable directly. When we do particle scattering experiments admixtures to this tensor of the form $R_{\mu\nu}$ and $Rg_{\mu\nu}$ are both necessary for renormalization and undetectable experimentally. Note that these admixtures are of higher order in κ and therefore infinitesimal. There is as yet no clash with causality.

If matter fields are added such admixtures do become visible when compared to similar admixtures in the other fields. Consequently these theories are not even one-loop renormalizable.

Introduction to Classical N-Particle Cosmology in $2+1$ Dimensions [8.2]

The previous paper treated quantum gravity in a way I would now call "conventional", which means that we assumed the problem to be to find a consistent prescription for calculating elements of the scattering matrix, under circumstances where we can still speak of a simply connected space-time, locally smooth and asymptotically flat.

But gravity raises a couple of conceptual problems when one tries to do better. One simple demonstration of this is to consider an extremely simplified version of gravity: gravity in 2 space- and one-time dimension. In such a world there are no gravitons, but the gravitational force does exist, and even though the Newtonian attraction between stationary objects vanishes the problems raised by the curvature of space-time are considerable.

My aim was originally to obtain a theory that can be formulated exactly and has two limits: one is a weak coupling limit in which ordinary (scalar) fields are weakly coupled to gravity and the perturbative approach makes sense; the other is a limit where Planck's constant vanishes such that we have a finite number of massive point particles gravitating in a closed universe.

But the system of classical point particles gravitating in a universe that is closed due to this gravitational force exhibits a number of features that predict evil for attempts to "quantize" it. For one, it may not be possible to define an S matrix. The reason is that under a wide class of initial conditions the universe does not end up in an asymptotic state where particles fly away from each other in classical orbits. Instead of this the universe may end up in a "big crunch". I discovered this in attempting to disentangle a paradox presented in a paper by J. R. Gott. He had argued that under certain conditions a pair of particles in this 2+1 dimensional universe may be surrounded by a region where a test particle may follow a closed timelike trajectory in space-time, and even move backwards in time. I could not believe that a sound physical system should admit such an event, and my intuition turned out to be correct: before the closed timelike curve gets a chance to come into being a "big crunch" ends it all. In a complete universe the region described by Gott is an illegal analytic extension; it is shut off from the physical world by a shower of particles crunching together.

The paper reproduced here explains what happens. I should add that I later found the transitions described in Figure 8 to be strictly speaking not complete. Other transitions are possible where polygons, as described in the paper, disappear altogether, and also sometimes an edge shrinks to zero but the system continues without the topological transition. These new possible transitions do not affect any of the conclusions however.[19]

Introduction to On the Quantum Structure of a Black Hole [8.3]

The previous paper demonstrates some of the difficulties that will form considerable obstacles when one wishes to formulate "quantum cosmology". The big crunch as an asymptotic state is extremely complex. Numerical experiments later showed that this crunch is in some sense "chaotic", and so we will not have an easy time enumerating the possible asymptotic states in a quantum mechanical "measuring process".

But one could argue that "quantum cosmology" is not our most urgent project. I think it is legitimate to ask first for a theoretical scheme describing scattering processes in regions of the universe that are surrounded by asymptotically flat space-time. This may not be possible in 2+1 dimensional worlds, but should be an option in 3+1 dimensions. Any finite number of particles with limited total energy will be surrounded by asymptotically flat space-time sufficiently far away. So should a scattering matrix exist there?

This seems to be plausible, at first sight. But now a new problem emerges, and it is an enormous one. We have to have at least four space-time dimensions. We better avoid taking more than four because then none of our ordinary models is even renormalizable. But in four dimensions the gravitational force, already before quantization, exhibits a fundamental instability: gravitational collapse can occur, and black holes can form.

Black holes will be a new state of matter, and it will be impossible to avoid them. They are regular solutions of the gravitational equations applied exclusively in those regions of space-time where we may expect that we know exactly how to formulate what happens. An event horizon may open up long before matter fields or particles here had a chance to interact in a way not covered by the Standard Model. Therefore it doesn't help to postulate that perhaps "black holes do not exist". They exist as legitimate solutions of the classical theory, and the real problem is to ask what they have to be replaced with in a completely quantized theory.

[19]G. 't Hooft, *Class. Quantum Gravity* **10** (1993) 1023.

At first sight a black hole seems to be just an "extended solution", of the kind I could have covered in my Banff lectures, Chapter 4.1. But if that were so it should have been possible to construct all their excited quantum modes just by performing standard field theory in a black hole background. This calculation has been done, or rather, that is what the authors thought. The outcome is a big surprise: not at all a reasonable spectrum of black hole states seems to emerge. What one finds is a continuous spectrum, as if black holes were infinitely degenerate. An immediate clash arises with the determination of the black hole entropy, which can be derived from its thermodynamical properties and it *is* finite.

So the background field calculations for black holes are incorrect. It turns out not to be hard to guess why they were incorrect: the gravitational interactions between ingoing and outgoing objects were neglected, in the first approximation. Under normal circumstances, as in conventional theories, one may always presume that such interactions can be postponed to later refinements, but gravity is not a conventional theory. We argue that these mutual interactions become infinitely strong and therefore may not be neglected even in the zeroth approximation. The big question is how to do things correctly then. One can do better, but a completely satisfactory approach is still missing.

I do not expect that a "completely satisfactory" formalism will be found very soon, because it can probably not be given without a completely satisfactory theory of quantum gravity itself. But what we can do is ask very detailed questions, insist on sensible answers, and hope that these answers, even if incomplete, also furnish a better insight in the problem of quantizing gravity itself. Another way of phrasing our point is this: if the gravitational force gives us so many problems when it becomes strong, why not concentrate entirely upon the question on how to formulate the theory where it is as strong as possible, which is near black holes?

The quantum black hole paradox began to intrigue me around 1984 and I began to study it in great detail. I am quite convinced that I made substantial progress in understanding its nature and its relevance for the entire theory of quantum gravity. I first learned that the real paradox already occurs at *our side* of the horizon, and that the "back reaction", i.e. the mutual (mostly gravitational) interactions between outcoming Hawking particles and objects falling into the hole somewhat later, is of crucial importance to resolve the problem. I found out that in contrast to a widespread misconception, the resolution of the problem will not require fundamental nonlocal interactions, but that it is strongly connected to very fundamental issues in the interpretation of quantum mechanics. I also found that one probably has to give up the idea that the quantum state of the black hole as it is seen by the outside observer can be described independently of the adventures of an observer who crosses the horizon to enter into the black hole. The inside region does not make any sense at all to the outside observer. One should never try to describe "super observers", a hypothetical simultaneous registration of what happens both inside and outside.

I was annoyed by the lack of interest in this problem and my work on it by most of the particle physics community and the general relativists, two groups who for a long time were unaware of the mutual incompatability of their views. But now that I am compiling my papers, early 1993, all this has changed. The literature is now being flooded with papers dealing with precisely the problem that kept me busy already for so long. I cannot say that these are already at a stage of affecting my present ideas, but an acceleration is certainly taking place, and the chances that new breakthroughs will occur (probably not by renown physicists but more likely by some young student), are now better than ever.

The following section is part of one of my first papers on black holes. The remainder of the paper is less relevant for the arguments that follow, and was therefore omitted. To understand the WKB approximation in Eq. (3.5) it is necessary to replace the coordinate r temporarily by σ according to

$$r - 2M = e^{\sigma} ,$$

because in this new coordinate the wave packets are much more regular than before and the WKB procedure much more accurate.

Introduction to the Gravitational Shock Wave of a Massless Particle [8.4]

So what is the gravitational interaction between ingoing and outgoing particles near a black hole horizon? This turns out to be sizeable even if the particles have very low energies as seen by a distant observer. It is large if the time interval between the ingoing and the outgoing object, as seen by the distant observer is of order $M \log M$ in Planck units or larger. That is still small compared to the total expected lifetime of the black hole. To find this result we need to compute the gravitational field of one of them. First take coordinates in which this is weak and therefore easy to compute. Then transform to any coordinate frame one wishes. It often becomes strong.

The calculation is done in the next paper, written together with Tevian Dray. The result is that there is a plane associated to the particle's trajectory in space-time (the "shock wave") such that geodesics of particles crossing this plane are shifted. The fact that the shift depends on the transverse distance makes it visible. The effect is much like a sonic boom. A fragile detector might break in pieces if it encounters such a shock wave.

Introduction to the S-Matrix Theory for Black Holes [8.5]

Because of the associated shock wave any object falling into a black hole has an effect upon the outgoing "Hawking" particles. The effect of these shifts may seem to be minute, until one realizes that the phases of the wave functions of particles coming out very late are extremely sensitive to precisely shifts of this nature. Thus we can now understand why information going into a black hole is always transmitted onto particles coming out, be it extremely scrambled. We can use this to construct a scheme for a unitary black hole S matrix.

As long as we concentrate on effects at moderately large distances across the horizon all interactions that contribute to this S matrix, gravitational and otherwise, are known. So many details of the S matrix we were searching for can be derived. Unfortunately, to see that this matrix will be unitary, and to determine the dimensionality of the Hilbert space in which it acts, we need more information concerning short distance interactions. Our hope was that there are still elements of the theory of general relativity that we have not yet used, and that they would be sufficient to remove most of our present uncertainties. Our feeling is that a lot of further inprovements should be possible along these lines, but they have not (yet?) been realized. The paper gives a good impression of the way I think about the problem at present. The algebra at the end should be seen as speculative and somewhat mysterious in its lack of unitarity. The idea is that it should inspire others to find improved methods along similar lines.

CHAPTER 8.1

QUANTUM GRAVITY

G. 't Hooft
University of Utrecht, The Netherlands

1. Introduction

The gravitational force is by far the weakest elementary interaction between particles. It is so weak that only collective forces between large quantities of matter are observable at present, and it is elementary because it appears to obey a new symmetry principle in nature: the invariance under general coordinate transformations.

Ever since the invention of quantum mechanics and general relativity, physicists have tried to "quantize gravity"[1)], and the first thing they realized is that the theory contains natural units of length (L), time (T) and mass (M). If

$$\kappa = 6.67 \cdot 10^{-11} \; m^3/kg \; sec^2$$

is the gravitational constant, then

$$\begin{aligned} L &= \sqrt{\kappa\hbar/c^3} = 1.616 \cdot 10^{-35} \; m \\ T &= \sqrt{\kappa\hbar/c^5} = 5.39 \cdot 10^{-44} \; sec \\ \text{and} \quad M &= \sqrt{\hbar c/\kappa} = 1.221 \cdot 10^{28} \; eV/c^2 \\ &= 2.177 \cdot 10^{-5} \; g \end{aligned}$$

But then the theory contains a number of obstacles.First there are the conceptual difficulties: the meaning of space and time in Einstein's general relativity as arbitrary coordinates, is very different from that of space and time in quantum mechanics. The metric tensor $g_{\mu\nu}$, which used to be always fixed and flat in quantum field theory, now becomes a local dynamical variable.

Advances have been made, from different directions[2,3,4)], to devise a language to formulate quantum gravity, but then the next problem arises: the theory contains essential infinities such that a field theorist would say: it is not renormalizable. This problem, discussed in detail in section 12, may be very serious. It may very well imply that there exists no well determined, logical, way to combine

Trends in Elementary Particle Theory, Edited by H. Rollnik and K. Dietz (Springer-Verlag, 1975).

gravity with quantum mechanics from first principles. And then one is led to the question: should gravity be quantized at all? After all, such quantum effects would be small, too small perhaps to be ever measurable. Perhaps the truth is very different, both from quantum theory and from general relativity.

Whatever one should, or should not do, our present picture of what happens at a length scale L and a time scale T is incomplete, and we would like to improve it. We claim that it is very worthwhile to try and improve our picture step by step, as a perturbation expansion in κ. In the following it is shown how to apply the techniques of gauge field theory and gain some remarkable results.

The sections 2-4 deal with the conventional theory of general relativity, seen from the viewpoint of a gauge field theorist. In section 5 it is indicated how quantization could be carried out in principle, but in practice we need a more sophisticated formalism to ease calculations.

This formalism, the background field method[2,5], is explained in section 6-11 . In these sections we mainly discuss gauge theories, and gravity is hardly mentioned; gravity is just a special case here.

Back to gravity in section 12, where we discuss numerical results. It is shown there why only pure gravity is finite up to the one-loop corrections.

2. Gauge Transformations

The underlying principle of the theory of general relativity is invariance under general coordinate transformations,

$$x'^{\mu} = f^{\mu}(x) \quad . \tag{2.1}$$

It is sufficient to consider infinitesimal transformations,

$$x'^{\mu} = x^{\mu} + \eta^{\mu}(x) \ , \ \eta \ \text{infinitesimal.} \tag{2.2}$$

Or, in other words, a function A(x) is transformed into

$$A'(x) = A(x + \eta(x)) = A(x) + \eta^{\lambda}(x)\partial_{\lambda}A(x) \ . \tag{2.3}$$

If A does not undergo any other change, then it is called a scalar. We call the transformation (2.3) simply a gauge transformation, generated by the (infinitesimal) gauge function $\eta^\lambda(x)$, to be compared with Yang-Mills isospin transformations, generated by gauge functions $\Lambda^a(x)$.

For the derivative of A(x) we have

$$\partial_\mu A'(x) = \partial_\mu A(x) + \eta^\lambda{}_{,\mu}\partial_\lambda A(x) + \eta^\lambda \partial_\lambda \partial_\mu A(x) , \qquad (2.4)$$

where $\eta^\lambda{}_{,\mu}$ stands for $\partial_\mu \eta^\lambda$, the usual convention. Any object A_μ transforming the same way, i. e.

$$A'_\mu(x) = A_\mu(x) + \eta^\lambda{}_{,\mu} A_\lambda(x) + \eta^\lambda \partial_\lambda A_\mu(x) , \qquad (2.5)$$

will be called a covector. We shall also have contravectors $B^\mu(x)$ (note that the distinction is made by putting the index upstairs), which transform like

$$B^{\mu\prime}(x) = B^\mu(x) - \eta^\mu{}_{,\lambda} B^\lambda(x) + \eta^\lambda \partial_\lambda B^\mu(x) , \qquad (2.6)$$

by construction such that

$$A_\mu(x) \, B^\mu(x)$$

transforms as a scalar. Similarly, one may have tensors with an arbitrary number of upper and lower indices.

Finally, there will be density functions $\omega(x)$ that transform like

$$\omega'(x) = \omega(x) + \partial_\lambda \left[\eta^\lambda(x) \, \omega(x)\right] . \qquad (2.7)$$

They enable us to write integrals of scalars

$$\int \omega(x) \, A(x) \, d_4(x) ,$$

which are completely invariant under local gauge transformations (under certain boundary conditions).

For the construction of a complete gauge theory it is of importance that the gauge transformations form a group. Of course they do, and hence we have a Jacobi identity. Let u(i) be the gauge transformations generated by $\eta^\mu(i,x)$. Then if

$$[u(1), u(2)] = u(3) ,$$

then

$$\eta^\mu(3,x) = \eta^\lambda(2,x)\,\partial_\lambda\eta^\mu(1,x) - \eta^\lambda(1,x)\partial_\lambda\eta^\mu(2,x). \qquad (2.8)$$

3. The Metric Tensor

In much the same way as in a gauge field theory[6], we ask for a dynamical field that fixes the gauge of the vacuum by having a non-vanishing vacuum expectation value. (Contrary to the Yang-Mills case it seems to be impossible to construct a reasonable "symmetric" theory.) To this end we choose a two-index field, $g_{\mu\nu}(x)$, which is symmetric in its indices,

$$g_{\mu\nu} = g_{\nu\mu} , \qquad (3.1)$$

and its vacuum expectation value is

$$\langle g_{\mu\nu}(x)\rangle_0 = \delta_{\mu\nu} \qquad (3.2)$$

(our metric corresponds to a purely imaginary time coordinate).

With $g_{\mu\nu}$, or its inverse, $g^{\mu\nu}$, we can now define lengths and time-intervals at each point in space-time:

$$|\ell|^2 = g_{\mu\nu}\ell^\mu\ell^\nu ,$$

and $g_{\mu\nu}$ can be used to raise or lower indices:

$$A^\mu = g^{\mu\nu}A_\nu , \quad A_\nu = g_{\nu\mu}A^\mu , \text{ etc.} \qquad (3.3)$$

Just as in the Yang-Mills case, we can now define covariant derivatives:

$D_\mu A = \partial_\mu A$ (the derivative of a scalar transforms as a vector),

$$\begin{aligned} D_\mu A_\nu &= \partial_\mu A_\nu - \Gamma^\alpha{}_{\mu\nu}A_\alpha , \\ D_\mu B^\nu &= \partial_\mu B^\nu + \Gamma^\nu{}_{\mu\alpha}B^\alpha . \end{aligned} \qquad (3.4)$$

The field $\Gamma^\alpha_{\mu\nu}$ is called the Christoffel symbol and is yet to be defined. First we write down how it should transform under a gauge transformation, such that the above-defined covariant derivatives be real tensors:

$$\Gamma^{\lambda\prime}_{\mu\nu} = \Gamma^\lambda_{\mu\nu} + \text{(ordinary terms for 3-index tensor)} + \partial_\mu\partial_\nu\eta^\lambda . \qquad (3.5)$$

We see that no harm is done by making the restriction that

$$\Gamma^\lambda_{\mu\nu} = \Gamma^\lambda_{\nu\mu} , \qquad (3.6)$$

because the symmetric part of Γ only is enough to make (3.4) covariant.

We now define the field Γ by requiring

$$D_\lambda g_{\mu\nu} = 0 \quad , \tag{3.7}$$

(from which follows: $D_\lambda g^{\mu\nu} = 0$). We see that we have exactly the right number of equations. By writing Eq. (3.7) in full we find that it is easy to solve

$$\Gamma^\lambda{}_{\mu\nu} = \frac{1}{2} g^{\lambda\alpha}(\partial_\mu g_{\alpha\nu} + \partial_\nu g_{\alpha\mu} - \partial_\alpha g_{\mu\nu}) \ . \tag{3.8}$$

Note that $\Gamma^\lambda_{\mu\nu}$ is not a covariant tensor.

The covariant derivative may be used just like ordinary derivatives when acting on a product:

$$D(XY) = (DX)Y + XDY \quad , \tag{3.9}$$

but two covariant differentiations do not necessarily commute:

$$D_\mu D_\nu A_\alpha \neq D_\nu D_\mu A_\alpha \quad . \tag{3.10}$$

Instead, we have:

$$D_\mu D_\nu A_\alpha - D_\nu D_\mu A_\alpha = R^\beta{}_{\alpha\nu\mu} A_\beta \tag{3.11}$$

with

$$R^\beta{}_{\alpha\nu\mu} = \Gamma^\beta{}_{\alpha\mu,\nu} - \Gamma^\beta{}_{\alpha\nu,\mu} + \Gamma^\beta{}_{\tau\nu}\Gamma^\tau{}_{\alpha\mu} - \Gamma^\beta{}_{\tau\mu}\Gamma^\tau{}_{\alpha\nu} \tag{3.12}$$

The comma denotes ordinary differentiation.
Since the l.h.s. of eq. (3.11) is clearly covariant and A_β is an arbitrary vector, $R^\beta{}_{\alpha\nu\mu}$ transforms as an ordinary tensor, in contrast with $\Gamma^\beta{}_{\alpha\mu}$. It is called the Riemann or curvature tensor (see the standard text books). Indices can be raised or lowered following (3.3). Without putting in any further dynamical equation, one finds the following identities,

$$\begin{aligned} R_{\alpha\beta\gamma\delta} &= R_{\gamma\delta\alpha\beta} = - R_{\alpha\beta\delta\gamma} \ , \\ R_{\alpha\beta\gamma\delta} + R_{\alpha\gamma\delta\beta} + R_{\alpha\delta\beta\gamma} &= 0 \ , \\ R_{\alpha\beta\gamma\delta;\mu} + R_{\alpha\beta\delta\mu;\gamma} + R_{\alpha\beta\mu\gamma;\delta} &= 0 \ . \end{aligned} \tag{3.13}$$

The semicolon denotes covariant differentiation. Further, we define*)

$$R_{\mu\nu} = R^{\alpha}{}_{\mu\alpha\nu} \quad , \quad R = R_{\mu\nu}\, g^{\mu\nu} \quad ,$$
$$G_{\mu\nu} = R_{\mu\nu} - \tfrac{1}{2} R g_{\mu\nu} \quad , \tag{3.14}$$

which satisfy, according to (3.13):

$$R_{\mu\nu} = R_{\nu\mu} \quad , \qquad D_\mu G_{\mu\nu} = 0 \ . \tag{3.15}$$

The metric tensor also enables us to define a density function [see (2.7)],

$$\omega(x) = \sqrt{\det\,(g_{\mu\nu}(x))} \quad . \tag{3.16}$$

In the quantum theory we shall encounter a fundamental problem: instead of $g_{\mu\nu}$ we could go over to a new metric $g'_{\mu\nu}$ with, for instance,

$$g'_{\mu\nu} = f_1(R) g_{\mu\nu} + f_2(R) R_{\mu\nu} \ . \tag{3.17}$$

So in a curved space there is some arbitrariness in the choice of metric (Section 12).We bypass this problem here.

4. Dynamics

The question now is whether we can make the fields $g_{\mu\nu}$ propagate. Indeed we can, because we can construct a gauge invariant action integral

$$S = \frac{c^2}{16\pi\kappa} \int \omega\, R\, d_4x \quad , \tag{4.1}$$

where κ is to be identified with the usual gravitational constant:

$$\kappa = 6{,}67 \cdot 10^{-11}\ \mathrm{m}^3\ \mathrm{kg}^{-1}\ \mathrm{sec}^{-2} \ . \tag{4.2}$$

For simplicity we shall take the units in which

$$\frac{c^2}{16\pi\kappa} = 1 \tag{4.3}$$

At a later stage one could put κ back in the expressions to find that the expansion in numbers of closed loops will correspond to an expansion

*) Note that there is a sign difference compared to some earlier papers.

with respect to κ.

One can also add other fields in the Lagrangian, for instance

$$\mathcal{L} = \omega \{ R - \frac{1}{2} g^{\mu\nu}\partial_\mu\phi\partial_\nu\phi - \frac{1}{2} m^2\phi^2 \} , \qquad (4.4)$$

where ϕ is a scalar field. We shall not repeat here the usual arguments to show that variation of the Lagrangian (4.4) really leads to the familiar gravitational interactions between masses, and to unfamiliar interactions between objects with a great velocity ("gravitational magnetism"). The equation for the gravitational field will be

$$G_{\mu\nu} = \frac{1}{2} T_{\mu\nu} , \qquad (4.5)$$

where $T_{\mu\nu}$ is the usual energy-momentum tensor (Einstein's equation). The action (4.1) has much in common with the action

$$- \frac{1}{4} G^a_{\mu\nu} \; G^a_{\mu\nu}$$

in Yang-Mills theories. As we shall indicate in the next section, a massless graviton with helicity $\pm$ 2 will propagate. Notice that we have been led to Einstein's theory of gravity almost automatically. It seems to be the simplest choice if we ask for a theory with invariance under general coordinate transformations.

5. Quantization

The first thing we must do is make a shift

$$g_{\mu\nu} = \delta_{\mu\nu} + A_{\mu\nu} , \qquad (5.1)$$

and consider $A_{\mu\nu}$ as the quantum fields. Here the problem mentioned in the introduction presents itself: what if we start with

$$g^{\mu\nu} = \delta^{\mu\nu} + B^{\mu\nu} \; ? \qquad (5.2)$$

The answer is that as long as we take as our elementary field any local function of the $g_{\mu\nu}$, the obtained physical amplitudes will be the same. The transformation from one function (for example, $g_{\mu\nu}$) to the other (for example, $g^{\mu\nu}$) will be accompanied by a Jacobian, or closed loops of fictitious particles (see the Zinn-Justin lectures on gauge theories).

But the propagators of these particles are constants or pure polynomials in k, because the transformation is local. If we now turn on the dimensional regularization procedure, which has to be used in order to get gauge invariant results, then the integrals over polynomials,

$$\int d^n k \quad \mathrm{Pol}\ (k) \ ,$$

vanish[7]. This is the reason why it makes no difference whether we start from Eq. (5.1) or Eq. (5.2). Non-local functions of $g_{\mu\nu}$ are not allowed. These non-local transformations would give rise to fictitious particles that do contribute in the cutting rules[8], and they are outlawed once unitarity has been established for the choice (5.1) or (5.2). By choosing a convenient gauge, comparable with the Coulomb gauge in QED, it is indeed not difficult to establish that the theory is unitary, in a Hilbert space with massless particles with helicity $\pm$ 2.

We can work out the bilinear part of the Lagrangian in Eq. (4.1) in terms of the fields $A_{\mu\nu}$:

$$\mathcal{L} = -\frac{1}{4}(\partial_\mu A_{\alpha\beta})^2 + \frac{1}{8}(\partial_\mu A_{\alpha\alpha})^2 + \frac{1}{2}L_\mu^2 + \text{higher orders}, \tag{5.3}$$

with

$$L_\mu = \frac{1}{2}\partial_\mu A_{\alpha\alpha} - \partial_\alpha A_{\mu\alpha} \ . \tag{5.4}$$

For practical calculations it seems to be convenient to choose the gauge

$$\mathcal{L}^c = -\frac{1}{2}L_\mu^2 \tag{5.5}$$

The Lagrangian in this gauge, $\mathcal{L} + \mathcal{L}^c$, has as a kinetic term

$$-\partial_\mu A_{\alpha\beta} W_{\alpha\beta|\gamma\delta} \partial_\mu A_{\gamma\delta} \tag{5.6}$$

where W is a matrix built from δ-functions.
The propagator is then

$$\frac{1}{k^2 - i\varepsilon}(W^{-1})_{\alpha\beta|\gamma\delta} \ . \tag{5.7}$$

Just in order to show the divergent character of the complications involved, we show here the Faddeev-Popov ghost[8] for this gauge, obtained by subjecting L_μ to an infinitesimal gauge transformation:

$$\mathcal{L}^{F.-P} = - \partial_\alpha \phi^*_\mu \partial_\alpha \phi^\mu + \phi^*_\mu [A_{\lambda\alpha,\alpha} \phi^\lambda_{,\mu} +$$

$$A_{\mu\lambda} \phi^\lambda_{,\alpha\alpha} + A_{\mu\lambda,\alpha} \phi^\lambda_{,\alpha} + A_{\mu\alpha,\alpha\lambda} \phi^\lambda + A_{\mu\alpha,\lambda} \phi^\lambda_{,\alpha}$$

$$- \tfrac{1}{2} A_{\alpha\alpha,\lambda} \phi^\lambda_{,\mu} - \tfrac{1}{2} A_{\alpha\alpha,\mu\lambda} \phi^\lambda - A_{\lambda\alpha,\mu} \phi^\lambda_{,\alpha}] . \tag{5.8}$$

The Lagrangian (4.1), expanded in powers of $A_{\mu\nu}$, with the gauge-fixing term (5.5) and the ghost term (5.8), form a perfect quantum theory. It is, however, more complicated than necessary, because gauge invariance is given up right in the beginning by adding the bad terms (5.5) and (5.8). The background field method, discussed in the next sections, is much more elegant because gauge invariance is exploited in all stages of the calculations, thus simplifying things a lot.

6. A Prelude for the Background Field Technique: Gauge Invariant Source Insertions

The methods described in this and the following sections are not only suitable for quantum gravity, but have a very wide applicability, in particular in gauge theories, for instance for calculations of renormalization group coefficients. First, it is convenient to introduce the concept of a gauge invariant source insertion. This is an artificial term in the Lagrangian of the form

$$J(x)\, R(x) \,,$$

where $J(x)$ is a c-number source function and $R(x)$ is some gauge invariant combination of fields, containing a linear part and quadratic or higher order corrections. Let us give some examples:

i) In a gauge theory with Higgs mechanism:

$$J_\mu \, \phi^* \cdot D_\mu \phi \,, \tag{6.1}$$

where ϕ is the Higgs field and $D_\mu \phi$ is its covariant derivative. We take $\partial_\mu J_\mu = 0$.

If

$$\langle \phi \rangle = F \;;$$
$$\phi = F + \psi \,,$$

then (6.1) becomes

$$J_\mu \left[gF^* A_\mu F + g\,\psi^* A_\mu F + gF^* A_\mu \psi + \psi^* D_\mu \psi\right] \tag{6.2}$$

The first term emits or absorbs single neutral vector particles. The other terms are higher order corrections, emitting two or three particles at once. In general one can find for all physical particles a similar gauge-invariant source that produces them predominantly and one by one.

ii) In a pure gauge theory (in momentum representation):

$$J^a_\mu(k)\, A^a_\mu(k) + g\, J^{ab}_{\mu\nu}\,(p,q)\, A^a_\mu(p)\, A^b_\nu(q) + \ldots \tag{6.3}$$

where $k_\mu J^a_\mu(k) \doteq 0$. Integration over k, p and/or q is understood. The higher terms are determined by requiring gauge invariance under

$$A^{a\prime}_\mu(x) = A^a_\mu(x) + g\, f_{abc}\, \Lambda^b(x)\, A^c_\mu(x) - \partial_\mu\, \Lambda^a(x)\ .$$

This implies

$$ip_\mu(J^{bc}_{\mu\nu}(p,q) + J^{cb}_{\nu\mu}\,(q,p)) = J^a_\nu\,(p+q)\, f_{abc}\ , \text{ etc.} \tag{6.4}$$

A source insertion that satisfies these conditions can be written in a closed form, in terms of an antisymmetric tensor source $J^a_{\mu\nu}(x)$:

$$J^a_{\mu\nu}(x)\quad T\left[\exp \int_{\text{path}}^{x} g\, A_\lambda(x')\, dx'^\lambda\right]_{ab} G^b_{\mu\nu}(x)\quad , \tag{6.5}$$

where $G^a_{\mu\nu} = \partial_\mu A^a_\nu - \partial_\nu A^a_\mu + g\, f_{abc}\, A^b_\mu\, A^c_\nu$, and A_λ stands for the matrix

$$A^{ac}_\lambda = f_{abc} A^b_\lambda\ .$$

The integral is along a path from infinity to x. The symbol T stands for time ordering along the path. Expanding (6.5) gives

$$J^a{}_{\mu\nu}(x)\ \left[G^a_{\mu\nu}(x) + g\, f_{abc} \int_{\text{path}}^{x} A^b_\lambda(x')\, dx'\ G^c_{\mu\nu}(x) + \ldots\right] \tag{6.6}$$

iii) In gravity one can do a similar thing:
a source $J^{\mu\nu}(x)$ satisfying

$$J^{\mu\nu} = J^{\nu\mu}\ ; \qquad \partial_\mu\, J^{\mu\nu} = 0 \tag{6.7}$$

can be coupled to $A_{\mu\nu}$ in eq. (5.1), and one can add higher order corrections to restore gauge invariance. The details are not very relevant for what follows, but note that we restrict ourselves to the *perturbative* theory.

The amplitude with which a gauge invariant source emits a single particle, obtains higher order corrections, see Fig. 1.

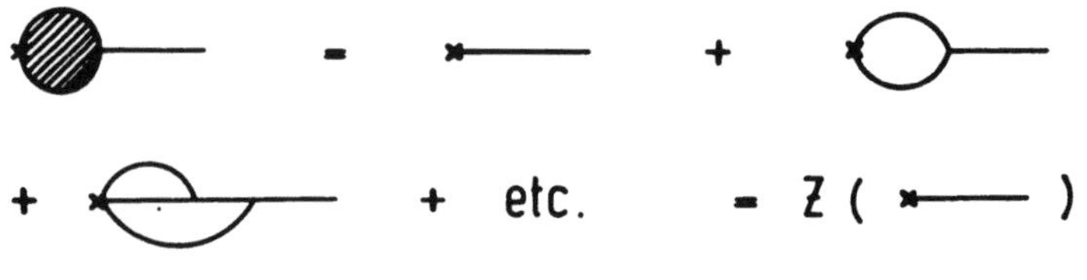

Fig. 1

A source can emit a single particle in different ways

The S-matrix can now readily be obtained in a gauge-independent way by considering vacuum-vacuum transitions in the presence of a gauge-invariant source. Then the external legs are amputated and put on mass-shell. From Fig. 1 it will be clear that in practice one can just as well calculate the amputated Green's functions directly and multiply them with some renormalization factor Z. This factor Z however may be gauge-dependent. The correct factor can be obtained by normalizing the imaginary part of the two-point function[8].

7. The Background Field Method

The background field method, useful for gauge theories, is practically indispensible for quantum gravity. Let the full Lagrangian be

$$\mathcal{L}(A) = \mathcal{L}^{inv}(A) + J\,R(A) - \tfrac{1}{2}\,C^2(A) + \mathcal{L}^{F.-P} \quad , \tag{7.1}$$

where $\mathcal{L}^{inv}(A)$ is the complete gauge invariant Lagrangian for all fields A_i. As described in the previous section, R(A) is a gauge invariant field combination, and J is a c-number source function. C(A) is the gauge-fixing function and $\mathcal{L}^{F.-P}$ describes the associated Feynman-deWitt-Faddeev-Popov ghost[1,2,8]. All irrelevant indices have been suppressed. The gauge transformation law will be written as

$$A_i' = A_i + s^a_{ij}\,\Lambda_a\,A_j + t^a_i\,\Lambda^a \tag{7.2}$$

where $\Lambda^a(x)$ is the infinitesimal generator. t and s are coefficients built from numbers and the space-time derivative ∂_μ.

We now introduce the notion of a classical field A^{cl}, which is a function of the c-number sources J. Usually[2,5] one defines its J dependence by requiring that A^{cl} satisfies the classical equations of motion. This is sufficient as long as we are interested in diagrams with at most one closed loop. In these notes we shall make that restriction also, but it must be kept in mind that for applications of these techniques at still higher orders it will be more convenient to add quantum corrections to the equation of motion. At this stage we require A^{cl} also to be in the gauge

$$C(A^{cl}) = 0 \quad . \tag{7.3}$$

Next, we perform a shift:

$$A = A^{cl} + \phi \quad , \tag{7.4}$$

where now ϕ is the new quantum field, and we rewrite the Lagrangian in terms of ϕ :

$$\mathcal{L}(A^{cl}+\phi) = \mathcal{L}_o(A^{cl},J) + \mathcal{L}_1(\phi,J,A^{cl}) + \mathcal{L}_2(\phi,J,A^{cl}) - \tfrac{1}{2}C^2(A^{cl}+\phi)$$
$$+ \mathcal{L}^{F.-P.} + \mathcal{O}(\phi^3). \tag{7.5}$$

Here $\mathcal{L}_1$ is linear in ϕ; $\mathcal{L}_2$ is quadratic in ϕ and $\mathcal{O}(\phi^3)$ is of higher order in ϕ.

Now, since A^{cl} satisfies the equation of motion

$$\frac{\delta\mathcal{L}}{\delta A^{cl}} = 0 \tag{7.6}$$

and $C(A^{cl}) = 0$, all terms linear in ϕ cancel:

$$\mathcal{L}_1(\phi) = 0\,. \tag{7.7}$$

So the source J is not coupled to terms linear in ϕ, and there are no vertices with only one ϕ-line. Therefore, the ϕ-lines can only go around in loops, and if we confine ourselves to one-loop diagrams then we can neglect the terms $\mathcal{O}(\phi^3)$. One loop diagrams now only consist of a ϕ-loop, with bilinear insertions of classical sources depending on A^{cl} and J. But remember that A^{cl} is a function of J, which can be obtained by solving the classical equations of motion by iteration. This iteration process corresponds to adding all possible trees to the single loop. Thus we reproduce the original Feynman rules, with the only change that loop lines are called ϕ-lines, and tree lines are called A^{cl}-lines.

8. The Background Field Gauge

The relevant part of the Lagrangian (7.5) is

$$\mathcal{L}(\phi) = \mathcal{L}^{\prime}_{inv}(\phi, J, A^{cl}) - \tfrac{1}{2}C^2(A^{cl}+\phi) + \mathcal{L}^{F.-P.} \quad . \tag{8.1}$$

This is just an ordinary gauge field Lagrangian where $\mathcal{L}^{\prime}_{inv} = \mathcal{L}_2 + \mathcal{O}(\phi^3)$ is invariant under what we shall call gauge transformations of type Q:

$$\phi^{\prime}_i = \phi_i + s^a_{ij}\,\Lambda^a(A^{cl}_j + \phi_j) + t^a_i\,\Lambda^a \quad ;$$

$$A^{cl\prime} = A^{cl} \tag{8.2}$$

The gauge is fixed by the function $C(A^{cl}+\phi)$.

Now there is another invariance of $\mathcal{L}^{\prime}_{inv}$, also broken by this C term. We call this a gauge transformation of type C:

$$\phi^{\prime}_i = \phi_i + s^a_{ij}\,\Lambda^a\,\phi_j \quad ;$$

$$A^{cl\prime}_i = A^{cl}_i + s^a_{ij}\Lambda^a\,A^{cl}_j + t^a_i\,\Lambda^a \quad . \tag{8.3}$$

The power of the present formulation is that we can go over to a different gauge function $C(A^{cl},\phi)$ which breaks the Q-gauge invariance (as is necessary) but preserves gauge invariance of type C. For example, if ϕ^a_μ is the quantum part of the vector field A^a_μ, then the choice

$$C_a = D_\mu \phi^a_\mu \quad , \tag{8.4}$$

where D_μ is the covariant derivative in the classical (C) sence:

$$D_\mu \phi^a_\mu = \partial_\mu \phi^a_\mu + g\, f_{abc}\, A^{cl}_{\mu b}\, \phi^c_\mu \quad ,$$

clearly preserves C-invariance. Such a gauge is also possible in the case of gravity. If

$$g_{\mu\nu} = g^{cl}_{\mu\nu} + A_{\mu\nu} \quad ,$$

we can take

$$C_\alpha = \sqrt[4]{g^{cl}} \cdot t^{\alpha\mu} \left(\tfrac{1}{2} g^{\kappa\lambda}_{cl} D_\mu A_{\kappa\lambda} - g^{\kappa\lambda}_{cl} D_\kappa A_{\mu\lambda}\right) , \qquad (8.5)$$

where

$$t^{\alpha\mu} t^{\alpha\nu} = g^{\mu\nu}_{cl} \quad .$$

Again, D_μ is the covariant derivative in the classical sence.

The one-loop (irreducible) vertex functions obtained in this gauge will be C-invariant also. It follows that they satisfy not only the Slavnov identities that describe the Q gauge symmetry, but in addition the much simpler Ward identities which are the direct generalizations of those in quantum electrodynamics.

Of course, the new Feynman rules in this background field gauge are independent of our original choice of the gauge C(A) in eq. (7.1). This implies that we can now also drop the gauge condition (7.3) for the classical fields since it can be replaced by another, arbitrary, gauge condition.

9. A Simple Example of the Background Field Gauge: Pure Yang-Mills Fields

Although we are mainly interested in quantum gravity, it is much more instructive to illustrate our methods in simpler field theories. Let us consider pure Yang-Mills fields. The invariant Lagrangian is

$$\mathcal{L}^{inv} = -\tfrac{1}{4}\, G^a_{\mu\nu}\, G^a_{\mu\nu} \, , \qquad (9.1)$$

with $$G^a_{\mu\nu} = \partial_\mu A^a_\nu - \partial_\nu A^a_\mu + g\, f_{abc} A^b_\mu A^c_\nu \quad .$$

The gauge invariance is

$$A_\mu^{a'} = A_\mu^a + g\, f_{abc} \Lambda^b A_\mu^c - \partial_\mu \Lambda^a \quad . \tag{9.2}$$

We leave aside the gauge invariance sources (sect. 6).

We shift

$$\begin{aligned} A_\mu^a &= A_{\mu a}^{cl} + \phi_\mu^a \;, \\ G_{\mu\nu}^a &= G_{\mu\nu}^{a\ cl} + D_\mu \phi_\nu^a - D_\nu \phi_\mu^a + g\, f_{abc} \phi_\mu^b \phi_\nu^c \;, \\ &(\text{where } D = \partial + g\, f\, A^{cl}) \end{aligned} \tag{9.3}$$

Using

$$(D_\mu D_\nu)^{ac} - (D_\nu D_\mu)^{ac} = g\, f_{abc} G_{\mu\nu}^{b\ cl} \quad , \tag{9.4}$$

we get

$$\begin{aligned} \mathcal{L}^{inv} = \mathcal{L}^{inv}(A^{cl}) &- \tfrac{1}{2} (D_\mu \phi_\nu)^2 + \tfrac{1}{2} (D_\mu \phi_\mu)^2 \\ &- g\, G_{\mu\nu}^{c\ cl} f_{abc} \phi_\mu^a\, \phi_\nu^b + \mathcal{O}(\phi^3) \\ &+ \text{total derivative.} \end{aligned} \tag{9.5}$$

A convenient background field gauge is

$$\mathcal{L}^c = -\tfrac{1}{2} C^2 = -\tfrac{1}{2} (D_\mu \phi_\mu^a)^2 \quad . \tag{9.6}$$

The ghost Lagrangian is then

$$\mathcal{L}^{F.-P.} = - D_\mu \psi_a^* D_\mu \psi_a \quad , \tag{9.7}$$

up to irrelevant interactions with ϕ_μ^a.

Of course the ghost is also C-invariant. Note that C-invariance permits us to write the interactions of ϕ_μ^a and ψ_a in a very condensed way. It is this feature that prevents overpopulation of indices in the case of gravity.

The one-loop infinities can be subtracted by a C-invariant counter term in the Lagrangian. The only candidate is

$$\alpha \cdot G^{a\,cl}_{\mu\nu} G^{a\,cl}_{\mu\nu} \quad .$$

The index α simultaneously governs the infinities of the two, three and four point vertices, and is therefore directly proportional to the Callan - Symanzik β-function[9]. To find this β-function one therefore only needs to investigate the two point function, contrary to the conventional formulation where also three point functions had to be calculated[10] in order to eliminate the Callan-Symanzik γ-function.

10. A Master Formula for all One-Loop Infinities

From the preceeding it will be clear that any one-loop amplitude* can be obtained from a Lagrangian bilinear in a set of fields ϕ_i. One can then rearrange the coefficients in such a way that

$$\mathcal{L} = \sqrt{g}\,\{-\tfrac{1}{2}(\partial_\mu \phi_i + N^{ij}_\mu \phi_j) g^{\mu\nu} (\partial_\nu \phi_i + N^{ik}_\nu \phi_k) + \tfrac{1}{2}\phi_i X_{ij} \phi_j\} \ , \qquad (10.1)$$

where N, g and X are arbitrary functions of space time x. Further,

$$N^{ij}_\mu = - N^{ji}_\mu \ ; \qquad X_{ij} = X_{ji} \quad .$$

In dealing with gravity[3] we found it very convenient first to calculate the infinities of this general Lagrangian, in terms of N, g and X . Of course, the background metric is allowed to be curved. Afterwards one may substitute the details such as: the way N and X depend on the background fields; and the fact that the indices i, j actually stand for pairs of Lorentz-indices. In gauge theories the objects N are mostly background gauge fields and in gravity the N contain the Christoffel symbols Γ.

In ref. 3), we used dimensional regularization and considered the poles at $n \to 4$ (n is the number of space-time dimensions). From power counting arguments one easily deduces that the residues of these poles can consist only of a limited number of terms. This number is even more restricted if we use the observation that, whatever N, g or X are, there is a C-gauge symmetry:

$\mathcal{L}$ is invariant under

---

* provided a Feynman-like background field gauge is chosen.

$$\phi'(x) = \phi(x) + \Lambda(x)\phi(x)$$

$$N'_\mu(x) = N_\mu - \partial_\mu\Lambda + \Lambda N_\mu - N_\mu\Lambda \qquad (10.2)$$

$$X'(x) = X + \Lambda X - X\Lambda$$

where Λ is an infinitesimal, antisymmetric matrix.

A straightforward calculation yields, that all one-loop infinities as $n \to 4$ are absorbed by the counter-Lagrangian[3]

$$\Delta\mathcal{L} = \frac{1}{8\pi^2(n-4)} \sqrt{g}\, \mathrm{Tr}\, \{ \frac{1}{24} Y^{\mu\nu}Y_{\mu\nu} + \frac{1}{4} X^2 - \frac{1}{12} RX$$
$$+ \frac{1}{120} R_{\mu\nu}R^{\mu\nu}\, I + \frac{1}{240} R^2\, I \} \qquad (10.3)$$

where

$$Y_{\mu\nu} = \partial_\mu N_\nu - \partial_\nu N_\mu + N_\mu N_\nu - N_\nu N_\mu \qquad (10.4)$$

and

Tr I = number of fields.
$R_{\mu\nu}$ and R are the Riemann curvature tensors for the background metric.
Raising and lowering indices by use of the background $g^{\mu\nu}$ is understood.

The formula (10.3) can also be used to calculate a similar master formula in the case of Fermions[9].

11. Substituting Equations of Motion.

In some cases the resulting formula (10.3) is not the end of the story. One may still make use of the information that the background fields are not arbitrary, but satisfy equations of motion. If, for instance, one finds a contribution in $\Delta\mathcal{L}$ proportional to

$$(D_\mu A^{cl})^2 \quad , \qquad (11.1)$$

and the equation of motion is

$$D^2 A^{cl} = V(A^{cl}) \quad , \qquad (11.2)$$

then one may replace (10.1) by

$$- A^{cl}\ V(A^{cl})\ . \qquad (11.3)$$

Addition of infinitesimal terms in the Lagrangian that vanish if the equation of motion is fulfilled, corresponds exactly to making an infinitesimal field renormalization: if the equation of motion is

$$\frac{\delta \mathcal{L}}{\delta A} = 0\ , \qquad (11.4)$$

then the terms in question must be

$$\varepsilon \cdot B \cdot \frac{\delta \mathcal{L}}{\delta A}\ , \qquad (11.5)$$

and we have

$$\mathcal{L}(A + \varepsilon B) = \mathcal{L}(A) + \varepsilon\ B\ \frac{\delta \mathcal{L}}{\delta A}\ . \qquad (11.6)$$

$A' = A + \varepsilon B$ is a field renormalization.
Note that all terms in $\Delta\mathcal{L}$ are always infinitesimal, because we neglect everything that comes from two-loop diagrams.

12. Some Numerical Results. Conclusions.

The master formula (10.3) has been applied to calculate the infinity structure in different cases. For pure gravitation we used[3] the gauge (8.5) and found from the gravitons

$$\Delta \mathcal{L} = \frac{1}{\varepsilon} \sqrt{g}\ \left(\frac{7}{24} R^2 + \frac{7}{12} R_{\mu\nu}R^{\mu\nu}\right)\ , \qquad (12.1)$$

where

$$\frac{1}{\varepsilon} = 1/8\pi^2(n-4)\ ,$$

and from the ghosts

$$\Delta \mathcal{L} = \frac{1}{\varepsilon} \sqrt{g}\ \left(- \frac{17}{60} R^2 - \frac{7}{30} R_{\mu\nu}R^{\mu\nu}\right) \qquad (12.2)$$

Here $g_{\mu\nu}$ and $R_{\mu\nu}$ are the classical metric and curvature .(At this level it is never necessary to consider quantum fields in the counter terms. We are also not interested in the renormalization of the source terms). From power counting one would expect a third possible term of the form

$$\sqrt{g}\ R_{\alpha\beta\gamma\delta}\ R^{\alpha\beta\gamma\delta} \quad . \tag{12.3}$$

It indeed occurs, but we made use of the identity[2,3,11]

$$\sqrt{g}\ (R_{\alpha\beta\gamma\delta}\ R^{\alpha\beta\gamma\delta} - 4\ R_{\mu\nu}R^{\mu\nu} + R^2) = \text{total derivative} \quad , \tag{12.4}$$

which implies that terms of the form (12.3) can be eliminated.

So all together we have

$$\Delta\mathcal{L} = \frac{1}{\varepsilon}\sqrt{g}\ (\frac{1}{120}\ R^2 + \frac{7}{20}\ R_{\mu\nu}R^{\mu\nu}) \ . \tag{12.5}$$

However, we still have the equations of motion of the background field (see previous section), which is

$$R_{\mu\nu} = 0 \quad ; \qquad R = 0 \tag{12.6}$$

This means that the infinity vanishes in pure gravity. It is interesting to observe that the field renormalization that corresponds to the elimination of this infinity (see previous section) is of an unusual type:

$$g_{\mu\nu} \rightarrow g_{\mu\nu} + \alpha\ R_{\mu\nu} + \beta\ R\ g_{\mu\nu} \quad . \tag{12.7}$$

This was the reason for our remark in the end of sect. 3. Note that eq. (12.4) was essential for this result.

Next we studied pure gravity with a (massless) Klein-Gordon field ϕ. The classical equations of motion are

$$\begin{aligned} D_\mu D^\mu \phi &= 0 \ ; \\ R_{\mu\nu} &= \frac{1}{2}\ D_\mu\phi D_\nu\phi \\ R &= \frac{1}{2}\ (D\phi)^2 \end{aligned} \tag{12.8}$$

There is one type of counterterm which is allowed by power counting and cannot be eliminated by substitution of the equations of motion:

$$\Delta\mathcal{L} = \frac{\sqrt{g}}{\varepsilon} \cdot \frac{203}{80} \cdot R^2 \tag{12.9}$$

The next thing that has been tried is pure gravity with in addition

Maxwell fields[12].
The Lagrangian is

$$\mathcal{L} = \sqrt{g}\ (R - \frac{1}{4} F_{\mu\nu} F_{\alpha\beta}\ g^{\mu\alpha} g^{\nu\beta}).$$

$$F_{\mu\nu} = \partial_\mu A_\nu - \partial_\nu A_\mu \ . \tag{12.10}$$

The equations of motion are

$$G_{\mu\nu} \equiv R_{\mu\nu} - \frac{1}{2} g_{\mu\nu} R = \frac{1}{2} T_{\mu\nu} \ ;$$

$$T_{\mu\nu} \equiv F_{\mu\alpha} F^{\alpha}{}_{\nu} - \frac{1}{4} g_{\mu\nu} F^{\alpha\beta} F_{\alpha\beta} \ ,$$

$$D_\alpha F^{\alpha\beta} = 0 \ ,$$

$$D_\alpha F_{\beta\gamma} + D_\beta F_{\gamma\alpha} + D_\gamma F_{\alpha\beta} = 0 \ . \tag{12.11}$$

From which

$$R = - \frac{1}{2} T^{\alpha}{}_{\alpha} = 0 \ .$$

In principle one may expect

$$\Delta\mathcal{L} = \frac{1}{\varepsilon} \sqrt{g} \left[\alpha_1 R_{\mu\nu} R^{\mu\nu} + \alpha_2 (F_{\alpha\beta} F^{\alpha\beta})^2 + \alpha_3 R_{\mu\nu\alpha\beta} F^{\mu\nu} F^{\alpha\beta} \right] \ . \tag{12.12}$$

Explicit calculation shows however:

$$\Delta\mathcal{L} = \frac{1}{\varepsilon} \sqrt{g} \cdot \frac{137}{60} R_{\mu\nu} R^{\mu\nu}. \tag{12.13}$$

So we see that there are some cancellations:

$$\alpha_2 = \alpha_3 = 0 \ . \tag{12.14}$$

Indeed, they occur in a miraculous way during the calculation and have not yet been explained.
In the coupled Einstein-Yang-Mills system [13] these cancellations persist and many new cancellations occur. Starting from the obvious generalization of the Abelian Lagrangian (12.10), one finds

$$\Delta\mathcal{L} = \frac{1}{\varepsilon} \sqrt{g} \left\{ \left[\frac{137}{60} + \frac{r-1}{10} \right] R_{\mu\nu} R^{\mu\nu} + f^2 C_2 \frac{11}{12} F_{\alpha\beta} F^{\alpha\beta} \right\} \tag{12.15}$$

where f is the gauge coupling constant,

$$C_2\delta_{ab} = f_{apq}f_{bpq} ,$$

$$C_2 r = f_{apq}f_{apq} \qquad (12.16)$$

Five other coefficients each happen to vanish. The second term in (12.15) is of the renormalizable type. It renormalizes the gauge coupling constant and fixes the Callan-Symanzik β-function.

Finally, also the combined Einstein-Dirac system has been investigated and nonrenormalizable infinities have been found also[14].

The fact that in all these systems where matter in some form is added to pure gravity infinities of the nonrenormalizable dimension survive really means that these theories cannot be renormalized in the perturbation expansion. In the case of pure gravity the infinities have been shown to be non-physical up to the one-loop level. No calculations have been performed to investigate renormalizability in the order of two loops. The calculations of Nieuwenhuizen and Deser show that "miraculous" cancellations often occur . Perhaps this is an indication of a new sort of symmetry that we are not aware of. Investigation of this symmetry could reveal new renormalizable models with gravity.

Even so, a renormalized perturbation expansion would only be a small step forward. At very small distances the gravitational effects must be large, because of the dimension of the gravitational constant, so the expansion would break down at small distances anyhow. We have the impression that not only a better mathematical analysis is needed, but also new physics. What we learned (see eq. (12.7) and the remarks in the end of sect.3) is that in such a theory the metric tensor might not at all be such a fundamental concept. In any case, its definition is not unambiguous.

REFERENCES

1. R. P. Feynman, Acta Phys. Polon. 24, 697 (1963)
   S. Mandelstam, Phys. Rev. 175, 1580, 1604 (1968)
   E. S. Fradkin and I. V. Tyutin, Phys. Rev. D2, 2841 (1970)
   L. D. Faddeev and V. N. Popov, Phys. Letters 25B, 29 (1967)

2. B. S. DeWitt, in Relativity, Groups and Topology, Summerschool of Theor. Physics, Les Houches, France, 1963 (Gordon and Breach, New York, London)
B. S. DeWitt, Phys. Rev. 162, 1195, 1239 (1967)
3. G.'t Hooft and M. Veltman, CERN preprint TH 1723 (1973), to be publ. in Annales de l'Institut Henri Poincaré
4. D. Christodoulou, proceedings of the Academy of Athens 47 (20 January 1972)
D. Christodoulou, CERN preprint TH 1894 (1974)
5. J. Honerkamp, Nucl. Phys. B48, 269 (1972);
J. Honerkamp, Proc. Marseille Conf.,19-23 June 1972, CERN preprint TH 1558
6. G. 't Hooft, Nucl. Phys. B35, 167 (1971)
7. G. 't Hooft and M. Veltman, Nucl. Phys. B44, 189 (1972)
8. G. 't Hooft and M. Veltman, DIAGRAMMAR, CERN report 73-9 (1973)
9. G. 't Hooft, Nucl. Phys. B61, 455 (1973)
G. 't Hooft, Nucl. Phys. B62, 444 (1973)
10. H. D. Politzer, Phys. Rev. Letters 30, 1346 (1973)
D. J. Gross and F. Wilczek, Phys. Rev. Letters 30, 1343 (1973)
11. R. Bach, Math. Z. 9, 110 (1921)
C. Lanczos, Ann. Math. 39, 842 (1938)
12. S. Deser and P. van Nieuwenhuizen, Phys.Rev.Letters ,32,245(1974); Phys. Rev. D10, 401 (1974)
13. S. Deser, Hung-Sheng Tsao and P. van Nieuwenhuizen, Phys. Rev. D10, 3337 (1974)
14. S. Deser and P. Van Nieuwenhuizen, Phys. Rev. D10, 411 (1974)

Class. Quantum Grav. 10 (1993) S79–S91.

CHAPTER 8.2

Classical N-particle cosmology in 2 + 1 dimensions

G 't Hooft

Institute for Theoretical Physics, University of Utrecht, PO Box 80 006, NL 3508 TA Utrecht, The Netherlands

Abstract. In 2 + 1 dimensional cosmology particles are topological defects in a universe that is nearly everywhere flat. We use a time dependent triangulation procedure of space to formulate the laws of evolution and to construct phase space of this system. A theorem is derived from which it follows that many configurations may have a big bang in their past or a big crunch in their future. The dimensionality of phase space for N particles is $4N - 11 + 12g$, where g is the genus of 2-space. This suggests to consider $2N - 6 + 6g$ pairs of canonical variables and one time variable. The quantum version of this model is speculated about.

1. Introduction

In 2 + 1 dimensional General Relativity [1,2] Einstein's equations for the vacuum,

$$R_{\mu\nu} = 0 \tag{1.1}$$

imply also that the entire Riemann tensor $R^{\alpha}_{\beta\mu\nu}$ vanishes. Hence the vacuum is flat. A particle at rest, at the position $\boldsymbol{x} = \boldsymbol{a}$, produces a curvature proportional to $\delta^2(\boldsymbol{x} - \boldsymbol{a})$. From that the global structure of space-time surrounding the particle can easily be seen to be a cone, and it is described by excising a wedge out of the plane (see figure 1), after which one identifies points at each side of the wedge (for instance the two arrows in figure 1 are identified). The angle of the wedge, the conical deficiency angle β, is proportional to the mass M of the particle:

$$\beta = 2\pi GM \tag{1.2}$$

where G is Newton's constant, which we will choose to be one from now on.

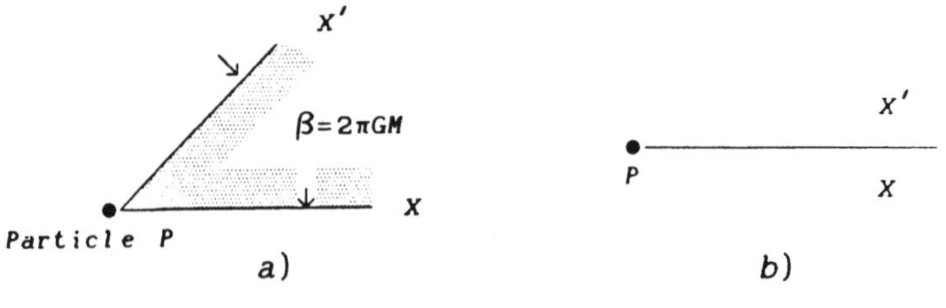

Figure 1. (*a*) Excised wedge near a particle P (shaded). (*b*) Diagrammatic notation.

If we wish to use Cartesian coordinates to describe locations in the neighbourhood of this particle we have to attach strings to each of these locations, and then the coordinates depend on how a string attached as indicated in figure 1 can also be described by the coordinates of $\boldsymbol{x}'$ with string as drawn, provided that

$$\boldsymbol{x}' = \boldsymbol{a} + \Omega(\boldsymbol{x} - \boldsymbol{a}) \tag{1.3}$$

where $\boldsymbol{a}$ is the location of P and Ω is the rotation

$$\Omega = \begin{pmatrix} \cos\beta & \sin\beta \\ -\sin\beta & \cos\beta \end{pmatrix} \qquad \beta = 2\pi M. \tag{1.4}$$

To describe a *moving* particle, one has to Lorentz transform this space-time. The relationship between the two coordinate frames then becomes

$$\begin{aligned} \begin{pmatrix} \boldsymbol{x}' \\ t' \end{pmatrix} &= L\left\{\begin{pmatrix} \Omega & 0 \\ 0 & 1 \end{pmatrix}\begin{pmatrix} \boldsymbol{x}-\boldsymbol{a} \\ t \end{pmatrix} + \begin{pmatrix} \boldsymbol{a} \\ 0 \end{pmatrix}\right\} L^{-1} \\ &= A\begin{pmatrix} \boldsymbol{x} \\ t \end{pmatrix} + \begin{pmatrix} \boldsymbol{b} \\ b^0 \end{pmatrix} \end{aligned} \tag{1.5}$$

where L is the Lorentz transformation that gives the particle the velocity $\boldsymbol{v}$ from its rest frame. A is a new Lorentz transformation and $\boldsymbol{b}$ and b^o are some shift vectors. Note that now the particle, its mass and its velocity, are determined by just giving an element of the Poincaré group. However not all elements of the Poincaré group specify a particle because the particle's space-time trajectory is given by the points satisfying

$$(\boldsymbol{x}', t') = (\boldsymbol{x}, t) \tag{1.6}$$

an equation that for generic elements of the Poincaré group has no solutions, as we will see.

Consider now two particles, moving with respect to each other. Following a path around both we find that points x and x' are identified by the product of two transformations of the type (1.5) (see figure 2).

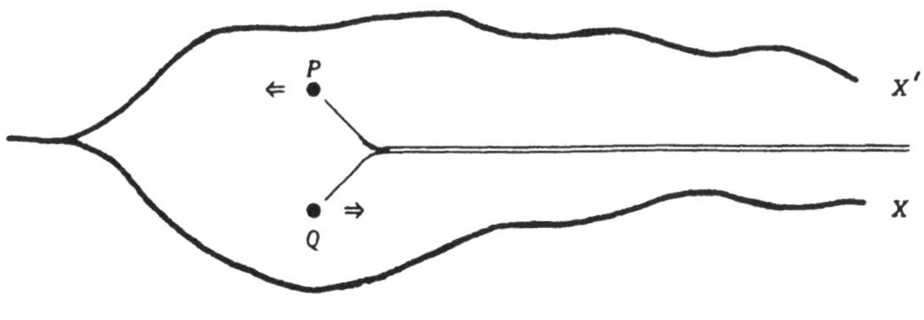

Figure 2. Two moving particles, P and Q.

The lines in figure 2 indicate the separation between the different coordinate frames used. The relation between the different coordinate frames for the point x, indicated as x and x', is obtained by the product of the transformations (1.5) for the two particles P and Q. Let us now consider the *center of mass frame*. This is the coordinate frame that is chosen such that

$$\begin{pmatrix} \boldsymbol{x}' \\ t' \end{pmatrix} = A_{cm} \begin{pmatrix} \boldsymbol{x} \\ t \end{pmatrix} + \begin{pmatrix} \boldsymbol{b} \\ b^o \end{pmatrix}_{cm} = \begin{pmatrix} \Omega(\boldsymbol{x} - \boldsymbol{a}) \\ t \end{pmatrix} + \begin{pmatrix} \boldsymbol{a} \\ \delta t \end{pmatrix} . \tag{1.7}$$

In general one can indeed find a Lorentz frame such that A_{cm} has only non-trivial elements in the first 2×2 block, as indicated, so that it is a pure rotation. It is natural to define the corresponding rotation angle to be the center of mass energy. Also one can find a vector $\boldsymbol{a}$ such that it acts as the origin of this rotation so that it can be interpreted as the location of the particle. But the new thing is the additional *time shift* δt as we encircle the two particles. It is not hard to verify that δt can be identified as being the *total angular momentum* [2]. The angles β over which one rotates are additive, and so are the time shifts δt. Thus one recovers the conservation laws of energy and angular momentum.

A complication is now that one can no longer define the trajectory of the center of mass as being the set of points invariant under (1.7). If the time shift is non-zero there are no invariant points. One has to go to the center of mass frame to locate the center of mass.

It is not difficult to derive the center of mass energy. Since the trace of a Lorentz matrix is Lorentz invariant one finds the cosine of the deficiency angle for the center of mass by taking the trace of the product of the Lorentz matrices corresponding to the two individual particles. The outcome of that (simple) calculation is [2]:

$$\cos \pi m_{cm} = \cos \pi m_1 \ \cos \pi m_2 - \sin \pi m_1 \ \sin \pi m_2 \ \cosh \gamma \tag{1.8}$$

where $m_{1,2}$ are the masses of the particles 1 and 2; m_{cm} is the center of mass energy; γ is the Lorentz boost parameter connecting the rest frame of 1 with that of 2. Note that in this expression we have the trigonometric functions of half the deficiency angles in stead of the angles themselves. This is because it is more convenient to work with the $SL(2, R)$ representation of the Lorentz group than it is with the $SO(2,1)$ representation.

We see that there only is a center of mass if the condition

$$\cosh \gamma \leqslant \frac{\cos \pi m_1 \ \cos \pi m_2 + 1}{\sin \pi m_1 \ \sin \pi m_2} \tag{1.9}$$

is satisfied. If this condition is *not* met P and Q are said to be a Gott pair [3]. The identification of the Cartesian coordinate frames connected by a loop around the particles is a Lorentz boost rather than a rotation. Such a pair can therefore be surrounded by a closed timelike curve (CTC). If no other particles occur it cannot be avoided that the universe is surrounded by an unphysical boundary condition [4, 5], even though CTC will *not* occur if we go sufficiently far to the past or to the future [6].

An open universe has an acceptable boundary if the total energy (corresponding to the sum of all deficiency angles in one coordinate frame divided by 2π) is less than 1. A universe closes if its energy is exactly 2. It then has the topology of a sphere (S_2, $g = 0$). But a torus ($g = 1$) or even 2-dimensional spaces with higher genus are also possible. The total energy vanishes for $g = 1$ and is negative in higher genus spaces. This is not in contradiction with positivity of the rest masses of the particles because when they move there can be negative energy in the surrounding gravitational fields as we shall see.

S M Carroll *et al* [5] proposed to consider a universe where the following happens. At $t = 0$ both M_1 and M_2 decay simultaneously each into a pair of light particles with masses:

$$M_1 \to m_1\, m_2 \qquad M_2 \to m_3\, m_4. \tag{1.10}$$

One easily finds that the two particles m_1 and m_2 created by the decay of M_1 move away from each other at some angle $\pi(1 - M_1)$ (see figure 3). With respect to each other they can never violate condition (1.9) because their center of mass energy is M_1. But m_1 and m_3 can meet each other head-on, and thus their relative boost parameter γ can exceed the value (1.9). Thus they may form a Gott pair. It turns out that this may only happen if

$$\text{tg}^2 \tfrac{1}{2}\pi M_1 > \text{tg}^2 \pi m + 1 \tag{1.11}$$

so that we must have $\frac{1}{2} < M_1 < 1$. This implies that the universe must be closed. So we add an 'antipode particle' X with mass $M_X = 2 - 2M_1$. The problem raised by these authors was the question whether closed timelike curves can then also arise in this universe.

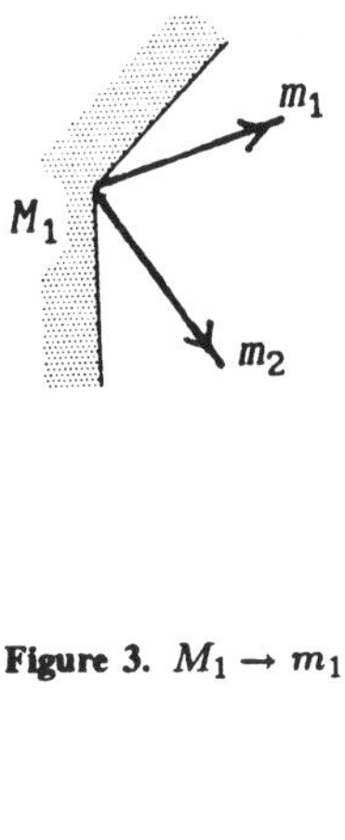

Figure 3. $M_1 \to m_1\, m_2$.

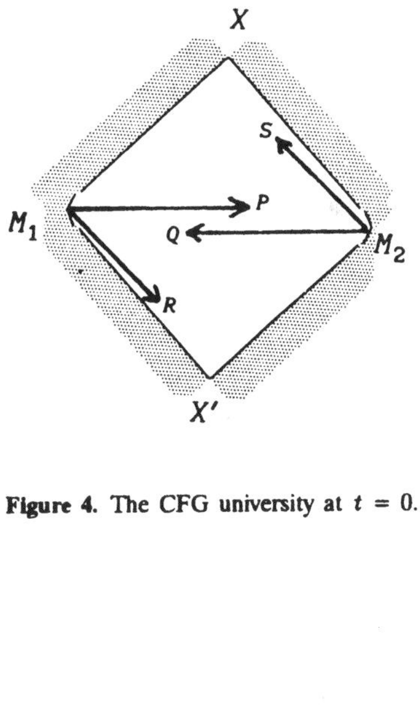

Figure 4. The CFG university at $t = 0$.

2. Triangulation and Cauchy surfaces

We now propose an approach to 2+1 dimensional gravity/cosmology that starts with construction of Cauchy cross sections of space-time. These are purely spacelike surfaces that are time-ordered, and together span the entire universe. We make optimal use of the fact that in between the particles space-time is flat. The ideal method is 'triangulation', which implies that we consider Cauchy surfaces that are built from entirely flat triangular 2-simplexes, glued together [7]. In practice however it is easier to take polygons rather than triangles, because in the generic case the vertex points will not connect more than three simplexes together. The general picture of such a Cauchy surface is sketched in figure 5, although in practice often the polygons will look somewhat more complex. Indeed, ultimately the most economic description will be using just one polygon, with rather elaborate prescriptions as to how the edges are to be sewn together.

All particles in our theory must be at vertex points, but vertex points where no particles sit are of course also allowed. In general we will put a particle at a 1-vertex; at such a point the adjacent sides of the corresponding polygon are glued together leaving the appropriate deficiency angle.

If all particles are at rest with respect to each other then the situation is easy to visualize. At all vertices we can take the angles to add up to exactly 2π, unless there is a particle present, and then the angles add up to $2\pi - 2\pi m$. If the total mass exceeds 1, then the Cauchy surface closes, so that there must be further particles at the 'antipodes'. The total mass then automatically adds up to 2.

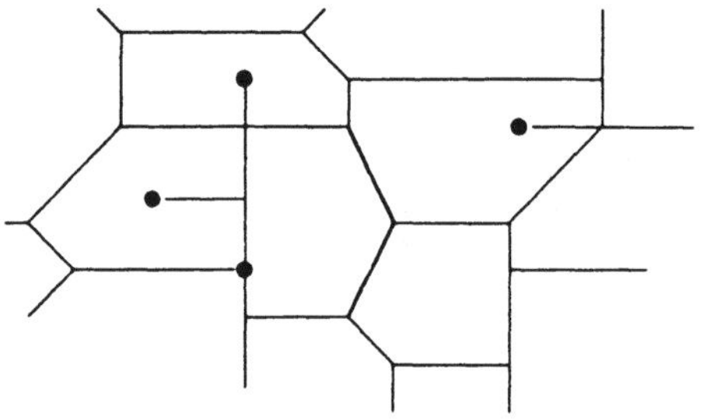

Figure 5. Cauchy surface built from polygons. The heavy dots are particles.

But if the particles move the situation is more complex. The polygons become deformed as a function of time, and in general the boundary between one polygon and another will involve a Lorentz transformation. It is now very instructive to formulate the rules that the moving polygons have to obey, if we insist that at the seams between all pairs of polygons space-time remains locally flat.

We consider in each polygon (which is actually a 3-simplex in space-time) a Cartesian coordinate frame. In general the sides all move in this frame, and their lengths change. Now in principle we could allow that on each polygon time runs at different speeds (as long as time runs forward), but we will impose the simplifying extra requirement that time runs equally fast at each polygon. In that case the evolution is uniquely determined, and furthermore the rules at the seams [7] become very simple:

1) The lengths of matching edges of two adjacent polygons, as measured in the coordinate frame of each, are equal. This is not a completely trivial statement because the matching goes with a Lorentz transformation.

2) The velocities with which the matching edges of two adjacent polygons move (always in a direction orthogonal to the orientation of the edge) in the coordinate frame of each, are the same, but the signs may differ. In general the signs are such that if the edge of one polygon recedes, the matching edge in the other one recedes also, because in the other case the matching becomes trivial.

3) The identification of points at two matching edges is such that a point moving in an orthogonal direction on one edge, remains identified with a point moving orthogonally on the matching edge, as seen in the corresponding frames.

4) The vertices between three polygons move in such way that rules 1, 2 and 3 remain valid. When new vertices are created (one vertex may split into several), special attention should be paid to whether they represent flat space or particles. Often this comes out all right automatically because of energy-momentum conservation at such space-time points.

The proof that polygons describing locally flat space can only move according to the above rules is not difficult. We will not here repeat the arguments already given in [7].

It is important to note that when a particle moves relatively to the coordinate frame chosen for the polygon that it is in, its velocity vector is only allowed to be in the direction of the bisector of the cusp, see figure 6.

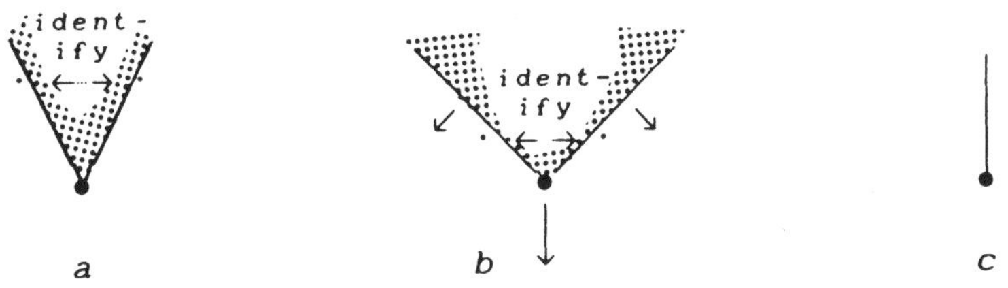

Figure 6. (*a*) Particle at rest. The shaded region is excised from flat space and the two edges are glued together. The two dots are points to be identified. (*b*) Particle moving downwards (arrow). We get the Lorentz contraction of picture (*a*), therefore the angle widens. Since the two dots in (*a*) are at the same height, they still will be at the same height in (*b*), and after the Lorentz transformation there will be no relative time shift. (*c*) Diagrammatic notation for a particle.

If ξ is the particle's rapidity, we see that a moving particle is Lorentz contracted by a factor $\cosh\xi$. Therefore it gets a widened deficiency angle β:

$$\mathrm{tg}\,\tfrac{1}{2}\beta = \mathrm{tg}(\pi m)\cosh\xi. \tag{2.1}$$

The boost parameter η describing the velocity $\mathrm{th}\eta$ with which the adjacent edges of the polygon widen or retract, is given by

$$\mathrm{th}\eta = \sin\tfrac{1}{2}\beta\ \mathrm{th}\xi. \tag{2.2}$$

Later it will turn out to be useful to eliminate ξ or β out of these expressions. One then obtains

$$\cosh\eta\ \cos\tfrac{1}{2}\beta = \cos\pi m \tag{2.3a}$$

$$\sinh\xi\ \sin\pi m = \sinh\eta. \tag{2.3b}$$

When three polygons meet at a vertex, see figure 7(*a*), and if there is no particle present precisely at this vertex (in the generic case there is none), then we can use the fact that space-time is locally flat here (note that although we talk about three different polygons, they could equally well be different regions of just one polygon, meeting at this particular vertex). The three coordinate frames can be seen as three points in a hyperbolic space (figure 7(*b*)). Since the sides of two adjacent polygons recede or widen with the same boost factor η in each coordinate frame, the distance between these two frames is the boost given by 2η. Hence in hyperbolic space the lengths of the sides of the triangle formed by the frames I, II and III are $2\eta_i$. The angles A_i in this triangle correspond to the angles α_i of the three polygons at the vertex, as follows:

$$A_i = \pi - \alpha_i. \tag{2.4}$$

These observations make it easy to understand the relations at each vertex as they were reported in ref. 7: writing

$$\sin\alpha_i = s_i \qquad \cos\alpha_i = c_i \qquad \sinh 2\eta_i = \sigma_i \qquad \cosh 2\eta_i = \gamma_i. \tag{2.5}$$

It was found [7] that:

$$s_1 : s_2 : s_3 = \sigma_1 : \sigma_2 : \sigma_3; \tag{2.6}$$

$$\gamma_2 s_3 + s_1 c_2 + c_1 s_2 \gamma_3 = 0; \tag{2.7}$$

$$c_1 = c_2 c_3 - \gamma_1 s_2 s_3; \tag{2.8}$$

$$\gamma_1 = \gamma_2\gamma_3 + \sigma_2\sigma_3 c_1; \tag{2.9}$$

$$\cot\alpha_2 = -\cot\alpha_1 \cosh 2\eta_3 - \coth 2\eta_2 \sinh 2\eta_3/\sin\alpha_1 \tag{2.10}$$

and all cyclic permutations. These are nothing but the trigonometric properties of the triangle of figure 7(*b*) in hyperbolic space.

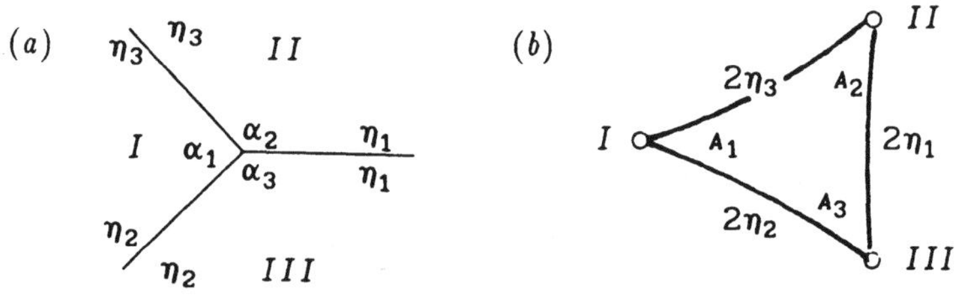

Figure 7.

It will be important to observe the ranges of values the angles can take: not more than one of the three angles α_i is allowed to exceed π. It is not hard to convince oneself that if there would be two angles $\alpha_i > \pi$ then there would be points in space-time that occur at two different spots on our Cauchy surface, which is not allowed. Now suppose one only knew the three boost parameters η_i then from (2.9) the three angles can be determined, except for an ambiguity $\alpha \Leftrightarrow 2\pi - \alpha$. But because the relative signs of $\sin\alpha_i$ are fixed by (2.6), and no more than one α is allowed to exceed π, this ambiguity can be resolved completely. *So the angles are completely determined by boosts η_i.*

3. Evolution

The polygons of the previous section must be seen as fixed time cross sections of 3-simplexes. At any given time t they together form a Cauchy surface. We now may ask how such a Cauchy surface evolves. There are several kinds of mutations that can take place at a given moment. First, an edge connecting two polygons (or different regions of a single polygon) may shrink to zero and be replaced by another edge, see figure 8(a). It is also possible that a particle hits an edge of the polygon it is in. It then crosses over into the next polygon (figure 8(b)). When the edge a particle is on shrinks to zero it hops into the adjacent polygon (figure 8(c). Finally, if one of the angles of a polygon is greater than π then that vertex may hit another edge and cause a 'vacuum cross-over' (see figure 8(d)).

In all these cases the angles and boost parameters of the newly opened edges are all fixed by the trigonometric equations, in particular (2.9) and (2.10). These fix all properties of the edge #1, if the boosts η_2, η_3 and the relative angle α_1 of the other edges are known. It is easily understood why all these properties are fixed: if the boosts and the angles of the transformations between frames I and II, frames II and IV, frames II and III, and frames III and IV are well known, then of course the transformation leading from I to IV (see figure 8(a)) is determined by that. And so are all other new edges in figure 8(a)–(d).

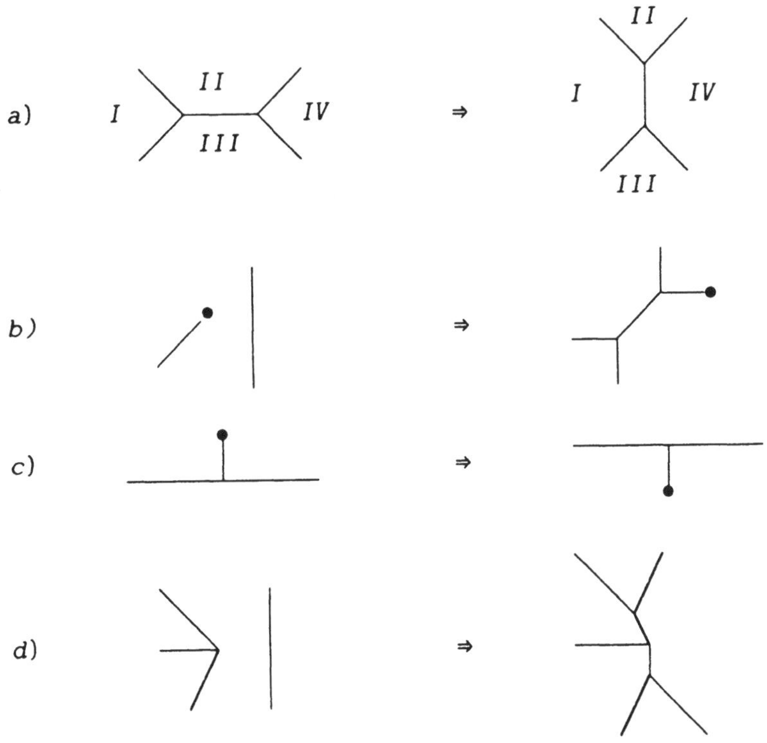

Figure 8. (a) Exchange; (b) cross-over; (c) hop; (d) vacuum cross-over.

In practice one needs no more than one single polygon, with identification rules at its edges. An example of a five-particle universe is pictured in figure 9(*a*). The identifications among the various edges are indicated by arrows.

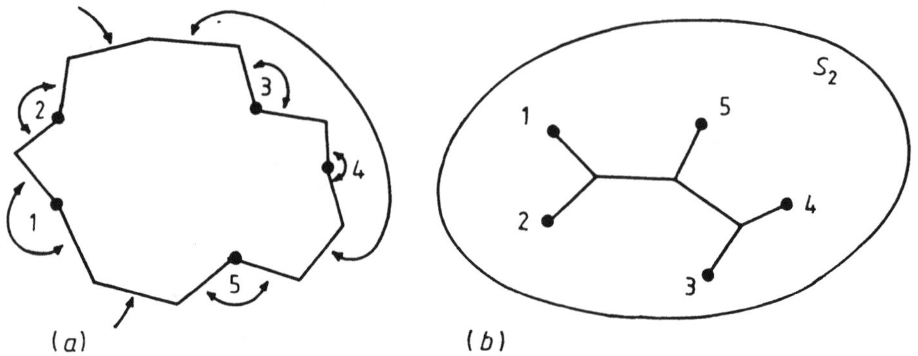

Figure 9. (*a*) Polygon representing a universe at given time *t*. (*b*) Diagrammatic notation of the same universe.

It is convenient to indicate the identification rules for the edges by drawing a diagram, obtained by drawing the polygon on an S_2 sphere. One can then glue the edges together. The seams obtained are shown in the diagram (figure 9(*b*)).

4. Crunch and bang theorem

An important property of 2+1 dimensional universes was indicated in [7]. It can actually be formulated in terms of a theorem:

Theorem. Let at one moment $t = t_1$ all edges of all polygons describing a closed or open universe be contracting, which means that all Lorentz boosts η at all edges are negative. (We will refer to this condition as the *Crunch condition*), then this condition will remain satisfied at all times $t > t_1$.

Note that the crunch condition implies that at all vacuum vertices (the 3-vertices containing no particle, so that (2.6)–(2.10) apply), the angles α_i are convex: $\alpha_i < \pi$, because due to (2.6) all $\sin\alpha_i$ have the same sign and because of the remark at the end of section 2. Where the particles are, the polygons have angles $\alpha = 2\pi - \beta$, where β is given by (2.1) or (2.3*a*). Thus $\alpha > \pi$ if $m < \frac{1}{2}$, and $\alpha < \pi$ if $m > \frac{1}{2}$. So where we have particles with mass $m < \frac{1}{2}$ the polygons are not convex. See figure 9(*a*). Evidently, we will not need figure 8(*d*).

The proof of the theorem can be given by checking all possible mutations, figure 8(*a*)–(*c*), one by one. Consider figure 8(*a*). The angles all start out being convex ($< \pi$). If η at the newly opened edge were positive all four adjacent angles would have to be $> \pi$, which is clearly not allowed. We see right away that that is geometrically impossible. So this transition keeps the Crunch situation as it was; the newly opened edge has $\eta < 0$ just like the others.

Next consider the cross-over, figure 8(*b*). This transition is only possible for a particle with mass $< \frac{1}{2}$ (otherwise it could only be involved with the hop transition, figure 8(*c*)). Again the angles start out being all convex. First look at the lower vertex that appears. If the new η that arises there would not be negative the two adjacent angles would come out larger than π, which is not allowed. We can then repeat the argument for the η of the edge attached to the particle after it entered the other polygon. One could argue alternatively that the particle as it enters gets an additional boost in the direction of the new polygon it is entering. This way also one finds unambiguously that the η of the particle (see equation (2.2)) must again be negative. Thus at this transition also the Crunch condition remains valid.

The case of the hop transition, figure 8(*c*), is again easy to check if the mass is $\geqslant \frac{1}{2}$. Then again it is the same geometrical argument that tells us that it is impossible for the two new angles at the vertex both to exceed π. If the mass is $< \frac{1}{2}$ this argument does not work. But careful analysis shows the theorem still to hold there. It is again geometry. The particle is entering a polygon at a convex point, where the two edges are contracting. The particle must outrun the edges and hence it must be receding also. It must have a negative η. We herewith checked all possible transitions. The crunch condition remains valid forever.

Of course the theorem can be time-inverted. Reversing the sign of all η coefficients we see that if one has at $t = t_1$ a situation such that all polygons are expanding at all sides, then this expansion must have existed at all times $t < t_1$. This is why we call this theorem the 'Crunch and Bang theorem'. If the Crunch condition is satisfied the universe must end in a Big Crunch. If the converse or 'Bang condition' is satisfied there must be a Big Bang in the past.

The theorem applies to the CFG universe. At the moment that the two particles M_i decay simultaneously into particles m_i all polygons satisfy the Crunch condition. The theorem tells us that the universe will continue to shrink forever. The particles will all approach each other with smaller and smaller impact parameters until they crunch. Before the crunch occurs there cannot possibly be a closed timelike curve simply because we were dealing with Cauchy surfaces all along. The purported CTC only is obtained when analytic extensions of space-time are considered beyond the Big Crunch. But we showed that the crunch singularity is an essential one; all particles meet each other there. Analytic extensions beyond that point are illegal.

The question could be asked whether perhaps the occurrence of a crunch is frame-dependent. But since the particles approach each other with ever decreasing impact parameters, and since in practice we found that at each mutation the crunch accelerates, we found that this question must be answered in the negative. There is no timelike path for which the moment of the crunch, in terms of its eigen time, can be postponed.

5. Degrees of freedom

An important question arises in these constructions. We would like to know exactly how to characterize the 'physical' degrees of freedom of 2+1 dimensional cosmologies. This is particularly of importance if one wishes to 'quantize' such models. Several quantization schemes have been proposed. It would be illustrative if a scheme could be devised that starts off with the degrees of freedom of a universe containing a finite number of particles, replacing the Poisson brackets of this finite dimensional phase space by commutators, and it would be of importance to consider the large distance, low mass limit of such a model where it should approach an ordinary Fock space consisting of a small number of non-interacting scalar particles, which we certainly know how to quantize. This question calls for the most economic way to represent a 2+1 dimensional universe with N particles at a given time.

Consider a universe of topology S_2, and containing N particles. Take coordinates in the rest frame of one of the particles, say M_0. We now construct a Cauchy surface at time t, corresponding to a moment τ in the eigen time of M_0. We spread its borders further and further out until points would be included that can be connected by timelike curves. A prescription can be worked out corresponding to moving the borders of our polygon about until a situation is reached that the edges touch each other without any time shifts. A typical situation is then shown in figure 9. The diagrammatic description of figure 9(b) is most appropriate. The construction allowed us furthermore to postulate that one particle is at rest: $\eta_1 = 0$. The direction in which the cusp of this particle is rotated is then also immaterial.

The remaining $N - 1$ particles are connected by a tree graph having $L = 2(N-1) - 3$ lines and $V = (N-1) - 2$ vertices. All diagrams of this sort satisfy

$$2L = 3V + (N-1). \tag{5.1}$$

There is one boost parameter at each line, and these fix all angles at the vertices via the identities (2.6)–(2.10). This gives us

$$L = 2N - 5 \tag{5.2}$$

dimensionless degrees of freedom. Now these also fix the cusp angles β where the particles are, via equation (2.3a), and all other angles including the cusp of particle M_0 were determined. Now since the polygon must close into itself, all external angles, $\pi - \alpha_i$, must add up to 2π. This gives us one more constraint, and so we are left with

$$P = 2N - 6 \tag{5.3}$$

degrees of freedom that determine the *angles* between the *velocities* of the particles. They can be compared with the *momentum* variables of the particles involved.

Then the lengths of all lines are arbitrary, giving us again $2N - 5$ degrees of freedom with the dimension of a length. Now not only the velocity of particle M_0 was fixed to be zero, also its position with respect to the other particles is completely determined now by the geometrical exercise of closing the polygon. Therefore we find

$$Q = 2N - 5 \tag{5.4}$$

degrees of freedom with the dimension of a length. It may seem to be surprising that Q is one bigger than P and that it is odd. This surprise disappears if one realizes that *time* t was introduced completely arbitrarily. If we look at the total number of degrees of freedom for *cosmologies*, where time is an undetermined parameter running over a certain domain of real numbers, we find that cosmologies span a

$$P + Q - 1 = 4N - 12 \tag{5.5}$$

dimensional phase space.

In an earlier stage of this research the author thought that 2-spaces with non-vanishing genus should not be allowed, because Euler's theorem would limit the total energy to be zero or negative. Now the total energy is obtained by adding up all deficiency angles of all particles *and* the angular deficits at all vertices where polygons meet. Indeed, moving particles all contribute here by amounts β_i, all having the same sign. But the deficits at the vertices count in the other direction. If for instance the Crunch condition of the previous section is satisfied then it is easily seen that equation (2.8), with $\gamma > 1$, gives

$$\cos\alpha_1 < \cos(\alpha_2 + \alpha_3) \Rightarrow \sum \alpha_i > 2\pi. \tag{5.6}$$

Thus, the vacuum vertices in general have negative curvature and hence they tend to cancel and reverse the contributions of the particles. We may interpret this as a negative contribution to the energy by the gravitational field. Due to this phenomenon total curvature can easily be negative, so that we can have a torus ($g = 1$, total deficit angles zero) or even higher genus surfaces. In these spaces the particles cannot be at rest; they have to move and the situation is far from stationary.

Careful study of the torus gave us that here

$$P + Q = 4N + 1. \tag{5.7}$$

In the general case the number of degrees of freedom is

$$P + Q = 4N + 12g - 11. \tag{5.8}$$

The one extra positional degree of freedom is again the time parameter. In these less trivial space parameters with dimensions of angles and boosts on the one hand and those with dimensions of lengths and times are not so easy to distinguish anymore.

An open question seems to be how exactly to characterize the topological shape of the spaces (5.3)–(5.8). Our difficulty here is that the transitions of figures 8(a)–(d) are mappings of these spaces into themselves; we would like to have definitions of Poisson brackets that are continuous under these mappings. Only then Poisson brackets may be replaced by commutators if one wishes to quantize the theory.

It is natural to view the parameters counted by equation (5.3) as being the 'momenta', they would turn into ordinary momenta of particles in Fock space in the large N limit, an ordinary non-interacting scalar particle model. The parameters counted by (5.4) are the coordinates plus a time variable. Considering the complicated topology of phase space it is however unclear to the present author whether and how a self consistent Fock space can be set up describing a cosmology with a fixed number of particles. An open question is also whether or not these particles, interacting only gravitationally can be pair created or destroyed. Classically (in the $\hbar \Rightarrow 0$ limit) as well as in the large N limit (where we have ordinary Fock space of free scalar particles), this does not happen. Existing quantization schemes [8] suggest that topology change occurs but are inconclusive as to whether particle pair creation and annihilation take place.

Acknowledgments

The author thanks R Jackiw, S Deser and A Guth for extensive correspondence leading to this work. He thanks H B Nielsen and G Gibbons for discussions concerning the last section.

References

[1] Staruszkiewicz A 1963 *Acta Phys. Polon.* **24** 734
Gott J R and Alpert M 1984 *Gen. Rel. Grav.* **16** 243
Giddings S, Abbot J and Kuchar K 1984 *Gen. Rel. Grav.* **16** 751

[2] Deser S, Jackiw R and 't Hooft G 1984 *Ann. Phys.* **152** 220

[3] Gott J R 1991 *Phys. Rev. Lett.* **66** 1126

[4] Deser S, Jackiw R and 't Hooft G 1992 *Phys. Rev. Lett.* **68** 267

[5] Carroll S M, Farhi E and Guth A 1992 *Phys. Rev. Lett.* **68** 263 (and A Guth personal communication)

[6] Cutler C 1992 *Phys. Rev.* D **45** 487 (see also Ori A 1991 *Phys. Rev.* D **44** R2214)

[7] 't Hooft G 1992 *Class. Quantum Grav.* **9** 1335

[8] Witten E 1988 *Nucl. Phys.* B **311** 46
Carlip S 1989 *Nucl. Phys.* B **324** 106; 1990 *Physics, Geometry and Topology (NATO ASI Series B, Physics)* **238** ed H C Lee (New York: Plenum) p 541

Nuclear Physics B256 (1985) 727–736

CHAPTER 8.3

ON THE QUANTUM STRUCTURE OF A BLACK HOLE

Gerard 't HOOFT

Institute for Theoretical Physics, Princetonplein 5, PO Box 80.006, 3508 TA Utrecht, The Netherlands

Received 26 October 1984
(Revised 30 November 1984)

The assumption is made that black holes should be subject to the same rules of quantum mechanics as ordinary elementary particles or composite systems. Although a complete theory for reconciling this requirement with that of general coordinate transformation invariance is not yet in sight, a number of observations can be made and a general framework is suggested.

1. Introduction

In view of the fundamental nature of both the theory of general relativity and that of quantum mechanics there seems to be little need to justify any attempt to reconcile the two theories. Yet the essential academic interest of the problem may be not our only motive. It seems to become more and more clear that "ordinary" elementary particle physics, in energy regions that will be accessible to machines in the near future, is plagued by mysteries that may require a more drastic approach than usually considered: the so-called hierarchy problems, and the freedom to choose coupling constants and masses. A variant of these problems also occurs in quantum gravity. Here it is the mystery of the vanishing cosmological constant. It would not be the first time in history if a solution to these mysteries could be found by first contemplating gravity theory; after all, the very notion of Yang–Mills fields was inspired by general relativity. In this paper we will present an approach in which the cosmological constant problem is acute and so, any progress made might help us in particle physics also.

We are not yet that far. The aim of the present paper is to set the scene, to provide a new battery of formulas that may become useful one day.

To many practitioners of quantum gravity the black hole plays the role of a soliton, a non-perturbative field configuration that is added to the spectrum of particle-like objects only after the basic equations of their theory have been put down, much like what is done in gauge theories of elementary particles, where Yang–Mills equations with small coupling constants determine the small-distance structure, and solitons and instantons govern the large-distance behavior.

Such an attitude however is probably not correct in quantum gravity. The coupling constant increases with decreasing distance scale which implies that the smaller the distance scale, the stronger the influences of "solitons". At the Planck scale it may well be impossible to disentangle black holes from elementary particles. There simply is no fundamental difference. Both carry a finite Schwarzschild radius and both show certain types of interactions. It is natural to assume that at the Planck length these objects merge and that the same set of physical laws should cover all of them.

Now in spite of the fact that the properties of larger black holes appear to be determined by well-known laws of physics there are some tantalizing paradoxes as we will explain further. Understanding these problems may well be crucial before one can proceed to the Planck scale.

At present a black hole is only (more or less) understood as long as it is in a quantum mechanically mixed state. The standard picture is that of the Hawking school [1, 3], but a competing description exists [2] which this author was unable to rule out entirely, and which predicts a different radiation temperature. The question which of these pictures is right will not be considered in this paper; we leave it open by admitting a free parameter λ in the first sections.

Whatever the value of λ, the picture is incomplete. If black holes show any resemblance with ordinary particles it should be possible to describe them as *pure* states, even while they are being born from an implosion of ordinary matter, or while the opposite process, evaporation by Hawking radiation, takes place. As we will argue, any attempt to "unmix" the black-hole configuration (that is, produce a density matrix with eigenvalues closer to 1 and 0) produces matter in the view of a freely falling observer. It is this "matter" that we will be considering in this paper.

We start with the postulate that there exists an extension of Hilbert space comprising black holes, and that a hamiltonian can be precisely defined in this Hilbert space, although a certain amount of ambiguity due to scheme (gauge) dependence is to be expected. Although at first sight this postulate may seem to be nearly empty, it is directly opposed to conclusions of the Hawking school [1]. These authors apply the rule of invariance under general coordinate transformations in the conventional way. We believe that, although our present dogma may well prove wrong ultimately, it stands a good chance of being correct if we apply general coordinate transformations more delicately, and in particular if the distinction between "vacuum" and "matter" may be assumed to be observer dependent when coordinate transformations with a horizon are considered.

Another point where the conventional derivations could be greeted with some scepticism is the role of the infalling observer. If the infalling observer sees a pure state an outside observer sees a mixed state. In our view this could be due simply to the fact that the "infalling observer" makes part of the system seen by the outsider: he himself is included in the outside Hilbert space. Clearly the very foundations of quantum mechanics are touched here and we leave continuation of this intriguing subject to philosophers, rather than accepting the simple-minded conclusion that transitions will take place between pure states and mixed states [1].

Nowhere a distinction may be made between "primordial" black holes and black holes that have been formed by collapse. Now we will make use of space–time metrics that contain a future and a past horizon, and at first sight this seems to be a misrepresentation of the history of a black hole with a collapse in its past. However, just because no distinction is made one is free to choose whichever metric is most suitable for a description of the present state of a black hole. It may be expected that in a pure-state description of a black hole a collapsing past will be very difficult to describe if this collapse took place much longer ago than at time $t = -M \log M$ in Planck units.

From now on a black hole will be defined to be any particle which is considerably heavier than the Planck mass and which shows only the minimal amount of structure (nothing much besides its Hawking radiation [3]) outside its Schwarzschild horizon.

The first part of this paper gives a number of pedestrian arguments indicating what kind of quantum structure one might expect in a black hole. Although they should not be considered as airtight mathematical proofs this author finds it extremely hard to imagine how the conclusions of sect. 2 of this paper could be avoided:

(i) The spectrum of black holes states is discrete. The density of states for heavy black holes can be computed up to an (unknown but finite) overall constant.

(ii) Baryon number conservation, like all other additive quantum numbers to which no local gauge field is coupled, must be violated.

The first of these conclusions is related to the expected thermodynamic properties first proposed by Bekenstein [4]. The second has been discussed at length by him and also by Zeldovich [5].

In sect. 3 a naive model ("brick wall model") is constructed that roughly reproduces these features. Being a model rather than a theory it violates the fundamental requirement of coordinate invariance at the horizon so it cannot represent a satisfactory solution to our problem, but in its simplicity it does show some important facts: it is the horizon itself, rather than the black hole as a whole that determines its quantum properties.

In the second part of the paper*) we formulate a much more precise and satisfying approach. A set of coordinate frames and transformation laws is proposed. One simplification is made that presumably is fairly harmless: we assume that, averaged over a certain amount of time, ingoing and outgoing particles near a black hole are smeared equally over all angles θ, φ. Certainly this is true for a Hawking-radiating Schwarzschild black hole. Furthermore, ingoing things may be *chosen* to be spherically smeared. As a result we may take the gravitational interactions between ingoing and outgoing matter (our most crucial problem) to be rotationally invariant. This assumption appears to be quite suitable for a first attempt to obtain an improved theory, and in fact it makes our whole approach quite powerful.

As stated before, how to interpret the coordinate transformations physically is yet another matter. The unorthodox interpretation that we proposed in ref. [2] is not ruled out (in fact it fits quite well in our description) but we do not insist on it. Rather, we allow the reader to draw his own conclusions here.

*) not reproduced in this book.

It is important to note that never space–times are considered that contain a "conical singularity" at the origin or elsewhere. Even in the formalism with $\lambda = 2$ there is no such singularity. This is because the "identification" of the points x with the points $-x$ is Minkowski space of ref. [2] only is made in the *classical limit*, not in the quantum theory.

Quantization of the space–time metric $g_{\mu\nu}$ itself is not considered explicitly, since we attempt to deduce the black-hole's properties from known laws of physics much beyond the Planck length. This may be incorrect: it could be that *only* by considering the full Hilbert space of all metrics a workable picture emerges. Our attitude is to first keep everything as simple as possible and only accept such complications when they clearly become unavoidable.

2. The black hole spectrum

A black hole can absorb particles according to the well-known laws of general relativity: the geodesics of particles with an impact parameter below a certain threshold will disappear into the horizon.

Conversely, black holes may emit particles as derived by Hawking. They radiate as black bodies with a certain temperature:

$$T_{\mathrm{H}} = \lambda / 8\pi M, \tag{2.1}$$

in units where the gravitational constant, the speed of light, Planck's constant and Boltzmann's constant are put equal to one. As is explained in sect. 1 there may be reasons to put into question the usual argument that $\lambda = 1$. There is an alternative theory with $\lambda = 2$, but the precise value of λ is of little relevance to the following argument.

If a black hole is to be compared with any ordinary quantum mechanical system such as a heavy atomic nucleus or any "black box" containing a number of particles and possessing a certain set of energy levels, which furthermore can absorb and emit particles in a similar fashion, then a conclusion on the density of its energy levels, $\rho(E)$, can readily be drawn.

Imagine an object with energy ΔE being dropped into a black hole with mass E, so that the final mass is $E + \Delta E$, where in Planck units

$$\Delta E \ll 1 \ll E. \tag{2.2}$$

Let R be the bound on the impact parameter; usually

$$R \simeq 2E. \tag{2.3}$$

Then the absorption cross section σ is

$$\sigma = \pi R^2. \tag{2.4}$$

From Hawking's result we conclude that the emission probability W is

$$W \simeq \pi R^2 \rho_{\Delta E}\, e^{-\beta_H \Delta E}, \tag{2.5}$$

where $\rho_{\Delta E}$ is the density of states for a particle with energy ΔE per volume element, and β_H is the inverse of the temperature T_H.

Now if the same processes can be described by a hamiltonian acting in Hilbert space, then we should have in the first case

$$\sigma = |\langle E+\Delta E|T|E, \Delta E\rangle|^2 \rho(E+\Delta E), \tag{2.6}$$

where T is the scattering matrix, and in the second case

$$W = |\langle E, \Delta E|T|E+\Delta E\rangle|^2 \rho(E)\rho_{\Delta E}, \tag{2.7}$$

by virtue of the "golden rule".

The matrix elements in both cases should be equal to each other if PCT invariance is to be respected (the expressions hold for particles and antiparticles equally). So dividing the two expressions we find

$$\frac{\rho(E+\Delta E)}{\rho(E)} = e^{+\beta_H \Delta E}, \tag{2.8}$$

$$\rho(E) = C\, e^{4\pi\lambda^{-1}E^2}. \tag{2.9}$$

Now this result could also have been concluded from the usual thermodynamical arguments [4]. The entropy S is

$$S = 4\pi\lambda^{-1}E^2 + \text{const}, \tag{2.10}$$

from which indeed (2.9) follows. We note that E^2 in (2.9) and (2.10) measures the total area of the horizon.

The constant in (2.9) and (2.10) is not known. Could it be infinite? We claim that this can only be the case if there exists a "lightest" *stable* black hole. Just compare eqs. (2.6) and (2.7) if $E+\Delta E$ represents the lightest black hole, and E and ΔE are ordinary elementary particles. If $\rho(E+\Delta E)$ is infinite but $\rho(E)$ finite then, since σ must be finite, W vanishes. Since this object fails to obey the classical laws of physics (it ought to emit Hawking radiation), it cannot be much heavier than the Planck mass. Now it is very difficult to conceive of any quantum theory that can admit the presence of such infinitely degenerate "particles". They are coupled to the graviton in the usual way, so their contributions to graviton scattering amplitudes and propagators would diverge with the constant C, basically because the probability for pair creation of this object in any channel with total energy exceeding the mass threshold is proportional to C.

Clearly the above arguments assume that several familiar concepts from particle field theory such as unitarity, causality, positivity, etc. also apply to these extreme energy end length scales, and one might object against making such assumptions. But we considered this to be a reasonable starting point and henceforth assume C

to be finite. An interesting but as yet not much explored possibility is that C, though finite, might be extremely large. After all, large numbers such as $M_{\mathrm{Planck}}/m_{\mathrm{proton}}$ are unavoidable in this area of physics. One might find interesting links with the $1/N$ expansion suggested by Weinberg [6], who however finds physically unacceptable poles in the $N \to \infty$ limit.

It is rather unlikely that C is exactly constant. One expects subdominant terms in the exponent. Also, we have not taken into account the other degrees of freedom such as angular momentum and electric (possibly also magnetic) charge. These can be taken into account by rather straightforward extrapolations.

Other additive quantum numbers cannot possibly be conserved. The reason is that the larger black hole may absorb baryons or any other such objects in unlimited quantities. If indeed it allows only a finite number of quantum states then any assignment of baryon number will fail sooner or later. Indeed one of the paradoxes we will have to face is that starting with a theory that is invariant under rotations in baryon number space one must end up with a theory where this invariance is broken [5].

3. The brick wall model

When one considers the number of energy levels a particle can occupy in the vicinity of a black hole one finds a rather alarming divergence at the horizon. Indeed, the usual claim that a black hole is an infinite sink of information (and a source of an ideally random black body radiation of particles [3, 7]) can be traced back to this infinity. It is inherent to the arguments in the previous sections that this infinity is physically unacceptable. Later in this section we will see that, as is often the case with such infinities, the classical treatment of the infinite parts of these expressions is physically incorrect. The particle wave functions extremely close to the horizon must be modified in a complicated way by gravitational interactions between ingoing and outgoing particles (sects. 5 and 6). Before attempting to consider these interactions properly we investigate the consequences of a simple-minded cut-off in this section.

It turns out to be a good exercise to see what happens if we assume that the wave functions must all vanish within some fixed distance h from the horizon:

$$\varphi(x) = 0 \quad \text{if} \quad x \leq 2M + h, \tag{3.1}$$

where M is the black hole mass. For simplicity we take $\varphi(x)$ to be a scalar wave function for a light ($m \ll 1 \ll M$) spinless particle. Later we will give them a multiplicity Z as a first attempt to mimic more closely the real world.

In the view of a freely falling observer, condition (3.1) corresponds to a uniformly accelerated mirror which in fact will create its own energy-momentum tensor due to excitation of the vacuum. As in sect. 1 we stress that this presence of matter and energy may be observer dependent, but above all this model should be seen as an elementary exercise, rather than an attempt to describe physical black holes accurately.

Let the metric of a Schwarzschild black hole be given by

$$ds^2 = -\left(1-\frac{2M}{r}\right)dt^2 + \left(1-\frac{2M}{r}\right)^{-1} dr^2 + r^2\, d\Omega^2 . \tag{3.2}$$

Furthermore, we need an "infrared cutoff" in the form of a large box with radius L:

$$\varphi(x) = 0 \quad \text{if} \quad x = L . \tag{3.3}$$

The quantum numbers are l, l_3 and n, standing for total angular momentum, its z-component and the radial excitations. The energy levels $E(l, l_3, n)$ can then be found from the wave equation

$$\left(1-\frac{2M}{r}\right)^{-1} E^2\varphi + \frac{1}{r^2}\partial_r r(r-2M)\partial_r\varphi - \left(\frac{l(l+1)}{r^2} + m^2\right)\varphi = 0 . \tag{3.4}$$

As long as $M \gg 1$ (in Planck units) we can rely on a WKB approximation. Defining a radial wave number $k(r, l, E)$ by

$$k^2 = \frac{r^2}{r(r-2M)}\left(\left(1-\frac{2M}{r}\right)^{-1} E^2 - r^{-2} l(l+1) - m^2\right), \tag{3.5}$$

as long as the r.h.s. is non-negative, and $k^2 = 0$ otherwise, the number of radial modes n is given by

$$\pi n = \int_{2M+h}^{L} dr\, k(r, l, E) . \tag{3.6}$$

The total number N of wave solutions with energy not exceeding E is then given by

$$\pi N = \int (2l+1)\, dl\, \pi n \stackrel{\text{def}}{=} g(E)$$

$$= \int_{2M+h}^{L} dr \left(1-\frac{2M}{r}\right)^{-1} \int (2l+1)\, dl \sqrt{E^2 - \left(1-\frac{2M}{r}\right)\left(m^2 + \frac{l(l+1)}{r^2}\right)} , \tag{3.7}$$

where the l-integration goes over those values of l for which the argument of the square root is positive.

What we have counted in (3.7) is the number of classical eigenmodes of a scalar field in the vicinity of a black hole. We now wish to find the thermodynamic properties of this system such as specific heat etc. Every wave solution may be occupied by any integer number of quanta. Thus we get for the free energy F at some inverse temperature β,

$$e^{-\beta F} = \sum e^{-\beta E} = \prod_{n,l,l_3} \frac{1}{1-e^{-\beta E}} , \tag{3.8}$$

or

$$\beta F = \sum_N \log\left(1 - e^{-\beta E}\right); \tag{3.9}$$

and, using (3.7),

$$\begin{aligned}\pi\beta F &= \int dg(E)\log\left(1-e^{-\beta E}\right)\\ &= -\int_0^\infty dE \frac{\beta g(E)}{e^{\beta E}-1}\\ &= -\beta\int_0^\infty dE \int_{2M+h}^{L} dr\left(1-\frac{2M}{r}\right)^{-1}\int\left(2l+1\right)dl\\ &\quad\times\left(e^{\beta E}-1\right)^{-1}\sqrt{E^2-\left(1-\frac{2M}{r}\right)\left(m^2+\frac{l(l+1)}{r^2}\right)}\,.\end{aligned} \tag{3.10}$$

Again the integral is taken only over those values for which the square root exists. In the approximation

$$m^2 \ll 2M/\beta^2 h, \qquad L \gg 2M, \tag{3.11}$$

we find that the main contributions are

$$F \simeq -\frac{2\pi^3}{45h}\left(\frac{2M}{\beta}\right)^4 - \frac{2}{9\pi}L^3\int_m^\infty \frac{dE(E^2-m^2)^{3/2}}{e^{\beta E}-1}\,. \tag{3.12}$$

The second part is the usual contribution from the vacuum surrounding the system at large distances and is of little relevance here. The first part is an intrinsic contribution from the horizon and it is seen to diverge linearly as $h \to 0$.

The contribution of the horizon to the total energy U and the entropy S are

$$U = \frac{\partial}{\partial\beta}(\beta F) = \frac{2\pi^3}{15h}\left(\frac{2M}{\beta}\right)^4 Z, \tag{3.13}$$

$$S = \beta(U-F) = \frac{8\pi^3}{45h}2M\left(\frac{2M}{\beta}\right)^3 Z. \tag{3.14}$$

We added a factor Z denoting the total number of particle types.

Let us now adjust the parameters of our model such that the total entropy is

$$S = 4\lambda^{-1}M^2, \tag{3.15}$$

as in eq. (2.10), and the inverse temperature is

$$\beta = 8\pi\lambda^{-1}M. \tag{3.16}$$

This is seen to correspond to

$$h = \frac{Z\lambda^4}{720\pi M}\,. \tag{3.17}$$

Note also that the total energy is

$$U = \tfrac{3}{8}M, \tag{3.18}$$

independent of Z, and indeed a sizeable fraction of the total mass M of the black hole! We see that it does not make much sense to let h decrease much below the critical value (3.17) because then more than the black hole mass would be concentrated *at our side* of the horizon.

Eq. (3.17) suggests that the distance of the "brick wall" from the horizon depends on M, but this is merely a coordinate artifact. The invariant distance is

$$\int_{r=2M}^{r=2M+h} \mathrm{d}s = \int \frac{\mathrm{d}r}{\sqrt{1-2M/r}} = 2\sqrt{2Mh} = \sqrt{\frac{Z\lambda^4}{90\pi}}. \tag{3.19}$$

Thus, the brick wall may be seen as a property of the horizon independent of the size of the black hole.

The conclusion of this section is that not only the infinity of the modes near the horizon should be cut-off, but also the value for the cut-off parameter is determined by nature, and a property of the horizon only. The model described here should be a reasonable description of a black hole as long as the particles near the horizon are kept at a temperature as given by (3.16) and all chemical potentials are kept close to zero. The reader is invited to investigate further properties of the model such as the average time spent by one particle near the horizon, etc.

The model automatically preserves quantum coherence completely, but it is also unsatisfactory: there might be several conserved quantum numbers, such as baryon number*. What is wrong, clearly, is that we abandoned the principle of invariance under coordinate transformations at the horizon. The question that we should really address is how to keep not only the quantum coherence but also general invariance, while dropping all global conservation laws.

References

[1] S.W. Hawking, Breakdown of predictability in gravitational collapse, Phys. Rev. D14 (1976) 2460; The unpredictability of quantum gravity, Cambridge Univ. preprint (May 1982)

[2] G. 't Hooft, Ambiguity of the equivalence principle and Hawking's temperature, J. Geom. Phys. 1 (1984) 45

[3] S.W. Hawking, Particle creation by black holes, Comm. Math. Phys. 43 (1975) 199;
J.B. Hartle and S.W. Hawking, Path-integral derivation of black hole radiance, Phys. Rev. D13 (1976) 2188;
W.G. Unruh, Notes on black-hole evaporation, Phys. Rev. D14 (1976) 870

[4] J.D. Bekenstein, Black holes and the second law, Nuovo Cim. Lett. 4 (1972) 737; Black holes and entropy, Phys. Rev. D7 (1973) 2333; Generalized second law of thermodynamics in black-hole physics, D9 (1974) 3292

[5] J.D. Bekenstein, Non existence of baryon number for static black holes, Phys. Rev. D5 (1972) 1239, 2403;
Ya.B. Zeldovich, A new type of radioactive decay: gravitational annihilation of baryons, Sov. Phys. JETP 45 (1977) 9;
A.D. Dolgor and Ya.B. Zeldovich, Cosmology and elementary particles, Rev. Mod. Phys. 53 (1981) 1

*) One may postpone this difficulty by inserting explicitly baryon number violating interactions near the horizon.

[6] S. Weinberg, in General relativity, an Einstein centenary survey, ed. S.W. Hawking and W. Israel (Cambridge Univ. Press, 1979) 795
[7] W.G. Unruh and R.M. Wald, What happens when an accelerating observer detects a Rindler particle, Berkeley preprint

Nuclear Physics B253 (1985) 173-188

CHAPTER 8.4

THE GRAVITATIONAL SHOCK WAVE OF A MASSLESS PARTICLE

Tevian DRAY[1] and Gerard 't HOOFT

Instituut voor Theoretische Fysica, Princetonplein 5, Postbus 80.006, 3508 TA Utrecht, The Netherlands

Received 20 August 1984

The (spherical) gravitational shock wave due to a massless particle moving at the speed of light along the horizon of the Schwarzschild black hole is obtained. Special cases of our procedure yield previous results by Aichelburg and Sexl [1] for a photon in Minkowski space and by Penrose [2] for sourceless shock waves in Minkowski space. A new derivation of the (plane) shock wave of a photon in Minkowski space [1] involving explicit calculation of geodesics crossing the shock wave is also given in order to clarify the underlying physics. Applications to quantum gravity, specifically the possible effect on the Hawking temperature, are briefly discussed.

1. Introduction

There are various reasons why one may be interested in exact expressions for the gravitational field surrounding a particle whose mass is dominated by kinetic energy rather than rest mass. For instance the first non-trivial gravitational effects to be seen in particle-particle interactions at extreme energies may be due to such fields. Our understanding of quantum gravity may be helped by considering these field configurations. A specific case of interest is the gravitational back-reaction and self-interaction of matter entering or leaving a black hole (Hawking radiation). At the black hole horizon the relative velocity of these particles approach that of light.

Aichelburg and Sexl [1] considered the gravitational field of a massless particle in Minkowski space, and showed that the resulting space-time is a special case of a *gravitational impulsive wave** [2] which is also an *asymmetric plane-fronted gravitational wave* [3]. Penrose [2] also gives explicit examples of *sourceless* gravitational impulsive waves in Minkowski space.

In this paper we first summarize the properties of the shock wave due to a massless particle in Minkowski space. We do this by presenting a new derivation of the results of Aichelburg and Sexl [1] involving explicit calculation of (null) geodesics crossing the shock wave. This enables the physical properties of such shock waves to be easily exhibited.

We then determine, for a particular class of vacuum solutions to the Einstein field equations, the (necessary and sufficient) conditions for being able to introduce

[1] Supported by the Stichting voor Fundamenteel Onderzoek der Materie.

* Note that Penrose [2] reserves the term *gravitational shock wave* for a metric which is C^1 whereas the metrics we consider are only C^0. We will nevertheless use the term "shock wave" for what are, in the terminology of [2], impulsive waves.

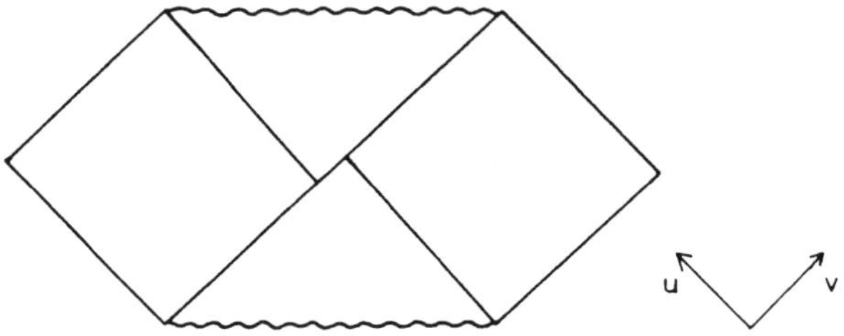

Fig. 1. The horizon shift (eq. (15)) due to the field of a massless particle moving in the v-direction along the horizon of the Schwarzschild black hole. The amount of the shift depends on θ.

a gravitational shock wave via a coordinate shift*. These conditions include both constraints on the metric coefficients and on the form of the shift. In Minkowski space they reduce to the plane-fronted wave of Aichelburg and Sexl [1] and, of course, to Penrose's results [2] for sourceless waves. However, for Schwarzschild black holes we obtain something new: there is a (spherical) shock wave at the horizon due to a massless particle at the horizon. (See fig. 1.)

Throughout this paper we think of the massless particle as the limit of a fast-moving particle with negligible rest mass**; this limit is given explicitly for the Minkowski case. Fig. 1 can thus be interpreted as describing an ordinary particle with small mass falling into the black hole from the left, as seen by an outside observer (on the left) at very late times; the particle is then seen close to the horizon and boosted to high energies.

The paper is organized as follows: in sect. 2 we summarize the situation for the (plane) shock wave due to a massless particle in Minkowski space and discuss the general physical features of such a wave. These results are based on a calculation of the null geodesics in such a space-time, which is given explicitly in appendix A. In sect. 3 we give the conditions, derived in appendix B, for a shock wave to be possible starting from a given "background" space-time. After showing that these conditions reduce to the correct ones [1, 2] in Minkowski space we then obtain the (spherical) shock wave at the horizon of the Schwarzschild black hole due to a massless particle there. In sect. 4 we discuss our results.

2. Shock waves: an example

Aichelburg and Sexl [1] (cf. eq. (A.37)) have shown that the gravitational field of a massless particle in Minkowski space is described by the metric

$$ds^2 = -d\hat{u}(d\hat{v} + 4p\ln(\rho^2)\delta(\hat{u})\,d\hat{u}) + dx^2 + dy^2\,, \tag{1}$$

* This is just the scissors-and-paste approach of Penrose [2] applied to more general space-times.

** We assume that the particle has no electric charge and no angular momentum. However, for an elementary particle for example we do not expect the results to differ significantly from those we derive here.

where $\rho^2 = x^2 + y^2$. The particle moves in the $\hat{v}$ direction with momentum p. By calculating geodesics which cross the shock wave, which is located at $u = 0$, we obtain the following two physical effects of such a shock wave (see appendix A): geodesics have a discontinuity $\Delta\hat{v}$ at $u = 0$ and are refracted in the transverse direction. The shift $\Delta\hat{v}$ is given by (cf. eq. (A.26))

$$\Delta\hat{v} = -\frac{4Gp}{c^3} \ln\frac{\rho_0^2}{l_{\rm Pl}^2}, \tag{2a}$$

which, for a photon, is

$$\Delta\hat{v} = -\frac{4l_{\rm Pl}^2\nu}{c} \ln\frac{\rho_0^2}{l_{\rm Pl}^2}, \tag{2b}$$

where we have put the units back in and where we have used $E = pc = \hbar\nu$, where ν is the frequency of the photon and $l_{\rm Pl}$ is the Planck length. ρ_0 is the value of ρ when the geodesic reaches $u = 0$. This shift is illustrated (for *nonzero* m and $x = 0$) in fig. 2; for $m = 0$ the shift occurs as a discontinuity at $u = 0$.

Note that the presence of a length scale in the *argument* of the logarithm is merely a reflection of our choice of units and has no physical meaning. It represents a *constant* shift in $\hat{v}$ which can be transformed away by a suitable redefinition of $\hat{v}$ (eq. (A.11)). Furthermore, by the same procedure, the value of ρ_0 for which $\Delta\hat{v} = 0$ (here $\rho_0 = l_{\rm pl}$) can be chosen arbitrarily far from the photon (ρ_0 large). In any case, only the *difference* in $\Delta\hat{v}$ for nearby geodesics is physically relevant.

There is also a refraction effect described by (cf. eq. (A.36))

$$\cot\alpha + \cot\beta = \frac{4Gp}{c^3\rho_0}, \tag{3a}$$

which, for a photon, is

$$\cot\alpha + \cot\beta = \frac{4l_{\rm Pl}^2\nu}{c\rho_0}. \tag{3b}$$

This is illustrated (for $x = 0$) in fig. 3, where the angles α and β are defined.

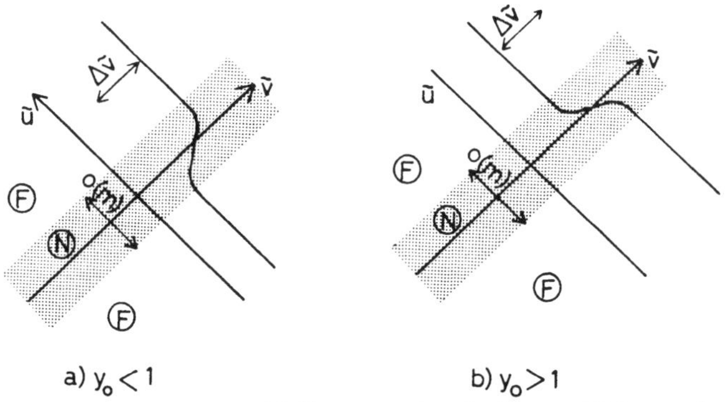

Fig. 2. The path of a null geodesic in the $(\hat{u}, \hat{v})$ plane as described by eq. (2) for $m \ll 1$, $\rho_0 \gg m$, and (a) $\rho_0 < 1$, (b) $\rho_0 > 1$. The near region N and the far region F, as well as the shift $\Delta\hat{v}$, are indicated.

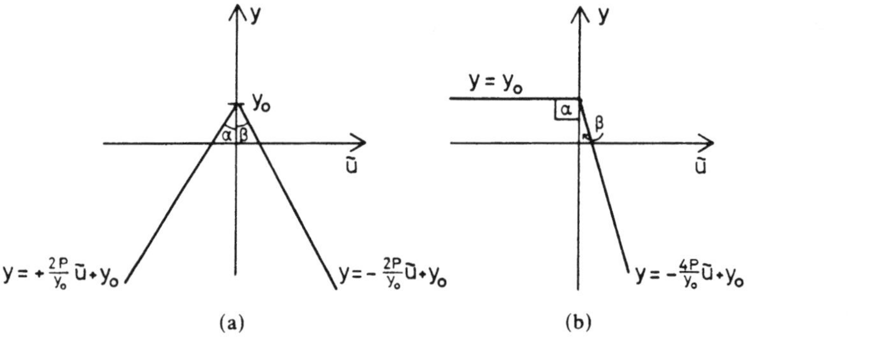

Fig. 3. The "spatial refraction" of null geodesics as described by eq. (3) for the two special cases (a) $\alpha = \beta$, and (b) $\alpha = \frac{1}{2}\pi$.

Eqs. (2) and (3) are the central results for these shock waves and describe physical effects which should also occur in more general situations. Note that if the shift (2) were *constant* it could be removed by a coordinate translation and would therefore not be physically observable (cf. the discussion after eq. (2)). Also, a shift *linear* in the transverse distance ρ would not be observable since it could be removed by a Lorentz rotation of one of the flat half-spaces with respect to the other. However the shift (2) is *logarithmic* in ρ and leads to physically observable effects. The *relative* shift for nearby observers goes as the first derivative $(1/\rho)$ while the relative refraction goes as the second derivative $(1/\rho^2)$ of the shift*. See fig. 4.

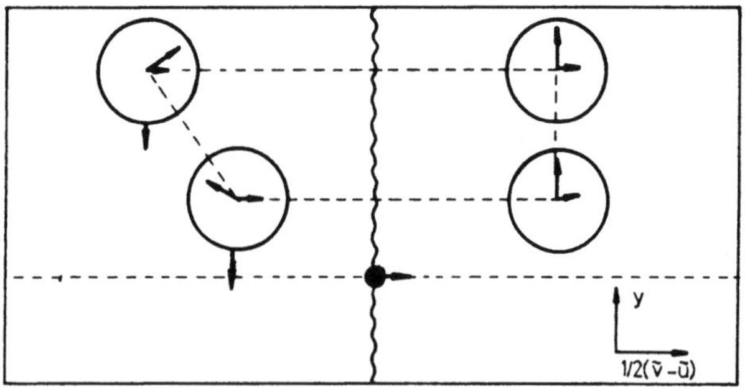

Fig. 4. Four synchronized clocks were originally situated at rest at the corners of a rectangle. A fast particle approaches from the left. The situation is shown when the shock wave has passed two of the clocks. The one closest to the trajectory of the particles has been shifted to the right with respect to the other; its clock now runs behind the other. They are also moving towards the trajectory of the fast particle at different speeds (arrows). Only their *relative* velocity, which is always away from each other, is locally observable.

* Note that a *local* observer can only detect the second derivative $(1/\rho^2)$ of the shift.

3. General result

Consider a solution of the vacuum Einstein field equations of the form

$$d\hat{s}^2 = 2A(u, v)\,du\,dv + g(u, v)h_{ij}(x^i)\,dx^i\,dx^j\,. \tag{4}$$

Under what conditions can we introduce a shift in v at $u = 0$ so that the resulting space-time solves the field equation with a photon at the origin $\rho = 0$ of the (x^i) 2-surface and $u = 0$? As shown in appendix B the answer is that at $u = 0$ we must have

$$A_{,v} = 0 = g_{,v}\,,$$
$$\frac{A}{g}\Delta f - \frac{g_{,uv}}{g} f = 32\pi p A^2 \delta(\rho)\,, \tag{5}$$

where $f = f(x^i)$ represents the shift in v, Δf is the laplacian of f with respect to the 2-metric h_{ij}, and the resulting metric is described by (B.2) or (B.4). Eqs. (5) represent our main result. We now turn to specific examples.

For a plane wave due to a photon in Minkowski space we have

$$d\hat{s}^2 = -du\,dv + dx^2 + dy^2\,, \tag{6a}$$

and thus

$$A = -\tfrac{1}{2}\,,$$
$$g = 1\,. \tag{6b}$$

The conditions on the metric are trivially satisfied, and the condition on the shift f is

$$\Delta f = -16\pi p \delta(\rho)\,, \tag{7}$$

where $\rho^2 = x^2 + y^2$. The solution of this equation, unique up to solutions of the homogeneous equation, is

$$f = -4p \ln \rho^2\,, \tag{8}$$

which agrees precisely with Aichelburg and Sexl [1] (cf. eqs. (2) and (A.26)).

For a *sourceless* plane wave in Minkowski space we set $p = 0$ to obtain

$$\Delta f = 0\,, \tag{9}$$

which agrees with Penrose [2].

For a *spherical* wave in Minkowski space we write the metric in the form

$$d\hat{s}^2 = -du\,dv + \tfrac{1}{4}(v-u)^2(d\theta^2 + \sin^2\theta\,d\varphi^2)\,, \tag{10a}$$

so that

$$A = -\tfrac{1}{2}\,,$$
$$g = r^2 = \tfrac{1}{4}(v-u)^2. \tag{10b}$$

But the derivatives of g are not identically zero at $u=0$. Thus, there are *no* spherical waves (of this form) in Minkowski space.

Physically this might seem mysterious because one expects spherical shock waves to arise in, e.g., the debris of a violent explosion. On closer inspection one concludes that there must be non-zero curvature behind such shock waves. However, note that Penrose [2] does exhibit the existence of sourceless spherical shock waves in Minkowski space but having a different form than our ansatz (eq. (B.2)).

We now turn to a more interesting example, namely the Schwarzschild metric which in (null) Kruskal-Szekeres coordinates takes the form

$$d\hat{s}^2 = -\frac{32m^3}{r}\,e^{-r/2m}\,du\,dv + r^2(d\theta^2 + \sin^2\theta\,d\varphi^2)\,, \tag{11a}$$

so that

$$A = -\frac{16m^3}{r}\,e^{-r/2m}$$

$$g = r^2\,. \tag{11b}$$

r is given implicitly as a function of u and v by

$$uv = -\left(\frac{r}{2m} - 1\right)e^{r/2m}\,, \tag{11c}$$

so that all v-derivatives of r are proportional to u. Thus, the conditions on the metric coefficients A and g are satisfied at $u=0$.

Furthermore, since $g_{,uv} \equiv A$ the condition on f becomes

$$\begin{aligned}\Delta f - f &= 32\pi p g A|_{u=0}\delta(\theta)\\ &= -2\pi\kappa\delta(\theta)\,,\end{aligned} \tag{12}$$

where $\kappa = 2^9 m^4 p\,e^{-1}$ and where we have arranged the coordinates so that the photon is at $\theta = 0 = u$.

We now solve eq. (12) by expanding f in terms of spherical harmonics $Y_{lm}(\theta, \varphi)$. We see immediately that only spherical harmonics with $m=0$ contribute; expressing these in terms of the Legendre polynomials $P_l(x)$ leads to

$$f = \kappa \sum_l \frac{l+\frac{1}{2}}{l(l+1)+1}\,P_l(\cos\theta)\,. \tag{13}$$

We can obtain an integral expression for f by using the generating function for the Legendre polynomials, namely

$$\sum_{l=0}^{\infty} P_l(x)t^l = (1 - 2xt + t^2)^{-1/2}\,, \tag{14a}$$

and the fact that

$$\int_{-\infty}^{0} t^{l}\, e^{s/2} \cos\left(\tfrac{1}{2}\sqrt{3}s\right) ds = \frac{l+\frac{1}{2}}{l(l+1)+1}, \tag{14b}$$

where $t = e^{s}$, to finally obtain

$$f = \kappa \int_{0}^{\infty} \frac{\sqrt{\frac{1}{2}} \cos\left(\frac{1}{2}\sqrt{3}s\right)}{(\cosh s - \cos\theta)^{1/2}} ds. \tag{15}$$

We have not attempted to perform the integration explicitly. We note that the homogeneous equation (eq. (12) with $p = 0$) has no solution. In the limit of small θ eq. (15) in appropriate coordinates reduces to eq. (8), with a well-determined value of the integration constant.

4. Discussion

The surprisingly simple geometric shape of a gravitational shock wave of massless particles in flat space can help us obtain a better understanding of gravitational interactions among particles at extreme energies. It is easy to argue that at extremely high energies interactions due to this shock wave will dominate over all quantum field theoretic interactions, simply because the latter will be postponed by an infinite time shift (due to the logarithmic singularitity in eq. (2), see fig. 4). This implies that cross sections at such energies will be entirely predictable.

A problem arises if *two* such particles are considered, both accompanied by their shock waves, that meet and collide. The result of such a collision will be curved shock waves which obey the vacuum Einstein field equations only if space-time after the collision in the region between both shock waves is curved, so that we then have to deal with the full complexity of general relativity. We have here a limiting case of the general problem of black hole encounters which has been studied in detail by D'Eath [4] and Curtis [5].

On physical grounds the Schwarzschild result, eq. (13), should not be surprising. The flat space result (e.g. eq. (8)) can be obtained [1] by infinitely boosting a (massive) source particle. Now take an $r = \text{const}$ observer in the usual Schwarzschild coordinates. Put a (nearly) massless particle at the horizon and wait. The observer will see a particle with an increasingly large boost! It is only natural to expect a similar result in both cases. The spherical nature of the wave in the Schwarzschild case (as opposed to the plane wave in Minkowski space) is merely a reflection of the spherically symmetric nature of the "boost" relating an $r = \text{const}$ observer to Kruskal-Szekeres coordinates. Physically, one expects any (weak) plane wave approaching the black hole to become gradually more spherical, as seen by an outside observer, as it comes closer to the horizon.

Returning to the picture of a particle of small mass falling into the black hole (see discussion after fig. 1) one expects a small increase of the Schwarzschild radius of the black hole, together with a slight expansion of its furture horizon. This expansion then grows exponentially with the Schwarzschild time coordinate. This is what our "shock wave" here actually describes. Eq. (15) is in closed form the extent of the horizon expansion. One might speculate what effect this expansion has on the quantum nature of the vacuum and, in particular, Hawking radiation. We believe that the gravitational interaction between infalling matter and Hawking radiation, crucial for a deeper understanding of the quantum properties of black holes themselves [6], should be described using our expression for the horizon expansion.

Finally, we are aware of the analogy with the electric field of a charged particle moving at the speed of light, which is similar to the gravitational field described here. Our gravitational shock wave can be compared to a limiting case of Cherenkov radiation.

We thank Paul Shellard for bringing the work of D'Eath [4] to our attention, which then led us to the previous work of Aichelburg and Sexl [1] and Penrose [2].

The computer calculations of the Ricci tenser were performed while one of us (T.D.) was a visitor at Queen Mary College, London. He is deeply indebted to Malcolm MacCallum and Gordon Joly for hospitality and assistance.

Appendix A

A PHOTON IN MINKOWSKI SPACE

Consider the linearized field of a point mass in Lorentz gauge:

$$ds^2 = -\left(1 - \frac{2m}{R}\right) dT^2 + \left(1 + \frac{2m}{R}\right)(dx^2 + dy^2 + dZ^2), \tag{A.1}$$

with $m \ll R$. This is the field of the particle as seen in its rest frame. Boost this rest frame with respect to coordinates (t, x, y, z) via

$$\begin{aligned} T &= t \cosh \beta - z \sinh \beta, \\ Z &= -t \sinh \beta + z \cosh \beta, \end{aligned} \tag{A.2}$$

and simultaneously set

$$m = 2p\, e^{-\beta} \tag{A.3}$$

for some constant $p > 0$.

Introduce null coordinates

$$\begin{aligned} u &= t - z, \\ v &= t + z. \end{aligned} \tag{A.4}$$

The momentum of the particle is

$$p^a = m[(\cosh\beta)\delta_t^a + (\sinh\beta)\delta_z^a], \tag{A.5}$$

and thus

$$\lim_{\beta\to\infty} p^a = p(\delta_t^a + \delta_z^a) = 2p\delta_v^a\,; \tag{A.6}$$

in the limit the particle is massless and moves (at the speed of light) in the v-direction; p is its momentum and is kept finite (possibly large).

Writing the metric in (u, v, x, y) coordinates we obtain

$$\mathrm{d}s^2 = \left(1+\frac{2m}{R}\right)[-\mathrm{d}u\,\mathrm{d}v + \mathrm{d}x^2 + \mathrm{d}y^2] + \frac{4m}{R}\left[\frac{p}{m}\mathrm{d}u + \frac{m}{4p}\mathrm{d}v\right]^2, \tag{A.7}$$

with

$$R^2 = x^2 + y^2 + \left(\frac{p}{m}u - \frac{m}{4p}v\right)^2. \tag{A.8}$$

The key idea is to notice that

$$\lim_{\substack{m\to 0\\ (u\neq 0,v,x,y)\ \text{fixed}}} \mathrm{d}s^2 = -\mathrm{d}u\left(\mathrm{d}v - 4p\frac{\mathrm{d}u}{|u|}\right) + \mathrm{d}x^2 + \mathrm{d}y^2, \tag{A.9}$$

which *is* flat although the coordinate v' satisfying

$$\mathrm{d}v' = \mathrm{d}v - \frac{4p\,\mathrm{d}u}{|u|} \tag{A.10}$$

suffers a discontinuity at $u = 0$ due to the absolute value sign. To make this somewhat more precise, introduce coordinates $(\hat{u}, \hat{v}, x, y)$ by*

$$\hat{u} = u + \frac{m^2 Z \ln(2R)}{pR},$$

$$\hat{v} = v + \frac{4pZ \ln(2R)}{R}. \tag{A.11}$$

Note that $(\hat{u}, \hat{v}, x, y)$ is obtained from (u, v, x, y) by adding $(4Z \ln(2R)/R)p^a$. Then

$$R^2 = x^2 + y^2 + \left(\frac{p}{m}\hat{u} - \frac{m}{4p}\hat{v}\right)^2, \tag{A.12}$$

$$\lim_{\substack{m\to 0\\ (u\neq 0,v,x,y)\ \text{fixed}}} \mathrm{d}s^2 = -\mathrm{d}\hat{u}\,\mathrm{d}\hat{v} + \mathrm{d}x^2 + \mathrm{d}y^2. \tag{A.13}$$

* The motivation for these coordinates is as follows: in the limit, Z/R acts like a θ-function and reproduces the effect of the absolute value sign, while $\mathrm{d}R/R = \mathrm{d}u/u$. Furthermore, $\ln R$ is finite at $u = 0$. The factor 2 is chosen for convenience. We use geometric units in which $G = c = \hbar = 1$; *all* quantities are dimensionless.

The metric (A.13) is flat. It remains to investigate its behavior near $u=0$, which, in the limit, is just $\hat{u}=0$. To do this we consider the behavior of (null) geodesics crossing $u=0$ for $m\neq 0$ and then take the limit as m goes to zero*. We do this both in a "near" region, which collapses to $\hat{u}=0$ in the limit - this is just the rest frame of the particle - and a "far" region, where $\hat{u}$ remains non-zero in the limit.

The (linearized) geodesics of the metric (1) are given by

$$\dot{T}=E\left(1+\frac{2m}{R}\right),$$

$$y\dot{Z}-Z\dot{y}=L\left(1-\frac{2m}{R}\right),$$

$$\dot{y}^2+\dot{Z}^2=-M^2\left(1-\frac{2m}{R}\right)+E^2, \tag{A.14}$$

where the dot denotes derivatives with respect to the affine parameter λ along the geodesic. We have assumed $x=0$ without loss of generality; the constants E, L, M denote the energy, angular momentum, and rest mass of the test particle, respectively.

In what follows we consider only "null" geodescis; i.e. we set $M=O(m^2)$. Expanding y, Z and T in powers of m and considering only the terms linear in m we have

$$y=y_0+my_1,$$

$$Z=Z_0+mZ_1,$$

$$T=T_0+mT_1, \tag{A.15}$$

and eqs. (A.14) now become

$$\dot{T}_0=E,$$

$$\dot{y}_0^2+\dot{Z}_0^2=E^2,$$

$$y_0\dot{Z}_0-Z_0\dot{y}_0=L,$$

$$\dot{T}_1=2E/R_0,$$

$$\dot{y}_0\dot{y}_1+\dot{Z}_0\dot{Z}_1=0,$$

$$y_0\dot{Z}_1-Z_1\dot{y}_0+y_1\dot{Z}_0-Z_0\dot{y}_1=-\frac{2L}{R_0}, \tag{A.16}$$

where $R_0^2=y_0^2+Z_0^2$.

* Penrose and MacCallum [7] describe some properties of such geodesics without actually calculating them. Penrose and Curtis have performed similar calculations of null geodesics crossing a shock wave, but as far as we know these have not been published [8].

However, since

$$u=\frac{m}{2p}(T-Z)\,,$$
$$v=\frac{2p}{m}(T+Z)\,, \tag{A.17}$$

we must require

$$\dot{Z}_0=-\dot{T}_0\equiv -E \tag{A.18}$$

if $\dot{v}$, and thus $\dot{\hat{v}}$, is to remain finite in the limit as m goes to zero. The second and third of eqs. (A.16) now yield

$$\dot{y}_0=0\,,$$
$$y_0=-L/E\,, \tag{A.19}$$

and the fifth of eqs. (A.16) implies

$$\dot{Z}_1=0\,. \tag{A.20}$$

Thus

$$\dot{u}=\frac{m}{p}E+\frac{m^2}{p}\frac{E}{R_0}\,,$$
$$\dot{v}=4p\frac{E}{R_0}\,. \tag{A.21}$$

Using eq. (A.18) these can be integrated directly to give

$$u=\frac{mE}{p}\lambda-\frac{m^2}{p}\ln(Z_0+R_0)\,,$$
$$v=-4p\ln(Z_0+R_0)\,, \tag{A.22}$$

where we have ignored an irrelevant integration constant, and thus

$$\hat{u}=\frac{mE}{p}\lambda+\frac{m^2}{p}\left[\frac{Z_0\ln(2R_0)}{R_0}-\ln(Z_0+R_0)\right],$$
$$\hat{v}=4p\left[\frac{Z_0\ln(2R_0)}{R_0}-\ln(Z_0+R_0)\right]. \tag{A.23}$$

We now separate into a near region N and a far region F as follows:

$$\mathrm{N}=\{|\lambda|<1/\sqrt{m}\}\,,$$
$$\mathrm{F}=\{\sqrt{m}\leq m|\lambda|<\infty\}\,. \tag{A.24}$$

Note further that

$$\lim_{\lambda\to-\infty} \hat{v}=0,$$

$$\lim_{\lambda\to+\infty} \hat{v}=-4p\ln y_0^2,$$

$$\lim_{\lambda\to\pm\infty} \hat{u}=\frac{mE}{p}\lambda. \tag{A.25}$$

Thus, there is a total shift in $\hat{v}$ given by

$$\Delta\hat{v}=-4p\ln y_0^2. \tag{A.26}$$

Note that, in the limit as m goes to zero, λ is infinite everywhere in F. Furthermore, in this limit $\hat{u}$ is identically zero in N, whereas $\hat{u}$ is a good affine parameter in F along the geodesic.

The shift (A.26) thus occurs, for small m, "essentially" only in N! Thus, in the limit as m goes to zero, the shift (A.26) occurs at $\hat{u}=0$ and represents a finite discontinuity in $\hat{v}$ along null geodesics! This can also be seen by calculating

$$\lim_{\lambda\to\pm\infty} \dot{\hat{v}}=0, \tag{A.27}$$

thus showing that in the limit as m goes to zero $\hat{v}$ is constant in F, i.e. for non-zero $\hat{u}$. This is just a reflection of the fact that, in the limit, F is flat. This is illustrated in fig. 2.

We now turn to the behaviour of y. We must solve the last of eqs. (A.16), which, on inserting eqs. (A.19) and (A.20) becomes

$$y_1\dot{Z}_0-Z_0\dot{y}_1=-2L/R_0. \tag{A.28}$$

The homogeneous equation clearly has the solution

$$y_1^h=AZ_0 \tag{A.29}$$

for any constant A; it remains to find a particular solution.

Multiplying eq. (A.28) by $\dot{Z}_0$ yields

$$E^2y_1-R_0\dot{R}_0\dot{y}_1=2LE/R_0. \tag{A.30}$$

But noticing that

$$\dot{R}_0^2=\frac{Z_0^2E^2}{R_0^2}\equiv E^2\left(1-\frac{y_0^2}{R_0^2}\right) \tag{A.31}$$

suggests an ansatz for y_1 as a power series in R_0. We thus obtain the particular solution

$$y_1^p=\frac{2L}{y_0^2E}\dot{R}_0\equiv-\frac{2R_0}{y_0}. \tag{A.32}$$

The general solution to eq. (A.28) is thus

$$y_1 = -\frac{2R_0}{y_0} + AZ_0 , \tag{A.33}$$

and therefore

$$y = -\frac{L}{E} + m\left[-\frac{2R_0}{y_0} + AZ_0 \right] . \tag{A.34}$$

We are interested in the behaviour of y in the far field F for m small. We obtain

$$\lim_{\substack{m\to 0\\ \hat{u}\neq 0}} \frac{\partial y}{\partial \hat{u}} = -\frac{2p}{y_0} \operatorname{sgn} \hat{u} - pA . \tag{A.35}$$

This behaviour is illustrated in fig. 3. In general we have

$$\cot \alpha + \cot \beta = \frac{4p}{y_0} \tag{A.36}$$

for the angles α and β as defined in fig. 3.

At this point several comments are in order. We have *not* considered all geodesics which cross the shock wave, but only a sufficient number to determine how to glue the two flat half-space together. That this is sufficient follows from the existence of coordinates in which the metric is in fact continuous (see appendix B).

Using the results of appendix B we see that

$$\begin{aligned}\lim_{m\to 0} \mathrm{d}s^2 &= -\mathrm{d}\hat{u}(\mathrm{d}\hat{v} + 4p \ln y_0^2\, \delta(\hat{u})\, \mathrm{d}\hat{u}) + \mathrm{d}x^2 + \mathrm{d}y^2 \\ &= -\mathrm{d}u\left(\mathrm{d}v + \frac{4p\, \mathrm{d}u}{u}(1 - 2\theta(u)) + 4p \ln 4\, \delta(u)\, \mathrm{d}u \right) + \mathrm{d}x^2 + \mathrm{d}y^2 ,\end{aligned} \tag{A.37}$$

which of course reduce to (A.13) and (A.9) respectively for $u \neq 0$. The first of (A.37) is just the result of Aichelburg and Sexl [1] (their eq. (3.9)), but the second of (A.37) disagrees with their eq. (3.10). Although this is at first disconcerting, a more careful analysis reveals the source of the discrepancy: we have taken a limit different from theirs. This can be seen by noting that for $m \to 0$, $u \neq 0$, our original coordinates (t, z) are related to their coordinates $(\bar{t}, \bar{x})$ by *infinite* scale factors.

Equivalently, note that the original Minkowski space given in (u, v, x, y) coordinates is "pushed to infinity" in the resulting space-time given in $(\hat{u}, \hat{v}, x, y)$ coordinates. Specifically, $\{u \neq 0; |v| < \infty\}$ corresponds to $(\hat{u} \neq 0;\ \hat{v} = -(\operatorname{sgn} \hat{u})\, \infty\}$, although the source located at $\{u = 0; |v| < \infty\}$, corresponds to $\{\hat{u} = 0; |\hat{v}| < \infty\}$. The corresponding statement for the $(\bar{t}, \bar{x}, \bar{y}, \bar{z})$ coordinates of Aichelburg and Sexl [1] would be somewhat different.

Finally, note that although we have linearized both the metric (A.1) and the geodesics (A.14) the result is in fact exact. Had we begun (as in [1]) with the *exact*

Schwarzschild metric in isotropic coordinates and expanded in powers of m only the linear terms we consider would have survived.

Appendix B

CALCULATION OF THE RICCI TENSOR

We start with the metric

$$d\hat{s}^2 = 2A(u, v)\, du\, dv + g(u, v) h_{ij}(x^i)\, dx^i\, dx^j , \tag{B.1}$$

which is assumed to satisfy the Einstein vacuum equations. We introduce a shock wave by keeping (B.1) for $u < 0$ but replacing v by $v + f(x^i)$ for $u > 0$*:

$$ds^2 = 2A(u, v + \theta f)\, du(dv + \theta f_{,i}\, dx^i) + g(u, v + \theta f) h_{ij}\, dx^i\, dx^j , \tag{B.2}$$

where $\theta = \theta(u)$ is the usual step function. Changing to coordinates $(\hat{u}, \hat{v}, \hat{x}^i)$ defined by

$$\begin{aligned} \hat{u} &= u , \\ \hat{v} &= v + \theta f , \\ \hat{x}^i &= x^i , \end{aligned} \tag{B.3}$$

we obtain

$$ds^2 = 2A(\hat{u}, \hat{v})\, d\hat{u}(d\hat{v} - \delta(\hat{u}) f\, d\hat{u}) + g(\hat{u}, \hat{v}) h_{ij}\, d\hat{x}^i\, d\hat{x}^j , \tag{B.4}$$

where $\delta = \delta(u)$ is the Dirac delta "function".

We note that the metric ds^2 given in (B.2) and (B.4) is in fact *continuous*, i.e. there *exist* coordinates $(\bar{u}, \bar{v}, \bar{x}^i)$ such that the metric coefficients are continuous. A possible choice is given implicitly by

$$\begin{aligned} \hat{u} &= \bar{u} , \\ \hat{v} &= \bar{v} + \theta \bar{f} - \tfrac{1}{2} \bar{u} \theta^2 \frac{\hat{A}}{\hat{g}} h^{mn} \bar{f}_{,m} \bar{f}_{,n} , \\ \hat{x}^i &= \bar{x}^i - u\theta \frac{\hat{A}}{\hat{g}} h^{im} \bar{f}_{,m} , \end{aligned} \tag{B.5}$$

where

$$\begin{aligned} \bar{f} &= f(\bar{x}^i) , \\ \hat{A} &= A(\hat{u}, \hat{v}) , \\ \hat{g} &= g(\hat{u}, \hat{v}) , \\ h^{ij} &= h^{ij}(x^i) .. \end{aligned} \tag{B.6}$$

* This is not quite the standard ansatz $g_{ab} = (1-\theta) g^-_{ab} + \theta g^+_{ab}$. However, in the examples considered here the corresponding Ricci tensors differ at worst by a term in R_{uu} proportional to $\theta(1-\theta)$, which is not physically relevant.

However, the coordinates (B.5) are extremely unwieldy both for mathematical computations and for preserving physical intuition. We will thus proceed as follows. Noting that the coordinate transformation between $(\bar{u}, \bar{v}, \bar{x}^i)$ and (u, v, x^i) coordinates is continuous, we will calculate the Ricci tensor for the metric (B.2). However since the metric (B.4) is much easier to work with, we will *formally* transform to $(\hat{u}, \hat{v}, \hat{x}^i)$ coordinates, calculate $R_{\hat{a}\hat{b}}$, and then transform back to obtain R_{ab}. Direct calculation yields*

$$R_{\hat{u}\hat{\imath}} = -\frac{\hat{A}_{,\hat{v}}}{\hat{A}} f_{,i}\delta\,,$$

$$R_{\hat{v}\hat{\imath}} = 0\,,$$

$$R_{\hat{\imath}\hat{\jmath}} = R^{(2)}_{ij} - h_{ij}\left[\frac{\hat{g}_{,\hat{u}\hat{v}}}{\hat{A}} + \frac{\hat{g}_{,\hat{v}\hat{v}}}{\hat{A}} f\delta\right],$$

$$R_{\hat{u}\hat{v}} = \left(\frac{\hat{A}_{,\hat{u}}\hat{A}_{,\hat{v}}}{\hat{A}^2} - \frac{\hat{A}_{,\hat{u}\hat{v}}}{\hat{A}} + \frac{1}{2}\frac{\hat{g}_{,\hat{u}}\hat{g}_{,\hat{v}}}{\hat{g}^2} - \frac{\hat{g}_{,\hat{u}\hat{v}}}{\hat{g}}\right) + \left(\frac{\hat{A}^2_{,\hat{v}}}{\hat{A}^2} - \frac{\hat{A}_{,\hat{v}\hat{v}}}{\hat{A}} - \frac{\hat{g}_{,\hat{v}}\hat{A}_{,\hat{v}}}{\hat{g}\hat{A}}\right) f\delta\,,$$

$$R_{\hat{v}\hat{v}} = -\frac{\hat{g}_{,\hat{v}\hat{v}}}{\hat{g}} + \frac{1}{2}\frac{\hat{g}^2_{,\hat{v}}}{\hat{g}^2} + \frac{\hat{g}_{,\hat{v}}\hat{A}_{,\hat{v}}}{\hat{g}\hat{A}}\,,$$

$$R_{\hat{u}\hat{u}} = \left(-\frac{\hat{g}_{,\hat{u}\hat{u}}}{\hat{g}} + \frac{1}{2}\frac{\hat{g}^2_{,\hat{u}}}{\hat{g}^2} + \frac{\hat{g}_{,\hat{u}}\hat{A}_{,\hat{u}}}{\hat{g}\hat{A}}\right) + \left(\frac{2\hat{A}_{,\hat{u}\hat{v}}}{\hat{A}} - \frac{2\hat{A}_{,\hat{u}}\hat{A}_{,\hat{v}}}{\hat{A}^2} + \frac{\hat{g}_{,\hat{v}}\hat{A}_{,\hat{u}}}{\hat{g}\hat{A}} + \frac{\hat{g}_{,\hat{u}}\hat{A}_{,\hat{v}}}{\hat{g}\hat{A}}\right) f\delta$$
$$+ 2\left(\frac{\hat{A}_{,\hat{v}\hat{v}}}{\hat{A}} - \frac{\hat{A}^2_{,\hat{v}}}{\hat{A}^2} + \frac{\hat{g}_{,\hat{v}}\hat{A}_{,\hat{v}}}{\hat{g}\hat{A}}\right) f^2\delta^2 + \frac{\hat{A}}{\hat{g}}\Delta f\delta - \frac{\hat{g}_{,\hat{v}}}{\hat{g}} f\delta'\,, \tag{B.7}$$

where $R^{(2)}_{ij}$ is the Ricci tensor derived from h_{ij}, Δ is the Laplace operator associated with h_{ij}, and $\delta = \delta(\hat{u})$.

Blithely ignoring the δ^2 term we transform to (u, v, x^i) coordinates and insert the vacuum equations (obtained by setting $f = 0$) to get

$$R_{vv} = R_{\hat{v}\hat{v}} = 0\,,$$

$$R_{vi} = R_{\hat{v}\hat{\imath}} = 0\,,$$

$$R_{ij} = R_{\hat{\imath}\hat{\jmath}} = -h_{ij}\frac{\hat{g}_{,\hat{v}\hat{v}}}{\hat{A}} f\delta\,,$$

$$R_{uv} = R_{\hat{u}\hat{v}} = \left(\frac{\hat{A}^2_{,\hat{v}}}{\hat{A}^2} - \frac{\hat{A}_{,\hat{v}\hat{v}}}{\hat{A}} - \frac{\hat{g}_{,\hat{v}}\hat{A}_{,\hat{v}}}{\hat{g}\hat{A}}\right) f\delta\,,$$

$$R_{ui} = R_{\hat{u}\hat{\imath}} + R_{\hat{u}\hat{v}}\theta f_{,i}$$
$$= f_{,i}\delta\left[-\frac{\hat{A}_{,\hat{v}}}{\hat{A}} + \theta f\left(\frac{\hat{A}^2_{,\hat{v}}}{\hat{A}^2} - \frac{\hat{A}_{,\hat{v}\hat{v}}}{\hat{A}} - \frac{\hat{g}_{,\hat{v}}\hat{A}_{,\hat{v}}}{\hat{g}\hat{A}}\right)\right],$$

* We note that Taub [9] has given a systematic presentation of space-times with distribution-valued curvature tensors.

$$R_{uu} = R_{\hat{u}\hat{u}} + 2R_{\hat{u}\hat{v}} f\delta$$

$$= \frac{\hat{A}}{\hat{g}} \Delta f\delta - \frac{\hat{g}_{,\hat{v}}}{\hat{g}} f\delta' + \left(\frac{\hat{g}_{,\hat{u}}\hat{g}_{,\hat{v}}}{\hat{g}^2} - \frac{2\hat{g}_{,\hat{u}\hat{v}}}{\hat{g}} + \frac{\hat{g}_{,\hat{v}}\hat{A}_{,\hat{u}}}{\hat{g}\hat{A}} + \frac{\hat{g}_{,\hat{u}}\hat{A}_{,\hat{v}}}{\hat{g}\hat{A}} \right) f\delta . \tag{B.8}$$

The stress-energy tensor for a massless particle located at the origin $\rho = 0$ of the (x^i) 2-surface and at $u = 0$ is

$$T^{ab} = 4p\delta(\rho)\delta(u)\delta^a_v\delta^b_v , \tag{B.9}$$

where p is the momentum of the particle. Thus, the only non-zero component is

$$T_{uu} = 4pA^2\delta(\rho)\delta(u) . \tag{B.10}$$

Inserting (B.8) and (B.10) into the Einstein field equations, partially integrating the δ' term, noting that e.g. $\hat{A}_{,\hat{v}}(\hat{u} = 0) = 0 \Leftrightarrow A_{,v}(u = 0) = 0$ yields precisely eqs. (5).

The calculation above was first done by hand and then checked using the algebraic manipulation computer system SHEEP. As a further check on the validity of working in the singular coordinates $(\hat{u}, \hat{v}, \hat{x}^i)$ (eq. (B.4)) SHEEP was also used to calculate the Ricci tensor directly in (u, v, x^i) coordinates (eq. (B.8)), thus checking the original calculation of 't Hooft for the Schwarzschild case. The same answer, namely eqs. (5), was of course obtained in all cases.

References

[1] P.C. Aichelburg and R.U. Sexl, J. Gen. Rel. Grav. 2 (1971) 303.

[2] R. Penrose, in General relativity: papers in honour of J.L. Synge, ed. L. O'Raifeartaigh (Clarendon, Oxford, 1972) 101;
in Battelle rencontres, ed. C.M. De Witt and J.A. Wheeler (Benjamin, New York, 1968) 198; in Differential geometry and relativity, ed. M. Cahen and M. Flato (Reidel, Dordrecht, 1976) 271

[3] J. Ehlers and W. Kundt, in Gravitation: an introduction to current research, ed. L. Witten (Wiley, New York, 1962) 85f;
H.W. Brinkman, Proc. Natl. Acad. Sci. (US) 9 (1923) 1

[4] P.D. D'Eath, Phys. Rev. D18 (1978) 990

[5] G.E. Curtis, J. Gen. Rel. Grav. 9 (1978) 987, 999

[6] G. 't Hooft, J. Geom. Phys. 1 (1984) 45

[7] R. Penrose and M.A.H. MacCallum, Phys. Reports 6 (1973) 270

[8] P.D. D'Eath, private communication

[9] A.H. Taub, J. Math. Phys. 21 (1980) 1423

CHAPTER 8.5

S-MATRIX THEORY FOR BLACK HOLES

G. 't Hooft
Institute for Theoretical Physics
Princetonplein 5, P.O. Box 80.006
3508 TA Utrecht, The Netherlands

ABSTRACT

We explain the principles of the laws of physics that we believe to be applicable for the quantum theory of black holes. In particular, black hole formation and evolution should be described in terms of a scattering matrix. This way black holes at the Planck scale become indistinguishable from other particles. This *S*-matrix can be derived from known laws of physics. Arguments are put forward in favor of a discrete algebra generating the Hilbert space of a black hole with its surrounding space-time including surrounding particles.

1. INTRODUCTION

There is quite a bit of controversy (and confusion) regarding the nature of physical law governing a black hole. Some of the difficulties have their origin in the deceptively clean picture given by the "classical" (here this means "non-quantum mechanical") solutions of Einstein's equations of gravity in the case of gravitational collapse. The metric tensor describing the fabric of space-time appears to be smooth and well-behaved in the vicinity of a region we call the "horizon", a surface beyond which there are space-time points from which no information can reach the outside world. It seems that one should be able to apply standard techniques from particle theory here to derive what a distant observer can perceive and, naturally, this exercise has been done[1].

There is one innocent-looking assumption that most practitioners then make. One observes that clouds of particles may venture into the "forbidden region", from which they can no longer escape or even emit any signal towards the outside world, and so one *assumes* that the corresponding states in Hilbert space may be treated the way one always does in quantum mechanics: the unseen modes are averaged over. Operators describing observations in the outside world are assumed to be diagonal in the sector of Hilbert space that is not seen, and hence in all computations one is obliged to sum over all unseen modes.

The immediate consequence of this practice is that the outside world alone is not anymore described by a single wave function but by a density matrix[2]. Even if one starts with a "pure" wave function, sooner or later one finds the system to be in a mixed state. It is as if part

New Symmetry Principles in Quantum Field Theory, Edited by
J. Fröhlich et al., Plenum Press, New York, 1992

of the wave functions "disappeared into the wormhole"; information escaped, as if the system were linked to a heat bath.

With large black holes one can perform Gedanken experiments, and consider observers who move semiclassically in the neighborhood of a black hole horizon. If we attach some sense of "reality" to these observers the correctness of the above assumptions seems to be an inescapable conclusion.

But what if the hole is small, so that classical observers are too bulky to enter? Or let us ask a question that is probably equivalent to this: suppose one keeps track of *all* possible states a black hole can be in, is it then still impossible to describe the hole in terms of pure quantum states alone? Will the very tiny black holes evolve according to conventional evolution equations in quantum physics or is the loss of information a fundamental new feature, even for them?

There is a big problem with any theory in which the loss of "quantum information" is accepted as a fundamental item. This is the fact that all effective laws become fuzzy. It is not difficult to construct an example of a theory in which pure states evolve into mixed states. Consider a system with a Hamiltonian that depends on a free physical parameter α (for instance the fine structure constant). A state $|\psi\rangle_0$ at $t=0$ evolves into the state

$$|\psi\rangle_t = e^{-iH(\alpha)t}|\psi\rangle_0 \tag{1.1}$$

at time t. The expectation value for an operator $\mathcal{O}$ evolves into

$$\langle\mathcal{O}\rangle_{t,\alpha} = {}_0\langle\psi|e^{iH(\alpha)t}\mathcal{O}\, e^{-iH(\alpha)t}|\psi\rangle_0 \ , \tag{1.2}$$

and from the ψ dependence one can recover the information that the system remained in a pure quantum state. But now assume that there is an *uncertainty* in α. We only know the first few decimal places. There is a distribution of values for α, each with a probability $P(\alpha)$. Our theory now predicts for the "expectation value" of $\mathcal{O}$:

$$\langle\mathcal{O}\rangle_t = \int d\alpha\, \langle\mathcal{O}\rangle_{t,\alpha}\, P(\alpha) \equiv \mathrm{Tr}\, \rho(t)\mathcal{O} \ ; \tag{1.3}$$

$$\rho(t) = \int d\alpha\, P(\alpha)\, e^{-iH(\alpha)t}|\psi\rangle_0\, {}_0\langle\psi|e^{iH(\alpha)t} \ . \tag{1.4}$$

This is an impure density matrix, of the kind one obtains in doing calculations with black holes. The outcome of a by now standard calculation is a thermal distribution of outgoing particles. A thermal distribution is always a mixed state.

In our example we clearly see what the remedy is. The extra uncertainty had nothing to do with quantum mechanics; the Hamiltonian was not yet known because of our incomplete knowledge of the laws of physics (in this case the value of α). By doing extra experiments or by working harder on the theory we can establish a more precise value for α, and thus obtain a more precise prediction for $\langle\mathcal{O}\rangle_t$.

Returning with this wisdom to the black hole, what knowledge was incomplete? Here I think one has a situation that is common to all macroscopic systems: because of the large number of quantum mechanical states it was hopelessly difficult to follow the evolution of just one such state precisely. One was forced to apply thermodynamics. The outcome of our calculations with black holes got the form of thermodynamic expressions because of the impossibility, in practice, to follow in detail the evolution of any particular quantum state.

But this does mean that our basic understanding of black holes at present is incomplete. In a statistical system such as a vessel containing an ideal gas, we have *in principle* a quantum theory that is precise enough to study pure quantum states. In particular, if we dilute the gas so much that a single atom remains, the thermodynamic description will no longer be correct, and we must use the real quantum theory. Similarly, if we want to understand how a black hole behaves when it reaches the Planck mass, we expect the thermodynamic expressions to break down.

The importance of a good quantum mechanical description is that it would enable us to link black holes with ordinary particles. The Planck region may well be populated by a lot of different types of fundamental particles. their "high energy limit" will probably consist of particles small enough and heavy enough to possess a horizon and thus be indistinguishable from black holes. What we want is a consistent theory that covers all of this region. If we had a "conventional" Schrödinger equation in this region, it would be relatively straightforward (at least conceptually) to extrapolate to large distance scales using renormalization group techniques, and recover the "standard model" (or more!)

There have been many proposals concerning the nature of our physical world near the Planck length. we have seen "supergravity", "string theory", "heterotic strings", et cetera. My problem with these ideas is that they seem to be ad hoc. The models are "postulated" and then afterwards the authors try to argue why things have to be this way (basically the argument is that the new model is "more beautiful" than anything else known).

It would be a lot safer if we could *derive* the only possible correct setting of variables and forces, directly from the presently established laws of physics. In these lectures we will argue that it is possible to do this, or at least to make a good start, by doing Gedanken experiments with black holes. The reason why black holes should be used as a starting point in a theory of elementary particles is that *anything* that is tiny enough and heavy enough to be considered an entry in the spectrum of ultra heavy elementary particles (beyond the Planck mass), must be essentially a black hole.

Black holes are defined as solutions of the classical, i.e. unquantized, Einstein equations of General Relativity[3]. This implies that we only know how to describe them reliably when they are considerably bigger than the Planck length and heavier than the Planck mass. What was discovered by Hawking[1] in 1975 is that these objects radiate and therefore must decrease in size. It is obvious that they will sooner or later enter the domain that we presently do not understand.

Curiously, it is not easy to see why Hawking's derivation of the thermal black hole radiation would not be exactly correct. Even in a functional integral expression for this calculation one might still expect wormhole configurations through which quantum information leaks towards a mystical "other universe"[4]. We will now decide to be merciless: topologically non-trivial space-times are forbidden (until further notice) so that, at least at the microscopic level, pure quantum mechanics can be restored. More precisely, what we require is first of all some quantum mechanically pure evolution operator, and secondly that this operator be consistent with all we know of large scale physics, in particular general relativity.

At first sight these requirements are in conflict with each other. General relativity predicts unequivocally that gravitational collapse is possible, and this produces a horizon with all its difficulties. However, we claim that a pure quantum prediction that naturally blends into thermodynamic behavior in the large scale limit is not at all impossible, but it is true that the requirement for this to happen is extremely restrictive. Combining it with all we already know about large scale physics may well yield an unambiguous theory. Anyway, we know for sure that the amendments needed at the horizon all refer to Planck scale physics. As long as this physics is not completely understood it will also be impossible to refute our theories on the ground of inconsistencies with known physics.

So this is our program. We *assume* that, as for all spatially confined systems, there exists such a thing as a "scattering matrix". One then tries to reconcile this scattering matrix with the laws of physics already known. We will find that this scattering matrix, to some extent, can be derived. More precisely: *the exact quantum behavior at large distance scales* (the distance scales reached in present particle experiments) *can be derived uniquely.*

The problem is an apparent acausality. If we apply linearized quantum field theory in the black hole background it seems ununderstandable how information that is thrown into the black hole can reemerge as information in the outgoing states. This is because the outgoing radiation originates at $t = -\infty$ and the ingoing matter proceeds until $t = +\infty$, so the information had to go backwards in time. We simply claim that precisely for this reason linearized quantum field theory is inappropriate here. One *must* take gravitational (if not other) interactions between in- and outgoing matter into account. One way to interpret what happens then is to assume that there is a fundamental *symmetry principle,* because matter inside the horizon is unobservable. One can then perform a transformation that transforms away the singularity at $t=+\infty$, and produces one at $t=-\infty$.

It is of crucial importance to note that what we are deriving is not only the (quantum) behavior of the black hole itself. It is the entire system, black hole *plus* all surrounding particles, that we are talking about. Using our (assumed) knowledge of physics at large distance scales we derive the properties of the black hole *and all other forms of matter* at energies larger than the Planck energy.

In ordinary quantum field systems behavior at small distance, or equivalently, at high energies, determines the behavior at large distances and low energies. In the present case the interdependence goes both ways, or, in other words, the whole construction will be over-determined. We expect stringent constraints of consistency, which, one might hope, may lead to a single unique theory. The point is that the symmetry principle just mentioned affects matter in an essential way, and thus may perhaps continue to be of relevance at the low energy domain.

This is the motivation of this work. It may lead to "the unique theory". Even though our work is far from finished, we will be able to show that there will be a remarkable role for the old string theory[5]. The mathematical expressions we derive are so similar to those of string theory that perhaps some of its results will apply without any change. But both the physical interpretation and the derivations will be very different. As a consequence, the mathematics is not identical. One important difference is the string constant (determining the masses of the excitations), which in our case turns out to be imaginary[6].

In the usual string theory one uses the obvious requirements of unitarity and causality to derive that the string is governed by a local Lagrangean on the string world sheet. To derive similar requirements for the strings born from black holes is far from easy. This is presently what is holding us back from considerations such as tachyon elimination and anomaly cancellation that so successfully seem to have given us the superstring scenario. What we advertise is a careful though slow process establishing the correct demands for a full black hole/string theory. If successful, one will know exactly the rules of the game and the ways how to select good from false scenarios and models.

2. QUANTUM HAIR

Classical black holes are characterized by exactly three parameters[3]:

the *mass* M, the *angular momentum* L, and the *electric charge* Q. If magnetic monopoles exist in nature then there will be a fourth parameter, namely magnetic charge Q_m, and if besides electromagnetism there are other long range $U(1)$ gauge fields then also their charges correspond to parameters for the black hole.

However, the existence of long range $U(1)$ gauge fields other than electromagnetism seems to be rather unlikely. Then, since L, Q (and Q_m) are all quantized, the number of different values they can take is limited, and indeed one can argue convincingly (more about this in Ref[6]) that the black hole can be in much more different quantum states than the ones labeled by L and Q (and Q_m), or in other words, the mass M must be a function of much more variables than these quantum numbers alone.

An interesting attempt to formulate new quantum numbers for black holes was initiated by Preskill, Krauss, Wilczek and others[7]. They took as a model field theory a $U(1)$ gauge theory in which the local symmetry undergoes a Higgs mechanism *via* a Higgs field with charge Ne. In addition one postulates the presence of particles with charge e. In such a theory there exist vortices, much like the Abrikosov vortex in a super conductor. These vortices can be constructed as classical solutions with cylindrical symmetry, at which the Higgs field makes one full rotation if one follows it around the vortex.

The behavior near the vortex of particles whose charge is only e is more complicated. One finds that because of the magnetic flux in the Abrikosov vortex the fields of these particles undergo a phase rotation when they flow around the vortex, in such a way that an Aharonov-Bohm effect is seen. The Aharonov-Bohm phase is $2\pi/N$, or, if we take a particle with charge ne, this phase will be $2\pi n/N$.

The importance of this Aharonov-Bohm phase is that it will be detectable for any charged particle, at any distance from the vortex, in such a way that we will detect its charge *modulo* N. This is surprising because *there is no long range gauge field present*!

An observer who can only detect large scale phenomena may not be able to uncover the chemical composition of the particle, but he can determine its charge modulo N. All he needs is a vortex, which to him will look just like a Nambu-Goto string.

Even if a particle were absorbed by a black hole, its electric charge would still reveal itself. Thus, charge modulo N is a quantum number that will survive even for black holes. It must be a strictly conserved charge.

One can then formalize the argument using only strings and charges modulo N, without ever referring to the original gauge field. Then there may exist many kinds of strings/vortices, so that the black hole may have a rich spectrum of these pseudo-invisible but absolutely conserved charges.

Will this argument allow us to specify all quantum numbers for a black hole? There are several reasons to doubt this. One is that an extremely large number of different kinds of strings must be postulated, which seems to be a substantial departure from the Standard Model at large distance scales.

Secondly, it is not at all obvious that it will be possible to do Aharonov-Bohm experiments with black holes. One then has to assume *first* that black holes indeed occur in well-defined quantum states, just like atoms and molecules. So this argument that black holes have quantum hair is rather circular.

In my lectures there is no need for the mechanism advertised by Preskill et al. It is neither necessary nor likely that all quantum states can be distinguished by means of some conserved quantum number(s). In my other lectures I use just the assumption that quantum states exist, and nothing else. No large-scale strings are needed.

3. DECAY INTO SMALL BLACK HOLES

Due to Hawking radiation the black hole looses energy, hence also mass. The intensity of the radiation will be proportional to T^4 , where T is the temperature, and the total area of the horizon, which for the Schwarzschild black hole is $4\pi R^2$; $R = 2M$. Since one expects#

$$T = 1/8\pi M \quad , \tag{3.1}$$

the mass loss should obey

$$\frac{\mathrm{d}M}{\mathrm{d}t} = - C\, T^4\, R^2 = - C'/\, M^2 \quad . \tag{3.2}$$

The constants C , C' depend on the number of independent particle types at the corresponding mass scale, and this will vary slightly with temperature; the coefficients will however stay of order one (as long as M stays considerably larger than the Planck mass).

Ignoring this slight mass dependence of C' , one finds

$$M(t) = C'' \, (t_o - t)^{\frac{1}{3}} \quad , \tag{3.3}$$

where t_o is a moment where the thing explodes violently. Conversely, the lifetime of any given Schwarzschild black hole with mass M can be estimated to be

$$t_1 = M^3/3C' \quad . \tag{3.4}$$

Now this is the time needed for the complete disappearance of the black hole. One may also ask for the average lifetime of a black hole in a given quantum mechanical state, i.e. the average time between two Hawking emissions.

A rough estimate reveals that the wavelength of the average Hawking particle is of the order of the black hole radius R , and that this is also the expected average spatial distance between two Hawking particles. Therefore the lifetime of a given quantum state is of order R , i.e. of order $1/M$ in Planck units.

In the language of particle physics this implies that the radiating black hole is a resonance state that in an S matrix would produce a pole at the complex energy value

$$E = M - C_3\, i/M \quad , \tag{3.4}$$

where C_3 is again a constant of order one. This corresponds to the value

$$E^2 = M^2 - 2C_3\, i M_{Pl}{}^2 \tag{3.5}$$

for the Mandelstam variable s . We see that all black hole poles are expected to be below the real axis of s at a universal average distance of order one in units of the Planck mass squared.

It is not altogether unreasonable to assume that a black hole is just a pole in the S matrix like any other tiny physical object.

#As was pointed out by this author[8], the derivation of this formula requires an assumption concerning the interpretation of quantum wave functions for particles disappearing into the black hole. Though plausible, one can imagine this assumption to be wrong, in which case the black hole temperature will be different from (3.1).

4. THE S-MATRIX ANSATZ AND THE SHIFTING HORIZON

The problem with linearised quantum field theory in the black hole background is that the ingoing particles then seem to be independent of the outgoing ones. Hilbert space is then a *product* space, $|\psi\rangle = |\psi\rangle_{in} \times |\psi\rangle_{out}$. If we were to describe a black hole that obeys an overall Schrödinger equation then these in- and out-spaces cannot be allowed to be independent of each other. In contrast, one would expect the existence of an S-matrix:

$$|\psi\rangle_{out} = S\ |\psi\rangle_{in} \ , \tag{4.1}$$

and with this mapping of in- to out-states the degrees of freedom pictured in Fig. 1*a* are replaced by the ones of Fig. 1*b* or Fig. 1*c*.

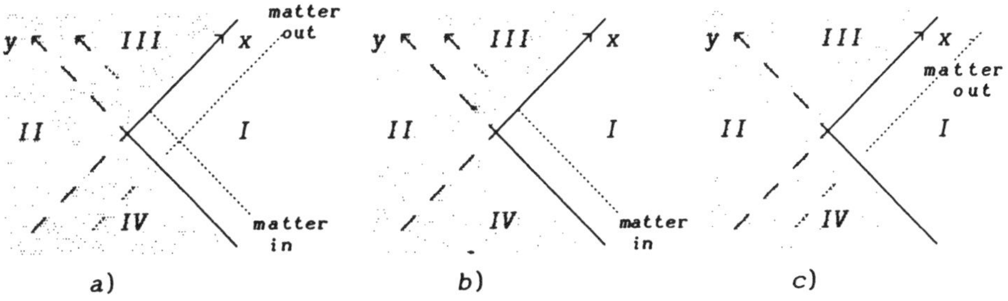

Fig. 1

a) Linearised quantum field theory produces a Hilbert space that is the product of two factors: $|\psi\rangle = |\psi\rangle_{in} \times |\psi\rangle_{out}$ Ingoing particles form the factor space $\{|\psi\rangle_{in}\}$ (*b*), and outgoing ones form $\{|\psi\rangle_{out}\}$ (*c*). x and y are the Kruskal coordinates for the Schwarzschild metric.

As stated earlier, the reason why superimposing in- and out-particles as in Fig. 1*a* is incorrect is the breakdown of linearised quantum field theory at distances closer than a Planck length from the horizon. Gravitational interactions there become super strong. We can obtain the black hole representations of Fig. 1*b* and Fig. 1*c* by adopting the following elementary procedure:

i) *Postulate* the existence of an S matrix, and

ii) take interactions between in- and out-states into account, in particular the gravitational ones.

We can look upon this procedure as a new and more precise formulation of the general coordinate transformation from Kruskal coordinates[3] to Schwarzschild coordinates, or from flat space-time to Rindler[9] space-time. There is no direct contradiction with anything we know about general relativity or quantum mechanics, but because of the crucial role attributed to the interactions the picture is only somewhat more complicated.

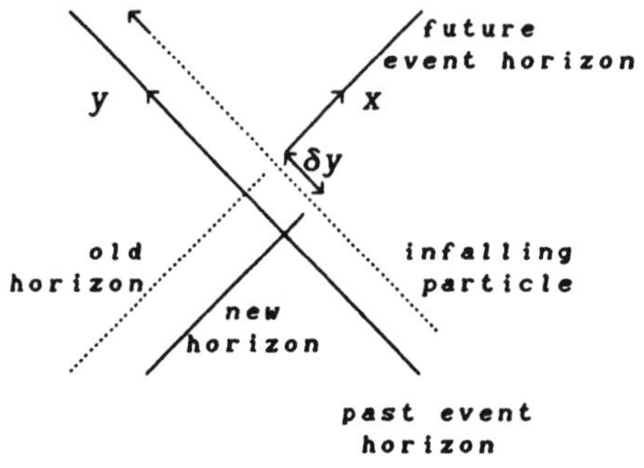

Fig. 2. The horizon displacement.

The most important ingredient of the gravitational interactions is the *horizon shift*[10]. Consider any particle falling into the black hole. Its gravitational field is assumed to be so weak that a linearised description of it, during the infall, is reasonable. The curvatures induced in the Kruskal frame are initially much *smaller* than the Planck length. Now perform a time translation (for the external observer). At the origin of Kruskal space this corresponds to a Lorentz transformation:

$$x \rightarrow \gamma^{-1}x \; ; \qquad y \rightarrow \gamma\, y \; . \tag{4.2}$$

The γ factors here can grow very quickly, exponentially with external time. The very tiny initial curvatures soon become substantial, but are only seen as shifts in the y coordinate, because y has expanded such a lot.

The result is a representation of the space-time metric where the ingoing particle enters along the y axis, with a velocity that has been boosted to become very close to the speed of light. Its energy in terms of the boosted coordinates has become so huge that the curvature became sizable. It is described completely by saying that two halves of the conventional Kruskal space are glued together along the y axis with a *shift* δy, depending explicitly on the angular coordinates ϑ and φ.

The calculation of the function $\delta y(\vartheta,\varphi)$ is elementary. In Rindler space, $\delta y(\tilde{x})$ is simply[11]

$$\delta y(\tilde{x}) \;=\; -\, 4G_N p\, \log(\tilde{x}^2) + C \;, \tag{4.3}$$

where G_N is Newton's constant, p is the ingoing particle momentum, and C is an arbitrary constant.

In Kruskal space the angular dependence of δy is a bit more complicated than the $\tilde{x}$ dependence in Rindler space. It is found by inserting Einstein's equation, which is $R_{\mu\nu} = 0$, everywhere except where the particle comes in. Starting with an arbitrary δy as an Ansatz, one finds Einstein's equation to correspond to

$$(1 - \Delta_{\vartheta,\varphi})\, \delta y(\vartheta,\varphi) \;=\; 0 \;, \tag{4.4}$$

where $\Delta_{\vartheta,\varphi}$ is the angular Laplacian. At the angles ϑ_o, φ_o where the particle enters we simply compare with the Rindler result (4.3) to obtain

$$(1 - \Delta_{\vartheta,\varphi})\, \delta y(\vartheta,\varphi) \;=\; \kappa\, p_{in}\, \delta^2(\vartheta,\varphi;\vartheta_o,\varphi_o) \;, \tag{4.5}$$

where κ is a numerical constant related to Newton's constant.

This equation can be solved:

$$\delta y(\vartheta,\varphi) = f(\vartheta,\varphi;\vartheta_o,\varphi_o)\, p_{in}(\vartheta_o,\varphi_o) \quad ;$$

$$f = \kappa_1 \int_0^\infty \frac{\cos\left(\frac{\sqrt{3}}{2}\, s\right)\, ds}{\left(\cosh s - \cos\vartheta_1\right)^{\frac{1}{2}}} \quad , \tag{4.6}$$

where ϑ_1 is the angular separation between (ϑ,φ) and (ϑ_0,φ_o) .

Other expressions for f are

$$f = \kappa_2 \int_{\vartheta_1}^{2\pi-\vartheta_1} dz\, \left(\cos\vartheta_1 - \cos z\right)^{-\frac{1}{2}} e^{-\frac{1}{2}\sqrt{3}\, z} \quad ; \tag{4.7}$$

and[12]

$$f = \frac{\pi\kappa_1}{\sqrt{2}} \frac{P_{-\frac{1}{2}+\frac{1}{2}i\sqrt{3}}\left(-\cos\vartheta_1\right)}{\cosh\left(\frac{1}{2}\pi\sqrt{3}\right)} \quad , \tag{4.8}$$

where P is a Legendre function with complex index (conical function). From (4.7) one sees directly that for all angles ϑ_1 f is positive.

5. SPACE-TIME SURROUNDING THE BLACK HOLE

The horizon shift discussed in the previous section is an essential ingredient in the S-matrix construction. Without it we would not be able to perform this task. Now we are. We will discuss this construction in the next chapter. First one has to understand what the relevant degrees of freedom are and where in space-time they live. Here, partly anticipating on our results, we observe that the outgoing configurations will depend on what goes in, and with a sensitivity that depends exponentially with $\delta t/4M$, where δt is the time interval as seen by the distant observer. However, the particles going out later than a

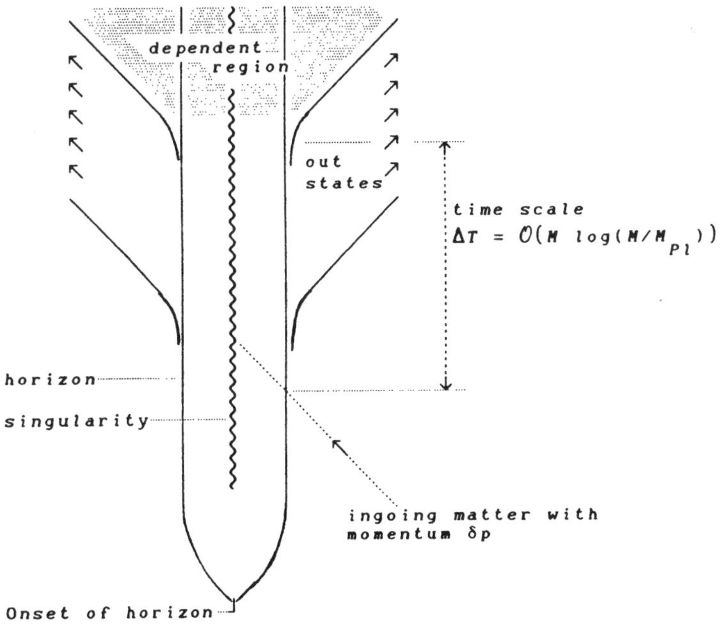

Fig. 3

lapse of time of the order of $\Delta t = 4M \log (M/M_{Pl})$, can no longer be described at all. We will assume, for simplicity, that these particles will be completely determined (in the quantum mechanical sense) by earlier events, or in other words, one is not allowed to choose these states any way one pleases. Simply counting states (as one can derive from the finite black hole entropy[14]) one notes that, given all ingoing particles, the outgoing particles can be chosen freely only during an amount of time Δt, or *vise versa*.

Any ingoing particle will not only affect the outgoing particles after a time lapse of order Δt, but also all that come out later than that. These limitations in choosing in- and outgoing states are just what they would be for any macroscopic system such as a finite size box containing a gas or liquid, connected to the outside world for instance via a tiny hole.

6. CONSTRUCTION OF THE S-MATRIX

The argument now goes as follows[6].

1. Consider one particular in-state and one particular out-state. Assume that someone gave us the amplitude defined by sandwiching the S-matrix between these two states:

$$\langle in | out \rangle \quad . \tag{6.1}$$

Both the in- and the out-state are described by giving all particles in some conveniently chosen wave packets. The ingoing wave packets look like

$$e^{-ip^i_{in} x} \, f^i_{in}(x,\vartheta,\varphi) \quad , \tag{6.2}$$

where i runs over all particles involved, and f^i_{in} are smooth functions. We assume them to be sharply peaked in the angular coordinates so that we know exactly where the particles enter into the horizon (so the *angular coordinates* and the *radial momenta* of all particles are sharply defined). Similarly the outgoing wave packets are

$$e^{-ip^i_{out} y} \, f^i_{out}(y,\vartheta,\varphi) \quad . \tag{6.3}$$

Now let us consider a small change in the ingoing state: $|in\rangle \rightarrow |in'\rangle$. This brings about a sharply defined small change in the distribution of the radial momenta $p_{in}(\vartheta,\varphi)$ on the horizon:

$$p_{in} \rightarrow p_{in} + \delta p_{in}(\vartheta,\varphi) \quad . \tag{6.4}$$

This δp_{in} now produces an (extra) horizon shift,

$$\delta y(\Omega) = \int f(\Omega-\Omega') \, \delta p_{in}(\Omega') \quad , \tag{6.5}$$

where f is the Green function computed in the previous section and Ω stands short for (ϑ,φ) ; $\Omega-\Omega'$ stands for the angle ϑ_1 between Ω and Ω' .

The horizon shift (6.5) does *not* affect the thermal nature of the Hawking radiation, but it *does* change the quantum states. All out-wave functions are shifted. (6.3) is replaced by

$$e^{-ip^i_{out}(y + \delta y(\Omega))} \, f^i_{out}(y,\Omega) \tag{6.6}$$

(the effect of the shift on f is of lesser importance). The shift δy

is assumed to be so small that the outgoing particle is not thrown over the horizon. This requires that we consider only those outgoing particles that are already sufficiently far separated from the horizon, or: they are not the ones that emerge later than the time interval Δt , as defined in the previous section, see Fig. 3. This is why the time interval Δt was necessary. Now such a restriction will also imply that we will have to reconsider the definition of inner products in our Hilbert space, and this will imply that the operator $\exp(-ip^i_{out}\delta y)$ might not be unitary. We will temporarily ignore this important observation.

We observe that the S-matrix element (6.1) is replaced:

$$\begin{aligned} \langle in'\,|\,out\rangle &= e^{-i\int p_{out}(\Omega)\ \delta y(\Omega) d^2\Omega}\ \langle in\,|\,out\rangle \\ &= e^{-i\int\int p_{out}(\Omega)\ f(\Omega-\Omega')\ \delta p_{in}(\Omega')\ d^2\Omega d^2\Omega'}\ \langle in\,|\,out\rangle \ . \end{aligned} \tag{6.7}$$

Here, $p_{out}(\Omega)$ is the total outgoing momentum at the angular coordinates Ω . What we have achieved is that we have been able to compute *another* matrix element of S . Now simply repeat this procedure many times. We then find *all* matrix elements of S to be equal to

$$\langle in\,|\,out\rangle = \mathcal{N}\ e^{-i\int\int p_{out}(\Omega)\ f(\Omega-\Omega')\ p_{in}(\Omega')d^2\Omega d^2\Omega'} \ , \tag{6.8}$$

where $\mathcal{N}$ is one common unknown factor. Apart from an overall phase, $\mathcal{N}$ should follow from unitarity.

The derivation of (6.8) ignores all interactions other than the gravitational ones. We will be able to do better than that, but let us first analyze this expression.

What is unconventional in the S-matrix (6.8) is the fact that the in- and out-states must have been characterized *exclusively* by specifying the *total* radial momentum distribution over the angular coordinates on the horizon. If there are more parameters necessary to characterize these states, these extra parameters will not figure in the S-matrix. But this would mean that two different states $|A\rangle$ and $|B\rangle$ could evolve into the same state $|\psi_{out}\rangle$, so these extra parameters will not be consistent with unitarity. *We cannot allow for other parameters than the total momentum distributions* (unless more kinds of interactions are taken into account).

Thus, if for the time being we only consider gravitational interactions, the in-states can be given as $|p_{in}(\Omega)\rangle$ and the out-states as $|p_{out}(\Omega)\rangle$. The operators $p_{in}(\vartheta,\varphi)$ commute for different values of ϑ and φ , and their representations span the entire Hilbert space; the same for $p_{out}(\vartheta,\varphi)$.

The canonically conjugated operators $u_{in}(\Omega)$, $u_{out}(\Omega)$ are defined by the commutation rules

$$[p_{in}(\Omega),u_{in}(\Omega')] = -i\delta^2(\Omega-\Omega') \tag{6.9}$$

(and similarly for the *out* operators), or,

$$\langle u_{in}(\Omega)\,|\,p_{in}(\Omega)\rangle = C\ \exp\ i\int d^2\Omega\ p_{in}(\Omega)u_{in}(\Omega) \ , \tag{6.10}$$

where C is a normalization constant.

In terms of the u operators the S-matrix is

$$\langle u_{out}(\Omega)\,|\,u_{in}(\Omega)\rangle = \int \mathcal{D}p_{out}\mathcal{D}p_{in}\ \exp\left(-ip_{in}u_{in} + ip_{out}u_{out} - ip_{out}fp_{in}\right) \ , \tag{6.11}$$

which is a Gaussian functional integral over the functions p_{out} and p_{in} . Since the inverse f^{-1} of f is $\kappa^{-1}(1 - \Delta_\Omega)$, the outcome of this functional integral is

$$\langle u_{out}(\Omega)|u_{in}(\Omega)\rangle = C\exp\left[-i\kappa^{-1}\int d^2\Omega\ u_{in}(\Omega)\ (1-\Delta_\Omega)\ u_{out}(\Omega)\right] =$$

$$C\exp\left[-i\kappa^{-1}\int d^2\Omega\ (\partial_\Omega u_{in}(\Omega)\ \partial_\Omega u_{out}(\Omega) + u_{in}(\Omega)\ u_{out}(\Omega))\right] . \tag{6.12}$$

The last term in the brackets is something like a mass term and becomes subdominant if we concentrate on small subsections of the horizon. Therefore it will be ignored from now on. Eq. (6.12) seems to be more fundamental than (6.8) because it is local in Ω.

Fourier transforming back we get

$$\langle p_{out}(\Omega)|p_{in}(\Omega)\rangle =$$

$$\mathcal{N}\int\mathcal{D}u_{in}\mathcal{D}u_{out}\ \exp\int d^2\Omega\left[ip_{in}u_{in} - ip_{out}u_{out} - i\kappa^{-1}\partial_\Omega u_{in}(\Omega)\ \partial_\Omega u_{out}(\Omega)\right] . \tag{6.13}$$

We reobtain (6.8) written as a functional integral.

It is illuminating to redefine

$$p_{out} = p_- \ ; \quad p_{in} = p_+ \ ; \quad u_{out} = x^- \ ; \quad u_{in} = -x^+ , \tag{6.14}$$

and to replace the angular coordinate Ω by a transverse coordinate $\tilde{x}$, so that if we define the transverse momentum components $\tilde{p}\cong 0$ one can write

$$\langle p_{out}(\tilde{x})|p_{in}(\tilde{x})\rangle = \mathcal{N}\int\mathcal{D}x^\mu(\tilde{x})\ \exp\int d^2\tilde{x}\left[ip_\mu(\tilde{x})x^\mu(\tilde{x}) - i\kappa^{-1}(\partial_\Omega x^\mu(\tilde{x}))^2\right] . \tag{6.15}$$

Suppose now that both the in-state and the out-state are written as sets containing a finite number of particles, having not only fixed longitudinal momenta but also transverse momenta $\tilde{p}^i$. We then have to convolute the amplitude (6.15) with transverse wave functions $\exp(i\tilde{p}^i\tilde{x}^i)$ for each particle i. It becomes

$$\mathcal{N}\prod_i\int d\tilde{x}^i\int\mathcal{D}x^\mu(\tilde{x})\ \exp\left[i\sum_i p_\mu{}^i x^\mu(\tilde{x}^i) + \int d^2\tilde{x}\left[-i\kappa^{-1}(\partial_\Omega x^\mu(\tilde{x}))^2\right]\right] . \tag{6.16}$$

This functional integral is very similar to the functional integral for a string amplitude, including the integration over Koba-Nielsen variables[13], except for the unusual imaginary value for the string constant:

$$T = 8\pi G_N i . \tag{6.17}$$

7. ELECTROMAGNETISM

What happens if more interactions are included? The simplest to handle turn out to be the electromagnetic forces. Suppose that the particles that collapsed to form the black hole carried electric charges. The angular charge distribution was

$$\rho_{in}(\Omega) . \tag{7.1}$$

As in the previous section, we consider a small change in this setting, so

$$\rho_{in}(\Omega) \rightarrow \rho_{in}(\Omega) + \delta\rho_{in}(\Omega) . \tag{7.2}$$

The $\delta\rho_{in}(\Omega)$ produces an extra contribution to the vector potential at the horizon which is not difficult to compute[♩].

$$\delta A_\mu = \frac{1}{r_o^2} \delta_{\mu x} \delta(x) A(\Omega) , \tag{7.3}$$

where r_o is the radius of the horizon, and $A(\Omega)$ must satisfy

$$\Delta_\Omega A(\Omega) = \delta\rho_{in}(\Omega) . \tag{7.4}$$

The field (7.3) is only non-vanishing on the plane $x=0$, where it causes a sudden phase rotation for all wave packets that go through. An outgoing wave undergoes a phase rotation

$$e^{iQ\Lambda(\Omega)} \; ; \; \Lambda(\Omega) = r_o^{-1} A(\Omega) . \tag{7.5}$$

This rotation must be performed for all outgoing particles with charge Q. All together the outgoing wave is rotated as follows:

$$|p_{out}(\Omega),\rho_{out}(\Omega)\rangle \rightarrow \exp -i \int d^2\Omega \int d^2\Omega' \; f_1(\Omega-\Omega')\; \rho_{out}(\Omega)\delta\rho_{in}(\Omega') \times |p_{out}(\Omega),\rho_{out}(\Omega)\rangle , \tag{7.6}$$

where $f_1(\Omega-\Omega')$ is a Green function that satisfies

$$\Delta_\Omega\, f_1(\Omega-\Omega') = -\kappa_e\delta^2(\Omega-\Omega') . \tag{7.7}$$

κ_e is a numerical constant.

And using arguments identical to the ones of the previous section we repeat the infinitesimal changes to obtain the S-matrix dependence on $\rho_{in}(\Omega)$ and $\rho_{out}(\Omega)$:

$$\langle p_{out}(\Omega),\rho_{out}(\Omega)|p_{in}(\Omega),\rho_{in}(\Omega)\rangle = \mathcal{N}\, e^{-i\int\int p_{out}(\Omega)\; f(\Omega-\Omega')\; p_{in}(\Omega')d^2\Omega d^2\Omega'} \times e^{-i\int\int \rho_{out}(\Omega)\; f_1(\Omega-\Omega')\; \rho_{in}(\Omega')d^2\Omega d^2\Omega'} . \tag{7.8}$$

Now let us replace $\rho_{out}(\Omega)\rho_{in}(\Omega')$ by

$$-\tfrac{1}{2}\big(\rho_{out}(\Omega)-\rho_{in}(\Omega)\big)\big(\rho_{out}(\Omega')-\rho_{in}(\Omega')\big) . \tag{7.9}$$

This differs from the previous expression by two extra terms in the exponent, one depending on $\rho_{out}(\Omega)$ only and the other depending on $\rho_{in}(\Omega)$ only. These would correspond to external "wave function renormalization factors" that do not describe interaction between the in- and the out-state. So we ignore them.

The electromagnetic contribution in (7.8) can then be written as a functional integral of the form

$$\int \mathcal{D}\Phi(\Omega) \exp \int d^2\Omega \Big(\frac{-i}{2\kappa_e}(\partial_\Omega\Phi)^2 + i\Phi(\rho_{out}-\rho_{in})\Big) . \tag{7.10}$$

Now it may also be observed that the charge distribution ρ is actually a combination of Dirac delta distributions,

---

[♩]The unit e of electric charge of the ingoing particle is here included in ρ_{in} .

$$\rho(\Omega) = \sum Q_i \delta^2(\Omega-\Omega_i) \quad ; \quad Q_i = n_i e \quad . \tag{7.11}$$

Therefore, if we add an integer multiple of $2\pi/e$ to the field Φ the integrand does not change. In other words: Φ is a periodic variable.

Adding (7.10) to (6.15) we notice that the field Φ acts exactly as a fifth, periodic dimension. Hence, electromagnetism emerges naturally as a Kaluza-Klein theory.

8. HILBERT SPACE

In this section we briefly recapitulate the nature of the Hilbert space in which these S matrix elements are defined. As explained in Section 4, the states whose momentum and charge distribution over the horizon were given by $p(\Omega)$ and $\rho(\Omega)$ *include* all particles in the black hole's vicinity. But (if for simplicity we ignore electromagnetism) we can also form a complete basis in terms of states for which the canonical operators $x(\Omega)$ are given. These $x(\Omega)$ (eq. 6.12) may be interpreted as the coordinates of the horizon. Apparently, *the precise shape of the horizon determines the state of the surrounding particles!*

Furthermore, the *in*-horizon and the *out*-horizon do not commute. Therefore, the positions of the future event horizon and the past event horizon do not commute with each other. If we define a "black hole" as an object for which the location in space-time of the future event horizon is precisely determined, we can define a "white hole" as a state for which the past event horizon is precisely determined. *The white hole is a linear superposition of black holes* (and vice versa); operators for white holes do not commute with the ones for black holes. In our opinion this resolves the issue of white holes in general relativity.

Obviously, it is important that the horizon of the quantized black hole is not taken to be simply spherically symmetric. In a black hole with a history that is not spherically symmetric, the onset of the horizon, i.e. the point(s) in space-time (at the bottom of Fig. 3) where for the first time a region of space-time emerges from which no timelike geodesic can escape to $\mathcal{I}^+$, has a complicated geometrical structure. Its mathematical construction has the characteristics of a caustic. One might conjecture that the topological details of this caustic specify the quantum state a black hole may be in.

The fact that the geometry of the (future or past) horizon should determine the quantum state of the surrounding particles gives rise to interesting questions and problems. In ordinary quantum field theory the Hilbert space describing particles in a region of space-time is Fock space; an arbitrary, finite, number of particles with specified positions or momenta together define a state. But now, close to the horizon, a state must be defined by specifying the *total* momentum entering (or leaving) the horizon at a given solid angle Ω . Apparently we are not allowed to specify further how many particles there were, and what their other quantum numbers were. Together all these possibilities form just one state. So, our Hilbert space is set up differently from Fock space. The difference comes about of course because we have strong gravitational interactions that we are not allowed to ignore.

The best way to formulate the specifications of our basis elements here is to assume a lattice cut-off in the space of solid angles (one "lattice point" for each unit of horizon surface area somewhat bigger than $\delta\Sigma$ (the Planck distance squared), and then to specify that there should be *exactly one ingoing and one outgoing particle* at each $\delta\Sigma$. The momenta are given by the operators $p_{in}(\Omega)$ and $p_{out}(\Omega)$ (and the charges by $\rho_{in}(\Omega)$ and $\rho_{out}(\Omega)$). The in- and out-operators of course do not commute.

One may speculate that since $\delta\Sigma$ is extremely small, the totality of all these particles may be indistinguishable from an ordinary Dirac sea for the large-scale observers.

Also one may notice that the way conventional string theory deals with in- and outgoing particles is remarkably similar. Before integrating over the Koba-Nielsen variables the string amplitudes also depend exclusively on the distribution of total in- and outgoing momenta (see concluding remarks in Sect. 6).

If Hilbert space is constructed entirely from the operators $p(\tilde{x})$ and $x(\tilde{x})$ then these operators are hermitean by construction. But we have also seen that in terms of ordinary Fock space $p(\tilde{x})$, and hence also $x(\tilde{x})$ are probably not hermitean. There are different ways to approach this hermiticity problem, but we shall not elaborate here on this point.

9. RELATION BETWEEN TERMS IN THE HORIZON FUNCTIONAL INTEGRAL AND BASIC INTERACTIONS IN 4 DIMENSIONS

In principle one can pursue our doctrine to obtain more precise expressions for our black hole S matrix by including more and more interactions that we actually know to exist from ordinary particle theory. We should be certain to obtain a result that is accurate apart from a limitation in the angular resolution, because particle interactions are known only up to a certain energy. In this section we indicate some qualitative results.

The details of our "presently favored Standard Model" may well change in due time. We will denote anything used as an input regarding the fundamental interactions among in- and outgoing particles near the horizon, at whatever scale, by the words "standard model".

Suppose the standard model contains a scalar field. The effects of this field will be felt by slowly moving particles at some distance from the horizon. But at the horizon itself these effects are negligible. Consider namely a particle such as a nucleon, surrounded by a scalar field such as a pion field. Close to the horizon this particle will be Lorentz boosted to tremendous energies. The scalar field configuration will become more and more flattened. But unlike vector or tensor fields, its intensity will not be enhanced (it is Lorentz invariant). So the cumulated effect on particles traversing it will tend to zero.

However, one effect due to the scalar field will not go away. Suppose our standard model contains a Higgs field, rendering a $U(1)$ gauge boson massive. This means that the electromagnetic field surrounding a fast electrically charged particle will be of short range only. One can derive that the field equation (7.4) will change into

$$(\Delta_\Omega - M_A^2)A(\Omega) \quad = \quad \delta\rho_{in}(\Omega) \quad . \tag{9.1}$$

One may say that the ingoing charge density $\rho_{in}(\Omega)$ is screened by charges coming from the Higgs particles.

This implies that the equations for the Φ field in Sect. 7 will obtain a mass term:

$$\int \mathcal{D}\Phi(\Omega)\ \exp \int d^2\Omega \Big(\frac{-i}{2\kappa_e}[(\partial_\Omega\Phi)^2 + M_A{}^2\Phi^2] \ + \ i\Phi\ \rho\ \Big) \quad . \tag{9.2}$$

Note that this mass term breaks explicitly the symmetry $\Phi \to \Phi + \Lambda$. This explicit symmetry breaking may be seen as a result of the finite and constant value of the Higgs field at the origin of Kruskal space-time.

Next, we may ask what happens if our standard model exhibits confinement. This means that at long distance scales no effect of the gauge field is seen and all allowed particles are neutral.

Confinement is usually considered to be the *dually opposite* of the Higgs mechanism: Bose condensation of magnetic monopoles. A magnetic monopole is an object to which the end point of a Dirac string is attached. A Dirac string is a singularity in a gauge transformation such that the gauge transformation makes one full rotation if we follow a loop around the string.

We must know how to describe the operator field of a monopole at the horizon. Suppose a monopole entered at the solid angle Ω_1 . This means that a Dirac string connects to the black hole at that point. The outgoing charged particles undergo a gauge rotation that rotates a full cycle if we follow a closed curve around Ω_1 (an anti-monopole may neutralize this elsewhere on the horizon).

The gauge jump for the vector potential field A can be identified with the periodic field Φ of Sect 7. So adding an entering monopole to the in-state implies that this field Φ is shifted by an amount $\Lambda(\Omega)$ where Λ makes a full cycle when followed over a loop around Ω_1. This is an operation that is called *disorder operator* in statistical physics and field theory. This operator, Φ_D , is dual to the original field Φ . We find that the dual transformation electricity $\leftrightarrow$ magnetism corresponds to the duality between Φ and Φ_D .

Thus, if we have confinement, a mass term will result in the equations for Φ_D . It explicitly breaks the symmetry $\Phi_D \to \Phi_D + C$. And this bars the transformation back to Φ . Therefore, *if confinement occurs, the field* Φ *is no longer well-defined, we have only* Φ_D . Its mass will be the glueball mass.

In Table 1 we list peculiarities of the mapping from 4 to 2 dimensions. The *generators* of local symmetry transformations in 4 dimensions correspond to the dynamic variables in 2 dimensions. Thus one expects that if the standard model includes a gravitino (requiring a supersymmetry generator of spin $\frac{1}{2}$) then a fermionic field variable will emerge in 2 dimensions.

But the above are merely qualitative features. They should be turned into precise quantitative rules and principles, for which further work is needed.

10. OPERATOR ALGEBRA ON THE HORIZON

A fundamental shortcoming of the procedure described above is that the dimensionality of Hilbert space is infinite from the start. The functions $p(\tilde{x})$ and $x(\tilde{x})$ generate an infinite set of basis elements. Yet the black hole entropy, as calculated from Hawking radiation, is finite. Indeed, we have not yet been able to reproduce Hawking radiation from our S-matrix. This is because we have ignored the *transverse* components of the gravitational shifts, and the string functionals we produced thus far only allow for infinitesimal string excitations. We shall now try to improve our description of the basis elements of Hilbert space. This we do by setting up an operator algebra. First we consider the algebra generated by the amplitudes we have.

Our starting point here is that states in Hilbert space are uniquely determined by specifying any one of the following four functions: the distribution of ingoing momenta $p_+(\tilde{x})$, the outgoing momenta $p_-(\tilde{x})$, the conjugated operators $x^+(\tilde{x})$, or $x^-(\tilde{x})$. They obey the algebra

$$[p_+(\tilde{x}),\ p_+(\tilde{x}')] = 0 \quad ; \quad [p_+(\tilde{x}),\ x^+(\tilde{x}')] = -i\delta^2(\tilde{x},\tilde{x}') \quad ; \tag{10.1}$$

$$[p_-(\tilde{x}),\ p_-(\tilde{x}')] = 0 \quad ; \quad [p_-(\tilde{x}),\ x^-(\tilde{x}')] = -i\delta^2(\tilde{x},\tilde{x}') \quad , \tag{10.2}$$

and we have the relation

$$x^-(\tilde{x}) = 4\pi G \int d^2\tilde{x}' \; f(\tilde{x},\tilde{x}') \; p_+(\tilde{x}') \quad . \tag{10.3}$$

This implies

$$[x^-(\tilde{x}),\; x^+(\tilde{x}')] = -4\pi i G \; f(\tilde{x},\tilde{x}') \quad , \tag{10.4}$$

so that we have also

$$x^+(\tilde{x}) = -4\pi G \int d^2\tilde{x}' \; f(\tilde{x},\tilde{x}') \; p_-(\tilde{x}') \quad . \tag{10.5}$$

Table 1

STANDARD MODEL IN 3+1 DIMENSIONS	INDUCED 2 DIMENSIONAL FIELD THEORY ON BLACK HOLE HORIZON
• *Spin 2:* $g_{\mu\nu}(\mathbf{x},t)$ local gauge generator: $u^\mu(\mathbf{x},t)$	String variables (*spin 1*): $x^\mu(\Omega)$
• *Spin 1:* $A_\mu(\mathbf{x},t)$ local gauge generator: $\Lambda(\mathbf{x},t)$ *mod* $2\pi/e$	Scalar variable (*spin 0*): $\Phi(\Omega)$ *mod* $2\pi/e$
• *Spin 0:* $\phi(\mathbf{x},t)$	No field at all
• Higgs mechanism: "spontaneous" mass M_A for vector field	explicit symmetry breaking; $\Phi(\Omega)$ gets same mass M_A .
• Confinement in vector field A_μ	Φ must be replaced by disorder op. Φ_D ; its symmetry broken.
• Non-Abelian gauge theory	only scalars Φ_i corresponding to Cartan subalgebra
• *Spin* $\frac{1}{2}$*: fermions*	no field at all
• *Spin* $\frac{3}{2}$*:* gravitino local gauge generator spin $\frac{1}{2}$	Spin $\frac{1}{2}$ fermion (?)

The algebraic relations among $\rho(\tilde{x})$ and $\phi(\tilde{x}')$ are slightly more subtle because of the quantization of electric charge and the ensuing periodic boundary conditions on ϕ . We will disregard these from here on.

The relations (10.1-5) are not infinitely accurate. This is because we neglected any gravitational curvature in the sideways directions. This is fine as long as transverse distance scales are kept considerably larger than the Planck scale. One may convince oneself that this implies neglecting higher orders in the derivatives $\partial x^{\pm}/\partial\tilde{x}$. Is there any way to obtain a more precise algebra? It is natural to search for an algebra that is invariant under Lorentz transformations. One might hope that such an algebra could generate the correct degrees of freedom at the Planck scale (in particular *quantized* degrees of freedom).

It was proposed in Ref[6] that Hilbert space on the horizon may be generated by the operator algebra of fundamental surface elements,

$$W^{\mu\nu}(\tilde{\sigma}) = \varepsilon^{ab} \frac{\partial x^{\mu}}{\partial \sigma^{a}} \frac{\partial x^{\nu}}{\partial \sigma^{b}} \quad . \tag{10.6}$$

where the transverse coordinates $\tilde{x}$ were replaced by more arbitrary surface coordinates σ^1, σ^2 . The relations (10.1-5) may be used in the case $\tilde{\sigma} = \tilde{x}$, when the derivatives are small. This means

$$W^{12} = 1 \quad ; \quad W^{1\mu} = \frac{\partial x^{\mu}}{\partial \sigma^{2}} \quad ; \quad W^{2\mu} = -\frac{\partial x^{\mu}}{\partial \sigma^{1}} \quad ; \quad W^{34} = O(\partial x^{\mu})^2 \; . \tag{10.7}$$

The commutation rules can then be rewritten in the form

$$\sum_{\lambda} [W^{\lambda\mu}(\tilde{\sigma}), W^{\lambda\nu}(\tilde{\sigma}')] = \tfrac{1}{2}T \, \varepsilon^{\mu\nu\kappa\lambda} W^{\kappa\lambda}(\tilde{\sigma})\delta^2(\tilde{\sigma}-\tilde{\sigma}') \quad , \tag{10.8}$$

which is written in such a way that it remains true in all coordinate frames. T is a constant ('string constant') equal to 8π in Planck units. In stead of (10.1-5) we can take this to be the equation that generalizes to arbitrary surfaces. It has the advantage of being linear in W .

Now (10.8) is not a closed algebra, because the left hand side still contains a summation. A complete algebra is obtained as follows.

Let K be i times the self dual part of W :

$$K^{\mu\nu} = i(W^{\mu\nu} + \tfrac{1}{2}\varepsilon^{\mu\nu\kappa\lambda}W^{\kappa\lambda}) \quad . \tag{10.9}$$

It has three independent components:

$$K_1 = i(W^{23} + W^{14}) \quad ; \quad K_2 = i(W^{31} + W^{24}) \quad ; \quad K_3 = i(W^{12} + W^{34}) \quad . \tag{10.10}$$

Now from (10.8) we derive that these obey a complete commutator algebra,

$$[K_a(\tilde{\sigma}), K_b(\tilde{\sigma}')] = iT\varepsilon_{abc}K_c(\tilde{\sigma})\delta^2(\tilde{\sigma}-\tilde{\sigma}') \quad . \tag{10.11}$$

Apart from a complication to be mentioned shortly, this is a local and complete algebra of the kind we were looking for. At first sight it seems to generate an infinite dimensional Hilbert space because the operators K , like the W , are distributions . But let us introduce test functions $f(\sigma)$, $g(\sigma)$ and define operators

$$L_a^{(f)} = T^{-1}\int K_a(\tilde{\sigma})f(\tilde{\sigma})\mathrm{d}^2\tilde{\sigma} \quad , \tag{10.12}$$

then these satisfy commutation rules:

$$[L_a^{(f)}, L_b^{(g)}] = i\varepsilon_{abc}L_c^{(fg)} \quad . \tag{10.13}$$

Let us now restrict to test functions $f(\tilde{\sigma})$ that can only take the values 0 or 1 . Then $L_a^{(f)}$ satisfy the commutation rules of ordinary angular momentum operators. Note that for such an f the integral (10.12) is nothing but a boundary integral:

$$L_1^{(f)} = iT^{-1} \oint_{\delta f} (x^2\mathrm{d}x^3 + x^1\mathrm{d}x^4) \quad , \quad \text{etc.}, \tag{10.14}$$

where δf stands for the boundary of the support of f . We conclude that for every closed curve δf on $\tilde{\sigma}$ space we have three 'angular momentum' operators $L_a^{(f)}$ that satisfy the usual commutation rules and addition rules for angular momenta. Given such a bunch of closed curves f_i we can characterize the contribution of that part of the horizon to Hilbert space by the usual quantum numbers l_i and m_i. These are discrete and so, in some sense, we seem to come close to our aim of realizing a discrete Hilbert space for black holes. We note an important resemblance with the loop variable approach to quantum gravity[15].

Unfortunately, there is a snag. The operators L_a are not hermitean. If we take x^i to be hermitean and x^4 anti-hermitean then in the definition (10.6) W^{ij} are hermitean and W^{i4} anti-hermitean. Therefore, $L_a^\dagger$ correspond to the *anti*-self dual parts of $W^{\mu\nu}$. The commutation rules between L_a and $L_a^\dagger$ are non-local (they follow from (10.1-5)). The operators L^2 are hermitean, but not necessarily positive (they are only nonnegative for *time-like* surface elements). If we may assume the smallest surface elements to be timelike we can still build our surface using quantum numbers l_i and m_i but the states we get are *not properly normalized* (it is for finding the norms of the states that we need hermitean conjugation). If

$$\psi\{l_i, m_i\}$$

are the basis elements constructed using the self dual operators L_i, and

$$\phi\{l_i, m_i\}$$

the basis elements generated by the anti-self dual $L_i^\dagger$, then we have

$$\langle \phi\{l_i', m_i'\} | \psi\{l_i, m_i\} \rangle = \prod_i \delta_{l_i, l_i'} \delta_{m_i, m_i'} , \tag{10.15}$$

but the ψ themselves, or the ϕ themselves, are not orthonormal.

Now remember our realization earlier that actually the operators $x^+(\tilde{x})$ and $x^-(\tilde{x})$ are *not* hermitean, when we pass from the "horizon Hilbert space" to ordinary Fock space, because the shift operators may move particles behind the horizon. It is conceivable that this will lead to hermiticity conditions altogether different from (10.15).

But it is far from clear whether or not we actually obtained a complete representation of our Hilbert space.

REFERENCES

1. S.W. Hawking, Commun. Math. Phys. **43** (1975) 199; J.B. Hartle and S.W. Hawking, Phys.Rev. **D13** (1976) 2188; W.G. Unruh, Phys. Rev. **D14** (1976) 870; R.M. Wald, Commun. Math. Phys. **45** (1975) 9
2. S.W. Hawking, Phys. Rev. **D14** (1976) 2460; Commun. Math. Phys. **87** (1982) 395; S.W. Hawking and R. Laflamme, Phys. Lett. **B209** (1988) 39; D.N. Page, Phys. Rev. Lett. **44** (1980) 301, Gen. Rel. Grav. **14** (1987) 299; D.J. Gross, Nucl. Phys. **B236** (1984) 349
3. C.W. Misner, K.S. Thorne and J.A. Wheeler, "Gravitation", Freeman, San Francisco, 1973; S.W. Hawking and G.F.R. Ellis, "The Large Scale Structure of Space-time", Cambridge: Cambridge Univ. Press, 1973; E.T. Newman et al, J. Math. Phys. **6** (1965) 918; B. Carter, Phys. Rev. **174** (1968) 1559; K.S. Thorne, "Black Holes: the Membrane Paradigm", Yale Univ. press, New Haven, 1986; S. Chandrasekhar, "The Mathematical Theory of Black Holes", Clarendon Press, Oxford University Press
4. S. Coleman, Nucl. Phys. **B310** (1988) 643; S.B. Giddings and A. Strominger, Nucl. Phys. **B321** (1989) 481; *ibid.* **B306** (1988) 890
5. P. Goddard, J. Goldstone, C. Rebbi and C.B. Thorn, Nucl. Phys. **B56** (1973) 109; M.B. Green, J.H. Schwarz and E. Witten, "Superstring

Theory", Cambridge Univ. Press; D.J. Gross, et al, Nucl. Phys. **B 256** (1985) 253
6. G. 't Hooft, Phys. Scripta **T15** (1987) 143; *ibid.* **T36** (1991) 247; Nucl. Phys. **B335** (1990) 138; G. 't Hooft, "Black Hole Quantization and a Connection to String Theory" 1989 Lectures, Banff NATO ASI, Part 1, "Physics, Geometry and Topology, Series B: Physics Vol. 238. Ed. H.C. Lee, Plenum Press, New York (1990) 105-128; G. 't Hooft, "Quantum gravity and black holes", in: Proceedings of a NATO Advanced Study Institute on Nonperturbative Quantum Field Theory, Cargèse, July 1987, Eds. G. 't Hooft et al, Plenum Press, New York. 201-226
7. L. Kraus and F. Wilczek, Phys. Rev. Lett. **62** (1989) 1221; J. Preskill, L.M. Krauss, Nucl. Phys. **B341** (1990) 50; L.M. Krauss, Gen. Rel. Grav. **22** (1990); S. Coleman, J.Preskill and F. Wilczek, preprint IASSNS-91/17 CALT-68-1717/ HUTP-91-A016
8. G. 't Hooft, J. Geom. and Phys. **1** (1984) 45
9. W. Rindler, Am.J. Phys. **34** (1966) 1174
10. T. Dray and G. 't Hooft, Nucl Phys. **B253** (1985) 173
11. W.B. Bonner, Commun. Math. Phys. **13** (1969) 163; P.C. Aichelburg and R.U. Sexl, Gen.Rel. and Gravitation **2** (1971) 303
12. C. Lousto, private communication
13 Z. Koba and H.B. Nielsen, Nucl. Phys. **B10** (1969) 633 , *ibid.* **B12** (1969) 517; **B17** (1970) 206; Z. Phys. **229** (1969) 243
14. J.D. Bekenstein, Phys. Rev. **D7** (1973) 2333; R.M. Wald, Phys. Rev. **D20** (1979) 1271; G. 't Hooft, Nucl. Phys. **B256** (1985) 727; V.F. Mukhanov, "The Entropy of Black Holes", in "Complexity, Entropy and the Physics of Information, SFI Studies in the Sciences of Complexity, vol IX, Ed. W. Zurek, Addison-Wesley, 1990. *See also* M. Schiffer, "Black Hole Spectroscopy", São Paolo preprint IFT/P - 38/89 (1989)
15. A. Ashtekar, *Phys. Rev.* **D36** (1987) 1587; A. Ashtekar *et al*, *Class. Quantum Grav.* **6** (1989) L185; C. Rovelli, Class. Quant. Grav. **8** (1991) 297, *ibid.* **8** (1991) 317

CHAPTER 9

EPILOGUE

CHAPTER 9.1

Can the Ultimate Laws of Nature be Found ?

Celsius/Linné Lecture, Uppsala University, Sweden

by

Gerard 't Hooft

Institute for Theoretical Physics
Princetonplein 5, P.O. Box 80.006
3508 TA UTRECHT, The Netherlands

Summary: Physicists are probing ever smaller structures in the world of fundamental particles. Each time we dig deeper, we find different kinds of particles ruled by different types of forces. Will there be an end to this, can there be an "ultimate law?" One thing we know for sure: at the distance scale of 10^{-33} cm (the "Planck length") particles will start feeling each others gravitational force, and this force is stranger and probably more fundamental than any other. Tiny "black holes" will make it impossible to consider distances smaller than the Planck length. If there is an ultimate law of physics, it should probably be formulated at this distance scale.

This lecture is commemorated to two great scientists from Uppsala. Both of them have left their marks in history. The first time I encountered the name of Carl von Linné was when I lived with my parents in the Hague, close to the sea shore. I used to stroll along the beach and collect shells. Back home I tried to identify my finds, and found that many of them had beautiful Latin names with a mysterious "L" behind them. The books did not even bother to explain what the "L" meant. It was supposed to be obvious. More than half of all shell species ever found on our beach were named and classified by the Linnaeus school and still carry the names given by him.

Anders Celsius not only rationalized the temperature scale, but is also known for quite a few important researches in astronomy. He managed to check by observations Newton's prediction that the earth is flattened, the length of one arc unit of the meridian being larger near the pole than near the equator. Like Linnaeus, he went to Lapland to do his research.

Before talking about "The Ultimate Laws of Physics" I have to explain what possibly can be meant with such a phrase. It is probably not only in my country that physicists who talk about such things are accused of being arrogant, if not a bit crazy, by colleagues in other fields. Those who introduced the pretentious words "Theory of Everything" probably deserve such criticism. We are now giving an impression of searching for a Stone of Wisdom, and we seem to pretend that Physics might provide for ultimate answers to all questions.

But of course this is not at all what we are trying to do. What we are after is most easily explained by giving a simple example of the kind of theories one might stumble upon when doing particle physics. In the early seventies it became fashionable among some physicists and mathematicians to play simple games on computers. Shortly after that the desk top computers made these games accessible to anyone. One such game was "Conway's Game of Life"[1]. It went as follows (see Fig. 1):

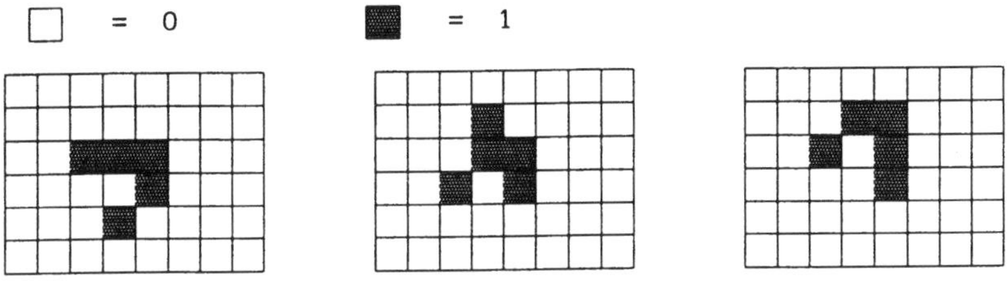

Fig. 1. Evolution of a particular pattern in Conway's Game of Life, at three consecutive times. It will propagate diagonally over the lattice.

On a rectangular infinite lattice we have "cells", each of which carries one bit of information: the cell is said to be "alive" if this bit of information is a one; it is "dead" if it is a zero. Then there is a clock. At every tick of the clock the contents of each cell is being updated. For each cell the new status, at time $t+1$, depends on the contents of itself and its nearest eight neighbors at time t (Fig. 2).

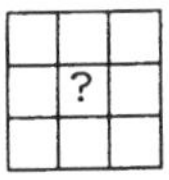

Fig. 2. The cell in the center is updated depending on what was there before, and on what was in the eight surrounding cells drawn here

The rule is as follows:

- *If exactly* 2 *neighbors are alive, the cell in the center will stay as it was.*
- *If exactly* 3 *neighbors are alive, the cell in the center will live.*
- *In all other cases the cell in the center will die.*

In Fig. 1, I show a pattern that, after a while, returns into itself, but at a different place: it moves! One can start with an initial configuration where several of these moving patterns are sent towards each other so that they will collide, and then study what happens. The results may become rather complex, and without a computer it is hopeless to calculate.

The point I want to make is that this system is a "model universe". Imagine that, given a large enough lattice and a patient enough computer, one can have "intelligent" creatures built out of these building blocks. They will investigate the world they are in, and perhaps ultimately discover the three fundamental "laws of physics" on which their universe is based.

The question I wish to address now is: could it possibly be that the universe we are in ourselves has such a simple structure, based on such a simple universal Law, a Law that is invariable and absolute? Is it conceivable that we will ever be able to discover that Law if it exists?

Theoretical physicists are studying the universal laws that govern our world, but it seems that we are still very far away from the "smallest possible structures", i.e. anything like the unit cells in Conway's model. The laws of physics that we did uncover could possibly be considered as "effective laws", laws that describe regularities, correlation phenomena, of patterns at very much larger distance scales and time scales than the smallest possible. It should not come as a surprise that these laws seem to be much more complicated than the Conway laws. If there is a smallest distance and time scale, what could the laws of physics there be like? One might suspect that these will be something much more clever and beautiful than Conway's game. What can be said about them?

In a certain way physicists have already come close to finding universal laws that govern the tiniest known particles of matter. These laws are known as "The Standard Model"[2]. The model is fairly complicated, and it cannot predict the reaction of particles and fields under all conceivable circumstances, but it seems to be a very good description of what is happening nearly everywhere in our universe. I will now give a rough description of this model and its laws.

The basic entities are "Dirac fermions", elementary particles that show a certain amount of spinning motion. This amount of spinning motion is measured in multiples of a fundamental constant of nature, Planck's constant divided by 2π : These particles are said to have "spin $\frac{1}{2}$". We start with particles that can only move with the speed of light. For these particles the *axis of rotation* can be seen to be always parallel to their velocity vector, and one can derive that this statement is unique and independent of the velocity of the observer. This is why one can distinguish unambiguously particles that spin "to the left" from particles that spin "to the right", with respect to this axis.

The forces these fermions exert onto each other are now described by introducing another set of particles, the *gauge bosons*. These are particles with spin one. they can either be considered as the "energy quanta" of various kinds of electric and magnetic fields, or one can view these particles themselves as the transmitters of these forces. When that picture is used one describes the force as being the consequence of an *exchange* of a gauge boson between two fermions: one fermion emits a boson and the other fermion absorbs it. If the *mass* of the boson is negligible then the efficiency of this exchange process is inversely proportional to the square of the distance: the Coulomb force law. If the boson has a certain amount of mass when at rest, then the force it transmits will range only up to a distance inversely proportional to that mass, and decrease very rapidly beyond that distance.

An important feature of the gauge boson force between two fermions is that before and after the emission of a gauge boson the fermion keeps the *same helicity*. This means that a left rotating particle remains left rotating, and a right rotating particle remains right rotating. This is of special importance for the neutrinos: neutrinos *only* exist as left rotating objects. Right handed neutrinos have never been observed (at least not for sure). In contrast, *anti*-neutrinos only come rotating towards the right, not left. By emitting a gauge boson a neutrino may turn into an electron, but then the electron must rotate towards the left.

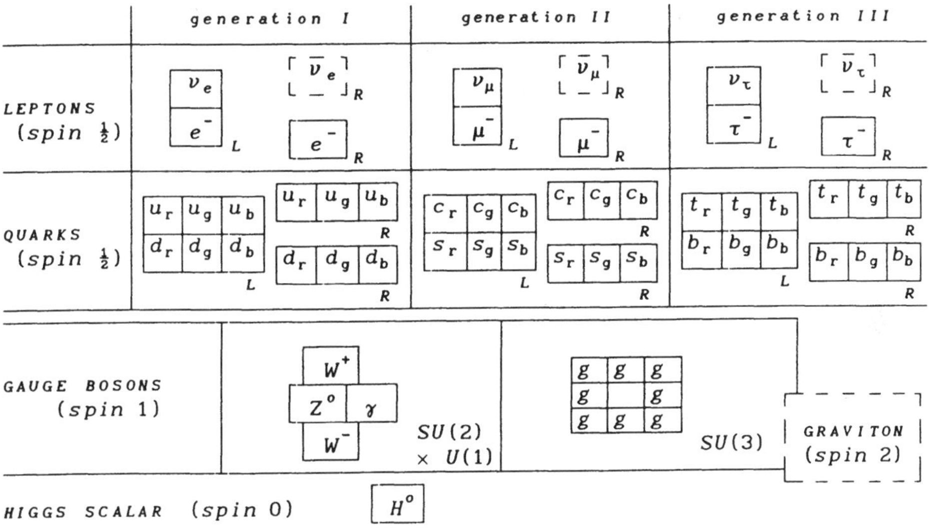

Fig. 3. THE STANDARD MODEL based on $SU(2)_{weak} \times U(1)_{em} \times SU(3)_{strong}$

Right handed neutrinos, which may exist, are indicated in dotted boxes.

We can now make a listing of all fermions and gauge particles. See Fig. 3. Here we see that the fermions are divided into *leptons* and *quarks*. The left rotating objects are indicated by an "*L*" and the right rotating ones by an "*R*". For the anti-leptons and anti-quarks, which are not shown, the *L* and *R* are interchanged. Of all fermions only the quarks are sensitive to the forces of the eight gauge bosons in the box called "$SU(3)$". This force is very strong and so all objects containing quarks will be strongly interacting with other objects.

All *left* rotating fermions are sensitive to the forces from the $SU(2)$ gauge bosons. Finally, all left rotating, and all electrically charged right rotating fermions feel the $U(1)$ force fields.

Now the standard model also contains a spin 0 particle, the "Higgs particle". It also transmits a force, but an important difference with the gauge boson forces is that if a fermion exchanges a Higgs boson it has to flip from left rotating to right rotating or *vice versa*. Another difference is that the spin 0 particle can disappear straight into the vacuum. This implies that some fermions can make spontaneous transitions from left to right and *vice versa*. Technically, this is the way one may introduce *mass* for these fermions. A particle with mass moves slower than the speed of light and it turns out that for such particles the rotation axis cannot be kept parallel to the velocity vector. So this mass can only be there if the particle has the ability to change its spin direction relative to its velocity vector. One finds that mass of a fermion is proportional to the coupling strength of that fermion to the field of the Higgs particle.

The *gravitational force* could be added to our description of the Standard Model by postulating a spin 2 particle, the *graviton*. The rules for this force are very strictly prescribed by Einstein's Theory of General Relativity, but only in as far as it acts collectively on many particles. The details of multiple graviton exchange between individual particles are not completely understood, mainly because any effects due to such exchanges would be so tremend- ously weak that no experimental verification will be possible in any foreseeable future.

The above is a qualitative description of the rules according to which the particles in the Standard Model move. To formulate these rules in more precise mathematical terms, as we were able to do for Conway's Game of Life, would lead us way beyond the scope of this lecture. In some sense the rules are nearly as precise.

Much of this information became known in the early 70's. A dramatic new prediction of this scheme at that time were the effects due to exchange of the newly predicted Z^0 component of the gauge fields. One of the first pictures made of an electron hit by a muon-neutrino is shown in Fig. 4.

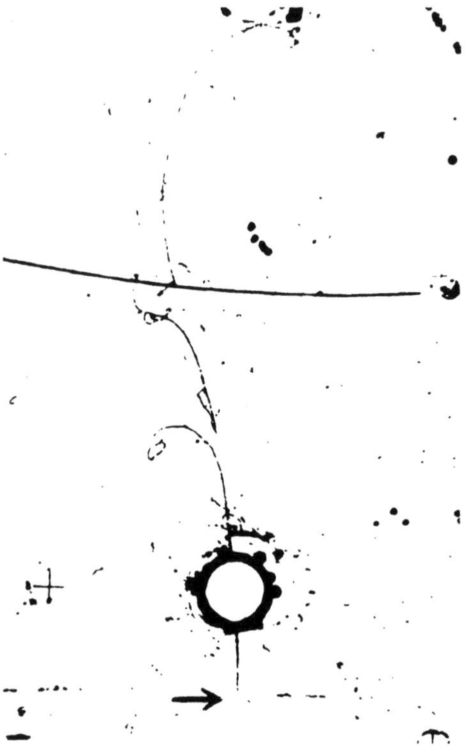

Fig. 4. "Neutral current event":
$e^- \nu_\mu \Rightarrow e^- Z^0 \nu_\mu \Rightarrow e^- \nu_\mu$
In a beam of neutrinos moving upwards, an electron (arrow) is thrown in an upwards direction, leaving a trace made visible in a bubble chamber.
(photo from H.Faissner, Aachen)

I would like to point out the similarities between the Standard model and an ideal "Theory of Everything": all material objects in the universe have to move according to its rules. But of course the model is far from perfect. One reason is that we expect the model not to provide the equations of motion for matter under all conceivable circumstances. One can imagine energies, temperatures and matter densities that are so large that the situation cannot be mimicked in any accelerator. Although our model could in principle be used to calculate what will happen, there are several indications that one should not take such predictions seriously. A second reason is that the *mathematics* of the model is less than perfect. Even if we had infinitely powerful computers to our disposal, some phenomena cannot be computed with any reasonable precision. One has to rely on certain perturbative approximation schemes that sometimes fail catastrophically.

Finally, in contrast with models such as Conway's, our interactions depend on a number of "constants of Nature" that are to be given as real numbers. The precision with which these real numbers are known will always be limited. There are essentially 20 independent numbers:

• 3 gauge coupling constants, corresponding to the strength with which the various kinds of gauge bosons couple to the fermions. The $SU(3)$ coupling constant is much larger than the one for $SU(2)$ and $U(1)$.

Then there are several interaction terms between the Higgs field and the fermionic fields (Yukawa terms). Many of them correspond to the masses of the various fermions:

• 3 lepton masses: m_e , m_μ and m_τ .
• 6 quark masses: m_u , m_d , m_c , m_s , m_t and m_b .
• 4 quark mixing angles, determining further details of the decay of exotic particles.
• 1 topological angle θ_s , a peculiarity relevant only for the strong interactions; as far as is known it is very close to zero.
• 2 self-interaction parameters for the Higgs field. One of these determines the Higgs mass M_H , the other determines the Higgs-to-vacuum transitions. In combinations with the other constants it produces the gauge boson masses.

This adds up to 19 "constants of Nature", which are incalculable; they have to be determined by experiment. We could then add Newton's gravitational constant G_N , but this could be used to fix the as yet arbitrary scale for mass, length and time. Strictly speaking there is also the so-called *cosmological coupling constant* which is also incalculable, but it may perhaps be set to be identically zero; it would be the 20th parameter.

In principle, the Standard Model can be applied at any length and time scale. Indeed, we have nearly scale-*invariance*. This means that when we study particles and their scattering properties under a magnification glass, we see very nearly the same as without the magnification glass. Only the *masses* of particles seem to become much *smaller* if we magnify them[3]. but also the (non-gravitational) coupling strengths will not be exactly the same. This is a subtle effect. When we consider a certain exchange process, for example the emission of a photon by a fermion, one might discover in the magnification glass that this process actually took place in several steps: there could have been a very short-lived particle created temporarily, to be reabsorbed immediately after. This implies that the effective coupling strength one experiences without the magnification glass, is actually the result of a more complicated effect of various couplings at the microscopic level. One finds that any coupling constant g is changed by a small amount, typically of order g^3 , when viewed at different scales.

And so it may happen that a coupling constant changes gradually when we go to widely different time- and length scales. Calculations show that the $U(1)$ force becomes gradually stronger at very small distance scales, the $SU(2)$ force becomes somewhat weaker and the $SU(3)$

force goes down more rapidly. Since these changes are fairly slow, one has to go to tremendously small length scales before all three couplings become about equal in strength. In the simplest version of the Standard Model this happens at about 10^{-30} cm, way beyond anything one can see with whatever microscope or accelerator.

A consequence of the variations of the coupling strengths at very small length scales is that Coulomb's Law of the electric attraction between opposite charges is no longer exactly valid. In stead of

$$F = C\, Q_1\, Q_2 \,/\, r^2 ,$$

one gets a very tiny modification of the power 2.

For the *gravitational coupling constant* the story is very different. Gravity acts upon the mass of a particle, not its charge. If we go to shorter distance scales, one would have to locate the position of a particle with increasing precision. Due to the quantum mechanical uncertainty relation

$$\Delta x\, \Delta p \approx \hbar \quad (= \text{Planck's constant}/2\pi) ,$$

such particles will tend to move around with very large momentum $p = mv$ and since the velocity v is limited to that of light, the masses m of our particles will increase. Therefore, the gravitational force will no longer behave as a $1/r^2$ force but rather become a $1/r^4$ phenomenon.

There will be a scale where the gravitational force *between individual particles* will become stronger than any of the other forces. At this scale the gravitational force will be the dominant one. It is easy to calculate where this happens. Using the three fundamental constants of nature, namely

- the speed of light, $c = 2\ 997\ 924$ km sec^{-1},
- Planck's constant, $\hbar = h/2\pi = 1.054\ 588 \times 10^{-27}$ erg sec,
- Newton's constant, $G = 6.672 \times 10^{-8}\ \text{cm}^3\ \text{g}^{-1}\ \text{sec}^{-2}$,

we find a fundamental unit of length,

$$\sqrt{\frac{G\,\hbar}{c^3}} = 1.6 \times 10^{-33}\ \text{cm} ;$$

one unit for time,

$$\sqrt{\frac{G\,\hbar}{c^5}} = 5.4 \times 10^{-44}\ \text{sec} ;$$

and a unit for mass:

$$\sqrt{\frac{\hbar\, c}{G}} = 22\ \mu\text{g} = 1.2 \times 10^{22}\ \text{MeV} = 1.2 \times 10^{16}\ \text{TeV} .$$

These units are called the Planck length, Planck time and Planck mass or energy.

It is for structures at these distance, time and mass scales that our Standard Model needs to be thoroughly revised. Fluctuations of the gravitational force here become uncontrollable. In any perturbative approach the higher order corrections become larger than the lower order ones. Technically what this implies is that any conventional candidate theory of "Quantum Gravity" becomes unrenormalizable: ill-defined effects from the higher order corrections cannot be absorbed into redefinitions of the interaction terms. This unrenormalizability problem is something one can try to cure and attempts of all sorts have been made. But the real problem of quantum gravity is not the unrenormalizability; it is the fact that the perturbative expansion has a dimensionful expansion parameter (Newton's constant) and that henceforth perturbative corrections at length scales much smaller than the Planck length will be divergent. The only cure to *that* would have to be a theory in which no time and length scales *exist* smaller than the Planck units. Certainly the space-time continuum will have to be either drastically different from what we are used to, or be abandoned altogether.

Although I do not believe that curing the unrenormalizability problem will help us much in understanding quantum gravity, let us look where some of those attempts led us anyway.

The first observation was that theories with higher symmetries are usually better convergent than others. A symmetry that was considered extremely promising was called "super symmetry[4]": a relationship between fermions (particles with spin integer + $\frac{1}{2}$) on the one hand and bosons (integer spin) on the other. If one tries to construct an interaction with this symmetry as a basis one finds as a force carrying particle the *gravitino*, a particle with spin $\frac{3}{2}$. Its supersymmetric partner must be the graviton, and so one obtains a theory in which the gravitational force has a natural place. Super symmetry greatly reduced the renormalization ambiguities, but did not eliminate the problems completely.

The most curious idea is called "String Theory[5]". According to this theory all point like particles must be replaced by string like objects, either open strings having two end points each, or little closed loops. Such objects can interact for instance by joining end points together followed by a rupture somewhere else, or by exchanging loops of string. The big advantage of the scheme as it was much advertised was that in a quantum theory of strings one did not have to specify where and when exactly the joining and rupture take place. The transitions are gradual, and the resulting expressions for the amplitudes, in particular the higher order corrections, seemed to become much less singular and divergent than corresponding expressions in conventional particle theory. The mathematics of these exchanges is very strict, and it turned out that the exchange of the lightest closed string loops reproduces all by itself the gravitational force. It was a discovery by Michael Green in England and John Schwarz in Caltech that calculations in string theory could be made consistent to a high degree provided that at those very small distance scales space-time exhibits 26 dimensions.

To describe also particles with half-integer spin such as the electron and the quarks it was necessary to introduce, again, super symmetry, but only on the string itself. A very ingenuous idea came from the Princeton school, namely that all fermionic excitations (all those that cause half-odd-integer contributions to the spin) run along the string only in one preferred direction. In that direction space-time must be 10-dimensional; in the other directions excitations can take place in 26 dimensions. This string is called the "heterotic string" and the nice thing about it is that it has a cork-screw nature, making a distinction between left and right, just like in the real world.

To explain that the real world only has 4 "visible" dimensions one would have to assume that the 22 (or 6) residual dimensions are "compactified". This means that in those directions space is rolled up very tightly like a tube. That there could be such compactified dimensions had already be proposed by Theodore Kaluza and Oscar Klein in the early 20's and is hence called the "Kaluza-Klein Theory". In this theory the momentum of a particle in one of the compactified directions corresponds to electric charge, and gravitational fields in space-time in those directions are to be interpreted as the electro-magnetic and other similar fields.

The most intriguing aspect of this theory was that it seems to admit no free parameters as constants of nature at all. This means that everything should be calculable and it was suggested that its laws are as precisely defined as in Conway's model. But this was not quite true. The theory, as it was formulated, requires the application of perturbative expansions just like the older particle field theories, and here, like always, one has the problem that the calculations will not converge ultimately. The real problem, as I see it, was that the theory is formulated in terms of *continuous* string coordinates as fundamental variables. They form a highly infinite manifold, and are vastly different from the nice, discrete zeros and ones in Conway's cellular automaton. Maybe one should be able to calculate everything with string theory, but presently we can't.

It is conceivable that string theory can be saved. Some suspect that it is a *topological* theory, which means that not the details of the continuum are relevant, but only the ways boundaries are connected together. Maybe so, but we do not understand how to turn such ideas into practical prescriptions. At the very best the theory is incomplete. Important pages of the manual are missing.

Rather than trying to postulate and guess what our world should be like at the Planck length, one could try to *deduce* this from more rigorous observations and arguments. One such observation is the following: *One cannot measure any distance with more precision than the Planck length*, 1.6×10^{-33} cm. Suppose namely that one tries to detect structures of such a small size. Now in particle theory we have the well-known "uncertainty principle":

$$\Delta x \cdot \Delta p \geq \tfrac{1}{2}\hbar ,$$

where Δx is the uncertainty in position, Δp the uncertainty in the momentum mv of a particle, and $\hbar$ Planck's constant divided by 2π. If Δx must be as small as the Planck length, then Δp must be as big as the Planck mass times the velocity of light c. Both the measuring device and the thing measured must possess structures as small as Δx and therefore contain momenta as large as Δp. Now if two objects with this much momentum collide against each other they carry a total amount of energy which in the center of mass frame exceeds the Planck energy, 10^{16} TeV. If they approach each other at a distance smaller than the Planck length the mutual gravitational interaction will become so strong that a tiny black hole will be formed. The size of a black hole is proportional to its mass, and this one will therefore have to be again larger than the Planck length. A black hole[6] can be viewed as being a conglomeration of pure gravitational energy and its attraction to matter is so strong that whatever object enters its immediate neighborhood will be swallowed without leaving a trace. In particular, the "thing we wanted to measure" will disappear into the hole.

Thus, whenever we wish to describe potential theories concerning structures with sizes smaller than the Planck length we run into the fundamental problem of black hole formation. This problem is a very special and unique one in theoretical physics, and therefore I have advocated for some time now that we should make theoretical studies of tiny black holes as best as we can[7]. By doing *Gedankenexperimenten* with these black holes we should be able to find out at least how to formulate purely logical restrictions of self consistency of whatever theory one might imagine to apply at the Planck scale. Some beautiful results have already been obtained in doing this.

In the by now conventional theory of gravity (Einstein's theory of general relativity) any conglomeration of a certain minimal amount of matter within a certain small enough volume contracts due to its own gravitational field, and this contraction must lead to an implosion that cannot be averted. Whatever kind of information there was inside the imploding cloud of matter, the residual black hole will carry exactly three numbers that enlist all properties that can ever be measured or transmitted: its total *mass* (or energy), its total *electric charge* (positive or negative), and its total *angular momentum*. Nothing can ever come *out* of the black hole, not even light, which is why it is black[8].

But that was the conventional or "classical" theory. It was Stephen Hawking's great discovery[9] early 1975 that one can say more than that if one applies what is known about *quantum theory*. He found that black holes do indeed emit matter, and one can compute precisely the probability for the black hole to emit just *anything*. The emission goes exactly as if the surface of the black hole is hot: it is the radiation emitted by any object with a certain definite temperature. Hawking computed the temperature and found it to be inversely proportional to the black hole mass. The tinier the black hole, the lighter it is and therefore the hotter it is.

Hawking's discovery seems to tell us that black holes are much more like ordinary physical objects than the ideal black holes as described in Einstein's general relativity. They emit radiation and therefore they *decay*, just like any heavy radio-active nucleus. This gave us the idea that the tiniest black holes may perhaps become indistinguishable from the heaviest "elementary particles". After all, particles carry gravitational fields just like black holes do, they decay radio actively just like black holes do, and if one tries to describe the gravitational field surrounding an elementary particle one finds the same expression as for a black hole.

The idea that particles and black holes may ultimately be the same things seems to be so natural that the consequences of such an important assumption will come as a surprise. The assumption namely *does not fit* with the technical details of Hawking's calculation. He found a very important difference between black holes and ordinary quantum mechanical objects. The laws of quantum mechanics and particle theory can be applied perfectly to any small region in the neighborhood of a black hole, but *not at all* to the black hole itself! We have become used to the fact that there are uncertainty relations in quantum mechanics that render it impossible for a physicist to predict exactly how a particle will behave under given circumstances. in stead, the particle physicist can give very precise *satistical* predictions: given a large number of particles and an experiment that is repeated many times, he can predict with ever increasing precision what the distribution of the outcome will be. For this he uses parameters called "constants of nature". The worrisome thing with black holes is that these parameters themselves will be shrouded by uncertainty relations. Even our predictions of the statistics will be fuzzy.

To me it is clear that this is a consequence of our understanding of the physics of black holes being incomplete. Not only incomplete but also wrong! The prediction of an experiment repeated many times should be sharp! Now here we have a perfect example of a paradox in our understanding of physical law, a paradox of the kind that in the past led to the discovery of fundamental new theories such as quantum mechanics and relativity.

Not every physicist agrees with this view. Hawking, like several others, claim that the fuzziness comes about because there is an infinite class of "universes" and we simply do not know which universe we are in. Each of these universes has slightly different constants of nature.

But this is an agnostic argument I find difficult to accept. In trying to devise an "improved" theory I hit upon very strange theories indeed. Requiring the black hole to behave as a whole in accordance with the laws of quantum mechanics I found striking similarities between black holes and strings. Their difference in appearance is merely superficial. Mathematically these objects are startlingly similar.

Conceptually both theoretical constructs are prohibitively difficult. It seems to be absolutely necessary that the theory to be constructed should be of a "topological" kind, as explained earlier. One can argue that the Hawking black hole can only support a limited amount of information. Its internal structure must, somehow, be discrete. As if it contained an array of zeros and ones, exactly like Conway's "game of life". But then we must realize that if you look at a black hole from close you just see empty space. Therefore one might suspect that indeed space and time themselves can carry only limited amounts of information, and as such may show much more similarity with Conway's game of life than ever expected.

Unfortunately it is impossible - and it will continue to be impossible for any foreseeable amount of time - to perform experiments that directly reveal the structure of space-time at the Planck length. We are condemned to do only thought experiments and are asked to sort out all possible theories by logical reasoning alone. This is a long and painful process without any guarantee of success. A poem by Chr. Morgenstern, as noted earlier by M. Veltman, should serve as a warning here (with English translation):

Es gibt viele Theorien
Die sich jedem Check entziehen,
Diese aber kann mann checken:
Elend wird sie dann verrecken.

Many theories are presented
For which no test can be invented.
But one test you will appreciate:
In boredom they'll disintegrate.

References

[1] J.H. Conway, 1970, unpublished;
M. Gardner, Sci. Amer. **224** (1971), Feb, 112; March, 106; April, 114; ibid. **226** (1972), Jan., 104;
S. Wolfram, Rev. Mod. Physics 55 (1983) 601.

[2] There are many text books on the Standard Model. See for instance:
I.J.R. Aitchison and A.J.G. Hey, "Gauge Theories in Particle Physics", Adam Hilger, 1989. More pedestrian:
N. Calder, "The Key to the Universe", BBC, London 1977.

[3] So this is opposite to what one should expect from daily life experiences: the mass of a sand grain will seem to be much larger when inspected through a magnification glass.

[4] See for instance P.C. West, Introduction to Supersymmetry and Supergravity, World Scientific 1986.

[5] See for instance M.B. Green, J.H. Schwarz and E. Witten, "Superstring Theory", Cambridge University Press 1987.

[6] See for instance "Black Holes: the Membrane Paradigm, ed. by K.S. Thorne, R.H. Price and d.A. Macdonald, Yale University Press, 1986.

[7] G. 't Hooft, Nucl. Phys. **B335** (1990) 138
G. 't Hooft, Physica Scripta **T36** (1991) 247
G. 't Hooft, *in* "EPS-8, Trends in Physics 1991, Proceedings of the 8th General Conference of the European Phys. Soc., Sept. 4-6, 1990, part I, p. 187.

[8] S.W. Hawking, Proc. Roy. Soc. London **A 300** (1967) 187
S.W. Hawking and R. Penrose, Proc. Roy. Soc. London **A 314** (1970) 529
R. Penrose, Phys. Rev. Lett **14** (1965) 57

[9] S.W. Hawking, Commun. Math. Phys. **43** (1975) 199J.B. Hartle and S.W. Hawking, Phys.Rev. **D13** (1976) 2188
W.G. Unruh, Phys. Rev. **D14** (1976) 870
S.W. Hawking, Phys. Rev. **D14** (1976) 2460
S.W. Hawking and G. Gibbons, Phys. Rev. **D15** (1977) 2738
S.W. Hawking, Commun. Math. Phys. **87** (1982) 395

INDEX